Smart Innovation, Systems and Technologies

Volume 79

Series editors

The Smart Innovation, Systems and Technologies book series encompasses the topics of knowledge, intelligence, innovation and sustainability. The aim of the series is to make available a platform for the publication of books on all aspects of single and multi-disciplinary research on these themes in order to make the latest results available in a readily-accessible form. Volumes on interdisciplinary research combining two or more of these areas is particularly sought.
The series covers systems and paradigms that employ knowledge and intelligence in a broad sense. Its scope is systems having embedded knowledge and intelligence, which may be applied to the solution of world problems in industry, the environment and the community. It also focusses on the knowledge-transfer methodologies and innovation strategies employed to make this happen effectively. The combination of intelligent systems tools and a broad range of applications introduces a need for a synergy of disciplines from science, technology, business and the humanities. The series will include conference proceedings, edited collections, monographs, handbooks, reference books, and other relevant types of book in areas of science and technology where smart systems and technologies can offer innovative solutions.

High quality content is an essential feature for all book proposals accepted for the series. It is expected that editors of all accepted volumes will ensure that contributions are subjected to an appropriate level of reviewing process and adhere to KES quality principles.

More information about this series at http://www.springer.com/series/8767

Arun K. Somani · Sumit Srivastava
Ankit Mundra · Sanyog Rawat
Editors

Proceedings of First International Conference on Smart System, Innovations and Computing

SSIC 2017, Jaipur, India

Editors
Arun K. Somani
College of Engineering
Iowa State University
Ames, IA
USA

Ankit Mundra
Department of Information Technology
Manipal University Jaipur
Jaipur, Rajasthan
India

Sumit Srivastava
Department of Information Technology
Manipal University Jaipur
Jaipur, Rajasthan
India

Sanyog Rawat
Department of Electronics and Communication Engineering
Manipal University Jaipur
Jaipur, Rajasthan
India

ISSN 2190-3018 ISSN 2190-3026 (electronic)
Smart Innovation, Systems and Technologies
ISBN 978-981-13-5502-8 ISBN 978-981-10-5828-8 (eBook)
https://doi.org/10.1007/978-981-10-5828-8

Printed on acid-free paper

This Springer imprint is published by Springer Nature
The registered company is Springer Nature Singapore Pte Ltd.
The registered company address is: 152 Beach Road, #21-01/04 Gateway East, Singapore 189721, Singapore

Preface

The 1st International Conference on Smart System, Innovations and Computing (SSIC 2017) is going to be held at Manipal University Jaipur situated in the outskirts of Pink City Jaipur during 15–16 April 2017.

The 84 selected papers would maintain the high promise suggested by the written abstracts and the program would be chaired in a professional and efficient way by the session chair, who have been selected for their international standing in this theme. The participants from 12 countries would make the conference truly international. The 84 manuscripts would be split between the three main conference areas, i.e., Smart Systems, Innovations and Computing, and the poster sessions along with the Ph.D. forum on the theme would be distributed across the days of the conference. A proceeding will be published by Springer in the form of book series "Smart Innovations, Systems & Technology", which would include the presented papers. There would be 11 keynote lectures covering different areas of the conference. In addition, there are 15 progress reports to be presented in the Ph.D. forum on ongoing current research, selected from the submitted abstracts based on their relevance and importance. The keynote lectures and the progress reports would bridge the gap between the different fields of Smart Systems, making it possible for non-experts to gain insight into new areas. Also, included among the speakers are eminent industry personals, who would bring new perspectives to the conference theme.

Finally, it is appropriate that we accord our thanks to our fellow members of the Technical Organizing Committee and Advisory Committee for their work in securing a substantial input of papers from within and outside India. We are also indebted to those who would serve as chairs at various positions during the conference.

We expect that SSIC 2017 conference will be a stimulating one, and the first of a very long chain of conference series.

Ames, USA — Arun K. Somani
Jaipur, India — Sumit Srivastava
Jaipur, India — Ankit Mundra
Jaipur, India — Sanyog Rawat

Contents

Contents

About the Editors

Arun K. Somani is currently Anson Marston Distinguished Professor and Jerry R. Junkins Endowed Chair Professor of Electrical and Computer Engineering at Iowa State University. He worked as Scientific Officer for Govt. of India, New Delhi and as a faculty member at the University of Washington, Seattle, WA. Professor Somani's research interests are in the area of fault tolerant computing, computer interconnection networks, WDM-based optical networking, and reconfigurable and parallel computer system architecture. He architected, designed, and implemented a 46-node multi-computer cluster-based system, Proteus. He has served as IEEE distinguished visitor, IEEE distinguished tutorial speaker, and has delivered several key note speeches. He has been elected a Fellow of IEEE for his contributions to "theory and applications of computer networks." He has been awarded a Distinguished Scientist member grade of ACM in 2006.

Prof. Sumit Srivastava is Professor, Head Department of Information Technology and the Convener of SSIC 2017 Conference held at Manipal University Jaipur. He has expertise in the domain of Data Analytics and Image Processing. He is also a senior member of IEEE with more than 14 years of teaching experience. He has organized many national and international conference under IEEE and Springer banner in Rajasthan. He has also served as publication chair of two consecutive editions of IEEE conference on MOOC, Innovation and

Computing (MITE). Prof. Srivastava has coordinated and mentored the Bilateral Exchange between the HEIG-VD University and Manipal University Jaipur for 2 consecutive years. He has published more than 50 research papers in peer-reviewed International Journals, Book series and IEEE/Springer/Elsevier conference proceedings.

Mr. Ankit Mundra is working as an Assistant Professor in the Department of Information Technology, School of Computer Science and IT, Manipal University Jaipur. Previously, he has worked at Department of CSE, Central University of Rajasthan for two years. He has completed his M.Tech in Computer Science and Engineering from Jaypee University of Information Technology (JUIT), Waknaghat, India and awarded with Gold Medal for scoring highest grades during M.Tech. program. His research interest includes Online Fraud Detection, Distributed Computer Networks, Parallel Computing, Cyber Physical Systems, Sensor Networks, Network Algorithms, and Parallel Algorithms. He has supervised eight M.Tech. Scholars and published 21 research papers in peer-reviewed International Journals, Book series and IEEE/Springer/Elsevier conference.

Dr. Sanyog Rawat is presently associated as Associate Professor with Electronics and Communication Engineering Department, Manipal University Jaipur. He has been into teaching and research for more than 12 years. He graduated with Bachelor of Engineering (B.E.) in Electronics and Communication, Master of Technology (M.Tech.) in Microwave Engineering and Ph.D. in the field of Planar Antennas. He is engaged actively in the research areas related to planar antennas for wireless and satellite communication systems. He has published more than 50 research papers in peer-reviewed International Journals, Book series and IEEE/Springer/Elsevier conference. He has organized several workshops, seminars, national and international conferences. He has been empaneled in the editorial and reviewer board of various national and International Journals. He has also edited the Proceedings of the International

Conference on Soft Computing—Theories and Applications (SoCTA 2016) for Springer publication (India). His current research interests include RF and Microwave devices, nature-inspired antennas and biological effects of microwave. He is an active member of various professional societies including IET (UK), ISLE, and IACSIT.

Cluster-Based Energy-Efficient Communication in Underwater Wireless Sensor Networks

Jyoti Mangla and Nitin Rakesh

Abstract There are various challenges in underwater wireless communication such as limited bandwidth, limited energy of sensor nodes, node mobility, high error rate, and high propagation delay. The main problem in underwater wireless sensor networks has limited power of battery as underwater sensor nodes cannot get charged due to deficiency of solar energy. Another main issue is the reliability. In this paper, an approach is proposed for energy-efficient communication among sensor nodes based upon the energy efficiency, reliability, and throughput of sensor node in UWSN (Underwater wireless sensor networks). It uses two step methods. First, it uses a method to identify node called as head node from a cluster which is required for effective communication. This node is responsible for communication with other nodes and further it sends the sensed data to surface station. Second, it finds path between the head node and underwater sink calculated by using Euclidean distance.

Keywords Underwater wireless communication · Node mobility
Underwater sink · Euclidean distance

1 Introduction

Almost 70% of the earth's surface is covered by water. So, we need to explore the underwater environment. The appearance of UWSNs provides new research areas to explore the ocean.

Underwater wireless communication networks (UWCNs) is a network consist of multiple sensing devices and autonomous vehicles that are known as acoustic

J. Mangla (✉) · N. Rakesh
Department of Computer Science and Engineering,
Amity University, Noida, Uttar Pradesh, India
e-mail: thejyotimangla@gmail.com

N. Rakesh
e-mail: nitin.rakesh@gmail.com

A.K. Somani et al. (eds.), *Proceedings of First International Conference on Smart System, Innovations and Computing*, Smart Innovation, Systems and Technologies 79, https://doi.org/10.1007/978-981-10-5828-8_1

underwater vehicles (AUV) which are used to perform various applications such as monitoring, collection of data from undersea, recording of climate, sampling for ocean, prevention from pollution and disaster, etc. [1]. Underwater networks use acoustic communication as radio wave does not propagate well in underwater and also optical waves are affected by some scattering. That is why underwater networks use acoustic communications [2].

The main problem in UWSN is power conservation as sensor nodes are not able to recharge or revive in underwater [3]. So, energy awareness should be considered for efficient utilization of battery power so that lifetime network nodes can be extended. To overcome the problem of battery power, we propose a method for energy-efficient communication in UWSN. This method uses the selection process based upon some optimization matrix.

There are two main issues related to underwater networks. They are energy efficiency and reliability. Reliability needs less error rate, i.e., correction of error, whereas error correction needs energy. So there is a need to work on efficient communication of nodes in underwater wireless sensor networks. The remaining paper is organized as follows: Section 2 includes some related works done in UWSN as a literature review. Section 3 consists of the 2D architecture of UWSN. Section 4 includes a detailed explanation of our proposed approach. Section 5 shows the analysis of our approach. Section 6 concludes the paper and discusses the future work.

2 Literature Review

Prasan et al. [4] presented the network architecture, research challenges, and its potential applications to offshore oil fields and identified research direction in underwater acoustic network used for short-range acoustic communication, high latency network protocol, and long duration network sleep.

Han et al. [5] presented localization algorithms for node mobility in underwater wireless sensor network. This algorithm is categorized into stationary, mobile, and hybrid localization, and then make a comparison among localization algorithm on the basis of time synchronization, location accuracy, localization time, computational complexity, and energy consumption.

Sehem et al. [6] presented two clustering algorithms for mobile underwater wireless sensor networks, i.e., KEER (k-means cluster-based energy-efficient routing algorithm) and EKEER (Extended KEER) which improve energy consumption of sensor node. Total energy consumption is depicted in both low mobility scenario and high mobility scenario.

Kantarci et al. [7] presented UASN architecture and localization techniques. Localization is the main task for UASN which is used for detection of target, tagging of data, and tracking of node. In this, properties of localization algorithms are discussed based on the anchor properties, ranging properties, and messaging properties. Localization is used to improve the performance of MAC and network protocols.

Zhu et al. [8] presented an algorithm for multihop localization which is used to solve the problem of isolated unknown nodes which appears due to flowing of water and touching of aquatic creatures. In this, a greedy approach is used to find the shortest path between anchor and unknown node which can improve the position prediction of sensor nodes.

Bhuvaneswari et al. [9] presented Efficient Mobility-Based Localization (EMBL) which predicts the location information of sensor node in UWSN. This approach discussed ordinary node localization and anchor node localization. In this, every node predicts its location of future mobility through its previous known location information. This scheme can reduce communication cost of the network.

Ayaz et al. [10] proposed dynamically reconfigurable routing protocol in which sensor nodes can re-route their communication if network configuration changes. This protocol provides reliable wireless communication. It reconfigures strategy within the protocol that provides an alternative path for efficient delivery of data. The protocol was demonstrated from sector-based routing and focused beam routing protocol. This provides reliable communication and successful transmission of data.

Ayaz et al. [11] presented a model for reliable data delivery in multihop underwater sensor networks. The paper proposed an algorithm for determining the size of data packet delivery. It uses two-hop acknowledgement model which is used to generate two copies of the same data packet. The paper also presented the connection held among the nodes of the network by packet size, throughput, bit error rate, and distance.

3 Architecture for 2D UWSN

The 2D architecture of UWSN consists of multiple sensor nodes which are anchored to the bottom of sea and further underwater sinks (uw-sinks) are linked to sensor nodes through acoustic links. Sensor nodes are connected to uw-sinks through multihop path or through direct links. These uw-sinks are responsible for sending and receiving information from undersea to surface station (Fig. 1).

Uw-sinks device consists of two transceivers. They are named as vertical and horizontal.

1. In horizontal transceiver, uw-sinks communicate with sensor nodes. Uw-sink sends commands and configured data to sensors and then sensors collect monitored data and send to uw-sinks.
2. In vertical transceiver, uw-sinks reply to surface station.

It is used for ocean-bottom monitoring and monitoring of environment.

In 2D architecture, each sensor node in a cluster sends gathered data to uw-sinks and then uw-sink nodes send it to surface station. There are multiple clusters that consist of sensor nodes and each cluster is having one uw-sink. All the sensor nodes

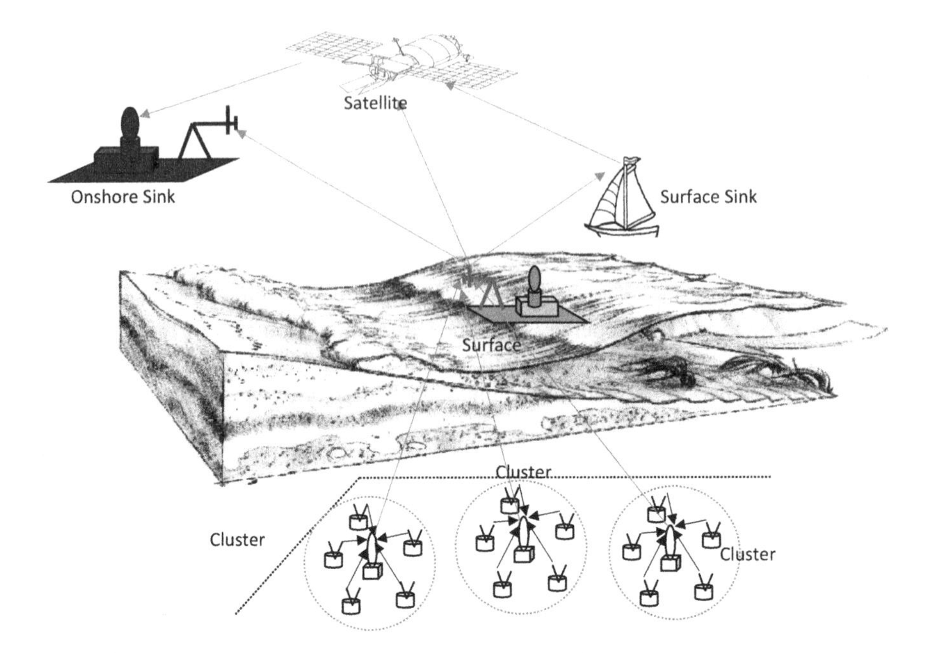

Fig. 1 2D architecture of UWSN [12]

in the cluster are responsible for sending data to uw-sink. This is not energy-efficient way as it is difficult for a single uw-sink to handle multiple nodes in a cluster and power required is also very high.

So, in my proposed approach mentioned in the next section, we can be able to solve this problem by considering some of the parameters of sensor node. It is not dependent on single uw-sink node and it is more energy efficient than previous one.

4 Proposed Approach

An approach is proposed for energy-efficient communication of sensor nodes based on some of the parameters of sensor network such as the energy efficiency, reliability, and throughput of sensor node in UWSN. These parameters are used to check the performance of a particular sensor node.

For efficient communication between sensor nodes, we have used two steps:

1. Selection of head node.
2. Calculate the distance between selected head node and multiple uw-sink nodes and then selects the shortest path among them.

Step 1. Selection of head node

For selecting a head node from cluster, we have taken an assumption that at least one node from a cluster is energy efficient; otherwise, this process cannot work, i.e., if all the nodes from a cluster go down, then cluster head node cannot be found out.

For selecting the head node, energy efficiency is selected as the optimization matrix which takes both energy and reliability constraints. Energy efficiency shows the portion of the total energy spends in communication links between sensors. It is defined as

$$E = T \times R,$$

where E is the energy efficiency, T is the throughput of energy, and R is the packet acceptance rate which explains the data reliability [13].

Assume a cluster has a list of nodes in underwater. It is used to select only that node having maximum energy and then which can perform efficient communication and can be able to send sensed data to uw-sinks (Fig. 2).

Algorithm:

1. Consider a cluster having multiple nodes.
2. Energy of each sensor node is calculated based upon the throughput and reliability of each node.
3. Then the network node having maximum energy efficient is responsible to communicates with other nodes.

Step 2. Find shortest path between head node and underwater sink

Here, underwater sinks (uw-sink) are responsible to collect sensed data from head node and then it is sent to surface station. So we need to find the minimum distance between head node (selected in the previous step) and uw-sink.

Minimum distance can be found out using Euclidean distance. The Euclidean distance is denoted by d_i. It is calculated using x and y coordinates of two points.

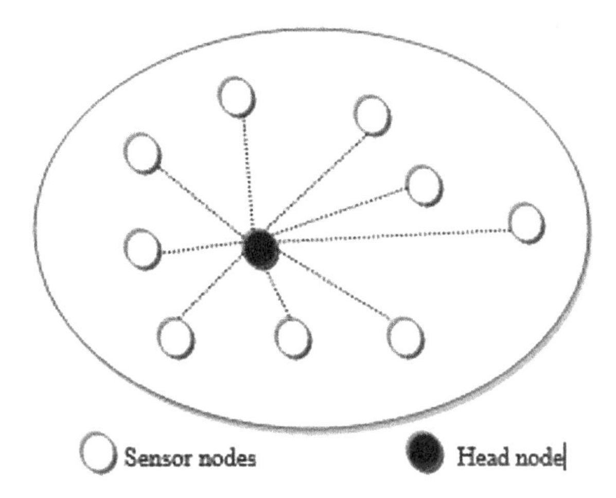

Fig. 2 Communication between head node and sensor node

For example, head node (A) has (x_1, y_1) coordinates and underwater sink (B) has (x_2, y_2) coordinates in two-dimensional network; then Euclidean distance can be calculated as

$$d_i = \sqrt{(x_2 - x_1)^2 + (y_2 - y_1)^2}.$$

5 Analysis of Proposed Approach

We consider that each sensor node knows their energy throughput and reliability. Energy efficiency can be calculated as

$$E = T \times R.$$

This function is calculated for all other nodes of that particular region. The network node that is having maximum energy is selected as a head node. The selected head node is responsible for communication with other nodes.

Let us consider a cluster having sensor nodes A to H which are having some energy values. We find that throughput of sensor nodes can be up to 100 kbps in underwater acoustic communication [14] and reliability, i.e., packet acceptance rate is between 0 and 1.

In Table 1, node E is more energy efficient. So we selected node E as a head node.

In another step, we take an example to find the minimum distance between head node and uw-sink, so we use the concept of Euclidean distance.

In Fig. 3, let us consider A as the selected head node in one of the clusters. Assume A has coordinates (x_1, x_2) and D has coordinates (y_1, y_2); then find the distance AD by using Euclidean and then by same procedure, we find the distance of AE. Then we compare the both distances, and uw-sink node which is closer to this cluster is chosen for communication between head node and uw-sink. Same procedure is followed for all other clusters. In another cluster, B is selected as head node; so distance between BD and BE is calculated and so on.

Table 1 Analysis of energy efficiency at each network node

Node	Throughput (T)	Reliability (R)	Energy efficiency (E)
A	40	0.1	4.0
B	60	0.5	30.0
C	50	0.4	20.0
D	100	0.3	30.0
E	90	0.6	54.0
F	90	0.2	18.0
G	60	0.8	48.0

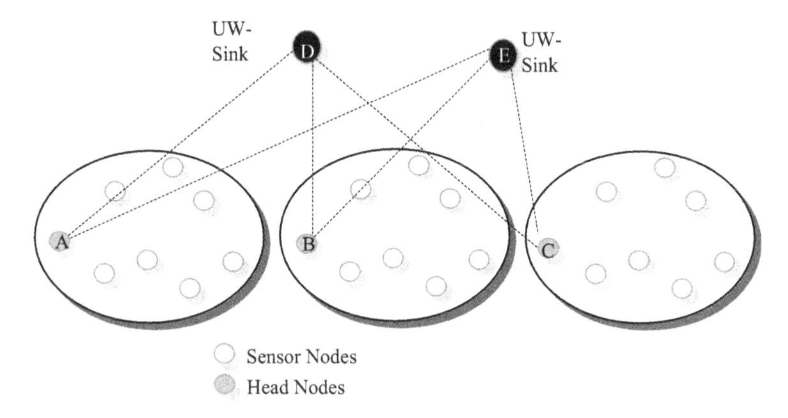

Fig. 3 Shows distance between uw-sink and head node

6 Conclusion and Future Work

In this paper, we discussed an approach for cluster-based efficient communication of sensor nodes in underwater wireless sensor network. It can be possible to some extent through our proposed approach to find the head node from each cluster which performs communication with other nodes so that reliable and efficient communication can be done. Communication cost can be decreased and network capability can be increased to a certain extent. Node mobility is one of the other challenges during communication as it results in improper communication. So, in future work, we improve this method in case of availability of mobile nodes.

References

1. Ian F. Akylidiz, Dario Pompili, Tommaso Melodia,"Underwater acoustic sensor networks: research challenges", Ad Hoc Networks (Elsevier), Volume 3, Issue 3, pp. 257–279, March 2005.
2. Ian F. Akylidiz, Dario Pompili, TommasoMelodia, "Challenges for Efficient Communication in Underwater Acoustic Sensor Network", Special issue on embedded sensor networks and wireless computing, Volume 1, Issue 2, pp. 3–8, ACM New York (2004).
3. K. Ovaliadis, N. Savage and V. Kanakaris, "Energy Efficiency in Underwater Sensor Networks: a Research Review," Journal of Engineering Science and Technology Review (Jestr), Volume 3, Issue 1, pp. 151–156 (2010).
4. U. DeveePrasan and Dr. S. Murugappan, "Underwater Sensor Networks: Architecture, Research Challenges and Potential Applications" International Journal of Engineering Research and Applications(IJERA), Vol. 2, Issue 2, pp. 251–256 (2012).
5. Guangjie Han, Jinfanq Jiang, Lei Shu, Yongjun Xu and Feng Wang, "Localization Algorithms of Underwater Wireless Sensor Network: A Surveys", Volume 12, Issue 2, pp. 2026–2061, Sensors, (2012).

6. SouikiSihem, HadjilaMourad, Feham Mohammed, "Energy Efficient Routing for Mobile Underwater Wireless Sensor Networks", 12th International Symposium on Programming and System, pp. 1–6, IEEE, April 2015.
7. MelikeErol-Kantarci, Hussein T. Mouftah, and SemaOktug, "A Survey of Architectures and Localization Techniques for Underwater Acoustic Sensor Networks", IEEE Communication of Surveys and Tutorials, Volume 13, Issue 3, pp. 487–502, IEEE Comunication society (2011).
8. Zhiwen Zhu, Wenbing Guan, Linfeng Liu, Sheng Li, Shanshan Kong, Yudao Yan, "A Multi-hop Localization Algorithm in Underwater Wireless Sensor Networks", International Conference on Wireless Communication and Signal Processing, pp. 1–6, IEEE Hefei (2014).
9. P.T.V. Bhuvaneswari, S. Karthikeyan, B. Jeeva and M. ArunPrasath, "An Efficient Mobility Based Localization in underwater sensor networks", International conference on Computational Intelligence and Communication Network (CICN), pp. 90–94, IEEE (2012).
10. BeenishAyaz, Alastair Allen, Marian and Wiercigroch, "Dynamically Reconfigurable Routing Protocol Design for Underwater Wireless Sensor Network", 8th International Conference on Sensing Technology, Sep, pp 2–4, Liverpool, UK (2014).
11. Muhammad Ayaz, Low Tang Jung, Azween Abdullah, IftikharAhmad,"Reliable data deliveries using packet optimization in multi-hop underwater Sensor networks", Journal of King Saud University - Computer and Information Sciences, Volume 24, Issue 1, pp 41–48, Jan 2012.
12. Kanika Agarwal, Nitin Rakesh, "Node mobility issues in underwater wireless sensor networks", International Conference on Computer, Communication and Computational Sciences (ICCCCS- 2016), Springer, accepted in press (2016).
13. Ammar Babikar, Nordin Zakaria, "Energy Efficient Communication for Underwater Wireless Sensor Networks, http://www.intechopen.com/book.
14. Alak Roy, Nityananda Sarma, "Selection of communication Carrier for Underwater wireless sensor networks", International Symposium on Advanced Computing and Communication (ISACC), pp. 334–340, IEEE (2015).

IoT-Based Solution for Food Adulteration

Karan Gupta and Nitin Rakesh

Abstract Food acts as an energy source for the organisms that help them to grow and sustain life. In order to maintain the proper hygiene and the safe supply of food products, the food quality should be checked and monitored regularly. To quench the thirst of greed, people add adulterants in the food products to get the monetary benefits by selling the low-quality food at the higher price. So, to avoid any compromise to the human health, food adulteration monitoring system can be used to detect the presence of adulterants in the food product. This system is governed by the Raspberry pi which controls the use of sensors in the system. The recorded data is transferred using the ZigBee module and results are displayed. The IoT technology has been introduced in the system for the purpose of making the system as smart device. With the use of this system, consumption of poor-quality food can be avoided. Moreover, the simplicity of the system can help everyone (commoner, food inspectors, and shop owners) to use food adulteration monitoring system.

Keywords Food adulteration monitoring system · IoT · ZigBee · Raspberry pi

1 Introduction

Food is one of the basic necessities of the human life and without it there will be no scope of human origin. Moreover, if the energy providing food is adulterated, then it will lead trauma to human existence. Furthermore, the present equipment used for monitoring the food adulterations is beyond the scope of a commoner as he is

---

K. Gupta (✉) · N. Rakesh
Department of Computer Science and Engineering, ASET, Amity University, Noida, Uttar Pradesh, India
e-mail: reachguptakaran@gmail.com

N. Rakesh
e-mail: nitin.rakesh@gmail.com

A.K. Somani et al. (eds.), *Proceedings of First International Conference on Smart System, Innovations and Computing*, Smart Innovation, Systems and Technologies 79, https://doi.org/10.1007/978-981-10-5828-8_2

unable to understand the basic principle of them. Moreover, methods used for monitoring are of orthodox level which cannot be used every time. For the present generation, food hygiene problem is the major issue of the concern. With the increase in the health problems, due to the presence of adulterants in food products, there has been huge loss to the immunity and to the health of a personage. Plus, food inspectors are not equipped with proper resources to check the food quality and to inform the public about it. As a result, there is complete negligence toward the health of the people and as a result, human race is unable to grow.

Food adulteration starts from the initial step of the food chain that is from the fields. In the fields, the fertilizers and pesticides are overused. There are some fertilizers and pesticides whose residue still remain and lead to entering in the food chain. The pesticides bioaccumulates in the human body which can lead to rise in toxic levels, thus leading to several diseases in the human body [1]. Diseases such as stomach ache, food poisoning, nausea, and appendix problems are some of the negative aspects due to the food adulteration. Moreover, people adulterate the food products for their monetary needs which reduce the food quality but increase the seller profit. It becomes tough for the buyer to hand-pick a food product due to misleading ads, inappropriate media prominence, and food adulteration. As a result of these malpractices, the ultimate victim is a consumer, who innocently takes adulterated foods and suffers. Although the seller is able to earn huge profit by selling the low-quality food at high cost, the health of humanity is compromised. This paper provides the solution which can be used to resolve the problems related to the adulterants present in the food product. In order to protect from the adulterated food, the food adulteration monitoring system will act as savior. This monitoring system will be able to check whether the given food product is adulterated or not. If the food product is adulterated, then it will prompt the warning to avoid consumption of the food product. The monitoring system will be within the approach of everyone (commoner, food inspectors, and shop owners) and they can use the system to check the food product. In this modern world, none is having enough time to keep regular check on the food products. So keeping that in mind IoT has been used in the system. The food product will be brought near the sensors and after sensing is done the notification is sent to the user. This will allow the user to formulate the notification to others accordingly.

The paper is structured into five sections as follows. Section 1 is the Introduction which introduces various food adulteration approaches and problems prevailing due to the food adulteration. The Sect. 2 is Problem Statement and Motivation which explains why there is a need of food adulteration monitoring system. In the Sect. 3, a complete model of the system is explained, i.e., food adulteration monitoring system is used for describing. The Sect. 4 talks about the classification of working that is the elements used in the system. And the Sect. 5 discusses about the future work and conclusion of the paper which includes the preventive measures for the food safety and the implementation of food adulterant monitoring system.

2 Problem Statement and Motivation

In a report issued by Public Health Foundation of India, more than 80% of the premature deaths are the result of poor food quality [2]. Kids and the mature individuals are the utmost susceptible to the problems caused by the adulterated food. However, the food monitoring system in the nation is neither uniform nor wise enough to keep a regular check of the adulterated food. With the rise in unemployment, individuals are curious to purchase inexpensive food products; as a result, they mostly end up by purchasing the adulterated food.

Moreover, during the festive seasons, there is a huge increase in the amount of adulterant used. In order to make maximum profit, shopkeepers add adulterants in the food products thus leading to several diseases in the human race [3]. Some of the well-known food adulteration shames that shocked our nation and prompted distress in everyone's hearts are: (i) Poisonous Alcohol in West Bengal (Sang Rampur), India—In this incident more than 156 people lost their life after consuming the poisonous alcohol which was distributed at the cheap rates that have been adulterated with methanol (2011). (ii) Midday Meal School Poisoning, Bihar (2013)—This scandal shocked the nation as the government initiated program didnot go well. In the official report more than 48 students were affected and 23 were died. Later on, it was found that the oil that has been used for the cooking purposes was adulterated with the pesticide content. (iii) Unaccepted levels of Lead in the Maggi, India (2015)—It was the most recent and popular scandal that shocked the entire youth. It was found that there was an excess amount of lead present which could damage the human health seriously. In these scandals, not only life of people was lost but also they harmed the coming generations. The most common vulnerable food products are as follows: "Saffron, milk, olive oil, apple juice, coffee, honey, and orange juice are the most vulnerable food products that can be targeted for economically or intentional driven adulteration of diet."

The FSSAI defines food adulteration as the addition or subtraction of any substance to or from food, so that the natural composition and quality of food substance is affected [4, 5]. *In nation like India generally, the adulteration in food products is prepared either for monetary advantage or due to negligence in appropriate sterile situation of processing, keeping, transportation in addition selling.* Food adulteration has quenched away the pleasure of food life. So, food adulteration monitoring system will not only protect the pleasure of life but also help in the buildup human health.

3 Proposed Model and Methodology

The food adulteration monitoring system helps us to determine the adulterants present in the system. First, the sample food product is selected and placed near the sensors so that signals are easy to transmit. This paper represents the list of some of the adulterants that can be found in the food products [6]. These adulterants sustain in the

Table 1 Food products and adulterants present

Food article	Adulterant	Harmful effects
Tea	Used tea leaves processed and colored	Liver disorder
Milk	Starch, vanaspati, urea	Stomach disorder
Khoa	Starch and less fat content	Less-nutritive value
Chilies powder	Brick powder, salt powder	Stomach disorder
Edible oils	Argemone oil	Loss of eyesight, heart diseases
Turmeric powder	Yellow aniline dyes	Carcinogenic
Sweets, juices, jam	Non-permitted coal tar dye	Metanil yellow is toxic
Honey	Molasses sugar	Stomach disorder

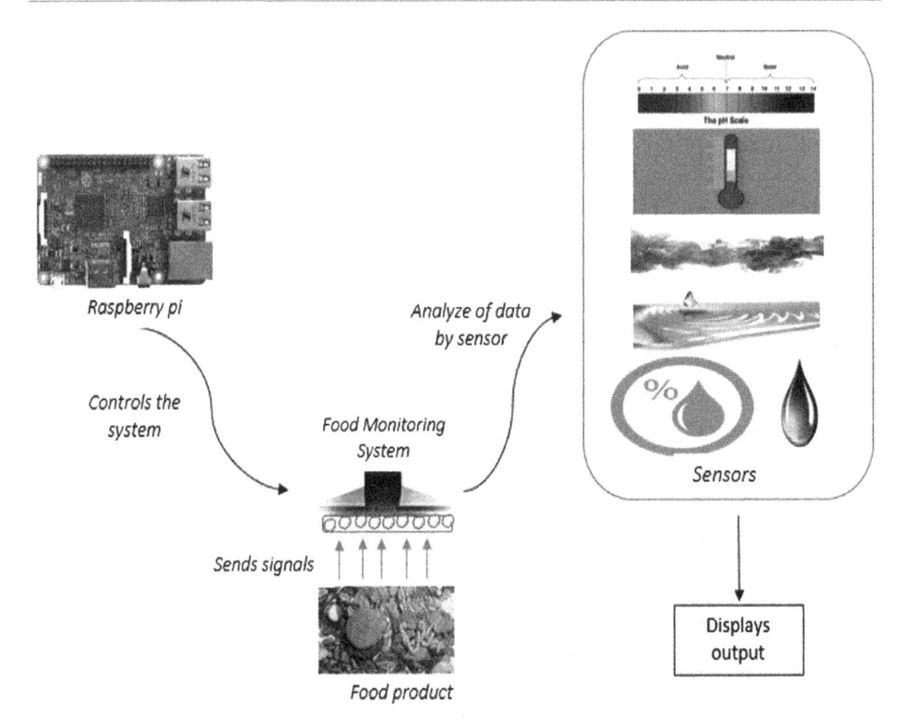

Fig. 1 Methodology

human body and can have several harmful effects on the body which are discussed in Table 1. Furthermore, the core methodology has been discussed in Fig. 1.

This system is administered by the Raspberry Pi. The Raspberry pi acts as the brain of the monitoring system. Raspberry pi handles all the operations taking place from sending command to the sensors to the displaying output. When the system is provided with power supply, the Raspberry pi gets switched on and then asks for the entire details related to the food product. Later on, Raspberry pi sends signal to the sensors to analyze the food product. Raspberry pi is low power consuming device, so it can be operated easily and more efficiently. After receiving the command from the raspberry pi, the sensors get activated and gets ready to perform the task. Moreover, the sensors sense the signals from the food product in various forms, for instance, temperature and oil sensor receive the signal in the form of radiation from the food product, whereas the metal detector receives the signals by electromagnetic property from the food product. Thereafter, the sensed data is transmitted as analog signals by the sensors which have to be converted into digital signals using the interface. After converting to digital signals, the signals are passed to the microcontroller where they are made linear and amplified. The motive of passing the signals through the microcontroller is to have linear and clear signals that make easy to transmit the signals by removing the other noises from the signals. Following it, the signals get transmitted to Raspberry pi and this task is achieved by using ZigBee module.

The ZigBee module is used to transmit the signals to the receiver end. ZigBee helps to transfer the signals more efficiently and easily with low power consumption. Then the Raspberry pi displays the output. In addition to this, IoT has been introduced in the system. After completing all the processes and recording the results, notification will send to the user using IoT. In this way, user will get to know about the food product. Figure 2 shows the explanation of complete working of the food adulteration monitoring system. The diagram includes all details about the process that has been taking place in the monitoring system. The IoT technology has been introduced in the system because after starting the monitoring system keeping the food product near the sensor, the user can indulge in other tasks. Furthermore, the notification received by the user can be formulated among the people related to the food product or the seller company. The notification can be used as evidence against the food processing company. Moreover, the user can upload the notification as Google Review which will help the entire world to know about the company whose product has been used. By the introduction of the IoT, this food adulteration monitoring system can be used both as personal and commercial devices. By keeping the result information to oneself system can be personal and formulating the results to others it can be used as commercial device. The displayed output can either show that food product is adulterated or food product is good for consumption. With the optimal use of this system, everyone will have affair chance to keep their sides and decide accordingly the best quality of food. This system will not only help in getting the good quality of food but also help in eradicating the negative aspects of the society.

In a nutshell, it will be right to say that by the introduction of this system many negative aspects of the food industry can be avoided and food adulteration can be

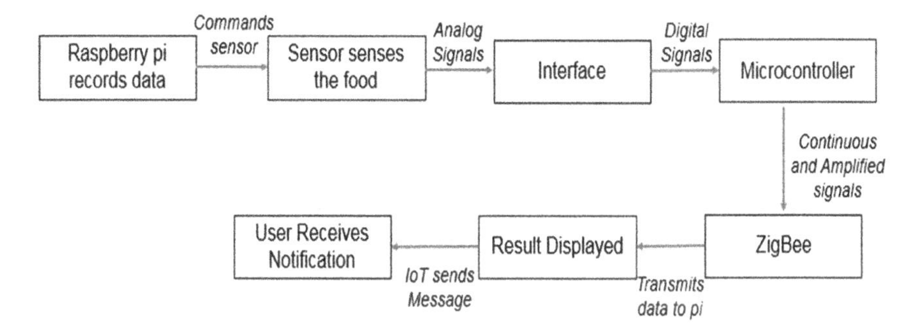

Fig. 2 Working

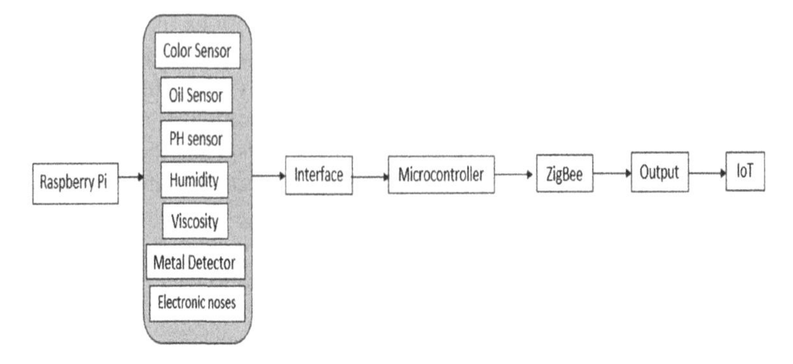

Fig. 3 Integration of the proposed model

controlled to a higher limit. Furthermore, using the IoT in an efficient manner, everyone will have the idea of the company products and this will bring uniformity in the quality monitoring system. This food adulteration monitoring system can be frequently used and easily managed by everyone (the food inspectors, commoner, other shop owners). Figure 3 shows all the equipment that has been used at a particular position during the process of monitoring the food product.

4 Classification of Working

This section of the paper explains the use of several instruments in the food adulterant monitoring system. This will provide all the necessary details of the particular technology that has been used in the model.

Raspberry pi: [7, 8] The Raspberry pi performs similar to a computer when plugs hooked on a computer display or TV, and applies a usual keyboard besides mouse. The raspberry pi board is a portable as well as low cost. In this system, the raspberry pi acts as a brain of computer which commands the sensor to work accordingly. Later on, raspberry pi displays the output of the certain food product accordingly. Every step in the system is monitored by the raspberry pi. The other equipments in the project revolve around the raspberry pi itself.

IoT: [9, 10] Internet of Things signifies a notion in which network equipments have capability to accumulate as well as sense information from the entire world; moreover, it shares that information throughout the web where that information can be operated and administered for several resolutions. In the food adulterant monitoring system, the IoT has a key role to play. The resulted information will be sent to the user using IoT technology. During the monitoring time, the user can engage in different works and when the monitoring is complete, then using IoT notification will be sent to the user providing the result. Furthermore, the user can use the notification as both personal and commercial.

ZigBee: [11–13] ZigBee is a less priced, small power, wireless network standard. The low price permits the tools to be broadly arranged in wireless controller plus observing applications. In the food adulterant monitoring system, the ZigBee protocol enabled us to transfer the data from the sensors to the raspberry pi using wireless technology efficiently and at low cost. The ZigBee protocol has enabled us to achieve great results after the proper quality of signals sent over the network.

Sensors: These will detect the adulterant in the food product, i.e., all the results will be based on the conditions information provided by the sensors. Table 2 shows

Table 2 Sensors

Sensor	Detection	Name of sensors	Attributes	Applications
Temperature	Thermal radiation and measuring electrical output relation with temperature	PT 1000, LM35, NTC sensor	•High accuracy •Robust and impact resistant	• Catering • Butchers • Restaurant • Cold stores
Oil	Thermal radiation, capacitance due to TPCs	Instruments like FOM310 and TESTO 270	•Exact determination frying oil quality •Adjustable to several oil types	• Catering • Sweet shops • Restaurant • Kitchens
Humidity	Electrical resistance proportional to humidity	DHT11 and DHT22	•Reliable and precise • Impact resistant	• Cold stores • Butchers
Salt meter	Salinity ions, electrical resistance	Instruments like SSX 210	• Easy operation • Handy and robust	•Meat, ham sausages, cheese, salads
Metal detector	Electromagnetic radiation	Using transmitter coil and reference coil	•Exact determination metals in food	• Cold stores • Shops
Color	Infrared, Silicon photodiodes	TCS3210 and TCS3200	• Handy and robust • Impact resistant	• Catering • Sweet shops • Restaurant • Kitchens
pH	Electronic ions, Hydrogen ion concentration	SEN0161 and pHT810	• Factory calibration certificate • Handy and robust	• Meat • Fluids • Butchers • Dairy
Viscosity	Fluidity using electromagnetic theory	Viscosity sensor 440–443 by the PAC industry	• Reliable and precise	• Fluids • Catering • Sweet shops
Electronic noses	Olfactrometry using MOSFET, conducting polymers, piezoelectric sensors	Devices like Cyranose 320, SensorFreshQ and JPL electronic nose	• Factory calibration certificate • Easy operation	• Catering • Butchers • Restaurant • Sweet shops

all the sensors that have been used in the food adulteration monitoring system with the knowledge of how they sense the food product. Moreover, the table lists the range of the sensors used and the attributes of the sensors. The applications of the sensors have all been in the table below [14–18]. The different sensors used in the system are given in Table 2.

5 Conclusion and Future Work

With the increasing amount of adulterants in the food products, there is a need to maintain a proper quality of food product without affecting the health of the humans. Improvement in the food adulterant monitoring system can provide a simple solution to maintain proper food quality. As the system will be affordable and easily available, anyone can make full use of it. This system can be installed at every food shop so that they get their customer satisfied. This will help in building stronger relationship between the consumer and the company. In addition to this, quality of the food products available in the market will increase, thus benefitting the personage. Moreover, food adulterants scandals can be avoided in the world.

So in a nutshell, it will be right to say that besides the use of food adulteration monitoring system, there should be proper implementation of the food laws, optimal use of advanced technology in the monitoring of food products, and reduce in the amount of pesticide used and proper public awareness. With the help of implementation of food adulterant monitoring system, everyone could avail the nutritious food and check the quality of their food product. And every commoner has the proper knowledge of the adulterants that can be used in the particular food products. There should be perfect execution of Public Food Safety Policy, Food safety laws, and Food adulterant laws. The future work is enhancing and improving the technology prevailing in the food industry so that everyone can make full use of the available food resources and maintain their proper health and stay fit [19]. Table 3 shows the rate of adulteration within the subsequent years and Fig. 4 shows

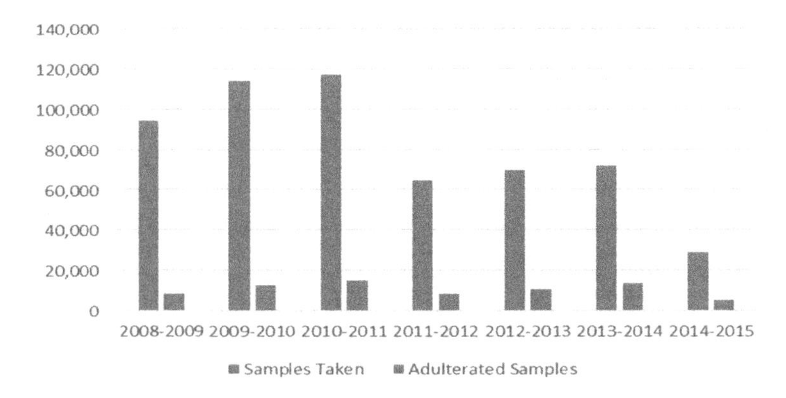

Fig. 4 Adulteration rate

Table 3 Adulteration rate

Year	Samples taken	Adulterated samples
2008–2009	94,470	8,304
2009–2010	113,969	12,692
2010–2011	117,061	14,806
2011–2012	64,593	8,247
2012–2013	69,949	10,380
2013–2014	72,200	13,557
2014–2015	28,731	4,861

the consequent details of Table 3. The table also includes the total samples taken and the adulterated samples found in total samples.

References

1. Dr. Shafqat Alauddin: Food Adulteration and Society: Department: Volume: 1 I Issue: 7 I Dec 2012: Global Research Analysis.
2. The National Food Security Act: http://dfpd.nic.in/nfsa-act.htm (2013).
3. Urban Cocktail—Grab house: https://grabhouse.com/urbancocktail/food-scandals-that-will-shock-you/.
4. FSSAI Manual on General Guidelines: Sampling. Food Safety and Standards Authority of India.
5. FSSAI Manual of Methods of Analysis of Foods, Food Additives: Food Safety and Standards Authority of India.
6. Quick Tests for some adulterants in the food: Food and Drug Administration, Government of Uttar Pradesh.
7. Harshada Chaudhari: Raspberry Pi Technology: A Review: International Journal of Innovative and Emerging Research in Engineering Volume 2, Issue 3, (2015).
8. Raspberry pi education manual, Version 1.0(2012).
9. Cheah Wai Zhao, Jayanand Jegatheesan and Son Chee Loon: Exploring IOT Application Using Raspberry Pi: International Journal of Computer Networks and Applications Volume 2, Issue 1(2015).
10. Hao Chen, XueqinJia, Heng Li: A Brief Introduction to IOT Gateway: Proceedings of ICCTA, Network Technology Research Center (2011).
11. ZigBee Alliance, ZigBee Specification Version 1.0: http://www.ZigBee.org.
12. Shizhuang Lin, Jingyu Liu, Yanjun Fang: ZigBee Based Wireless Sensor Networks and Its Applications in Industrial: IEEE International Conference on Automation and Logistics, page(s):1979–1983 (2007).
13. Nisha Ashok Somani and Yask Patel: Zigbee: a low power wireless technology for industrial applications: International Journal of Control Theory and Computer Modelling (IJCTCM) Vol.2, No.3 (2012).
14. E-bro, a xylem brand: Instruments for the Food Technology.
15. SENSIT, Sensors of temperatures: http://www.sensit.cz/sensors-of-temperature-humidity-and-flowin-food-processing-applications-1404043103.html.
16. PAC, Viscometer sensor 440/443 high pressure: http://www.paclp.com/lab_instruments/viscometer_sensor_440-443:_high-pressure/.

17. Christine Thuen: Sensing odour with e-nose: the past and future trends of odour detection.
18. Fritzsche, Wolfgang, Popp, Jürgen (Eds.): Chapter—Color Sensors and Their Applications: Optical Nano- and Microsystems for Bio analytics.
19. Rediff, "In India, 3 of 4 companies adulterating, misbranding food goes unpunished." Retrieved from http://www.rediff.com/business/special/special-indiaspend-in-india-3-of-4-companies-adulterating-food-go-unpunished/20150708.htm.

SEE THROUGH Approach for the Solution to Node Mobility Issue in Underwater Sensor Network (UWSN)

Nishit Walter and Nitin Rakesh

Abstract One of the most crucial and anticipated problems of underwater sensor network (UWSN) is of node mobility issue. It is a problem that arises due to mobile nature of nodes. In a communication, there is a situation when either of the source or the destination nodes displaces from their original position, thus resulting in a condition of communication failure. We have proposed an approach named SEE THROUGH to overcome this problem. To explain the same, few cases have also been considered. The objective of this paper is to present a new approach to node mobility issue.

Keywords UWSN · Mobility issue · Communication network · Distance

1 Introduction

Wireless Sensor Network (WSN) is a network that consists of a group of sensing nodes that are connected to anchors deep under the ocean. For communication, these sensor nodes use acoustic links and by using these links the nodes can get connected to one or more underwater sinks [1, 2]. These stations play the role of passing and receiving information from underwater to the on-surface stations. Horizontal transceivers and vertical transceivers are the two kinds of transceivers that are consisted of underwater sinks. Vertical transceivers are the ones that are used for long range for applications [3], whereas horizontal transceivers are used by underwater sinks for communication with sensing nodes that are there in the network (Fig. 1).

N. Walter (✉) · N. Rakesh
Department of Computer Science and Engineering, Amity University, Noida, Uttar Pradesh, India
e-mail: nishitwalter1@hotmail.com

N. Rakesh
e-mail: nitin.rakesh@gmail.com

A.K. Somani et al. (eds.), *Proceedings of First International Conference on Smart System, Innovations and Computing*, Smart Innovation, Systems and Technologies 79, https://doi.org/10.1007/978-981-10-5828-8_3

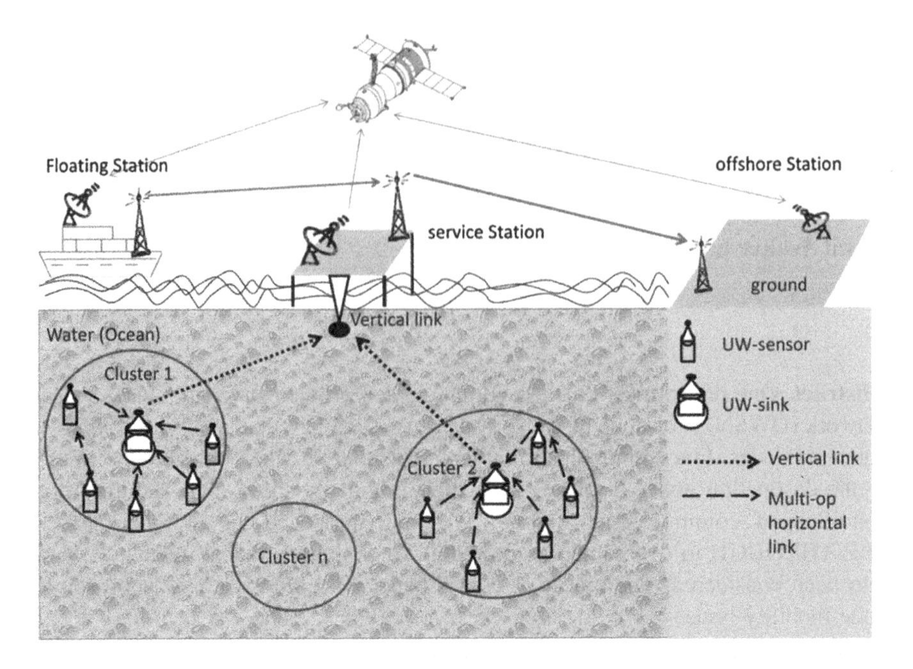

Fig. 1 Underwater sensor network

Surface stations are also allotted with transceivers for the management of multiple communication lines from underwater sinks and are also provided with transmitters for communication with satellite. Underwater sinks are connected with sensor nodes via direct links or sometimes by multihop path. Underwater sinks receive information directly from nodes. Due to large distances, this communication is not very energy efficient [4–7].

Node mobility is a condition where the nodes get drifted from their initial position under various circumstances, like environmental conditions, etc. [8–10].

Here, we are considering 2-D representation of underwater communication system [11] and propose a solution named SEE THROUGH to provide an effective approach for managing and maintaining communication irrespective of the fact that source node or the destination node might get drifted from their original position, resulting in node mobility issue and thus establishing a communication network that can withstand the node mobility issue and can pass and receive information seamlessly, leading a working communication network (Figs. 2, 3, 4, and 5).

Δ—Surface buoy
S—Source Node
D—Destination node

There are three cases or source to drift from its quadrant, three cases for destination node, and nine cases (when considering both, node mobility = NM) (Tables 1, 2, and 3).

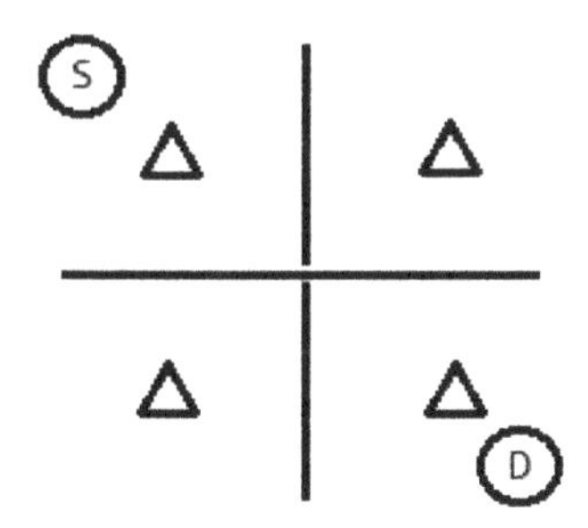

Fig. 2 General depiction of network

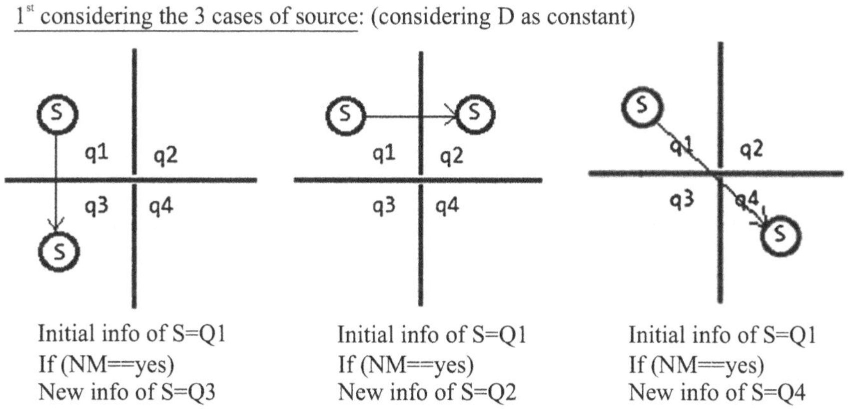

1st considering the 3 cases of source: (considering D as constant)

q1 q2 q3 q4

Initial info of S=Q1
If (NM==yes)
New info of S=Q3

Initial info of S=Q1
If (NM==yes)
New info of S=Q2

Initial info of S=Q1
If (NM==yes)
New info of S=Q4

Fig. 3 Source cases

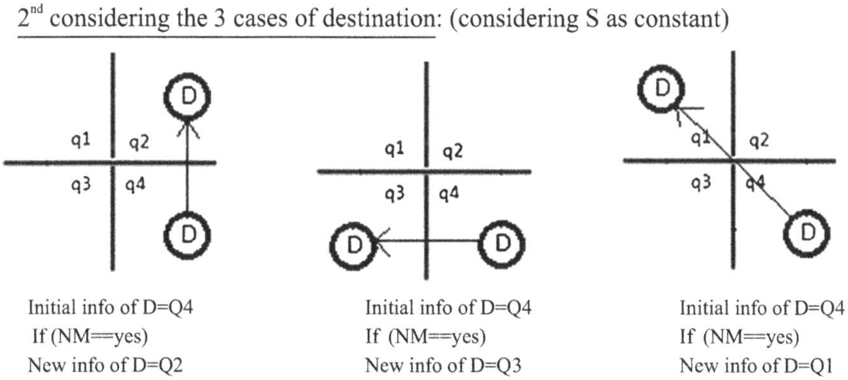

2nd considering the 3 cases of destination: (considering S as constant)

q1 q2 q3 q4

Initial info of D=Q4
If (NM==yes)
New info of D=Q2

Initial info of D=Q4
If (NM==yes)
New info of D=Q3

Initial info of D=Q4
If (NM==yes)
New info of D=Q1

Fig. 4 Destination cases

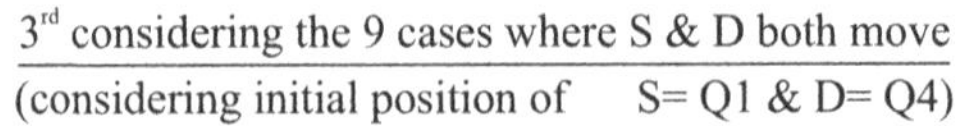

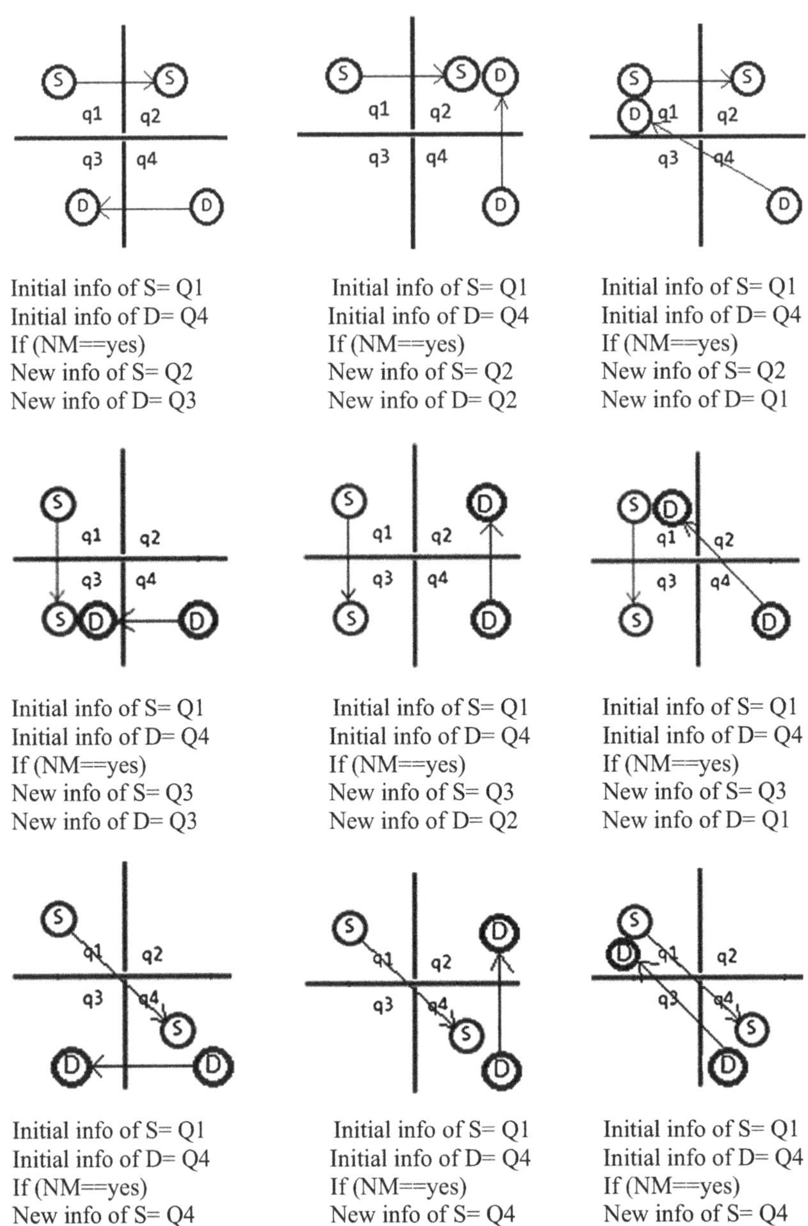

Fig. 5 Source and destination cases (combined)

Table 1 Possibility of S node to move

S. no.	Initial quadrant of S	Final quadrant of S
1	1	2
2	1	3
3	1	4
4	2	1
5	2	3
6	2	4
7	3	1
8	3	2
9	3	4
10	4	1
11	4	2
12	4	3

Table 2 Possibility of D node to move

S. no.	Initial quadrant of D	Final quadrant of D
1	4	1
2	4	2
3	4	3
4	3	1
5	3	2
6	3	4
7	2	1
8	2	3
9	2	4
10	1	2
11	1	3
12	1	4

Table 3 Possibility for S and D nodes to move, when communication is between 1 and 4

S. no.	Initial quadrant of S	Initial quadrant of D	Final quadrant of S	Final quadrant of D
1	1	4	2	2
2	1	4	2	3
3	1	4	2	1
4	1	4	3	2
5	1	4	3	3
6	1	4	3	1
7	1	4	4	2
8	1	4	4	3
9	1	4	4	1

2 Proposed Approach

The idea for solution of given problem makes use of surface buoys. Now let each sensing quadrant has one surface buoy and four nodes as shown in Fig. 6.

Here, we are considering that the surface buoy of source node quadrant and surface buoy of destination node quadrant contain information about the nodes between which the communication is taking place along with their quad information and the information that is to be passed is already present on the network formed by nodes that are being used for the communication purpose.

If any node other than source and destination node gets drifted from its place, the communication network will be maintained by routing algorithm, i.e., it is used in case of node mobility issue like in our previous work, but whenever source or destination node moves from their location, they will send signal to the nearest surface buoy determining its quadrant zone and notifying other quadrants about its present location.

Proposed approach has been designed in such a way that it can turn the disadvantage of node mobility issue into an advantage. If now source or destination node changes its location, then it is a dead lock condition, in the sense that it will terminate the communication.

Surface buoy will have the information about quadrants. Let us say the communication is going to take place between quads 1 and 4. Now, this information is transmitted to nearby buoys that a communication is going to be established between 1 and 4, where S is the source node and D is the destination node.

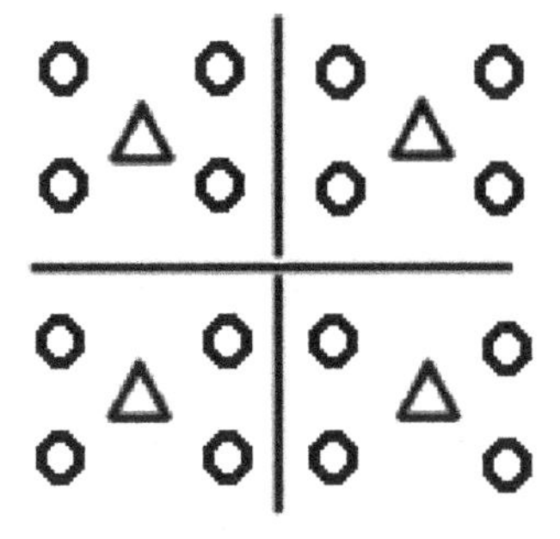

Fig. 6 Four nodes and one surface buoy in each quad

If S node or D node moves from their place, they will contact nearby buoy to check in which quad they are in and using the buoy of that quad send the information to its new location to all nearby buoys resulting in "the nodes used in communication from different quads will get notified by the surface buoy so that the new routing will take place from the node, i.e., nearest to that quad using any routing algorithm" [12, 13].

3 Proposed Algorithm

Define quads among which communication is going to take place. Send the quad information along with the details of source node and destination node to the surface buoys. If node mobility occurs, then S and D nodes will contact nearest buoys and determine their location and share this information with other buoys. Now, buoys send this information in their quads. If there is a node, i.e., involved in the communication path (that was formed after finding Euclidean distance [14] and later using routing algorithm determining the nodes that will be involved in communication) falling in the respective quad. Node is selected that gives minimum distance between communication node and D node; then we will use that node for further routing from that point.

It is not necessary that nearest node is necessarily the same as D node quad. So in that case, when the information about quad, S and D nodes, is regulated, then routing algorithm is used from the communication node, i.e., nearest to the quad of D node or the node itself and further completing the communication network.

If communication is not established, due to node mobility of S node or D node, then repeat the process till communication is successful. Once communication is successful, then stop (Fig. 7).

Let,

T be the threshold of distance
Ts be the threshold of source node
Td be the threshold of destination node information about quads among which communication is going to be established, i.e., the quad of S and D nodes.
S be the source node
D be the destination node
M be the mobility issue
SB be the surface buoy
Q be the quadrant
CN be the communication network.

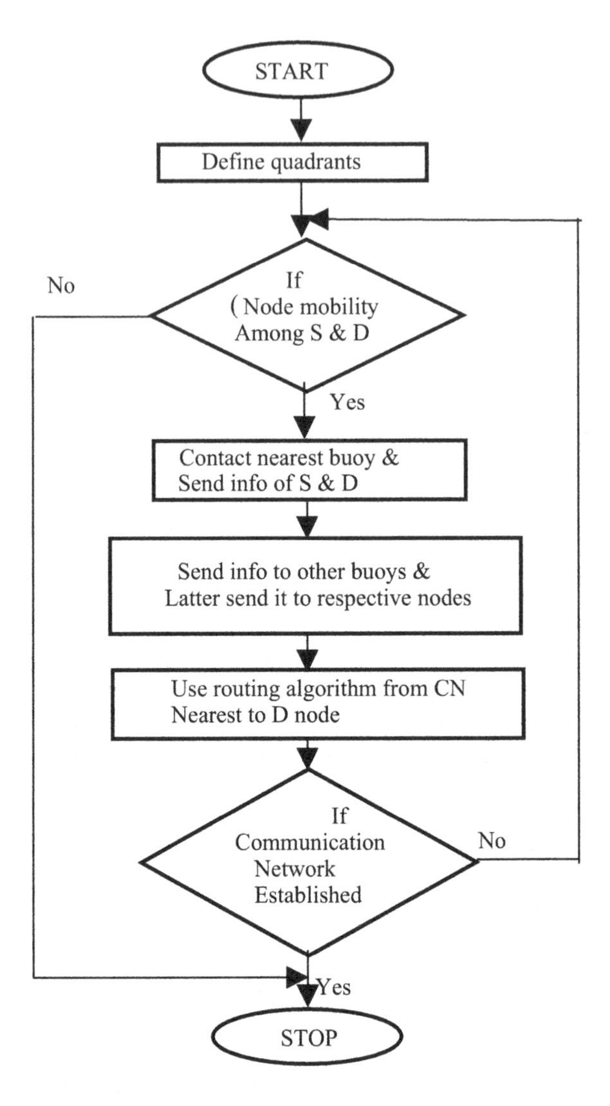

Fig. 7 Flow chart of proposed algorithm

```
1.  Define the Q's for communication
2.  Send info of S & D to SB
3.  If (M>Ts || Td)
            {
            S &&B contact nearest SB
            Get new info
            Update new info to SB's
            }
    Else
            {
            Follow normal procedure of routing
            }
4.  Send info to Q node
    If (node==communication node)  //nearest to D
            {
            Check (nearest CN)
            Use CN
            }
    Else
            {
            Search for CN
            Check (nearest CN)
            Use CN
            }
5.  If (communication establish==No)
            {
            Goto step 3
            }
    Else
            Exit
6.  Stop
```

Note

(x1, y1) are coordinates of D
(x2, y2) are coordinates of CN
Check ()
{
$d = \sqrt{(x_2 - x_1)^2 + (y_2 - y_1)^2}$; //for all nearby CN
Min (d);
Use CN where d = min;
}

4 Results and Future Work

Considering two-dimensional area for our work, node mobility issue is a well-known topic but this subcategory of node mobility that either source or destination moves from its position is unique and new and by using our proposed algorithm we observe that even if source node or the destination node moves from their location, we can deliver the information to the destination node, keeping in mind we have to transfer our information (that we want to send) to all the communication nodes.

In future, we will use this approach and extend its work to 3-D environment and implement it in UWSN and do real-time analysis of it.

Merits are as follows: approach is lucid and flow chart of approach is also given.

Demerit is that the approach is in initial phase.

References

1. ChinnaDurai, M; MohanKumar, S; Sharmila, "Underwater wireless sensor networks", Compusoft journal, volume- IV, ISSUE- VII, 1899–1902, 2015.
2. Jan Erik Faugstadmo; Subsea, Kongsberg Maritime, Horten, Norway; Magne Pettersen; Jens M. Hovem; Arne Lie, "Underwater wireless sensor network", IEE, 422–427, 2010.
3. J. Heidemann; Inst. of Inf. Sci., Southern California Univ., CA; Wei Ye; J. Wills; A. Syed, "Research challenges and applications for underwater sensor networking, IEEE, 228–235, 2006.
4. Ian F. Akylidiz, Dario Pompili, Tommaso Melodia, "Challenges For Efficient Communication in Underwater Acoustic Sensor Network", ACM, 2004.
5. U. Devee Prasan and Dr. S. Murugappan, "Underwater Sensor Networks: Architecture, Research Challenges and Potential Applications" International Journal of Engineering Research and Applications, Vol. 2, Issue 2, pp. 251–256, Mar-Apr 2012.
6. Jun-Hong Cui; Connecticut Univ., Storrs, CT, USA; Jiejun Kong; M. Gerla; Shengli Zhou, "The challenges of building mobile underwater wireless ntworks for aquatic applications", IEEE, vol 20, issue 3, 12–18, 2006.
7. Jim Partan, Jim Kurose, Brian Neil Levine, "A survey of practical issues in underwater networks", ACM, VOL-11, ISSUE-4, 23–33, 2007.
8. Djauhari, M., & Gan, S. Optimality problem of network topology in stocks market analysis. Physica A: Statistical Mechanics and Its Applications, 419, 108–114, 2015.
9. Dario Pompili, Rutgers and Ian F. Akyildiz, "Overview of Networking Protocols for Underwater Wireless Communications", IEEE Communications Magazine, January, pp 97–102, 2009.
10. Mohamed K. Watfa, Tala Nsouli, Maya Al-Ayache and Omar Ayyash "Reactive localization in underwater wireless sensor Networks", University of Wollongong in Dubai—Papers, 2010.
11. S. Naik and Manisha J. Nene, "Realization of 3d underwater wireless sensor networks and influence of ocean parameters on node location estimation", International Journal of Wireless & Mobile Networks (IJWMN) Vol. 4, No. 2, April 2012.

12. Nishit Walter, Nitin Rakesh "KRUSH-D approach for the solution to node mobility issue in Underwater sensor network (UWSN)", International Conference on Recent Advancements in Computer Communication and Computational Sciences (ICRAC3S-2016), Springer 2016,, *accepted, in press.*
13. Kanika Agarwal, Nitin Rakesh "Node Mobility Issues in Underwater Wireless Sensor Network", International Conference on Computer, Communication, and Computational Sciences (IC4S-2016), Springer 2016, *accepted, in press.*
14. Udit Agarwal, "Algorithms Design and Analysis", Dhanpat Rai & Co., 6th Edition, 2008.

Analysis of Mobility Aware Routing Protocol for Underwater Wireless Sensor Network

Jyoti Mangla, Nitin Rakesh and Rakesh Matam

Abstract Underwater wireless sensor network (UWSN) finds various issues such as low bandwidth, node mobility, high error rate, limited energy. UWSN have a dynamic topology due to the movement of sensor nodes which results in improper communication between sensor nodes and underwater-sinks (uw-sinks). In this paper, we have studied various routing protocols which has been proposed for underwater wireless sensor based on some of the parameter, i.e., packet delivery ratio (PDR), end-to-end delay (E2E delay) and energy efficiency. A comparison among these protocols is shown under different network scenario. The overall performance of routing protocol is evaluated by using these parameters and best possible protocols among them are to be identified.

Keywords Underwater wireless sensor networks · Node mobility
Dynamic topology · Underwater-sinks

1 Introduction

Underwater sensor network comprises of multiple sensor nodes used for oceanographic data collection. Underwater wireless sensor network has various applications such as ocean sampling, environment monitoring, undersea exploration for detecting oilfields, disaster prevention, and tactical surveillance application [1]. To design Underwater Acoustic Networks, there are various issues

---

J. Mangla (✉) · N. Rakesh
Department of Computer Science and Engineering, Amity University, Noida, Uttar Pradesh, India
e-mail: thejyotimangla@gmail.com

N. Rakesh
e-mail: nitin.rakesh@gmail.com

R. Matam
Indian Institute of Information Technology, Guwahati, India
e-mail: rakesh@iiitg.ac.in

A.K. Somani et al. (eds.), *Proceedings of First International Conference on Smart System, Innovations and Computing*, Smart Innovation, Systems and Technologies 79, https://doi.org/10.1007/978-981-10-5828-8_4

- Limited available bandwidth.
- Limited battery power due to the unavailability of solar energy in underwater.
- High bit error rates.
- Due to fouling, corrosion, there is high probability for failures of sensors node [2].

There are so many routing protocols proposed for terrestrial wireless sensor network but they cannot be used for underwater wireless sensor network due to different underwater properties. Due to the complex underwater environment such as node mobility, it is difficult to transmit collected data from source to sink. So, to overcome the problem of node mobility, there are various existing routing protocols has been proposed which can be efficient for node mobility to some extent. So, a comparison among various routing protocols based on some of the parameter is presented in this paper. We take the following parameters to study the performance of different routing protocol which are efficient for node mobility:

(1) Packet Delivery Ratio (PDR)—rate of packets received to the packets generated at the source node.
(2) End-to-end Delay—time taken by a packet to move from the source to sink node.
(3) Total Energy Consumption—consumption of total energy including delivery of packets, receiving, transmitting and the energy consumed by all nodes.

These parameters are the main parameters which can be used to check the overall performance of routing protocol and identify best possible protocols among them.

First of all, a comparison is made among various protocols on the basis of above-mentioned protocol, and second these parameters are used to show the performance of these protocols in different network scenarios. The remaining paper is arranged as follows: Sect. 2 includes related work done which consists of existing routing protocol used for underwater network. Section 3 shows the comparison among exiting routing protocol. Section 4 includes performance evaluation of routing protocol. Section 5 shows the node mobility effect on the performance of underwater routing protocols. Section 6 includes the conclusion of the paper and discusses the future work.

2 Related Work

Xie et al. [3] proposed the first geographical routing protocol and was named as Vector-based forwarding (VBF) protocol. It is an energy-efficient, scalable, robust routing protocol and it is also depending on the location of the sensor node. In this, packets from source to destination are forwarded along interleaved and redundant path. Source node computes the vector pipe to sink. Then it finds the forwarder within the pipe. If forwarder found, then packet transmitted to destination otherwise source node finds new vector pipe. Node selection in VBF is done by using the self-adaptation algorithm. The most desirable nodes are selected as forwarders which depends on the value of $T_{adaption}$ given by

$$T_{adaption} = \sqrt{\propto} \times T_{delay} + \frac{R-d}{v_0},$$

where $T_{adaption}$ is the packet holding time before forwarding it, $\propto$ is the desirableness factor, v_0 is the propagation speed (1500 m/s), d is the distance between close nodes, R is the transmission range, and T_{delay} is a predefined maximum delay. Each packet in underwater network is handled by the RANGE field to overcome the node mobility problem. Every packet also has a predefined threshold, i.e., RADIUS field, to determine the closeness to the routing vector.

Xie et al. [4] proposed the enhancement of VBF, i.e., Hop-by-hop vector-based forwarding (HH-VBF) protocol. Instead of unique pipe, it defines the per hope virtual pipe from source to sink. After receiving the packet from the source, it computes the vector from source to sink according to each hop in the network. That is why it is known as hop by hop VBF. If the distance calculated to that vector have a small value than the predefined threshold, it forwards the packet.

Yan et al. [5] proposed a protocol which uses depth information of the underwater nodes named as Depth-based routing (DBR). When the packet reaches to the destination, the depth of forwarding nodes decreases. So, on receiving a packet, it computes the depth d_1 of previous hop packet. Then it compares its depth d_2 with d_1. If $(d_2 < d_1)$, it transmits the packet. Otherwise, packet is dropped.

As shown in Fig. 1 node A is a sink node, and s_1, s_2, s_3 are the neighboring nodes. As s_1 and s_2 are closer to sink node so they are chosen as candidate forwarding nodes. Also, node s_1 is chosen as compared to s_2 to forward the packets as s_1 forward packets before the predefined sending time for the packet.

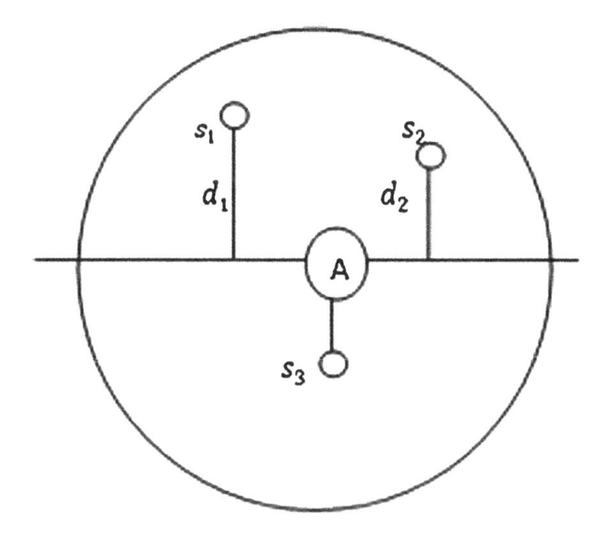

Fig. 1 Selection of forwarding node

Ayaz et al. [6, 7] proposed Hop-by-hop dynamic addressing based (HH-DAB) protocol which uses dynamic Hop ID (can be change with node movement) for routing decision and Node-ID (unique ID for every node) for node identifier. It uses two phases to complete its task. In first phase, a route is created by assigning dynamic hop ids to every node. In second phase, forwarding of data packets to sink are done by using these hop ids. In this protocol, hello packet is sent by every sink node. Hello packet consists of sink ID (used to identify floating node) and hop ID (used to find out the number of hops away from the sinks).

Ayaz et al. [8] proposed a Temporary Cluster Based Routing (TCBR). It makes different clusters based to their locations. It is suitable for static, dynamic networks as well as for the hybrid networks. It worked well for reliability and energy consumption (less nodes are required for E2E process). Also, no location information of sensor nodes is required.

Li et al. [9] presented Depth-Based Routing aware MAC protocol (DBR-MAC), which integrates a DBR protocol and handshaking-based. In this, a depth-based transmission scheduling algorithm is proposed based on the depth, and angle information, and overhead transmissions of one-hop neighboring nodes to access the channel of high priority key nodes than other nodes. It is a cross-layer scheme that forward packets to the floating sink node by the least hops. It improves the energy, throughput, and time efficiency.

3 Comparison Among Routing Protocol

In this paper, a comparison among various routing protocol VBF [3], HH-VBF [4], DBR [5], HH-DAB [6, 7] is described with their advantages and disadvantages. It is found that these protocols can be efficient for node mobility to some extent (Table 1).

4 Impact of Node Mobility

The impact of static and mobile network is explored on the performance of these protocols. We compare the performance of these protocols in different environment, i.e., in static environment/network (we assumed that all nodes are stable, i.e., no effect of water current) as well as in mobile network (where one or more nodes get mobile due to different underwater activities) (Table 2).

In VBF, it is identified that the ratio of packet delivery increases during node mobility as compared to static nodes. Also, average delay is stable for both the network. It is not energy-efficient for mobile network as it only suitable for dense network and having only single sink architecture.

In HH-VBF, packet delivery ratio is high for stable nodes but it decreases approx. 50% during node mobility. We observe there is no impact of the E2E delay,

Table 1 Comparison among protocols for UWSN with advantages and disadvantages

Routing protocol for UWSN	Advantages	Disadvantages
Vector based forwarding	• Scalable to size of network as state information required • Energy efficient as node along the routing pipe are involved in routing • Handle node mobility • Handle packet loss and node failures • Suitable for dense network	• End-to-end delay is high • PDR is less for stable network • Not suitable for sparse network • Need full localization information as it is localization based protocol
Hop by hop vector based forwarding	• Suitable for sparse network • Handle node mobility to some extent • End-to-end delay is less • PDR is high	• It is not energy efficient for static network
Depth-based routing	• Full dimensional location information is not required • Handle node mobility well • Suitable for multi-sink architecture • Suitable for dense network due to high packet delivery ratio	• It is not as much energy efficient
Hop by hop dynamic addressing based routing	• Complex routing tables are not required • No need of specialized hardware • Full dimensional location information is not required • High packet delivery ratio is achieved in both dense and sparse network • Handle node mobility easily	• Complex architecture

Table 2 Comparison of performance between static and dynamic network

Routing protocols	Parameters	Static network	Mobile network
VBF	Packet delivery ratio	Low	High
	Delay efficiency	Stable	Stable
	Energy efficient	High	Low
HH-VBF	Packet delivery ratio	High	Low
	Delay efficiency	Stable	Stable
	Energy efficient	Low	High
DBR	Packet delivery ratio	Stable	Stable
	Delay efficiency	High	High
	Energy efficient	High	Low
HH-DAB	Packet delivery ratio	High	Low
	Delay efficiency	High	Low
	Energy efficient	High	High

it is more energy-efficient for mobile network rather than stable network as it is suitable for sparse and dense network both.

In DBR, it is observed that the ratio of delivery does not change with node mobility because routing decisions uses depth information of sensor node in DBR. Also, no need of topology or route information is required among neighboring nodes. So, DBR can handle node mobility. Also it is not as much energy efficient for mobile network.

In HH-DAB, there is a minor difference with mobile nodes as compared to static nodes. So, there is no serious effect of node mobility on the rate of data delivery as well as energy consumption, as it does not require any location information and also no complex routing tables maintained. Also it can easily handle vertical node movement efficiently (ignoring horizontal movement of node).

5 Performance Evaluation

In the table, the performance of protocols VBF [3], HH-VBF [4], DBR [5], HH-DAB [6, 7] are shown based on the different parameters. In the table given below, we have given the comparative results among these protocols on the basis of the performance metrics, i.e., parameters in case of mobile network (Table 3).

It is find out that VBR is more suitable for dense network, whereas HH-VBR is suitable for both (dense and sparse) network and PDR is high in sparse network. In DBR, high packet delivery ratios are achieved for dense networks. But in case of HH-DAB, high packet delivery ratio is high in both dense and sparse network. So, packet delivery ratio is high in case of HH-DAB.

Average E2E delay is low in case of multi-sink DBR. But delay is high for VBF and one sink DBR. Also E2E delay in HH-DAB is lower than that of DBR. So, HH-DAB is delay efficient.

VBF and HH-DAB is more energy-efficient than that of HH-VBF and DBR. Because in case of HH-DAB, it is not required to maintain location information of underwater nodes and no need of complex routing tables.

The overall performance of HH-DAB is expected to be high than all other protocols in case of both static network and dynamic network.

Table 3 Overall performance of routing protocols using different parameters

Routing protocols	Packet delivery ratio (for sparse network)	Delay efficient	Energy efficient	Overall performance (expected)
VBF	Low	Low	Average	low
HH-VBF	High	Average	Low	Average
DBR	High	High	Low	Average
HH-DAB	High	Average	Average	High

6 Conclusion and Future Work

In this paper, we have shown the comparison among various routing protocols based on some of the parameters and also explored the impact of static and mobile network on the performance of these protocols. From the comparison, we have found that the HH-DAB shows better results than other protocols in terms of packet delivery ratio (efficient for both sparse and dense network), average E2E delay (due to multi-sink architecture), and more energy-efficient (no requirement of complex routing tables). Also it is identified that HH-DAB can handle node mobility more efficiently than other protocols as it dynamically allocates addresses to the nodes when nodes float. In this paper, we have shown expected results. So, in future, we will analyze these protocols with actual results.

References

1. Ian F. Akyildiz, Dario Pompili, Tommaso Melodia,: Challenges for Efficient Communication in Underwater Acoustic Sensor Networks, In: Special issue on embedded sensor networks and wireless computing, Volume1, Issue2, pp. 3–8, ACM New York (2004).
2. Kanika Agarwal, Nitin Rakesh,: Node mobility issues in underwater wireless sensor networks", In: International Conference on Computer, Communication and Computational Sciences (ICCCCS—2016), Springer, accepted in press (2016).
3. Peng Xie, Jun-Hong Cui, Li Lao: VBF- Vector-Based Forwarding Protocol for Underwater Sensor Networks", proceeding of IFIP Networking, Coimbra, Portugal, pp. 1216–1221, May 2006.
4. Peng Xie, Zhong Zhou, Nicolas Nicolaou, Andrew See, Jun-Hong Cui, and Zhijie Shi, "Efficient Vector-Based Forwarding for Underwater Sensor Networks", EURASIP Journal on Wireless Communications and Networking, 2010.
5. Hai Yan, Zhijie Jerry Shi, and Jun-Hong Cui. "DBR: Depth-Based Routing for Underwater Sensor Networks", Proceeding of Networking, Singapore, vol. 4982, pp. 72–86, 2008.
6. Muhammad Ayaz, Azween Abdullah, "Hop-by-Hop Dynamic Addressing Based (H2-DAB) Routing Protocol for Underwater Wireless Sensor Networks", International Conference on Information and Multimedia Technology, 2009, ICIMT '09, pp. 436–441, IEEE, 2009.
7. Muhammad Ayaz, Azween Addullah, Ibrahima Faye, Yasir Batira, "An Efficient Dynamic Addressing Based Routing Protocol for Underwater Wireless Sensor Networks," Computer Communication., vol. 35, no. 4, pp. 475–486, 2012.
8. Muhammad Ayaz, Azween Abdullah, Low Tang Jung, Temporary Cluster Based Routing for Underwater Wireless Sensor Networks", IEEE International Symposium in Information Technology (ITSim), 2010.
9. Chao Li, Yongjun Xu, Boyu Diao, Qi Wang, and Zhulin An, "DBR-MAC: A Depth-Based Routing Aware MAC Protocol for Data Collection in Underwater Acoustic Sensor Networks", IEEE Sensors Journal, vol. 16, issue. 10, pp. 3904–3913, IEEE (2016).

Arduino Controlled Chessboard

Soikat Chakrabarty, Rupanshu Goyal and Nitin Rakesh

Abstract The major findings of this paper are as follows. After completion, it is evident that an Arduino is fully capable of controlling an automated chessboard using a permanent magnet. The MEGA used for this purpose does so by analyzing the incoming Bluetooth signals. A java code is used to design a virtual chess playing program, and the Bluetooth signals are sent through a laptop. The Arduino program uses the concept of "Least Hindrance Path" in order to move and capture the pieces in the board.

Keywords Arduino MEGA · Automated chessboard · Bluetooth
Least hindrance path

1 Introduction

Arduino is a very interesting area to work on. The portability and versatility of the device has influenced many young researchers and engineering aspirants. Prior to me, many others have designed similar projects on chess programs. One such project is the Arduino plus Raspberry PI controlled chessboard [1]. In this project, the Arduino controls the board which is connected to Raspberry PI monitoring the chess engine Stockfish. Piece positions are recognized using reed switches which signals its move using LEDs on each square.

Now, let us discuss a bit about Raspberry PI and its applications. It is basically a fully operational computer, with a dedicated processor, memory, and a graphics

S. Chakrabarty (✉) · R. Goyal
Amity School of Engineering and Technology, Noida, India
e-mail: soikat.chakrabarty@student.amity.edu

R. Goyal
e-mail: rupanshu.goyal@student.amity.edu

N. Rakesh
Amity University, Noida, Uttar Pradesh, India
e-mail: nitin.rakesh@gmail.com

A.K. Somani et al. (eds.), *Proceedings of First International Conference on Smart System, Innovations and Computing*, Smart Innovation, Systems and Technologies 79, https://doi.org/10.1007/978-981-10-5828-8_5

driver customized for HDMI output. It requires a constant supply of 5 v power, and undergoes shut down like a classic computer [2]. The PI can be utilized for an array of real-time applications like robotics, game development, app development, efficient way of computing and analyzing large volumes of data, weather forecast, etc. The points above only highlight a very small spectrum of applications for Raspberry PI.

The main purpose of this project is to illustrate the necessary details related to the development of this project such as, the methodology adopted, the design criteria, the equipment used and the experiments conducted in order to make this project possible. All these points will be discussed thoroughly in this paper. Arduino's compatibility, complexity and networking capabilities are also an issue here which needs to be addressed, since we are using a wireless peer-to-peer Bluetooth connection in our working model. These question needs to be answered that—How much can one achieve using Arduino? What is the extent to which we can rely on Arduino? Is there any limitation corresponding to its processing and flexibility criteria? Coding for this model has been done on a java platform. Therefore, the quality of the solution, completeness, and space and time complexity of the code will also be discussed in this paper. The purpose of this paper has already been discussed. This paper is divided into various modules/sections. This paper will be divided into five sections. The first section will comprise of the *Proposed Model* in detail. The second section will contain the *Methodology* adopted for the model. *Results and Analysis* of the model will be included in the third section. The fourth section will contain the *Conclusion*. And the final section will contain the list of the sources used in this paper, i.e., *References*.

2 Proposed Model

The main components of the model includes an Arduino MEGA, Ultrasonic Distance Sensors, HC-05 Bluetooth Adapter, Neodymium Magnet, Gear motor and pulleys, Servo Motor. The base idea is that the moves generated by the java code will be transferred to the Arduino via a wireless Bluetooth connection. The Arduino on receiving the data will move the magnet to its desired position, which will then be activated to hold the piece and move it to its desired location on the board.

2.1 Working

Full Algebraic Notation is used to record the moves in a chess game. A chessboard is divided into ranks and files. The rows are called ranks and the columns are files. Rows are numbered from 1–8 and files from a–h (Fig. 1).

The working principle of this model is actually the wireless Bluetooth connection established between the software and Arduino. A Bluetooth connection is developed

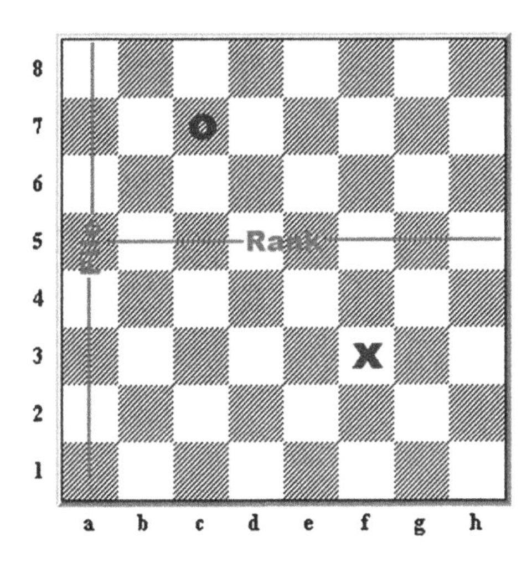

Fig. 1 Ranks and files on a chessboard

to function in a noisy radio frequency environment. It makes the link robust by using a fast acknowledged, frequency hopping spread spectrum (FH-SS). Bluetooth radios mostly hops faster and uses shorter packets (Table 1), compared to systems operating in the similar frequency band. This makes a wireless Bluetooth transmission absolutely robust against quintessential interference phenomenon [3, 4].

The moves generated by the software are transferred to Arduino via Bluetooth. The data transferred through this process is in the form of an auto-generated String, i.e., *(f2–f4—on the first move the pawn on the f2 square moved to the f4 square. fxe5 d6—white pawn on the f file takes the pawn on e5; the black pawn moves to d6)* [5].

These auto-generated Strings are analyzed by Arduino to process the correct position of a particular piece on the board. This is done by calculating the distance between *(i) the origin and the initial position and (ii) the initial position and the final position* of the piece using ultrasonic distance sensors. On accessing this information, the permanent magnet is moved to its desired position with the help of gear motors and pulleys, which then activates to hold the chess piece and move it to

Table 1 Bluetooth technical information

Frequency band	2.4 GHz ISM band
Output power	1 mW/2.5 mW/100 mW
Range	10 –100 m (dependent on Power Class)
Modulation	2-GFSK
Frequency hopping	Up to 1600 hops/s; 79 (23) Hopping Channels; (1 MHz)
No of devices	8 per Piconet ~ 10 Piconets in one environment
Speed	~ Asynchronous: 2.1 Mbps/160 kbps; ~ Synchronous: 1 Mbps/1 Mbps
Speech-codecs	~ LogPCM/CSVD
Module (HC-05)	~ 26.9 mm × 13 mm × 2.2 mm; ~ + 3.3VDC 50 mA
Protocol	Bluetooth Specification v2.0 + EDR

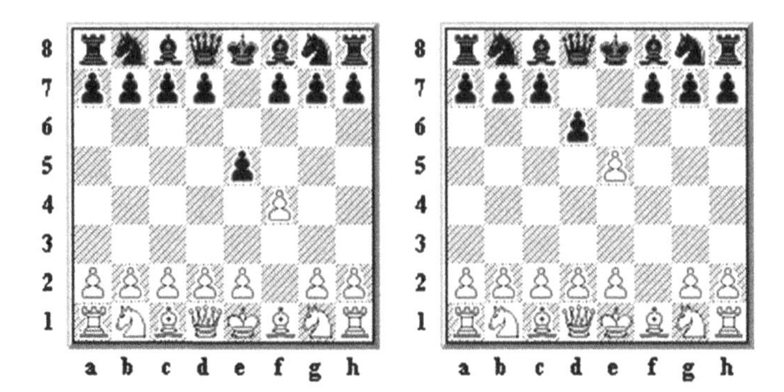

Fig. 2 f2-f4 e7-e5/f4xe5 d7-d6

its desired position. The model consists of three sets of gear motors and pulleys. Two of them are parallel to each other and the third set is perpendicular to the other two forming an "I". The permanent magnet is mounted on the servo motor attached to the third set of gear and pulleys which allows it to move freely, covering the entire chess board. It moves the pieces by traversing through the outlines of the squares on the chessboard. Another crucial point to notice is that, while moving and capturing the pieces it is programmed to follow a path of least hindrance on the board, i.e., *(it avoids taking the piece through crowded places where too many pieces are concentrated at the same time; thus allowing efficient use of memory and fast processing time).* After taking a piece to its desired position, the permanent magnet moves to its initial position, i.e., the origin on the board represented by a1. This concludes the working of the model being discussed (Fig. 2).

2.2 Model Description

The necessary equipments required for this model have already been listed above. In this section, we will discuss the design specifications and the equipments of the model in detail. The basic structural framework of the model comprises of a rectangular wooden framework having a cross section of 55 cm × 70 cm. A 4 mm thick plywood will be installed on top of this framework. The chessboard will be drawn on this ply wood. The reason behind using a rectangular framework for this model is that, the chessboard consists of a matrix of 64 equal sized squares on grids of 8. The extra area on the two sides of ply will be used for positioning the captured pieces from the chessboard. Assembling of gear motor and pulleys has already been explained. With each pair of gear motor and pulley, an ultrasonic sensor will be involved acting as an auxiliary. These ultrasonic sensors will calculate the distances in each step of the movement and feed it to the Arduino. Using the cumulative distance gathered from three ultrasonic sensors, the Arduino will be able to position the electromagnet efficiently.

Arduino Mega 2560 is a microcontroller board established on the ATmega2560. It comprises of 54 I/O pins, 16 analog inputs, 4 hardware serial ports, a 16 MHz crystal oscillator, a USB connection, a power jack, an ICSP header, and a reset button. This piece of equipment easily interacts with its environment in real time. It can be powered using a battery, an AC-to-DC adapter or a USB cable connected to a computer [6]. Designed for transparent wireless connection setup, the HC-05 is basically a *Serial Port Protocol* module. This module is totally qualified for Bluetooth *Enhanced Data Rate* supporting 3Mbps modulation along with 2.4 GHz radio transceiver and baseband. The working principle of this model is a CSR *Bluecore 04-External single chip* Bluetooth system with CMOS technology and *Adaptive Frequency Hopping Feature* [7]. Ultrasonic Sensors are characterized by their reliability and outstanding versatility, because their measuring methods work reliably under almost all conditions. It sends out a high-frequency sound wave and then estimates the time taken for the echo to reflect back. The sensor consists of two openings on the front side; one opening transmits ultrasonic waves, and the other receives them. One of them acts as a miniature speaker while the other as a miniature microphone [8]. Neodymium magnet is the most widely used rare earth element. A permanent magnet with a tetragonal crystalline structure of $Nd_2Fe_{14}B$ is formed from an alloy of neodymium, iron, and boron [9]. This magnet is attached to a servo motor which holds it and positions it accordingly. Gear motor corresponds to a combination of motor plus a reduction gear train. The gear reduction helps in reducing the speed of the motor, significantly increasing the torque. In addition, the gear reducer can be used as a means to reorient the output shaft in a different direction. This is only possible in case of a small transmission ratio N, which can be accomplished only with a single gear pair [10]. Suitable for use in a closed-loop control system, it consists of a motor coupled to a sensor and position feedback. The magnet attached to this motor helps in moving the pieces on the board to their desired position. After every move the motor rotates the magnet by 200° so that the magnet cannot dislocate any other piece on the board [11] (Figs. 3 and 4).

Fig. 3 Arduino MEGA 2560

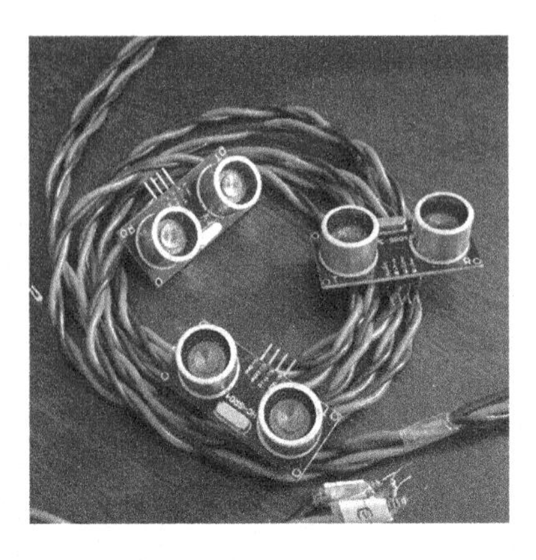

Fig. 4 Ultrasonic sensors

2.3 Functionality and Applicability

Swing was developed as an integral part of Java so as to provide a more sophisticated set of *Graphical User Interface* (GUI) components compared to the earlier *Abstract Window Toolkit* (AWT). Thus, a basic definition of Swing would be a simple *GUI widget toolkit* for Java. A major part of the chess playing software is programmed using Swing. The reason behind this is actually the architectural framework of Swing.

It is platform-independent as it is completely written in Java, highly modular based architecture, lightweight UI reflecting high level of flexibility. These features of Swing help us in building a dynamic framework of the software. It is basically the software which controls the entire model's operations. The Bluetooth connection established between the two transfers the strings generated into the Arduino, which is then analyzed and the model is activated accordingly.

In order to establish the Bluetooth connection between the java application and Arduino; BlueCove [12] is implemented as a back door into the software. It is a java library for Bluetooth having the capacity to interface with the Mac OS X, WIDCOMM, Bluesoleil and Microsoft Bluetooth stack found in Windows XP SP2 or Windows Vista. BlueCove JSR-82 Emulator module [13] is an additional module to simulate Bluetooth stack. BlueCove can be used with J2SE 1.1 or newer.

It has been tested on various JVMs. Some of them are as follows:

- SUN Java 1.1.8 on Windows XP/X86
- SUN Java 2 Platform Standard Edition 6.0 on Windows XP/X86 and Windows Vista.
- WebSphere Everyplace Micro Environment 6.1.1, CLDC 1.1, MIDP 2.0 on Linux/X86
- Java for Mac OS X 10.4 and 10.5, Release 5 on PowerPC and Intel processors

Bluecove provides JSR-82 Java interface for the following Bluetooth profiles:

- SDAP—Service Discovery Application Profile

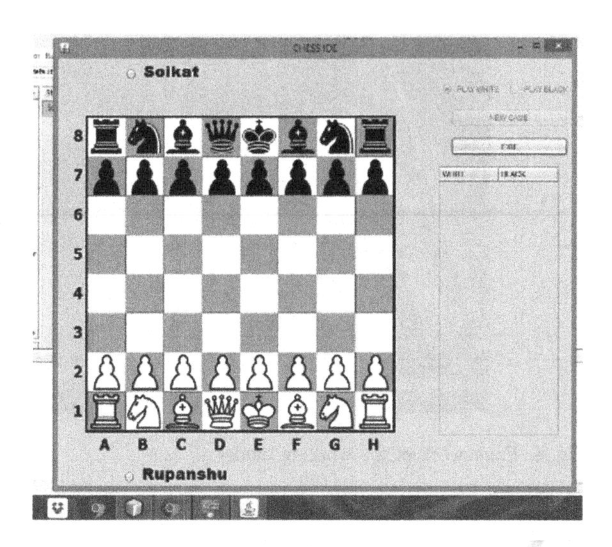

Fig. 5 Chess IDE developed using Java

- RFCOMM—Serial Cable Emulation Protocol
- L2CAP—Logical Link Control and Adaptation Protocol
- OBEX—Generic Object Exchange Profile (GOEP) on top of RFCOMM and TCP

Therefore, it is evident from this discussion that the software is acting as a backbone for the model. If the software does not process accordingly and there is difficulty in establishing a wireless communication, then the model will not work at all. Also, I have already mentioned that we need a strong magnet. This is because the magnet should be able to hold the piece and move it properly through a 6 mm wooden board, since we do not want the magnet to disrupt other pieces on the board. Therefore, the magnet will be attached to a servo motor. The servo motor provided an angular movement of 200° approximately. The servo motor rotates the magnet by 200° after every move, so that the magnet does not disturb the pieces on the board.

Regarding the applicability of this model, this project illuminates the dynamic feature of Arduino. This microcontroller is one of the greatest inventions of the last decade and we are yet to explore its boundaries. Thus, I think that if an Arduino can be used to control an automated chessboard, then its possibilities are endless (Fig. 5).

3 Methodology Adopted

Initially the idea was to create an android app for chess instead of a java application, and to embed a chess engine into the android application so as to analyze the moves on the board. But the apk file of a chess engine was not available. Therefore, the idea shifted from its origin and a java application for the same was programmed

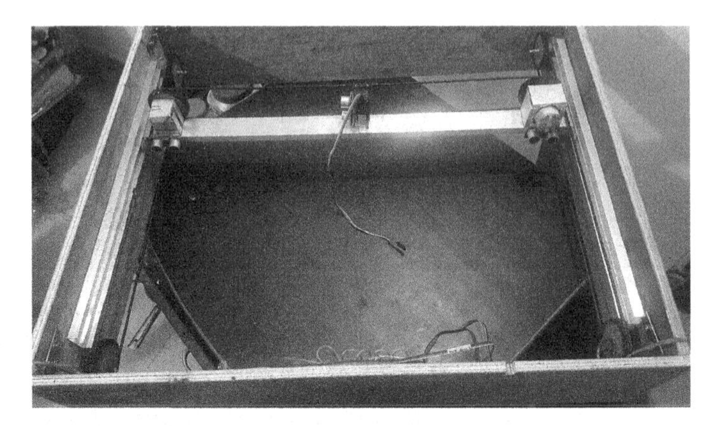

Fig. 6 Framework of the working model

along with a chess engine. The reason behind embedding a chess engine was that we wanted to create a single player automated chess machine.

Stockfish 8 [14] and Komodo 9 [15] are two of the most effective chess engines currently in play. The java software makes use of both these chess engines in order to estimate the moves appropriately. Any chess engines can be embedded into the software by simply downloading it in the same directory of the software after renaming it to "chess-engine". Method *exec ()* of Runtime class is used to start the chess engine in the background for the GUI and *printwriter* command is used to write the commands and the string is retrieved. The chess engines use *UNIVERSAL CHESS INTERFACE (UCI) PROTOCOL* [16] to communicate with the Graphical User Interfaces (GUI's). For example, *ucinewgame* command starts a new game that the engine analyses. The engines use either the whole concatenated string of moves or *FEN Strings* [17]. We use concatenated strings to communicate with the engine.

As mentioned earlier, the structural framework of the model is a rectangular wooden framework. All the necessary equipments are mounted on this wooden framework. It uses three sets of ultrasonic sensors to calculate the correct position of the pieces on the board. Two sensors are parallel to each other while the third is perpendicular to the other two. This orientation helps by estimating the distance between the ranks and files separately. This information is then utilized by the Arduino to position the magnet attached to the servo motor.

Metal pieces are attached below the chess pieces so that the magnet can hold them through the wooden board. A permanent magnet is used for this purpose along with a servo motor so as to position the pieces on the board accordingly (Fig. 6).

3.1 Sample Algorithm

This section explains the working of the chessboard with the help of a simple algorithm

Step 1: Two tables are maintained at each time for both black and white pieces.
Table 1: contains name, type, and position of each piece.
Table 2: contains legal moves of each piece on the board.
Step 2: Initially both tables gets hard-coded positions, i.e., start position of each piece.
Step 3: As the game progresses, Table 1 is updated with the new position of each piece. Table 2 is updated with new legal positions.
Step 4: When a piece is captured, its entry is deleted from both the tables.
Step 5: If a pawn is promoted, its type is changed to the desired promoted piece and Table 2 is updated with new legal moves.
Step 6: Table 2 is updated with respect to all the rules existing in the chess game.

4 Result and Analysis

A simple analysis would suggest that the model is working like it was supposed to. The Bluetooth connection is maintained perfectly, Arduino is able to interpret the incoming string commands and position the permanent magnet, the ultrasonic sensors estimate the final position of the pieces and the permanent magnet is able to move them accordingly. This is just an initial assessment of the work done. The final assessment of the experiments conducted is illustrated in the next paragraph.

A vital task in building the model was of using a magnet with just the right amount of strength. The reason being in the game of chess, the pawn is the smallest piece and the king and queen are the largest piece on the board. Also each piece had metal pieces attached to their base which made them even heavier. Therefore, if the magnet is not strong enough, then it will not be able to hold the pieces through a wooden board. On the other hand, if the strength of magnet exceeds the required property, then it would be difficult to detach the magnet from the piece. This means that the servo motor will not be able to rotate the magnet accordingly. This simplifies the problem only to a limited percentage. It does not eliminate the problem entirely. I have mentioned above that we will be using a wooden board on which a chessboard will be drawn. The magnet will be moving the pieces positioned on this board. This requires the board to provide the least possible opposing resistance. If there is a significant amount of resistance being encountered by the pieces, then it would not be able to move freely on the board. This requires the board to be polished thoroughly so as to eliminate unwanted resistance. Fixing the gear motors and pulleys was a task that required utmost accuracy, as they need to be in perfect alignment with each other. Finally, a major portion of our time was spent in debugging and resolving bugs in our software.

5 Conclusion

A report from Google Trends regarding the word "Arduino" illuminates the rise and popularity of the Arduino microcontroller on the Internet [18]. The use of Arduino is actually an enactment of *Design-Based Learning (DBL)* [19]. This project is still under development phase and a number of modifications can be included in this model to make this interesting. For example, we could develop an android/IOS application for the software instead of a java application. This would resolve portability issues to a significant extent. Further improvements would include modifications like voice modulation, remote access, local server access, size modifications, etc.

References

1. "Wooden Chessboard with piece recognition." A Raspberry pi based chess computer that uses Stockfish and reed switches to recognize piece and signals the move. https://create.arduino.cc/projecthub/Maxchess/wooden-chess-board-with-piece-recognition-872ffb?ref=platform&ref_id=424_trending___&offset=26.
2. Comparison of Raspberry Pi and Arduino Boards. Available: http://www.digitaltrends.com/computing/arduino-vs-raspberry-pi/.
3. "Specification of the Bluetooth System", Core, V1.0 B, Dec. 1999.
4. "Specification of the Bluetooth System", Profiles, V1.0 B, Dec. 1999.
5. Chess notations according to a game played on a chessboard. Available: http://www.chesscorner.com/tutorial/basic/notation/mont.gif.
6. Analysis of Arduino Boards – MEGA 2560. Available: https://www.arduino.cc/en/Main/ArduinoBoardMega2560.
7. Analysis of Bluetooth Module – HC-05. Available: https://www.itead.cc/wiki/Serial_Port_Bluetooth_Module_(Master/Slave)_:_HC-05.
8. C-SR04 Ultrasonic Sensors. Available: http://arduino-info.wikispaces.com/Ultrasonic+Distance+Sensor.
9. Chemical composition of Neodymium Magnets. http://www.ndfeb-info.com/neodymium_magnets_made.aspx.
10. Analysis and Classification of gear motors. http://hades.mech.northwestern.edu/index.php/Gear_Motor http://www.globalspec.com/learnmore/motion_controls/motors/gearmotors.
11. Working principle of Servo motors. https://en.wikipedia.org/wiki/Servomotor.
12. BlueCove java library information. http://bluecove.org/.
13. JSR-82 Analysis. http://bluecove.org/bluecove-emu/.
14. Stockfish chess engine Analysis. https://stockfishchess.org/ and http://support.stockfishchess.org/kb/getting-started/using-the-stockfish-engine.
15. Komodo chess engine analysis. https://komodochess.com/Komodo9-43a.htm.
16. UCI protocols. http://wbec-ridderkerk.nl/html/UCIProtocol.html.

17. Forsyth-Edwards Notation. https://chessprogramming.wikispaces.com/Forsyth-Edwards+Notation.
18. (2015). Google Trends - Web Search interest - Worldwide, 2004 - present. Available: https://www.google.com/trends/explore.
19. S. Chandrasekaran, G. Littlefair, M. Joordens, and A. Stojcevski, "A Comparative Study of Staff Perspectives on Design Based Learning in Engineering Education," 2014.

Development of Effective Technique for Integration of Hybrid Energy System to Microgrid

Sheeraz Kirmani, Majid Jamil and Iram Akhtar

Abstract Due to lack of energy resources, renewable energy sources have increased fabulous consideration and settled rapidly in recent years. It is a very decisive issue to integrate hybrid energy system to the microgrid. In this paper, a control model for integration of hybrid energy sources to the microgrid and the effectiveness of the control algorithm for DC/DC converter under variation of different load demands are presented. A nearly constant dc voltage at the output of hybrid energy system is preferred for high efficiency So, there is a necessity for developing control techniques for a grid integration hybrid system including a method for output voltage control that stabilizes the voltage and dc-link capacitance. The simulation results of MATLAB/SIMULINK model report that the proposed control algorithm has a good performance. Therefore an idea of an effective technique for integration of solar/wind system to microgrid has been given.

Keywords Microgrid · Renewable energy sources · DC-DC converters
PI controller · Signal generator

1 Introduction

Nowadays, PV technology and wind technology developed very rapidly, because it has many advantages. Basically, MI cuk converter is used for the integration of renewable energy sources to the microgrid. Speed control method for a permanent magnet synchronous generator is chosen to extract the maximum wind power below

S. Kirmani (✉) · M. Jamil · I. Akhtar
Faculty of Engineering & Technology, Department of Electrical Engineering, Jamia Millia Islamia, New Delhi 110025, India
e-mail: sheerazkirmani@gmail.com

M. Jamil
e-mail: majidjamil@hotmail.com

I. Akhtar
e-mail: iram1208@gmail.com

A.K. Somani et al. (eds.), *Proceedings of First International Conference on Smart System, Innovations and Computing*, Smart Innovation, Systems and Technologies 79, https://doi.org/10.1007/978-981-10-5828-8_6

and above the rated wind speed [1]. An intelligent mining of finest power due to the fuzzy logic controller and the dispatch of maximum power from a grid connected system consists of permanent magnet synchronous generator, which is based on wind turbine system and a little concentration photovoltaic generation system. For the photovoltaic system, maximum power point tracking control is developed by using a fuzzy logic controller and taking the effect of solar irradiance changes. The fuzzy logic controller is also used for power extracting from wind turbine without affecting output voltage. This can decrease high-frequency oscillations due to wind energy system. A 1:1 delta Y-grounded transformers are taken at inverter output to remove the triple harmonic [2]. The boost converter is used to check the result of the wind, the main function of this method to take the maximum Power Point Tracking. P and O are one of the famous methods for maximum power point tracking particularly for both the wind and solar system. Furthermore, this converter raises the voltage generated by the photovoltaic and wind generator to connect this system to inverter circuit [3]. To achieve superior harmonic benefit performance using distributed generation, an adaptive HCM controlled method is very effective. An adaptive HCM can reduce the harmonic from the output voltage, so filter requirement can be reduced. The numbers of filter reduction give the cost-effective solution for this, so the adaptive method is a viable solution. Additionally, a phase-locked loop is not essential as the microgrid-based frequency, automatically recognized by the control loop the voltage system and current control method, improved harmonic compensation scheme for power electronics based on renewable energy sources. The voltage system and current control method improve renewable energy sources' steady-state power management performance, when there are different frequency deviations. By using hybrid voltage system and current control method, the renewable energy sources also understand superior system harmonic recompense presentation under variable microgrid frequency circumstances [4]. A dual-based voltage source inverter scheme is also used to improve the power quality of the entire system, as well as the reliability of the system. The scheme consists of two inverters, which allows the microgrid to interchange the power generated by the distributed energy resources and to recompense any type of load. ISCT scheme is used to change the mode of the grid, i.e., grid sharing mode or grid connected mode. ISCT increases the reliability of the system, lower bandwidth requirement for any inverter, lower cost solution because of reduction of filter size [5].

To decrease the intermittencies of the distribution grid and to improve the power quality of the grid, ultracapacitor power conditioners are used as it has active power capability, which is very useful in tackling the grid intermittencies [6]. An adaptive hybrid voltage scheme and current control method are extremely useful to larger harmonic compensation using renewable energy source. The adaptive hybrid voltage scheme and current control method can reduce the numbers of filters in the renewable energy sources based microgrid system. Furthermore, here is no need of phase-locked loops system for the microgrid frequency deviance. It can be spontaneously identified by the power control system.

Therefore, the hybrid voltage scheme and current control method can provide chances to reduce Renewable energy sources control difficulty. For steady-state condition, this technique can reduce the alteration of input current difference; the moment continuous power change occurs the voltage protection circuit can easily be activated. Therefore, this technique is used to stop the false alarm and capacitance range. It can affect the harmonic compensation enactment. Maintaining power quality is an additional important portion which is an imperative concern while we work on the integration of renewable energy sources to the microgrid. The propagation of power electronics devices and electrical loads with nonlinear currents has tarnished the power quality of the system, so different power quality improvement techniques have been used for this purpose. The wind power plants and solar energy system give the huge part to the production of renewable energy. The contribution of wind power plants and solar system faces firm challenges like voltage regulation and integration of these sources to microgrids. In an electrical power system, microgrid is the group of different electric loads and power generation sources like wind, solar, diesel, etc. A controllable operation is needed for better performance and for improving power quality of the system.

The intermittent environment of renewable energy resources joined with the unpredictable changes in the wind speed, or solar irradiance is the matter of concern nowadays. The sharing of the battery-based supercapacitor storage system to grip sudden change or average variations in power surges provides the fast DC-link voltage regulation [7]. The voltage direct current link connected with a damping-based controller based on the adaptive network which is based on fuzzy implication system by using this damping development of an addition of the wind, solar, and naval current systems is achieved [8]. It is observed that variations of frequency are commonly seen in a microgrid [9, 10], because of main grid voltage disturbances or effects of sudden load changes.

In this paper, hybrid microgrid energy system is described in Sect. 2. In Sect. 3, the control strategy for DC/DC converter with power flow control scheme is presented. Results and discussion are described in Sect. 4. Finally, concluding statements are presented in Sect. 5.

2 Modeling of a Hybrid Energy System to Microgrid

In a microgrid, power developed from different renewable energy sources are added to grid AC or dc and loads by using different power electronic devices as shown in Fig. 1. A grid side inverter is needed for swapping power from the microgrid to the main grid and the different linked load. The dc-link side voltage of the auxiliary inverter should be maintained as constant for proper operation.

The energy extract from wind turbine can be calculated as

$$W_{wind} = V_a \frac{1}{2} \rho \left(V_1^2 - V_3^2\right) \quad (1)$$

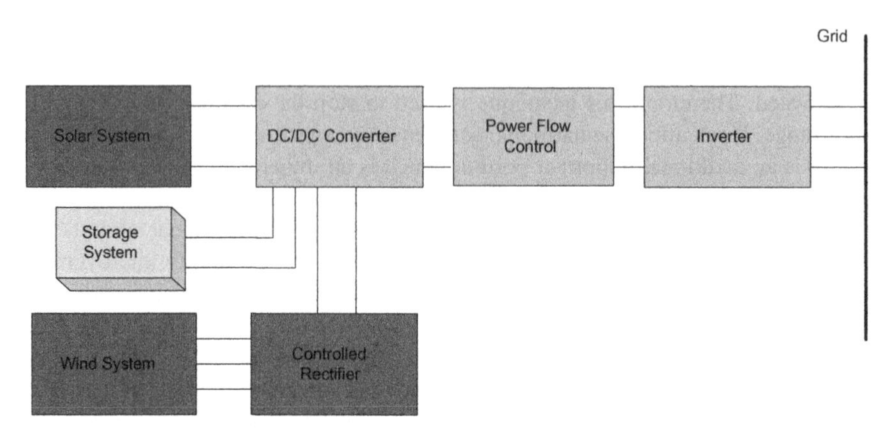

Fig. 1 Microgrid hybrid energy system

Wind power available

$$P_a = \frac{d(V_a \frac{1}{2}\rho(V_1^2 - V_3^2)}{dt} \tag{2}$$

$$P_a = \frac{1}{2}\rho A_r(V_1^2 - V_3^2)V_2 \tag{3}$$

According to Betz, maximum wind power output when $V_2 = 2/3V_1$ and $V_3 = 1/3V_1$

$$P_{max} = \frac{16}{54}A_r\rho V_1^3 \tag{4}$$

$$A_r = A$$

Available wind power is calculated as follows:

$$\mathrm{P_a} = \frac{1}{2}\rho \mathrm{AV}^3 \tag{5}$$

$$\mathrm{A} = \frac{\pi}{4}\mathrm{D}^2 \tag{6}$$

$$\mathrm{P_a} = \frac{1}{8}\rho\pi \mathrm{D}^2\mathrm{V}^3 \tag{7}$$

where,

V_1: Upstream wind speed [m/s]
V_2: Wind speed at turbine [m/s]
V_3: Downstream turbine speed [m/s]

V: Wind velocity [m/s]
A_r: Rotor shaft area [m^2]
P_a: Wind power [W]
A: Swept area [m^2]
ρ: Air density [kg/m^3]
D: Rotor diameter []

This shows that extreme power from wind energy system is the square of the diameter of the capture area. $(D_1 < D_2 < D_3 < D_4)$. By the use of power coefficient of wind system, the output energy is attained, if swept area, air density and, wind velocity are constant. Wind system gives 8 kW power when MPPT control algorithm is used with wind system.

PV system is 55 kW connected to microgrid network via DC-DC converters. The I-V relationship of the model photovoltaic,

$$I = I_{pv} - I_0 \left[exp\left(\frac{V + IR_s}{AkT_w} \right) - 1 \right] - 1 \frac{V + IR_s}{R_{sh}} \tag{8}$$

$$I_{pv} = G\left[I_{sc} + k\left(T_w - T_{ref}\right)\right] \tag{9}$$

$$I_0 = I_s \left(\frac{T_w}{T_{ref}} \right)^3 \exp\left\{ \frac{qEG\left(\frac{1}{T_{ref}} - \frac{1}{T_w}\right)}{kA} \right\} \tag{10}$$

I_{pv} is the photo current, and I_0 is the saturation current R_s is the series resistance, A is the diode identity factor, T_w is the working temperature, T_{ref} is the reference temperature, G is the solar insulation, I_s is the reverse saturation current, A hybrid microgrid model using dc/dc converter is developed with MATLAB Simulink/Sim power systems. This model has provided the good performance to integrate renewable energy sources to microgrid. This is simple, trustworthy, cost-effective, stable system for hybrid microgrid.

3 Control Strategy of Microgrid Connected Hybrid Energy System

For controlling the output voltage of dc/dc converter, control algorithm is developed. The two voltage signals V_{ref} and V_{dc} are related. V_{ref} is a reference voltage which can be calculated by the input current calculation and output power calculation. V_{dc} is also the average feedback voltage of the dc/dc converter, and this voltage is sampled by using different modules for the execution of the control algorithm. Figure 2 shows the control strategy for dc/dc converter.

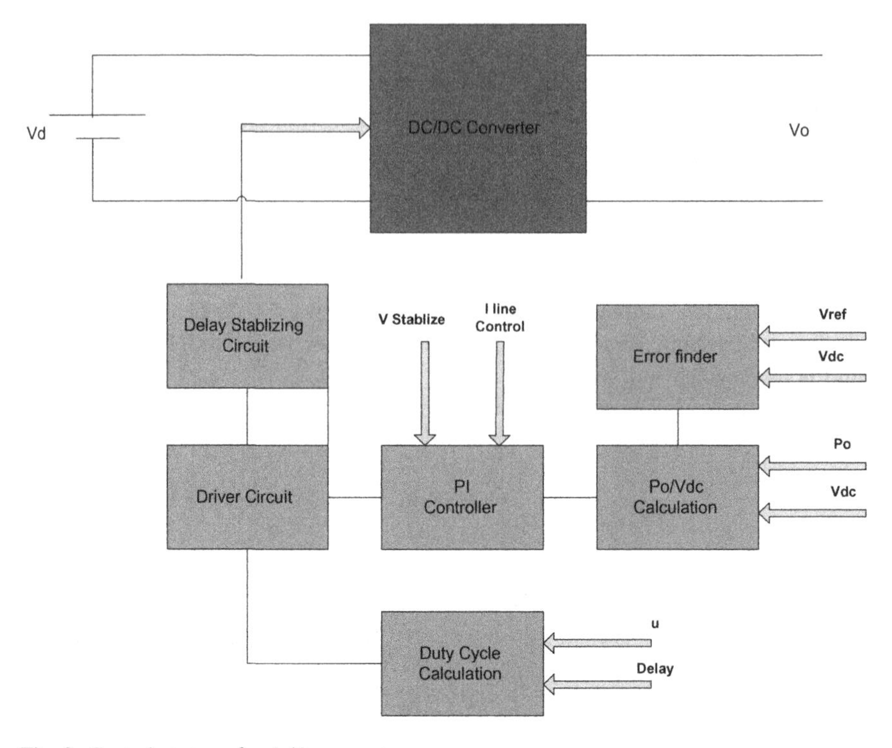

Fig. 2 Control strategy for dc/dc converter

An error signal which is proportional to $V_{ref} - V_{dc}$ is used with PI controller for integrating this and transfer this error to driver based circuit, where PWM techniques are used to generate the pulse signals for switches of a dual boost converter. The pulse generator circuit can easily control the switch; when the error is different from zero driver circuit it sends the pulses to switches, so output voltage get changed, and error is returned to again zero. This driver circuit is very useful to control the output of dc/dc converter and to execute the control algorithm. Power flow can be controlled by control algorithm when there is surplus power; then this power is used to charge the battery and when power is needed, the battery provides the extra power. Therefore, this control technique is very effective for integration of renewable energy sources to the microgrid.

4 Simulation Studies

The simulation model of the hybrid energy system with control algorithm for dc/dc converter is developed in MATLAB Simulink/Sim power systems.

Wind system gives 6 kW and the solar system gives 54 kW output to feed the microgrid.

4.1 Control Performance of the PV System

Figure 3 shows the PV characteristics of PV system. This characteristic shows that maximum power is 55 W, which is obtained by PV system at the voltage of 18 V. These values can be modified according to output power requirement. It can be controlled easily by the controller to get maximum output. Maximum power tracking is done by the planned controller so that maximum power can be extracted most of the time.

4.2 Control Performance of the Wind Energy System

Permanent Magnet Synchronous Generator is used in wind system, because it has many advantages in comparison with other machines. A proper conversation between renewable energy sources and control system handling communities is supposed to be done. A discussion between these two communities can improve the system performance. An example effort is the combined panels at the IEEE Power and Energy Society General Meeting.

There is an all time need for controlling the output voltage of different renewable energy sources. This kind of analysis is more useful in the control area, but it helps the different loads to operate safely. There is a need for high-voltage microgrid system, because there are so many customers such as residential, industrial,

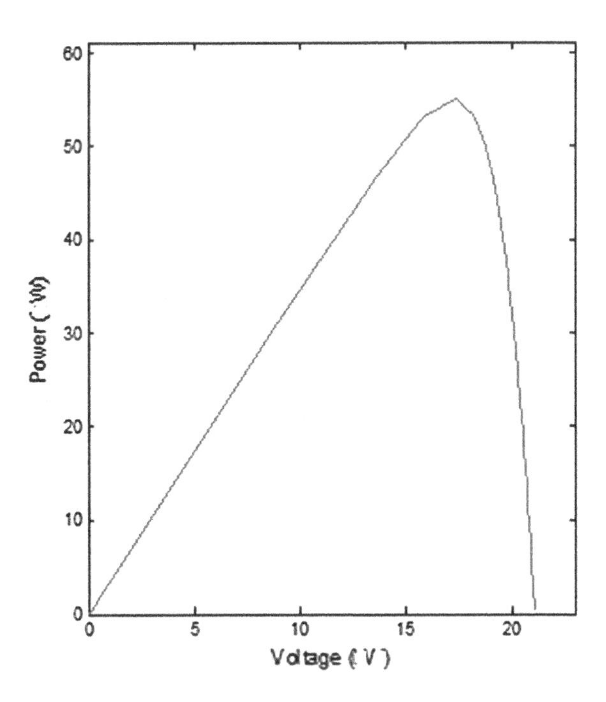

Fig. 3 P-V characteristics of PV module with MPP

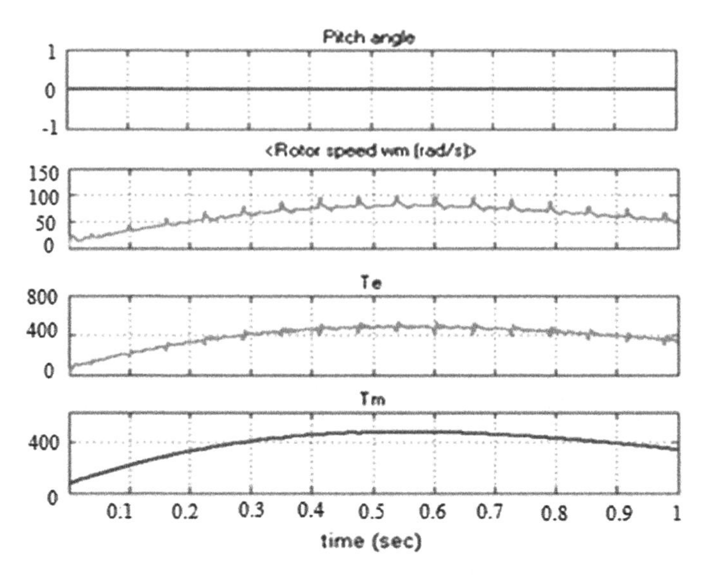

Fig. 4 Wind turbine control performance. Pitch angle, turbine rotor speed, electromagnetic torque, and shaft torque

commercial, etc., so high voltage is needed for different costumers. So control system designs the separate controller for this purpose and power electronics controllers can enhance the voltage profile.

Figure 4 shows the wind turbine control performance, i.e., generator speed (per unit), Turbine rotor speed, Pitch Angle, Electromagnetic Torque, and Shaft Torque. All these parameters are efficiently controlled to get the maximum output and lower harmonic distortion. Rotor speed is 86.12 rad/s to get the distortionless output. Figure 5 shows the wind energy conversion performance, where different parameters have been shown which can be controlled to get the smooth and efficient output.

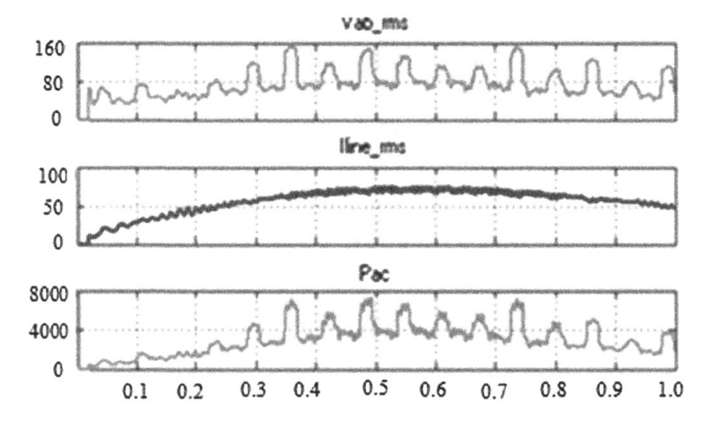

Fig. 5 Wind energy system performance

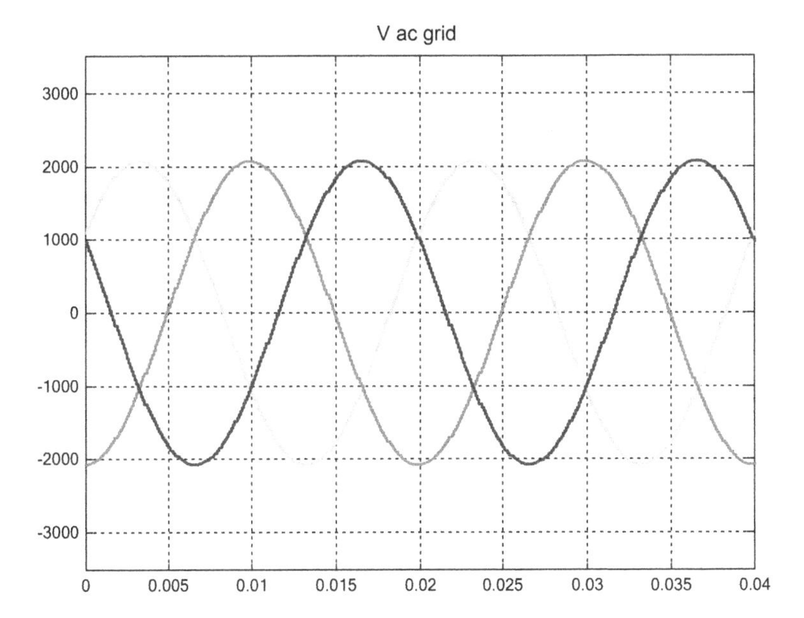

Fig. 6 Inverter output voltage

4.3 Inverter Output Voltage

Figure 6 shows the inverter output with less distortion. The inverter is connected to the output of dc/dc converter. The output of inverter has low harmonic distortion, so power quality is also improved by using this control strategy.

4.4 Output Voltage of Dc/Dc Converter with Control Algorithm

The distorted voltages and currents due to power electronics devices have harmonics in the system, and this may disturb the other equipments which are connected to the system. PI controller is used with this system to control the output of dc/dc converter. The value of $K_i = 2.5$ and $K_p = 40$. The PI controller can easily control the output voltage ranging from 77 to 540 V as shown in Figs. 7 and 8. When the reference voltage changes outside this limit, output voltage becomes irrepressible. Table 1 shows the output voltage of dc/dc converter at different reference voltages.

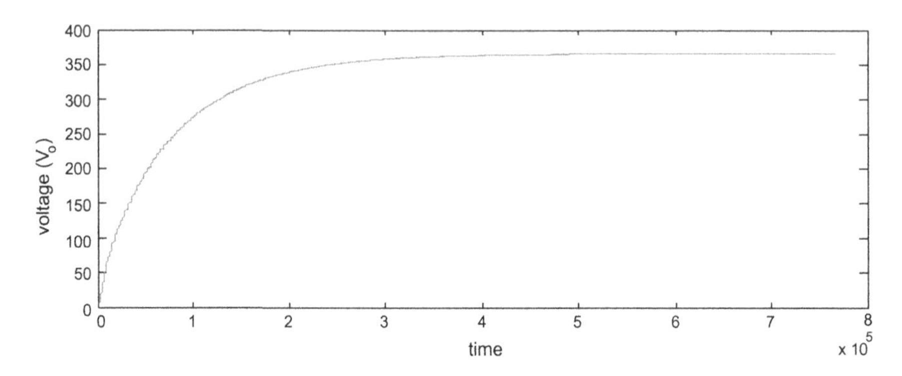

Fig. 7 Output voltage of dc/dc converter

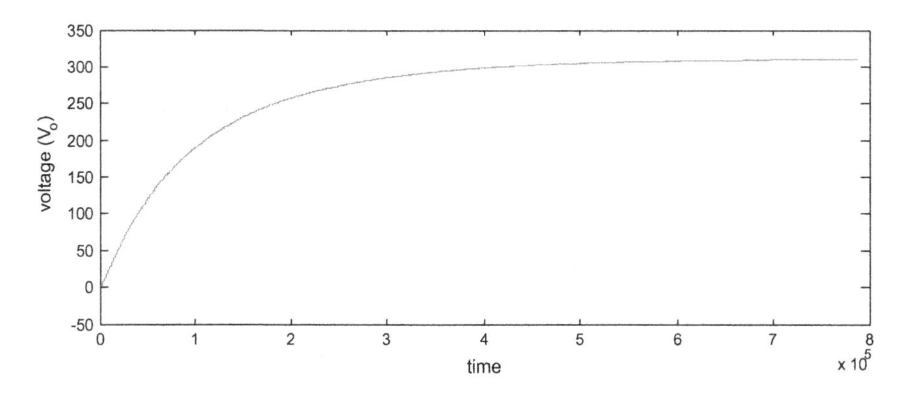

Fig. 8 Output voltage at reference voltage 310 V

Table 1 Output voltage at different reference voltages

Reference voltage (V)	Output voltage (V)
110	110
230	230.1
310	310.4
420	420.1
500	500.3

5 Conclusion

Due to lack of fossil fuels for the upcoming year, renewable energy sources are developed very rapidly. In this paper, MATLAB/Simulink model is developed for integration of renewable energy sources to the microgrid. A control algorithm is used to regulate the output voltage of cuk converter, so whenever there is a change in wind speed or solar irradiance, the output voltage remains constant; hence system stability improves. This is cost-effective solution, because PI controller controls the duty cycle of dc/dc converter switches. The excess power can be controlled effectively when there is surplus power in the system; then this power is used to charge the battery. Therefore, this control technique is very effective for integration of renewable energy sources to the microgrid. Therefore the simulation results of MATLAB/SIMULINK model show the effectiveness of this method.

References

1. S. Bae, S. Member, and A. Kwasinski, "Dynamic Modeling and Operation Strategy for a Microgrid With Wind and Photovoltaic Resources," vol. 3, no. 4, (2012), 1867–1876.
2. C. Bhattacharjee and B. K. Roy, "Advanced fuzzy power extraction control of wind energy conversion system for power quality improvement in a grid tied hybrid generation system," IET Gener. Transm. Distrib., vol. 10, no. 5, (2016), 1179–1189.
3. A. Costa De Souza, F. Cardoso Melo, T. Lima Oliveira, and C. Eduardo Tavares, "Performance Analysis of the Computational Implementation of a Simplified PV Model and MPPT Algorithm," IEEE Lat. Am. Trans., vol. 14, no. 2, (2016), 792–798.
4. J. He, Y. W. Li, and F. Blaabjerg, "Flexible microgrid power quality enhancement using adaptive hybrid voltage and current controller," IEEE Trans. Ind. Electron., vol. 61, no. 6, (2014), 2784–2794, 2014.
5. M. V. Manoj Kumar, M. K. Mishra, and C. Kumar, "A Grid-Connected Dual Voltage Source Inverter with Power Quality Improvement Features," IEEE Trans. Sustain. Energy, vol. 6, no. 2, (2015), 482–490.
6. D. Somayajula and M. L. Crow, "An ultracapacitor integrated power conditioner for intermittency smoothing and improving power quality of distribution grid," IEEE Trans. Sustain. Energy, vol. 5, no. 4, (2014), 1145–1155, 2014.
7. N. R. Tummuru, M. K. Mishra, and S. Srinivas, "Dynamic Energy Management of Renewable Grid Integrated Hybrid Energy Storage System," IEEE Trans. Ind. Electron., vol. 62, no. 12, (2015), 7728–7737.
8. L. Wang and M. S. N. Thi, "Stability enhancement of large-scale integration of wind, solar, and marine-current power generation fed to an sg-based power system through an lcc-hvdc link," IEEE Trans. Sustain. Energy, vol. 5, no. 1, (2014), 160–170.
9. A. V. Timbus, M. Ciobotaru, R. Teodorescu, and F. Blaabjerg, "Adaptive resonant controller for grid-connected converters in distributed power generation systems," in Proc. IEEE APEC Expo., (2006), 6.
10. J. Dong, T. Xia, Y. Zhang, T. Weekes, J. S. Thorp, and Y. Liu, "Monitoring power system disturbances at the distribution level," in Proc. IEEE Power Energy Soc. Gen.Meet.–Del. Elect. Energy 21st Century, (2008), 1–8.

Purple Fringing Aberration Detection Based on Content-Adaptable Thresholds

Kannan Karthik and Parveen Malik

Abstract Purple fringe aberration (PFA) patterns stem from specific defects in certain camera sensor grids, leading to the fraying of edges near high-contrast regions. Much of the literature deploys predefined, absolute, experimentally determined static thresholds for detecting purple fringes. Given the potential diversity in the spectral signature of the local light source, these fringes may not have a static distribution. It is therefore important to make the detection procedure content-adaptable and in tune with the environmental settings. In this paper, we propose a PFA detection procedure in the Y-Cb-Cr CHROMATIC space, by first using a global relativistic Y-channel-gradient threshold for detecting the high-contrast regions and then use the fact that PURPLE and GREEN are antipodes of each other to segregate PURPLE fringes reliably. Comparisons with the state-of-the-art detection approaches are presented. The advantage with the proposed approach rests with the fact that the threshold is content-adaptable and non-static and can therefore be used to pick up diverse fringe patterns (not necessarily confined to the seat of PURPLE).

Keywords Purple fringing aberration · Content-adaptable threshold
Purple and green antipodes

1 Introduction

Defects in image acquisition induced either due to camera sensor noise or lens aberrations, prove extremely useful towards performing a forensic analysis of captured images, for the images bear the noise signature of the camera device.

K. Karthik (✉) · P. Malik
Department of Electronics and Electrical Engineering, Indian Institute of Technology Guwahati, Guwahati 781039, Assam, India
e-mail: k.karthik@iitg.ernet.in

P. Malik
e-mail: parveen@iitg.ernet.in

A.K. Somani et al. (eds.), *Proceedings of First International Conference on Smart System, Innovations and Computing*, Smart Innovation, Systems and Technologies 79, https://doi.org/10.1007/978-981-10-5828-8_7

One such sensor defect stems from a process called sensor blooming in cameras [1, 2], wherein certain cameras leave behind traces of PURPLE fringes near high-contrast edges, at the time of image acquisition. This defect is called a PURPLE fringing aberration (PFA). Attempts have been made to derive secondary statistics from these fringes and use them as a forensic tool to detect splicing in images [3]. However, as a precursor to any forensic process, the primary goal is to ensure stability of the PFA detection procedure and guarantee some form of reliability of the detected PURPLE fringes. The reason why PURPLE fringes lend themselves as a forensic tool is because the spectral diversity of the fringes, witnessed in the CHROMATIC space, is a function of the local illumination profile and the sensor grid residing in the camera. This binds the environmental setting with the camera device, while remaining insensitive to the content in the scene (including the arrangement of objects relative to the camera and the light source). This form of quasi-binding of the fringes without being overly dependent on the micro-details opens up PFA as a viable forensic tool. While the actual application of PFA to forensic analysis is beyond the scope of this paper, we focus on the RELIABILITY of the PFA detection procedure.

The PFA detection procedure entails two fundamental steps

- Short-listing pixels in and around high-contrast regions. In some papers, these regions are identified as near saturation regions [4]; while in others the gradient profile is used in conjunction with the saturation constraint to isolate these regions [5, 6]. Since the sensor blooming phenomenon involves a charge spill over, it is only natural to search for these patterns in areas where the intensity differentials are very high.
- Once these high-contrast neighborhoods have been identified, coloration constraints [5, 6] are applied and then the PURPLE fringe pixels are detected.

The coloration constraint in the case of Kang [4] and Chung et al. [7] happened to be a comparison across the R, G, and B channels associated with a single pixel. Since the seat of PURPLE was not clearly defined at the time of fringe detection, the definition of PURPLE was fuzzy. This was likely to lead to several false positives. These coloration constraints could be deployed in any of the color spaces, not necessarily confined to R-G-B. One of the issues with the RGB space was the drift associated with intensity variations about the high-contrast zones. This is expected to induce a shift in the fringe coloration patterns. To normalize this effect and to further mitigate the effect of light source compositional variations, Kim and Park [8] deployed the coloration constraint in the CIExy space. However this luminance normalization does not work fully. Complete independence from the luminance space was achieved in Ju and Park [6], by transforming RGB to Y-Cb-Cr, wherein the chrominance channels carried exclusive color-bias information, opening up the idea of specifying a zone in the chrominance space, where the PURPLE concentration is maximum. While Ju and Park [6] deployed angular constraints to correct fringing aberrations, they did not use this for detecting fringe pixels. Plain near saturation along with gradient constraints were used to detect fringe pixels.

In almost all the existing PFA detection approaches in the literature, the intermediate thresholds used to isolate the PFA region are absolute [4–7]. While these authors must have indulged in a considerable amount of experimentation to sediment a set of standardized thresholds, given the diversity of camera devices and its defects, it is unlikely for these systems to cover all forms of PURPLE fringing.

The rest of the paper is organized as follows: Given the observation from the literature that almost all papers deploy ABSOLUTE, experimentally determined thresholds for detecting PURPLE fringes, we devise a novel method which uses a relativistic and content-adaptable threshold for detecting PURPLE fringe patterns. The CORE of the proposed PFA detection algorithm is discussed in Sect. 2. Finally experimental results and comparisons with the state-of-the-art PFA detection procedures are presented in Sect. 3.

2 Proposed PFA Detection Algorithm

While the exact mechanism and model for the generation of a PURPLE fringe remains unknown, it is clear from numerous observations and experiments that several low-mid cameras produce these fringes around high-contrast areas. The spectral base for all these fringes happens to be PURPLE. We first make the following critical observation:

The SEAT of the PURPLE COLOR is a COMPOSITION OF ONLY RED and BLUE. There is no GREEN in PURPLE.

This crucial observation eventually leads to another interesting observation in the CHROMATIC space, wherein the SEAT of GREEN color is found to be an ANTIPODE to the SEAT of PURPLE. We exploit this antipodal property towards STABLE detection of PURPLE FRINGES without involving absolute thresholds. The first step towards PFA detection involves separation of the LUMINANCE component from the CHROMATIC space. This is done by transforming the RGB palette to YCbCr. The normalized Y_n, C_{bn} and C_{rn} values are given by Eq. (1) [9],

$$\begin{pmatrix} Y_n \\ C_{bn} \\ C_{rn} \end{pmatrix} = \begin{pmatrix} 0.299 & 0.587 & 0.114 \\ -0.169 & -0.331 & 0.5 \\ 0.5 & -0.419 & -0.0813 \end{pmatrix} \begin{pmatrix} R_n \\ G_n \\ B_n \end{pmatrix}, \tag{1}$$

where $R_n, G_n, B_n \in [0, 1]$; $Y_n \in [0, 1]$, $C_{bn} \in [-0.5, 0.5]$ and $C_{rn} \in [-0.5, 0.5]$. Once the LUMINANCE component Y_n is segregated, it is used to detect the HIGH-CONTRAST NEAR saturation regions with the help of the GRADIENT PROFILE. Gaussian weighted X and Y gradients are computed at every pixel in Y_n. If the GRAY-SCALE image in question is represented by the function, $Y_n(x, y)$, the results of the convolution with respective horizontal and vertical GAUSSIAN WEIGHTED gradient kernels are given by,

$$Y_h(x,y) = Y_n(x,y)*G_x(x,y)$$

$$Y_v(x,y) = Y_n(x,y)*G_v(x,y),$$

where, the * operator indicates a two-dimensional convolution of the image with a KERNEL. The gradient magnitude evaluated at a point (x_i, y_i) is given by the equation

$$M_Y(x_i,y_i) = \sqrt{[Y_h(x_i,y_i)]^2 + [Y_v(x_i,y_i)]^2} \tag{2}$$

The MEAN gradient magnitude over all the points $(x_i, y_i)(\mu_{grad})$ and its standard deviation (σ_{grad}) are computed. Let $\alpha \in [0,3]$ be a parameter for setting the threshold for picking the largest of gradients. The global gradient threshold magnitude is set as,

$$T_{grad} = \mu_{grad} + \alpha\sigma \tag{3}$$

A reasonable choice of α is $\alpha \geq 0.5$ to pick the largest among the gradient magnitudes which correspond to the high-contrast regions. The GAUSSIAN weighting process ensures that striations and textural patterns are left out of the gradient selection, by computing a smoothened discrete derivative. Every pixel at location (x_i, y_i) which satisfies,

$$M_Y(x_i,y_i) > T_{grad} \tag{4}$$

is selected as a high-contrast pixel. Let the locations of such pixels be represented by $(x_{h(i)}, y_{h(i)})$; $i = 1, 2, \ldots N_h$, where N_h is the total number of high-contrast pixels detected. Around each of these high-contrast pixels construct a $w \times w$ window with $w = 5$, just large enough to pick up the entire girth of the fringe. For a window positioned at location $(x_{h(i)}, y_{h(i)})$, all the pixels within that window are scanned. Let these pixels be represented by coordinates $(x_{h(i)} + p, y_{h(i)} + q)$ with $p, q \in 0, \pm1, \pm2, \cdots, \pm(w-1)/2 \cdot$ The chrominance values are computed at all these w^2 locations as,

$$C_b(p,q) = C_{bn}(x_{h(i)} + p, y_{h(i)} + q)$$

$$C_r(p,q) = C_{rn}(x_{h(i)} + p, y_{h(i)} + q) \tag{5}$$

Form a sequence of w^2 vectors comprising of these CHROMINANCE components.

$$\bar{X}_{p,q} = \begin{bmatrix} C_b(p,q) \\ C_r(p,q) \end{bmatrix} \tag{6}$$

Now we define the following ANTIPODAL COLOR SEATS:

$$\bar{X}_{purple} = [0.331, 0.42]^T$$

$$\bar{X}_{green} = [-0.331, -0.42]^T \tag{7}$$

For each vector $X_{p,q}$, compute the ratio of NORMS,

$$\rho_{p,q} = \frac{||\bar{X}_{p,q} - \bar{X}_{purple}||_2}{||\bar{X}_{p,q} - \bar{X}_{green}||_2} \tag{8}$$

This process is illustrated in Fig. 1. The decision region for the constraint

$$\rho_{p,q} < T_{PFA} \tag{9}$$

With $T_{PFA} < 1$ is the inside of the circle shown in Fig. 1. Since this threshold is also relativistic (just as the gradient threshold T_{grad}), the fringe detection process is robust to compression and scaling operations.

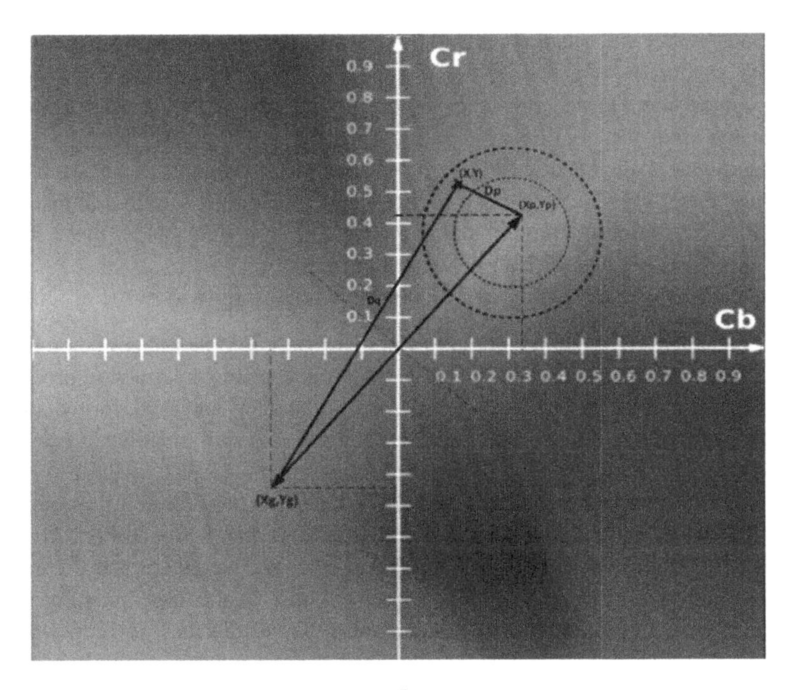

Fig. 1 The purple seat $\bar{X}_{purple}$ and Green point $\bar{X}_{green}$ in the Cb-Cr space are set as $(0.331, 0.42)$ and $(-0.331, -0.42)$ respectively at $Y_n = 1$ plane. For a given point $(\bar{X}_{p,q})$ inside a HIGH-CONTRAST WINDOW of size $w \times w$, the norms $\bar{X}_{\mathrm{p,q}} - \bar{X}_{\mathrm{purple}}$ and $\bar{X}_{\mathrm{p,q}} - \bar{X}_{\mathrm{green}}$ are computed and compared. The decision region for selecting a FRINGE POINT is the circle shown in the above figure

3 Experimental Results

The proposed PFA detection algorithm based on RELATIVISTIC THRESHOLDS was compared with several existing approaches through many PFA-effected images. Since there was no formal database of PFA-images to date, most of these test images were obtained either by GOOGLING for PURPLE FRINGING ABERRATIONS and searching for portals which hosted such ABERRATED IMAGES or by manually taking snapshots using low-end cameras. The images which have been displayed in Fig. 2 have been obtained from EXTERNAL PORTALS. Broadly the test images belonged to the following categories:

- Natural images carrying leaves and branches with plenty of BACK-LIGHT (correspond to sub-figures (a, o, v) in Fig. 2). It has been observed that when there is a strong back-light because of the sun or another artificial source, contrast changes at the periphery of the leaves and branches are extreme.
- Natural images with strong shadows (corresponding to sub-figure (aq) in Fig. 2). Unlike images with back-light, here the light source is either on the side of the object and front of the camera or behind the camera with the object being in front. Either way shadows are cast on the object or on the surfaces by its side. These shadow-regions are regions of high-contrast and near saturation towards the lighter side. Hence, fringes are produced towards the darker side because of a charge spill over.
- Images with artificial back-lighting (correspond to sub-figures (ac, aj) in Fig. 2). In one case, Fig. 2(ac), a false-ceiling lamp introduces a high-contrast zone leading to a string of fringe patterns. In another case (Fig. 2(aj)), these fringes are produced because of back-light at the panels.
- Artificially spliced images carrying objects with strong shadows (Fig. 2(h)).

The test images in the first column of Fig. 2 are only sample images belonging to one of the above categories. On the whole the proposed fringe extraction procedure was tested on more than 100 authentic/natural images and 100 forged/spliced images carrying PFA effects. Much of the database was self-constructed using our own cameras and the splicing operations were done using ADOBE PHOTOSHOP. The proposed PFA detection algorithm was compared with five existing PFA detection algorithms, viz. Kang [4], Kim and Park [5], Kim and Park [8], Chung et al. [7] and Ju and Park [6]. We recognize that there is no formal basis for comparing PFA-detected images, since quality assessment becomes a BLIND AFFAIR. There is also an uncertainty regarding the extent to which a fringe can be regarded as a PURPLE fringe. Some attempt was made in Ju and Park [6], to impose angular constraints towards quantifying and segregating PURPLE in a specific color space. However, in environments where the light sources can have diverse spectral signatures, fringe profiles in CHROMATIC SPACES are expected to show some variations. So a question arises as to whether we should treat the FRINGE as any FRINGE, or focus only on picking up fringes which have a PURPLE TINGE to them. It is also obvious that if a PFA detection procedure is not

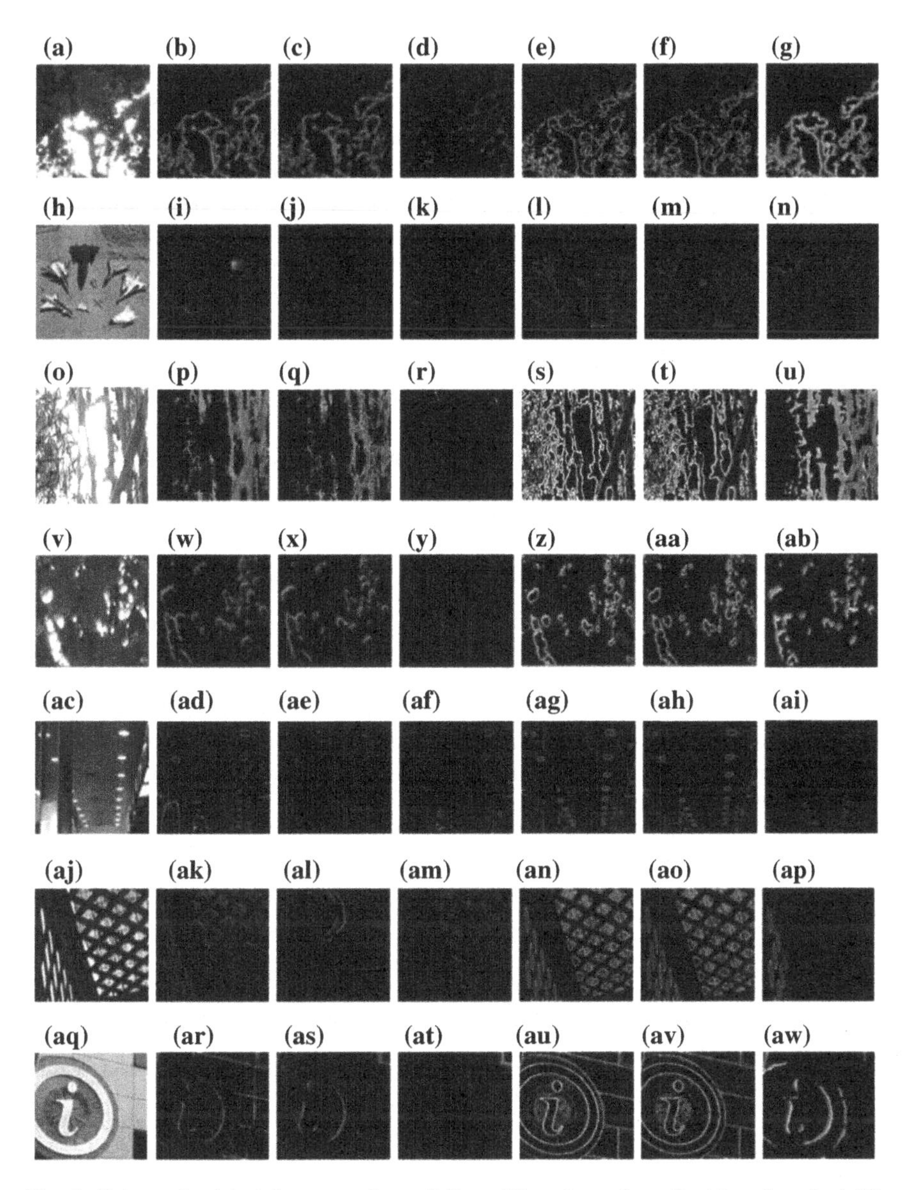

Fig. 2 Column 1-original images, column 2-Kang [4], column 3-gradient based method [5], column 4-xy chromaticity [8], column 5-Chung et al. [7], column 6-Ju and Park [6], column 7-proposed method

ATTUNED to pick up fringes perfectly, it will incur several false positives, by picking up a piece of regular scene content as a fringe. This problem is witnessed in Chung et al. [7] and Ju and Park [6], wherein the GREEN portions of leaves have been picked up as false positives because of an imperfect PURPLE extraction procedure (Fig. 2 with sub-figures (e, f), (s, t) and (z, aa) respectively).

Since most algorithms may deploy morphological operations to impart region continuity, there is another concern as to whether the PURPLE fringe is excessive and the so-called detection process is OVER-DONE or the PURPLE region is OVER-GROWN. This is precisely why we have limited the size of the structuring element in our system to ONE UNIT RADIUS. When the test image used was Fig. 2(aq), one may observe that almost all the existing approaches including Kang [4] and Kim and Park [5] (except the proposed one) incur false positives. The PURPLE fringes are a lot more precise with minimal false positives with the proposed algorithm (Fig. 2(aw)).

On the whole, the performance of the proposed procedure based on RELATIVISTIC thresholds is much better compared to most of the recent approaches. The advantage with the proposed approach rests with the fact that the threshold is content-adaptable, non-static and can therefore be used to pick up diverse fringe patterns (not necessarily confined to the seat of PURPLE).

4 Conclusions

In this paper, we have presented a content-adaptable threshold based PFA detection procedure which is reliable and incurs minimal false positives. Scope for its application as a forensic tool is promising, as the PFA detection procedure is virtually content-independent and stable.

References

1. Nakamura, J.: Image sensors and signal processing for digital still cameras. CRC Press (2005).
2. CCD versus CMOS (http://www.dallmeier.com/fr/produits-electronic/cameras/bon-a-savoir/ccd-vs-cmos.html).
3. Yerushalmy, I., Hel-Or, H.: Digital image forgery detection based on lens and sensor aberration. Volume 92., Springer (2011) 71–91.
4. Kang, S.: Automatic removal of purple fringing from images (July 5 2007) US Patent App. 11/322,736.
5. Kim, B., Park, R.: Automatic detection and correction of purple fringing using the gradient information and desaturation. In Proceedings of the 16th European Signal Processing Conference. Volume 4. (2008).
6. Ju, H.J., Park, R.H.: Colour fringe detection and correction in ycbcr colour space. Image Processing, IET 7(4) (June 2013) 300–309.
7. Chung, S.W., Kim, B.K., Song, W.J.: Detecting and eliminating chromatic aberration in digital images. In: Image Processing (ICIP), 2009 16th IEEE International Conference on. (Nov 2009) 3905–3908.

8. Kim, B.K., Park, R.H.: Detection and correction of purple fringing using colour desaturation in the xy chromaticity diagram and the gradient information. Image and Vision Computing 28(6) (2010) 952–964.
9. YCbCr (https://en.wikipedia.org/wiki/YCbCr).

Comparative Analysis of Feature Extraction Techniques in Motor Imagery EEG Signal Classification

Rajdeep Chatterjee, Tathagata Bandyopadhyay, Debarshi Kumar Sanyal and Dibyajyoti Guha

Abstract Hand movement (both physical and imaginary) is linked to the motor cortex region of human brain. This paper aims to compare the left–right hand movement classification performance of different classifiers with respect to different feature extraction techniques. We have deployed four types of feature extraction techniques—wavelet-based energy–entropy, wavelet-based root mean square, power spectral density-based average power, and power spectral density-based band power. Elliptic bandpass filters are used to discard noise and to extract alpha and beta rhythm which corresponds to limb movement. The classifiers used are Bayesian logistic regression, naive Bayes, logistic, variants of support vector machine, and variants of multilayered perceptron. Classifier performance is evaluated using area under ROC curve, recall, precision, and accuracy.

Keywords EEG · BCI · Motor imagery · Signal processing
Feature extraction · Classification · ROC · Sensitivity · Precision

1 Introduction

Human activities especially limb movements and thoughts are controlled by brain. Each and every activity exhibits a different signature encoded in microvolt electric signal generated inside the brain. This signal can be measured in many ways. One such way is *electroencephalography* which is widely used due to its noninvasiveness,

R. Chatterjee (✉) · T. Bandyopadhyay · D.K. Sanyal · D. Guha
School of Computer Engineering, KIIT University, Bhubaneswar 751024, India
e-mail: cse.rajdeep@gmail.com

T. Bandyopadhyay
e-mail: gata.tatha14@gmail.com

D.K. Sanyal
e-mail: debarshisanyal@gmail.com

D. Guha
e-mail: dibya.guha@gmail.com

A.K. Somani et al. (eds.), *Proceedings of First International Conference on Smart System, Innovations and Computing*, Smart Innovation, Systems and Technologies 79, https://doi.org/10.1007/978-981-10-5828-8_8

portability, and cost-effectiveness. In this method, electrodes are placed in different regions on the scalp of the human brain. The spatial positioning of the electrodes is done according to International 10–20 system [25]. The signal obtained in this procedure is known as *encephalogram (EEG)* [4].

Now let us understand the brain activities during hand movement. Right portion of human body is controlled by left hemisphere of human brain and vice versa. So right hand movement can be detected through the EEG activity reading in C3 electrode which is placed in the motor cortex region of left hemisphere and the opposite is also true; left hand movement is connected to the C4 electrode in motor cortex region of right hemisphere of human brain and so left hand movement can be detected through the EEG activity reading in C4 electrode. It is generally agreed that out of all the brain rhythms, only alpha and beta bands are sufficient to distinguish the type of hand movement [13]. Existing literature also states that, during right hand movement, the alpha band in *C*3 will be diminished and beta band will be increased. The common understanding is that alpha band signifies relaxed state of human brain, whereas beta band suggests muscle movements [19, 23, 27].

In this paper, we experiment with different feature extraction and classification techniques on EEG data for imagined hand movement. Raw EEG signal is very noise prone. So we first de-noise the signal using digital filters. Then we generate feature vectors from the filtered signal with the help of multiple feature extraction techniques. Finally, a variety of classifiers are run on the feature vectors to build left–right hand movement prediction model. We measure and report the performance of the classifiers.

Paper organization: This paper is divided into five sections. Section 1 is the overview of EEG-based motor imagery classification. Section 2 briefly describes related work, problem statement, and our contribution. Data preparation and feature extraction techniques used in the study are explained in Sect. 3. Section 4 presents results and discussion. Finally, we conclude our work in Sect. 5.

2 Related Works, Problem Statement, and Contributions

A. Related works on feature extraction: Classification of EEG signals has attracted a lot of attention from scientists and engineers working in theoretical and applied neuro-computing due to its tremendous potential in understanding the basic brain functions and its applications in medical field, gaming, virtual reality, etc. We focus on the techniques used for feature extraction and classification of motor imagery. A brief overview of related work on various feature extraction techniques is provided in Table 1.

B. Problem statement: Given a dataset of motor imagery EEG signals known to contain left hand or right hand movements, our goal is to identify which set of features lead us to better left–right hand movement classification. The prominent questions that we encounter in this paper are as follows:

Table 1 Related work on feature extraction

Sl. No.	Feature extraction techniques	References
1	Fast fourier transformation (FFT)	[5, 15]
2	Short-time fourier transformation (STFT)	[22, 26]
3	Power spectral density (PSD)	[3, 6, 7, 24]
4	Discrete wavelet transform (DST)	[2, 7, 17, 18]
5	Continuous wavelet transform (CWT)	[22]
6	Band power	[7, 16]

1. Which features of EEG data should be used for classification?
2. Given these features, which classifiers should be used?

C. Contributions: We conducted experiments to answer the questions posed in the previous section. Our major contributions from the study are as follows:

1. We found that wavelet-based features are best for EEG-type nonstationary signals.
2. We found logistic and linear Support Vector Machine (SVM1) are best performer classifiers.
3. We were able to characterize the performance of these classifiers on the data in a precise way using widely used metrics like recall, precision, and ROC.

3 Data Preparation and Feature Extraction

For our analysis, BCI competition II 2003 (dataset III) has been taken from the Department of Medical Informatics, Institute for Biomedical Engineering, University of Technology, Graz. The data has been used widely as a benchmark data in this field of EEG-based motor imagery classification problem [1].

A. Data Preprocessing: An elliptic bandpass filter is used to extract the frequency band of 0.5–30 Hz which covers both the alpha (8–12 Hz) and beta (18–25 Hz) bands. Also, we have used sampling frequency at 128 Hz.

B. Hardware and Software configuration: MATLAB *R*2014*a* is used for feature extraction and WEKA 3.6 classification tool is used to classify the obtained feature vectors with an Intel Core i5 computer.

C. Feature Extraction Techniques used: From past literatures, we find various types of feature extraction techniques which were used for left–right hand movement classification. For comparison, all the popular and mostly used feature extraction techniques are employed in this paper and in addition their combinations are

Table 2 Details of feature sets used

Feature sets name	Actual extracted feature sets	Features length
Ap	Average power	4
Bp	Average band Power	4
EngEnt	Wavelet-based energy–entropy	4
RMS	Wavelet-based root mean square	2
BpAp	Combining (2) & (1)	8
EngEntAp	Combining (3) & (1)	8
EngEntBp	Combining (3) & (2)	8
EngEntRms	Combining (3) & (4)	6
RmsAp	Combining (4) & (1)	6
RmsBp	Combining (4) & (2)	4

also used to explore their discriminating effectiveness. These are the Power Spectral Density (PSD)-based average power (Ap), band power (Bp), wavelet-based energy–entropy (EngEnt), and RMS values (RMS). Detailed description of the feature extraction techniques can be found in our previous work in [8]. We provide in Table 2 a brief description of the feature sets used in our experiments. The maximum feature set length is eight and the minimum is two. In machine learning, feature set length is a critical performance parameter.

4 Results and Discussion

The primary purpose of the paper is to compare the performance of different feature sets and whether multiple classifiers behave consistently on the same feature set. In our previous works, it is observed that both the classifiers, support vector machine (SVM) and multilayered perceptron (MLP), perform better than other popular classifiers (e.g., Naive Bayes, Logistic, etc.) in motor imagery classification [8, 9, 11]. However, in this study, we have incorporated probabilistic to nonlinear classifiers so that in total, five primary types of classifiers with three variants of SVM and two variants of MLP (Table 3) are used. In Table 3, C and Nu are the regularization parameters, used to give relaxation to the training error in SVM. After reasonable tuning, the C value is taken as 1.0 and Nu value as 0.5. If this value is close to 0, it signifies a large margin, i.e., it simply ignores those points which are violating the margin constraint. On the other hand, high value indicates narrow margin where constraints are hard to ignore. The length of feature sets is the number of inputs in MLPs. The average of feature set length and class labels is used as number of hidden layers in our MLP variants.

In the past, many papers used the said feature extraction techniques separately. This paper uses most of the prominent extraction methods independently and in a

Table 3 Types and variants of classifiers

Classifiers	Variants	Kernel functions
Bayesian logistic regression	–	NA
Naive Bayes	–	NA
Logistic	–	NA
SVM1	C-SVC (C = 1.0)	Linear
SVM2	Nu-SVC (Nu = 0.5)	Linear
SVM3	Nu-SVC (Nu = 0.5)	Polynomial (degree = 3)
	Learning rate	Momentum
MLP1	0.3	0.20
MLP2	0.7	0.29

combined way. This provides a comparative analysis of the quality of the feature sets [10, 12, 14, 20]. It is well known in machine learning that the classification is as good as the data set is. Most of the modern-day classifiers are not robust and suffer data-dependent predictions. Now our feature extraction techniques are fundamentally based on two principles: one, wavelet-based and two, power-based. All types are explained in the earlier section of this paper. These are common digital signal processing (DSP) techniques widely discussed and explained in any book on DSP or in related papers [21].

So the power spectral-based feature extraction technique deals with the power distributions of alpha and beta bands. Again, average band power talks about the concentration of power for the individual bands. Our results obtained from power-based extraction methods are performing well in terms of our measures but they are not the best. The possible explanation could be the power exhibition for both the right and left hands are more or less same with minimal distinctive information. Another extraction principle is wavelet-based feature sets. Wavelet is most appropriate for any nonstationary signal like EEG. It considers both the frequency range and temporal range at the time of signal processing. Unlike power-based features, wavelet-based feature extraction techniques are applied on the filtered (0.5–30 Hz) EEG signal without considering the alpha and beta bands specifically. Wavelet-based energy–entropy provides the best consistent results irrespective of the classifiers. Another interesting finding is that wavelet-based RMS feature set with only two features gives performance as good as four energy–entropy features, particularly for the Naive Bayes, logistic, and SVM2 classifiers.

Except area under ROC, SVM with linear kernel outperforms others in all other performance measures. Energy–entropy combining with AP, BP, and RMS is also used as feature sets. It is suggested that the combined results do not include any additional distinctive features and thus give similar result what is demonstrated by the energy–entropy individually. However, energy–entropy with BP loses its significance in combination. RMS with other combinations is as good as RMS itself rather it loses in few cases. BLR, Naive Bayes, and logistic classifiers exhibit improvement

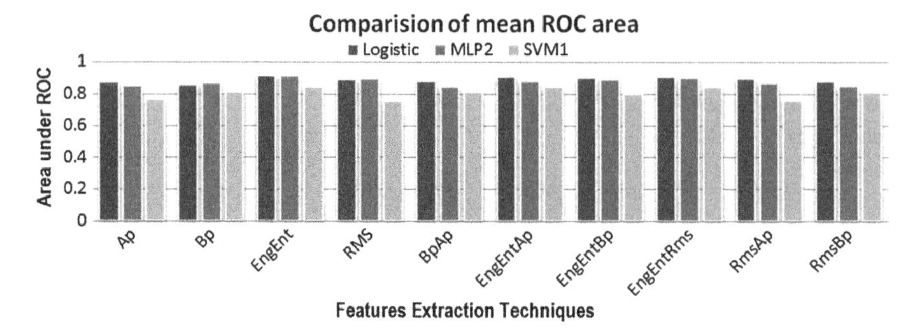

Fig. 1 ROC area chart for best three classifiers

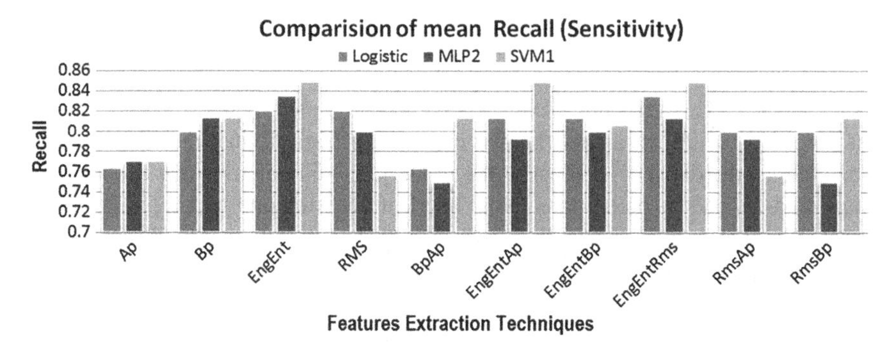

Fig. 2 Recall chart for best three classifiers

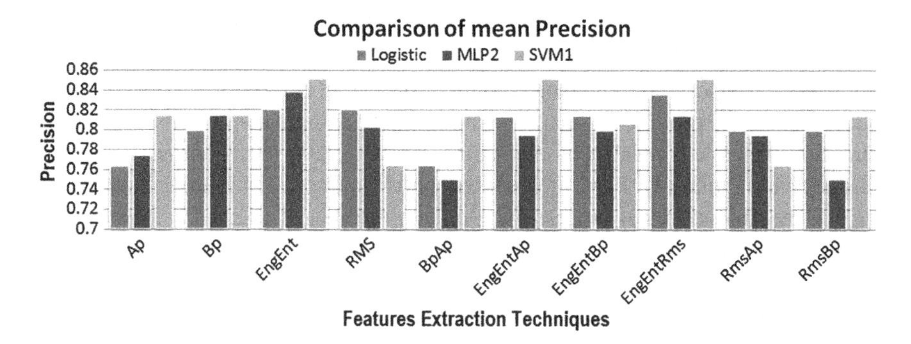

Fig. 3 Precision chart for best three classifiers

in performance when they use energy–entropy with RMS combination over energy–entropy or RMS independently.

In Table 4, the observations of our study are represented in a concise manner. Performances of all the classifiers as per the measures are given in the tabular formats in Tables 5, 6, 7, and 8 in Appendix A. Performance plots for best three classifiers (Logistic, MLP2, and SVM1) are shown in Figs. 1, 2, 3, and 4.

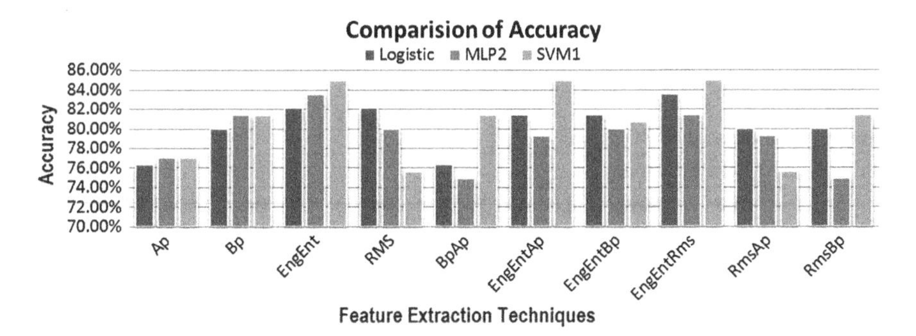

Fig. 4 Accuracy chart for best three classifiers

Table 4 Observations obtained from results

Sl. No.	Empirical findings	Significance
1	Logistic gives highest ROC area and MLP2 gives second highest ROC area	1. Large area under ROC ensures higher degree of confidence in the prediction model
2	SVM1 provides best sensitivity, precision, and accuracy	1. It indicates the quality of prediction (test) is consistent
3	Wavelet-based energy–entropy features perform better than other feature sets	1. It gives same results as its combination with others. Sometimes even better than its combinations 2. No information gain by adding the average power and average band power as additional features
4	RMS gives competitive results to energy–entropy	1. Having feature set length of two, it has as good discriminating information as others 2. Moreover, wavelet-based variants comprise more distinctive information over features obtained from power spectral-based variants on alpha and beta bands

5 Conclusion

In this paper, we have used and discussed a variety of feature extraction techniques and their combinations. Summarizing our findings, we can state that the feature extraction technique based on joint time–frequency-based wavelet transform is the most suitable one for classification of EEG signals for left–right hand movement discrimination. However, our work is limited to offline analysis of EEG signals. In future, our focus will be on implementing the best extraction algorithms in real-time systems for brain-state discrimination.

Appendix: A

Table 5 Results based on ROC area

For ROC area (under the curve)

	BLR	Naive Bayes	Logistic	MLP1	MLP2	SVM1	SVM2	SVM3
Ap	0.729	0.817	0.875	0.867	0.854	0.771	0.664	0.564
Bp	0.793	0.858	0.860	0.87	0.874	0.814	0.807	0.814
EngEnt	0.800	0.897	0.918	0.894	0.917	0.850	0.821	0.757
RMS	0.793	0.876	0.892	0.898	0.899	0.757	0.821	0.514
BpAp	0.793	0.885	0.880	0.869	0.848	0.814	0.807	0.814
EngEntAp	0.636	0.886	0.909	0.879	0.880	0.850	0.821	0.757
EngEntBp	0.786	0.883	0.906	0.874	0.896	0.807	0.807	0.814
EngEntRms	0.807	0.893	0.910	0.890	0.907	0.850	0.821	0.757
RmsAp	0.779	0.881	0.900	0.884	0.870	0.757	0.821	0.514
RmsBp	0.793	0.872	0.880	0.876	0.855	0.814	0.807	0.814

Table 6 Results based on sensitivity (recall)

For recall (sensitivity)

	BLR	Naive Bayes	Logistic	MLP1	MLP2	SVM1	SVM2	SVM3
Ap	0.729	0.757	0.764	0.771	0.771	0.771	0.664	0.564
Bp	0.793	0.786	0.800	0.771	0.814	0.814	0.807	0.814
EngEnt	0.800	0.800	0.821	0.814	0.836	0.850	0.821	0.757
RMS	0.793	0.800	0.821	0.807	0.800	0.757	0.821	0.514
BpAp	0.793	0.771	0.764	0.779	0.75	0.814	0.807	0.814
EngEntAp	0.636	0.793	0.814	0.836	0.793	0.850	0.821	0.757
EngEntBp	0.786	0.793	0.814	0.786	0.800	0.807	0.807	0.814
EngEntRms	0.807	0.807	0.836	0.814	0.814	0.850	0.821	0.757
RmsAp	0.779	0.786	0.800	0.764	0.793	0.757	0.821	0.514
RmsBp	0.793	0.800	0.800	0.750	0.750	0.814	0.807	0.814

Table 7 Results based on precision

For mean precision

	BLR	Naive Bayes	Logistic	MLP1	MLP2	SVM1	SVM2	SVM3
Ap	0.765	0.759	0.764	0.772	0.775	0.815	0.666	0.569
Bp	0.793	0.787	0.800	0.772	0.815	0.815	0.808	0.815
EngEnt	0.802	0.802	0.821	0.817	0.839	0.852	0.822	0.780
RMS	0.793	0.801	0.821	0.812	0.804	0.765	0.822	0.515
BpAp	0.793	0.772	0.765	0.779	0.751	0.815	0.808	0.815
EngEntAp	0.789	0.793	0.814	0.837	0.796	0.852	0.822	0.780
EngEntBp	0.786	0.793	0.815	0.786	0.800	0.807	0.809	0.817
EngEntRms	0.812	0.810	0.836	0.817	0.815	0.852	0.822	0.780
RmsAp	0.781	0.786	0.800	0.765	0.796	0.765	0.821	0.515
RmsBp	0.793	0.800	0.800	0.750	0.751	0.815	0.808	0.815

Table 8 Results based on accuracy

For accuracy (%)

	BLR	Naive Bayes	Logistic	MLP1	MLP2	SVM1	SVM2	SVM3
Ap	72.86	75.71	76.43	77.14	77.14	77.14	66.43	56.43
Bp	79.29	78.57	80.00	77.14	81.43	81.43	80.71	81.43
EngEnt	80.00	80.00	82.14	81.43	83.57	85.00	82.14	75.71
RMS	79.29	80.00	82.14	80.71	80.00	75.71	82.14	51.43
BpAp	79.29	77.14	76.43	77.86	75	81.43	80.71	81.43
EngEntAp	63.57	79.29	81.43	83.57	79.29	85.00	82.14	75.71
EngEntBp	78.57	79.29	81.43	78.57	80.00	80.71	80.71	81.43
EngEntRms	80.71	80.71	83.57	81.43	81.43	85.00	82.14	75.71
RmsAp	77.86	78.57	80.00	76.43	79.29	75.71	82.14	51.43
RmsBp	79.29	80.00	80.00	75.00	75.00	81.43	80.71	81.00

References

1. Brain Computer Interface Competition II, Department of Medical Informatics, Institute for Biomedical Engineering, University of Technology Graz (Jan 2004 (accessed June 6, 2015)), http://www.bbci.de/competition/ii/
2. Adeli, H., Ghosh-Dastidar, S., Dadmehr, N.: A wavelet-chaos methodology for analysis of eegs and eeg subbands to detect seizure and epilepsy. IEEE Transactions on Biomedical Engineering 54(2), 205–211 (2007)
3. AlZoubi, O., Calvo, R.A., Stevens, R.H.: Classification of eeg for affect recognition: an adaptive approach. In: Australasian Joint Conference on Artificial Intelligence. pp. 52–61. Springer (2009)
4. Andersen, R.A., Musallam, S., Pesaran, B.: Selecting the signals for a brain–machine interface. Current opinion in neurobiology 14(6), 720–726 (2004)
5. Aris, S.A.M., Taib, M.N.: Brain asymmetry classification in alpha band using knn during relax and non-relax condition. ICCIT (2012)
6. Aris, S.A.M., Taib, M.N., Lias, S., Sulaiman, N.: Feature extraction of eeg signals and classification using fcm. In: Computer Modelling and Simulation (UKSim), 2011 UkSim 13th International Conference on. pp. 54–58. IEEE (2011)
7. Bhattacharyya, S., Khasnobish, A., Chatterjee, S., Konar, A., Tibarewala, D.: Performance analysis of lda, qda and knn algorithms in left-right limb movement classification from eeg data. In: Systems in Medicine and Biology (ICSMB), 2010 International Conference on. pp. 126–131. IEEE (2010)
8. Chatterjee, R., Bandyopadhyay, T.: Eeg based motor imagery classification using svm and mlp. In: Proceedings of the 2nd International Conference on Computational Intelligence and Networks (CINE). pp. 84–89 (2016)
9. Cristianini, N., Shawe-Taylor, J.: An introduction to support vector machines and other kernel-based learning methods. Cambridge university press (2000)
10. Fawcett, T.: An introduction to roc analysis. Pattern recognition letters 27(8), 861–874 (2006)
11. Hagan, M.T., Demuth, H.B., Beale, M.H., De Jesús, O.: Neural network design, vol. 20. PWS publishing company Boston (1996)
12. Hernández-Orallo, J., Flach, P., Ferri, C.: A unified view of performance metrics: Translating threshold choice into expected classification loss. Journal of Machine Learning Research 13(Oct), 2813–2869 (2012)
13. Jeannerod, M.: Mental imagery in the motor context. Neuropsychologia 33(11), 1419–1432 (1995)
14. Lasko, T.A., Bhagwat, J.G., Zou, K.H., Ohno-Machado, L.: The use of receiver operating characteristic curves in biomedical informatics. Journal of biomedical informatics 38(5), 404–415 (2005)
15. Li, K., Sun, G., Zhang, B., Wu, S., Wu, G.: Correlation between forehead eeg and sensorimotor area eeg in motor imagery task. In: Dependable, Autonomic and Secure Computing, 2009. DASC'09. Eighth IEEE International Conference on. pp. 430–435. IEEE (2009)
16. Meinel, A., Castaño-Candamil, J.S., Dähne, S., Reis, J., Tangermann, M.: Eeg band power predicts single-trial reaction time in a hand motor task. In: 2015 7th International IEEE/EMBS Conference on Neural Engineering (NER). pp. 182–185. IEEE (2015)
17. Murugappan, M., Rizon, M., Nagarajan, R., Yaacob, S., Zunaidi, I., Hazry, D.: Eeg feature extraction for classifying emotions using fcm and fkm. International Journal of Computers and Communications 1(2), 21–25 (2007)
18. Murugappan, M., Ramachandran, N., Sazali, Y., et al.: Classification of human emotion from eeg using discrete wavelet transform. Journal of Biomedical Science and Engineering 3(04), 390 (2010)
19. Pfurtscheller, G., Neuper, C.: Motor imagery and direct brain-computer communication. Proceedings of the IEEE 89(7), 1123–1134 (2001)
20. Powers, D.M.: Evaluation: from precision, recall and f-measure to roc, informedness, markedness and correlation (2011)

21. Proakis, J.: Dg manolakis on digital signal processing (2006)
22. RamaRaju, P., AnogjnaAurora, N., Rao, V.M.: Relevance of wavelet transform for taxonomy of eeg signals. In: Electronics Computer Technology (ICECT), 2011 3rd International Conference on. vol. 3, pp. 466–470. IEEE (2011)
23. Schwartz, A.B., Cui, X.T., Weber, D.J., Moran, D.W.: Brain-controlled interfaces: movement restoration with neural prosthetics. Neuron 52(1), 205–220 (2006)
24. Sulaiman, N., Hau, C.C., Hadi, A.A., Mustafa, M., Jadin, S.: Interpretation of human thought using eeg signals and labview. In: Control System, Computing and Engineering (ICCSCE), 2014 IEEE International Conference on. pp. 384–388. IEEE (2014)
25. Tanner, A., et al.: Automatic seizure detection using a two-dimensional eeg feature space (2011)
26. Valenzi, S., Islam, T., Jurica, P., Cichocki, A.: Individual classification of emotions using eeg. Journal of Biomedical Science and Engineering 7(8), 604 (2014)
27. Xu, H., Lou, J., Su, R., Zhang, E.: Feature extraction and classification of eeg for imaging left-right hands movement. In: Computer Science and Information Technology, 2009. ICCSIT 2009. 2nd IEEE International Conference on. pp. 56–59. IEEE (2009)

Block Matching Algorithm Based on Hybridization of Artificial Bee Colony and Differential Evolution for Motion Estimation in Video Compression

Kamanasish Bhattacharjee, Arti Tiwari and Nitin Rakesh

Abstract Block matching is the most efficient technique for motion estimation (ME) in video compression and there are many algorithms to implement block matching. This paper discusses the block matching algorithms based on differential evolution (DE) and artificial bee colony (ABC) and proposes a new algorithm hybridizing these two algorithms aiming to get better results in block matching than the individual algorithms. In the proposed algorithm, food source generation operation of ABC is replaced by mutation and crossover operations of DE with the objective to utilize the search space exploration ability of DE and the solution exploitation ability of ABC.

Keywords Block matching · Artificial Bee Colony (ABC)
Differential Evolution (DE)

1 Introduction

In video compression, block matching (BM) for motion estimation is an essential technique. In block matching, video frames are split into macroblocks. Search space of a frame is used to find the best matching block for each block of next frame to minimize the objective function as sum of absolute differences (SAD) or mean absolute difference (MAD) or mean squared error (MSE) between two blocks to optimize the solution.

K. Bhattacharjee (✉) · A. Tiwari · N. Rakesh
Department of Computer Science, Amity School of Engineering & Technology,
Amity University, Noida, Uttar Pradesh, India
e-mail: kamanasish_b@live.com

A. Tiwari
e-mail: arti.tiwari94@gmail.com

N. Rakesh
e-mail: nitin.rakesh@gmail.com

A.K. Somani et al. (eds.), *Proceedings of First International Conference on Smart System, Innovations and Computing*, Smart Innovation, Systems and Technologies 79, https://doi.org/10.1007/978-981-10-5828-8_9

In block matching technique, this SAD/MAD/MSE calculation is the most time-consuming operation. Thus, block matching for motion estimation is considered as an optimization problem having an objective to search the best matching block for target block from the search space. Full search algorithm (FSA) provides the most precise motion vector through a comprehensive computation for all aspects of the search window. Various block matching approaches have been introduced to speed up the calculation by calculating only a fixed subset of search areas at the cost of deficient accuracy. So, there must be a trade-off between accuracy and speed.

The metaheuristic methods (evolutionary techniques and swarm intelligence techniques) have shown good optimization of this trade-off. Many such techniques have been used by researchers for block matching such as genetic algorithm (GA) [1, 2], harmony search [3], artificial bee colony [4], particle swarm optimization [5], and differential evolution [6]. Empirical studies have shown that hybridization of these evolutionary algorithms gives better result than the individual algorithms. Hence, this paper gives a novel idea for block matching by hybridizing DE and ABC.

The paper is further extended into six sections. Second section focuses on the basic concept of block matching. Third and fourth sections discuss the block matching algorithm based on DE and ABC, respectively. The proposed hybridized algorithm is explained in fifth section. Sixth section presents the conclusion and future work.

2 Block Matching

Block matching is the technique to find the matching macroblocks in a sequence of video frames. It is used to find the temporal redundancy in the video which is utilized in video compression.

In Fig. 1, a is the previous frame or target frame and b is the anchor frame or current frame. The current block under consideration is marked with green border,

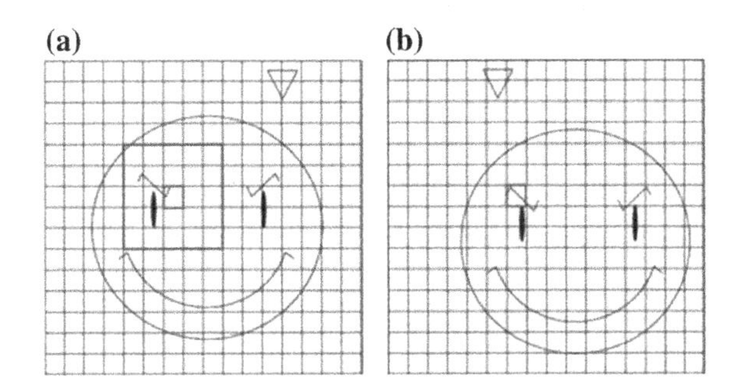

Fig. 1 **a** Previous frame **b** Current frame

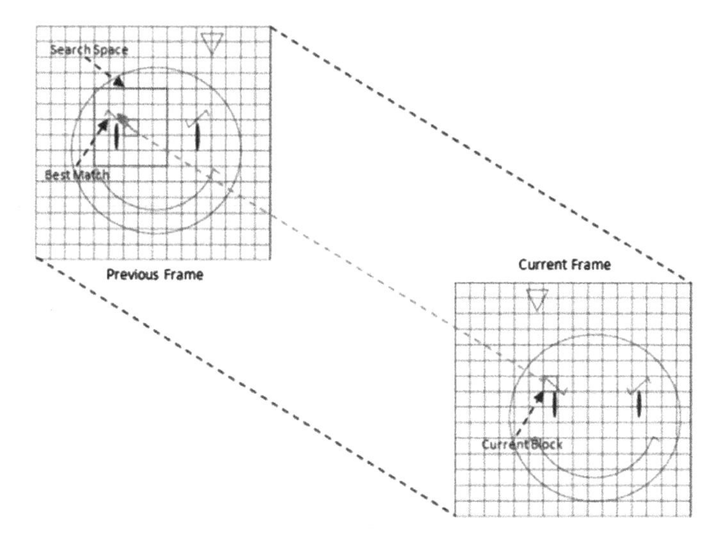

Fig. 2 Block matching, green block shows the target block and blue block is the entire search space in previous frame, red arrow shows the best matching block in previous frame

shown in Fig. 1b Current Frame. The same block shown in Fig. 1a Previous Frame is also marked with green border. Search space is specified by the displacement parameter or maximum allowed displacement (W). In this case W = 2. The search space is of $(2W + 1)(2W + 1)$ dimension, i.e., 5 × 5 in this case. The search space is marked with blue border in the previous frame containing the target block in center. In case of ABC, DE and the hybrid algorithm, the macroblocks are selected from this search space of 25 blocks and each of these macroblocks is compared with the current block under consideration in the current frame.

Here, the objective is to search the best matching block, i.e., the minimum difference between the two blocks. Figure 2 represents the method. Sum of absolute difference (SAD) is used as the difference measure.

$$SAD(h, v) = \sum_{j=0}^{N-1} \sum_{i=0}^{N-1} |I_t(x+i, y+j) - I_{t-1}(x+h+i, y+v+j)|, \quad (1)$$

where $I_t(.)$ is the current block's pixel intensity and $I_{t-1}(.)$ is the pixel intensity a block from previous frame. Therefore, the motion vector in (h, v) is defined as

$$(h, v) = \min SAD(h, v),$$

where (h,v) ϵ Search Space and-$W \leq h, v \leq W$.

3 Block Matching Using Differential Evolution

In [6], Cuevas et al. proposed an algorithm for block matching based on DE [11]. It reduces the search locations in block matching. DE is an evolutionary algorithm which tries to improve the solution vector iteratively and optimizes the problem by initializing a large population and then through—the mutation, crossover and selection operations. The following steps are applied to optimize the problem through DE.

3.1 Population Generation

In this step, a random population of D dimensional vector is generated up to NP candidates. These are called the parent vectors where X_i is the ith parent vector.

3.2 Mutation

To exchange information among several solutions, mutation is used. For mutation, three random parent candidates are chosen whose indexes are not equal to each other and the iteration number. Weighted difference vector (X_{n2} – X_{n3}) is added to the third parent candidate X_{n1}:

$$V = X_{n_1} + F^*(X_{n_2} - X_{n_3}), \tag{2}$$

where F is the Mutation probability and V is the mutant vector.

3.3 Crossover

Crossover between original vector and mutant vector is used to expand the diversity of the argument vector. Here, uniform crossover between parent vector (X_i) and mutant vector (V) is used to generate utility vector or child vector (U) with CP as the crossover probability. If the value of rand (0, 1) is less than CR then attribute value is chosen from mutant vector, otherwise from parent vector. Where j_{rand}є (1,2.….D)…

$$U = \begin{cases} V_{j,i} & if\ rand(0,1) \leq CP\ or\ j = j_{rand}, \\ X_{j,i} & otherwise \end{cases}, \tag{3}$$

where j, j_{rand} є (1,2.….D).

3.4 Selection

Fitness value is calculated through objective function for each parent vector–utility vector pair. If utility vector is superior to corresponding parent vector then it replaces the parent in the population otherwise parent remains same. Through this operation, population is generated for next generation.

$$X_i = \begin{cases} V_{j,i} & if\, f(u_i) \leq f(x_i), \\ X_{j,i} & otherwise \end{cases} \tag{4}$$

DE based Block Matching Algorithm

```
begin
  1.  Initialize the DE parameters F = 0.5 and CP = 0.5 D =2.
  2.  Select the random initial population NP of D dimensional from the search space i.e.
      (2W+1)* (2W+1) and NP = 5.
  3.  Calculate SAD value between current target block B(x, y) and each block from NP (X_i
      where i ϵ 1 to NP) using eq. 1.
  4.  while termination criteria is not satisfied do
  5.   for i = 1 to NP
      /* Mutation */
           Select three random numbers n_1, n_2 and n_3 where i ≠ n_1 ≠ n_2 ≠ n_3
       V_i = X_{n_1} + F * (X_{n_2} - X_{n_3})
      for j = 1 to D
      /* Crossover */
                if rand(0,1) ≤ CR then,
                     trial vector U_{j,i} = V_{j,i}
                else
                     trial vector U_{j,i} = X_{j,i}
                end if
      end for
      end for
      /* Selection */
  6.  for i = 1 to NP
          Calculate SAD value between current    block (b) and each trial vector block (U_i)
          using eq. 1.
          if SAD(U_i)< SAD(X_i)
                X_i =U_i
         end if
      end for
      take the best solution till now.
  7.  end while
end
Select block with the minimum SAD value for Motion Vector calculation.
```

4 Block Matching Using Artificial Bee Colony (ABC)

In [4], Cuevas et al. proposed an algorithm for block matching based on ABC [12]. It reduces the number of search place in the block matching. The following steps are implemented to optimize the problem through ABC.

4.1 Initial Food Source Generation

In this step, a random population of D dimensional vector is generated up to NP candidates. These are called food sources where X_i is the ith food source. Now, calculate the fitness value for each food source as:

$$fitness = \begin{cases} 1/(1+f(x)) & if\, f(x) \geq 0 \\ 1+abs(f(x)) & otherwise \end{cases} \tag{5}$$

4.2 New Food Source Generation by Employed Bees

Each employed bee (equal number of food sources) will generate new food source V in neighborhood by

$$V_{j,i} = X_{j,i} + r^*(X_{j,i} - X_{j,k}) \tag{6}$$

Here, r is a random number in the range of [−1,1] and i, j, k are the index parameters where i, k ϵ (1,2…NP) with constraint i $\neq$ k. Calculate the fitness value for each food sources (V), these fitness values will be compared with fitness values of initial food sources (X) and best food sources will be updated.

4.3 Selection of Food Sources by Onlooker Bees

Onlooker bee will select the food source based on their probabilities. Probability of the food source is calculated as:

$$\Pr ob_i = fitness_i / \sum_{i=1}^{NP} fitness_i \tag{7}$$

Onlooker bee will select the food sources according to their probabilities, go to those locations and generate new food sources in the neighborhood and if generated candidate has better fitness value, it will replace the old food source otherwise old will remain same.

4.4 Determine Scout Bees

After completing the repetitive process up to specified limit, if there is no improvement in any food source then it is determined as abandoned and the corresponding onlooker bee will become scout bee. These scout bees will generate the new population and continue with the same process as the above sections:

ABC based Block Matching Algorithm

begin

1. *Initialize the ABC parameter, set the limit to stop the process.*
2. *Set the D dimensional random initial population NP from the search space (2W+1)*(2W+1) and NP = 5.*
3. *while the terminating criteria is not satisfied do*
4. *Calculate the SAD between current block B(x, y) and each block of NP (X_i where $i \in 1$ to NP)using eq.1*
5. *Calculate the fitness value for each individual of NP using eq. 5*

   / new food source generation by employed bees */*
6. *for i = 1 to NP*

   r = rand(-1,1) and select a number k where k ≠ i

   for j = 1 to D

   $V_{j,i} = X_{j,i} + r*(X_{j,i} - X_{j,k})$

   end for

   end for

   Now, calculate the SAD value of each newly generated food source (V_i) using eq.1 followed by calculating fitness value using eq. 5.

   if fitness(V_i) > fitness(X_i) then

   update the food source by new food source(V_i)

   end if

   */*selection of food source by onlooker bee*/*
7. *onlooker bee will select the food source based on their probability calculate probability of each food source using eq. 7*

   for i = 1 to NP

   $Prob_i = fitness_i / \Sigma_{i=1}^{NP} fitness_i$

   if rand(0,1) < $Prob_i$

   follow step 6 for this block

   end if

   end for

   / Determining the scout bee */*
8. *if the fitness(X_{l-1}) ≥ fitness(X_l) where l is the limit, then*

   block is abandoned and a new block is randomly selected.

   end if
9. *end while*

end

Final is considered as the best to calculate the Motion Vector (MV)

5 Proposed Algorithm for Block Matching

Hybridization of DE and ABC has already been proposed [7–10]. But these algorithms are not generalized. We cannot use these algorithms directly for every problem. Every problem has its own set of parameters and to take care of those parameters, we have to customize the algorithm according to the problem. This paper has proposed a customized approach of DE-ABC hybrid algorithm to fit the goal to minimize the number of SAD evaluations with the acceptable solution for block matching. The proposed hybrid method is presented below:

Hybrid ABC-DE based Block Matching Algorithm

begin

1. *Initialize the parameters $F = 0.5$, $CP = 0.5$ for DE and limit for ABC.*
2. *Randomly select the initial population NP of D dimensional from the search space i.e. $(2W+1)*(2W+1)$, $D=2$ and $NP = 5$.*
3. *while the terminating criteria is not satisfied do*
4. *Calculate the SAD between current block $B(x, y)$ and each block of NP (X_i where $i \in 1$ to NP) using eq. 1*
5. *Calculate the fitness value for each individual of NP using eq. 5.*
6. *New population of NP blocks is generated using mutation and crossover using eq. 2 and 3.*
   for i = 1 to NP
   Select three random numbers p, q and r where $p \neq q \neq r \neq i$ $\quad V_i = X_p + F*(X_q - X_r)$
   for j = 1 to D
   if rand(0,1) $\leq$ CP then,
   trial vector $U_{j,i} = V_{j,i}$
   else
   trial vector $U_{j,i} = X_{j,i}$
   end if
   end for
   end for
   Now, calculate the SAD value of each newly generated food source (V_i) using eq. 1 followed by calculating fitness value using eq. 5.
   if fitness(U_i) > fitness(X_i) then
   update the food source by new food source(V_i)
   end if
7. *calculate probability of each selected food source using eq. 7*
   for i = 1 to NP
   $Prob_i = fitness_i / \sum_{i=1}^{NP} fitness_i$
   if rand(0,1) < $Prob_i$
   follow step 6 for this block
   end if
   end for
8. *if the fitness(X_{l-1}) $\geq$ fitness(X_l) where l is the limit, then*
   block is abandoned and a new block is randomly selected.
   end if
9. *end while*

end

Final is considered as the best to calculate Motion Vector (MV)

5.1 Advantages of the Hybridization (ABC-DE)

In the proposed algorithm, search space exploration ability of DE is combined with the solution exploitation ability of ABC. Search space exploration is needed in ABC in two steps—new food source generation by employed bees and new food source generation by onlooker bees. In both these cases, the proposed algorithm uses the mutation and crossover operations of DE. Through mutation and crossover operations of DE, the proposed method explores the search space better while ABC overcomes the solution exploitation problem of DE.

5.2 Justification

In [10], Abraham et al. have proposed a hybridized algorithm of ABC and DE and have done a comparative study between standard DE, standard ABC and their proposed hybrid algorithm using the standard benchmark functions. All the benchmark functions used are standard minimization functions. The benchmark functions used in [10] are listed in Table 1 and the graphs shown in Figs. 3 and 4 are formulated from the results of [10]. In Fig. 3, the blue, red, and green lines correspond to DE, ABC, and hybrid ABC-DE, respectively. It can be clearly seen that ABC and ABC-DE have better mean best value for every minimization function than DE. Now, for a comparative analysis of performance between ABC and ABC-DE, we refer Fig. 4. Here, the blue and red lines correspond to ABC and hybrid ABC-DE, respectively. It can be inferred from the figure that for some cases, ABC-DE performs as good as ABC while for other cases it gives better mean best value than ABC. Hence, it can be seen that their proposed hybrid ABC-DE algorithm performs better than both DE and ABC.

Hybridization of ABC and DE is applied for solving other problems also in [7–9]. The results of all these papers establish the superiority of hybrid ABC-DE algorithm. After reviewing various proposed hybrid ABC-DE algorithm, it is seen that the hybrid algorithm performs better for minimization problems. In this paper, the problem is also a minimization problem. Hence, it can be inferred that the hybridization of ABC and DE will perform better than the individual algorithms for Block Matching.

Table 1 Benchmark functions

Function	Mathematical representation	Theoretical optimum f_{min}
Sphere function (f_1)	$f_1(\vec{x}) = \sum_{i=1}^{D} x_i^2$	$f_1(\vec{0}) = 0$
Rosenbrock (f_2)	$f_2(\vec{x}) = \sum_{i=1}^{D-1} [100(x_{i+1} - x_i^2)^2 + (x_i - 1)^2]$	$f_2(\vec{1}) = 0$
Rastrigin (f_3)	$f_3(\vec{x}) = \sum_{i=1}^{D} [x_i^2 - 10\cos(2\pi x_i) + 10]x$	$f_3(\vec{0}) = 0$
Grienwank (f_4)	$f_4(\vec{x}) = \frac{1}{4000} \sum_{i=1}^{D} x_i^2 - \prod_{i=1}^{D} \cos(\frac{x_i}{\sqrt{i}}) + 1$	$f_4(\vec{0}) = 0$
Ackley (f_5)	$f_5(\vec{x}) = -20\exp(-0.2\sqrt{(\frac{1}{D}\sum_{i=1}^{D} x_i^2)}) - \exp(\frac{1}{D}\sum_{i=1}^{D} \cos(2\pi x_i) + 20 = e$	$f_5(\vec{0}) = 0$
Step (f_6)	$f_6(\vec{x}) = \sum_{i=1}^{D} (\lfloor x_i + 0.5 \rfloor)^2$	$f_6(\vec{p}) = 0$ $\frac{-1}{2} \le p < \frac{1}{2}$
Schwefel's problem (f_7)	$f_7(\vec{x}) = \sum_{i=1}^{D} \|x_i\| + \prod_{i=1}^{D} \|x_i\|$	$f_7(\vec{0}) = 0$
Schaffer's F6 function (f_8)	$f_8(\vec{x}) = 0.5 + \frac{\sin^2(\sqrt{(x_1^2 + x_2^2)}) - 0.5}{(1 + 0.001(x_1^2 + x_2^2))^2}$	$f_8(\vec{0}) = 0$
Six-Hump Camel back function (f_9)	$f_9(\vec{x}) = (4x_1^2 - 2.1x_1^4 + \frac{1}{3}x_1^6 + x_1x_2 - 4x_2^2 + 4x_2^6)$	$f_9(0.089, -0.71) = f_9(-0.089, 0.71) = -1.0316285$
Goldstein-Price function (f_{10})	$f_{10}(\vec{x}) = \{1 + (x_0 + x_1 + 1)^2(19 - 14x_0 + 3x_0^2 - 14x_1 - 6x_0x_1 + 3x_1^2)\}$ $\{30 + (2x_0 - 3x_1)^2$ $(18 - 32x_0 + 12x_0^2 + 48x_1 - 36x_0x_1 + 27x_1^2)\}$	$f_{10}(0, -1) = 3$

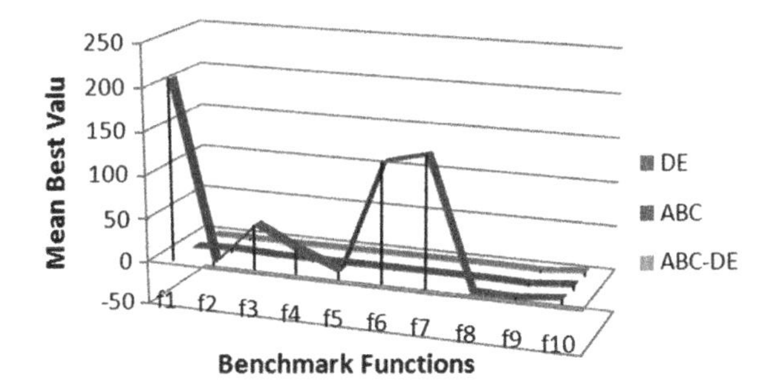

Fig. 3 Performance analysis of DE, ABC, and Hybrid ABC-DE

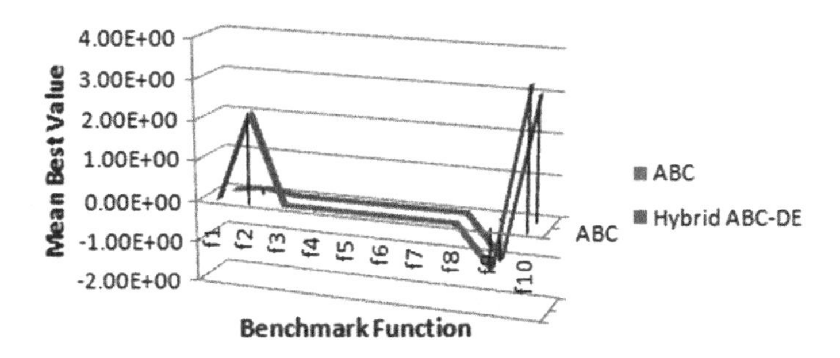

Fig. 4 Performance analysis of ABC and hybrid ABC-DE

6 Conclusion and Future Work

In this paper, the emphasis is given on the theoretical formulation of the hybrid algorithm for block matching. Various papers [7, 8, 9, 10] have proposed the hybridization of ABC and DE for global optimization or solving different problems. But such hybridization has not been attempted yet to optimize the problem of block matching. These papers [7, 8, 9, 10] are referred to understand the basic concepts behind the hybridization with respect to various parameters and inferences are drawn from the papers to justify why the hybrid algorithm will provide better results than the individual algorithms in case of block matching. Practical implementation of the algorithm and comparison with other algorithms based on various parameters like PSNR, degradation ratio, number of visited search points, etc., are the next phase of this work.

References

1. Chun-Hung, L., Ja-Ling W.: A Lightweight Genetic Block-Matching Algorithm for Video Coding, IEEE Transactions on Circuits and Systems for Video Technology, 8(4), (1998), 386–392.
2. Wu, A., So, S.: VLSI Implementation of Genetic Four-Step Search for Block Matching Algorithm, IEEE Transactions on Consumer Electronics, 49(4), (2003), 1474–1481.
3. Cuevas, E.: Block-matching algorithm based on harmony search optimization for motion estimation, Applied Intelligence, 39 (1), 165–183 (2013) W. (eds.) Euro-Par 2006. LNCS, vol. 4128, pp. 1148–1158. Springer, Heidelberg (2006)
4. Cuevas, E., Zaldívar, D., Pérez-Cisneros, M., Sossa, H., Osuna, V.: Block matching algorithm for motion estimation based on Artificial Bee Colony (ABC). Applied Soft Computing Journal 13 (6), 3047–3059 (2013)
5. Yuan, X., Shen, X.: Block Matching Algorithm Based on Particle Swarm Optimization, International Conference on Embedded Software and Systems (ICESS2008), 2008.
6. Cuevas, E., Zaldívar, D., Pérez-Cisneros, M., Oliva, D.: Block-matching algorithm based on differential evolution for motion estimation, Engineering Applications of Artificial Intelligence, 26 (1), 488–498 (2013)
7. Li, X., Yin, M.: Hybrid differential evolution with artificial bee colony and its application for design of a reconfigurable antenna array with discrete phase shifters, IET Microwaves, Antennas & Propagation. 6(14), (2012)
8. Yang, J., Li, W., Shi, X., Xin, L., Yu, J.: A Hybrid ABC-DE Algorithm and Its Application for Time-Modulated Arrays Pattern Synthesis, IEEE Transactions on Antennas and Propagation. 61 (11), (2013)
9. Worasucheep, C.: A Hybrid Artificial Bee Colony with Differential Evolution, International Journal of Machine Learning and Computing. 5 (2015)
10. Abraham, A., Jatoth, R.K., Rajasekhar, A.: Hybrid Differential Artificial Bee Colony Algorithm, Journal of Computational and Theoretical Nanoscience. 9, 1–9 (2012)
11. Storn, R., Price, K.: Differential evolution-a simple and efficient heuristic for global optimization over continuous spaces. Journal of Global Optimization. 11, 341–359 (1997)
12. Karaboga, D.: An idea based on honey bee swarm for numerical optimization, Technical Report-tr06, Erciyes University, Engineering faculty, Computer Engineering Department, Vol. 2000 (2005)

Enhanced Online Hybrid Model for Online Fraud Prevention and Detection

Harsh Tyagi and Nitin Rakesh

Abstract With the advent of technology, the business process, worldwide, has shifted from a slow—person-to-person physical interaction to a better and faster service platform—the Internet. The biggest advantage of this platform is that it allows the people to undergo their business transactions without meeting with the other party/parties and still making the benefits out of the agreement between them. Though this approach has a major edge over the conventional methods, still has many drawbacks to be pondered upon. The Internet has also emerged as a platform to commit crimes at a very high rate. Frauds committed online has put a question on this very approach of revolutionized business since no proposed model yet has given a sound and complete way to handle these issues. This paper is an extension of the previously generated OHM model, and enhances the parameters and flow of the entire OHM system, to prevent online frauds. The key frauds where Enhanced Online Hybrid Model (EOHM) can be used would be auction frauds, no-delivery frauds, and identity theft frauds.

Keywords Auction fraud · No-delivery fraud · Identity theft fraud · HMM

1 Introduction

The trend of the modern era to undergo most of the trades is through online business approach. This online business approach is concocted based on e-commerce and e-business, which have been the prominent results of the Internet. The e-commerce refers to all types of electronic transactions between organization and stakeholders whether they are financial transactions or exchange of any service

H. Tyagi (✉) · N. Rakesh
Department of Computer Science and Engineering,
Amity University, Noida, Uttar Pradesh, India
e-mail: 23ht11tyagi@gmail.com

N. Rakesh
e-mail: nitin.rakesh@gmail.com

A.K. Somani et al. (eds.), *Proceedings of First International Conference on Smart System, Innovations and Computing*, Smart Innovation, Systems and Technologies 79, https://doi.org/10.1007/978-981-10-5828-8_10

while e-business is the transformation of key business processes using various Internet technologies [1]. As per a report of dazeinfo.com [2], there has been a rapid increase in the use of e-commerce website over the past decade. The figure below depicts the increase in the e-commerce sale in India over the past few years and predicts the future statistics of the same (Fig. 1).

Also, an analysis of eMarketer [3], which is one of the most renowned market research companies, shows the increase in retail e-commerce sales worldwide reaching over $2 trillion in 2016 and is a forecast of the future years (Fig. 2).

The growth in this methodology of doing trades and business has apparently opened a gateway to illegal and malicious activities [4]. To identify crimes and frauds on Internet is an arduous and difficult task, eventually increasing the number of such violations over the internet in the past years. The National White Collar Crime center, bifurcate these frauds into various categories, out of which, auction, nondelivery and identity theft, are the most common ones.

The Auction process over the Internet works as follows:

- Online users visit auction websites to buy and sell various items
- Interested buyers bid on the auction item
- Buyer, after winning, then pays for the auctioned item
- Product is then delivered

This paper enhances the original Online Hybrid Model, which primarily worked on online-in-auction fraud and nondelivery frauds and was sectioned into two stages: verifying an authentic user and web server and informing the victim on fraud detection; the EOHM increases the efficacy of the authentication process and

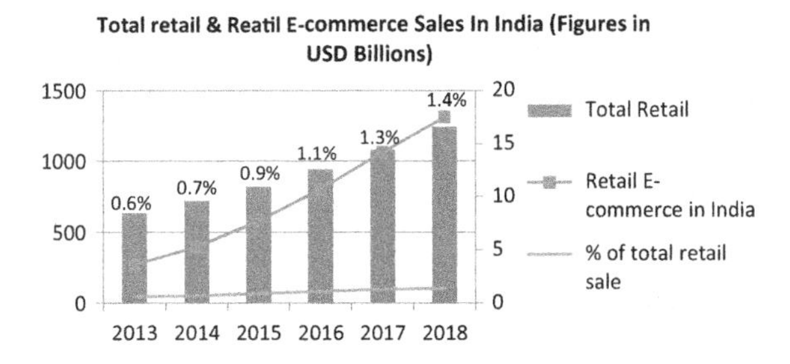

Fig. 1 Retail sale analysis of e-commerce in India. *Source* eMarketer.com analysis

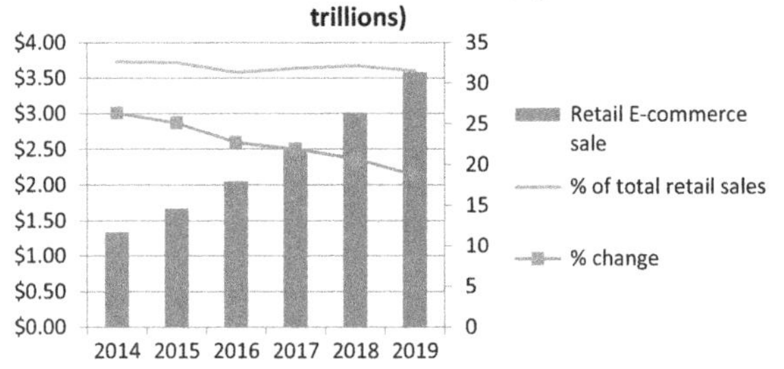

Fig. 2 Ecommerce growth. *Source* eMarketer.com analysis

introduces a theoretical approach to regularly monitor each user, to prevent frauds. It also introduces a process to reduce the no-delivery frauds by creating a monitored cash transaction system on every sale/purchase.

This paper is bifurcated into five sections. The introduction of the various frauds which is already discussed above, is the first section, the following section will include the overview of the existing system and the formulation of the problem. The third section is the entire description of EOHM named under Sect. 3. Section 4 tells how the model would be implemented and the final section is the conclusion and the future work.

2 Literature Review

This paper is an extension of the Online Hybrid Model [5] which aimed at reducing Internet frauds. The model uses the attributes such as users, the model and the servers to detect and prevent the frauds. This model in the registration window asks for the credentials of the user so that it can store the information for any verification purposes, however, there is no step to verify the information being collected.

This model also verifies the web servers, which is a pretty procedure of verification being adopted. It evaluates the authenticity of the digital signature certificate and verifies the past transactions and converges on the ranking of the system, based on which the OC is issued.

The **auction fraud** [6] occurs when, there are false bids placed on the auction products, or there is **no-delivery** [7] made or the product delivered is not what was promised on the portal. Such frauds are dealt using the EOHM, which will try to curb all these possibilities.

The **identity theft** is simply a person, using the credentials of someone else using information such as credit card credentials [8], etc., and thus committing fraud.

The impediment to the evaluation and detection of such frauds is prominent because of the following two reasons—the variety of frauds and changes in fraud behavior to avoid detection [9]. Thus, there was a need for a feedback mechanism to keep the detection methods aware of the new frauds.

3 Existing System and Problem Formulation

3.1 Existing Model

The OHM [5], as mentioned above, already considers factors such as collusive bidding, multiple bidding, etc., which were not checked in conventional methodologies used such as Shill-deterrent fee [10] schedule mechanism and the hidden Markov model [11]. The OHM consisted of two models: mobile authentication model and HMM model.

3.2 Problem Formulation

The increase in the online auction fraud has been the reason for continuing this research. The problem primarily focuses on—"Analyzing the efficiency of parameters of the OHM."

4 Enhanced Online Hybrid Model

The initial approach of OHM was to prevent or detect the detrimental activities on the online transactions by authenticating the parties involved in the transactions. This validation process had three attributes: Users, OHM, and Web Servers. These three assets are still considered in the EOHM, but the process of validating them has been modified to make the system more efficient. The process assets are defined below:

1. Users: This majorly contains buyers or sellers. Verification of the OHM certificate is mandatory for them. This attribute is the sole fragment of the system that interacts with any electronic-commerce or electronic-business platform [5].
2. OHM: This approach, as defined in the previous paper [10], is to authenticate the users and the server. The OC (OHM Certificate) is issued after the validation process [5].
3. Web Servers: As defined in the previous paper [10], a platform and management approach is provided by the web servers providing auction and retail services to be focused upon. This is also provided with an OC which is digitally signed [5].

4.1 EOHM Approach

As the original OHM classifies the process into two tasks: Prevention and Detection; the EOHM follows the parent concept but has enhanced both sub-agendas.

4.1.1 EOHM for Prevention

The Enhanced OHM has shortened the registration process of its parent method. Though, if the criteria for this new approach are not met, the user is provided the OC but is not an "assured user". The OC without an assured user simply depicts that the person's information is stored in the database, as provided by the person himself, but not verified by any authority. However, the server authentication remains the same.

Attributes needed for User Authentication:

1. Username: The unique username to identify the user.
2. Email: Email address for verification. It can act as a username too.
3. Password: Password for the account.
4. Contact: Contact details.

The above four attributes are the prime factors needed for the registration phase. This is then followed by the email verification system. The email sent by the server to the email id provided in phase one has a link to it. This link is appended with a unique token id generated by the server for that user, to avoid phishing attacks. The link when opened asks for a unique identification code number, such as Aadhaar Card (India), Social Security Number (USA), etc., which is sent to the server which then issues OC to the user. The user's account is activated but the user is not titled as an "OC Assured User". The unique identification proof provided by the user in the email verification process is validated by the server. If the user is defined as an authentic user per the government's database, then he is titled as an "OC Assured User" and is notified of the same. The buyers must put their trust on assured users as compared to nonassured user.

The diagram below shows the original OHM approach (Fig. 3).

EOHM Approach Diagrammatic Representation:

The EOHM model, however, bifurcates into two phases: Registration and Verification (Fig. 4).

REGISTRATION:

The above diagram displays the various phases of the registration process of the enhanced online hybrid model (Fig. 5).

VERIFICATION:

This process is divided into two stages:

1. OC: Issuing the OHM certificate after taking in the details of the user along with the nationally accepted unique id proof number.
2. OHM Assured User: This is when the user is authenticated using the identification proof provided by him/her (Fig. 6).

Fig. 3 Old OHM approach

Fig. 4 EOHM approach

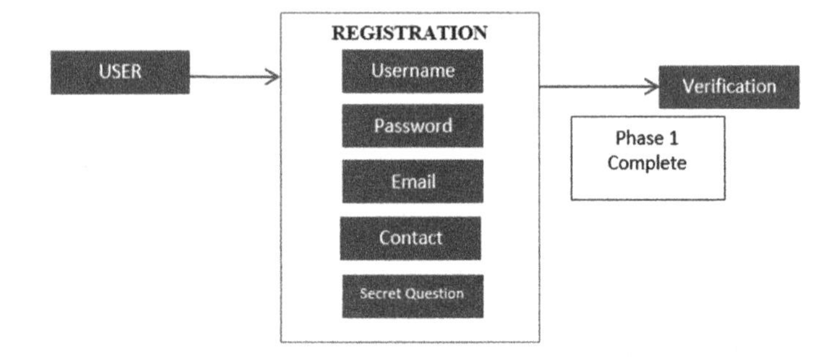

Fig. 5 Registration

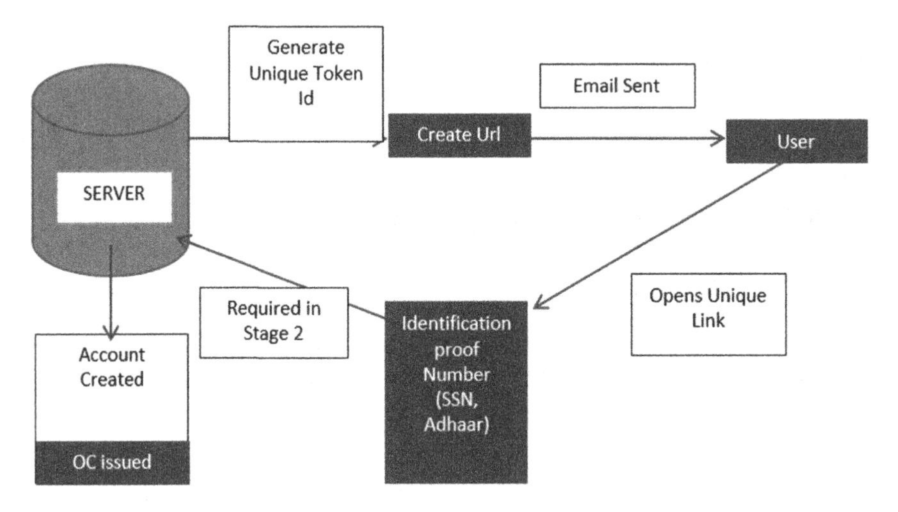

Fig. 6 EOHM certificate issue

Stage 2: OHM-Assured User (OAU):

This signifies that the user is an authenticated user thus the buyers can trust them more than OC sellers. This is more of a server-side task where the server validates the user against the identification proof provided to it in the registration process. The user must enter the OTP number for authentication purposes.

The user, if has OAU, will get a check mark in front of his/her name when other buyers or sellers see their profile or products.

The users can be differentiated by a simple check mark in front of the user's name, depicting it is a verified user.

Example

Name: John Doe✓; Contact: +91-9*******; Place: Noida, India**

The tick verifies the OAU certification (Fig. 7).

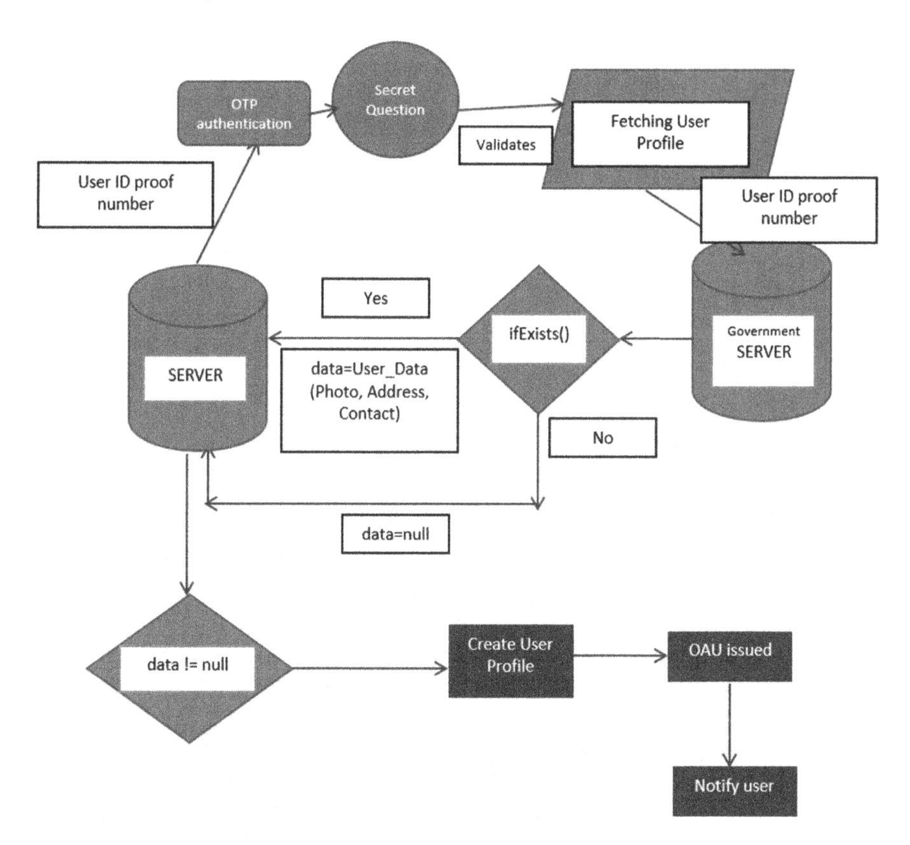

Fig. 7 OAU flow diagram

4.1.2 Alternative Approach

However, there is a possibility that the user might not have a mobile registered or the country in which one lives in does not allow to fetch data from the database and only allows for validation; in such cases the user must fill the complete form details as in the former OHM registration process, but the information will then be parceled into a packet and validated against the government database. The information entered in the mandatory sections should match the information of the government identification that is being used to verify the details. Any discrepancy would lead to the failure in the authentication process and the user will not be marked under the category of OHM assured user (OAU).

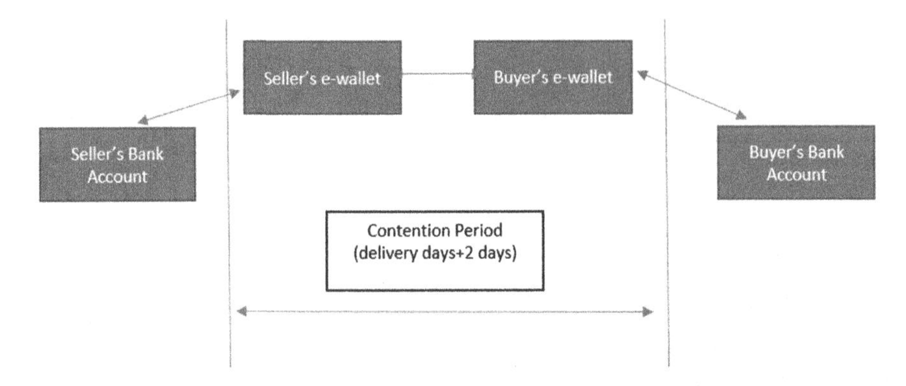

Fig. 8 Transaction system middleware

4.2 EOHM for Nondelivery Fraud

The increase in the nondelivery fraud in online auction is a nightmare to the people doing online business. The victim count as per the report of IC3 of 2015, has reached to more than 67,000 [4]. As discussed, the nondelivery auction fraud includes, the no-delivery of the product and even the wrong delivery, or if the product delivered is not what the product was promised.

This problem is dealt with creating a middle ware in the transaction window while doing the purchase on both ends—sellers and the buyers as well. This protocol will reduce the fraudulent behavior in this domain of frauds to minimal.

Creation of the middle ware:

The prime aim is to augment an e-wallet with each user's account. This e-account will hold the transaction money, in the seller's account, until the product is delivered to the buyer allowing a 2-day contention window to the buyer, in which the buyer can report any issue, if any, related to the problem. After the contention window period expires, the money will be automatically transferred from the e-wallet of the seller to his bank account; thus, making a safe transaction circumventing a high probability of fraudulent behavior. The contention period includes—the number of days in which the product is delivered to the buyer along with an extension of two days for reporting any issues (Fig. 8).

Workflow of the framework:

See Fig. 9.

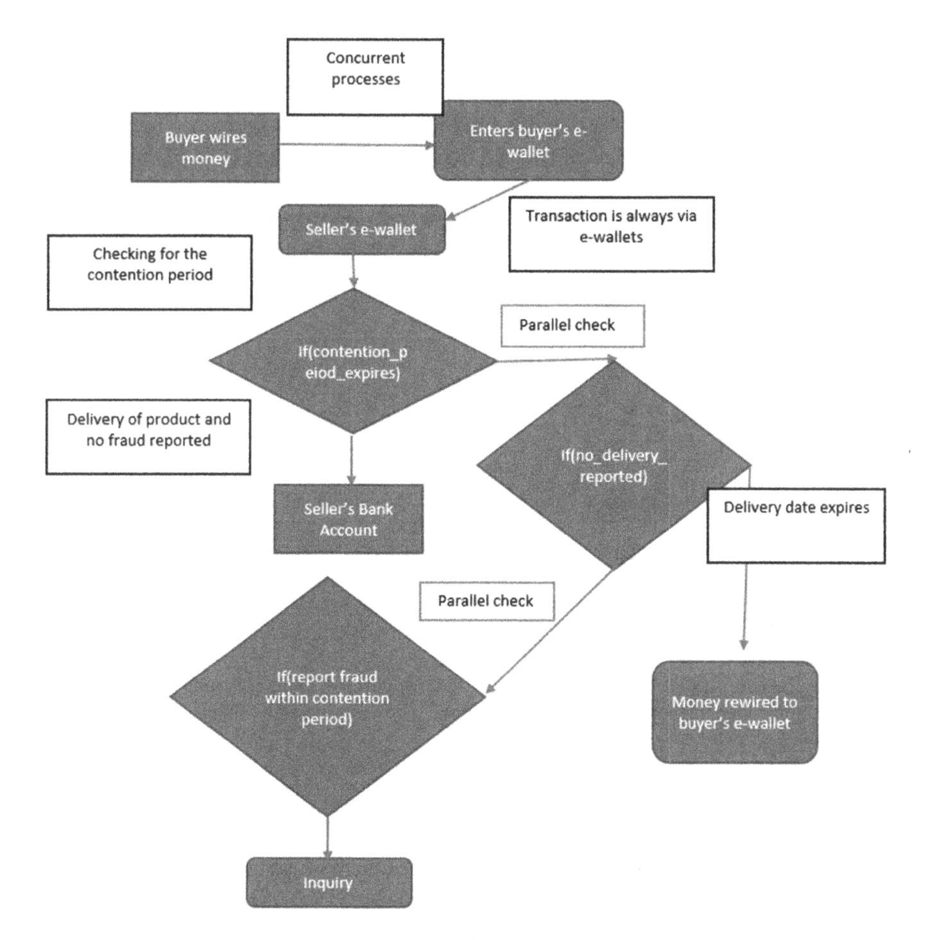

Fig. 9 Flowchart of the transaction system: (Seller sells to highest bidder)

4.3 EOHM Feedback Mechanism

This system also encourages the users to **rate** each other based on their interaction and the transaction they went through. The average rank of any user would also speak of its credibility and past stature on the portal.

As the technology grows, so do the frauds. The EOHM includes a special section of reporting a fraud. In this, any user can report a fraud and categorize it under the given fraud categories or even add a new category, which can be later evaluated and a proper analysis can be done.

5 Future Scope

The research is intended to be extended in the "Bidding" mechanism. The previous paper—OHM already inculcates the use of SDFS (Shill-deterrent fee schedule) by Wang et al. [10] inculcating the Markov model [11], but EOHM will include an anomaly detection, which would be a progressive and dynamic way of learning fraud patterns and prevent users from indulging in fraudulent activities. Also, the feedback mechanism may include subjects such as data mining and text mining for the analysis of user feedback.

References

1. Chaffey, D. (2009). E-business and E-commerce Management. Essex: Pearson Education Limited
2. Mahajan, A. C. (2015). Retail Ecommerce Sales In India 2014–2018. dazeinfo.com.
3. eMarketer. (2015). WORLDWIDE RETAIL ECOMMERCE SALES: EMARKETER'S UPDATED ESTIMATES AND FORECAST THROUGH 2019. New York: eMarketer, Inc.
4. Internet Crime Complain Center: Internet Crime Report, (2015); http://www.ic3.gov/media/annualreport/2015_IC3Report.pdf
5. Nitin Rakesh, Ankit Mundra, Online Hybrid Model for Online Fraud Prevention and Detection, Department of Computer Science and Engineering, Jaypee University of Information Technology, Solan, 2014
6. Chui, K., Xwick, R.: Auction on the Internet: A Preliminary Study, http://repository.ust.hk/dspace/handle/1783.1/1035, July 2008
7. Chu W,Zinkevich M, Li L, Thomas A, Tseng B. Unbiased online active learning in data streams. In ACM conference on Data Mining, 2011
8. Singh, S.P., Shukla, S.S.P., Rakesh, N., Tyagi, V.: Problem reduction in online payment system using hybrid model. Int. J. Manag. Inf. Technol. 3(3), 62–71 (2011)
9. Donga, F., Shatza, S.M., Xub, H.: Combating online in-auction fraud: clues, techniques and challenges. Comput. Sci. Rev. 3(4), 245–258 (2009)
10. Wang, W.L., Hidvègi, Z., Whinston, A.B.: Shill Bidding in English Auctions, Technical report, Emory Uni., http://oz.stern.nyu.edu/seminar/fa01/1108.pdf (2001)
11. Srivastava, A., Kundu, A., Sural, S., Majumdar, A.K.: Credit card fraud detection using hidden Markov model. IEEE Trans. Dependable Secure Comput. 5(1), 1062–1066 (2008)

Robust and Efficient Routing in Three-Dimensional Underwater Sensor Network

Rakshit Jain and Nitin Rakesh

Abstract Three-Dimensional (3D) Underwater Wireless Sensor Networks (UWSN) different from the terrestrial Network forms various challenges like propagation delay, battery, packet loss. Sensor deployed in water is mobile and hence is the prominent challenge while transferring the data packet. Various routing protocols have emerged but still finding a static and reliable route remains a foremost challenge. We propose Robust and Efficient Routing Protocol (RERP) attaining a more static route for the communication between two nodes. In RERP, we assign a rank to each node by formulating important aspects of the sensor in order to form a stable network. So, by selecting least rank node for communication, a more efficient and reliable routing path is constructed. 2-Hop Acknowledgment along with the mobility factor also ensures high packet delivery ratio.

Keywords Routing protocol · Three-dimensional underwater wireless sensor network · Node mobility · Robustness

1 Introduction

Majority portion (71%) of the Earth consists of water. Underwater Wireless Acoustic Sensor Networks (UW-ASN) has been emerging area of research in recent time. Acoustic underwater vehicles (AUVs) which interact with the various sensors deployed underwater for performing a wide range of applications such as monitoring, data collection from underwater, recording of climate, sampling for the ocean, prevention from pollution and disaster, etc. [1–3]. For making these above applications viable, there is a need for communication protocols among underwater devices.

R. Jain (✉) · N. Rakesh
Department of Computer Science and Engineering, Amity University,
Noida, Uttar Pradesh, India
e-mail: rjain.jain444@gmail.com

N. Rakesh
e-mail: nitin.rakesh@gmail.com

A.K. Somani et al. (eds.), *Proceedings of First International Conference on Smart System, Innovations and Computing*, Smart Innovation, Systems and Technologies 79, https://doi.org/10.1007/978-981-10-5828-8_11

Over the years, many routing protocols have been discovered but still, it faces many challenges. Primarily challenges faced are high propagation delay due to various factors such as the propagation speed underwater is 1.5×10^3 m/s (acoustic signals) while the propagation speed of radio is 3×10^8 m/s (radio signals), low battery power, the mobility of the nodes, limited bandwidth, error rate, etc. [4, 5]. Thus, packets can be delivered in a number of ways and examining these challenges, Routing protocol is to be developed for more efficient and robust transfer of the data packet. As the radio waves does not compliment well with conducting medium such as water and also due to high attenuation of the signal, the acoustic medium is preferred. With the distinction in the communication for a terrestrial domain that uses radio waves to the underwater that uses acoustic medium, various challenges are faced while developing a routing algorithm.

In the Three-Dimensional UW-ASN, the sensors are deployed in varying depths and are also the reason for divergent mobility of the sensors. In this paper, we propose to solve the following:

(1) Routing path which is composed of less mobile sensors.
(2) Efficient consumption of the battery of the sensors.
(3) Increase in probability of the delivery of data packet to the destination.

The paper is structured as follows—Sect. 2 is a brief description of related researches done in previous years, Sect. 3 determines the proposed protocol, Sect. 3.2 provides the simulation of the protocol along with the comparisons among existing protocols and Sect. 4 includes the conclusion of the research.

2 Literature Review

In [3] Hop-by-hop dynamic addressing-based (HH-2DAB) approach, sensors are deployed in various depth with a unique ID. This unique ID is the result of the how much deep they are deployed from the surface. With the help of these IDs, sensors determine appropriate forwarding node without any reliable delivery of the packet or acknowledgment of the packet from the forwarding node. It also has more energy consumption near the sink. 2H-ACK [4] approach is suggested as an improvement to the HH-2DAB with the more reliable delivery of the data packet. In this approach, acknowledgment for the delivery of packet is sent after 2-hops transfer of the data including three nodes at a time, which guarantees delivery of the packet [5]. In Vector-based forwarding, an imaginary pipe of a certain width is considered from source to the destination and only the nodes in that pipe can be considered for forwarding of the packet. This approach would collapse in a network of low-density nodes as it is sensitive to routing pipe [6]. Depth-based routing can handle dense network efficiently. Sensor nodes are equipped with the information of their respective location and next forwarding node is chosen with less depth but this approach is not so efficient for the sparse network of nodes.

Amit and David [7] proposed Epidemic routing protocol which forwards packets to every node within transmission range. This approach ensures reliable delivery of the packet but consuming a large amount of energy which is a challenging factor of UW-ASN. Kanika and Nitin [8] proposed an arc movement to solve the mobility issue. It considers a 2D topology and lacks the possibility of solving real problems based on 3D topology. In [2] greedy routing approach, the author presented KRUSH-D an enhanced approach of the arc movement for the solution to node mobility issue in an underwater sensor network. It attempts to solve the mobility issue but in a small environment. In an enormous network of sensors, it would be very challenging to carry out this approach. Robust Routing Protocol [9] aims to achieve robustness and reliable delivery of the packet with the help of the data structure called backup bin. In the backup bin, alternate links are stored in the case of any broken link due to some abnormality. This protocol is not storage efficient due to storage of large volume of backup links.

John Heidemann, Milica Stojanovic, and Michele Zorzi [10] presented reliable and energy-balanced algorithm (REBAR) which is a location-based routing focused on the efficient energy utilization of the nodes. It assumes that each node has a defined radius relating to the distance with the sink. As the energy of the nodes near the sink is used up very early which causes a separation between the nodes and the sink. In sector-based approach [11], a Sector based on destination location prediction for an underwater sensor network is presented which assumes that it knows the position of the destination node and choose the next possible node which is nearer to the destination node along with the priority of the corresponding sector. In [12] Direction Flooding-Based Routing (DFR), the reliable delivery of the packet is proposed. It controls the flooding of the packet to the nearby node by determining the link with the destination. If the link is poor, more nodes are included in the flooding process (vice versa) to establish a good link. In [13] Void-Aware Pressure Routing, a beacon is used that carries the information relating hop count, depth and sequence number so that next hop could be estimated. When the beacon is received from the shallow depth, the direction is set to upward direction otherwise downwards direction. This direction helps in locating sink which is situated at the top by selecting nodes with direction pointing upwards. In [14] Focused beam routing (FBR), less energy consumption of the sensor is achieved. In this, only the location of the nodes is required. Forwarding node is selected which is within the transmitting radius and having least power level. As many of the terrestrial clustering techniques cannot be implemented in underwater [15], Location-based Clustering Algorithm for Data Gathering is proposed. In this network is divides in 3D different Grids each having their own cluster heads which are responsible for the forwarding of the packets from the grid to the destination.

3 Proposed Solution

In this section, we propose our Reliable and Efficient routing protocol (RERP) in detail. The Three-Dimensional (3D) underwater sensor network architecture has been considered for the working of the routing protocol. In this paper, we consider a freely mobile environment where there is no restriction on the mobility of the node. So, it becomes difficult to transfer packets from source to destination. Nodes are deployed considering the three dimension UWSN topology. Further, in this section, we have discussed Protocol Overview, Protocol Design, and also detail working of protocol with flowchart of the routing protocol.

3.1 Overview of Energy Efficient Fitness-Based Routing Protocol for UWSNs

We assume that a node-deployed underwater has a basic information about the location, residual energy, and is allocated a unique ID. Our main aim is to calculate rank for every node which is the deciding factor for selecting forwarding node in order to pass the packet. Rank is calculated considering following elements:

- More priority is given to the nodes those are closer to the line connecting the source with the destination.
- Less mobile nodes are chosen by two factors:

(i) Their total number of previous successful forwarding of packet
(ii) If they are moving toward or away from source node

Above described constituents along with the residual energy helps in the formation of an efficient routing algorithm. The 2-Hop Acknowledgment provides reliability factor in the protocol by ensuring that packet is successfully forwarded and the copy of the packet can be removed also ensuring optimal usage of memory.

3.2 Protocol Design

Protocol design describes different constituents required for the routing protocol which includes the Packet Format, Mobility Factor and the distance vector used to calculate the distance between two nodes in the Three-Dimensional Environment and lastly, formulation of flow diagram by assembling constituents is presented.

3.2.1 Packet Format

First of all, the node sends a "Hello packet" to all the nodes within the transmission range. It would consist of Senders ID and Location Coordinates. In reply, of the Hello packet, Acknowledgment packet will be sent consisting of Senders ID, Location Coordinates and Residual Energy (Figs. 1 and 2).

3.2.2 Calculating Distance Between Two Nodes

Considering the Three-Dimensional deployment of the nodes, they are assigned (x, y, z) coordinates which represent their position in underwater. Euclidean 3D distance formula is used to calculate the distance between two sensors.

$$D_{sd} = \sqrt{(x_2 - x_1)^2 + (y_2 - y_1)^2 + (z_2 - z_1)^2} \tag{1}$$

D_{sd} is the distance between source and destination node. Source node has (x_1, y_1, and z_1) coordinates and destination node has (x_2, y_2, and z_2) coordinates, respectively. Distance of the nodes within transmission range is calculated which acts as a deciding factor while calculating the rank of the node.

3.2.3 Waiting Time

For the reliable delivery of the packet, we have introduced a constrained called the "Waiting Time". This constraint notifies node about successful delivery of the ejected packet ahead in the routing path and gives approval for deletion of the copy of packet maintained in the buffer. If an acknowledgement is not received in that specific time, then current forwarding node is removed from the routing table and next minimum rank node is chosen for forwarding of the packet, thus, reducing the storage requirement along with a decrease in a number of multiple copies to multiple neighbor nodes for reliable delivery.

$$W = 4R/V + 2t_p \tag{2}$$

Fig. 1 Hello packet

Senders Id	Location Coordinates

Fig. 2 Acknowledgment packet

Senders Id	Location Coordinates	Energy

R is the maximum transmission range of a node-deployed underwater, t_p is the processing time required and V is the propagation speed in the medium (Figs. 3 and 4).

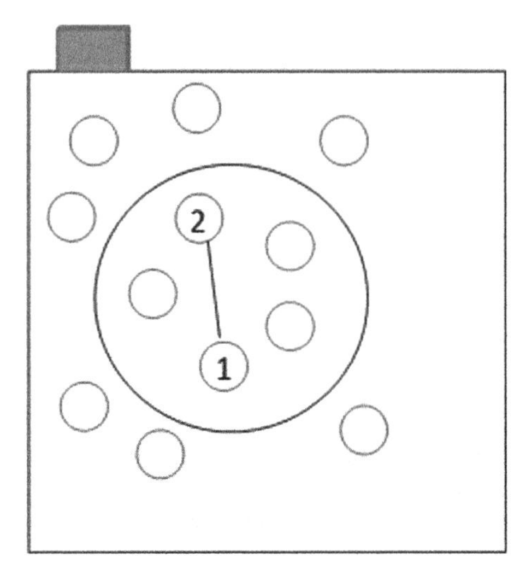

Fig. 3 Packet transfer between 1 and 2

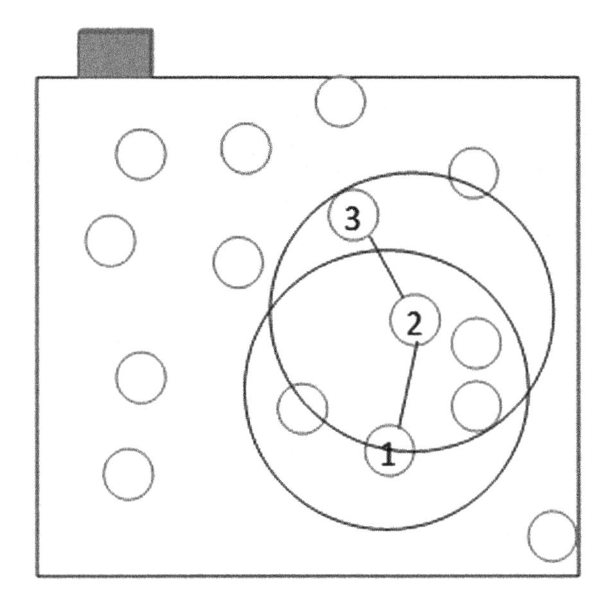

Fig. 4 Packet transfer between 2 and 3

3.2.4 Mobility Factor

In this, we use the 2-hop Acknowledgment component as a deciding factor for reliable delivery of the packet from the node. If the intermediate node can transfer the packet from the source node to the next forwarding node with no error and receive an acknowledgement, then its mobility will be decreased. From Fig. 2 we can deduce that if second node can transfer packet to third node successfully then an acknowledgment would be sent back to second node which it will pass on to first node. This method helps us to find the less mobile and reliable node for the forwarding of the packet.

Case I: Acknowledgment is not received. In this case, the 1st node waits for a period, i.e., "Waiting Time" and then a copy of the packet is passed to other node based on the ranks of the neighboring nodes and the mobility of the intermediate node is decreased by 0.5 factor.

$$M = M * 0.5 \text{ (M is the mobility of the intermediate node)} \quad (3)$$

Case II: Acknowledgment is received. In this case, the 1st node gets the acknowledgment before the Waiting Time is evaded. After this, we delete the copy of the packet in a buffer and increase the mobility by 0.5 factor.

$$M = M \div 0.5 \quad (4)$$

The format of the acknowledgment packet received would consist of the sender's ID along with the packet Sequence number (Fig. 5).

With the mobility factor, we get an insight into the sensor movement and probability of the successful passing of packet if included in the routing path. It acts as a major factor in the determination of the rank of a node (Fig. 6).

3.2.5 Rank

Rank of the node determines the probability of delivery of the packet. Rank of the node is calculated by aggregating the important factors discussed above. Here mobility of the sensor is the dominant factor in determining the rank of the sensor and also predicts the possible delivery of the packet.

$$Rank = \frac{(D^2 \times M \times (1 - cos(A)))}{E} \quad (5)$$

E is the residual energy, M is the mobility factor, D is the distance between the sensors and A is the angle between the destination sensor and the forwarding

Fig. 5 Acknowledgment packet

Senders ID	Packet Sequence Number

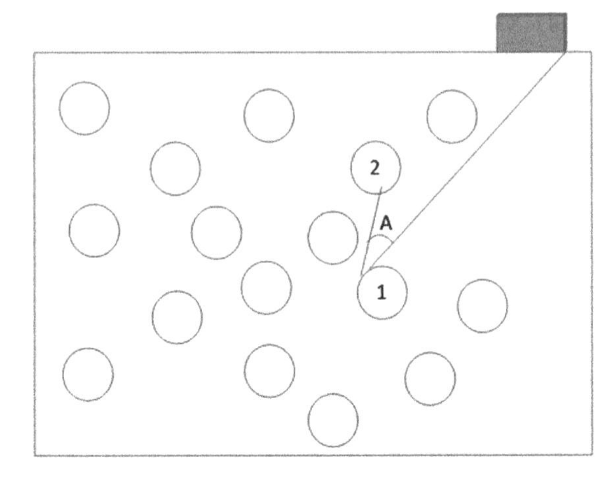

Fig. 6 Angle between the destination node and sink Node

sensor. For selecting the forwarding sensor, the rank of all the neighbor sensors are calculated and the minimum rank sensor is selected for the routing of the packet. If the ranks of two or more packets are same than the packet with least distance is chosen.

3.2.6 Flow Diagram

From the Flow diagram described above, the routing protocol can be described in detail. First of all, the node which wants to send the packet sends the hello packets to the nodes in the transmitting range. Flag is set to 0 upon delivery of the packet to the intermediate node. After that, sensors receive the hello packet, it replies with the acknowledgment packet consisting of the basic information required to calculate the rank. Node with the minimum rank is selected and the packet is forwarded to that node but the acknowledgment will only be returned after it is successfully forwarded to the next relay node. This acknowledgment from the relay node with the deciding factor for improving the mobility of the intermediate node. But if the packet is not forwarded or acknowledgment is not received from the relay node then sender node will again send the packet to another node after Waiting Time of the node gets to 0. Another routing path will be followed by the same process. The main advantage of the protocol is to determine a static path for the reliable and efficient delivery of the packet between nodes (Fig. 7).

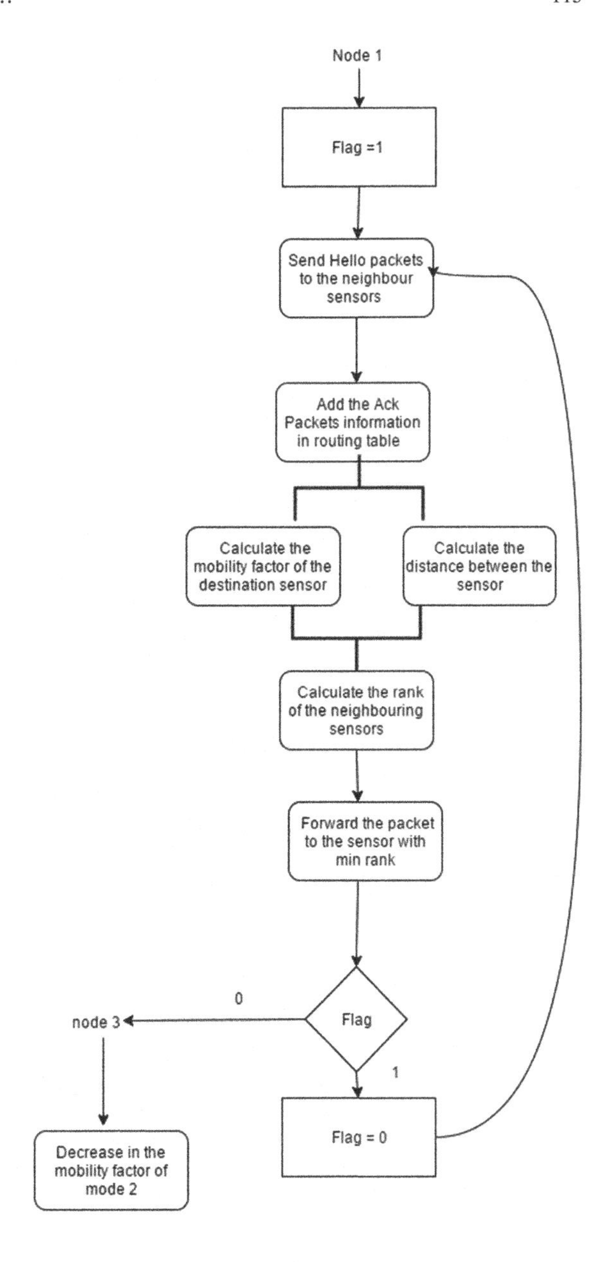

Fig. 7 Flow diagram

4 Conclusion and Future Works

In UWSNs, acoustic communication faces mobility issue which results in various adversities in the delivery of the data packets between the nodes. In order to deal with this issue, we have proposed a Robust and Efficient Routing Protocol (RERP)—an enhancement to the previously proposed KRUSH-D approach. RERP is an adaptive protocol the dynamically forms route by determining the rank of the nodes. It also includes reliable delivery of the data packet with the help of 2-hop Acknowledgments. RERP aims to form a route with less mobile nodes and hence more stable path. In future, we will consider improving the data structure used to store neighboring nodes information.

References

1. R. B. Manjula and S. M. Sunilkumar, "Issues in underwater acoustic sensor networks,"International Journal of Computer and Electrical Engineering, vol.3, no.1, pp. 101–110, 2011.
2. M. T. Kheirabadi and M. M. Mohamad, "Greedy routing in underwater acoustic sensor networks: a survey," International Journal of Distributed Sensor Networks, vol.2013, ArticleID 701834, 21 pages, 2013.
3. C. Detweiler, M. Doniec, I. Vasilescu, and D. Rus, "Autonomous depth adjustment for underwater sensor networks: design and applications," IEEE/ASME Transactions on Mechatronics, vol.17, no. 1, pp. 16–24, 2012.
4. S. Basagni, C. Petrioli, R. Petroccia, and M. Stojanovic, "Opti- mized packet size selection in underwater wireless sensor net- work communications," IEEE Journal of Oceanic Engineering, vol.37, no.3, pp. 321–337, 2012.
5. J. M. Jornet, M. Stojanovic, and M. Zorzi, "On joint frequency and power allocation in a cross-layer protocol for underwater acoustic networks," IEEE Journal of Oceanic Engineering, vol.35, no. 4, pp. 936–947, 2010.
6. Guangjie Han, Jinfanq Jiang, Lei Shu, Yongjun Xu and Feng Wang, "Localization Algorithms of Underwater Wireless Sensor Network: A Survey", february, 2012.
7. Muhammad Ayaz, Low Tang Jung, Azween Abdullah, Iftikhar Ahmad, "Reliable data deliveries using packet optimization in multi-hop underwater Sensor networks", pp 41–48,vol 24, 2012.
8. Muhamad Felemban, Basem Shihaday and Kamran Jamshaidy, "Optimal Node Placement in Underwater Wireless Sensor Networks", 2013.
9. Mohamed K. Watfa, Tala Nsouli, Maya Al-Ayache and Omar Ayyash "Reactive localization in underwater wireless sensor Networks", University of Wollongong in Dubai – Papers, 2010.
10. John heidemann, Milica stojanovic and Michele zorzi "underwater sensor networks: applications, Advances and challenges", pp 158–175, 2015.
11. S. Naik and Manisha J. Nene, "Realization of 3d underwater wireless sensor networks and influence of ocean parameters on node location estimation", International Journal of Wireless & Mobile Networks (IJWMN) Vol. 4, No. 2, April 2012.
12. Kashif Ali and Hossam Hassanein, "Underwater Wireless Hybrid Sensor Networks", pp. 1166–1171, IEEE, 2008.
13. Yi Zhou, Kai Chen, Jianhua He, Jianbo Chen and Alei Liang, "A Hierarchical Localization Scheme for Large Scale Underwater Wireless Sensor Networks" 11th IEEE International Conference on High Performance Computing and Communications, 2009.
14. Davide Anguita, Davide Brizzolara and Giancarlo Parodi, "Building an Underwater Wireless Sensor Network based on Optical Communication: Research Challenges and Current Results", International Conference on Sensor Technologies and Applications, pp. 476–479, 2009.
15. Geetangali Rathee, Ankit Mundra, Nitin Rakesh, and S.P. Ghera, "Buffered Based Routing and Resiliency Approach for WMN", IEEE International Conference of Human Computer Interaction, Chennai, India, 1–7, 2013.

Limiting Route Request Flooding Using Velocity Constraint in Multipath Routing Protocol

Rajshree Soni, Anil Kumar Dahiya and Sourabh Singh Verma

Abstract Mobile ad hoc network (MANET) diverges from the conventional wireless Internet infrastructure. MANETs are deployed in an environment which has no preexisting infrastructure along with the irregular movement of nodes. Because of the inconsistent movement of nodes, routes break frequently and the reestablishment of routes utilizes more network resources and energy. So, it is essential to provide a routing protocol which handles the mobility of nodes and helps in reducing packet drops. Ad Hoc On-Demand Multipath Distance Vector (AOMDV) is a preexisting routing protocol for MANET whose performance may degrade due to high velocity of mobile nodes. Furthermore, the AOMDV protocol incorporates excessive flooding of RREQ packets at the time of route discovery phase. So to deal with these issues, we are introducing a new routing protocol named Velocity Constrained Multipath Routing Protocol (VC-AOMDV). And it also results in better QoS performance simultaneously. The only routes that are velocity constrained and having better link reliability will get chosen for information transmission. The simulation results prove that the proposed protocol VC-AOMDV will perform better with delay, PDF, packet loss, etc., as compared to AOMDV protocol. It will also reduce the packet drop and delay with the increment of successfully delivered packets.

Keywords QoS · MANET · AOMDV · Disjoint · Multipath
Node velocity · VC-AOMDV

R. Soni (✉) · A.K. Dahiya · S.S. Verma
Department of CSE, Mody University of Science & Technology,
Lakshmangarh (Sikar), India
e-mail: rajshreesoni24@gmail.com

A.K. Dahiya
e-mail: anilkumar.cet@modyuniversity.ac.in

S.S. Verma
e-mail: ssverma80@gmail.com

A.K. Somani et al. (eds.), *Proceedings of First International Conference on Smart System, Innovations and Computing*, Smart Innovation, Systems and Technologies 79, https://doi.org/10.1007/978-981-10-5828-8_12

1 Introduction

Mobile ad hoc network (MANET) is a self-coordinated, infrastructure less, and self-structured wireless communication network. Similarly as the communication range of nodes is limited, multiple hops are generally essential for exchanging information among nodes. In MANET, each node is allowed to relocate its position dynamically, and thus the topology varies accordingly. In MANET, each and every node serves as a host and a router and therefore forwards data to rest of the nodes without any central server. The uni-path routing protocols provides single path from route discovery procedure, and so it is not suitable for QoS aware applications like real-time multimedia audio and video transmissions. Thus, the need for multipath routing aroused. It is likely to generate multiple routes for source-destination by means of multipath routing. It is typically suggested for providing load balancing and increasing the consistency of data transmission.

AOMDV allows finding many routes between source and destination during the route discovery procedure. The problem for managing these routes increases simultaneously with the node velocity. The route discovery procedure of AOMDV is initiated by the source. Upon receiving the RREQ by every intermediate node, RREQ broadcasting is performed. This route discovery leads to large number of flooded packets in the network [1]. In preexisting multipath routing protocol, network congestion is increased because of the flooding of RREQ packets initiated by the source. Moreover, if the data is being transmitted through the routes found from the route discovery of AOMDV, there is a great chance of route failure. To solve the problem of excessive flooding and frequent route breaks due to node velocity, a routing protocol named Velocity Constrained Multipath Routing Protocol (VC-AOMDV) is introduced in this work. VC-AOMDV handles the node velocity and dynamic route breaks at the time of route discovery. The performance of VC-AOMDV along with AOMDV is compared considering the packet delivery fraction (PDF), end-to-end delay, packet loss, and throughput. Thus, the prime focus of this proposal is on modifying the preexisting multipath routing protocol which will compute multiple paths along with velocity constraint to accomplish the QoS requirements. VC-AOMDV can be described according to the following architecture Fig. 1.

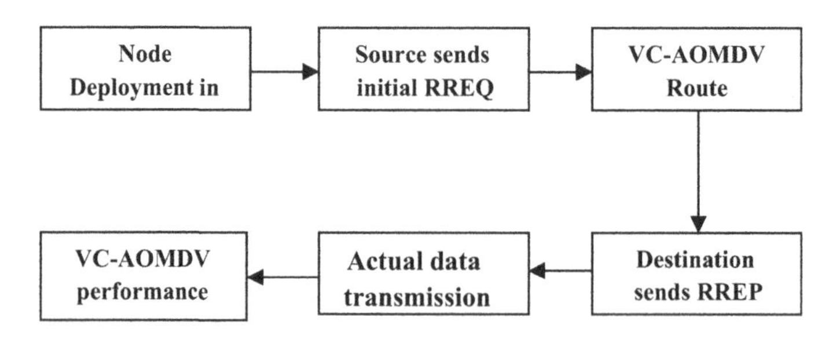

Fig. 1 VC-AOMDV architecture

The simulation is performed using ns-2.35 and the results justify that the introduced protocol provides higher performance comparing with the traditional multipath routing protocol.

2 Related Work

Various QoS depending routing protocols have been introduced for MANET. A large number of them depend on the well-known on-demand routing protocols, AODV and DSR. The important sections of the issues that are being dealt with finding the route are link stability, energy efficiency, node velocity, end-to-end delay, etc. A part of these issues is labeled in multipath routing whereas some of them are in single-path routing.

Multipath routing provides different routes between every source-destination pair. If the failure of first route arises, the reinforcement routes are utilized for proceeding with information transmission. In multipath routing protocols, several paths between source-destination match can be found in distinct fashion such as node disjoint, link disjoint, and zone disjoint. For providing good QoS support, a fraction of the above-listed issues is examined in VC-AOMDV. It is likely to change the route discovery procedure of AOMDV in such a way that it will provide multiple paths satisfying multiple constraints. These paths can be node disjoint or link disjoint. A proposal offered in this study is to find multiple paths in MANETs using node disjoint approach.

2.1 Link Disjoint Paths Based Routing Protocols

On-demand based routing protocol either AODV or AOMDV may be extended to provide velocity constrained multipath routing protocol. In split multipath routing (SMR) [2], RREQs are forwarded by intermediate nodes which they have accepted via distinct links and with the hop count shorter than the initially received RREQ. Route providing the initial RREP will get selected by destination node and at that point it holds up to get more RREQs. The maximally disjoint path from shortest delay path will be chosen by destination node. If more than a single such paths (i.e., maximally disjoint paths) are found then the path having shorter hop count will be chosen to break the tie [3, 15] (Fig. 2).

2.2 Review of Node Disjoint Paths Based Routing Protocols

AOMDV is an extension of AODV routing protocol for discovering the node disjoint routes between source-destination pair. The purpose of node disjoint

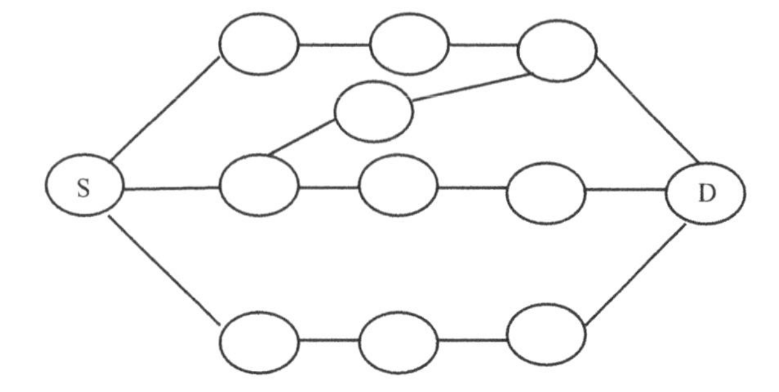

Fig. 2 Link disjoint paths

multipath routing is to find out node disjoint routes that are reliable and exhibits lower overhead in routing control [4]. When the route discovery begins, the link reliable routes having better life time are selected as initial and backup routes. While transmitting data through the primary route if any failure occurs, the backup routes will be utilized for continuing the transmission [5]. The selection of back up routes relies on the connection quality and life time of the link.

2.3 *Extension of AOMDV in Various QoS Related Aspects*

The proposed discussion in [2] is to expand the route discovery procedure of AOMDV routing protocol. This approach uses delay bound along the hop limit. The route discovery of AODMV is extended for finding multiple node disjoint paths along with node velocity constraint and it results in overall reduction of the RREQ packets dropped.

The work proposed in [6] extends AODMV and makes use of link life time. Since the link breaks frequently because of the dynamically changing network topology, it is feasible to have an estimation of the life time of the link dynamically and finding the most stable links. Thus, it is being concluded that finding a path with better stability and greater life time is important [7, 14]. Additionally, having an efficient use of energy is also desirable because of the limitation of battery power of mobile nodes. Thus, the introduced routing protocol, i.e., velocity constrained multipath routing protocol (VC-AOMDV) is an extension of the preexisting multipath routing protocol [8]. It deals with the node velocity and if the node's velocity deviates from a standard velocity value then the packet will get dropped. Thus, the overall performance will get increased as the unnecessary flooding of packets is reduced and thereby the successful packet delivery is increased.

3 Proposed Work

VC-AOMDV routing e protocol is an extension of preexisting AOMDV routing protocol which permits discovering many routes between source and destination during the VC-AOMDV route discovery procedure. The route discovery guarantees the freedom and disjointness of alternate routes, and only the one route will be utilized for information transmission. Two types of control messages are included in the route discovery procedure of VC-AOMDV. These messages include RREQ and RREP.

The VC-AOMDV route discovery is initiated by the source and RREQ gets broadcasted in the network. All the neighboring nodes receive and forward the RREQ. Upon receiving the RREQ by any intermediate node, the packet forwarding decision will be taken considering the node velocity constraint. The result of this discovery process indicates that overall packet drop is reduced to a great extent and so the flooding of RREQ is reduced.

The flow chart of the introduced protocol is given Fig. 3.

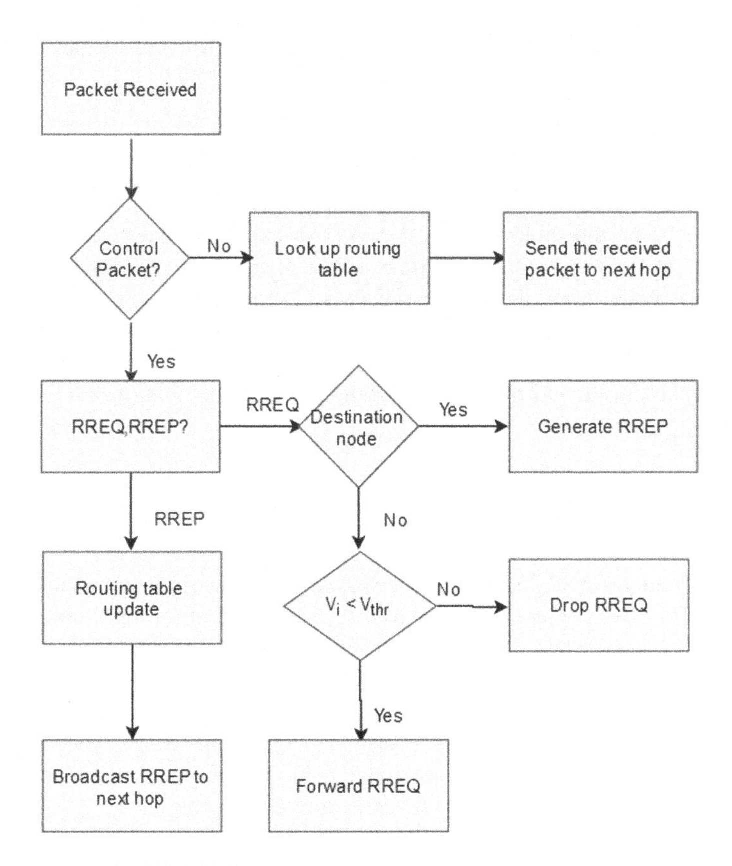

Fig. 3 Flow chart for VC-AOMDV

The performance of VC-AOMDV is measured using the matrices like throughput, pdf (Packet Delivery Fraction), delay, and packet drop. Let us say we have n number of nodes having m connections generated using cbr (Constant Bit Rate) traffic. A MANET can be graphically demonstrated as a set of nodes and edges using G = (V, E), where V represents the set of nodes and E represents the set of edges. A link between the nodes is represented by an edge, which lies within the transmission range of the nodes. The set of neighbors of a node v_i is denoted by $N(v_i)$. Let us assume that we have a path of length n between source S and destination D, represented by ($S = V_0, V_{1...}Vn = D$) where each $v_i \epsilon V$ and $v_i \epsilon N(v_i - 1)$. Let us say that the speed of two consecutive nodes is S1 and S2, and θ be the angle between them. Then according to the cosine theorem, the relative speed of S1 and S2 is

$$S = \sqrt{\left(S1^2 + S2^2 - 2S1S2\cos\theta\right)} \tag{1}$$

The speed of every node is calculated during route request phase and then gets compared against the threshold velocity, which is chosen from the range [Vmin, Vmax]. Vmin is always set to 1 m/s, and we vary Vmax from 5 to 25 m/s. Then, the decision of forwarding packet to next hop is demonstrated by the following equation.

$$\mathrm{IF}\begin{pmatrix} vi \leq Vthr & \text{Forward(Packet RREQ)} \\ \text{else} & \text{Drop(Packet RREQ)} \end{pmatrix}, \tag{2}$$

where Vthr is the threshold velocity a node can exhibit in the network, v_i is the velocity of a node at any instance of time say Δt. If any node exceeds this threshold velocity that node drops the packet otherwise it forwards that packet. So using VAOMDV, we can scale down the unnecessary flooding of packets and so the network congestion. When the threshold velocity is increased, the packet delivery fraction gets reduced, and the overall throughput also gets minimized.

4 Simulation Setup and Results

Network Simulator ns-2.35 is used for simulation. This simulation tool depends upon an event-driven approach and provides a great support for multi-hop networks [9, 10]. For various test conditions, simulation is performed using cbr traffic. Simulation runs for a number of scenarios (e.g., Varying range and simulation time) and cbr traffic (e.g., Varying node's number). Results of simulation are recorded using different traffic patterns, node density, connections, range, and simulation time. The following matrices are considered in this proposed work to perform comparison of performance among several routing protocols:

(1) Throughput: It is described as the rate at which data packets are delivered productively from one node to another. The unit for measuring the throughput [11] is bits per second (bits/sec).

$$\text{Throughput} = \frac{(\text{number of packets delivered} * \text{size of packet})}{\text{total simulation time}} \quad (3)$$

(2) PDF: It is defined as the proportion of effectively delivered packets to the total emissions of packets from the source [12].

(3) Delay: End-to-end delay is the total elapsed time for a packet to reach destination node from source node. The unit for delay measurement is milliseconds [13].

Figure 4 shows the packet delivery fraction against varying simulation time. It is illustrated that as the simulation time increases, the PDF of VC-AOMDV increases over AOMDV (Table 1).

Figure 5 shows the packet drop against simulation time. The total number of packet loss is less for VC-AOMDV as compared to AOMDV.

Figure 6 shows that as the network density increases, VC-AOMDV provides good PDF against AOMDV.

Figure 7 shows that as the network density increases, the packet drop in VC-AOMDV drastically reduces because of the prevention of packet sent from high mobility nodes.

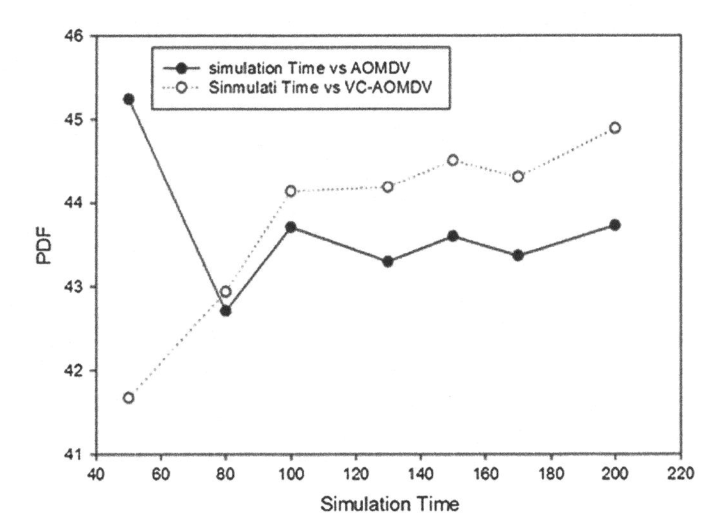

Fig. 4 Simulation time versus PDF

Table 1 Simulation parameters used in ns-2.35

Parameters	Value
Simulation time in seconds	50, 80, 100, 130, 150,170, 200
Number of nodes	20, 50
Map size	500*500
Number of connections	8,10
Rate of data transmission	1/2 to 1/10
Mobility model	Random way point
Traffic type	CBR
Channel type	Wireless channel
MAC type	802.11
Packet size	210, 512 up to 900

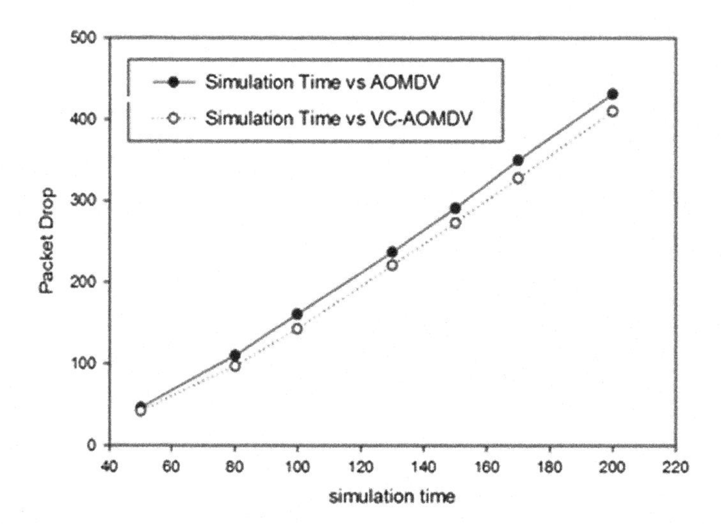

Fig. 5 Simulation time versus packet drop

Figure 8 shows that the end-to-end delay of VC-AOMDV is minimized with the varying packet size. The end-to-end delay in packet transmission of these protocols is nonuniform due to the limited battery power of nodes and node mobility.

Figure 9 shows that VC-AOMDV exhibits less number of packets dropped when scaled against packet size.

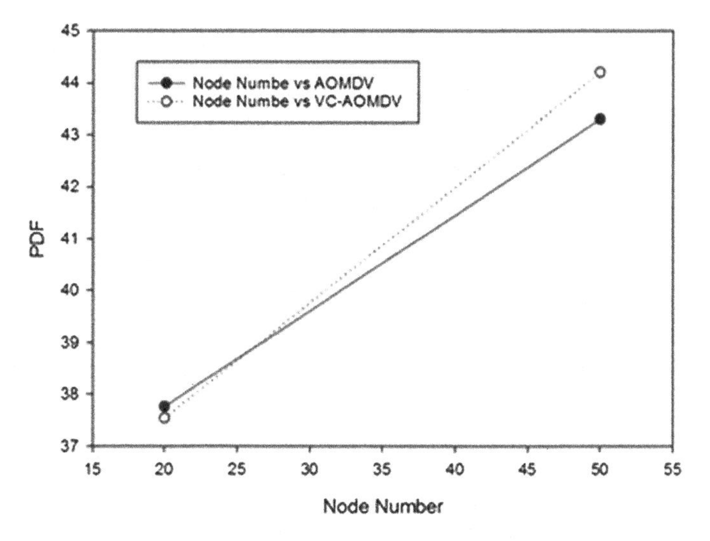

Fig. 6 Number of node versus PDF

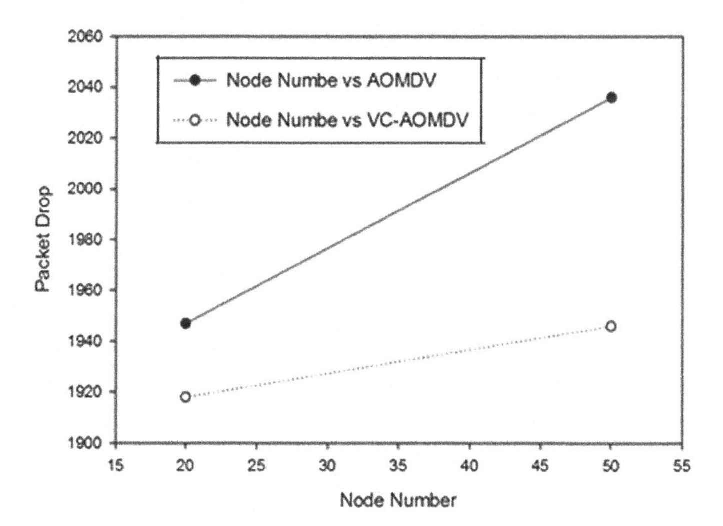

Fig. 7 Number of node versus packet drop

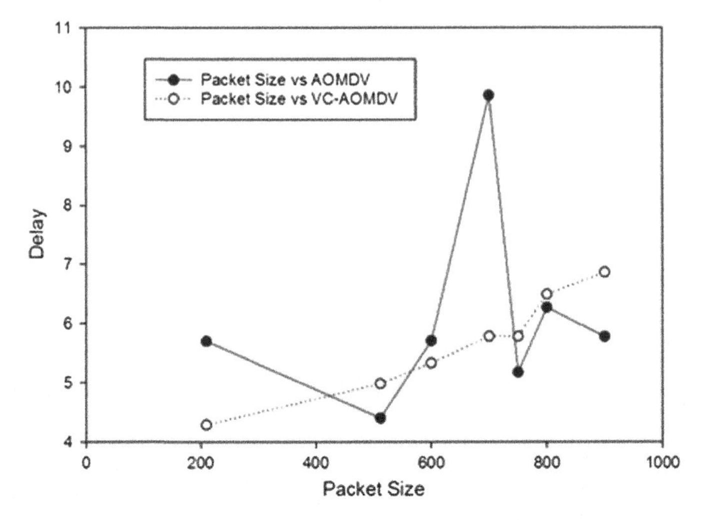

Fig. 8 Packet size versus delay

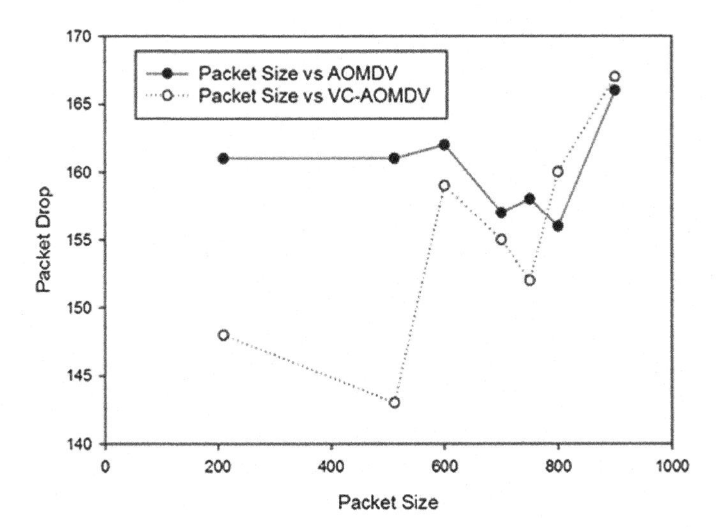

Fig. 9 Packet size versus packet drop

5 Conclusion

Here, we presented a modified multipath routing protocol for MANET which uses velocity as a constraint for its functioning, also it reduces RREQ flooding. In this work, our main focus was to examine and enhance the performance of MANET with multiple routes, under different scenarios. For comparison, both AOMDV and VC-AOMDV were evaluated considering constant bit rate. The comparison was evaluated using several performance matrices like end-to-end delay, throughput, packet delivery fraction, packet drop. Results show that VC-AOMDV performs better for these metrics against AOMDV. VC-AOMDV also results in reduction of network traffic by controlling the unnecessary flooding of RREQ packets. Since the route discovery process of VC-AOMDV uses velocity constraint, it provides reliable paths for information transmission with a less number of packet drops and reduced congestion in the network.

References

1. Verma S.S., Patel R.B. Lenka S.K.: Analyzing varying rate flood attack on real flow in MANET and solution proposal: real flow dynamic queue (RFDQ). Int. J. Inf. Commun. Technol. (in press) Inderscience, **7** (2015).
2. Mukherjee A. & Somprakash B.: Location management and routing in mobile wireless Networks. (2003) Amazon.
3. Er. Rubia Singla and Er. Jasvir Singh.: Node-Disjoint Multipath Routing -Based on AOMDV Protocol for MANETS. proceedings of (IJCSIT) International Journal of Computer Science and Information Technologies, Vol. 5 (4), 2014, 5491–5496.
4. V. Loscri and S. Marano.: A New Geographic Multipath Protocol for Ad Hoc Networks to Reduce the Route Coupling Phenomenon. Proceedings of the 63rd IEEE Vehicular Technology Conference, VTC 2006-Spring, Vol. 3, pp. 1102–1106, 2006.
5. M. Li, L. Zhang, V. O. K. Li, X. Shan and Y. Ren: An Energy- Aware Multipath Routing Protocol for Mobile Ad-Hoc Networks. Proceedings of ACM SIGCOMM Asia. International Journal of Wireless & Mobile Networks (IJWMN) Vol.2, No.4, November 2010.
6. Jamali, S., Safarzadeh, B., & Alimohammadi, H. SQRAODV: A stable QoS-aware reliable on-demand distance vector routing protocol for Mobile Ad Hoc networks. Scientific Research and Essays Academic Journals, 6 (2011).
7. Lim, G., Shin, K., Lee, S., & Yoon, H.: Link stability and Route life time in Ad Hoc Wireless networks. IEEE conference on parallel processing (2002).
8. S. Lee: Split Multi-path Routing with Maximally Dis-joint Paths in Ad-Hoc Networks. Proceedings of IEEE International Conference on Communications, Vol. 10, pp. 3201–3205, 2001.
9. Teerawat Issariyakul, Ekram Hossain: Introduction to Network Simulator NS2. 2009 Springer Science & Business Media, LLC.
10. Tutorial for network Simulator Available: http://www.isi.edu/nsnam/ns/tutorial.
11. Ibrahim Zagli, Min Song.: A Routing Protocol Improving Packet Delivery Ratio for MANET. WSEAS Transactions on Communications 5(6):1047–1054. June 2006.
12. Parulpreet Singh, Ekta Barkhodia, Gurleen Kaur Walia: Performance Study of Different Routing Protocols (OLSR, DSR, AODV) Under Different Traffic Loads and with Same Number of Nodes in MANET using OPNET. IJECT, Vol. 3, pp no.-155–157, March 2012.

13. Wang, Z.: QoS routing for supporting multimedia applications. IEEE Journal on Communication (1996).
14. Verma, S.S., Patel, R.B., Lenka, S.K.: Investigating variable time flood request impact over QOS in MANET. In: 3rd International Conference on Recent Trends in Computing 2015 (ICRTC-2015).
15. Balachandra, M., Prema K. V. & Makkithaya K.: Enhancing the efficiency of MANETs by improving the on demand routing protocols and QoS metrics. International Journal of Recent Patents on 082X (Online), ISSN: 2215–0811 (Print).

Forensic Writer Identification with Projection Profile Representation of Graphemes

Pallavi Pandey and K.R. Seeja

Abstract Handwriting is one of the behavioral biometric techniques for person identification. Our paper proposes a novel grapheme-based handwriting recognition system. First, the images of handwritings are segmented into graphemes and then graphemes are represented as projection profile. Then, dictionary is learnt from the images of handwriting using k-means clustering. Now using dictionary, writer feature vectors are generated and stored with writer label. Finally, query document features are matched with writer feature vectors using k-nearest neighbor. The outcome is the writer label with the highest similarity score.

Keywords Graphemes · Projection profile · Dictionary · Writer feature vector

1 Introduction

In forensics, writer identification is a major problem. Often, it is required to identify the writer of threatening letters, suicidal notes, historical documents, duplicate wills, etc. with their handwriting. Another application of writer identification is the detection of fraudulent alteration in ball point pen strokes. Writer identification problem comes in the domain of pattern recognition and is directly applicable in the field of forensic document analysis and historical document analysis. Images of handwritten samples will be used and various image processing techniques will be applied. The writer identification techniques are classified as grapheme-based and text-line based. Graphemes are the structural primitives which are obtained by the segmentation of the handwriting. Text line based features are based on those characteristics of the handwriting which is visible, for example, the width of characters and height of writing zones.

P. Pandey (✉) · K.R. Seeja
Indira Gandhi Delhi Technical University for Women, Delhi, India
e-mail: jipallavi@gmail.com

K.R. Seeja
e-mail: seeja@igdtuw.ac.in

A.K. Somani et al. (eds.), *Proceedings of First International Conference on Smart System, Innovations and Computing*, Smart Innovation, Systems and Technologies 79, https://doi.org/10.1007/978-981-10-5828-8_13

Writer identification is a challenging problem. Problem occurs with the variability in writing style of alphabets and numeric digits. For example, one character (letter or digit) can be written in different ways in a handwriting style and the same writer may write alphabets in a different manner.

In this paper, we will address the writer identification problem from scanned images of handwritten text document. Here, a new representation scheme of graphemes, which is a modified projection profile, is proposed.

2 Literature Survey

In the past years, various attempts have been made for the automation of the process of forensic document examination with an objective to achieve high writer identification rates. Schomaker et al. [1] worked on writer identification system and have used fragmented connected component contours. In this work, the contours of connected component are identified and then fragmented in the mixed style hand written samples of limited size to obtain the grapheme. Bulacu and Schomaker [2] have analyzed combinations of various features. He has given a fusion scheme of features in which the distance between two handwritten documents was calculated by computing the average of the distances obtained by individual features from the combination. He has worked on a large dataset which was containing 900 writers. By fusion of multiple features like grapheme directional, and run length PDFs, he has got performance improvement in writer identification and verification system. Kumar et al. [3] proposed a forensic writer identification based on sparse representation of graphemes. Due to the sparsity, feature representation could be more sensible to represent handwriting features and writer specific features could be captured more accurately. Chanda et al. [4] proposed a writer identification system for Oriya script based on curvature feature and support vector machine which could perform reasonably well when the amount of text is small. Chanda and Franke [5] proposed a text-independent writer identification for Bengali script by using gradient feature and discrete directional feature with SVM. Bulacu and Schomaker [6] worked on writing style that was from oriented edge fragments. A new version of the location-specific features was computed on the top half of the text line and then bottom half of text lines and finally combined. This new location-specific feature has given significant improvements in writer identification performance.

Marti et al. [7] proposed a writer identification and used text line based features. They have extracted 12 features from hand written text lines and these 12 features were used to identify the writer. Panagopoulos et al. [8] have proposed a method to classify ancient Greek inscriptions. These inscriptions are characterized by the writer who carved them. Brink et al. [9] worked for the amount of handwritten text that is required for text-independent approach. He stated that the performance for text-independent and offline writer verification and identification increases if the document contains larger amount of text. Srihari et al. [10] did a study of the handwriting of twins to describe the individuality of handwriting. They have found that handwriting is affected by training, psychology, and other behavioral factors.

Bensefia et al. [11] developed their writer identification and verification system by information retrieval model which was based on textual information. Brink et al. [12] proposed a writer identification method using directional ink trace width measurements. Zois and Anastassopoulos [13] used the feature vector which is derived by processing words horizontal projection functions morphologically. Zois and Anastassopoulos [13] have given automatic segmentation method for the IAM database. This segmentation procedure gives information about words too. Zimmermann and Bunke [14] worked on various feature integration and got promising results. Cha and Srihari [15] explored the procedure of handwriting identification and have given a forensic writer identification technique. Morris [16] proposed a 2-D Gabor filtering technique for feature extraction from the textures of the handwriting images and used a weighted Euclidean distance classifier for identification. Said et al. [17] developed handwriting identification which was based on run length coding. Arazi [18] divided handwriting in small fragments. They have computed texture descriptors from these obtained fragments (textures). Hannad et al. [19] have done word segmentation with quadratic assignment problem. In this, pair wise correlation between text gap is considered.

3 Proposed Methodology

In the proposed method, IAM offline database [20], which is a collection of scanned images of documents, is used. All the images are converted from grayscale to binary images using Otsu method. Then unwanted texts are removed. In all the forms, few lines are printed in English in each form and the same text is copied by writers. First of all, graphemes are extracted from the images and then dictionary is learnt. Then, labeled writer database is created with writer feature vectors. Now, writer feature vector of query document is matched with the database to find the writer label. Implementation details are organized in the five steps which are explained in the following subsections:

3.1 Grapheme Extraction

The proposed method is based on graphemes. To extract the graphemes, after image preprocessing, thinning of the documents images is done. With the help of thinning, we will get a line which is one pixel thick, through the middle of an image object and the topology of that object will be preserved. Thinning is a preprocessing step for various image processing algorithms. Thinned images reduce processing time for other image processing operations. To perform thinning, Zhang-Suen thinning algorithm [21] is used. Then, the thinned images of handwritings are segmented into graphemes. For this purpose, check the number of neighbors of each pixel. If the neighborhood pixels are greater than two for any pixel that pixel will be deleted. Graphemes are obtained after deletion of these pixels. Some of the graphemes are very small in size and useless. So they need to be cleaned. For cleaning, the number of pixels, which is called the area, was counted in each connected component and the threshold was calculated using mean and standard error. But before performing these operations, connected component needs to be labeled. Labeling is done as 1 for first connected component, 2 for second connected component, and so on. Entries are zero in the connected component labeling matrix, where the pixel value is zero that is black (background). The graphemes whose size are less than the threshold in the document are removed. Above process is done for all the documents of training set to get the final set of graphemes.

3.2 Representation of Graphemes

In the proposed solution, graphemes are represented using horizontal projection profile. The general horizontal projection profile gives the total number of black (or white) pixels in each row. That is, it gives pixel frequency in each row for a particular pixel value. In this paper, a new projection profile technique is proposed. The proposed projection profile is defined as the position of the first white pixel in each row. To implement this, all the graphemes are inserted into a bounding box. Bounding box is a rectangle which tightly holds the grapheme image. The bounding box is of different sizes. So, it is necessary to fix the size. Here, it is fixed as 43×43 pixels bounding box to represent graphemes. This size is obtained by varying the size from 30 to 50 and better result is obtained taking 43. After that image will be resized so that actual image of grapheme fits into the bounding box of size 43×43. Few obtained graphemes are shown in the Fig. 1.

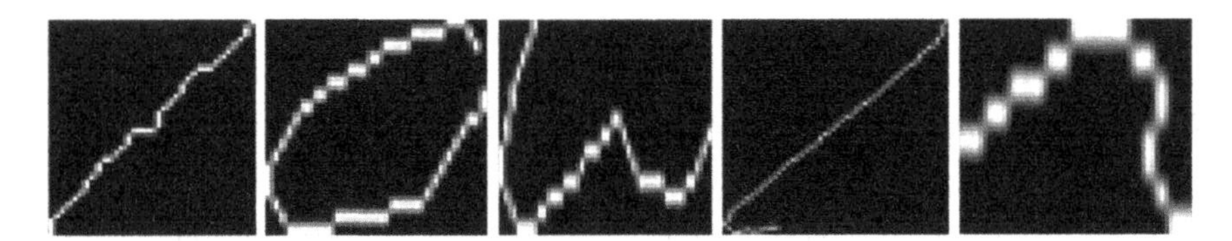

Fig. 1 Graphemes

In this representation of graphemes, the vector contains 43 entries, i.e., one for each row. The entry is the position of the first white pixel value in the image. Generally horizontal projection profile gives the number of pixels in each row but since images are thinned already, there will be single pixel in each row (not always) and the first white pixel value is recorded. Apply same procedure in all row of the matrix and record the values. A 1×43 matrix of the form [40, 30, 35, ……………….., 20] is obtained. This grapheme representation vector is then normalized. In this way, the graphemes feature vector for all the documents written by various writers are collected and stored in a matrix data structure.

3.3 Dictionary (Code Book) Learning

Dictionary learning is one of the major tasks in writer identification systems. For dictionary or codebook learning, clustering method is applied on the obtained matrix of graphemes. Dictionary of graphemes shows different types of graphemes possible to occur in English language handwritings. To prepare the dictionary (codebook) of graphemes, documents from 650 writers have been taken, which is nothing but the images of handwritings of different writers. The dictionary prepared from various documents gives the database of different possible graphemes in English language handwriting. For dictionary learning, K-means clustering algorithm is used. The value of k is varying from 100 to 250 to observe the results. Best results are obtained for the k equal to 240.

3.4 Writer Feature Generation

In our forensic writer identification, we believe on the hypothesis that each writer produces a specific kind of curve in his/her handwriting. So working on this hypothesis, we have tried to extract the features of handwriting which is directly related to the curve produced by writers. Therefore to generate the writer feature vector, handwritten document is segmented into graphemes then each grapheme is matched with the dictionary, i.e., first grapheme goes to which cluster, second goes to which, and so on. In this way, we will get the vector of size $1 \times k$ where k is the value used in k-means clustering. Finally, all the writer feature vectors are stored with its writer label in a database.

3.5 Writer Identification

In writer identification phase, when a query document comes, its writer feature vectors are computed in the same way as given in Sect. 3.4. Then, query document features are matched with the database which contains the writer feature vectors with writer label which is obtained by the process given in Sect. 3.4. Nearest neighbor algorithm is used for this purpose. Output is the writer label with highest similarity measures.

4 Results and Discussion

The proposed methodology has been implemented in MATLAB and IAM Handwriting Database is used. This database contains 1539 scanned text pages contributed by 657 writers. The documents of 650 writers are used for dictionary learning. For writer feature vector generation, documents from 101 writers have been taken. Only those writers are taken who have contributed two or more documents. Single pages from all these writers have been taken for writer features.

The performance of the proposed writer identification system is evaluated by varying different parameters like cluster numbers and bounding box size. The number of clusters has taken the values 100 to 250 incrementing 10 at a time. Accuracy was varying from 84 to 88.54% where the best performance is obtained at 240. Similarly, the size of bounding box to hold the grapheme tightly has been varying from 30 × 30 to 50 × 50 and the highest accuracy for writer identification is obtained when the size of bounding box was taken as 43 × 43.

The accuracy of k-means classification method is 88.57% as out of 70 documents, it has classified 62 documents correctly. Table 1 shows the comparison of the proposed model with grapheme based models and have used IAM database.

It is found that the proposed model gives better accuracy than other reported grapheme-based methods.

Table 1 Performance comparison

S. No.	Model	Number of writers used for dictionary	Number of clusters	Accuracy (%)
1	Bulacu and Schomaker [22]	650	400	80.00
2	Siddiqi and Vincent [23]	650	100	84.00
3	Kumar et al. [3]	650	200	88.43
4	Proposed method	650	240 100	88.57 84.20

5 Conclusion

In this paper, a grapheme-based forensic writer identification method is proposed which follows the text-independent approach. Graphemes are represented using a modified horizontal projection profile in which the position of first foreground pixel of grapheme in each row of bounding box is recorded. K-means clustering is used to prepare the dictionary and K-nearest neighbor is used for writer identification. The performance evaluation of the proposed model shows better accuracy than other reported grapheme-based methods.

References

1. Schomaker, L., Bulacu, M., Franke, K.: Automatic writer identification using fragmented connected component contours. In: F. Kimura & H. Fujisawa, Proc of 9th IWFHR, Japan, Los Alamitos: IEEE computer society (2004) 185–190.
2. Bulacu, M., Schomaker, L.: Combining Multiple Features for Text Independent Writer Identification and Verification. Proc. of 10th International Workshop on Frontiers in Handwriting Recognition, IWFHR. La Baule, France (23–26 October 2006) 281– 286.
3. Kumar, R., Chanda, B., Sharma, J. D.: A novel sparse model based forensic writer identification. Special issue in Frontier in Handwriting Processing, Pattern Recognition Letters 35 (2014) 105–112.
4. Chanda, S., Franke, K., Pal, U.: Text Independent Writer Identification for Oriya Script. Document Analysis Systems (2012) 369–373.
5. Chanda, S., Franke, K.: Text Independent Writer Identification for Bengali Script. International Conference on Pattern Recognition (2010).
6. Bulacu, M., Schomaker, L.: Writer Style from Oriented Edge Fragments. In proc of the 10th international conference on computer analysis and pattern (2003).
7. Marti, U.-V., Messerli, R., Bunke, H.: Writer Identification using text line based features. IEEE (2001).
8. Panagopoulos, M., Papaodysseus, C., Rousopoulos, P., Dafi, D., Tracy, S.: Automatic Writer Identification of Ancient Greek Inscriptions. IEEE transactions on pattern analysis and machine intelligence, vol. 31, no. 8 (2009).
9. Brink, A., Bulacu, M., Schomaker, L.: How Much Handwritten Text Is Needed for Text-Independent Writer Verification and Identification. International Conference on Pattern Recognition (2008).
10. Srihari, S., Huang, C., Srinivasan, H.: On the Discriminability of the Handwriting of Twins. J Forensic Sci, Vol. 53, No. 2 (2008).
11. Bensefia, A., Paquet, T., Heutte, L.: A writer identification and verification system. Pattern Recognition Letters, 26(2005) 2080–2092.
12. Brink, A. A., Smit, J., Bulacu, M L., Schomaker, L.R.B.: Writer identification using directional ink-trace width measurements. Pattern Recognition 45.1 (2012) 162–171.
13. Zois, E. N., Anastassopoulos, V.: Morphological waveform coding for writer identification. Pattern Recognition, volume 33, issue 3 (2000) 385–398.
14. Zimmermann, M., Bunke, H.: Automatic segmentation of the IAM offline handwritten text {English} database. 16th international conference on Pattern Recognition, Canada, volume 4 (2002) 35–39.

15. Cha, S. H., Srihari, S.: Multiple Feature Integration for writer identification. 7th m international workshop on Frontiers in handwriting recognition, IWFHR VII, Amsterdam, The Netherlands. (2000) 333–342.
16. Morris, R. N.,: Forensic Handwriting Identification. Academic Press(2000).
17. Said, H.E.S., Tan, T. N., Baker, K.D.: Personal Identification Based on Handwriting. Pattern Recognition, vol. 33 (2000) 149–160.
18. Arazi, B.: Handwriting Identification by Means of Run-Length Measurements. IEEE Transactions on Systems, Man, and Cybernetics—TSMC 01/1977; 7(12):878–881. doi:10.1109/TSMC.4309648.
19. Hannad, Y., Siddiqui, I., Kettani.: Writer identification using texture descriptors of handwritten fragments. Expert Systems With Applications, 47(2016) 14–22.
20. Gole, K., Mulani, J., Kumar, G., Gosavi, D.: Writer identification for handwritten document based on structured learning. International Journal of Engineering Research, Volume 5 (2016) 54–55.
21. Marti, U., Bunke, H.: The IAM-database: An English Sentence Database for Off-line Handwriting Recognition. Int. Journal on Document Analysis and Recognition, Volume 5 (2002) 39 –46.
22. Bulacu M., Schomaker, L.: Text-Independent Writer Identification and Verification Using Textural and Allographic Features. IEEE Transactions on pattern Analysis and machine intelligence, volume 29, Issue 4 (April 2007) 701–707.
23. Siddiqi, Imran, Vincent, N.: Text independent writer recognition using redundant writing patterns with contour-based orientation and curvature features. Pattern Recognition. 43.11 (2010) 3853–3865.

Secure Public Auditing Using Batch Processing for Cloud Data Storage

G.L. Prakash, Manish Prateek and Inder Singh

Abstract Cloud computing enables the users to outsource and access data economically from the distributed cloud server. Cloud provider offers data storage as a service and share the data across multiple authorized users in a cost-effective manner. In the cloud-computing model, once the data leaves the owner premises, there is no control of data to the data owner. To provide security for the outsources, data is one of the challenging tasks in cloud computing. Public auditing of outsourced data to verify the integrity of the data provides the data security. In this paper, we proposed a multisector public auditing system by utilizing a linear combination of homomorphic linear authenticator tags of the file blocks. Our proposed method performs the multiple auditing tasks simultaneously, without retrieving the entire file, which reduces the computation and communication overheads in the auditing. The analysis of the proposed system shows that, multisector public auditing method is better than the existing auditing method in terms of data security, communication, and computation overheads by utilizing linear combination of homomorphic linear authenticator tags of the file blocks. Our proposed method performs the multiple auditing tasks simultaneously, without retrieving the entire file, which reduces the computation and communication overheads in the auditing. The analysis of the proposed system shows that, multisector public auditing method is better than the existing auditing method in terms of data security, communication, and computation overheads.

Keywords Security · Authenticator · Public auditing · Integrity
Storage as a service

G.L. Prakash (✉) · M. Prateek · I. Singh
School of Computer Science Engineering, University of Petroleum
and Energy Studies, Dehradun, India
e-mail: prakashgl@ddn.upes.ac.in; glprakash78@gmail.com

A.K. Somani et al. (eds.), *Proceedings of First International Conference on Smart System, Innovations and Computing*, Smart Innovation, Systems and Technologies 79, https://doi.org/10.1007/978-981-10-5828-8_14

1 Introduction

In cloud computing, storage as a service is provided by the cloud service provider such as Google Drive, iCloud, Dropbox, etc., to data users with lower cost than traditional storage service. The cloud service providers not only stored the owner's data in cloud but also shared to authorized users. Once the data leaves from the data owner premises, there is no control of outsourced data to data owner.

To provide the security to the outsourced data is a challenging task in cloud computing. The most important parameters for data security are confidentiality, integrity, and availability in cloud data storage service. Before utilizing the cloud data by authorized user, the data integrity should be verified using any verification method.

The traditional method of data verification proposed in RSA [1] MD5 [2] is not efficient in cloud-computing model because of the amount of data for data verification. To verify the correctness of the outsourced data in cloud computing, recently there are many public auditing mechanisms proposed in [3–8] without downloading the entire data from cloud server.

In the proposed method the data is divided into smaller chunks and each chunk identified by its authenticator (data owner) and the random combination of these chunks integrity is checked for the entire data before accessing. In this paper, we proposed a single and multiple public auditing to solve the security issues on shared data by utilizing the homomorphic authenticator and random masking to hide data, so that the public auditor verify the integrity of the shared data without retrieving the entire blocks. Further, we extend this work for batch auditing, which can perform multiple tasks simultaneously. To improve the efficiency of verification we introduce the multisector batch auditing techniques.

The rest of this paper is organized as follows; In Sect. 2, we present the related work to the proposed method, Sects. 3 and 4 define the statement of the problem and design goals and the traditional identity verification methods, respectively. Sections 5 and 6 explain the public auditing method for single and batch processing, respectively, Sect. 7 presents analysis of our proposed method, and finally conclusion is provided in Sect. 8.

2 Related Work

In cloud computing, auditing the outsourced owner data on the untrusted server is a challenging task. To design an efficient data auditing method, some of the data auditing works are proposed in [9–12] as follows.

Juels et al. [9] have presented the work Proof of Retrievability (PoRs) to check the correctness of outsourced data at cloud server using error correcting code. This method supports local auditing without supporting trusted third-party auditor.

The public auditing method a provable data possession (PDP) proposed Ateniese et al. in [10] and Boyang Wang et al. [13], to provide the integrity of remotely stored data by checking the few file blocks. As compared to local auditing, the public auditing provides data verification and also minimizes the user's overheads by introducing an intermediated trusted third-party auditor. Apart from public auditing, there are some other data security, privacy protection, batch and dynamic auditing are concerned for public auditing in cloud storage.

Erway et al. [11] have proposed a dynamic provable data possession (DPDP) by a rank-based authenticated skip list. They demonstrated a general pattern for dynamic data auditing using dynamic data structures with verification algorithms. With the continuation of this work, Wang et al. [12] presented a dynamic auditing using Merkle Hash Tree (MHT), which supports privacy-preserving and batch verification. These methods have more computation overhead at third-party auditor and communication overhead for updating and verification.

3 Problem Statement

3.1 System Model

Consider the proposed cloud data storage system model consisting of three entities such as data owner, auditor, and cloud server as shown in Fig. 1.

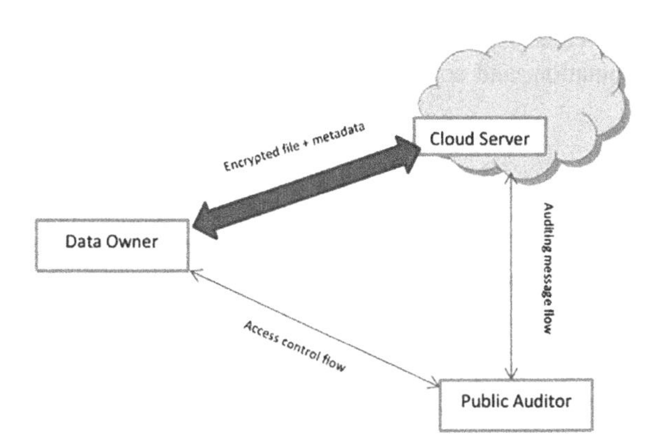

Fig. 1 Cloud system model

The data owner stores his data on a cloud server, and then there is no control for his data so that data owner is less important in this cloud model. Providing security and integrity of outsourced owner data is an important task in a cloud server.

3.2 Threat Model

The attacks for outsourced data can come from either internal or external attacks at cloud server. These attacks can originate from software problems, hardware failure, communication problem, etc., for the cloud server reputation, cloud server hides these data corruption threats from the data owners and users. Since the cloud server and auditor are untrusted entities, they can learn the outsourced data after auditing. These data integrity threats for owners data can come at server and auditor is more important in business rather than honest process.

3.3 Design Goals

To audit the outsourced data at untrusted cloud server in a secured and cost-effective method the following design goals are considered.

- **Verification**: Verify the correctness of the outsourced owner data using batch auditing without retrieving outsourced data.
- **Cost effective**: Design a cost-effective model for data correctness verification using batch auditing.
- **Data privacy**: Ensure that, the cloud server and auditor cannot derive outsourced user data using metadata collected during auditing.
- **Minimal overheads**: Data privacy auditing process is achieved with minimal storage, computation, and communication overhead.

4 Background

4.1 Message Authentication Code(MAC)-Based Method

Before outsourcing owner data to cloud server, owner split the data file to data blocks and generate the MAC for each block using the secrete key (sk) then send data blocks and MACs to the server and corresponding MAC secrete key to the auditor. For verification of data blocks, auditor retrieves data blocks from the server using (sk). This linear computation and communication cost for the auditor is the drawback for this method of verifications.

4.2 Restricted MAC Method

To overcome the drawbacks of the MAC method and privacy of the outsourced data, owner restricts the number of verification for equality checking. In this method, data owner generates a set of MAC secret keys $\{sk_i\}$ for entire file F *i.e.* $MAC_{sk_i}\{F\}\ for\ i = 1\ to\ s$. After preparing metadata for the file $\{MAC\ and\ sk\}$, owner outsource file to the server and publish metadata to the auditor. For each audit, auditor ask MAC of the file to server using MAC secret key *sk*.

4.3 Homomorphic Linear Authenticator Method

To improve the proof of data possession/irretrievability of outsourced data Ateiese et al. [10] proposed a homomorphic linear authenticate (HLA) code scheme, which is the efficient approach in cloud storage model.

The design technique to construct the HLA tags is as follows; for a file F is divided into n blocks *i.e.* $F = \{b_i\}$ where i = *1* to n and generates a homomorphic authenticator tag σi for each block bi with the help of authentication key.

The data verification prover (cloud service provider) can handle the verification challenges V_i using a pair of aggregated block $\prod_{i=1}^{n} \sigma_i^{V_i} mod\, p$ and tag $\sum_{i=1}^{n} V_i^{b_i}$. This aggregated block and tag, which enables the verifier (public auditor) to check the validity of the cloud service provider using authentication key. To generate a correct response message to the auditor, HLA scheme enables the CSP to generate the correct response message with one block–tag pair. Using this HLA scheme minimizes the computational and communication overheads as compared to the non-homomorphic scheme (MAC-based scheme).

5 Public Auditing for Single Owner

In this section, we proposed a public auditing for single data owners file by utilizing HLA scheme for proof of irretrievability of outsourced data file. The design steps for this method such as key generation, file setup, and public auditing are described as follows.

5.1 Key Generation

Generate the secret and public parameters for file, data owner follows the following steps.

- Select the random signing key pair (ssk, spk) to generate the secret and public parameters
- Select the random integer number x and u from the group G1, then compute v such that $v = g^x$.
- Secrete parameter $sk = (x, ssk)$ and public parameter $pk = (spk, v, g, u, e(u, v))$

5.2 File Storage on Cloud

The initial file setup for file outsourcing at cloud server the data owner prepares the metadata using the following steps

1. Prepare the metadata of the file F

- Split the file F into n blocks, i.e., $F_i = \{b_1, b_2, \ldots b_n\}$ of equal size
- Compute the authenticator (σ) for each block b_i as follows:
- Select the unique identifier for the file F and append block index to it i.e., $W_i = id||i$, $\sigma_i = \left(H(W_i).u^{b_i}\right)^x$
- Prepare the authenticator set $\phi = \{\sigma_i\}$ where $i = 1$ to n
- Generate the file tag t for the integrity of the file identifiers $t = id||signature_{ssk}(id)$

2. Send the file F and metadata $\{\phi, t\}$ to the cloud server.
3. Remove file and metadata from the local storage.

5.3 Auditing

To verify the correctness of the owner's data file stored on untrusted cloud server using public verifier without retrieving the original file as follows:

1. Auditor retrieves the file tag t from the server and verifies the *signature(id)* using public key *spk*
2. Generate a challenge message $chal = \{i, V_i\}$, Where i = *{s1, s2, …, sc}* and $i <= Sc < [1, n]$ and *Vi* is the random number.
3. Sends the challenge message to the server.
4. Server prepare the response message *{μ, σ, R}* using the following steps
   a. Select the random number r and compute; $R = e(u, v)^r \epsilon\ G_T$
   b. Compute the linear combination of sampled data blocks; $\mu' = V_i^{b_{ij}}$
   c. Blind μ with r using; $\mu = r + \Gamma\mu' mod\ p$ where $\Gamma = h(R)$
   d. Calculate the aggregated authenticator $\sigma = \prod_{i \epsilon I} \sigma^{V_i}$
5. Send the response message *{μ, σ, R}* to the auditor.
6. Auditor computes $\Gamma = h(R)$ and verifies $\{\mu, \sigma$ and $R\}$ using Eq. 1.

$$R.e\left(\sigma^{\Gamma}, g\right)? = e\left(\prod_{i=s1}^{s_c} H(W_i)^{V_i})^{\Gamma}.u^{\mu}, v\right) \quad (1)$$

6 Batch Auditing

Auditing multiple user's data file using single auditing method is more expensive in terms of computation and communication overheads for the server as well as an auditor. To overcome these overheads, the auditor groups the multiple files in the batches and submits each bath to cloud server for audit as shown in Fig. 2. The detailed procedure for this auditing explained in Algorithm 1.

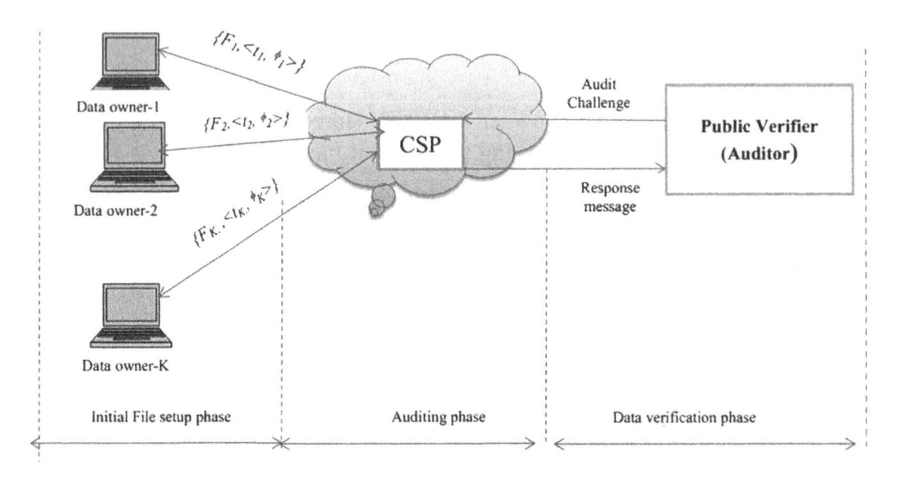

Fig. 2 Batch auditing model

6.1 System Model

Let k be the number of data owners in the system and each owner has data file

$F_k = \{b_{k1}, b_{k2}, \ldots, b_{kn,}\}$ to store on a cloud server and each file has n number of data blocks. To audit the outsourced files stored on cloud server using batch processing the detailed procedure is as follows.

6.2 Key Generation

For each data, owner choose the secret private and public keys (*ssk, psk*) and a random number x_k secret key $= (x_k, ssk)$ public parameter is $(spk_k, v_k, g, u_k, e(u_k, v_k))$
Where $v_k = g^{x_k}$

6.3 Metadata Preparation

1. generate the file tag t_k for the kth user data file $t_k = id_k||signature_{ssk_k}(id_k)$
   Where id_k is the file identifier
2. Calculate the authenticator (σ_{ki}) for each data block of the file F_k
   $\sigma_{ki} = \left(H(W_{ki}).u_k^{b_{ki}}\right)^{x_k}$ Where i = 1 to n and $\phi = \{\sigma_{k1}, \sigma_{k2}, \ldots \sigma_{kn}\}$
3. send $\{F_k, t_k, \phi_k\}$ to the cloud server
4. delete $\{F_k, t_k, \phi_k\}$ from the local storage.

Alg. 1 MULTI-SECTOR BATCH AUDITING

Data owner:

1. Split the file in to n equal sized data blocks $F_k = b_{ij}\ for\ i = 1\ to\ n, j = 1\ to\ s$
2. Generate the secret and public key parameters;
   Secret key = (x_k, ssk_k) and public key = $(spk_k, v_k, g, u_k, e(u_k, v_k))$
3. Generate the file fag t_k; $t_k = id_k || sig_{ssk_k}(id_k)$
4. Compute the authenticator Store $\{F_k, \sigma_{k_i}, t_k\}$ at server side

Auditor:

5. Retrieve and verify the file tag t_k for the auditing k^{th} user file.
6. Send the request $\{i, Vi\}$ to the server
7. Response for this request from the server, auditor verify the storage correctness equation.

Server:

8. After receiving request from auditor, for each request server prepares the response message as follows;
9. Compute $\{\mu_k, \sigma_k, R_k\}$ using the following equations.
   $$\mu_k = \sum_i^c V_i^{b_{ki}}, \quad \sigma_k = \prod_{i=1}^n \sigma^{V_i}, \quad R_k = e(u_k, v_k)$$
10. Compute $R = R_1 . R_2 . \ldots . R_k$.
    $L = v_{k1} || v_{k2} || v_{k3} || \ldots || v_{kK}$
    $\Gamma_k = h(R || v_k || L)$
11. Compute $\mu_k = r_k + \Gamma_k u_k \bmod p$
12. Send the response message $\{\mu_k, \sigma_k, R_k\}$ to the auditor.

6.4 Auditing

1. Auditor retrieves all file tags t_k from cloud server and verifies all the tags t_k are matched or not.
2. If tags verification fails, the auditor will discard those tags.
3. Once the tags are verified, the auditor prepares audit challenge message for each audit file and sends it to the server.
   $chal = \{(i, V_i)\}\ for\ i \in I$ Where i is the position of the data block bk, V is the random value, I is the subset of random elements chosen set $[1, n]$
4. After receiving challenge message from the data owners, server prepares a response message as follows:

- pick a random variable r_k and compute

$$R^{'} = R_1.R_2 \ \ldots \ R_k$$

$$L = v_{k1}||v_{k2}|| \ \ldots \ v_{kK}$$

$$\Gamma_k = h(R||v_k||L)$$

response message = $\{\mu_k, \sigma_k, R^{'}\}$
where, $\mu_k = \Gamma_k \sum_{i=1}^{n} (V_i^{b_{ki}}) + r_k \, mod \, p$, $\sigma_k = \prod_{i=1}^{c} \sigma_{ki}^{V_i}$

5. Server sends the response message to auditor
6. The auditor checks the response message is validity using the following equation.
   - Compute $\Gamma_k = h(R||v_k||L)$
   - Check

$$R^{'}.e\left(\prod_{k=1}^{K} \sigma_k^{\Gamma_k}, g\right)? = \prod_{k=1}^{K} e\left((\prod_{k=1}^{K} H(W_{ki})^{V_i})^{\Gamma}.u_k^{\mu_k}, v_k\right) \tag{2}$$

The verification proof for the above equation is as follows;

$$\begin{aligned} LHS &= R^{'}.e\left(\prod_{k=1}^{K} \sigma_k^{\Gamma_k}, g\right) \\ &= R_1.R_2 \ \ldots \ R_k\left(\prod_{k=1}^{K} e(\sigma_k^{\Gamma_k}, g)\right) \\ &= \left(\prod_{k=1}^{K} R_k e(\sigma_k^{\Gamma_k}, g)\right) \\ &= \left(\prod_{k=1}^{K} (u_k, v_k)^{r_k} e(\sigma_k^{\Gamma_k}, g)\right) \\ &= \left(\prod_{k=1}^{K} (u_k, v_k)^{r_k} e(\sigma_k^{\Gamma_k}, v_k^{r_k})\right) \end{aligned}$$

7 Performance Analysis

We analyze the performance analysis of our proposed cloud data auditing system in terms of storage, computation, and communication overhead.

Consider a file *Fi* consists of n blocks and each of s sectors of data. The storage overhead is analyzed based on the number of sectors that are used in each block to store data. For example, the data owner divides the file into *n* blocks, the size of each block is calculated as block size = $|F|/n$, which contains one or more sectors. For $s = 1$, i.e., every block contains only one sector. The storage overhead of file metadata contains signatures of the blocks and file tag. For smaller data block size, the size of the metadata also increases linearly, which increases the storage overhead. In order to reduce the storage overhead of metadata, the data owner will split the file into larger size blocks. As the number of blocks is reducing, the size of metadata also reduces, which reduces the data verification cost. Our proposed auditing method is more flexible to adjust the storage and computation overhead for selecting the variable number of sectors in a block.

Cost of Auditing: In our proposed method, the cost for data auditing is calculated based on the number of cryptographic operations that are used for verifying c number of blocks in the task. The computation overhead considered in terms of server computation, auditor computation, and communication between them for verification of outsourced data blocks.

The cost analysis of auditing c number of data blocks with one sector ($s = 1$) each is as follows:

Server side: The cost required for auditing c blocks at server side is the sum of the blinded linear operation (μ), bilinear pairing (R), and the aggregated authenticator (σ) computation cost. The response message includes $\{\mu, \sigma, R\}$ so that the communication cost for this response message is very small.

Auditor side: After receiving a response message from the server, the auditor verification cost is the sum of the cryptographic operations cost required to evaluate the Eq. 2.

For larger blocks, the auditor computation cost is more expensive than the server computation for auditing outsourced data blocks. So that, auditor side has little computation overhead than server side for single user auditing. To improve the computational efficiency at auditor side, batch auditing has designed for auditing tasks in a group. The average auditing task time is defined as the ratio of total auditing time and a total number of task. Using batch auditing, the auditor takes less computational operations to verify the outsourced data blocks, which reduces the average auditing task time.

8 Conclusions

In this paper, we propose a multisector public auditing for securing the outsourced user data on cloud storage model. In this method, we introduced a data block masking which hides the data from the public verifier. In specific, public verifier will not derive the user data during the auditing process. We also proposed a batch auditing to handle the multiuser auditing of outsourced data files. Further, we extend our design for auditing multisector public auditing for data verification and correctness, which minimizes the computation and storage overheads compared to the individual used data auditing method. The analysis of proposed system shows that multisector public auditing system is better than the existing single sector block auditing system. The main problem of this method is, during auditing if any one-user auditing fails then entire auditing will fail. To overcome this drawback we can extend this work by segregating the failed audit and successful audits.

References

1. Rivest, R. L. and Shamir, A. and Adleman, L., A method for obtaining digital signatures and public-key cryptosystems, Commun. ACM, vol. 21, no. 2, pp. 120–126, 1978.
2. Rivest, R., The md5 message-digest algorithm, 2014.
3. Yan Zhu, Gail-Joon Ahn, Hongxin Hu, Stephen S. Yau, Ho G. An, and Chang-Jun Hu, Dynamic audit services for outsourced storages in clouds, in IEEE TRANSACTIONS ON SERVICES COMPUTING, vol. 6, pp. 227–238, April-June 2013.
4. Jiang Deng, Chunxiang Xu, Huai Wu and Liju Dong, A new certificateless signature with enhanced security and aggregation version, Concurrency and Computation: Practice and Experience, vol. 28, no. 4, pp. 1124–1133, 2016.
5. Tao Jiang, Xiaofeng Chen, and Jianfeng Ma, Public integrity auditing for shared dynamic cloud data with group user revocation, in IEEE Transactions on Computers, 2015.
6. Cong Wang, Sherman S.M. Chow, Qian Wang, Kui Ren, and Wenjing Lou, Privacy-preserving public auditing for secure cloud storage, in IEEE Transactions on Computers, vol. 22, no. 2, 2013.
7. Jiawei Yuan; Shucheng Yu, Public integrity auditing for dynamic data sharing with multiuser modification, IEEE Transactions on Services Computing, vol. 10, pp. 1717–1726, 2015.
8. Hui Tian; Yuxiang Chen; Chin-Chen Chang; Hong Jiang; Yongfeng Huang; Yonghong Chen; Jin Liu, Dynamic-hash-table based public auditing for secure cloud storage, IEEE Transactions on Services Computing, vol. 99, 2016.
9. Juels, Ari and Kaliski, Jr., Burton S., Pors: Proofs of retrievability for large files, in Proceedings of the 14th ACM Conference on Computer and Communications Security, CCS 07, pp. 584–597, ACM, 2007.
10. Ateniese, Giuseppe and Burns, Randal and Curtmola, Reza and Herring, Joseph and Kissner, Lea and Peterson, Zachary and Song, Dawn, Provable data possession at untrusted stores, in Proceedings of the 14th ACM Conference on Computer and Communications Security, CCS 07, pp. 598–609, ACM, 2007.
11. Erway, C. Chris and Kupc u, Alptekin and Papamanthou, Charalampos and Tamassia, Roberto, Dynamic provable data possession, ACM Trans. Inf. Syst. Secur., vol. 17, no. 4, 2015.
12. Qian Wang, Cong Wang, Kui Ren, Wenjing Lou, and Jin Li, Enabling public auditability and data dynamics for storage security in cloud computing, in IEEE Proceedings of the 14th European Symposium on Research in Computer Security (ESORICS09), 2009.
13. Boyang Wang; Hui Li; Xuefeng Liu; Fenghua Li; Xiaoqing Li, Efficient public verification on the integrity of multi-owner data in the cloud, Journal of Communications and Networks, vol. 16, no. 6, pp. 592–599, 2014.

Workflow Scheduling in Cloud Computing Environment Using Bat Algorithm

Santwana Sagnika, Saurabh Bilgaiyan and Bhabani Shankar Prasad Mishra

Abstract The data handling and processing capabilities of current computing systems are increasing, owing to applications involving the bigger size of data. Hence, the services have become more expensive. To maintain the popularity of cloud environment due to less cost for such requirements, an appropriate scheduling technique is essential, which will decide what task will be executed on which resource in a manner that will optimize the overall costs. This paper presents an application of the Bat Algorithm (BA) for scheduling a workflow application (i.e., a data intensive application), in cloud computing environment. The algorithm is successfully implemented and the results compared with two popular existing algorithms, namely Particle Swarm Optimization (PSO) and Cat Swarm Optimization (CSO). The proposed BA algorithm gives an optimal processing cost with better convergence and fair load distribution.

Keywords Cloud computing · Task scheduling · Cat swarm optimization
Particle swarm optimization · Bat algorithm · Optimization

1 Introduction

The concept of cloud computing is the next step towards future computing technologies where the computing demand is much higher than the existing processing capabilities [1]. Cloud computing is a computing archetype where the service offerings (i.e., resources) are grouped together to form a resource hub/pool and a user is allowed into the hub through a virtual cloud network. Here a cloud service

S. Sagnika (✉) · S. Bilgaiyan · B.S.P. Mishra
School of Computer Engineering, KIIT University, Bhubaneswar, Odisha, India
e-mail: santwana.sagnika@gmail.com

S. Bilgaiyan
e-mail: saurabhbilgaiyan01@gmail.com

B.S.P. Mishra
e-mail: mishra.bsp@gmail.com

A.K. Somani et al. (eds.), *Proceedings of First International Conference on Smart System, Innovations and Computing*, Smart Innovation, Systems and Technologies 79, https://doi.org/10.1007/978-981-10-5828-8_15

broker (may be a third party or the cloud itself) ensures that the user attains optimal services with minimum charges and maximum quality of services. There are some big names behind these service offerings like Google, Amazon ec2/Web Services, Engine Yard, GoGrid, Microsoft, VMware, Salesforce, RackSpace, etc. [2]. In a cloud computing environment, the users need not purchase physical instances of any resources, but only hire their virtual instances on the basis of a pay-as-you-demand policy where processing costs are much less than purchasing the actual physical resources [3]. This overall computing scenario is transparent to the user in terms of configuration, location, instances, etc. The access to cloud services is open through a constant internet connection [4, 5]. Figure 1 shows a general cloud computing architecture.

The growing size of work to be processed on cloud platforms requires massive computing power; hence, such applications are termed under a special category, i.e., data intensive applications. Scientific workflow, astrophysics, neuroscience, turbulence, etc. are examples of some data intensive applications [3, 6]. They require special care for managing their execution to optimize the overall scenario [5, 7].

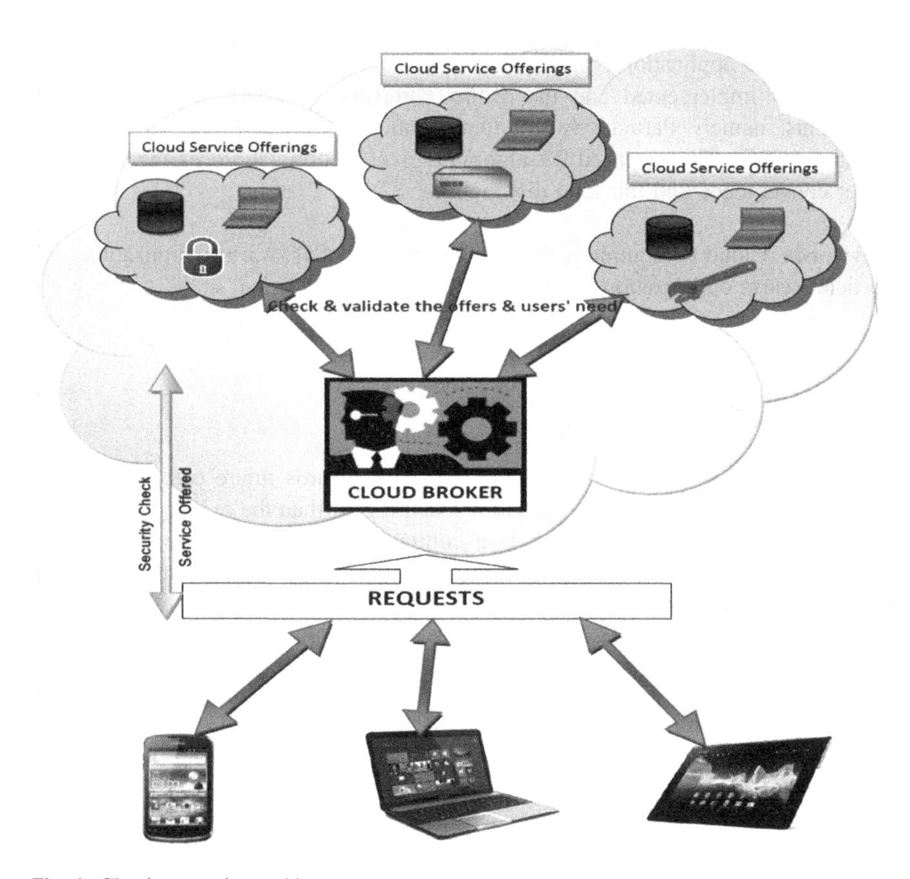

Fig. 1 Cloud computing architecture

PSO is one of the most popular and efficient soft computing approaches for optimization problems like task scheduling, software cost estimation, change detection, etc. PSO is able to find the best possible solution for a problem on the basis of stored historical information of its particles. However, when PSO is applied to solve the problem function with high dimensions, the algorithm becomes prone to being trapped and sinking into local minima. [8–10]. CSO is a latest optimization technique based on social activities of cats. CSO is becoming popular among existing optimization algorithms due to its ability to find the best outcome in minimum iterations. Execution time of the algorithm for larger problems, like change detection in images increases rapidly along with the number of iterations, which is a major drawback of this technique [11, 12].

Like CSO, BA is also one of the recent emerging meta-heuristic optimization algorithms. BA has been introduced by Yang in 2010 and is based on the echolocation process of bats that they implement while searching for the food [13]. It has been successfully applied in various fields like prediction, fuzzy cluster, optimization, artificial intelligence problems, etc. This technique performs intelligent movement to reach the solution, just like CSO, but uses simple steps and parameters, like PSO. Hence, it achieves the speed of PSO as well as the efficiency of CSO while overcoming their disadvantages. This makes the technique attractive for applying to data intensive optimization problems.

The main objective of this article is to get an optimal task scheduling solution that will minimize the overall processing cost in minimum time with a fast convergence rate by efficiently distributing the tasks among available resources.

2 Related Work

In a cloud computing system, the presence of resources distributed over various locations working in parallel leads to the requirement of a strategy to divide and distribute the different tasks for simultaneous and efficient execution. Thus, the need arises to focus on developing smarter mechanisms for tasks scheduling in cloud systems [14–16]. In this section, the authors discuss various approaches related to this problem (with a special focus on evolutionary computing techniques) that have been proposed till date.

Pandey et al. [17] introduced the use of particle swarm optimization (PSO) algorithm for scheduling jobs over a set of resources and compared it with best resource selection (BRS) algorithm. The objective was to minimize computational cost. The proposed method gave lower cost than BRS as well as better load distribution.

Wu et al. [18] modified the general PSO to a revised discrete PSO (RDPSO) and applied it to perform workflow scheduling, taking into account the computation and transmission cost. They compared it with standard PSO and BRS techniques. The results of the proposed work helped reduce the search space and give better performance.

Bilgaiyan et al. [19] applied a new approach, i.e., cat swarm optimization (CSO) for minimizing computational cost in a distributed cloud environment, by providing an effective mapping scheme for workflow scheduling. The proposed technique gave the effective results in minimum iterations as compared to PSO. Also, the CSO gave an effective load distribution among the available resources.

Bitam [20] modified the bee colony optimization (BCO) algorithm to make it suitable for task scheduling on cloud systems. The new technique, referred as bees life algorithm, was tested to distributing workloads on different systems so as to reduce the total execution time. It gave a better result than the existing GA system.

SundarRajan et al. [21] have implemented the firefly algorithm, an emerging technique, for workflow scheduling. The method efficiently reduces the makespan and ensures fair load distribution, as compared to PSO and CSO.

Among the existing mechanisms for workflow scheduling, it has been found that PSO takes the minimum time to reach the solution but the number of iterations required is more. On the other hand, the best convergence is observed through the CSO technique but it has a high execution time.

3 Swarm-Intelligence-Based Techniques: A General Idea

3.1 Social Behavior of Birds

Particle Swarm Optimization (PSO) is among the most popular swarm-intelligence-based mechanisms, based on the flocking and food quest of birds. Birds use a simple yet effective mechanism of social behavior while looking for food sources. They use a specific velocity for movement to update their current position to a new position. This velocity is influenced by their personal best position as well the best position of the whole flock. Starting from a random initial position, they use a fitness function as a performance measure using which they select or reject their next generated position. Using this movement system, PSO proves to be a fast and effective searching technique applicable for optimization problems and is hugely popular among researchers as a basic optimization technique [10, 17].

3.2 Behavioral Traits of Cats

Cats have a natural hunting talent, combined with a strong curiosity. They spend a majority of their time being physically dormant, all the while staying observant and analyzing their surroundings. This is known as the seeking mode. When they decide upon the next best move to be followed, they switch to the tracing mode, in which they move according to specific velocities. Cats in the tracing mode behave somewhat similar to the movement of birds as detailed in the PSO technique. Cats prefer to stay predominantly in the seeking mode and move in a calculated manner only when necessary. Hence, their mechanism is more intelligent compared to the random behavior of birds. Therefore, researchers have worked on an optimization technique Cat Swarm Optimization (CSO) based on such behavior of cats, which is a smarter technique than PSO and converges faster. On the flip side, it takes more time per iteration than PSO, owing to its calculative nature [22–24].

3.3 Hunting Behavior of Bats

Bats move about and hunt for their prey at night. Due to poor visibility, they make use of sound waves. They emit sound waves of specific loudness and frequencies and wait for their echo to come back. The difference in the loudness and frequency between the sent wave and received echo gives them the idea of the distance and location of the objects ahead of them, be it their residing place, obstacles or prey. This phenomenon is known as echolocation [13, 15, 25]. The sound pulses that are emitted are very short lived, i.e., up to 10 ms, generally in the frequency range of 25–100 kHz. In each second, bats can emit multiple pulses, starting from 10, sometimes reaching to 200 pulses, on nearing the prey. The loudness lies in the ultrasonic region, with a very high value of around 110 dB when they prey is being searched, and reducing to very low values on approaching the prey. By using echolocation, bats can determine the location, size, speed, and position of the prey ahead. By mapping this behavior of bats to computing systems, an efficient searching technique can be formulated, namely Bat Algorithm approach, which can be applied to solve optimization problems based on large search spaces [26, 27].

4 Representation of Workflows

This section represents the basic structure of a workflow. The workflow is represented as a set of jobs $J = \{J_1, J_2, \ldots, J_n\}$, and a set of resources $S = \{S_1, S_2, \ldots, S_n\}$, each having its own memory system $M = \{M_1, M_2, \ldots, M_n\}$.

Figure 2 represents a sample workflow model involving a set of 17 jobs or tasks. It follows a standard UML representation. Figure 3 shows a set of resources distributed over various geographical locations, each having their own storage and interaction ability with each other [28].

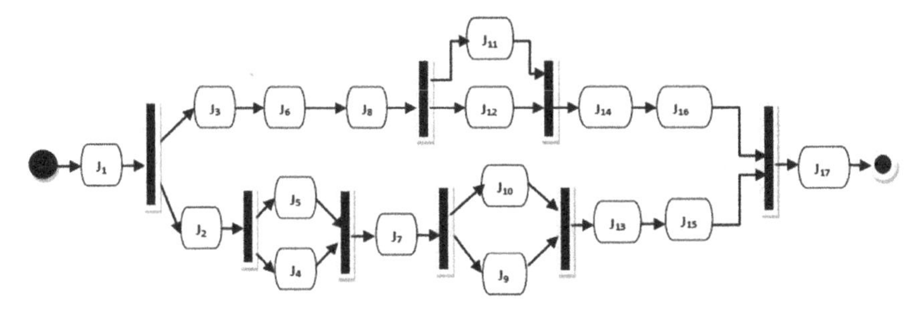

Fig. 2 Model for workflow

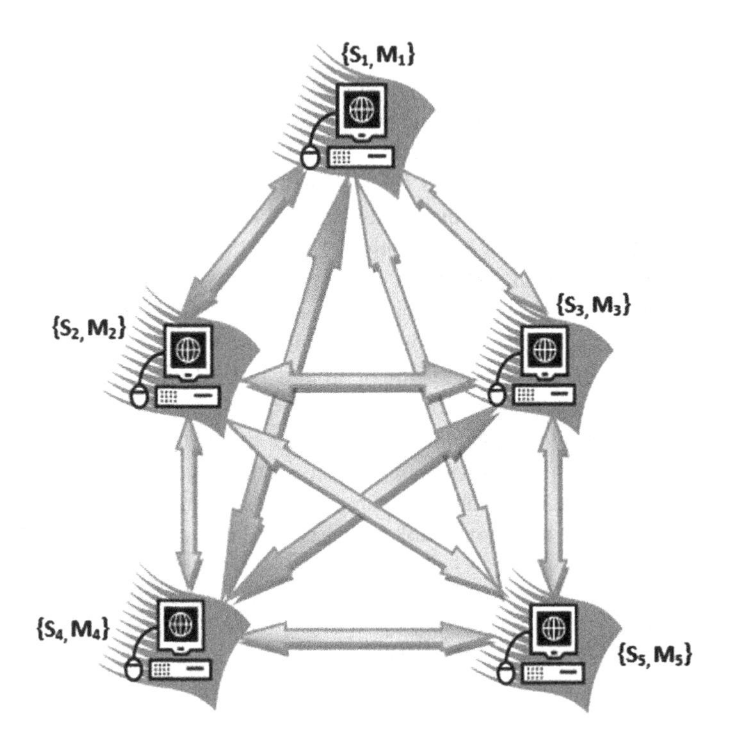

Fig. 3 Distribution of interacting resources

5 Proposed Bat Algorithm Technique

5.1 Problem Representation

The problem can be specified as the allotment of the various tasks of a workflow to the different resources, in such a manner that minimum makespan for tasks is ensured. An algorithm is needed which can perform the scheduling in minimum time and iterations while taking care that the load is evenly distributed over the available resources. The following mathematical model represents workflow scheduling as an optimization problem that can be solved using a meta-heuristic technique.

Let S_x represent a specific resource, on which various tasks are being executed, J_y being one of them. Mp represents a mapping of tasks to resources so that *Mp[Jy]* gives S_x. The computation cost of J_y on S_x is represented by *CC[y][x]* and the transmission cost from S_x to $S_{\times 1}$ is represented by *TC[x][x1]*. The transmission cost incurred for task J_y is the sum of the transmission costs from S_x to all such resources where those tasks which are dependent on J_y are being executed. Thus,

$$Cost\_trans[J_y] = \sum_z TC[x][x1], \quad (1)$$

where $Mp[J_y] = S_x$ and $Mp[J_z] = S_{x1}$, provided J_z is dependent upon J_y. *Cost_trans* $[J_y]$ gives the net transmission cost for an individual task.

So, the total cost of resource S_x is the sum of the total cost, i.e., computation cost and transmission cost, for all tasks. Hence,

$$Cost\_total[S_x] = \sum_y CC[y][x] + Cost\_trans[J_y] \quad (2)$$

To ensure proper load distribution, it is required to find the maximum value of the total cost among all resources for the mapping, i.e.

$$Cost\_max(Mp) = max(Cost\_total[S_x]), \forall x \quad (3)$$

This value needs to be minimized to find the most optimal solution. Hence,

$$Minimize(Cost\_max(Mp)) \quad (4)$$

Equation 4 gives the fitness function for the optimization problem.

5.2 PSO Approach

The specified problem has been solved using PSO by Pandey, Guru, and Buyya [17], as a simple and efficient method which can execute in real-time systems quite fast. The method employs a population of N particles. The number of dimensions is the total number of tasks to be performed, and the values assigned to them represent the resource which is to be assigned. Hence, each particle corresponds to a particular mapping of tasks to resources, which is randomly generated. The following algorithm shows the steps that PSO follows, with regard to the mentioned problem [29].

1. A flock of N birds is created.
2. The birds are initialized with random positions and velocities.
3. The fitness value of each bird is computed.

4. For every bird, the fitness is compared with its best till now. If found to be better, the current position is set to be the new personal best for that bird.
5. Among all the birds, the best solution is taken and compared with the global best till now. If found to be better, that position is set to be the new global best for the whole population.
6. The velocities and positions of the birds are updated according to Eqs. (5) and (6).
7. Repeat steps 3 to 6 till the terminating condition is obtained.
8. The following equations are used to change the velocity and position of the *kth* bird.

$$\begin{aligned} v\_kt+1 = & (w^*v\_kt) + (c1^*r1^*(x\_best\_k - x\_kt)) \\ & + (c2^*r2^*(g\_best - x\_kt)) \end{aligned} \tag{5}$$

$$x\_kt+1 = x\_kt + v\_kt+1 \tag{6}$$

where $v\_kt + 1$ is the new velocity, w is the inertial weight, $c1$ and $c2$ are acceleration constants, v_kt is the current velocity, $r1$ and $r2$ are random values ranging from 0 to 1, x_best_k is the personal best of the bird, g_best is the global best for all birds, x_kt is the current position and $x\_kt + 1$ is the new position.

5.3 CSO Approach

The given problem was again approached by Bilgaiyan, Sagnika, and Das [19] using a newer technique, i.e., CSO, which is proven to find the solution in a lesser number of iterations. The representation of the problem elements is same as described in the PSO approach [11]. The following algorithm outlines the procedure followed by CSO to solve the problem under discussion [30].

1. An initial population of N cats is considered.
2. Each cat is given a random position and velocity.
3. The mode of each cat is arbitrarily decided, i.e., seeking mode or tracing mode, according to Mixture Ratio (MR).
4. The fitness function is evaluated for each cat. Each cat retains its own best position and the global solution is updated to the best fitness value thus found.
5. On each cat, if it is in seeking mode then the seeking mode procedure is applied; else the tracing mode procedure is applied.
6. The new positions of the cats are re-evaluated for fitness. If found to be better, the new solutions replace the old solutions.
7. The cats are re-distributed into seeking and tracing modes according to MR.

8. Steps 5–7 are iterated till the stopping criterion is reached.

The Mixture Ratio (MR) denotes to the ratio of cats in the seeking mode to those in the tracing mode.

The following steps detail the seeking mode procedure.

1. The number of copies of each cat is created according to SMP.
2. The position of each copy is then changed randomly, for CDC number of dimensions.
3. The fitness of all such copies is evaluated.
4. The copy having best fitness is selected and used to replace the original cat.

Here, SMP (Seeking Memory Pool) defines the seeking memory size, i.e., the number of copies to be made for each cat in the seeking mode. It symbolizes different points in the search space which the cat analyses while in seeking mode and judges which position to move to among them. CDC (Count of Dimensions to Change) refers to the number of dimensions which are needed to be varied.

The following steps detail the tracing mode procedure.

1. The velocity of each cat is updated according to Eq. (7).

$$v\_m^{t+1} = w*v\_m^t + c*r*(p\_best - p\_m^t) \tag{7}$$

where v_m^{t+1} is the new velocity for m^{th} cat, w is the inertial weight, v_m^t is the old velocity, c is an acceleration constant, r is a random value between 0 and 1, p_best is the best position discovered till that time and p_m is the current position of the mth cat.

2. The position of each cat is updated according to Eq. (8).

$$p\_m^{t+1} = p\_m^t + v\_m^{t+1} \tag{8}$$

where p_m^{t+1} is the new position for the mth cat and p_m^t is its current position.

5.4 Bat Algorithm Approach

This paper shows a method to apply the Bat Algorithm for performing task scheduling in a cloud computing environment. The aim is to minimize the computation and transmission costs incurred during execution. Every bat is a representation of a specific mapping of tasks onto resources. A population of N bats is taken, who collectively work towards finding the most optimal solution, i.e., in this case the mapping that gives the minimal cost. The minimal cost can be obtained by finding the fitness of the individual bats. The fittest bat represents the best mapping that can be obtained [31]. The following algorithm shows the steps of Bat Algorithm, as applied to the scheduling problem [32].

1. An initial population of N bats is considered.
2. Velocities for each bat, v_i, are initialized with random values.
3. Frequencies of pulse, f_i are also initialized for all bats.
4. Loudness A and pulse rate r are given values.
5. Evaluation of fitness function for each bat is done.
6. Velocities and positions are updated using given Eqs.
7. A random solution is selected with some probability.
8. Around this solution, random local walk is used to find a new local solution.
9. If the new solution has a lower frequency and amplitude, the new solution is accepted, then A is reduced and r is increased.
10. The new solutions are compared with the existing solutions. If better, then old solutions are replaced with the new solutions.
11. Steps 5 to 10 are repeated till stopping criteria is attained.

The following equations are used in the algorithm for updating the velocity and position.

$$f\_i = (f\_max - f\_min)*r + f\_min \tag{9}$$

$$v\_i^{t+1} = v\_i^t + f\_i*(x\_i^t - x\_best) \tag{10}$$

$$x\_i^{t+1} = x\_i^t + v\_i^{t+1} \tag{11}$$

Here r is a random value between 0 and 1. f_max and f_min denote the maximum and minimum permitted values for the frequency, based on the problem specification. x_best in the current global best solution. The current velocity v_i^{t+1} is generated from the previous velocity v_i^t, which is added to the previous position x_i^t to reach the current position x_i^{t+1}. The local random walk of step 8 is performed using the equation

$$x\_new = x\_old + s*A^t \tag{12}$$

Here s is a randomly generated number between -1 and 1, and A^t gives the average amplitude value for the specific time value.

As it is required to decrease the amplitude and increase the rate of emitting pulses when the prey is near, both these tasks are done as per the following equations.

$$A\_i^t = p*A\_i^{t-1} \tag{13}$$

$$r\_i^t = r\_i^0*\left(1 - e^{-q*t}\right) \tag{14}$$

Here p and q are constants, where p lies in between 0 and 1, and q is a non-zero positive value. r_i^0 is the initial pulse rate emission and r_i^t is the current pulse rate emission. The current amplitude A_i^t is dependent on the previous amplitude A_i^{t-1}.

6 Experiments and Results

6.1 Experimental Setup

The experiment was performed on a hypothetical workflow consisting of 17 jobs to be mapped on five geographically distributed resources as shown in Sect. 4. For performing the experiment, the platform used was MATLAB version 7.14, on personal laptop having 4 GB RAM, Intel i3 core processor. The bat population size was set to 50 and the number of iterations was varied from 50 to 300. The input size of data was fixed at 128 MB for experimenting. The computing costs were based on some standard cloud offerings by popular service providers like Mosso, Amazon web services, GoGrid, Engineyard, Rackspace, etc. Tables 1 and 2 represent the computation cost of each task over each resource, and data transmission costs among resources, respectively.

Table 1 Computation Cost (CC) in cents

	S_1	S_2	S_3	S_4	S_5
T_1	1.19	1.09	1.13	1.16	1.20
T_2	1.18	1.17	1.23	1.15	1.19
T_3	1.13	1.14	1.10	1.16	1.16
T_4	1.27	1.13	1.11	1.16	1.15
T_5	1.21	1.13	1.22	1.17	1.18
T_6	1.23	1.10	1.14	1.15	1.16
T_7	1.12	1.10	1.09	1.14	1.11
T_8	1.27	1.11	1.15	1.16	1.14
T_9	1.10	1.08	1.14	1.17	1.13
T_{10}	1.29	1.11	1.15	1.13	1.17
T_{11}	1.17	1.18	1.26	1.13	1.18
T_{12}	1.26	1.11	1.15	1.14	1.15
T_{13}	1.24	1.12	1.15	1.14	1.13
T_{14}	1.25	1.12	1.13	1.16	1.17
T_{15}	1.26	1.11	1.15	1.15	1.17
T_{16}	1.10	1.11	1.13	1.16	1.12
T_{17}	1.16	1.15	1.29	1.13	1.15

Table 2 Transmission Cost (TC) in cents

	S_1	S_2	S_3	S_4	S_5
S_1	0	0.16	0.02	0.18	0.21
S_2	0.01	0	0.16	0.20	0.04
S_3	0.16	0.14	0	0.20	0.19
S_4	0.19	0.20	0.21	0	0.14
S_5	0.19	0.02	0.13	0.17	0

6.2 Results

The following graphs represent the results obtained from the experiment. Figure 4 represents the convergence obtained by the bat algorithm approach, in comparison to the convergence of PSO and CSO approaches for a given data size. Figure 5 gives the optimal load distribution obtained by the proposed technique. Figure 6 compares the execution times of BA, PSO and CSO techniques for the same data size.

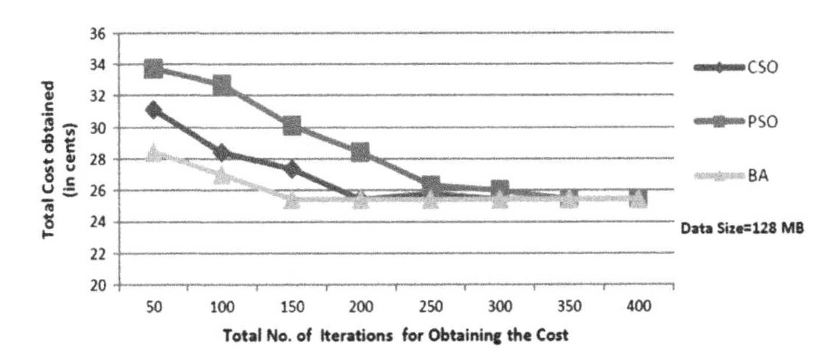

Fig. 4 Comparative graph of convergence for various techniques having data size 128 MB

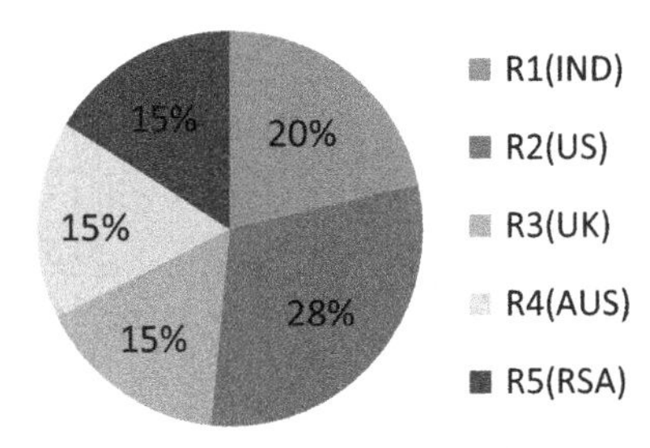

Fig. 5 Load distribution graph

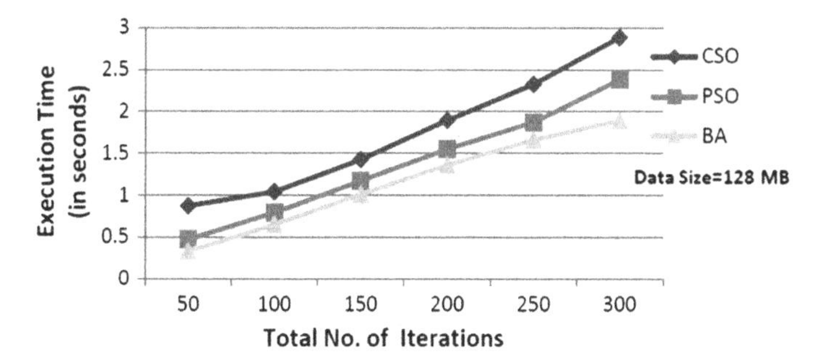

Fig. 6 Comparative graph of execution time for various techniques having data size 128 MB

7 Analysis

From Fig. 4, it is evident that the proposed bat algorithm approach gives a very fast rate of convergence, lesser than the CSO technique. Figure 5 shows that the proposed technique also ensures a uniform load distribution on the different resources without overloading any particular resource. Figure 6 also shows that the execution time of the algorithm is very less, even lesser than PSO. The proposed technique is fast because the bat algorithm follows a similar technique of updating position based on velocity, like the PSO approach. It also gives good convergence as it performs intelligent and informed updating of velocity, based on the distance from the solution, instead of randomly updating the values. Furthermore, the local random walk for specific solutions also makes sure that any iteration provides a larger step towards the solution. Both the efficiency and speed of the bat algorithm approach are provided by the echolocation behavior of bats, which performs simplistic yet smart movement towards the solution. The availability of parameters that have the flexibility of being adjusted during the running of the algorithm also provides ease of managing the convergence rate. Thus, the proposed technique is fast in execution as well as good in convergence, thereby overcoming the disadvantages of both PSO and CSO techniques. So, the proposed approach can be taken as an efficient mechanism to perform job scheduling in cloud environments using meta-heuristics.

8 Conclusion and Future Work

In this paper, the authors put forth the application of a new optimization technique, Bat Algorithm, to solve the task-resource mapping problem in a cloud computing environment. The objective was to minimize the net computation and transmission costs incurred during the execution of a workflow on a distributed cloud system. The applied method gave optimal results in minimum amount of time. It exhibited a high convergence rate and also ensured a proper load distribution over various resources without over-burdening any specific resource. It gave results in very less iterations, lesser than the CSO technique, and its execution time was lesser than the fastest known mechanism PSO. Hence, it can be said that Bat Algorithm is a very suitable mechanism for workflow scheduling in cloud systems. Future work in this field can involve extending this method to optimize scheduling taking multiple objectives into consideration. Fine-tuning and modifying the variable parameters can also be experimented with, to make the efficiency of this algorithm even better.

References

1. Truong-Huu, T., Tham, C.-K.: A Novel Model for Competition and Cooperation among Cloud Providers. IEEE Transactions on Cloud Computing, 2(3), 251–265, (2014).
2. Top 10 cloud computing providers of 2012, http://searchcloudcomputing.techtarget.com/photostory/2240149049/Top-10-cloudproviders-of-2012/11/1-Amazon-Web-Services#contentCompress.
3. Bilgaiyan, S., Sagnika, S., Sahu, S.S.: Cloud Computing: Concept, Terminologies, Issues, Recent Technologies. Research Journal of Applied Sciences, Medwell Journals, 9(9), 614–618 (2014).
4. Khalil, I.M., Khreishah, A., Azeem, M.A.: Cloud Computing Security: A Survey. Computers, 3(1), 1–35 (2014).
5. Bilgaiyan, S., Sagnika, S., Das, M.: A Multi-Objective Cat Swarm Optimization Algorithm For Workflow Scheduling In Cloud Computing Environment. In: International Conference on Intelligent Computing, Communication & Devices (ICCD), Proceedings of ICCD, Springer, 1, 73–84 (2014).
6. Zhao L., Li H.: Median-Oriented Bat Algorithm for Function Optimization. In: Huang DS., Bevilacqua V., Premaratne P. (eds) Intelligent Computing Theories and Application. ICIC 2016. Lecture Notes in Computer Science, vol 9771. Springer, Cham (2016).
7. Gil, Y., Deelman, E., Ellisman, M., Fahringer, T., Fox, G., Gannon, D., Goble, C., Livny, M., Moreau, L., Myers, J.: Examining the challenges of scientific workflows. IEEE Computer Society, 40(12), 24–32 (2007).
8. Shao, L., Bai, Y., Qiu, Y., Du, Z: Particle Swarm Optimization Algorithm Based on Semantic Relations and Its Engineering Applications. Systems Engineering Procedia, Elsevier, 5, 222–227 (2012).
9. Cheng, R., Jin, Y.: A Social Learning Particle Swarm Optimization Algorithm for Scalable Optimization. Information Sciences, Elsevier, 291, 43–60 (2014).
10. Kennedy, J., Eberhart, R.: Particle Swarm Optimization. In: Proceedings of IEEE International Conference on Neural Networks IV, 1942–1948 (1995).
11. Chu, S.C., Tsai, P.-W., Pan, J.S.: Cat Swarm Optimization. In: Proceedings of 9th Pacific Rim International Conference on Artificial Intelligence, Guilin, Springer, 4099, 854–858 (2006).
12. Pradhan, P.M., Panda, G.: Solving Multiobjective Problems using Cat Swarm Optimization. Expert Systems with Applications, Elsevier, 2956–2964 (2011).
13. Yang, X.-S.: Bat Algorithm: Literature Review and Applications. International Journal of Bio-Inspired Computation, 5(3), 141–149 (2013).
14. Gherbi Jaddi, N.S., Abdullah, S., Hamdan, A.R.: Optimization of neural network model using modified bat-inspired algorithm. Applied Soft Computing, 37, 71–86 (2015).
15. Yılmaz, S., Küçüksille, E.U.: A new modification approach on bat algorithm for solving optimization problems. Applied Soft Computing, 28, 259–275 (2015).
16. Wang, Y., Shi, W.: Budget-Driven Scheduling Algorithms for Batches of MapReduce Jobs in Heterogeneous Clouds. IEEE Transactions on Cloud Computing, 2(3), 306–319 (2014).

17. Pandey, S., Wu, L., Guru, S.M., Buyya, R. A Particle Swarm Optimization-based heuristic for scheduling workflow applications in cloud computing environments. In: 24th IEEE International Conference on Advanced Information Networking and Applications, 400–407 (2010).
18. Wu, Z., Ni, Z., Gu, L., Liu, X.: A Revised Discrete Particle Swarm Optimization for Cloud Workflow Scheduling. In: International Conference on Computational Intelligence and Security, IEEE Computer Society, 184–188 (2010).
19. Bilgaiyan, S., Sagnika, S., Das, M.: Workflow Scheduling in Cloud Computing Environment using Cat Swarm Optimization, In: IEEE International Advance Computing Conference (IACC), 680–685 (2014).
20. Bitam, S.: Bees life algorithm for job scheduling in cloud computing. International Conference on Computing and Information Technology (ICCIT), 186–191 (2012).
21. SundarRajan, R., Vasudevan, V., Mithya, S.: Workflow Scheduling in Cloud Computing Environment using Firefly Algorithm. In: Proceedings of International Conference on Electrical, Electronics, and Optimization Techniques (ICEEOT), 955–960 (2016).
22. Zhang, Y., Tian, Y. An Improved Cat Swarm Optimization Algorithm and Application Research. In: 7th IEEE International Conference on Advanced Computational Intelligence, 207–211 (2015).
23. Crawford B., Soto R., Berrios N., Olguín E., Misra S.: Cat Swarm Optimization with Different Transfer Functions for Solving Set Covering Problems. In: Gervasi O. et al. (eds) Computational Science and Its Applications—ICCSA 2016. ICCSA 2016. Lecture Notes in Computer Science, vol 9790. Springer, Cham (2016).
24. Razzaq S., Maqbool F., Hussain A.: Modified Cat Swarm Optimization for Clustering. In: Liu CL., Hussain A., Luo B., Tan K., Zeng Y., Zhang Z. (eds) Advances in Brain Inspired Cognitive Systems. BICS 2016. Lecture Notes in Computer Science, vol 10023. Springer, Cham (2016).
25. Yang, X-S.: A New Metaheuristic Bat-Inspired Algorithm. Nature Inspired Cooperative Strategies for Optimization (NICSO), Studies in Computational Intelligence, Springer, 284, 65–74 (2010).
26. Wang, G., Guo, L.: A Novel Hybrid Bat Algorithm with Harmony Search for Global Numerical Optimization. Journal of Applied Mathematics, Hindawi Publishing Corporation, 2013, 1–21 (2013).
27. Yang, X.-S., Gandomi, A.H.: Bat Algorithm: A Novel Approach for Global Engineering Optimization. Engineering Computations, 29(5), 464–483 (2012).
28. Liu, H., Sun, S., Abraham, A.: Particle Swarm Approach to Scheduling Work-Flow Applications in Distributed Data-Intensive Computing Environment. In: Proceedings of the Sixth International Conference on Intelligent Systems Design and Applications, IEEE Computer Society, 661–666 (2006).
29. Awada, A.I., El-Hefnawya, N.A., Abdel_kader, H.M.: Enhanced Particle Swarm Optimization for Task Scheduling in Cloud Computing Environments. Procedia Computer Science, Elsevier, 65, 920–929 (2015).
30. Crawford, B., Soto, R., Berríos, N., Johnson, F., Paredes, F., Castro, C., Norero, E.: A Binary Cat Swarm Optimization Algorithm for the Non-Unicost Set Covering Problem. Mathematical Problems in Engineering, Hindawi Publishing Corporation, 1–8 (2015).
31. Ye, Z.-W., Wang, M.-W., Liu, W., Chen, S.-B.: Fuzzy entropy based optimal thresholding using bat algorithm. Applied Soft Computing, 31, 381–395 (2015).
32. Gandomi, A.H., Yang, X.-S, Alavi, A.H., Talatahari, S.: Bat Algorithm for Constrained Optimization Tasks. Neural Computing and Applications, Springer, 22, 1239–1255 (2013).

Recognition of Telecom Customer's Behavior as Data Product in CRM Big Data Environment

Puja Shrivastava, Laxman Sahoo and Manjusha Pandey

Abstract This paper approaches toward the standardization of telecom customer's behavior by specifying the call activities like frequency, duration, time of calls with the type of calls like local, national, and international. In the same way specifying the SMS/MMS activities as behavior of customers plus the rate of data pack and talk-time recharge. It is an attempt to identify meaningful attributes to describe behavior of customer plus study the available call detail records in big data environment and recognize the procedure that uses customer behavior for the designing of data product which is tariff plan.

Keywords Behavior · Customer · CRM · Pattern recognition
Big data · Call activity

1 Introduction

Throat cut competition in the telecom market has opened new dimensions for the researchers to determine new strategy planning techniques through the big data analytics on data of customer relationship management systems (CRM) of telecom companies. Business Support System (BSS) and Operation Support System (OSS) are the existing systems in the CRM, which provide the customer details that can be called as content data and call plus data usage details known as context data, respectively. The customer life cycle in CRM consists of six phases including customer contact to customer service representative (CSR), services provided to the customer, use of services by a customer, bill generation, bill payment, and offer of

P. Shrivastava (✉) · L. Sahoo · M. Pandey
School of Computer Engineering, KIIT University, Bhubaneswar, Odisha, India
e-mail: pujashri@gmail.com

L. Sahoo
e-mail: laxmansahoo@yahoo.com

M. Pandey
e-mail: manjushapandey82@gmail.com

A.K. Somani et al. (eds.), *Proceedings of First International Conference on Smart System, Innovations and Computing*, Smart Innovation, Systems and Technologies 79, https://doi.org/10.1007/978-981-10-5828-8_16

new services to customers. The utilization of services by the customer is a most important phase of customer cycle since it provides the behavior of the customer. The idea is to first collect data from the existing systems and then apply pattern recognition algorithms on obtained data to find out patterns which are the behavior of customers. Once, patterns are identified then apply most suitable big data clustering method to attain clusters of customers who show similar behavior in their call activities. Discovered clusters are helpful in providing the count of no. of customers of similar behavior, which is further supportive in designing new tariff plan with the calculation of revenue. The existing literature regarding the presented study is not complete and still, a big scope is open to work. Since advertising and attracting new customer costs more than retaining old customer; so the current scenario of business is to provide customer oriented services to retain customer long and attract new customers too. The objective of retaining customers can be achieved by offering plans near to needs and desires of both customer and company. Customer needs and desires can be known by two ways: 1st is to do a survey either by interviewing each and every customer or using some online/paper form filling, which is neither feasible nor suitable in the fast growing circumstances; 2nd way is to study the call activities and data usage, since the data is easily available in the company's existing system, which is feasible and more suitable due to the available techniques and fast research outputs. This paper is organized into four sections: 1st is the introduction of the paper, 2nd part defines an attribute of the customer behavior in the telecom market, the 3rd segment defines behavior to be recognized as a pattern in telecom big data environment and the 4th last section discusses the conclusion and future scope.

2 Customer Behavior

Definition of customer behavior needs some terminologies to be known first such as: who is customer? The answer is any person/group of persons/business organization who procures goods or services from any person/group of persons/business organization for his/her/their end-use are known as customer. Types of customers and types of services, in the telecommunication industry, are tabulated in Table 1.

Taste, need, and ability to spend are three attributes which play important role in the behavior of the customer. Cultural factors create the ideology of customers and affect the service consumption by the customer; occupation, age, life style, personality, motivation, perception, beliefs, and attitude are additional attributes disturbing the behavior of the customer. The behavior of telecom market customers also gets affected by these attributes. Call Detail Records (CDRs)/Usage Detail Records (UDRs) are the base for the definition of customer behavior, since it contains all details about the phone call such as calling party, called party, start of call in terms of date and time, call duration, call type (voice call, SMS, Data) and a unique sequence number for the detection of record.

Table 1 Types of customers and services

S.N.	Mobile services	Fixed line services	Net-Pack services
1.	Pre-Paid individual customers	Pre-Paid individual customers	Pre-Paid individual customers
2.	Post-Paid individual customers	Post-Paid individual customers	Post-Paid individual customers
3.	Pre-Paid company customers	Pre-Paid company customers	Pre-Paid company customers
4.	Post-Paid company customers	Post-Paid company customers	Post-Paid company customers

Traditionally four types of customer behavior models are defined: Economic, learning, psychoanalytic and sociological model. Economic model is based on maximum utility, learning model is based on the change of customer behavior after the service use. Psychoanalytical model based on the personal interests and ideology of customer and sociological model based on the sociological condition of customer [1]. Author has canvassed the behavior of telecom customer in the form of calling activities, use of messaging services and data usage as shown in Table 2 named as details of customer behavior.

Objective of this work is to study the call activities and net usage of the customer as behavior to further generate tariff plans for the benefit of both customer and company. Customer behavior is defined here with the help of attributes according to the services such as calling, messaging, and net pack with additional value vouchers [2]. Table 3 describes the list of attributes for customer behavior.

Total number of calls made by per customer can be calculated as the sum of $Call_{local}$, $Call_{nationa}$, and $Call_{international}$:

Call	=	$Call_{local}$	+	$Call_{national}$	+	$Call_{international}$
SMS	=	SMS_{local}	+	$SMS_{national}$	+	$SMS_{international}$
MMS	=	MMS_{local}	+	$MMS_{national}$	+	$MMS_{international}$

To analyze the customer behavior two data sets are used first is set of attributes for the description of calling/messaging/net-usage activities and other set consist of attributes describing customer details; so both together can give some hidden insights for the further designing of tariff plans, most suitable for the customer and beneficial for the company too. Customer defining attributes are described in Table 4.

Table 2 Details of customer behavior

S.N.	Name of service	Kind of service	Count of use	Day of use	Time of use	Duration
1.	Calls	Local	No. of calls per day/week	Working days/weekends	Working hours/non-working hours	Short (5 mnts)/medium (5–10 mnts)/long (more than 10 mnts)
		National				
		International				
2.	SMS/MMS	Local	No. of SMS/MMS per day/week	Working days/weekends	Working hours/non-working hours	Short/long SMS/MMS
		National				
		International				
3.	Internet pack	Net use	No. of net begin	Working days/weekends	Working hours/non-working hours	Minutes or hours spent on internet

Table 3 Attributes for customer behavior

Attribute name	Description
$Call_{short}$	Call duration less than 10 min
$Call_{medium}$	10 min < Call duration <= 30 min
$Call_{long}$	30 min < Call duration
$Call_{local}$	Count of local calls per 24 h
$Call_{national}$	Count of long distance calls per 24 h
$Call_{international}$	Count of calls made out of country per 24 h
$Call_{workinghours}$	No. of calls in working hours
$Call_{non\text{-}workinghours}$	No. of calls in non-working hours
$Call_{workingdays}$	No. of calls in working days
$Call_{non\text{-}workingdays}$	No. of calls in non-working days
SMS_{short}	160 characters
SMS_{long}	More than 160 characters
SMS_{local}	No. of local SMS per 24 h
$SMS_{national}$	No. of national SMS per 24 h
$SMS_{international}$	No. of international SMS per 24 h
$SMS_{workinghours}$	No. of SMS in working hours
$SMS_{non\text{-}workinghours}$	No. of SMS in non-working hours
$SMS_{workingdays}$	No. of SMS in working days
$SMS_{non\text{-}workingdays}$	No. of SMS in non-working days
MMS_{short}	600 KB
MMS_{long}	More than 600 KB
MMS_{local}	No. of local MMS per 24 h
$MMS_{national}$	No. of national MMS per 24 h
$MMS_{international}$	No. of international MMS per 24 h
$MMS_{workinghours}$	No. of MMS in working hours
$MMS_{non\text{-}workinghours}$	No. of MMS in non-working hours
$MMS_{workingdays}$	No. of MMS in working days
$MMS_{non\text{-}workingdays}$	No. of MMS in non-working days
Net_{loging}	No. of times net connected per 24 h
$Net_{workingdays}$	Net usage in working days
$Net_{non\text{-}workingdays}$	Net usage in non-working days
$Net_{workinghours}$	Net usage in working hours
$Net_{non\text{-}workinghours}$	Net usage in non-working hours
Net_{time}	No. of hours spent on internet
VV_{call}	Value voucher for calling service
$VV_{message}$	Value voucher for messaging service

Table 4 Attributes for customer description

Attribute name	Description
Cust_Id	Identification of Customer
Gender	Female/Male
$Age_{teenage}$	13 = < Age < 18
Age_{Adult}	<= 18
$Age_{seniorcitizen}$	<= 65
Profession	Type of work
Social-Status	Annual income
Phone-Service	Mobile, Land-Line, Internet
Type-of-Service	Pre-Paid, Post-Paid
Payment-Method	Cash, E-Payment, Credit Card
Paperless-Bill	Yes/No
Bill-Range	Money spent on Phone Recharge

3 Behavior as Pattern

Extension of telecom CRM in big data environment provides a broader view of the customer including telecom service-usage behavior further useful for the designing of customer oriented tariff plans. Customer behavior can be studied as associations between:

Talk-Time-Recharge	→	Financial Status
Net-Pack	→	Educational Background
Net-Pack	→	Profession + Age
Call Duration	→	Gender + Age
Profession	→	Frequency-of-Call + Time-of-Call
Profession	→	Frequency-of-Recharge
Profession	→	Recharge-Pack
Profession	→	Roaming-Pack
2G/3G/4G Recharge	→	Social-Statuses + Mobile-Set
Age + Profession	→	SMS/MMS Pack
Age + Profession	→	Value Voucher
Festive-Season	→	Net-Pack/Talk-Time-Recharge
Psychology	→	Service-Usage
Gender	→	Service-Usage

Call detail records consists additional values other than defined attributes such as phone number of the customer who is making call (Calling Party), phone number of customer who is receiving the call (Called Party), billing phone number that is charged for the call, identification of telephone exchange arranging and recording the call, route of call entering the exchange, route of call leaving the exchange and fault conditions if encountered [3].

Let B = {b_1, b_2, b_3, … b_n}, be the set of defined behavior that is call, messaging, and net-usage activities. C is a subset of set B and |C| denotes the no. of defined behaviors included in C. *CDR* = {set of call detail records}, obtained from OSS (Operation Support System), where each call detail is a kind of behavior. |*CDR*| denotes the total no. of call detail records in a particular time period. Each behavior can be identified by behavior identification no (bid). Support of C is the proportion of behavior in *CDR* that contains C, i.e., $\phi(C) = |\{CDR|\ CDR \in \mathit{CDR}, C \subseteq CDR\}| / |\mathit{CDR}|$. The support count or frequency of C is the no. of call detail records in *CDR* that contains C [4].

CDR is set of call detail records taken for consideration and partitioned as CDR = {R_1, R_2, R_3,…R_n}. To mine pattern of behavior in the big data environment, a set of distributed memory works.

Let D = {D_1, D_2, D_3,…D_n} be the set of distributed memory, so the set of records are distributed as follows:

$$
\begin{array}{c}
R_1 \rightarrow D_1 \\
R_2 \rightarrow D_2 \\
R_3 \rightarrow D_3 \\
R_4 \rightarrow D_4 \\
. \\
. \\
R_n \rightarrow D_n
\end{array}
$$

Records are stored in memory blocks and analyses with the MapReduce. Figure 1 represents architecture for behavior/pattern identification as data product [5].

This architecture takes call detail records as input and distributes it to list of available data nodes for the processing and MapReduce function performs a modified parallel frequent-pattern growth algorithm to identify the patterns which are by default behavior of customer [6]. A parallel FP-Growth algorithm for MapReduce consists of five phases such as Sharding, Parallel Counting, Grouping Items, Parallel FP-Growth, and Aggregating. Figure 2 shows the FP-Growth algorithm in MapReduce [7].

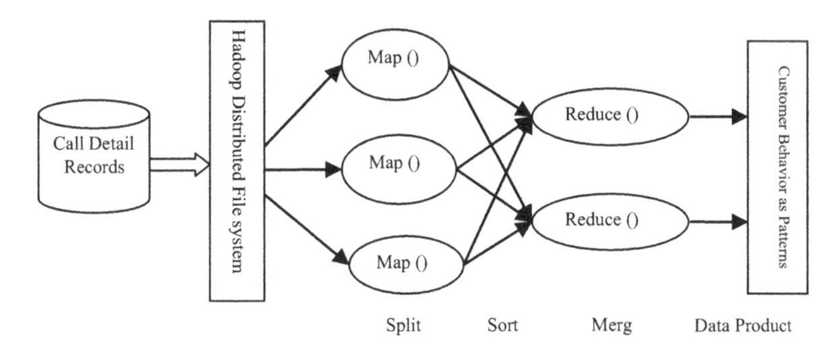

Fig. 1 Architecture for behavior/pattern identification

Phase 1	Sharding	Divides CDR into successive parts and stores the parts on different N computers. Division and distribution of data is called sharding, and each part is called a shard.
Phase 2	Parallel Counting	Perform a MapReduce pass to count the support values of all records that appear in CDR and result is stored in F-list (Frequent-List).
Phase 3	Grouping Items	Divides all the |B| items on F-list into M groups and the list of groups is called group list (G-list). F-list and G-list both are small and the time complexity is O(|B|).
Phase 4	Parallel FP-Growth	This is the key step with one MapReduce pass and map and reduces both perform different functions.
Phase 5	Aggregating	Aggregation of the results obtained in phase 4, which are the patterns of behaviors.

Fig. 2 Parallel FP-growth algorithm in MapReduce

Sharding and Parallel Counting: Mapper is fed with shards of CDR, its input key-value pair is like < key, value = R_i >, where $R_i \subset$ CDR is a record—set of behaviors. For each behavior, say $b_i \in R_i$, the mapper outputs a key-value pair < key = b_i, value = 1 >. The reducer collects the outputs in a key–value pair and provides the output in the form of a whole key–value pair. Figure 3 shows the algorithm which performs first two steps that are Sharding and Parallel Counting [7].

Parallel FP-Growth Algorithm: This is the key step which performs Mapper and Reducer two steps as follows:

Mapper—generates group—dependent behavior: it reads G-list and produces key—value pairs using R_1. Key is group-id and its related value is a group-dependent behavior.

Reducer—FP-Growth on group—dependent shards: When all mapper cases complete their work, it automatically groups all group-dependent behaviors for each group-id into a shard of group-dependent behavior. A local FP-tree is formed for each shard and recursively conditional growth in the tree which reveals frequent patterns [7].

Aggregating: Aggregates the outputs found from the step 4 by selecting the behavior with maximum support value or high-frequency value [7].

Fig. 3 Algorithm for Sharding and Parallel Counting

```
Procedure: Mapper (key, value = Ri)
foreach behavior bi in Ti do
        Output (<bi, 1>);
end
Procedure: Reducer(key = bi, value = S(bi))
C ←0;
foreach behavior 1 in Ri do
     C ← C + 1;
End
Output (<null, bi + C>);
```

4 Conclusion

Recognition of customer behavior as a pattern in the big data environment through the call detail records and customer information obtained from the customer relationship management system is important due to the do or die competition in the telecom market. Identification of customer behavior is further helpful in designing of tariff plans according to the needs of the customer and makes them feel more important by making it customer oriented. For example, if the customer is a student and studying far from home and, due to the classes and studies, makes calls to home at evening or late evening, so a tariff plan "Chhatra" can be offered to a student with less charges in that duration. It will make them feel important due to the plan according to their requirements. In the same way, if customers are housewives and make calls in mid of the day, due to the free time—since husband is in job and kids are in school, so a tariff plan called "Saheli" can be offered to the group of housewives who talk in free time together, with less charges. These offers will not reduce the revenue of the company, since no extra investment is needed, and company can make feel their customer important plus retain the customer and attract a new one. This paper identifies telecom customer behavior in the form of call activities through the described attributes containing customer information and call detail records. Storing the partitioned call detail records in Hadoop distributed file system and analyzing with MapReduce functions provides a big picture of customer telecom usage which is further helpful in the designing of customer oriented tariff plan with the calculation of revenue generation.

References

1. "Consumer_behavior_models" https://www.tutorialspoint.com/consumer_behavior/consumer_behavior_models_types.htm.
2. Yongbin Zhang, Ronghua Liang, Yeli li, "Behavior-Based Telecom Tariff Service Design with Neural Network Approach," 2011 Crown.
3. Cisco Unified Communications Manager Call Detail Records Administration Guide OL.
4. David C., Jeremy Iverson, Shaden Smith and George Karypis, "Big Data Frequent Pattern Mining," http://glaros.dtc.umn.edu/gkhome/node/1121.
5. Yen-hui Liang, and Shiow-yang Wu, "Sequence-Growth: A Scalable and Effective Frequent Itemset Mining Algorithm for Big Data Based on MapReduce Framework," IEEE International Congress on Big Data, 2015, pp 393–400.
6. Sankalp Mitra, Suchit Bande, Shreyas Kundale, Advait Kulkarni, Leena A. Deshpande, "Efficient FP-Growth using Hadoop—(Improved Parallel FP-Growth)," International Journal of Scientific and Research Publications, July 2014, Volume 4, Issue 7, pp 1–3.
7. Haoyuan Li et al., "PFP: Parallel FP-Growth for Query Recommendation," ACM conference on Recommendation Systems, 2008, pp 107–114.

A Perspective Analysis of Phonological Structure in Indian Sign Language

Vivek Kumar Verma and Sumit Srivastava

Abstract Linguists have been the interesting area for researcher from many years but sign language linguists become a new challenge. Contrasted with spoken language phonology, the field of sign-based phonology is quite young. In the fundamental work for sign-based communication, some researchers have shown the need of grammatical structure for the sign language as well. In this paper, we included the linguistic structure of Indian Sign Language (ISL) with phonological counterparts. ISL phonology has different aspects in which few are illustrated and discussed as phonemic list, allophones, and internal prosodic structure.

Keywords Indian Sign Language · Linguistics · Sign grammar
Phonology

1 Introduction

Phonology is all the more firmly attached to the generation and recognition frameworks than some other conceptual level of any language structure. Linguistic experts have been attracted to the investigation of sign language dialects for many years and found related difficulties to manage the grammatical characteristic of sign language [1, 2]. As communications through signing treated as an alternate physical methodology, but it has a phonology and considered particularly significant for language structure. It has been investigated that the signs in the vocabulary ISL are not comprehensive motions, but rather are involved a generally little number of insignificant units that may recombine to create a conceivably vast dictionary.

V.K. Verma (✉) · S. Srivastava
SCIT, Manipal University, Jaipur, India
e-mail: vermavivek123@gmail.com

S. Srivastava
e-mail: sumit.srivastava@jaipur.manipal.edu

A.K. Somani et al. (eds.), *Proceedings of First International Conference on Smart System, Innovations and Computing*, Smart Innovation, Systems and Technologies 79, https://doi.org/10.1007/978-981-10-5828-8_17

The phonological parts of the ISL are communicated through different signs, instead of vocal sound as in spoken language. As there are numerous sign-based communications in the world, which has no direct relationship between the prevailing spoken language and corresponding sign language in any group [3]. A man who is hard of hearing during childbirth and does not take in a sign-based language semantically and intellectually denied similarly as any listening to individual and falsely kept from taking in natural language dialects. It is critical to consider what the condition of our insight about ISL is, subsequent to sign-based language additionally offers one of a kind open doors for testing thoughts regarding the way of dialect itself, thoughts by and large detailed only from perceptions about natural language dialect. Our assignment as ISL phonologists is to discover which the negligible units of the framework are, which parts of this sign are contrastive, and how these units are obliged by the tactile frameworks that deliver and see them. The considerable number of things of the rundown of contrasts and likenesses amongst sign-based and natural language dialects the ranges that present the most striking divergences happen in phonology. The interface amongst phonology is without a doubt diverse, given the opportunities and limitations accessible to the system.

In linguistic concern, phonology is the study of precise association of sounds in natural language. It is the general investigation of phonemes which is the smallest part of any language. These small parts join in obliged approaches to make the expressions of the language. As for the reference of the phoneme in natural language 'P' pronounce as and 'B' pronounce as in the English word 'pat' and 'bat' respectively (Fig. 1).

In case of sign dialects, there are three noteworthy phonological classes: hand shape, Position of hand, and hand movement. Instead of sounds, the phonemes are considered as the different signs present in a row of hand signs [4, 5] (Fig. 2).

As in ISL two sign for 'GIRL' and 'NOT/NO' both has the same handshape but two different location. Similarly two sign for 'SIT' and 'TRAIN' has the same hand shape with two different hand movements.

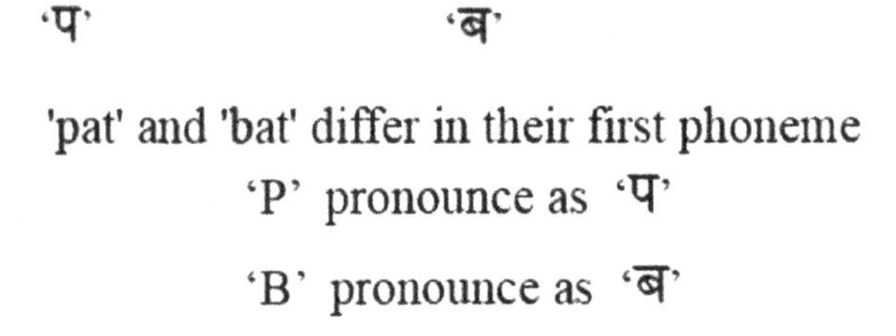

Fig. 1 Phoneme for natural language

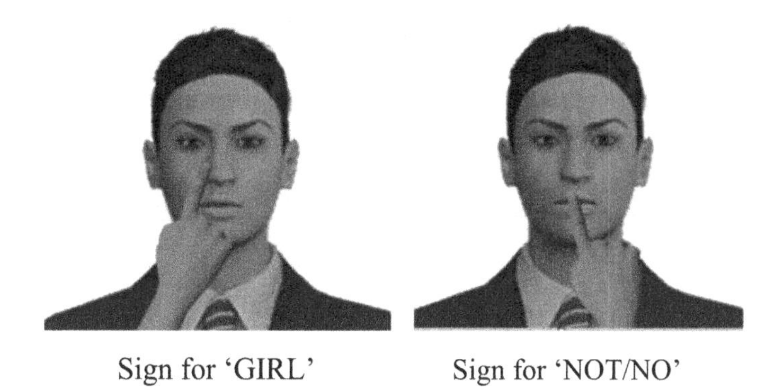

Sign for 'GIRL' Sign for 'NOT/NO'

Fig. 2 Phoneme for ISL [6]

2 Phonological Structure of ISL

There are various aspects of ISL phonology, including phonemic sign list, allophones, and internal prosodic structure.

2.1 *Phonemic Sign List*

It includes all the possible phonemes for the generation of a complete expression in sign language. A physical part of dialect is phonemic in the event that it serves to show lexical complexities (i.e., unusual from different parts of physical structure). This property of duality of designing is crucial to human phonology [7]. For accommodation, ISL handshape, for the most part, are named after the signs in which they show up, yet they are not themselves morphemes. Notwithstanding handshape contrast, there are additionally differentiates in position, way movement, neighborhood, and nonmanual components.

Phoneme based on movement: A pair of signs that contrast in path movement. Two signs move across in front of the chest with the handshape with the palm facing outward, but in second sign hand moves across horizontally in a straight line path.

Phoneme based on local movement: MAN and THANK is a pair of signs that contrast in the local movement. In both, the handshape is used, but in MAN, the hand rotates while in THANKS the palm bends back to chin.

Phoneme based on orientation: ABOUT and AROUND is a pair of signs which has the same phoneme with different hand orientations. Both use the handshape on a turn about the finger over another hand but the orientation of both turnabouts are different.

Phoneme based on nonmanual features: UP and DOWN is a pair of signs that uses a nonmanual marker features. They both involve the same handshape with different movement using looking upward and looking downward shows the exact meaning.

2.2 Allophones

It includes multiple possible signs for the generation of a single phoneme.

'WHAT TIME' and 'HOW BIG' two different sign with one common hand shape using a phoneme called 'WHAT'. They both sign has a common phoneme but the second phoneme is different as for TIME indicates through pointing the finger on the wrist while in other hand for BIG two open palm with large difference [8, 9].

2.3 Internal Prosodic Structure

There is variation in prosodic properties of language usually it includes intonation, tempo and stress, length, pausing, etc. In case of ISL includes the properties of changes in timing, including pause, and manifested by the signing hands. The analysis of internal prosodic structure is quite difficult to understand in sign phonology. Signs are best dissected as portion estimated entire units, as consecutive strings of section, measured units, as syllables in common dialect comparable yet unmistakeable from talked dialect prosody [10, 11]. In signs, completely still partitions are less resonating (fiery) than segments where the hand does not change position yet there is neighborhood development (handshape change), which is thusly are less vibrant than bits where the hand has way development with or without nearby development.

3 Challenges in ISL Phonology

- To train visual dataset to distinguish and recognize different phonemes
- Manifestation of two different handshapes is closely linked to the specific sign
- Loosely related different signs need to be further organized for next level grouping
- In ISL there are no defined phonemes which can be included in the automated system

- ISL has many ambiguities with the same combination of phonemes for different signs. As, for example, in WHAT and WHY has same phoneme structure
- Unavailability of standard dataset for the ISL
- Manifestation of two different handshapes is closely linked to the specific sign.

4 Transliteration Approach for ISL

The notation based translation system for sign languages is an alphabetic system describing signs on a mostly phonetic level. Homonyms are the words that have the same spelling as well as same pronunciation but have different meanings; Words with the same spelling but different meaning are called homographs. While for signing these words will have different signs but when used for translation there will be an issue because. While generating signs for words like 'May' the month may have different sign while as the verb 'may' have different sign so if the sign for 'may' has to be displayed the computer has to see which 'may' it is it the month or the verb thus a certain type of intelligence has to be integrated for such situations.

The approach we could use is like such words need to be finger spelled instead of signing the word. For this the all the possible homonyms have to be figured out and then these words need to be finger spelled as required. ISL is the composition of different sign categories based on the features of visual communication to make a sign for a single unit, e.g., letter, word, and sentences. These signs are further decomposed into manual and nonmanual as manual consist of hand shape, movement, location [8], and orientation as shown in Fig. 3, whereas nonmanual gestures consist of facial expression, head movement, shoulder movement, and mouthing. In general nonmanual signs are used along with manual markers.

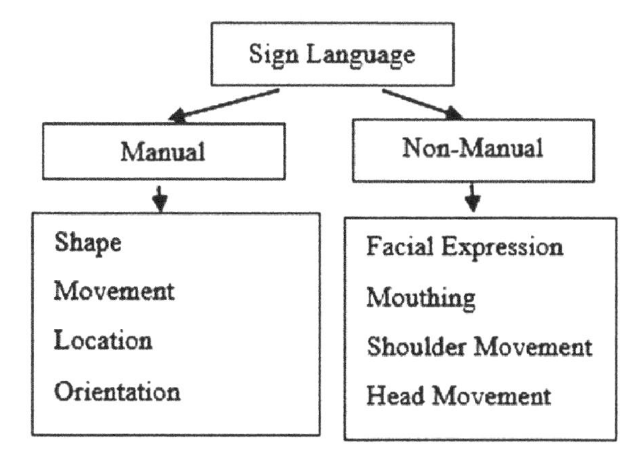

Fig. 3 Types of sign

5 Conclusion

As to conclude this paper described some of the important issues of areas of research in ISL phonology. We have investigated with a visual corpus of 1200 signs for ISL and analyses the phonological structure to generate the automated form ISL generation. As in ISL has many ambiguity need to be removed through proper integration of the automated sign and related grammatical structure. In ISL grammatical structure there are no defined rules which can be utilized for the processing of signs in a different application. It needs to be defined through analysis of the sign in a real environment and covers all constraints for sign automation.

References

1. Verma, V.K., Srivastava, S. K.: Towards linguistic issues of Indian Sign Language. CSI Convention, New Delhi, Dec. 2015 In. Springer, (2015).
2. Chiu, Ming Ming.: Statistical discourse analysis of an online discussion: Cognition and social metacognition. Productive multivocality in the analysis of group interactions Springer US, PP. 417–433, (2013).
3. Cooper, Helen, Brian Holt, Richard Bowden.: Sign language recognition. In: Visual Analysis of Humans. Springer London, (2011).
4. Sandler, Wendy.: The phonological organization of sign languages. In: Language and linguistics compass 6.3, PP. 162–182, (2012).
5. Dzikovska, Myroslava, et al.: BEETLE II: Deep natural language understanding and automatic feedback generation for intelligent tutoring in basic electricity and electronics. In: International Journal of Artificial Intelligence in Education 24.3, 284–332. (2014).
6. Kishore, P. V. V., and P. Rajesh Kumar.: A video based Indian sign language recognition system (INSLR) using wavelet transform and fuzzy logic. In: IACSIT International Journal of Engineering and Technology 4.5 (2012).
7. Stein, Daniel, Christoph Schmidt, and Hermann Ney.: Analysis, preparation, and optimization of statistical sign language machine translation. In: Machine Translation 26.4 PP. 325–357, (2012).
8. Curiel, Arturo, and Christophe Collet.: Sign language lexical recognition with propositional dynamic logic. In: arXiv preprint arXiv: PP.1403.6636 (2014).
9. Sako, Shinji, and Tadashi Kitamura.: Subunit modeling for japanese sign language recognition based on phonetically depend multi-stream hidden markov models. In: Universal Access in Human-Computer Interaction. Design Methods, Tools, and Interaction Techniques for EInclusion. Springer Berlin Heidelberg, PP. 548–555. (2013).
10. Caridakis, George, Stylianos Asteriadis, and Kostas Karpouzis.: Non-manual cues in automatic sign language recognition. In: Personal and ubiquitous computing 18.1, PP. 37–46, (2014).
11. Friginal, Eric, et al.: Linguistic characteristics of AAC discourse in the workplace. In: Discourse Studies (2013).

Air Pollution Prediction Using Extreme Learning Machine: A Case Study on Delhi (India)

Manisha Bisht and K.R. Seeja

Abstract Outdoor air pollution has emerged as a serious threat to public health across the globe. Air quality monitoring and forecasting are required to provide the policy makers a scientific basis for formulating a robust policy on abatement of air pollution. Moreover, if air pollution forecasts are issued to the public, they can take preventive measures to minimize their exposure to unsafe levels of air pollutants. In this paper, an intelligent air pollution prediction system using Extreme Learning Machine (ELM) has been proposed to predict the air quality index for five pollutants (PM_{10}, $PM_{2.5}$, NO_2, CO, O_3) for the next day. It is found that the prediction of ELM-based proposed system is better than the existing air pollution prediction systems.

Keywords Air pollution prediction · Extreme Learning Machine Regression · Artificial neural networks

1 Introduction

Air pollution is a condition under which the ambient concentration of some gaseous or particulate substances exceeds their prescribed safe limits in the environment. This happens due to both natural and man-made causes, the latter having more pronounced effects. This stems from the trend of rapid and haphazard urbanization and industrialization. Especially in metropolitan cities like Delhi, which have a huge burden of population to support, the economic development concerns often supersede the ecological considerations. Additionally, massive amounts of municipal solid waste are generated by the city each day. The emissions from these

M. Bisht (✉) · K.R. Seeja
Department of Computer Science & Engineering, Indira Gandhi Delhi Technical University for Women, Kashmere Gate, Delhi 110006, India
e-mail: manishabisht27@gmail.com

K.R. Seeja
e-mail: seeja@igdtuw.ac.in

A.K. Somani et al. (eds.), *Proceedings of First International Conference on Smart System, Innovations and Computing*, Smart Innovation, Systems and Technologies 79, https://doi.org/10.1007/978-981-10-5828-8_18

vehicles and industries as well as those generated from the burning of solid waste are the most significant sources of air pollution. Air pollution is, however, not independent of natural phenomena [1]. The meteorological conditions like relative humidity, wind speed, wind direction, temperature, and rainfall also affect the level of air pollutants at any time and place. For instance, air quality is found to be worst during winters due to the prevalent meteorological conditions in that climate. The adverse health impacts of air pollution cannot be emphasized enough. Pollutants like nitrogen dioxide, sulphur dioxide, particulate matter and carbon monoxide may cause respiratory and cardiovascular diseases. Lead may cause impairment of brain functions and ozone may affect normal liver function [2].

Various other methods for air pollution prediction can be found in the literature. These methods can be classified into two categories: traditional techniques and soft computing techniques. Traditional techniques use statistical models and soft computing uses artificial intelligence models for forecasting pollutants. Artificial neural networks can be trained to learn from dataset so that it can give accurate predictions when fed with new data. A neural network-based prediction of +1, +2 and +3 pollution levels of SO_2, PM_{10} and CO has been proposed [3]. This system used feed forward back propagation network. Back propagation, though widely used learning technique in artificial neural networks, has slow convergence due to iterative tuning and gradient descent. Statistical techniques [4] such as ARIMA (Auto Regressive Integrated Moving Averages) and SRS (single exponential smoothing) were also used for forecasting NO_2, SO_2, SPM and RSPM. They used the historical pollutant data to find patterns and predict solely on past values of pollutant, without taking into account the prevalent meteorological conditions. In another research, a SOSE (site optimized semi-empirical) model has been proposed to predict CO concentrations [5]. SOSE is a statistical model that estimates the pollutant concentrations based on wind speed and wind direction only. An Analytical dispersion model for prediction of PM_{10} has also been used by taking into account the dispersion of pollutants due to wind in the atmosphere [6]. Another system to forecast PM_{10} levels uses a three-layer feedforward back propagation neural network and a recurrent neural network called Elman network [7]. A comparison of the two proposed networks is done by the authors and it was found that Elman networks required more training time and hence FFNN was found to be better. A comparative study of statistical regression and techniques such as ARIMA (Auto Regressive Integrated Moving Averages), LRA (Linear Regression Analysis) and PCA (Principal Component Analysis) for the prediction of ozone levels has been done and the authors [1] have found classification techniques better for their data.

In this research, an intelligent air pollution prediction system is proposed for predicting the various pollutant levels in next day.

2 Materials and Methods

In this research, an existing air pollution monitoring and forecasting system SAFAR is used as the data source as well as for performance evaluation and ELM is used for prediction of air pollution in Delhi.

2.1 SAFAR

India is one of the most polluted countries in the world, with its capital, Delhi being ranked as the most polluted city in WHO's 2014 report 'Ambient air pollution in cities' [8]. SAFAR (System of Air Quality Forecasting and Research) [9], is a programme under the aegis of Ministry of Earth Sciences, India. It was set up by the government of India to assess the impact of emission reduction measures undertaken by it in preparation of the 2010 commonwealth games, which were hosted by Delhi [10]. SAFAR has 10 monitoring stations across Delhi, where the concentrations of five air pollutants (PM_{10}, $PM_{2.5}$, NO_2, CO, O_3) as well as weather parameters (relative humidity, rainfall, temperature, wind speed, wind direction) are monitored using specialized equipment. The collected data is then processed at IITM (Indian Institute of Tropical Meteorology), Pune, which uploads the current and next day's predicted air quality indices, as well as relevant health advisories corresponding to each of the monitored pollutants on its web portal (http://www.safar.tropmet.res.in). SAFAR uses WRF-Chem to make 1-day prediction of pollutants [11]. The present-day air quality status as well as the next day's prediction is given on its web portal not in terms of absolute concentrations, but rather, the raw concentration values of pollutants are converted into indicative numbers called air quality indexes. This is done because the interpretation of the air quality index is easier for the general public as the air quality index ranges from 0 to 500, with each band of 100 having a different air pollution label. The air quality index 0–100 is labelled as 'Good/satisfactory', 101–200 is labelled as 'moderate', 201–300 is labelled as 'poor', 301–400 is labelled as 'very poor' and 401–500 is labelled as 'severe'.

2.2 Extreme Learning Machine

Extreme Learning Machine is a learning algorithm for single hidden layer feed-forward neural networks [12]. In ELM, there is a single hidden layer and the input weights and biases of hidden layer neurons are assigned randomly. Moreover, no iterative tuning is required in ELM, resulting in much faster learning compared backpropagation-based FFNN. ELM has been found to have good generalization in a variety of applications. ELM can be used for a large variety of machine learning tasks such as regression, classification, clustering, feature selection, etc. [13].

The ELM algorithm is summarized below:

If the training set S, of N readings is S = $\{(x_i, t_i)\}$, for i = 1, 2,…N), activation function is g(x) and number of hidden layer neurons is m, the ELM algorithm is given as

Step 1: For each of the hidden layer neurons, assign input weight, w_i and bias, b_i randomly.

Step 2: Calculate the output matrix for the hidden layer, H.

Step 3: Use the matrix H computed in step 2 to calculate the output weight as: $\beta = H^{\dagger}T$ where $H^{\dagger}$ is the Moore–Penrose generalized inverse matrix for the hidden layer output matrix, H.

The ELM algorithm uses a random function to generate values for w_i and b_i and then calculates the corresponding β_i using Moore–Penrose generalized inverse matrix.

2.3 *Proposed Methodology*

The proposed methodology is summarized as follows:

Data Collection Daily meteorological and air pollution data for the Delhi University monitoring station was collected from SAFAR's web portal (http://safar.tropmet.res.in), which provides air quality index for PM_{10}, $PM_{2.5}$, NO_2, CO, O_3 along with the current values of weather conditions like relative humidity, wind speed, temperature, rainfall and wind direction. 37 days data has been collected from SAFAR. Out of this, first 30 days data was used to train the network and next 7 days data was used for testing it.

Variable Selection and Construction of ELM Model In the data collected from SAFAR portal, it was observed that except for CO, all other air pollution indices were continuous values in the range [0, 500] but CO was taking only two values of air quality index—59 and 104. Hence linear regression model of ELM is used to predict air quality index for PM_{10}, $PM_{2.5}$, NO_2, O_3 and binary classification model of ELM is used to predict air quality index for CO. After conducting various experiments, 8 input parameters were chosen: previous day values of PM_{10}, $PM_{2.5}$, NO_2, CO, O_3, temperature, humidity and wind speed. Thus, the ELM has 8 input neurons. For predicting each of the five different pollutants (PM_{10}, $PM_{2.5}$, NO_2, CO, O_3), five different ELM networks were constructed. The output layer of each of these networks consists of only one neuron corresponding to the predicted pollutant.

The network shown in Fig. 1 is used to predict PM_{10} values for the next day. Similarly, networks were created for prediction of $PM_{2.5}$, NO_2 and O_3.

Training and Testing of ELM Model Training experiments were carried using various activation functions and a variable number of hidden layer neurons for each

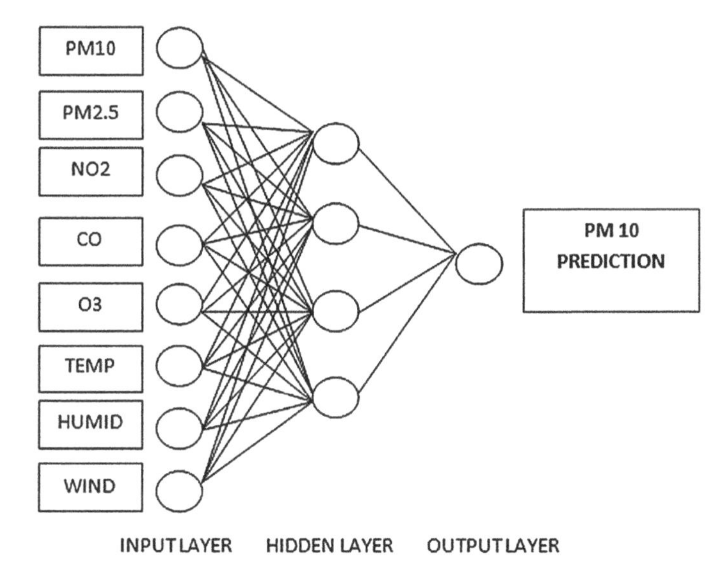

Fig. 1 Proposed ELM network for prediction of PM_{10}

pollutant. The parameters for which optimal results were obtained during training phase were used to perform testing.

Performance Evaluation of Proposed Model The forecasting performance for the proposed ELM model was evaluated with the prediction of SAFAR as well as with actual values. 'Root mean square error' was used as a measure for comparison of prediction performance of PM_{10}, $PM_{2.5}$, NO_2 and O_3, since they were predicted using the regression model of ELM. On the other hand, 'testing accuracy' was used to compare prediction performance of CO, since classification model of ELM is used to predict it.

3 Experiments

The data was collected from SAFAR portal during February and March 2016, for Delhi University monitoring station. The dataset was then curated to remove duplicate readings on consecutive days, especially for weekends, during which the portal did not regularly update the values. The value of all pollutants was in the range 0–500, as provided by the portal. All columns of the dataset, except for the target output was normalized between [0, 1], by dividing by suitable power of 10. The first 30 readings were given as training file and next 7 readings were given as testing file in.txt format to the ELM program. Table 1 summarizes the optimal values of ELM parameters like 'activation function' and 'number of hidden layer

Table 1 ELM parameters for different pollutants

Pollutant	No. of hidden layer neurons	Activation function
PM_{10}	4	radbas
$PM_{2.5}$	6	sig
NO_2	4	sin
CO	3	sig
O_3	3	sig

neurons' for each of the five pollutants, as concluded from the experiments performed.

4 Results and Discussion

The experimental results obtained from each of the five proposed ELM networks have been given below. In these bar graphs, y-axis represents the pollutant air quality index value and x-axis represents the day for which prediction is given. The graph shows a comparison between prediction of SAFAR and prediction of the proposed ELM network, with respect to the actual value for that day, as given on the SAFAR portal.

4.1 *Prediction of PM_{10}*

The seven days prediction of PM_{10} is shown in Fig. 2. It is found that the air quality index values predicted by ELM are close to the actual values for all the seven days in the testing data set.

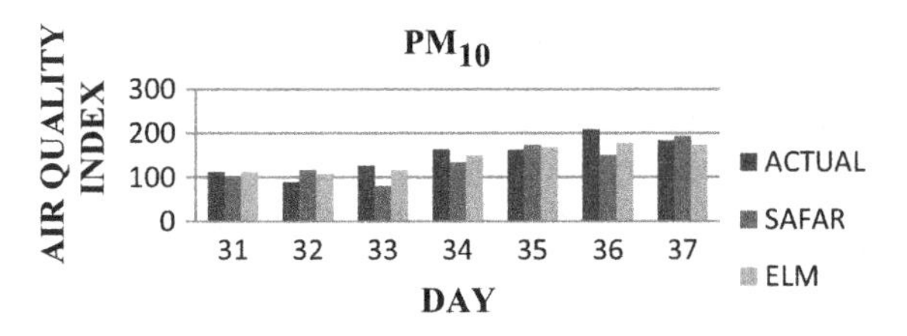

Fig. 2 PM_{10} prediction comparison

4.2 Prediction of $PM_{2.5}$

The seven days prediction of $PM_{2.5}$ is shown in Fig. 3. It is found that the air quality index values predicted by ELM are close to the actual values for all the seven days in the testing data set.

4.3 Prediction of NO_2

The seven days prediction of NO_2 is shown in Fig. 4. It is found that the air quality index values predicted by ELM are close to the actual values for six days out of seven days in the testing data set.

4.4 Prediction of Carbon Monoxide

The seven days prediction of CO is shown in Fig. 5. It is found that the air quality index values predicted by ELM are same as the actual values for six days out of the seven days in the testing data set.

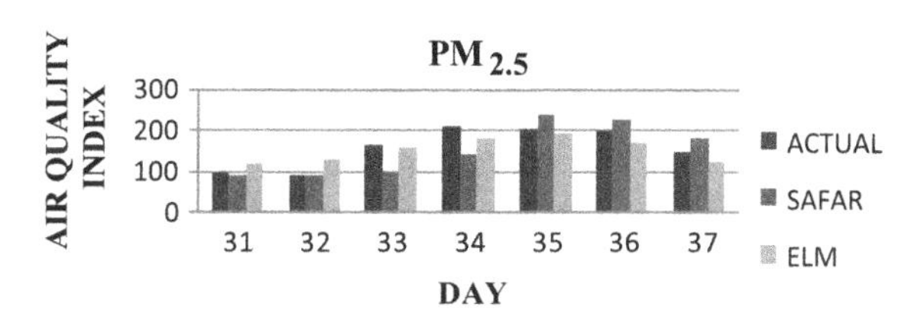

Fig. 3 $PM_{2.5}$ prediction comparison

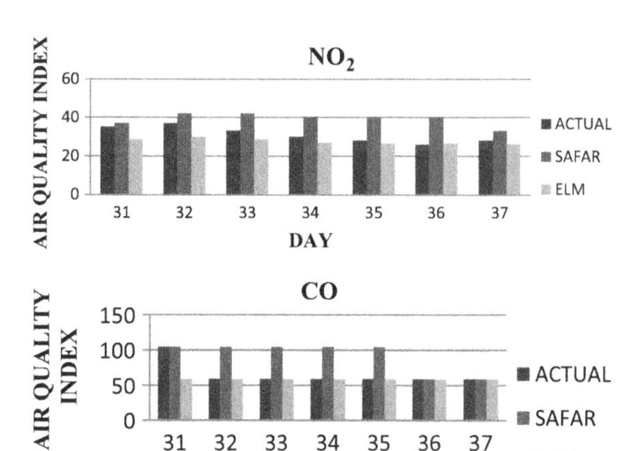

Fig. 4 NO_2 prediction comparison

Fig. 5 CO prediction comparison

4.5 Prediction of Ozone

The seven days prediction of O_3 is shown in Fig. 6. It is found that the air quality index values predicted by ELM are close to the actual values for five days out of seven days in the testing data set.

The performance evaluation measure, Root Mean Square Error (RMSE) is used for comparing the performance of existing system SAFAR and proposed ELM-based system and is shown in Fig. 7.

Classification accuracy comparison for CO is shown in Fig. 8.

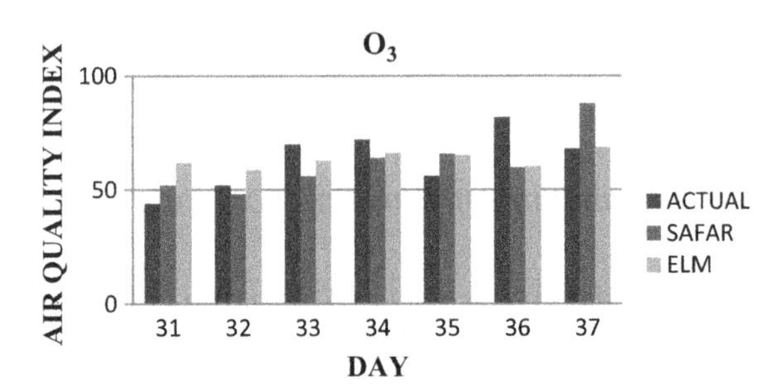

Fig. 6 O_3 prediction comparison

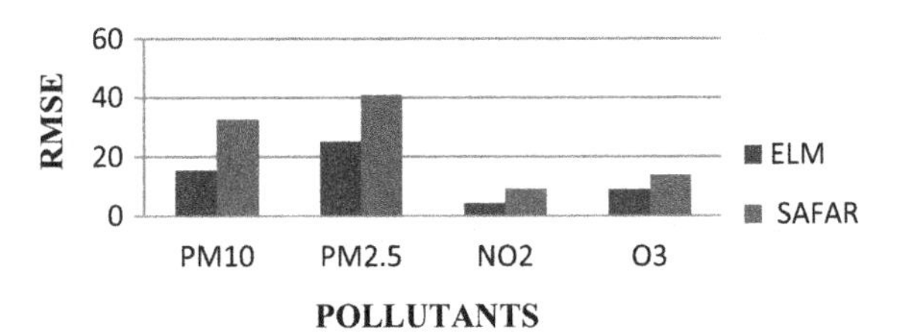

Fig. 7 RMSE comparison of ELM and SAFAR

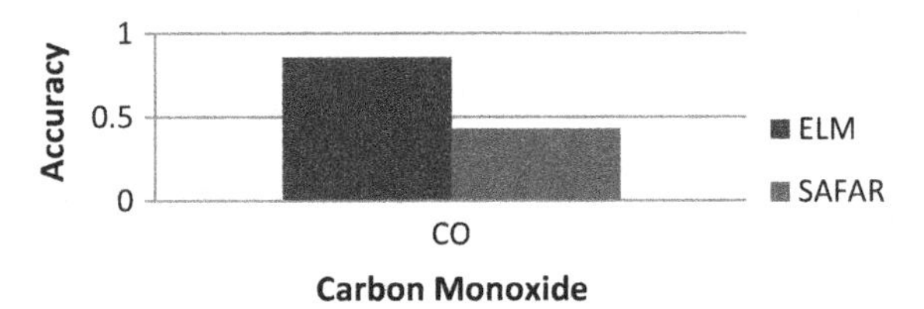

Fig. 8 Comparison of ELM and SAFAR for prediction of CO

5 Conclusion

In this work multi-variable linear regression model of ELM is used to predict air quality index for PM_{10}, $PM_{2.5}$, NO_2, CO, O_3. In the proposed model, the previous day air quality index of pollutants and meteorological conditions are used for prediction. Performance of the proposed model was compared with the prediction of an existing prediction system, SAFAR as well as with the actual values of next day. ELM-based prediction was found to have greater accuracy than the existing.

References

1. Athanasiadis, Ioannis N., Kostas D. Karatzas, Pericles A. Mitkas.: Classification techniques for air quality forecasting." Fifth ECAI Workshop on Binding Environmental Sciences and Artificial Intelligence, 17th European Conf on Artificial Intelligence. 2006.
2. Niharika, Venkatadri M., and Padma S. Rao. "A survey on Air Quality forecasting Techniques." International Journal of Computer Science and Information Technologies 5.1 (2014): 103–107.
3. Kurt, Atakan, Betul Gulbagci, Ferhat Karaca, and Omar Alagha. "An online air pollution forecasting system using neural networks." Environment International 34, no. 5 (2008): 592–598.
4. Kandya, Anurag, and Manju Mohan. "Forecasting the urban air quality using various statistical techniques." In Proceedings of the 7th International Conference on Urban Climate. 2009.
5. Peace, Margaret, Kim Dirks, and Geoff Austin. "The Prediction of Air Pollution Using A Site Optimised Model And Mesoscale Model Wind Forecasts." (2005).
6. Kumar, Anikender, and P. Goyal. "Air Quality Prediction of PM10 through an Analytical Dispersion Model for Delhi." Aerosol Air Qual. Res 14 (2014): 1487–1499.
7. Ababneh, Mohammad F., and Mohammad Hjouj Btoush. "PM10 Forecasting Using Soft Computing Techniques." Research Journal of Applied Sciences, Engineering and Technology 7, no. 16 (2014): 3253–3265.
8. WHO pollution statistics report. (accessed on 30.04.16): http://www.who.int/phe/health_topics/outdoorair/databases/cities/en/.
9. SAFAR web portal: http://safar.tropmet.res.in.
10. Beig, Gufran, Dilip M. Chate, Sachin D. Ghude, A. S. Mahajan, R. Srinivas, K. Ali, S. K. Sahu, N. Parkhi, D. Surendran, and H. R. Trimbake. "Quantifying the effect of air quality control measures during the 2010 Commonwealth Games at Delhi, India." Atmospheric Environment 80 (2013): 455–463.
11. Srinivas, Reka, Abhilash S. Panicker, Neha S. Parkhi, Sunil K. Peshin, and Gufran Beig. "Sensitivity of online coupled model to extreme pollution event over a mega city Delhi." Atmospheric Pollution Research 7, no. 1 (2016): 25–30.
12. Huang, Guang-Bin, Qin-Yu Zhu, and Chee-Kheong Siew. "Extreme learning machine: theory and applications." Neurocomputing 70, no. 1 (2006): 489–501.
13. Huang, Gao, Guang-Bin Huang, Shiji Song, and Keyou You. "Trends in extreme learning machines: a review." Neural Networks 61 (2015): 32–48.

Load-Balanced Energy-Enhanced Routing Protocol for Clustered Bee-Ad Hoc MANETs

Sasmita Mohapatra and M. Siddappa

Abstract To make MANETs energy efficient and to balance the load swarm intelligence along with clustering as Bee-Ad Hoc-C has been chosen as the best method in our previous work with improvized routing by the use of BCN. But in this routing technique optimum load balancing and energy efficiency have not been improved much. In the present paper, one new routing algorithm has been introduced which takes care of the above factors. Here, the energy efficiency is increased by taking into consideration the number of nodes in the cluster and remaining battery power of the nodes. Also to avoid any redundancy in the routing alternate router has been initiated if any shortest path is busy. The work is carried using the NS2 (Network Simulator-2). By the proposed method, the MANET routing can be properly balanced and improved in terms of Energy Efficiency, End-to-end delay, Throughput, Packet Delivery Ratio, Route Discovery Time.

Keywords MANET · Load balance · Energy efficiency · Residual energy Number of nodes · Alternate router

1 Introduction

The application of MANETs has gone beyond any limit starting from disaster and emergency relief to personal and industrial use as mobile conferencing to defense system. According to [1], dynamic directing is nearly the most vital issue in MANETs. For proper hierarchical architecture, clustering has been considered as the main protocol. In our research work for clustering swarm intelligence is taken as the best method which is naturally followed. In our previous papers [2–4] Bee-Ad Hoc-C is taken as the best method. In [3] the algorithm has been done which

S. Mohapatra (✉) · M. Siddappa
Sri Siddhartha Academy of Higher Education, Tumkur, India
e-mail: sasmitamohapatra0@gmail.com

M. Siddappa
e-mail: siddappa.p@gmail.com

A.K. Somani et al. (eds.), *Proceedings of First International Conference on Smart System, Innovations and Computing*, Smart Innovation, Systems and Technologies 79, https://doi.org/10.1007/978-981-10-5828-8_19

proposed a Stable Cluster Maintenance Scheme and focuses on minimizing the CH changing. In [4] the algorithm has been introduced where the routing has been made more systematic by making the use of BCN (Border Cluster Node) mainly for inter-cluster routing. But in these works load balancing in the cluster has not been considered. Though the energy efficiency in IBAC is more compared to BAC still it has not reached the optimum value. So in the proposed routing algorithm (LBEE-IBAC) care has been taken for the following aspects:

- Cluster Head and other cluster members are selected based on Received Signal Strength (RSS) and residual energy of the node.
- For cluster maintenance load-balanced re-clustering has been adopted, where there is a limit for minimum and maximum number of nodes in any cluster.
- For avoiding delay and balancing the load data transmission has been done by the alternate router if the shortest path is busy.

2 Related Work

As MANET has become a key improvement in wireless technology so according to [7–10] making MANET more energy efficient has become the most prominent research area for many researchers. According to [5] Bee Sensor-C has been introduced as a scalable and energy-efficient multipath routing protocol for sensor networks. Taking the algorithm into consideration, we have introduced two routing algorithms as Bee-Ad Hoc-C (BAC) and Improvised Bee-Ad Hoc-C (IBAC) for MANETs.

2.1 Working of Bee-Ad Hoc-C MANET

In [3], we have introduced a new routing protocol as Bee-Ad Hoc-C (BAC) which has been proved to be a multipath scalable routing protocol for MANET. In this regard, bee-inspired routing along with clustering has been introduced as the best routing method. Here as in a Bee Hive the work has been distributed between the bees similarly in the network the total work of the network is divided among the dynamic nodes. According to that, in any cluster nodes have been divided as cluster head, foragers, and scouts. When an event occurs according to the received signal strength the cluster is formed and in a cluster the nodes are divided as Cluster Head, Foragers, and Scouts. The duty of the nodes is different according to the category. Once a cluster is formed and cluster members are decided the next work is to decide the path from the source to destination. In this regard, the CH takes the help of forager first to find the destination within the cluster. If the forager can find the path then the data transfer takes place within the cluster by forager alone. However, if the forager cannot find the destination then it takes the help of scouts to search the destination within the cluster or else it sends them to search the destination outside the cluster. Thus, the scouts move randomly to find a path to the destination. In this regard, care has been taken to count the number of maximum hop for the scout and whether the scout visits the same node more than once. With respect to these parameters any scout can be disabled also. Thus, once the destination is found the forward scout is converted to backward scout and sent to the CH with hop information in the cache. After the route is found the CH sends the data to the destination through the shortest route. This algorithm supports to find multiple paths from source to destination for data transfer. Along with this, in this paper, we have introduced SCMS (Stable Cluster Maintenance Scheme) algorithm which minimizes the CH selection cost by delaying the process up to the acceptable time period to make the cluster more stable. But in this process, the traffic is more and the redundancy increases as the scouts move arbitrarily outside the cluster in search of the destination. Also, the energy efficiency was reduced as the nodes are used unnecessarily. To come over this problem the next algorithm was introduced in our next work where the routing protocol has been introduced as Improvised Bee-Ad Hoc-C (IBAC).

2.2 *Working of Improvised Bee-Ad Hoc-C MANET*

According to [4] in the new IBAC algorithm, the BAC algorithm is improved for inter-cluster communication, where the routing is established between the clusters by the scouts at the border of two clusters known as Border Cluster Node. In this algorithm, when an event is detected and CH sends a request to the forager to search the destination if the forager cannot find the destination it takes the help of the scouts to search the destination within the cluster. If it cannot find the destination within the cluster then it sends an RERR (error) signal to the CH. In that case, the CH sends a request to the BCNs to find the destination in the attached clusters. Again the request is carried by the other CHs to search the destination in their respective clusters. Thus, once the destination is found RREP (reply) signal is sent by the respective BCN showing that the route has been found for the destination and data transfer takes place (Fig. 1).

But in this regard, the main drawback is that though it is a systematic process because of the use of BCNs, the routing time has increased. Also in this paper care has not been taken to balance the load with respect to the number of cluster members or availability of shortest path with fewer loads and no alternate routing is introduced. Also, energy efficiency of the system has not been considered with respect to the residual energy of the nodes. In the proposed routing algorithm utmost care has been taken to still improve the routing by considering mainly three modules:

- Cluster head selection based on residual energy and RSS.
- Cluster Maintenance by load-balanced re-clustering.
- Data Transmission via alternate routers if the shortest path is busy.

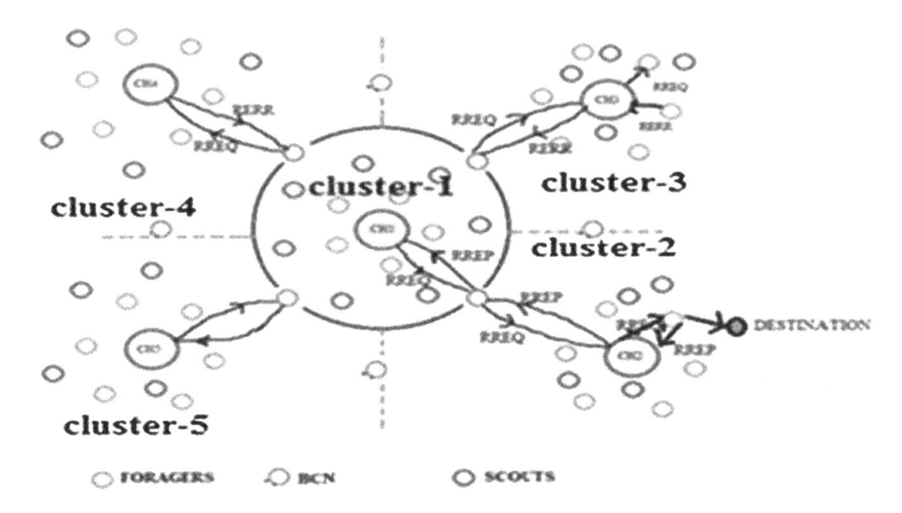

Fig. 1 Workflow of improvised Bee-Ad Hoc-C network

Finally, the performance comparison has been done for both proposed LBEE-IBAC and IBAC.

3 Algorithm for Working of Load-Balanced Energy-Enhanced Routing for Bee-Ad Hoc-C (LBEE)

This protocol works mainly in four modules in order to make the Bee-Ad Hoc-C MANET more energy efficient, for proper load balancing and to reduce the delay time. The algorithm for working on the protocol can be summarized as

3.1 Cluster Formation
3.2 Cluster head and cluster member selection
3.3 Cluster maintenance by load-balanced re-clustering

 3.31 Cluster Merging

  3.311 Cluster head selection after merging of clusters

 3.32 Cluster Split up

  3.321 Cluster head selection after cluster split up

3.4 Data transmission via alternate router after calculation of probability value of path.

3.1 Cluster Formation

The MANET we have designed is a linear network. Initially, the area of the network is decided. For calculating the number of cluster we take X-axis as a reference where:

$$\mathbf{No.of\ clusters} = \frac{\mathbf{val(x)}}{\mathbf{communication\ range}} \tag{1}$$

3.2 Cluster Head and Cluster Member Selection Based on Residual Energy and Received Strength Signal (RSS)

As the mobile nodes in a MANET are battery operated so there can be a serious problem if any of the nodes have a very low energy left. Basically, in a Clustered MANET, the maximum work is done by a CH. So it is obvious that the battery power of CH drains fast. Hence it is required to balance the energy consumption among mobile nodes to avoid node failure as any node failure in the clustered network may lead to network partition or interruption in transferring the data thus increasing the routing time.

- Implementation: In the earlier case we have given the ID to different nodes and categorized them according to the RSS (Received Signal Strength) but in this case, the nodes are decided to be given a virtual ID not only according to the RSS but also according to the remaining Residual Energy (RE). To remain inside the cluster a minimum RSS and RE is decided for the nodes based on certain factors.
- Deciding RSS: Here as an event occurs the first destination is decided. Then RSS is decided for the nodes depending on their distance with respect to the event. For any node: High distance indicates RSS is Low while Low distance indicates RSS is High.
- Deciding Residual Energy: The residual energy of the nodes is also calculated. As in C++, it is inbuilt each node maintains the current energy. Thus, we have extracted the residual energy for each node using energy model.
- Deciding Virtual ID (VID): Thus depending on RSS and RE the VID is decided for the nodes as

$$\mathbf{VID[index]} = (0.5 * \mathbf{Avgdis}) + (0.5 * \mathbf{residualenergy}) \tag{2}$$

Here average distance takes care of RSS which is calculated according to the distance of the respective node taking into consideration the X-axis and Y-axis placing of the node. Thus the node with highest VID is decided as the CH. According to the VID, the other nodes are decided to be Foragers, Scouts. The lowest VID node is considered as BCNs. Any node when drain out more and the residual energy is below the limit (in our case 5 J) and goes out of the range of cluster making RSS very low then its VID is changed to zero and thus neglected to be a member of the cluster. If any CH drains out and makes its VID to zero, then any other member of the cluster with next highest VID is selected as the CH.

3.3 Cluster Maintenance by Load-Balanced Re-clustering

As in case of a MANET the nodes are dynamic so the number of nodes in any cluster keeps varying. Thus the MANET may not be properly balanced with respect to load. A too large cluster may put too heavy load on the cluster heads causing cluster heads to become bottleneck of a MANET and this reduces system throughput. A too small cluster can increase the no. of clusters thus increasing the length of hierarchical routes resulting in longer end-to-end delay. Thus upper limit and lower limit has been fixed on the number of mobile nodes that a cluster can deal with. In this process, if a cluster size reduces or exceeds its predetermined limit then re-clustering can take place.

Implementation: This can be done using Adaptive Multi-hop Clustering where the nodes (Foragers, Scouts, and BCNs) periodically broadcast certain information like node VID, Cluster ID and status (whether a forager, scout or BCN). The nodes are equally distributed in the cluster. Where

$$\mathbf{int\,max} = \frac{\mathbf{Total\ no.of\ Nodes}}{\mathbf{initial\ no.of\ Clusters}} \tag{3}$$

$$\mathbf{int\,min} = \frac{\mathbf{int\,max}}{2} \tag{4}$$

Here two cases can take place-Cluster merging and Cluster split up.

3.3.1 Cluster Merging

If the number of nodes in the cluster goes below the minimum limit then the cluster tries to merge with other cluster, which has less number of cluster members. But here care has been taken that after merging the maximum number of cluster members should not exceed the predetermined limit. This is implemented with few steps:

Step1: Let the number of nodes in any cluster is fixed from Nmin to Nmax

Step2: If the number of members in a cluster Ci is less than Nmin, then the cluster Ci tries to find a cluster Cj where total number of nodes will be <= Nmax

Step3: If the total number of nodes in Ci and Cj is greater than Nmax for all neighboring clusters then it tries to find a Cj to minimize the sum value.

Cluster Head Selection

While merging the clusters for selecting the CH two parameters are considered. One is the remaining energy of the node and the number of nodes the cluster had previously. Thus CH is decided as:

$$\mathbf{CH_1 = W_1 * ResE(CH_1) + W_2 * nodes(CH_1)} \quad (5)$$

$$\mathbf{CH_2 = W_1 * ResE(CH_2) + W_2 * nodes(CH_2)} \quad (6)$$

$$\mathbf{CH = max(CH_1, CH_2)} \quad (7)$$

where,

ResE (CH_1)—Residual energy of current cluster head
Nodes (CH_1)—number of nodes of current cluster head
ResE (CH_2)—Residual energy of cluster head corresponding to joining cluster
Nodes (CH_2)—number of nodes of cluster head corresponding to joining cluster
W_1, W_2 are the weight factors.

The cluster head that has high CH value is selected as a new cluster head.

3.3.2 Cluster Split up

Step 3: If the total number of nodes in Ci and Cj is greater than Nmax for all neighboring clusters then it tries to find a Cj to minimize the sum value.
Step 4: If the number of nodes in Ci alone is greater than Nmax then the cluster undergoes division mechanism.

Cluster Head Selection

Similarly while dividing the cluster in each cluster the CH is selected according to the VID of the node as

$$\mathbf{CH = W_1 * RSS(node\,i) + W_2 * ResE(node\,i)} \quad (8)$$

CH value is calculated for every node i in the cluster. The node i that gives high CH value is selected as the cluster head.

3.4 Data Transmission via Alternate Router

At any time, if the current shortest path is busy, the alternate shortest path is utilized for data transmission to avoid delay. During data transfer from sources to destination each router maintains the traffic flow count in addition to hop count and residual energy. If router involved in any flow, that node flow count is increased. Thus waggle dance value decides the path for routing.

$$\mathbf{Waggle\ dance\ formula} = (\mathbf{W}_1 * \mathbf{hops}) + (\mathbf{W}_1 * \mathbf{ener}) + (\mathbf{W}_1 * \mathbf{fc}) \quad (9)$$

where $W_1 = 0.3$, fc = Flow count

All weights are equal. For finding the best routing path the hop probability is calculated. For example—If source node 3 contains 3 paths (3 next hop). Then waggle dance value is calculated for 3 next hops and then the probability is calculated for the hop. Where

$$\begin{aligned}\mathbf{Total\ waggle\ dance} = & \ \mathbf{nexthop1\ Waggle\ dance} \\ & + \mathbf{nexthop2\ Waggle\ dance} \\ & + \mathbf{nexthop3\ Waggle\ dance}\end{aligned} \quad (10)$$

$$\mathbf{Next\ hop1\ probability} = \frac{\mathbf{nexthop1\ Waggle\ dance}}{\mathbf{Total\ Waggle\ dance}} \quad (11)$$

$$\mathbf{Next\ hop2\ probability} = \frac{\mathbf{nexthop2\ Waggle\ dance}}{\mathbf{Total\ Waggle\ dance}} \quad (12)$$

$$\mathbf{Next\ hop3\ probability} = \frac{\mathbf{nexthop3\ Waggle\ dance}}{\mathbf{Total\ Waggle\ dance}} \quad (13)$$

Thus, the node which has a high probability that node is selected as a next hop.

4 Simulation Model

In MANET, nodes N are located in the network area of X * Y. X is the network width and Y is the network height. Here the area of the network is decided depending on the X-axis and Y-axis dimension. We have set: X-axis = 1000 m and Y-axis = 200 m. Each cluster is slit into 200 m communication range in the network area. Thus in our case no. of clusters = 1000/200 = 5. Thus each cluster operates over an area of $200 \times 200\ \text{m}^2$. The total number of nodes in the network is 50. Thus the minimum and maximum limit for nodes in any cluster is 5 and 10 respectively. Initially, each node has a residual energy of 10 J. The simulation has

Table 1 Simulation model

Simulator	Network simulator 2.35
Number of nodes	50–100
Area	1000 m × 200 m
Communication range	200 m
Interface type	Phy/wireless phy
MAC TYPE	IEEE 802.11
Queue type	Drop tail/priority queue
Queue length	50 packets
Antenna type	Omni antenna
Propagation type	Two ray ground
Routing agent	LBEE-IBAC, IBAC
Transport agent	UDP
Application agent	CBR
Initial energy	10 J
Transmission power	20 mW
Reception power	10 mW
Simulation time	50 s

been carried for the new methodology in the certain suitable scenario as shown in Table 1.

4.1 Simulation Scenario

The simulation for the proposed method is carried using Network Simulator 2.35. Various states of the simulation are mentioned below: As in Fig. 2a, nodes are dynamic in nature and seen in the simulation area where node 0 is found as the destination node. The Fig. 2b shows the scenario when the clusters are formed with a different number of nodes. Thus 5 clusters are created with different no. of cluster

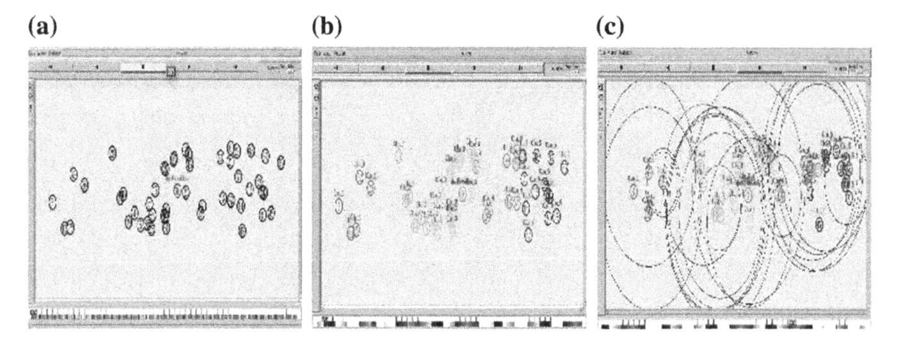

Fig. 2 **a** Dynamic nodes in the simulation area with identified destination **b** Formation of clusters **c** Routing from source to destination

members. Thus the routing takes place from the source to the destination through the best-suited path which is decided according to the number of waggle dance of the foragers as shown in Fig. 2c.

5 Simulation Results

The simulation results are considered both for Load-Balanced Energy-Enhanced Routing (LBEE) for Bee-Ad Hoc-C which is proposed protocol and Improvised Bee-Ad Hoc-Clustered (IBAC) Routing which is the existing protocol. The output trace files for both protocols are executed and the results are compared. Different parameters are considered for comparing and finding the best routing protocol.

Energy Efficiency: The energy efficiency of the MANET is calculated according to the total energy consumed by nodes and the remaining energy of the nodes as they are dynamic and battery operated. Usually, maximum energy is consumed for the CH. Energy efficiency is decreased in both protocols while increasing the traffic load as depicted in Fig. 3a. However, energy efficiency the proposed protocol is improved than existing due to the load-balanced clustering and selection of cluster head, routers based on residual energy for data transmission in the proposed protocol (LBEE).

End-to-End Delay: End-to-End delay is the time taken for a packet to reach the destination from the source node. As in Fig. 3b, the delay is increased in both protocols while increasing the traffic load. But the rate of increase in delay in the proposed method is less compared to the previous due to the load-balanced clustering and selection of routers based on traffic load and hop count for data transmission.

Throughput: It is the amount of data successfully received at the destination per second. For y-axis visibility, value is converted to 0.01 kbps. As in Fig. 3c, throughput is increased in both protocols while increasing the traffic load. Throughput in proposed protocol is better than existing due to the load-balanced clustering.

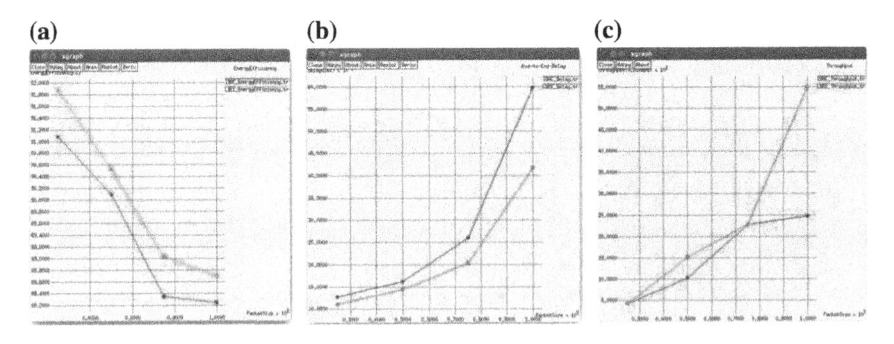

Fig. 3 **a** Energy efficiency, **b** End-to-end delay, **c** Throughput

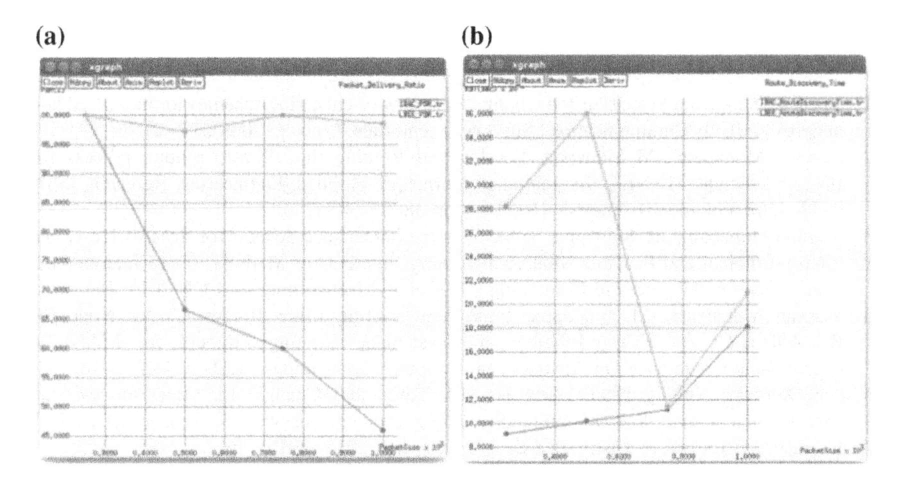

Fig. 4 **a** Packet Delivery Ratio, **b** Route discovery time in sec

Packet Delivery Ratio: This shows the ratio between the number of delivered packets to the destination and the generated packets in the CH.

As in Fig. 4a, PDR is decreased in both protocols while increasing the traffic load. PDR in proposed protocol is better than existing which does not change much with an increase in packet size.

Route discovery time: It is the time taken for a CH to find a route to the destination. It is found by finding the difference between sending time of forward scout and receiving time of backward scout at CH. As in Fig. 4b, route discovery time is increased in both protocols while increasing the traffic load. Route discovery time in the proposed protocol is increased than existing protocol due to the increased number of clusters while load balancing.

6 Conclusion and Future Scope

Thus a new routing protocol as LBEE-IBAC is established to make the MANET more balanced and energy efficient so that there should not be any wastage of energy by any node while searching the route to destination randomly. According to the simulation result found with respect to different parameters the performance of LBEE is found to be better to the existing protocol IBAC.

However, the protocol can be still analyzed by considering different parameters like routing overhead. The protocol can still be improved by introducing some other parameters for load balancing with respect to channel bandwidth to make the routing process more efficient.

References

1. Yu, Jane Yang, and Peter Han Joo Chong. "A survey of clustering schemes for mobile ad hoc networks." IEEE Communications Surveys & Tutorials 7, no. 1 (2005): 32–48.
2. Sasmita Mohapatra, M Siddappa, Bee-Inspired Routing the ultimate routing process for Energy Efficient MANET, International Journal of Applied Engineering Research ISSN 0973–4562 Volume 10, Number 18 (2015) pp 38855–38862.
3. Sasmita Mohapatra, M. Siddappa, Stable Cluster Maintenance Scheme for Bee-AdHoc-C: An Energy-Efficient and Scalable Multipath Routing Protocol for MANET, Third International Conference On Advances in Computing, Control and Networking - ACCN 2015, Bangkok.
4. Sasmita Mohapatra, M. Siddappa, Improvised routing using Border Cluster Node for Bee-AdHoc-C: An Energy-Efficient and systematic Routing Protocol for MANETs, International Conference On Advances in Computer Applications, IEEE ICACA-2016.
5. Cai, Xuelian, Yulong Duan, Ying He, Jin Yang, and Changle Li. "Bee-Sensor-C: an energy-efficient and scalable multipath routing protocol for wireless sensor net-works." International Journal of Distributed Sensor Networks 2015 (2015): 26.
6. Saleem, Muhammad, and MuddassarFarooq. "Beesensor: a bee-inspired power aware routing protocol for wireless sensor networks." In Workshops on Applications of Evolutionary Computation, pp. 81–90. Springer Berlin Heidelberg, 2007.
7. Maan, Fahim, and NaumanMazhar. "MANET routing protocols vs mobility models: A performance evaluation." In 2011 Third International Conference on Ubiquitous and Future Networks (ICUFN), pp. 179–184. IEEE, 2011.
8. Gopinath.S, Sureshkumar.N, Vijayalakshmi.G, Natraj.N.A, Senthil.T, Prabu.P, Energy Efficient Routing Protocol for MANET, IJCSI International Journal of Com-puter Science Issues, Vol. 9, Issue 2, No 1, March 2012.
9. M. Saleem, S. A. Khayam, and M. Farooq, A formal performance modeling framework for bio-inspired Ad Hoc routing protocols, in Proceeding of the 10th Annual Genetic and Evolutionary Computation Conference (GECCO'08), pp. 103–110, New York, NY, USA, July 2008.
10. Feeney, Laura Marie. "An energy consumption model for performance analysis of routing protocols for mobile ad hoc networks." Mobile Networks and Applications 6, no. 3 (2001): 239–249.

A Novel Self-transforming Image Encryption Algorithm Using Intrinsically Mutating PRNG

Soorya Annadurai, R. Manoj and Roshan David Jathanna

Abstract In this research paper, a new approach to image encryption has been proposed. The technique involves using a seed to generate a random list of prime numbers, which is then used to generate subsequent lists recursively. In this manner, a random number generation is created by recursive seeding, rather than one direct seed. This final set of lists is used to encrypt each pixel of the image. Moreover, each pixel is encrypted with its counterpart pixel in the original image, to reduce the possibility of deducing the original image without knowledge of the key. Statistical analysis tests are used to justify the strength of the algorithm.

Keywords Cryptography · PRNG · Mutation · AES · Entropy · PSNR
NPCR · Correlation algorithm

1 Introduction

When sensitive data is transmitted from one digital entity to another, mechanisms are required to secure the transmission. This data may be digitized as text, files, images, videos, or others. Image encryption has a diverse field of applications, such as private Internet communications, medical imaging systems, military or government communications, and so on. Cryptography is one of several techniques that may be employed to prevent such unauthorized access of sensitive information, and aims at achieving high standards of data confidentiality, data integrity, and authentication [1, 2]. Cryptographic transmissions can generally be classified into two broad

S. Annadurai (✉) · R. Manoj · R.D. Jathanna
Computer Science & Engineering, Manipal Institute of Technology,
Manipal University, Manipal, Karnataka, India
e-mail: soorya.annadurai@learner.manipal.edu

R. Manoj
e-mail: manoj.r@manipal.edu

R.D. Jathanna
e-mail: roshan.jathanna@manipal.edu

A.K. Somani et al. (eds.), *Proceedings of First International Conference on Smart System, Innovations and Computing*, Smart Innovation, Systems and Technologies 79, https://doi.org/10.1007/978-981-10-5828-8_20

categories: symmetric/private-key algorithms, and asymmetric, or public key algorithms Symmetric key algorithms are computationally less expensive (in general) to implement, when compared to asymmetric algorithms. These algorithms can be implemented in many ways, such as round-key substitution, chaotic maps, Feistel cipher mechanisms, and combinations of the above mentioned techniques and more. In the proposed algorithm, a symmetric schematic is implemented, tested with various indexes, and compared against other strong algorithms. A symmetric key algorithm has been chosen as this class of encryption algorithms is generally more suitable for bulk message transmissions [6], as opposed to the asymmetric class. Here, the images under consideration would qualify as bulk transmission of pixel data.

2 Literature Review

Unlike the encryption procedures of text or bytes in files, images have some unique properties that distinguish them from other kinds of media, such as the correlation between adjacent pixels, redundancy, noise, and others. This implies that the kinds of attacks and security measures to protect such images will differ from regular modes of encryption [3]. There are different approaches to encrypting an image. Stream encryption algorithms involve building a pseudo-random key sequence, and then influencing the original stream of pixels with the key sequence with the help of an exclusive-or operation [4, 5, 7, 8]. The A5/1 stream cipher algorithm [9] uses a 64-bit key for key production from a set of linear feedback shift registers. Similarly, the W7 stream cipher algorithm uses a 128-bit key. Block cipher cryptosystems can also be used, which involves encrypting text by applying an encryption key and a corresponding encryption algorithm to a block of data (continuous pixels) instead of transforming the pixels on at a time (as performed in stream-cipher algorithms). Encryption can also be broadly classified into three other categories of substitution, permutation, and combinations of substitution and permutation [10]. Substitution schemes change the actual pixel values by means of an externally applied algorithm and/or key, while permutation schemes merely shuffle the given set of pixel values as specified by the algorithm. In some cases, both of the methods are employed to improve security standards. A combination of Arnolds cat map and Chens chaotic system has been proposed in [11]. In [12], three permutation techniques have been implemented by shuffling the pixels at the bit level, pixel level, and block level at a pre-specified order. In [13], the AES algorithm has been significantly improved by adding the usage of a key stream generator. The implementation in [14] uses a bit-level permutation based on a chaotic crypto-system. This idea is especially effective, as permutation at the bit level has the dual advantage of changing both the order and value of the pixels in the image. In [15], another crypto-system based on a total shuffling scheme has been demonstrated. In [16, 17], multiple logistic maps and chaotic systems have been utilized to improve the encryption security. Another implementation [18] uses a key whose size is equal to the product of the dimensions of the original gray-scale image. The technique in [21] is based on Nearest Neighbouring

Coupled Map Lattices (NCML). In this manner, several methodologies have been used and/or combined to increase encryption standards. Here, a new algorithm based on substitution is proposed. Instead of direct permutation, a dependency of every original pixel is introduced into generation of all encrypted pixels. Similar work is done in recent works [22, 23]. But in [22], the image is xor-transposed before the introduction of a variable factor, i.e. the key. If this variable factor were to be brought in during the transposition itself, it would introduce a higher rate of NPCR as the key has a direct role to play in the xor-transposition stage. In [23], the key is being mutated sufficiently by generating the key with sufficient xor operations, but the original image itself can be processed in a more secure and efficient manner. Our algorithm patches these setbacks and presents a very high standard of encryption, as depicted by analysis results.

3 Methodology

Assume the dimensions of the image to be (m, n, o). Each pixel value has range [0, 255], and uses 1 byte of memory.

3.1 Encryption Procedure

PRNG Generation Let $\Omega(n)$ be a function to return the last 8 bits of n, $\Theta(n)$ be a function to return the largest prime factor, and $\oplus$ be the XOR operation.

- A random number, α is passed as a seed.
- μ is generated by manipulating the current system time.
- Let the array (to hold the PRNG) be rarr[1:256].
- For i = 1:256: $\mu = \mu + \alpha$, rarr[i] = $\Omega(\theta(\mu) * \mu)$
  The array with the new PRNG, rarr[1...256], is returned, with each element of the random array rarr[ ] using 1 byte of memory.

Self-XOR Transform

- Take the middle element of PRNG, x, and obtain $r = x\%(m * n * o)$.
- Arrange the pixels of the image in a cyclic, one-dimensional order. Start from the rth position in the linear array, and XOR it with the previous element in the logical circular array. Continue for all pixels in the image.

Transforming the Image with the PRNG Arrange all the pixel values of the image in a linear, one-dimensional order. Let this array be called encr[m*n*o].

- Initialize index i = 0 for the iterations, and c = 0 for the PRNG index.
  Calculate encr[i] = encr[i] $\oplus$ rarr[c] Modify rarr[c] = ((rarr[c]^2 rarr[c+1]) % 255) $\oplus$ encr[i], c = (c+1) % 255
  Perform this transformation for all pixel values by incrementing i.

- Initialize index i = m*n*o, and c = 0.
  Calculate encr[i] = encr[i] ⊕ rarr[c] Modify rarr[c] = ((rarr[c]2 * rarr[c+1])%255) ⊕ ((encr[i]+(i/(n*3))) * (encr[i] + (i%(n*3))/3) * (encr[i]+(i%(n*3))%3))
  Then increment c in a cyclic order, i.e. c = (c+1) % 255
  Perform this transformation for all pixel values by decrementing i.
- Store the transformed image as the final encrypted image.

3.2 Decryption Procedure

PRNG Generation The PRNG is generated in the same manner as in the encryption procedure. The same random number α is passed as a seed, along with the previously generated μ. The generated PRNG array, rarr[1...256], is returned, with each element using 1 byte of memory.

Transforming the encrypted image with the PRNG

- Arrange all the pixel values of the image in a linear, one-dimensional order.
- Initialize the index i = m*n*o, and c = 0. For every ith pixel value in the linear image, XOR it with the cth element in the generated PRNG, rarr[ ].
  Before continuing the iteration, modify rarr[c] = ((rarr[c]^2 rarr[c+1])%255) ⊕ ((encr[i]+(i/(n*3))) * (encr[i]+(i%(n*3))/3) * (encr[i]+(i%(n*3))%3))
  Then increment c in a cyclic order, i.e. c = (c+1) % 255
  Perform this transformation for all pixel values by decrementing i.
- Initialize i = 0 for the iterations, and c = 0 for the PRNG index. XOR the ith pixel value in the linear image with the cth element in rarr[ ]. Store the resultant value in another array encr[ ].
  Modify rarr[c] = ((rarr[c]^2 rarr[c+1])%255) ⊕ encr[i], c = (c+1) % 255
  Perform this transformation for all pixel values by incrementing i.

Self-XOR Transform

- Take the middle element of PRNG, x, and obtain $r = x\%(m * n * o)$.
- Arrange the pixels of the image in a cyclic, one-dimensional order. Start from the rth position in the linear array, and XOR it with the previous element in the logical circular array. Continue for all pixels in the image.
- Store the generated image as the final decrypted image.

4 Results and Analysis

In this section, the different tests that were performed to analyze the strength and security of the proposed image encryption algorithm are presented. Test images from the USC-SIPI Image Database [19], were used. An RGB image of size 512×512 (Lena, 4.2.02.tiff) was used to perform the tests on MATLAB r2015a.

Fig. 1 Original and encrypted forms of the Lena image

4.1 *Visual Test*

It must be ensured that there is no visual pattern, relationship, or similarity between the original and encrypted forms of the test image. Figure 1 verifies this.

4.2 *Histogram Analysis*

Statistical attacks can be used to derive conclusions from the encrypted forms of images. These attacks are especially effective when a known plaintext attack is executed. To prevent the leakage of information by such attacks, it must be ensured that the original and encrypted forms of the image do not exhibit any statistical similarity. A histogram analysis provides information about the distribution of the number of pixels in an image for each value of a pixels intensity. In this test, three histograms have been plotted for each color plane of the three-dimensional color image. If in the encrypted histogram, there is an equal probability of generating pixels of all intensities, there is a higher degree of encrypted symmetry, as there is no possibility of eliminating attack options. Moreover, if the histograms of the original and encrypted images show no statistical resemblance, an attack based on the histogram analysis of the encrypted image will not deterministically provide information about the original image. This idea has been exploited here [20], where Shannon suggests that several cryptographic ciphers can be broken by an appropriate statistical analysis. Here, the resistance to statistical attacks has been demonstrated by means of histogram and correlation tests. Figure 2a shows that the original image has a non-uniform histogram, implying that data can be visually extracted. However, in Fig. 2b, the histograms resemble a straight line. This shows that every pixel intensity in the encrypted image

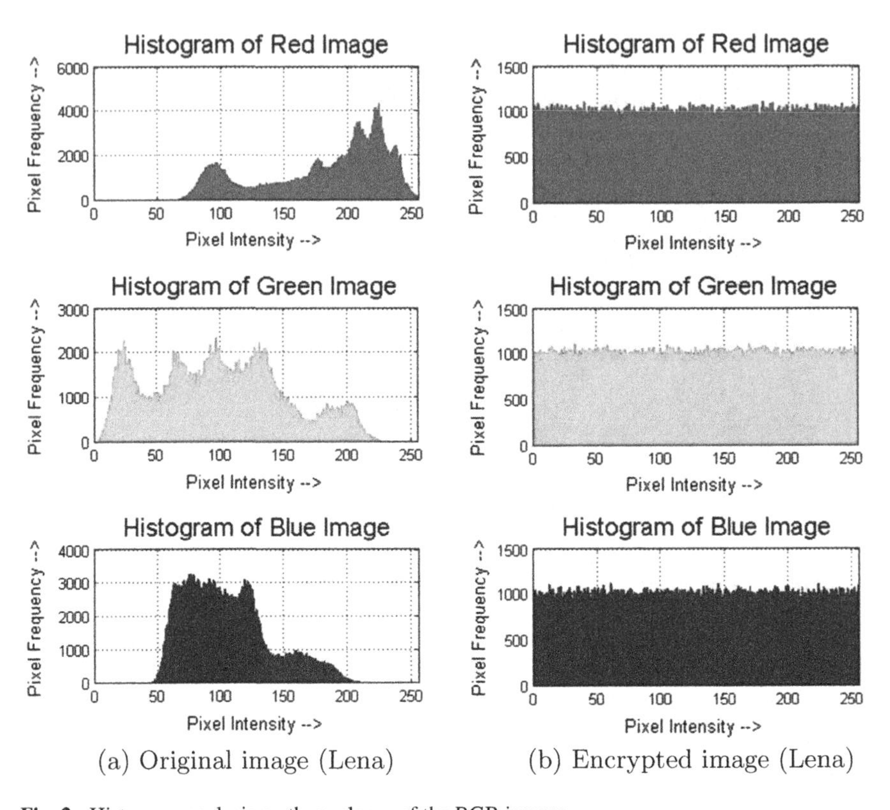

Fig. 2 Histogram analysis on three planes of the RGB images

has a nearly equal probability of generation. It can also be observed that there is no similarity between the histograms of the original and encrypted images, thus showing resistance to a known-plaintext attack.

4.3 Correlation Coefficient Analysis

In an ordinary image, adjacent pixels will generally be related to each other, in all directions; horizontally, vertically, and diagonally. However, the pixels of an encrypted image should show minimal correlation to its neighbors. The correlation test performed here considers all pairs of adjacent pixels in the horizontal, vertical, and diagonal axes.

$$E(x) = \frac{1}{n}\sum_{i=1}^{N} x_i, E(y) = \frac{1}{n}\sum_{i=1}^{N} y_i$$

Table 1 Correlation coefficients of two adjacent encrypted pixels

Lena (4.2.04.tiff)	Correlation coefficient of adjacent pixels		
	Horizontal	Vertical	Diagonal
Original image	0.9502767042	0.9654508326	0.9372732083
Method by Saeed and Majid [2]	−0.0074	0.0072	0.0105
DCT+AES [24]	0.0138	−0.0383	0.0171
Proposed algorithm	−0.0009676951	0.0009016908	0.0022480862

$$V(x) = \frac{1}{n}\sum_{i=1}^{N}(x_i - E(x)^2),\ V(y) = \frac{1}{n}\sum_{i=1}^{N}(y_i - E(y)^2)$$

$$cov(x,y) = \frac{1}{n}\sum_{i=1}^{N}(x_i - E(x))(y_i - E(y))$$

$$\gamma_{xy} = \frac{cov(x,y)}{\sqrt{V(x)}\sqrt{V(y)}},\ with V(x) \neq 0,\ V(y) \neq 0$$

Here, x_i and y_i are the values of two adjacent pixels in a particular color plane. N denotes the number of pixel pairs (x_i, y_i) in the test image. () and () are the expectation/mean values of all the x_i and y_i values. $V(x)$ and $V(y)$ are the individual variances of all of the x_i and y_i values. $cov(x, y)$ is the covariance of the distributions of all x_i and y_i values. γ_{xy} represents the correlation between the x_i and y_i distributions.

The correlation coefficient can range from $[-1, +1]$. When it is 0, the two variables are independent of each other. This implies that the covariance between x and y, $cov(x, y)$ is equal to zero. Here, the two variables considered are the intensities of any two adjacent pixels in a particular axis. In this test, all three axes were considered for the three color planes of the RGB image.

As in Table 1, all correlation coefficients for the image encrypted by the proposed algorithm tend to approach the ideal value of 0 faster than other algorithms, implying that there is minimal correlation between the pixels of the encrypted images. As in Fig. 3a, there is a very high degree of correlation in all axes, shown by the 45-degree pattern in the correlation plots. As in Fig. 3b, the correlation plots of the encrypted image in all axes and color planes show completely scattered patterns, indicating low correlation among adjacent encrypted pixels.

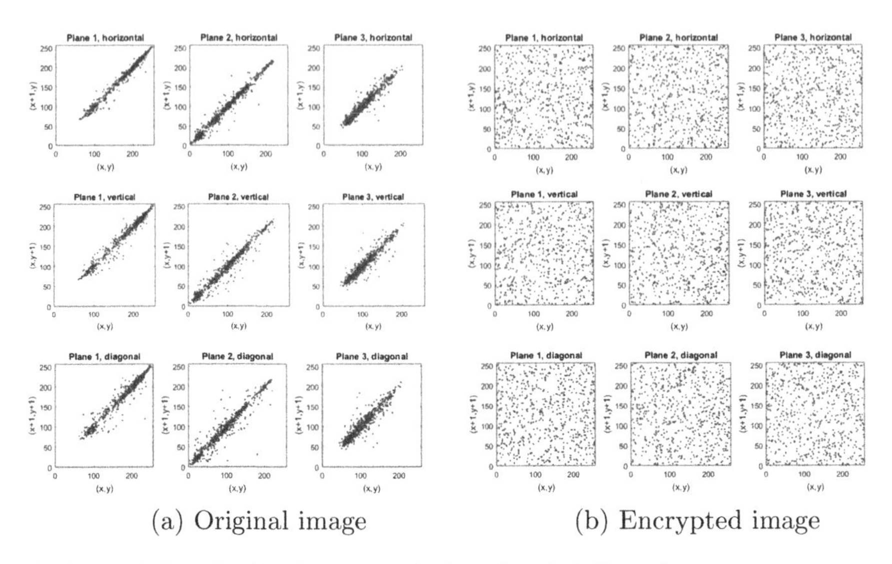

Fig. 3 Correlations plots (3 color planes) in horizontal, vertical, diagonal axes

4.4 Entropy Analysis

Entropy indicates the randomness of the occurrence of a particular pixel value. Shannon [20] introduced the concept of information entropy as a quantifiable measure of data in a source. It is defined as:

$$H(s) = -\sum_{i=1}^{3N} P(s_i) \times \log_2(P(s_i))$$

Here, $3N$ is the total number of symbols (color planes per pixel), and $P(s_i)$ is the probability of the occurrence of the symbol s_i. In this case, a color plane of a pixel is 8 bits in size. Thus, s_i can result in 2^8 different outcomes. If we assume these to be generated by the encrypting algorithm with equal probability, then the entropy of the set of symbols is 8. Entropy can range from 1 to 8. If the entropy value is less than 8, then the probability of predicting the original symbols increases, which compromises encryption strength.The near-ideal value of entropy by the proposed algorithm in Table 2 shows that the predictability of encrypted pixels by the proposed algorithm is negligible, and is thus capable of resisting against entropy-based attacks. This is significant, as the high entropy value indicates the pixel values in all three color channels cannot be determined in computationally feasible time.

Table 2 Entropy values of encrypted images

Lena (4.2.04.tiff)	Entropy value
Original image	7.7501974797
Method by Saeed and Majid [2]	7.9890
DCT+AES [24]	7.8683
Proposed algorithm	7.9997980997

4.5 *Differential Analysis*

The encrypted image should be sensitive to small changes introduced in the original image. If not, then many concurrences can be observed in the encrypted forms of two original images with small differences, and a series of known-plaintext attacks can relate the original and encrypted forms of an image. Moreover, this type of attack facilitates the determination of the keys used during the encryption process (μ and α). Two measures were considered: NPCR (Number of Pixels Change Rate), and UACI (Unified Average Changing Intensity). NPCR gives an indication of how many encrypted pixel values exhibit a change when exactly one pixel of original image is changed. UACI measures the average intensity of the differences between the original image and the encrypted image. If two plain images P_K and $\overline{P_K}$ differ in exactly one color plane of exactly one pixel, let their encrypted forms be C_K and $\overline{C_K}$. For an RGB image of height M, width N, and number of planes O, UACI and NPCR are:

$$UACI = \frac{1}{M \times N \times O} \sum_{X=1}^{M} \sum_{Y=1}^{N} \sum_{Z=1}^{O} \left[\frac{C_K(X,Y,Z) - \overline{C_K}(X,Y,Z)}{255} \right]$$

$$NPCR = \frac{1}{M \times N \times O} \sum_{X=1}^{M} \sum_{Y=1}^{N} \sum_{Z=1}^{O} D_K(X,Y,Z)$$

$$D_K(X,Y,Z) = \begin{cases} 0 & \text{if } C_K(X,Y,Z) = \overline{C_K}(X,Y,Z) \\ 1 & \text{if } C_K(X,Y,Z) \neq \overline{C_K}(X,Y,Z) \end{cases}$$

For this test, six separate encryptions were generated, where one encryption was made on the image in its intact form, and five others were made upon changing exactly one plane of exactly one pixel in random locations. High NPCR (ideally 100%) and UACI (ideally 33%) values indicate that the encryption algorithm will reflect large differences on minor changes in the original image.

Table 3 shows that the proposed algorithm is highly sensitive to changes in the original image.

Table 3 NPCR and UACI values of encrypted images

Lena (4.2.04.tiff)	NPCR value of images	UACI value of images
Method by Khaled and Berdai [2]	0.995850	0.286210
DCT+AES [24]	0.995941	0.337715
Proposed algorithm	0.9948659261	0.3344753151

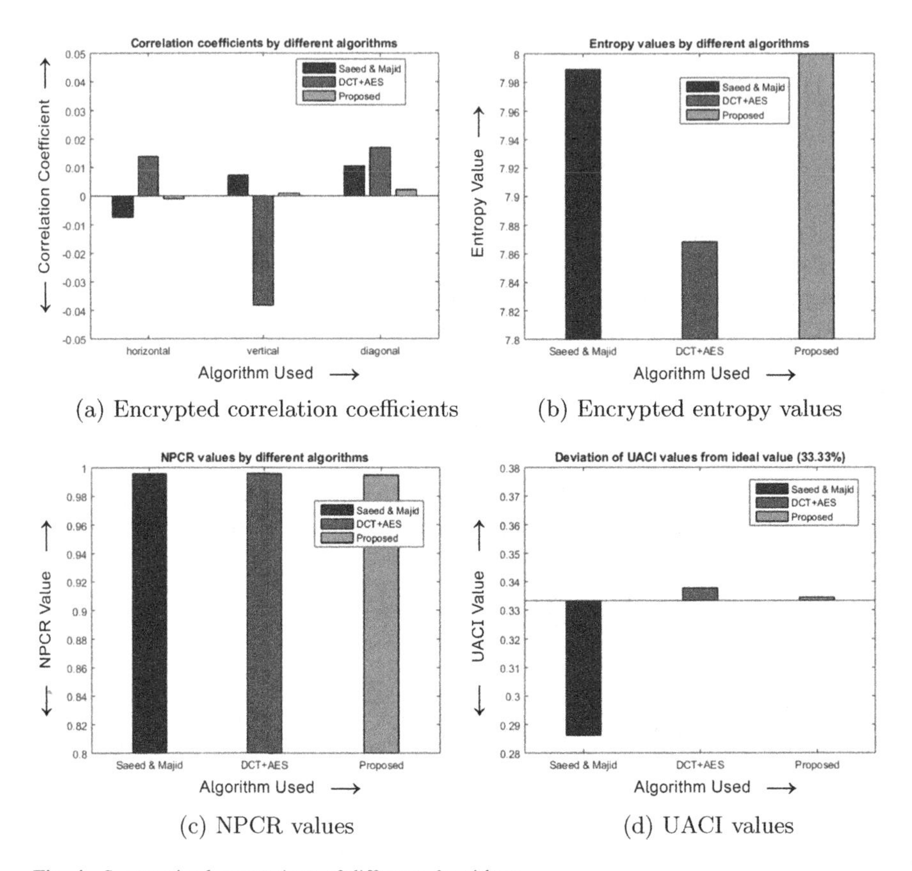

(a) Encrypted correlation coefficients (b) Encrypted entropy values

(c) NPCR values (d) UACI values

Fig. 4 Summarized comparison of different algorithms

4.6 Peak Signal-to-Noise Ratio Analysis

PSNR indicates the change between the original pixels and the distorted pixels, calculated in decibels. It can be used for either encryption or decryption tests. The ideal PSNR value between the original and encrypted image is 0 (implying complete introduction of noise into the image), and the ideal value between the original and the decrypted image is infinity (implying complete recovery of data). Here, the PSNR value was infinity for decrypted images, implying no loss in data, and the value for the encrypted image was significantly low (8.643746373387632).

5 Conclusion

Figure 4a shows that the proposed algorithm gives lower correlation among adjacent encrypted pixels. It shows little similarity between the original and encrypted images, and the histogram analysis verifies this assessment. Figure 4b shows a nearly ideal value of 8.00 for entropy by the proposed algorithm. Figure 4c shows nearly equal values of NPCR for all considered algorithms. Figure 4d shows negligible deviation from the ideal 33.33% for UACI. The proposed algorithm is stronger than several other candidate algorithms, and in conclusion, it can be expected to find many real-world applications.

References

1. Nigam Sangwan, Text Encryption with Huffman Compression, International Journal of Computer Applications (0975 8887) Volume 54 No.6, September 2012.
2. Ayushi, A Symmetric Key Cryptographic Algorithm International Journal of Computer Applications (09758887) Volume 1 No. 15.
3. S. Lian and X. Chen, On the design of partial encryption scheme for multimedia content, Mathematical and Computer Modelling.
4. F. Bao and R. H. Deng, Light-weight encryption schemes for multimedia data and high-speed networks, GLOBECOM 07, pp. 188192, November 2007.
5. A. Jolfaei and A. Mirghadri, Image Encryption Using A5/1 and W7, vol. 2, no. 8.
6. Cryptography & Network Security, 6th Edition - William Stallings.
7. N. Thomas, D. Redmill, and D. Bull, Secure transcoders for single layer video data, Signal Processing, vol. 25, no. 3, pp. 196207, 2010.
8. F. Liu and H. Koenig, A survey of video encryption algorithms, Computers and Security, vol. 29, no. 1, pp. 3–15, 2010.
9. S. Bahrami, M. Naderi, Image Encryption Using a Lightweight Stream Encryption Algorithm, Advances in Multimedia, Article ID 767364, 2012.
10. Secure Image Encryption Algorithms: A Review, International Journal of Scientific & Technology Research Volume 2, Issue 4, April 2013, ISSN 2277-8616.
11. Zhi-Hong Guan, Fangjun Huang, Wenjie Guan, Chaosbased image encryption algorithm, Physics Letters A 346, Elsevier, 2005.
12. Mitra, Y. V. Subba Rao and S. R. M. Prasanna, A New Image Encryption Approach using Combinational Permutation Techniques, IJECE, 2006.
13. M. Zeghid, M. Machhout, L. khriji, A modified AES based algorithm for image Encryption, World Academy of Science, Engineering and Technology 3, 2007.
14. Z. Zhu, W. Zhang, K. Wong, A chaos-based symmetric image encryption scheme using bit-level permutation, Information Sciences, 1171–1186 Elsevier, 2010.
15. G. Zhang and Q. Liu, A novel image encryption method based on total shuffling scheme, Optics Communications, vol. 284, no. 12, pp. 2775 2780, 2011.

16. Ismail, Amin, A Digital Image Encryption Algorithm On Composition Of Two Chaotic Logistic Maps, Proc. 27th IEEE Intl Conf. Signal Processing, 733–739, 2011.
17. H. Alsafasfeh, A. A. Arfoa, Image encryption based on the general approach for multiple chaotic system, Journal of Signal & Information Processing 2, 238–244, 2011.
18. Khaled Loukhaoukha, J. Chouinard, A. Berdai, A Secure Image Encryption Algorithm Based on Rubiks Cube Principle, Laval University, Canada G1K 7P4, 2011.
19. USC-SIPI Image Database, Signal & Image Processing Institute.
20. C. E. Shannon, Communication theory of secrecy systems, Bell System Technical Journal, vol. 28, no. 4, pp. 656715, 1949.
21. Yong Wanga, Kwok-Wo Wong, Xiaofeng Liaoc, Guanrong Chen, A new chaos-based fast image encryption algorithm, Applied Soft Computing, Elsevier, 2011.
22. P. Sree, M. Padmaja, An Improved Image Encryption Using Pixel XOR Transpose Technique Over Images, IJETTCS, Vol. 4, Issue 3, May–June 2015 ISSN 22786856
23. A. Dixit, P. Dhruve, Image encryption using permutation and rotational XOR technique (Eds): SIPM, FCST, ITCA, WSE, ACSIT pp. 01–09, 2012.
24. Belazi Akram, B. Oussama, H. Safya, Selective Image Encryption Using DCT with AES Cipher, (CS & IT), No. https://doi.org/10.5121/csit.2014.41306, pg. 69–77, 2014.

A Model of Computation of QoS of WSCDL Roles from Traces of Competing Web Services

Ravi Shankar Pandey

Abstract Roles in WSCDL choreography define the behavior of participants in collaboration to achieve business objectives. These roles may be realized by one or more web services. The quality of service (QoS) of roles and choreography are dependent on these QoS of competing web services. Service providers may or may not publish the values of attributes of the QoS of their services to enable their selection. Consequently, it is desirable to have an alternative strategy to compute values of attributes of QoS of web services. Further, WSDL descriptions of web services do not contain an order of invocation of operations. This invocation protocol may be needed by the client either for adaption or selection of web services. In this paper, I propose a method to discover invocation protocol of a web service and to capture values of attributes of QoS from its execution traces of web services. I have represented execution traces as Labeled Transition Systems (LTS) and merge these LTSs to obtain service invocation protocol. My LTS model defines a set of functions for attributes of QoS. These functions take a transition as input and return the value for a specific attribute of QoS. These functions are used to compute the range of values of QoS attributes. I have demonstrated my methodology through an example.

Keywords Web service · QoS · WSCDL · Roles

1 Introduction

Roles in WSCDL choreography define the behavior of participants in collaboration to achieve business objectives. These roles may be realized by one or more web services. The quality of service (QoS) of roles and choreography are dependent on these QoS of competing web services. Service providers may or may not publish

R.S. Pandey (✉)
Department of Computer Science, Birla Institute of Technology, Ranchi, India
e-mail: ravishankarbit@yahoo.com

R.S. Pandey
Extension Centre, Allahabad, UP, India

A.K. Somani et al. (eds.), *Proceedings of First International Conference on Smart System, Innovations and Computing*, Smart Innovation, Systems and Technologies 79, https://doi.org/10.1007/978-981-10-5828-8_21

the values of attributes of the QoS of their services to enable their selection. Consequently, it is desirable to have an alternative strategy to compute values of attributes of QoS of web services. Further, WSDL descriptions of web services do not contain an order of invocation of operations. This invocation protocol may be needed by the client either for adaption or selection of web services.

In this paper, I propose a method to discover invocation protocol of a web service and to capture values of attributes of QoS from its execution traces of web services. I have represented execution traces as Labeled Transition Systems (LTS) and merge these LTSs to obtain service invocation protocol. My LTS model defines a set of functions for attributes of QoS. These functions take a transition as input and return the value for a specific attribute of QoS. These functions are used to compute the range of values of QoS attributes. I have demonstrated my methodology through an example.

2 Related Work

Zhao et al. [1] considered the activity of choreography as a basic concept and associated cost and time as QoS parameters. They proposed a trace-based semantic model of choreography which included cost and time. Further, they also provided a model for sharing of cost of choreography amongst the participating roles. Yunni et al. [2] identified three QoS metrics for a choreography. These metrics included expected-process-normal-completion-time, process normal-completion-probability, and expected-overhead-of-normal-completion. They estimated these metrics using Petri net and continuous-time Markov chain. They used simulation to validate them using theoretical estimation. In [3], web services have been modeled as a finite state machine and relationship between them as invokability in the context of choreography. Using these models, they described an algorithm for dynamic selection of web services based on their reliability for reliable execution of choreography. D'Ambrogio in [4] proposed an extension to the metamodel of WSDL to include QoS attributes. These attributes included reliability, availability, and access policy at service level, bitrate, delay, jitter, packetloss as network-related QoS at port level, message encryption at message level, and operation demand and latency at operation level. Based on this extended model of WSDL and metamodel of web service choreography [5], I proposed additional QoS attributes of choreography [6]. These QoS attributes included dominant role, dominant relationship, and dominant interaction. Further, [7] I argued that it is desirable to specify the values of these attributes as a range instead of single value [8] has also proposed a model for computing QoS of a participating role of a choreography. However, my methodology to estimate the QoS of choreography was based on assumption that QoS attributes of participating web services are known a prior as an advertisement. As this assumption is not valid in all scenarios, it motivated us to look for estimation model based on execution traces of web service.

3 My Proposal

A web service is an LTS, which captures the invocation order of the operations and quality of service attributes. This LTS is generated after merging of the LTSs of web service execution traces. The behavior of client using web services is stored in the form of traces. My proposal consists of three steps as given in Fig. 1. These steps include collection of traces of different web services, merging of these traces and computation of attributes of QoS. To merge these execution traces, each trace is represented as Label Transition System (LTS). I model an LTS as five tuple < S; T; s0; sf; F >. In this model, S is the set of states. s0 and sf are the initial and final states. T is the set of transitions and F is the set of functions. A transition t ϵ T is three tuple (sh, opName, and st) where sh, st ϵ S and opName is the name of the invoked operation of web service. Sh and st represent head (source) and tail (destination) states of transition t. As stated earlier, I uses different notations for unsuccessful and successful invocations of operation of a web service. The unsuccessful is indicated by dash (-) in the name of operation. Consequently, a transition in LTS may be representing either successful invocation or unsuccessful invocation. The set of function F may contain one or more functions. Each function of this set takes an LTS and returns a value of specific attributes of QoS to demonstrate my proposal. I define the following four functions for this set F.

fLatmin: It returns the minimum value of latency of operation of a transition.
fLatmax: It returns the maximum value of latency of operation of a transition.
fScount: It returns the number of successful invocation of operation of a transition.
fFailure: It returns the number of unsuccessful invocation of operation of a transition.

This step of merging of traces is crucial for both protocol discovery and computation of QoS attribute range of values. It consists of two steps.

Initialization:

In this phase, I generate one LTS for each execution trace. It involves the creation of one state and a transition for every successful and unsuccessful

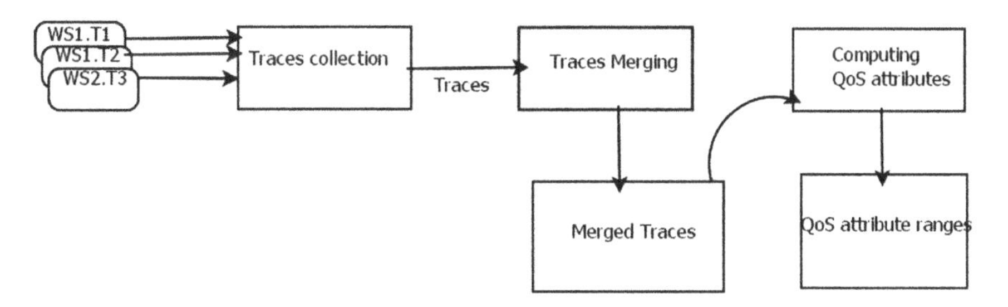

Fig. 1 Overview of proposal

invocations of operation and defining the value of the function, which is to be 3 returned for this transition.

Merge:

I have used k-tail algorithm to merge LTSs generated for different execution traces. This algorithm is based on concepts of k-paths and k-equivalence states. The definitions of these two concepts are reproduced below from [9] for ready reference. Definition (k-paths): Given a state s in LTS a set of paths with length k, called k-path(s), is defined as a set {path1...pathr}, where pathi is a sequence of method invocations, where pathi = st1; st1...stk such that there exists a sequence of transitions (s0; t1; s1) (s1; t2; s2)...(sk-1; tk; sk) in the LTS.

Definition (k-equivalence): Two states s1 and s2 in the LTS are k-equivalent if and only if k-paths(s1) = k-paths(s2). Two states are merged if they are k-equivalent. The merging of two LTSs is defined as a funtion M: LTS* LTS- > LTS.

Let lts1, lts2, and lts3 LTS are defined as

lts1 = (S1, s01, sf1, T1,F1), lts2 = (S2, s02, sf2, T2, F2) lts3 = (S3, s03, sf3, T3, F3) where S1, S2 $\subseteq$ S3,
s03, sf3 $\subseteq$ S3 T3 = T1 $\cup$ T2 and for every tk ϵ T3, the minimum and maximum latencies and success and failure count functions are defined as

fLatmin(tk):: = min {fLatmin(ti);fLatmin(tj)}
if for t ϵ T1 and tj ϵ T2 head state of ti and tj
are k-equivalent
else
fLatmin(ti) or fLatmin(tj) for tk = ti or tk = tj
fLatmax(tk):: = max {fLatmax(ti); fLatmax(tj)}
if for ti_T1 and tj_T2
head states of ti and tj are k-equivalent
else
fLatmax(ti) or fLatmax(tj) for tk = ti or tk = tj
fScount(tk):: = fScount(ti) + fScount(tj)
else
fScount(ti) or fScount(tj) for tk = ti or tk = tj
fFailure(tk):: = fFailure(ti) + fFailure(tj)
else

fFailure(ti) or fFailure(tj) for tk = ti or tk = tj

4 Example

In this example, I have taken same example as given in [4]. This is a virtual e-store which is an e-commerce application. It is an online book shopping web portal. It has front and back ends. The web portal is font end while in back end some e-stores are present. The function of e-stores is to provide facility to customer to find best offers using all these catalogs through e-stores. In back end e-store numbers may increase over a period of time.

I use example given in [6] to describe my proposal. The example application is called VirtualStore. It enables users to find best offers through a search of catalog e-stores and to buy one or more products from them. VirtualStore uses service discovery infrastructure to connect to set of e-stores. All of these e-stores provide the following operations: ItemSearch (I), CartCerate (C), CartAdd (A), and CartPurchase (P). Symbols given after names of the operations within parenthesis are going to be used as abbreviations for these operations. Even though the operations are same, their order of invocations may be different due to different protocols of e-stores. Since execution traces are likely to contain both successful and unsuccessful invocations of operations, I use distinct notations from them. A bar on the short name of the operation indicates unsuccessful invocation, whereas the absence of this horizontal bar represents successful invocation. Further, each successful invocation is suffixed with the duration of time taken to complete a request. In other words, the successful invocation has name of the operation and its latency. In the case of unsuccessful invocation, this latency is undefined and represented as a dash (-).

Example traces for two stores are given below. These traces are numbered as Tr1, Tr2, Tr3, and Tr4 and are prefixed with the name of e-stores. Trace Amazon. Tr1 contains successful invocation of ItemSearch, CartCerate, CartAdd again CartAdd, and CartPurchase. Their latencies are 8,12,15,18, and 30, respectively. This trace does not contain any unsuccessful invocation. The second trace contains both successful and unsuccessful invocations. Amazon.Tr2 contains unsuccessful invocations of ItemSearch, CartCerate, CartAdd, and CartPurchase. Similarly, PetStore traces Tr3 and Tr4 are shown below:

$$\text{Amazon.Tr1: } < I: 08; C: 12; A: 15; A: 18; P: 30 >$$

$$\text{Amazon.Tr2: } < I: -; \overline{I}: -; I: 20; \overline{C}: -; C: 20; A: 22; A: 30; P: -; \overline{P}: 10 >$$

$$\text{PetStore.Tr3: } < C: 14; I: 12; A: 15; I: 20; A: 18; I: 30; A: 20; P: 20 >$$

$$\text{PetStore.Tr4: } < \overline{C}: ; \overline{C}: -; C: 20; \overline{I}: -; I: 23; A: 15; \overline{I}: -; \overline{I}: -; I: 20; A: 18; \overline{I}: -; \overline{I}: -; I: 30; A: 20; \overline{P}: -; P: 20 >$$

The LTSs of the above traces are given in Fig. 2: The first LTS does not contain unsuccessful invocation of any operation. LTS in B contains unsuccessful invocation of three different operations. Further, it also indicates that after failed invocation of I(ItemSearch) or C(CartCreate) or A(CartAdd) or P(CartPurchase). These operations can be invoked again, whereas the LTS in C indicates that unsuccessful invocation of p negotiates the repeat of I, C, A, A, and P. Some example values of the attributes returned by the functions for some of the transitions belonging to LTS given below:

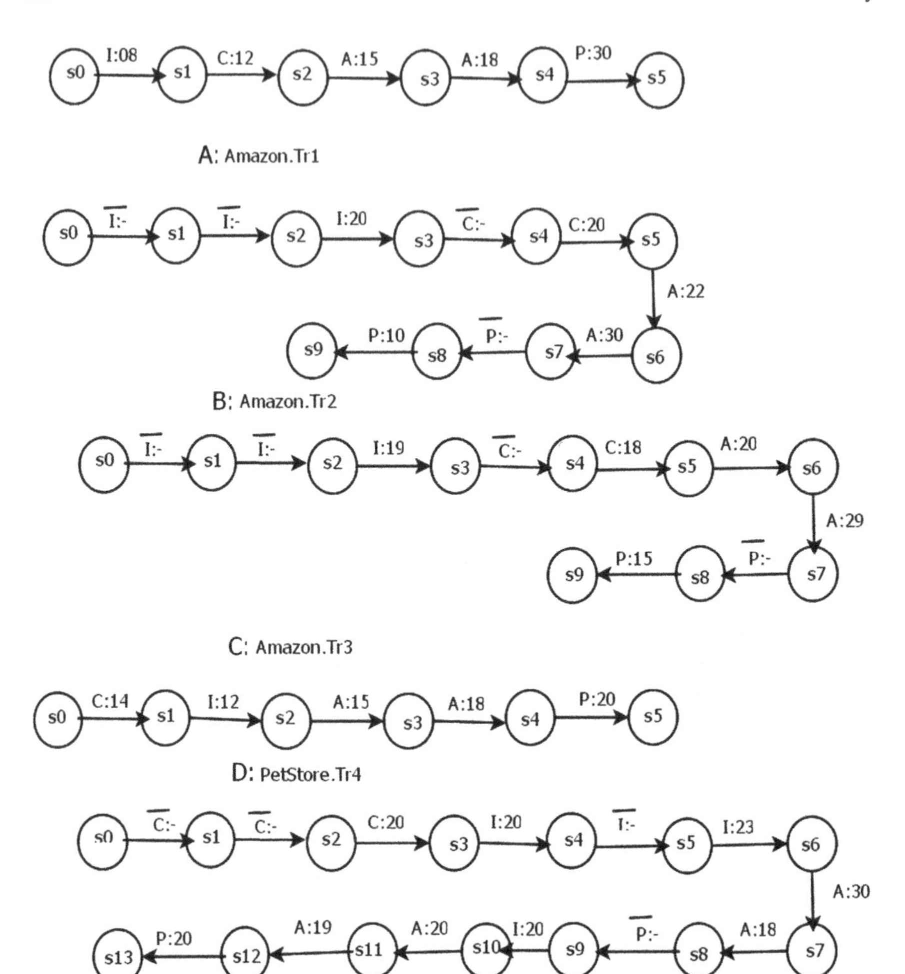

Fig. 2 LTSs of web services

Amazon.Tr1:
fLatmin(s0,s1,I) = 8, fLatmax(s0,s1,I) = 8
fScount(s0,s1,I) = 1, failure (s0,s1,I) = 0
fLatmin(s4,s5,P) = 8, fLatmax(s4,s5,P) = 8
fScount(s4,s5,P) = 1, failure (s4,s5,P) = 0

The example for functions is given below of Amazon.Tr2 trace which contains failed invocation:

fLatmin(s0,s1,I) = 20, fLatmax(s0,s1,I) = 20
fScount(s0,s1,I) = 1, failure (s0,s1,I) = 0

fLatmin(s0,s1; $\overline{I}$)=-, fLatmax(s0,s1; $\overline{I}$)=-
fScount(s0,s1; $\overline{I}$)=0, failure (s0,s1; $\overline{I}$)=1

I iteratively computed k-paths for every state of two LTSs to be merged for different values of k and checked for their k-equivalence between states for merging them. The results of merging LTSs are given in Fig. 2 A, B, and C and Fig. 3. The values returned by the functions for some of the transitions of this merged LTS are given below. It may be noted that this merge operation consists of higher order function, which takes functions associated with individual transition and generates a new set of functions for the merge transition. In other words, the functions defined for transition in Fig. 3 have been generated by higher order function taking values of these functions on respective traces before merge. For my convenient, I did not change the name of the functions, but only change the values returned by them.

k-paths are iteratively computed for every state of two LTSs merged for different values of k and checked for their k-equivalence between states for merging them. The result of merging LTSes given in Fig. 2 A, B and C is shown in Figs. 3 and 4. The values returned by the functions for some of the transitions of this merged LTS are given below.

fLatmin, fLatmax, and other attributes for the first transition of Figs. 3 and 4 are computed from values of these functions for the first transition of Fig. 2A, and third transitions of Fig. 2B, C.

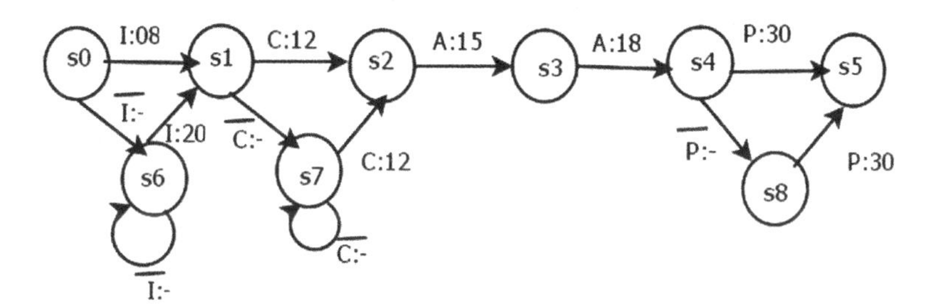

Fig. 3 fLatmin(s0,I,s1) = 8, fLatmax(s0,I,s1) = 20 fScount(s0,I,s1) = 3, fFailure(s0,I,s1) = 2

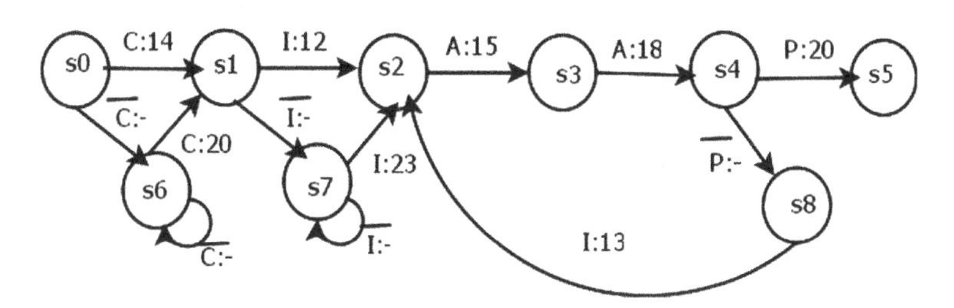

Fig. 4 fLatmin(s2,A,s3) = 15, fLatmax(s2,A,s3) = 30 fScount(s2,A,s3) = 2, fFailure(s2,A,s3) = 0

Algorithm to compute QoS attributes of a web service

The algorithm to compute the QoS attributes of a web service from the merged LTS obtained from its execution traces is described below:

Algorithm: compute QoS attributes of a web service

Input: merged LTS
Output: CS, CF, lmin, lmax

Variables CS and CF store the total number of successful and unsuccessful invocations of all operations. Similarly, lmin and lmax store minimum and maximum latencies for all operations

1. Repeat for each transition ti of an LTS
2. CS = CS + fScount(ti)
3. CF = CF + fFailure(ti)
4. if(lmin > fLatmin(ti)) then
   lmin = fLatmin(ti)
5. if(lmax < fLatmax(ti))
   lmax = fLatmax(ti)

The minimum and maximum latencies of web service Amazon.com are 08 and 30, respectively. A total number of successful and unsuccessful invocations of operations are 18 and 08, respectively. The PetStore.com web service minimum and maximum latencies are 12 and 23, respectively. The total numbers of successful and unsuccessful invocations of operations are 16 and 05, respectively.

The invocation protocols for two e-stores are given below

$$\text{Amazon: } (((\,(I+(I+.\,I).\,(C+(C+.\,C)).\,(A+:(P+P.P\,)))$$

$$\text{PetStore: } ((((C\ +(C+.\,C).\,(I+(I+.\,I)).\,(A+.\,(P+P.\,(\,I.A+.P)+)))$$

The protocols are regular expressions obtained from merged LTSs given in Figs. 4 and 5 using method suggested in [10].

5 Conclusion

In this paper, I have proposed a methodology to discover invocation protocol and to compute the values of attributes of QoS of competing web services for a role of WSCDL choreography. These protocols and values of QoS attributes are computed based on measurements and recording of execution traces of web services. These execution traces are available only with clients. Accordingly, the collection of traces by service discovery infrastructure is not feasible with current standards.

References

1. Xiangpeng, Zhao and Chao, Cai and Hongli, Yang and Zongyan, Qiu A QoS View of Web Service Choreography, in Proc. IEEE Int. Conf. on e-Business Engineering (ICEBE '07), Hong-Kong, 2007, pp. 607–611.
2. Yunni Xia, Jun Chen, Mingqiang, and Yu Huang A Petri-Net- Based Approach to QoS Estimation of Web Service Choreographies, in Proc. Springer-Verlag (LNCS) APWeb and WAIM 2009, Berlin Heidelberg, 2009, pp. 113–124.
3. San-Yih Hwang, Wen-Po Liao, Chien-Hsiang Lee Web Services Selection in Support of Reliable Web Service Choreography, in Proc. IEEE International Conference on Web Services, 2010, pp. 115–122.
4. A. DAmbrogio, A model-driven wsdl extension for describing the qos of web services, in Proc. IEEE Int. Conf. on Web Services(ICWS 06), Washington, DC, USA, 2006, pp. 789–796.
5. http://www.w3.org/TR/ws-chor-model/.
6. Ravi Shankar Pandey, B.D.Chaudhary A Meta-Model Based Proposal for QOS of WSCDL Choreography, in Proc. Int. Multi-Conf. of Engineers and Computer Scientists(IMECS-10), Hong-Kong, 2010, Vol I, pp. 973–982.
7. Ravi Shankar Pandey, B.D. Chaudhary An Estimation of Min- Max of QoS Attributes of a Choreography, Accepted in IEEE-International conference on advances in engineering, science and management(ICAESM-2012).
8. Tew_k Ziadi, Marcos Aurlio Almeida da Silva, Lom Messan Hillah, Mikal Ziane A Fully Dynamic Approach to the Reverse Engineering of UML Sequence Diagrams, in Proc. TAVWEB 06, July 17, Portland, Maine, USA. 2006, pp. 10–16.
9. Ravi Shankar Pandey, B. D. Chaudhary A cost model for participating roles based on choreography semantics, in Proc. IEEE Asia-Paci_c Services Computing Conference, Tiwan, 2008, pp. 277–283.
10. G. Denaro, M. Pezz e, D. Tos, Daniela Schilling Towards Self-Adaptive Service Oriented Architectures, in Proc. 16th IEEE International Conference on Engineering of Complex Computer Systems 2011, pp. 107–116.

A Model to Measure Readability of Captions with Temporal Dimension

Muralidhar Pantula and K.S. Kuppusamy

Abstract Video content has become one of the most noticeable online trends of recent times. It is no secret that videos have become a staple of everyone's life. WCAG 2.0 suggested that closed captioning is needed to provide accessible audio content for the audience who are Deaf and Hard of Hearing due to their disability in pursuing the auditory information. Caption files are plain text files with time codes information for each frame. Readability of caption is one major challenge, due to variation in speaking rate and reading rate. The time duration of caption frame affects the flow of text perceived by the reader. This paper proposes a novel model for caption readability based on time dimension. The proposed algorithm calculates the readability of each frame corresponds to the time duration. As a result, we analyzed the caption frames with respect to time, word count, and syllables and identified that the time gap enhances the caption readability.

Keywords Caption frame · Readability · Timecode · Deaf and Hard of Hearing

1 Introduction

More than 60% of all online traffic is videos and is expected to rise by 80% in next 5 years [1]. Audience interacts interestingly and gets entertained with video contents. Forrester's researchers specified if one picture paints 1000 words, then one-minute

M. Pantula · K.S. Kuppusamy (✉)
Department of Computer Science, Pondicherry University, Pondicherry 605014, India
e-mail: kskuppu@gmail.com

M. Pantula
e-mail: vijayamuralisarma@gmail.com

A.K. Somani et al. (eds.), *Proceedings of First International Conference on Smart System, Innovations and Computing*, Smart Innovation, Systems and Technologies 79, https://doi.org/10.1007/978-981-10-5828-8_22

video is worth 1.8 millions. Audio-visual media becomes closer to real life because of availability of both auditory and visibility clues and content. Research shows that use of online videos in marketing has doubled in last year [1].

W3C's (World Wide Web Consortium) internal multimedia accessibility policy is to ensure the audio and video content accessible for all, including Deaf and Hard of Hearing people, who cannot hear auditory information due to their physical disability. Multimedia accessibility is easy to establish/enable by providing a simple text for the auditory information also called as captions [2].

Caption files are simple plain-text files, having time code and text transcription of the frame. The time code specifies the starting and ending time of each frame. The text transcription is to be displayed, for particular time duration. The reader has to perceive the text within the frame duration. Captions are displayed based on the auditory information and most times the frame rate of the caption affects the readability. The readability ease measure is needed for captions.

Readability ease explains the measure of understanding the content in other writings. The caption is a simple text for auditory information in the multimedia content. Till now, Readability is calculated based on the text content and presentation of the text. We have not found any study on readability based on time factor.

2 WCAG

The WCAG (Web Content Accessibility Guidelines) were proposed by W3C in 1999 for the improvement of web accessibility. In 2008, WCAG again updated with four principles, perceivable, operable, understandable and robust, for making universally accessible web content [3]. WCAG is guiding the developers and academicians to make the web content more accessible to people including a wide range of disabilities. These guidelines are also usable by elderly persons and the people falling under situational disability. Perceivable principle explains the need of alternative for time-based media is mentioned in Guideline 1.2 [4]. Time-based media content is categorized as pre-recorded and live. This paper focuses on measuring the readability ease of caption based on time factor.

3 Caption

Caption is an explanation for an interactive information such as the dialog, sound effects, and relevant musical clues [5]. Captioning opens the key of information in media, accessible to deaf and hard of hearing people. Captions also aid for an audience in a noisy environment where audio is not discernible and for nonnative viewers [6]. Captioning can improve the effectiveness of audiovisual presentation and enhances viewer language skills [7]. A large number of students of oral language input use captions to enhance their listening comprehension in face-to-face interaction. Captions aid the phonological visualization of aural cues in minds of listeners [8].

As per caption guidelines, elements of quality captions are Accurate, Consistent, Clear, Readable, and Equal [9, 10]. There are no universally accepted guidelines, but some forums have developed guidelines for captions [11]. They specified that captions must be synchronized with the audio, equivalent with the content in the audio, and accessible to whoever needs it. Based on the visibility, captions are of two types: (1) Open Captions which are embedded into the video and (2) Closed Captions which can be turned on/ off based on user's interest [12]. MAGPIE, Softpedia's Subtitle Workshop, Captionate, and Universal subtitles are popular tools for closed captions [10, 13–15]. Caption files are simple files having time code and the text content to be displayed for the time. The files are stored in different formats supported by the media players. Some of the popular caption formats are CAP, CPT.XML, DFXP, SRT, WebVTT, QT, and SAMI (SMI).

4 Readability

Readability is a metric to measure the reader's understandability in written content. As per natural language processing, readability can be measured by text content and its presentation. Text content is the complexity of vocabulary and syntax. Presentation aspects explain about font size, line height, and line length. Nikolai A Rubakin stated that the main blocks of comprehension are unfamiliar words and long sentences [16]. Plain language in web content enhances readability for people with English as a second language. Various readability studies have done such as speech of perception, eye movement, rate of work, fatigue in reading, etc., using a number of factors such as direct and indirect statements, idea density, nominalization, structural cues, document age, etc. [16].

Out of all readability measures, the following are very popular to calculate the readability ease of the content. The Flesch–Kincaid readability test calculates the readability ease based on number of one-syllable words per 100 words and average sentences in words [17]. Dale Chall formula calculates based on a percentage of difficult words but not on Dale Chall word list and average sentence length. Gunning fog formula is based on average sentence length and percentage of hard words [18]. Hard words specify the words with more than two syllables. Fry reading graph is calculated by the average of x- and y-axes per hundred words [19]. Plot the average number of sentences and syllables in the graph. McLaughlin's SMOG formula is based on polysyllable count for a sample of 30 sentences [20]. Forecast formula calculates readability ease by number of single syllable words in a 150-word sample.

Readability ease measures explain the grade at which the document is comfortable to understand. Most researchers have done on calculating the readability ease measure on books [19], websites [18], online patient education materials [21], and medical reports [22]. Websites, books, and medical reports are static contents, and captions in videos are dynamic. Pausing the video content for reading captions may reduce viewers interest. This paper proposes a statistical model for caption readability based on temporal dimension. Using this model, we have analyzed the caption frames and results identify that the readability enhances with the time gap in captions.

5 Challenging Issues

ACCAN (Australian Communications Consumer Action Network) disability policy advisor, Wayne Hawkins, suggested that captions of audiovisual content must be readable and comprehensible [23]. Caption files are simple text files having time code and text. Each block is treated as a caption frame. The reader has to perceive the frame content within the duration.

Two factors affecting the readability of caption are (1) speaking rate vs reading rate and (2) time consumed by pauses in reading. Captions are nothing but auditory information of media content. Speaking rate varies from 175 to 275 wpm which is faster than the reading rate caption of around 100–150 wpm [24]. The continuous reading is impossible for anyone; reader will take pauses in the reading process [25]. Readability is analyzed by the flow of text perceived by the reader. Lasecki et. al. showed that DHH people fall behind spoken content especially when audio paired with video content [26].

Section 4 specifies popular readability formulas for calculating the readability ease but they are good for static text and calculate readability for overall text. Captions in videos are dynamic, disabled based on the time frame. The user has to grasp the content within the time frame. The existing formulae are not accurate for measuring the readability ease of captions. The purpose of the study is to explore the effect of caption rate on comprehension.

6 Algorithm

Algorithm 1. TempRead: Finding the Readability of the Caption

---

1: **procedure** TempRead(Captionfile)
2: **for** $i \in Frame$ **do**
3: $ReadFramescore \leftarrow TempReadFrame(Fi)$
4: $rScore \leftarrow rScore + ReadFramescore$
5: $rScore \leftarrow \frac{rScore}{Framelength}$
6: return $rScore$
7: **procedure** TempReadFrame(Fi)
8: $Tot_{ns} \leftarrow Noofsyllables$
9: $Tot_{sws} \leftarrow Noofstopwordsyllables$
10: $W_c \leftarrow NoofWords$
11: $SW_c \leftarrow NoofStopWords$
12: $T_{sec} \leftarrow TimedurationofFrame$
13: $ReadFramescore = 100 - \frac{(Tot_{ns} - Tot_{sws}) * (W_c - SW_c)}{T_{sec}}$
14: return $ReadFramescore$

This section explains about the algorithm TempRead (Method for readability in temporal dimension). Caption files are taken as input and produce the overall readability of caption. Caption files are having the time code and text. The above algorithm parses the time code and text of each caption frame, and calculates the time duration of the frame in seconds. The readability of each frame is calculated based on the formula in Eq. 2 so that we can easily identify the frame which is less readable to reader within the frame duration and the readability of overall caption is calculated through Eq. 1.

The algorithm returns a value of range from 0 to 100. The values nearer to zero are difficult to read and the values nearer to 100 are easy. The frames getting the values nearer to zero may need more time to comprehend:

$$RC = \sum_{i=1}^{n} \frac{Rcap(F_i)}{n}, \tag{1}$$

where

RC -> Readability of the Caption, Rcap(Fi)-> Readabaility of each Frame, n ->Number of frames in caption file

$$Rcap(F_i) = 100 - \frac{(Tot_{ns} - Tot_{sws}) * (W_c - SW_c)}{T_{sec}}, \tag{2}$$

where

$Rcap(F_i)$->Readability of each Frame, Tot_{ns}->Total number of syllables in the word,
Tot_{sws} ->Total number of syllables in stop words, W_c ->No of Words in Frame, SW_c ->No of stop words in Frame, T_{sec} ->Time of the Frame.

7 Experimental Results

NPTEL website is providing online courses of higher education [27]. With the increase in accessibility awareness, NPTEL had recently placed caption files for their videos [28]. We downloaded the caption files for measuring the readability ease based on the temporal dimension. A prototype implementation was developed using Python for evaluating the readability of the video file. Our system takes the SRT file of encoding UTF-8 as input and parses the data content in the caption file into time code and caption text for each frame. The time code provides the duration of the frame. Our program computes the number of words, number of stopwords, number of syllables in a frame, and number of stop word syllable, and derives the readability of each frame using Eq. 2; and using Eq. 1 we calculated the readability of overall caption as a mean of all frames and identified the variation of readability with various parameters.

We have downloaded an NPTEL SRT file which belonged to artificial intelligence course. A 20-min SRT is sliced based on time code in the file having average of 18 total words (nonstopwords +stopwords), 10 stopwords, 8 nonstopwords, and 25 syllables. The readability ease measure is calculated based on TempRead algorithm mentioned in Sect. 6 on the SRT file. The readability metric is analyzed based on the number of parameters such as the number of words, number of stop words, total number of words (sum of nonstopwords and stopwords), number of syllables, and time duration. The figures show readability analysis of each parameter.

Figure 1 explains the readability of each frame in the caption. The readability of overall caption is 72.302. We clearly observe that even though the readability of overall caption is good the frames 66, 68, 74, and 83 have very low readability. The reader may feel difficult to read within the time duration and lost his interest due to missing connectivity in frame. Figure 2 shows the readability with the number of words in the frame and identifies that frames having the same number of non-stopwords vary the readability and analyze that the frame having more than 20 words reduces the readability of the frame.

Figure 3 compares readability with number of syllables in the caption frame and we analyzed that the frames having same number of nonstopword syllables may not have same readability and also observed that the frame having lowest syllable rate is having high readability and the frame with high number of syllables having low readability. Figure 4 compares the readability with frame duration and identifies

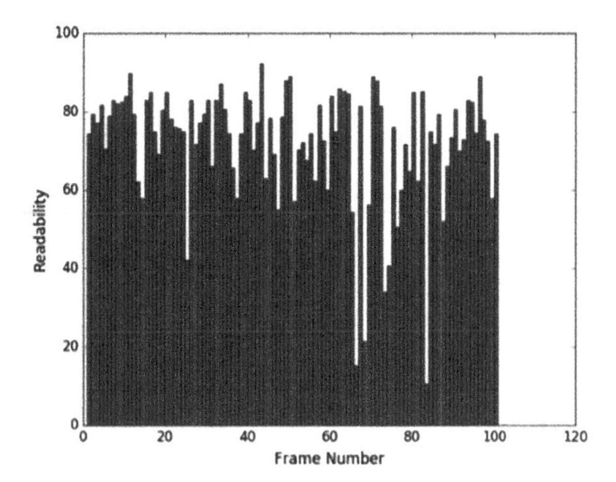

Fig. 1 Readability for each frame

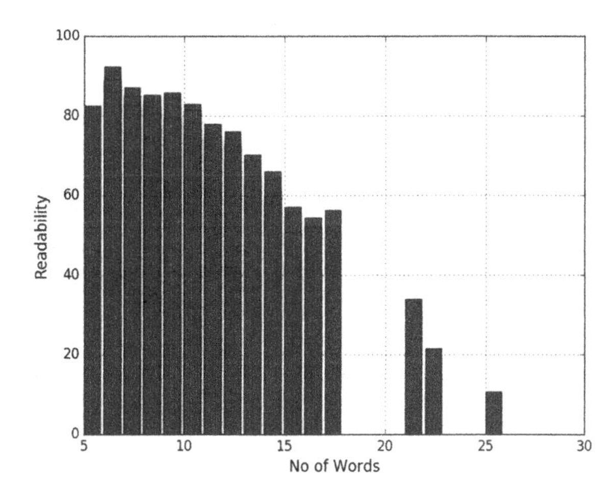

Fig. 2 Readability compared with number of words

that frames having same time duration may not have the same readability and figured out that the frame with maximum duration increases readability.

Figure 5 compares the readability with number of words and number of stop words in the frame and analyzes that the frame having more words and less stop words will affect the readability of the frame. The frame with number of words between 14 and 16 and number of stop words between 4 and 6 gaining maximum readability. Figure 6 compares readability with the total number of words (number of nonstopwords + number of stop words) and time duration of the frame, and analyzes that total number of words of ranges from 18 to 22 having frame time of 6–7 increasing readability.

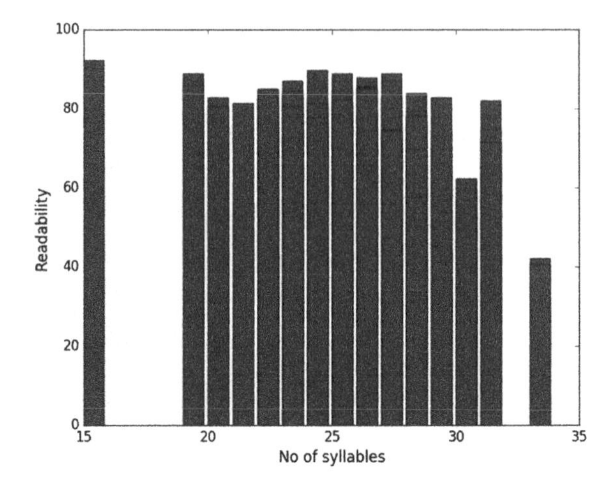

Fig. 3 Readability compared with number of syllables in caption frame

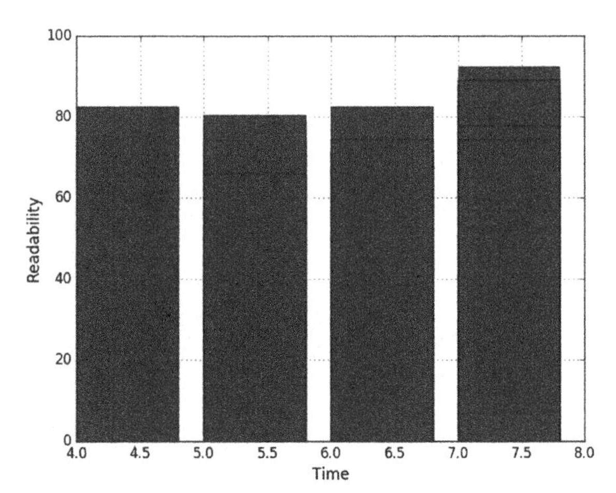

Fig. 4 Readability compared with time period of frame

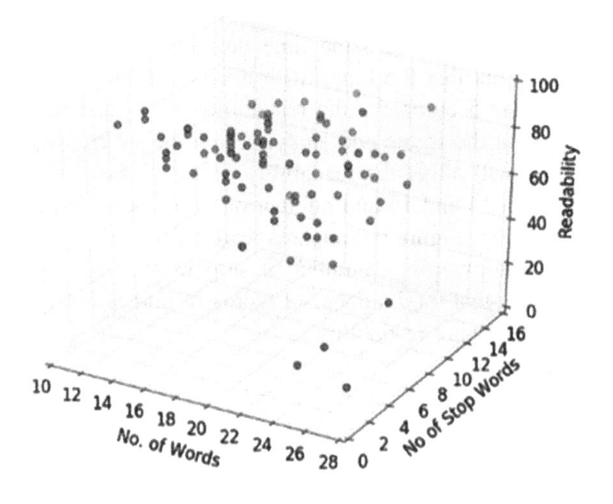

Fig. 5 Readability with number of words and number of stop words in caption frame

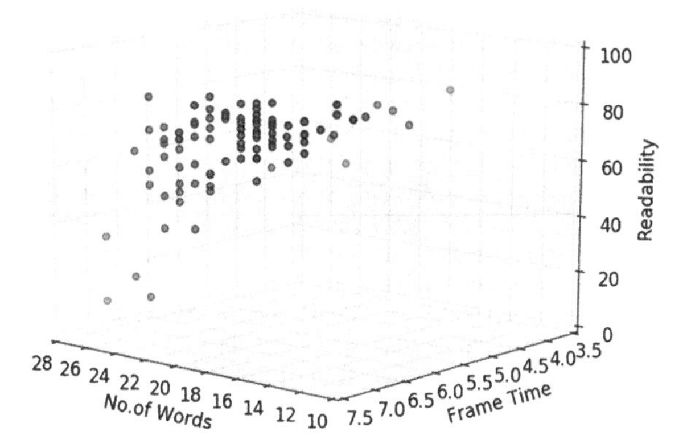

Fig. 6 Readability with total number of words and time taken by frame

8 Conclusion and Future Direction

Captions are displayed on the screen for auditory information accessibility. The caption file is having the timecode for each frame. The user has to perceive the text of caption frame within the time duration of display. Captions are created based on the auditory information of the video. Since there is huge variation in speaking rate with the reading rate, the readability of caption is to be evaluated based on the temporal dimension. The result shows that the frames with less number of syllables and more stop words are highly readable. The frame having more words requires more time for comprehension. So frame readability enhances with the time gap of the frame. Caption readability varies with time duration of the frame. Comparing TempRead algorithm on huge caption data will evaluate the optimum time duration with respect to number of words and compare with other readability mechanisms. We can also use TempRead readability technique for caption segmentation based on the time duration and compare with other segmentation methods as a future work.

References

1. Cisco VNI, White paper: Cisco VNI Forecast and Methodology, 2015–2020, 2016.
2. W3C, Multimedia Accessibility, 2008. https://www.w3.org/2008/06/video-notes.
3. Web accessibility for developers https://www.wuhcag.com/web-contentaccessibility-guidelines/.
4. Web Content Accessibility Guidelines (WCAG) 2.0, 2008. https://www.w3.org/TR/WCAG/.
5. Captioning. https://www.techsmith.com/tutorial-camtasia-mac-captions.html.
6. Closed Captioning, 2016. https://en.wikipedia.org/wiki/Closedcaptioning.
7. Martine Danan, Captioning and Subtitling: Undervalued Language Learning Strategies, Traduction audiovisuelle/ Audiovisual Translation, Volume 49n, p. 67–77, 2004. http://www.erudit.org/revue/Meta/2004/v49/n1/009021ar.html.

8. Paula Winke, Susan Gass, Tetyana Sydorenko, The Effects of Captioning Videos used for foreign language listening activities, Language Learning & Technology, Volume 14, Number 1, pp. 6586, 2010. http://llt.msu.edu/vol14num1/winkegasssydorenko.pdf.
9. Caption Key, National Association of the Deaf, 2011. http://www.captioningkey.org/qualitycaptioning.html.
10. Manitu Group, Captionate http://www.buraks.com/captionate/.
11. Diana Dee-Lucas, Reading Speed and Memory for Prose, Journal of Literacy Research, vol. 11 no. 3 221–233, 1979. http://jlr.sagepub.com/content/11/3/221.refs.
12. Universal subtitle: Add subtitle for any video in web, http://boingboing.net/2010/12/09/universal-subtitles.html.
13. Captioning Guidelines, Media Access Australia http://www.mediaaccess.org.au/practical-web-accessibility/media/captionguidelines.
14. Digital Captioning, High Tech Center Training Unit. http://www.htctu.net/trainings/manuals/web/DigitalCaptionMAGpie2.pdf.
15. http://www.softpedia.com/get/Multimedia/Video/Other-VIDEO-Tools/SubtitleWorkshop.shtml.
16. Readability, 2016. https://en.wikipedia.org/wiki/Readability.
17. James Hartley, Is time up for the Flesch measure of reading ease?, Scientometrics, 2016.
18. Muhammad Kunta Biddinikaa et al. Measuring the readability of Indonesian biomass websites: The ease of understanding biomass energy information on websites in the Indonesian language, Volume 59,1349–1357, 2016.
19. Hoke, Brenda Lynn, Comparison of Recreational Reading Books Levels Using the Fry Readability Graph and the Flesch-Kincaid Grade Level, Kean University, 1999.
20. Terry Smith, The problems with current readability methods and formulas: Missing that usability design, Professional Communication Conference (IPCC), IEEE, 2016.
21. Daniil L. Polishchuk, Readability of Online Patient Education Materials on Adult Reconstruction Web Sites, The Journal of Arthroplasty, Vol. 27, No. 5 2012.
22. Elena T. Carbone, Jamie M. Zoellner, Nutrition and Health Literacy: A Systematic Review to Inform Nutrition Research and Practice, Journal of the Academy Of Nutritions AND Dietetics, 2011.
23. ACCAN, Online video captions must be accurate and readable, 2015. https://accan.org.au/our-work/1125-online-video-captions-must-be-accurateand-readable.
24. Diana Dee-Lucas, Reading Speed and Memory for Prose, Journal of Literacy Research, vol. 11 no. 3 221–233, 1979. http://jlr.sagepub.com/content/11/3/221.refs.
25. Raja S. Kushalnagar, Walter S. Lasecki, Jeffrey P. Bigham, A Readability Evaluation of Real-Time Crowd Captions in the Classroom, ASSETS12, ACM, 2012. https://web.eecs.umich.edu/wlasecki/pubs/captionreadability.pdf.
26. Walter S. Lasecki, Raja Kushalnagar, Jeffrey P. Bigham, Helping students keep up with real-time captions by pausing and highlighting, W4A '14 Proceedings of the 11 th Web for All Conference, Article No. 39, 2014.
27. NPTEL, A Project by MHRD. http://nptel.ac.in/course.php.
28. NPTEL video lectures, Information and Communication Technology. http://textofvideo.nptel.iitm.ac.in/.

Readability Assessment-cum-Evaluation of Government Department Websites of Rajasthan

Pawan Kumar Ojha, Abid Ismail and K.S. Kuppusamy

Abstract Readability of text is an important factor to be considered during the design of contents which are aimed to transfer information to a large population. Texts which are easy to comprehend improves retention, reading persistence and reading speed. The paper presents evaluation result of 60 departmental websites of Rajasthan government by using various readability tools such as Flesch–Kincaid Reading Ease, Flesch–Kincaid Grade Level, Gunning Fog Index, Simple Measure of Gobbledygook, Coleman–Liau Index and Automatic Readability Index. Also, the classification of websites into four categories are presented such as Type-1 falls in very easy to understand, Type-2 falls in easy to understand, Type-3 falls in hard to understand and Type-4 falls in very hard to understand the text. By providing readability training and awareness programme to developers and designers in terms of standard guidelines, to achieve this readability-cum-accessibility in a better way. This analysis provides the feedback to website administrator to improve the readability-cum-accessibility and helps to achieve the accessibility for all.

Keywords Readability Indices · Classification · Accessibility · Rajasthan Government Departmental Websites · Readability assessment

P.K. Ojha (✉) · A. Ismail · K.S. Kuppusamy
Department of Computer Science, Pondicherry University, Pondicherry 605014, India
e-mail: pawanojhacs2@yahoo.in

A. Ismail
e-mail: abidpu2015@gmail.com

K.S. Kuppusamy
e-mail: kskuppu@gmail.com

A.K. Somani et al. (eds.), *Proceedings of First International Conference on Smart System, Innovations and Computing*, Smart Innovation, Systems and Technologies 79, https://doi.org/10.1007/978-981-10-5828-8_23

1 Introduction

Nowadays, Internet is one of the most important sources of information, Internet is an interconnection of different networks which provides services via websites and web pages. Because of easy access to Internet connection economically, ease of availability of services and launch of *Digital India Programme*, people are more aware and using the websites for their important tasks in their day-to-day life. As the readability of the web page increases, it will be barrier free for a large section of society.

World Wide Web Consortium (W3C) has framed various guidelines to make these resources universally accessible. The W3C proposed WCAG 2.0 guidelines which is a stable, referenceable technical standard. It consists of 12 guidelines categorized under four principles abbreviated as POUR. POUR stands for Perceivable, Operable, Understandable and Robust. Under each guideline, there are testable success criteria. Guideline 3.1 discusses readability, which intends to provide guidelines to allow text content to be read by users, to ensure that information that is required to be understand is available.[1]

Readability is a factor which gives an indication of understandability of written text and level of grade education needed to be understand. Different readability formulas consider count of words, sentences, complex words,syllables, etc. Websites of departments of government are the resources which people come across for many important tasks. The readability of these websites needs to be in accordance with the level of education and understanding of the people accessing the websites.

In this paper, we have focused to check the readability-cum-grade score of websites of a large and important state situated in the northwestern part of India known as Rajasthan. Departmental websites of Rajasthan state government are chosen as a source of data for our readability analysis, and this study is a part of a series of zone-wise analysis of web page accessibility-cum-readability of the entire nation. The primary objective of this study is to highlight the necessary steps that need to be taken to improve the readability of the websites so that it can be utilized by a large spectrum of users with the minimal barrier.

2 Background

Due to the huge impact of websites on their usability and accessibility over common people, the research is in accordance to assess the readability of websites of different domains has been an active area of research. A research aimed to examine the readability of websites related to providing information over biomass renewable energy was being conducted using readability standard formulas [1]. Another work carried out by David W. Ferguson [2] aimed to find that municipal websites in U.S. are creating digital inequalities of opportunities, and had performed a study assessing the grade-level readability of municipal websites.

[1] https://www.w3.org/WAI/intro/wcag.php

The availability of healthcare information on the Internet has increased the involvement of patients for health regarding tips, the reading level of average people should be taken into account. With this regard, a study of 138 online patient education resources from RadiologyInfo.Org was conducted by researchers [3]. Readability assessment of criteria for eligibility of clinical trial and Internet-based education materials for patients related to breast cancer was performed in [4, 5].

Readability of text is being used to predict the level of difficulty faced by people in understanding written text. A collaborative study from Microsoft and University of Washington [6] presented an article which used traditional readability formulas to study the language used on the Twitter website for a sample of tweets. Various studies were conducted for evaluating the accessibility of online government services in Saudi Arabia [7] and also, a similar study has been conducted to evaluate the accessibility of websites of Universities in Jordan [8] and Universities of India [9].

E-governance refers to provide governmental services electronically usually over the Internet [10], active participation of people in e-governance entails transparency and ease of access to services. E-governance is almost adopted in all the government organizations in India, a study has been done towards successful implementation of e-governance in government organizations [11].

3 Readability

The aim of readability assessment is to provide a quantifiable measure to text according to the understandability of human. There are various readability formulas we came across in the literature survey used for evaluation process, but for our study we used following formulas for website evaluation of Rajasthan Government (departments websites). The formulas used for evaluation are based on sentence length, count of syllables, percent of complex words and count of characters. The formulas are given by Eqs. 1, 2, 3, 4, 5 and 6.

1. **Flesch-Kincaid Reading Ease (FKRE)** [12]:

$$FKRE = 206.835 - (1.015 * \text{Average Sentence Length}) - (84.6 * \text{Average syllables per Word}) \quad (1)$$

2. **Flesch-Kincaid Grade Level (FKGL)** [13]:

$$FKGL = (0.39 * \text{Average Sentence Length}) + (11.8 * \text{Average Syllables per Word}) - 15.59. \quad (2)$$

3. **Gunning Fog Index (GFOG)** [13]:

$$GFOG = 0.4 * (\text{Average Sentence Length} + \text{Percent Complex Words}). \quad (3)$$

4. **Simple Measure of Gobbledygook (SMOG)** [13]:

$$SMOG = 3 + \sqrt{\text{Polysyllable Count}} \tag{4}$$

5. **Coleman-Liau Index (CLI)**[2]:

$$\begin{aligned} CLI = {} & 0.0588 * (\text{Average number of letters per 100 words}) - \\ & 0.296 * (\text{Average number of sentences per 100 words}) - 15.8 \end{aligned} \tag{5}$$

6. **Automatic Readability Index (ARI)** [14]:

$$ARI = 4.71 * \left(\frac{characters}{words}\right) + 0.5 * \left(\frac{words}{sentences}\right) \tag{6}$$

4 Methodology

We have collected 71 departmental websites' URLs of Rajasthan government from wikipedia and government portal site of Rajasthan.[3] Among 71 URLs only 60 URLs' (Table 1) readability information was available to be processed using a web tool [4] during the second week of December 2016.

The Readability Checker Tool takes the URL of website as input and outputs the following information.

1. Readability Indices uses the following readability formulas
   (a) *Flesch–Kincaid Reading Ease Readability formula:* The overall result of 60 websites with their readability score given by FKRE readability formula is shown in Fig. 1a. The graph shows how much score varies between selected websites from minimum reading score −89 to maximum reading score 114.7. This indicates that least score websites are difficult to understand and highly scored websites are easy to comprehend. The overall inference is explained in Sect. 5.
   (b) *Flesch–Kincaid Grade-Level Readability formula*: Fig. 1b represents scores using FKGL readability formula which varies from −1.5 to 27. Score −1.5 denotes that contents of the website are very easy to understand and contents of a website with score 27 are very difficult to understand.
   (c) *Gunning Fog Index Readability formula*: The scores using GFOG readability formula ranges from 1.2 to 13.8 as shown in Fig. 1c. As compared to FKRE and FKGL the distribution of scores is narrow, the website with less score is more readable compared to a website with more score.

[2]http://www.readabilityformulas.com/coleman-liau-readability-formula.php

[3]https://rajasthan.gov.in/Pages/default.aspx

[4]http://www.webpagefx.com/tools/read-able/

Table 1 List of websites with website ID

Website ID	URLs	Website ID	URLs
1	http://www.ard.rajasthan.gov.in/	31	http://www.rajliteracy.rajasthan.gov.in/index.htm
2	http://www.krishi.rajasthan.gov.in	32	http://www.lsgraj.org
3	http://animalhusbandry.rajasthan.gov.in	33	http://medicaleducation.rajasthan.gov.in
4	http://museumsrajasthan.gov.in	34	http://www.rajswasthya.nic.in
5	http://ayurved.rajasthan.gov.in	35	http://www.dmg-raj.org
6	http://www.biofuel.rajasthan.gov.in	36	http://www.minorityaffairs.rajasthan.gov.in
7	http://www.dce.rajasthan.gov.in	37	http://rajpension.nic.in
8	http://cad.rajasthan.gov.in	38	http://www.dop.rajasthan.gov.in
9	http://rajtax.gov.in	39	http://petroleum.rajasthan.gov.in
10	http://www.rajcooperatives.nic.in	40	http://www.planning.rajasthan.gov.in
11	http://devasthan.rajasthan.gov.in	41	http://prosecution.rajasthan.gov.in
12	http://www.rajrelief.nic.in	42	http://www.rajwater.gov.in
13	http://www.rajshiksha.gov.in	43	http://pwd.rajasthan.gov.in
14	http://www.ceorajasthan.nic.in	44	http://police.rajasthan.gov.in
15	http://esi.rajasthan.gov.in	45	http://www.rsad.rajasthan.gov.in
16	http://www.rajrojgar.nic.in	46	http://rpg.rajasthan.gov.in
17	http://www.rajenergy.com	47	http://igrs.rajasthan.gov.in
18	http://environment.rajasthan.gov.in	48	http://www.rdprd.gov.in
19	http://evaluation.rajasthan.gov.in	49	http://rajsanskrit.nic.in
29	https://rajexcise.gov.in	50	http://sje.rajasthan.gov.in
21	http://www.rajfab.nic.in	51	http://rajsainik.raj.nic.in

(continued)

Table 1 (continued)

Website ID	URLs	Website ID	URLs
22	http://fisheries.rajasthan.gov.in	52	http://sdri.rajasthan.gov.in
23	http://www.food.rajasthan.gov.in	53	http://sipf.rajasthan.gov.in
24	http://gad.rajasthan.gov.in	54	http://rajasthantourism.gov.in
25	http://home.rajasthan.gov.in	55	http://ctp.rajasthan.gov.in
26	http://horticulture.rajasthan.gov.in	56	http://transport.rajasthan.gov.in
27	http://rajind.rajasthan.gov.in	57	http://www.udh.rajasthan.gov.in/SitePages/Home.aspx
28	http://dipr.rajasthan.gov.in	58	http://waterresources.rajasthan.gov.in
29	http://doitc.rajasthan.gov.in	59	http://watershed.rajasthan.gov.in
30	http://www.rajlabour.nic.in	60	http://wcd.rajasthan.gov.in

(d) *SMOG Readability formula*: The variance in the scores of websites using SMOG readability formula ranges from 1.8 to 15.8 as shown in Figure 1d, the readability of the content increases with increase in score.

(e) *Coleman–Liau Index Readability formula*: The CLI readability scores for our data considered for evaluation varies from −14.7 to 41.3, refer Fig. 1e. The score −14.7 represents that readability of website is understandable to people with low age whereas to understand the content of the website of score 41.3 much higher age is required.

(f) *Automatic Readability Index Readability formula*: The ARI test evaluated and generated scores ranging from −18 to 27 as shown in Figure 1f, in which −18 denotes contents are ease to understand whereas difficulty increases with increase in score.

Table 2 represents overall statistics of readability indices, the bold representation of numerals (score) in the table shows the scores assigned to websites which are very easy to read in accordance with the formula used.

2. Text Statistics of the website contains the count of sentences, count of words, count of complex words, percent of complex words, average words per sentence and average syllables per word. Table 3 represents the overall analysis of text statistics of contents of departmental websites of Rajasthan government. The average count of sentences per website is 71.2 and the standard deviation in the number of sentences is 86.74 for the 60 websites, which infers that there is a high

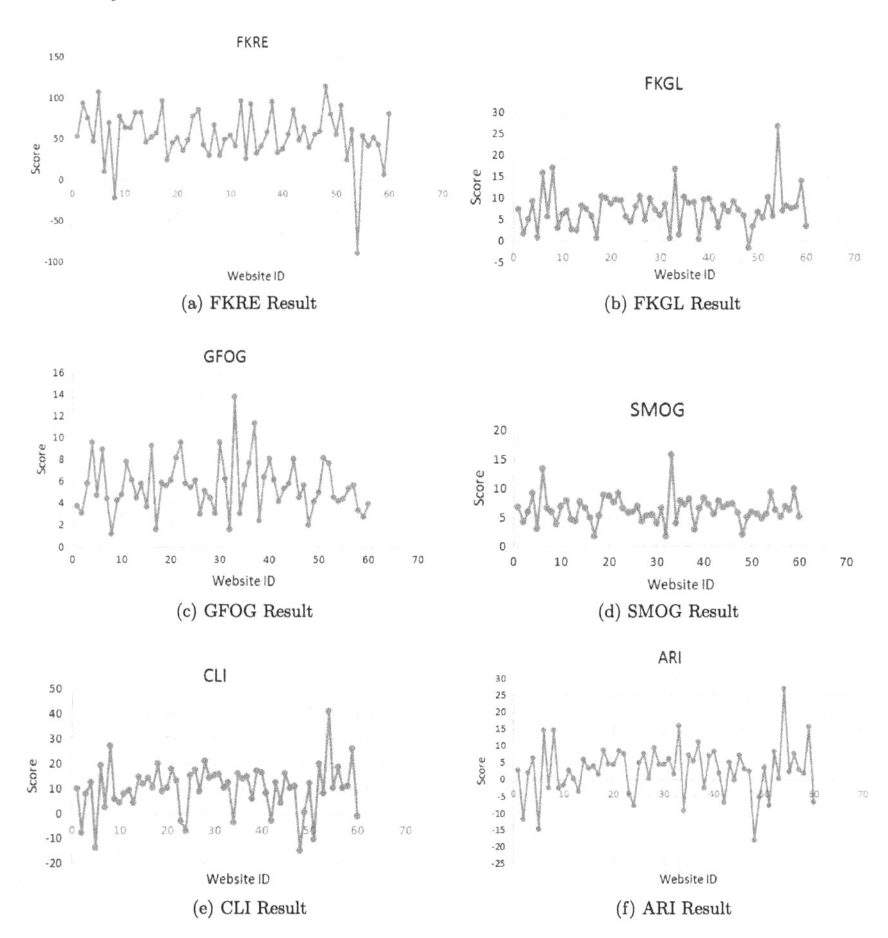

(a) FKRE Result

(b) FKGL Result

(c) GFOG Result

(d) SMOG Result

(e) CLI Result

(f) ARI Result

Fig. 1 Overall readability indices result of 60 departmental websites of Rajasthan government

Table 2 Scores statistics

S.No	Formula/Statistics	Min	Max	Mean	STDEV
1	FKRE	−89	**114.7**	54.67	32.14
2	FKGL	**−1.5**	27	7.36	4.60
3	GFOG	**1.2**	13.8	5.59	2.44
4	SMOG	**1.8**	15.8	6.46	2.40
5	CLI	**−14.7**	41.3	10.22	9.81
6	ARI	**−18**	27	2.94	7.61

Note Readability Tools (FKRE—Flesch-Kincaid Reading Ease, FKGL—Flesch-Kincaid Grade Level, GFOG—Gunning FOG, SMOG—Simple Measure of Gobbledygook, CLI—Coleman-Liau Index, ARI—Automatic Readability Index)

Table 3 Text statistics

S.No	Statistics	NOS	NOW	NOCW	PCW
1	Sum	4272	26239	4340	1179.17
2	Mean	71.2	437.31	72.33	19.65
3	Standard deviation	86.71	534.85	75.61	10.22

Table 4 Classification of websites based on grade level

S.No	Grade level	Passed websites	Status
1	<= 5	**22**	Very easy
2	6–10	**32**	Easy
3	11–15	**5**	Hard
4	16–20	**0**	Very hard
5	>= 20	**1**	Too complex

variance in the text content (sentences) from site to site. The results resemble the number of words (NOW) is similar to sentences in websites. The average number of complex words per website approximately to 72 and the deviation from site to site is approximately 76 with respect to number of complex words (NOCW) on websites.

The average of percent of complex words (PCW) on the websites is 19.65, that is an average 20% words are complex on every website.

3. Test result provides the information regarding average grade level and the age of person required to understand the content of a website. Table 4 represents a classification of websites based on grade-level output of websites, we categorized the overall result into five categories assigning status from very easy to too complex as shown in the table. The result shows that out of 60, 32 websites passed in the category of easy and 22 websites are very easy to understand whereas 5 websites are hard to comprehend and only one website passed the category where grade level is above 20 and too complex to understand.

5 Readability Classes

We have evaluated the sixty (60) departmental websites of Rajasthan Government using different readability formulas and categorized on the basis of scores available into very easy, easy, hard and very hard categories. For FKRE readability formula, the categorization on the basis of score was already available and our evaluation shows that 38.8% of websites fall in the category from *very easy* to *standard* and almost 61% of websites are in the category from *fairly difficult* to *very confusing*.

For the results of FKGL, GFOG, SMOG, CLI and ARI formulas we make classification of these selected websites on the basis of following formula in Eq. 7 which classify the websites into four categories namely *Type-1, Type-2, Type-3 and Type-4*.

$$A = AVERAGE(MinScore, MaxScore) \quad B = AVERAGE(MinScore, a) \quad C = AVERAGE(a, MaxScore) \tag{7}$$

The description of these Type categories are as follows:
Type-1 ranges from Min Score to B = Very Easy Status of Reading
Type-2 ranges from B to A = Easy Status of Reading
Type-3 ranges from A to C = Hard Status of Reading
Type-2 ranges from C to Max Score = Very hard Status of Reading

On evaluation with FKGL readability formula 30% of websites passed the criteria of very easy, 61.66% fall under easy category, 6.66% of websites are hard and 1.66% of websites are very hard.

GFOG readability formula resulted in 30% of websites into very easy, 46.6% of websites into easy, 20% of websites into the category of hard and 3.3% of websites into very hard category. The results of SMOG readability formula shows that 28.3% of websites are very easy to understand, 60% of websites are easy to understand, 8.3% of websites are hard to understand and 3.3% of websites are very hard to understand for readers.

CLI readability formula puts 15% of websites into very easy category, 48.33% of websites into the category of easy, 35% of websites falls in the category of hard and 1.6% of websites are very hard. On evaluating our data using ARI readability test 10% of websites are very easy, 45% of websites are easy, 41.66% of websites are hard and 3.33% of websites are very hard.

6 Conclusions and Future Directions

The formulas we used to predict readability of websites are specially designed for English text, so our study focus on the plain text of websites but there is a need to have readability formulas which are language independent. The formulas on assessment provide Grade Level which is US based and used almost all over the world. There has not been much research to get country specific Grade Level.

Formulas used to assess the readability of text is used to assess the readability of websites also, there is need to have readability formula which can be exclusively used for predicting the readability of websites. The readability formula should consider the effect of web components such as links, images, typeface (fonts), captions and contrast ratio along with text on the web page while predicting the readability of a website.

During our study we found that there are 22.95% Type-1 websites, 33.69% Type-2 websites, 21.16% Type-3 websites and 22.12% Type-4 websites. Hence, almost 43.28% websites are hard to understand the text, there is need to minimize the score so that the readability can be increased. We suggest to decrease the percentage of websites in hard to very hard category by introducing simple to read terms in the web content so that the information resources can be made more and more accessible.

References

1. Biddinika, Muhammad Kunta, et al. "Measuring the readability of Indonesian biomass websites: The ease of understanding biomass energy information on websites in the Indonesian language." Renewable and Sustainable Energy Reviews 59 (2016): 1349–1357.
2. Ferguson, David W. Grade-level readability of municipal websites: Are they creating digital inequalities of opportunities that perpetuate the digital divide?. Diss. The University of Akron, 2014.
3. Hansberry, David R., et al. "A critical review of the readability of online patient education resources from RadiologyInfo. Org." American Journal of Roentgenology 202.3 (2014): 566–575.
4. Kang, Tian, Noémie Elhadad, and Chunhua Weng. "Initial Readability Assessment of Clinical Trial Eligibility Criteria." AMIA Annual Symposium Proceedings. Vol. 2015. American Medical Informatics Association, 2015.
5. AlKhalili, Rend, et al. "Readability assessment of Internet-based patient education materials related to mammography for breast cancer screening." Academic radiology 22.3 (2015): 290–295.
6. Davenport, James RA, and Robert DeLine. "The readability of tweets and their geographic correlation with education." arXiv:1401.6058 (2014).
7. Al-Faries, Auhood, et al. "Evaluating the accessibility and usability of top Saudi e-government services." Proceedings of the 7th International Conference on Theory and Practice of Electronic Governance. ACM, 2013.
8. Kamal, Israa Wahbi, et al. "Evaluating Web Accessibility Metrics for Jordanian Universities." International Journal of Advanced Computer Science and Applications 1.7 (2016): 113–122.
9. Abid Ismail, and K. S. Kuppusamy. "Accessibility of Indian universities' homepages: An exploratory study." Journal of King Saud University-Computer and Information Sciences (2016).
10. Manoharan, Aroon, ed. Active Citizen Participation in E-Government: A Global Perspective: A Global Perspective. IGI Global, 2012.
11. Barua, Mithun. "E-governance adoption in government organization of India." International Journal of Managing Public Sector Information and Communication Technologies 3.1 (2012): 1.
12. James Hartley, *Is time up for the Flesch measure of reading ease?*, Scientometrics, 2016.
13. Begeny, John C., and Diana J. Greene. "Can readability formulas be used to successfully gauge difficulty of reading materials?." Psychology in the Schools 51.2 (2014): 198–215.
14. Al Tamimi, Abdel Karim, et al. "AARI: automatic arabic readability index." Int. Arab J. Inf. Technol. 11.4 (2014): 370–378.

A Novel Technique for Frequently Searching Items with Alarm Method

R. Priya and G.P. Rameshkumar

Abstract There are many technologies that are already available for searching items with mobile phone. But in this method, user should have a mobile phone in working condition. The proposed system is based on wireless sensor network. The wireless sensor network is a network of nodes that can sense and control the environment and also has interaction between the user, computer, and environment. This research paper has been developed to search the frequently used homely products during emergencies through alarm with transmitter and Receiver connection. If any user wants to search items, this application is very much useful during emergency.

Keywords Wireless network · Transmitter · Receiver · Searching items

1 Introduction

Nowadays, people's day-to-day life becomes more faster, which creates stress and tension. If users want to go out urgently, at that time, they may miss house key, car key, or bike key, etc. So users' minds are fully absent and the user has to spend their valuable time for searching such items. To avoid this problem, the proposed model has been developed for searching the items in that particular place like wireless personal area network. There are various types of method available for finding items such as radio frequency (RF) and so on. RF finder is used to track devices, while Bluetooth finder is used in smart phones. Depending on the device and condition is used the ranges allocated between 250 and 300 feet.

R. Priya (✉)
Department of Computer Science, Bharathiar University, Coimbatore 641046, India
e-mail: priyaramato2016@gmail.com

G.P. Rameshkumar
Government Arts College, AiyarMalai, Kulithalai, Karur, India
e-mail: gpr.snr@gmail.com

A.K. Somani et al. (eds.), *Proceedings of First International Conference on Smart System, Innovations and Computing*, Smart Innovation, Systems and Technologies 79, https://doi.org/10.1007/978-981-10-5828-8_24

In this method, the system has two sides of communication. One is for input and another one is for output. In the input side, the system has attached to switches and each switch is connected to the particular things. In the output side, the things are fixed with the buzzer alarm. When the input switches are ON, the search of particular item will happen.

1.1 Wireless Personal Area Network (WPAN) and Wireless Sensor Network (WSN)

This is a low range network. It serves to interconnect all the ordinary computing and communicating devices which cover a few dozen meters [1]. It is mainly used for connecting peripheral devices like printer, cell phone, computer, home appliance, etc., without using a wired connection. Several technologies are available for this network like Bluetooth, HomeRF, WiFi, zigbee, WiMAX, etc.

The wireless sensor network consists of nodes that are used to sense the object and control the environment and it also connects the user, computer, and environment [2]. The activity of sensing, working, and interaction [3] with specific energy will increase the power of cross-layer design method that requires the joint consideration of distributed data processing, communication protocols, and medium access control [4].

1.2 RF Module

The wireless system has two constraints such as distance and the amount of information transfer. RF module covers the small dimension, but wide voltage range from 3 to 12 v. It uses 433 MHz of transmitter and receiver. When transmitting logic is zero, it does not supply any power. When a carrier frequency is used to an extent, it consumes low battery consumption. The logic one is sent to the carrier frequency, its supplies 4.5 mA with 3 V. The transmitter sends the data in sequence as received by the tuned receiver (Fig. 1).

1.3 Features of RF Module

(1) 433 MHz frequency used in receiver, (2) 105Dbm typical frequency used in receiver, (3) Current 3.5 mA, (4) Low power consumption, (5) It receives about 5 V, (6) 433.92 MHz frequency used in transmitter, (7) Input voltage of 3–6 V for transmitter and (8) Output voltage of 4–12 V for transmitter.

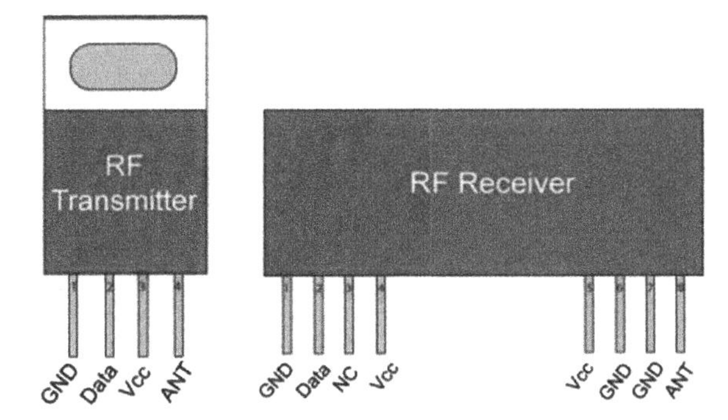

Fig. 1 RF Transmitter and RF Receiver

2 Existing Method

Already users have some technology for searching items with Bluetooth technology in mobile phone. These methods can be used in iPhone or Android phone only such on Tile, TrackR, and Duet Reviewed and so on. But in this method, user should have a mobile phone in his hand. This mobile phone already should have this software for searching items with Bluetooth. In this method, each item has sensor chips and mobile phone can connect with sensor chips. In this method, mobile phones need to be in working condition. And mobile should not be missed anywhere. This method is used in mobile phone with the help of Bluetooth connectivity. Existing methods are used to key-finding with Bluetooth technology and working with battery. The user should know how many days are working batteries charges and how long sensor batteries would last before it needs a new battery. These wireless key finders applications are used in public park and home also. It would be tested with various places in various distances, alarm or alerts were audible in barrier places also. In this case, two are more product work simultaneously in the same place, Bluetooth connectivity was disturbed by one of the sensors to others. The proposed system is mainly concentrated on sensor application for searching frequently used items through buzzer alarm. In this method, Zigbee is used to connect the sensor chip and switches with RF transmitter and receiver [5].

2.1 Zigbee

ZigBee is a wireless networking and aims at remote control and sensor applications which are suitable for any operation in harsh radio environments and in isolated locations [6]. Zigbee is one of the wireless protocols used in wireless personal area networks. Already zigbee was used to detect the leaking gas or fire like infrared detector, temperature sensor, smoke detector, and gas sensor with alarm message remotely. If user wants to search items, which items should have sensor chips like Tile, TrackR and Due Reviewed application, etc? In these items, Tile is a

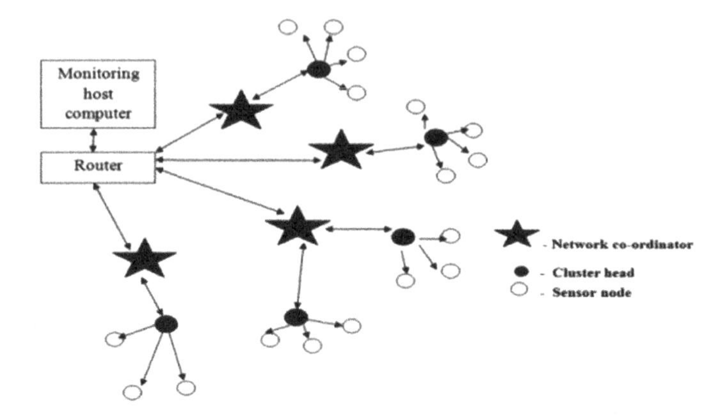

Fig. 2 Structure of wireless sensor network based on ZigBee technique

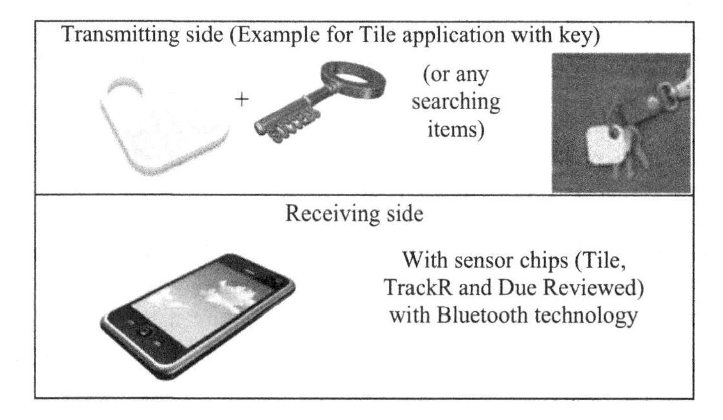

Fig. 3 Existing method of searching items

square-shaped tracker lost item product that attaches to bike key, home key, any bag, and more. Nowadays, Tile gives support very much for Android smart phone to find lost items or stolen items, and allows a larger group of peoples using their mobile phones [7] (Fig. 2).

If user using wants to search any items, activate wireless network within his mobile phone through Bluetooth, and then connect to searching items based on sensor chips through searching application (Fig. 3).

2.2 *Sensor Node*

The sensor node is a basic unit of the wireless sensor network and it is used for data analog/digital conversion. The processing module of sensor node is responsible for controlling the whole operation and binary information transmitted from other devices. It is mainly used for exchanging information and controls the receiving or transmitting data [8].

3 Literature Review

Nowadays, wireless sensor network works in different places such as a fire detection system, leaking gas detection, open/close status of gates and doors in large warehouses, industrial automation, traffic monitoring, air traffic control, robot control, monitoring weather conditions, monitoring medical devices, video surveillance, automated smart homes, etc. Some of the applications are given below:

- Forest fire detection system is one of the wireless sensor systems and it consists of monitoring nodes, base stations, communications systems, and Internet access. The forest fire detection system has a large number of the different types of sensors to monitor the value change, including temperature, humidity, smoke, or gas detector, etc. All sensor nodes are divided to some cluster. Each cluster is equivalent to a relatively fixed self-organizing network. All the data are stored in the particular database [9].
- Industrialization increases air pollution by releasing the unwanted gases in the environment. Sensor nodes are used to detect the percentage of pollution by using a Zigbee technique [10].
- The air traffic control (ATC) tower is located in or near the aerodrome and usually has visual contact with the aircraft. The tower may consist of a local controller, ground control, and clearance delivery. The local controller is responsible for all traffic within the control zone and on the runways, the ground controller for all traffic on the taxiways. It may provide all kinds of information and especially traffic information to the pilot [11].
- The traffic density is increasing in developing countries, it needs to implement advance intelligent traffic signals to replace the conventional manual and time-based traffic signal system. These advance intelligent traffic signals have a central microcontroller at every junction which receives data from tiny wireless sensor nodes place on the road. The sensor nodes have sensors that can detect the presence of vehicles and the transmitter wirelessly transmits the traffic density to the central programmable microcontroller. The microcontroller makes use of the proposed programmed algorithm to find ways to manage and regulate traffic in a systematic manner efficiently [12].
- A quantum mechanical algorithm is a factor algorithm than other classical algorithm, this paper says that the quantum mechanical algorithm is only polynomial faster than classical algorithm. Even though it is difficult for factorization problem. For example, unsorted database contains a number of items with the single items satisfies the given condition [13].

- In today's world, there is a greater opportunity to know about the physical environment using online information. It is helpful to find the information about car price, book reviews, how to rectify the jammed printer, etc. [14].

4 Proposed Model

This proposed model is mainly focused on sensor application connected with the frequently used items to perform searches through buzzer mechanism. In this block diagram, the proposed model has two sections such as transmitting section, receiving section. The communication is done between the transmitting section and receiving section (Fig. 4).

In this block diagram, the proposed model has two parts such as transmitting section and receiving section, in between the both sections of wireless sensor network. The transmitting section has electrical switches for mobile phone, TV remote, bike key, car key, etc. If user press any switch, the electrical switch connection with RF transmitter. The receiving section has searching items with alarm. RF receiver activates with RF transmitter, and find outs the items, and then alarm makes noise for detecting items. In this method, first, user should set electric switches for each item in his house like home switches (Figs. 5 and 6).

The transmitting section has 433mhdr transmitter, HD 12d IC chip, 1 K resistor, 330 amp resistor, push button, waver sender antenna, C 104 capacitors, 5 V power supply battery, IC power controller a472j, and LED light. The receiving section has 433mhdr receiver, HD 12e IC chip, 2 Nos. of 1 resistor, 330 amp resistor, C 104 capacitor, and buzzer alarm.

In a transmitting side, the push button is placed. At the same time, in the receiving part, the alarm has placed and the item which wants to be searched has fixed. According to Fig. 7, it shows the searching of misplaced key. To find the

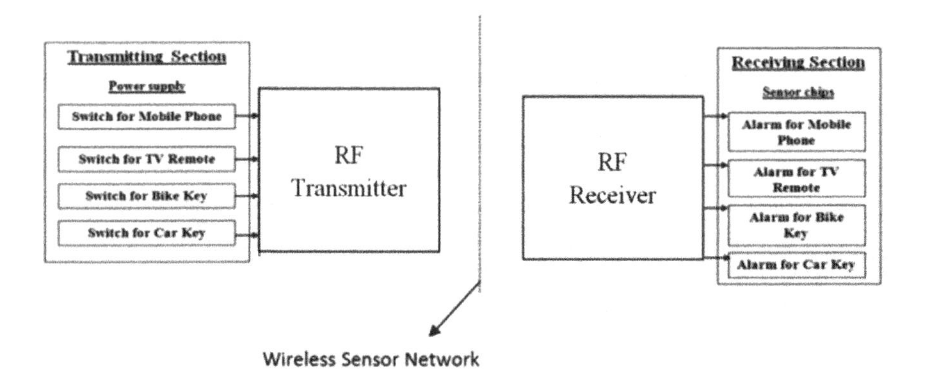

Fig. 4 Block diagram of proposed system

Fig. 5 Transmitting section

Fig. 6 Receiving section

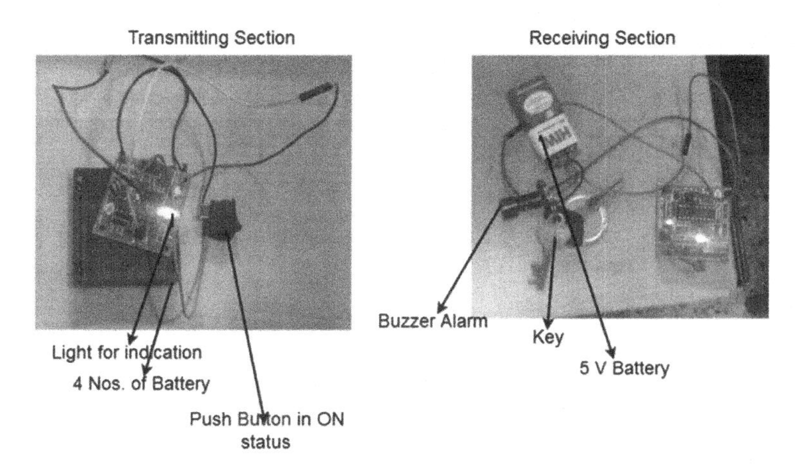

Fig. 7 The proposed system for finding items with transmitter and receiver

misplaced key, the push button which is placed on the transmitter side has to be switched ON, after the switch is ON, on the receiver side it raises the alarm. On the receiver side, the 5 V battery for power supply and buzzer alarm is set up on below the battery.

4.1 Flow Diagram of Proposed Model

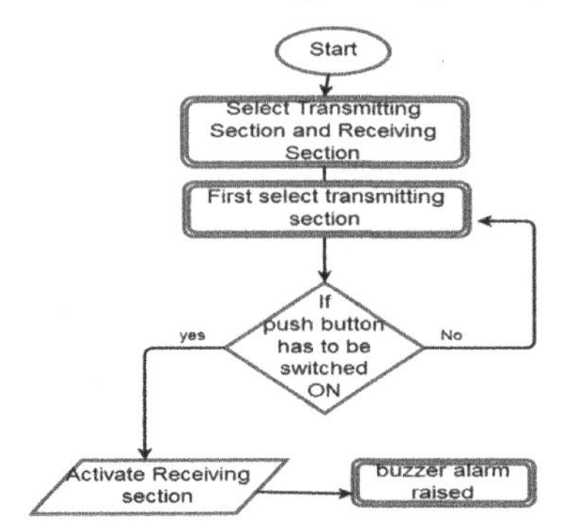

The communication channel consists of RF transmitter and RF receiver. In both, transmitting section and receiving section worked with wireless sensor network. The transmitting section has electrical switches for finding mobile phone, TV remote, bike key, car key, etc. If user presses any switch in transmitting section, the action has to be established for the specified item. The receiving section has alarm.

4.2 Algorithm of Proposed Model

```
create database "searching_items";
create table trans;
{433mhdr transmitter, hd 12d ic chip, 1 k resistor, 330
amp resistor, push button, waver sender antenna, c 104
capacitor, 5v power supply battery, ic power controller
a472j,led light, read pinmode =13}
create table receiv
{433mhdr receiver, hd 12e ic chip, 2 nos. of 1 k
resistor, 330 amp resistor, c 104 capacitor, buzzer
alarm,read pinmode =13}
sensordect( )
{ if(searching_items.trans ==searching_items.receiv)
{buzzer alarm raised;}
elseif(searching_items.trans not matched to
searching_items.receiv)
{buzzer alaram keep silent;}
end if
define searching()
{searching_items. Trans()
{read pushbutton
if (pushbutton ==high)
//activate receiving section and led switched on
enable searching_items.receiv( )
{sensordetect();
{alarm == activated}}
else
     sensordect( );
end if
```

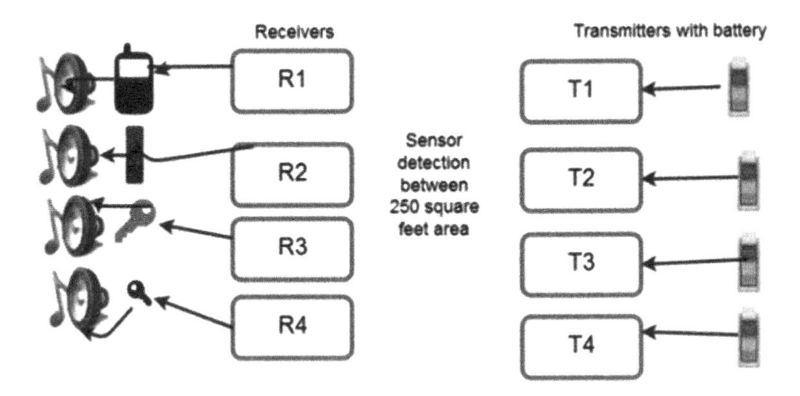

Fig. 8 Proposed system model

4.3 *Architecture of Proposed Model*

The architecture of proposed model consists of two sections such as transmitting section, receiving section and the connection is established through wireless network. It can be used within 250 square feet or room. It can be extended in the future development. The transmitting section consists of individual switches for the items —mobile phone, TV remote, Bike key and Car key. The receiving section consists of respective alarms for the same. If the system wants to find more than one item, the system should have the individual transmitter and receiver based on the number of items (Fig. 8).

But, in this paper, the connection is made with a single switch for finding key only. The proposed system has been explained the switch of the mobile phone is getting connected to its respective transmitter T1. Similarly, the same method can be used to connect the individual items to its Transmitter T2, T3, and T4. The output lines from the entire transmitter from T1 to T4 will get connected to the receivers through wireless network. There are four receivers R1, R2, R3, and R4 available in the receiving section. Each receiver receives the incoming signal from the transmitter and sends it to the respective alarm which is on the receiving section. Then, the receivers R1, R2, R3, and R4 are connected to its alarm. In the future enhancement, a single transmitter and a single receiver would help to find more than one item.

4.4 *Benefits of the Proposed Model*

- Time consuming: The main advantage of the proposed model is time consumption. However, it is easy to identify the frequently used accessories like mobile phone, TV remote, and so on. Whenever, the user is in need of those accessories, user time is consumed just by pressing the respective switch of that

accessory and on pressing the switch, the alarm will indicate where the device located. For example, when the mobile phone is misplaced somewhere inside the home and the mobile phone was switched off, then searching process will be difficult due to the disconnection. At the time, when the transmitter switch is ON, then the user can easily find the mobile using this alarm.
- Reducing stress and tension: The proposed model reduces the stress and tension of the users during their busy schedule, while they spend a lot of time searching the lost frequently used accessories.
- Easy to handle: Even small children can make use of it.

4.5 Limitations of the Proposed Model

Cost: This proposed model requires an individual transmitter and receiver for every accessory which are included in it. Though, it increases the amount to be spent for transmitters and receivers.

5 Further Enhancement

To overcome the only disadvantage of the proposed models, the cost involved in transmitters and receivers is reduced. Instead of using the individual transmitters and receivers for every individual item, as single transmitter and receiver has to be used as common for all the items. The block diagram explains the further enhanced model clearly. The architecture of enhancement model consists of three sections such as transmitting section and receiving. The transmitting section consists of individual switches for each item. The receiving section consists of respective alarms for the items. The output lines from the transmitter will get connected to a single receiver. Next the voice and playback circuit with the amplifier are connected to single receiver; it receives the incoming signal from the transmitter and sends it to the respective alarm which is on the receiving section.

6 Discussion About Proposal Model

In the existing method, if the mobile phone is missing, then the user can search it using the other mobile phone through making calls. At the time, both the mobile must be in the switch ON condition and also the network must be in coverage.

The proposed system will be work in the certain square feet distant. In this proposed system, the user does not need to check on the network condition and mobile conditions. Instead of that, the transmitter switch is turned ON, and then users can easily find the mobile by the rising of alarm. And also this method is very useful while the mobile phones are in silent mode and not in working condition.

7 Conclusion

This method mainly pointed out unique advantages of safety data transmission, flexibility in building the network and low cost. The topology structure of the system is an adaptation of a cluster tree. Compared with a reticular structure, a cluster tree structure can be built more easily and the information path takes less memory space. At the same time, the chain structure needs to be stable and its scale is limited, which needs to be improved in future investigations. People's day-to-day life becomes faster. Though, the user wants to save their time from spending it in searching the frequently used items. For example, if users want to go out urgently, at that time, they may miss house key, car key, or bike key, etc.. At this time, users spend their time for searching such items. To avoid such problem, the proposed model has been used for searching the lost items.

References

1. http://ccm.net/contents/834-wpan-wireless-personal-area-network
2. Broring, A. et al. New generation sensor web enablement. Sensors, 11, 2011, pp. 26522699. ISSN 1424-8220. Available from: doi:10.3390/s110302652
3. Sensai, "Integrating the physical with the digital world of the network of the future"
4. IEC, "Internet of Things: Wireless Sensor Networks", White paper, ISBN 978-8322-1834-1, 2014
5. R. Priya and G.P. Rameshkumar, "A Novel Architecture for Searching Items in Wireless Personal Area Networks (WPAN) with Zigbee Technology", Communications on Applied Electronics (CAE) – ISSN: 2394-4714, 2016
6. http://www.radio-electronics.com/info/wireless/zigbee/zigbee.php
7. https://www.thetileapp.com/press
8. Junguo ZHANG, Wenbin LI, Ning HAN, Jiangming KAN, "Forest fire detection system based on a ZigBee wireless sensor Network", Research Article, Front. For. China 2008, 3(3): 369–374, DOI 10.1007/s11461-008-0054-3
9. Nandini Ammanagi, Yogesh Prajapati, Hardeep Purswani, Milesh Janyani and Atharva Nandurdikar, "Automation in Air Traffic Control using ZigBee and Ultrasonic Radar", International Journal of Emerging Technology and Advanced Engineering, ISSN 2250-2459, ISO 9001:2008 Certified Journal, Volume 5, Issue 4, April 2015
10. P. Vijnatha Raju, R.V.R.S. Aravind, B Sangeeth Kumar, "Pollution Monitoring System using Wireless Sensor Network in Visakhapatnam", International Journal of Engineering Trends and Technology (IJETT) – Volume 4 Issue 4- April 2013, ISSN: 2231-5381
11. P.S. Jadhav, V.U. Deshmukh, Vidya Pratishthans College of Engineering, Baramati, Pune University, "Forest Fire Monitoring System Based On ZIG-BEE Wireless Sensor Network", International Journal of Emerging Technology and Advanced Engineering, ISSN 2250-2459, ISO 9001:2008 Certified Journal, Volume 2, Issue 12, Dec'2012

12. Rashid Hussian, Sandhya Sharma, Vinita Sharma, Sandhya Sharma, "WSN Applications: Automated Intelligent Traffic Control System Using Sensors", International Journal of Soft Computing and Engineering (IJSCE), ISSN: 2231-2307, Volume-3, Issue-3, July 2013
13. Lov K. Grover, "Quantum Mechanics helps in searching for a needle in a haystack", 1997
14. David Merrill and Pattie Maes, "Augmenting Looking, Pointing and Reaching Gestures to Enhance the Searching and Browsing of Physical Objects", International Conference on Pervasive Computing, 2007

A Novel System of Automated Wheel Bed with Android Phone and Arduino

R. Swathika and K. Geetha

Abstract The exigency of many people with ataxia can be fulfilled with hand operated or not automatic bed segment of ataxia people society chance on it hard or laborious to use bed. There is broad on computer-controlled bed where sensors and intellectual algorithms have been used to minimize the level of interference. This paper describes an innovative, motorized, voice controlled wheel bed for physically challenged people using the embedded system. Proposed design supports the voice activation system for physically challenged persons which also have manual operation. Arduino microcontroller and voice recognition processor have been used to support the guide of the wheel bed. The wheel bed does not respond to a wrong speech command. Microcontroller controls the wheel bed directions through the voice commands, at the same time it inactivates the wheel bed for the commands which are not trained and using ultrasonic sensors impediments are avoided.

Keywords Arduino uno • Bluetooth module • L293D motor driver
Gear motor • Wheel • Chasis • Bread board

1 Introduction

More than 0.06% of a people in India are physically challenged. There is a consistent growth in the number of physically challenged people in the past three decades [1]. They face more challenges in economic side, health concerns, and also to have recognition in the society. They need health care provider's skill for day-to-day activities which include transport, taking food, routine works. Voice-enabled control systems [2–4] control systems work based the eye movement

R. Swathika (✉) · K. Geetha
Department of Computer Science, Bharathiar University, Tamil Nadu, India
e-mail: Swathi19cs@gmail.com

K. Geetha
e-mail: geethakab@gmail.com

A.K. Somani et al. (eds.), *Proceedings of First International Conference on Smart System, Innovations and Computing*, Smart Innovation, Systems and Technologies 79, https://doi.org/10.1007/978-981-10-5828-8_25

has also been found in the literature. Applications [5] are controlled using ARM controllers were also developed for the disabilities.

These days, the subject of "web of things" is more well-known. Automation systems are widely used in applications like wheel chair, voice-enablled car driving, controlling robots through command, etc., which makes our lifestyle to a great extent. The main aim of the proposed automation system is used to reduce human effort, time, and easy access without the manpower to assist an ataxia-affected person to move to the direction as he/she desires these automated systems are extremely useful to lead a more sophisticated life with technology. Voice controlled wheel bed framework expects to empower the handicapped individual with no reliance on others.

With the aim to assist such people not to depend on care providers to some extent, the proposed model is designed to respond to the voice of the user to move the bed in four directions: right, left, forward, and backward. This system has been developed for huge amount of ataxia person in the world. The proposed automation voice recognizing wheel bed is very much used to regular life such as used in hospital, home, etc. Using this system, a patient can move in four directions by own without any assistance from the care taker by giving the directions as commands to the system. With the advent of new technologies, there is required a less need of training with the ability to use it in an easier way.

2 Literature Review

Bala Krishna and Nagendram [6] developed Wireless Home Automation System (WHAS) to provide voice command and control on/off status of electrical devices, such as lamps, fans, television, etc., in home environment using new communication technologies such as GSM/GPRS networks, wireless sensor networks, Bluetooth, power line carriers, and the Internet. They used the preprocessing stages such as capturing the command using a microphone, sampled at 8 kHz, filtered, and then the signals are converted to digital data using an analog-to-digital converter. The resultant data is then compressed and sent serially as packets to the Central Controller Module. A user interface has been developed using Visual Basic, running on the Personal Computer which uses Microsoft Speech API library for the voice recognition. Based upon the commands, home appliances will enact. They suggested to add more functions to improve the system versatility, integrating with smart phones with the help GSM.

Faisal Baig et al. [7] developed a hand gesture-based control command application using real-time video-based pointing method by integrating the technologies such as OCR analysis, smart phone, finger tracking, human–computer interface, and smart home–user interface. It has been designed in such a way to accept the commands through an installed camera to control the home appliances. They tested the application on the two test beds computer and mobile device and three types of communication links: dedicated wire, Bluetooth and Global System for Mobile (GSM). They suggested using the YCbCr color space to reduce the problem of recognitions in low impressive finger prints of old age users. Shabana Tadvi et al. [8] developed an android mobile application to interact with the wheel chair to work on four control operations to move right, left, forward, backward directions and to stop. They fitted a microcontroller in the wheelchair and enable communication from Android mobile the wheelchair via Bluetooth. Srishti et al. [9] developed a smart wheelchair which can be controlled by the head gesture as well as with the help of voice commands.

Sonali Sen et al. [10] implemented a mobile based application to instruct the microcontroller to switch on/off an appliance. They use Arduino Uno, HC-05 Bluetooth Module, Smartphone, and Voice Controller for the implementation of the microcontroller based system. The developed application Auto Home is an android based application which requires a Bluetooth module and relays circuit attached to the switch board so as to make an interaction between the smart phone and an appliance. Further, they suggested to include GSM module to extend the area to be covered. Mohammed Asgar et al. [11] proposed an Automated Innovative Wheelchair, moves according to the movement of the neck. For detecting the movement of the neck, they have used LEDs and photodiodes. They used IC HT12E and HT12D for encoding and decoding, TX-433 AM/ASK for wireless transmission and SM RX-433 AM/ASK for reception of signals, SL-HC-SR04 for detecting any obstacle coming in the path of wheelchair and Microcontroller 89V51RD2. They tested the system with four command start, stop, left and right.

Apsana et al. [12] designed and implemented a speaker dependent Voice Actuated Miniature Model of Wheelchair to assist physically challenged people. The developed system is trained with four commands. Microphone, microcontroller, motor driver, and two DC motors were used in the experiment.

Prathyusha et al. [13] Wheelchair driving is difficult to work even with normal person hands. The disabled persons cannot operate wheel chair with their hands due to lack of force such as orientation, mobility, etc., so robotic wheel chair has been developed to overcome this problem. Nikam et al. [14] the system design aims to reactive shared-controlled system to perform with unknown contours and terrains. The reactive and shared controlled system is used to move the wheel chair without manpower. It is based on the fuzzy-controller system. It has been designed with intelligent obstacle avoidance, collision detection, and contour avoidance through voice commands or joystick. Abed [15] developed a smart wheel chair using Arabic commands to operate the wheel chair.

3 Proposed System for Voice Controlled Automation System

The architecture of the proposed system is shown in Fig. 1, in which it has three major components: Smart phone used to give the instruction, microcontroller with arduino to analyze the voice input and to pass signals to the blue tooth receiver and a wheel bed with Bluetooth receiver. Other side of arduino cable is connected with

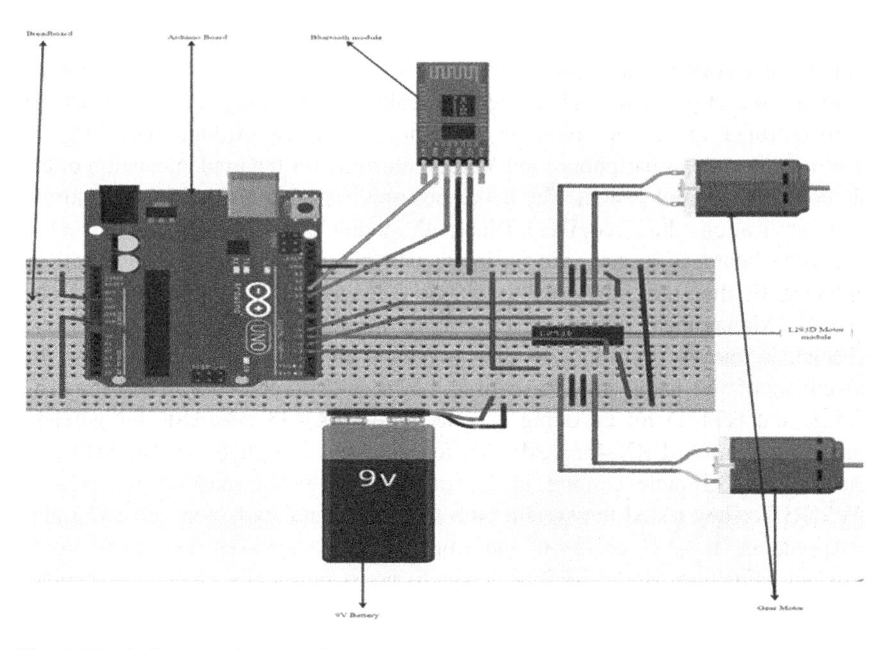

Fig. 1 Block diagram of proposed system

Fig. 2 The proposed system for automated wheel bed

a ground of breadboard from the ground of breadboard, the Bluetooth module is connected through cables. On the other hand, L293D motor module chip is connected on the breadboard. Pin 2 and Pin 5 of chip is connected to gear motor on both side, respectively. A 9 V battery is used in the experiment to provide power supply to system testing. Figure 2 shows the arrangement of wheel bed used in this proposed system.

3.1 Components of the Proposed System

A. **Arduino Uno R3**

The Arduino Uno R3 is a microcontroller board and it has 14 digital input/output pins, 6 analog inputs, a USB connection, a power jack, and a reset button. Arduino is an open source platform based on I/O board and it to view is shown in Fig. 3. It supports the microcontroller fully. The proposed system can be either connected through system or operated using a 9 V battery. The Arduino circuit acts as an interface between software and hardware [1]. Arduino uno have pins to adopt the power supply from 9 V battery through the breadboard. Arduino can be used to develop systems such command and control application in which it requires human machine interaction.

B. **Bluetooth module HC-05**

This Bluetooth module is used to design the setup for transparent wireless serial connection. In other words, this module can be used with Master or Slave configuration to enable wireless communication. These modules can be configured by commands. In these modules, the slave part does not not communicate with other

Fig. 3 Arduino uno R3

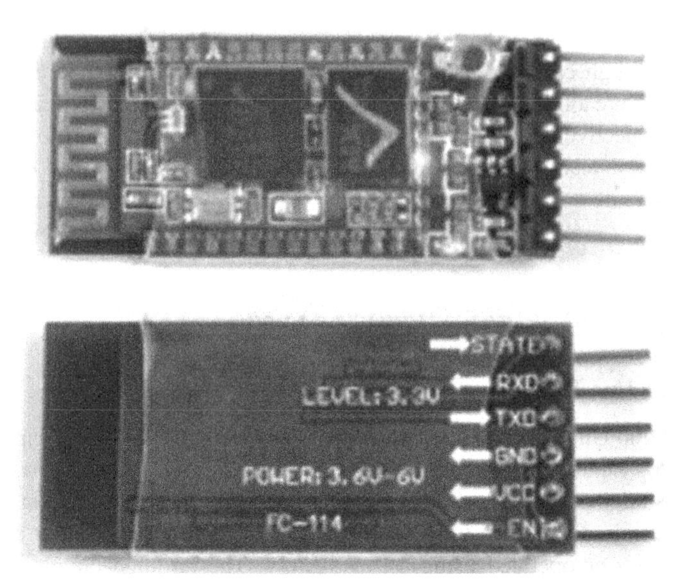

Fig. 4 Front and back pose of bluetooth module

Fig. 5 L293D motor module chip

devices, but accept all connections. Similarly, the master part of module is to initiate and establish the connection between devices. Figure 4 shows the top and bottom view of Bluetooth module.

C. **L293D motor module chip**

L293D is a typical Motor driver or Motor Driver IC which allows DC motor to move in either direction. L293D is a 16-pin IC which can control a set of two DC motors simultaneously in any direction. It is a dual H-bridge Motor Driver integrated circuit (IC). By setting the appropriate Voltage Specification, a L293D IC can be used to drive a small to quite big motors.

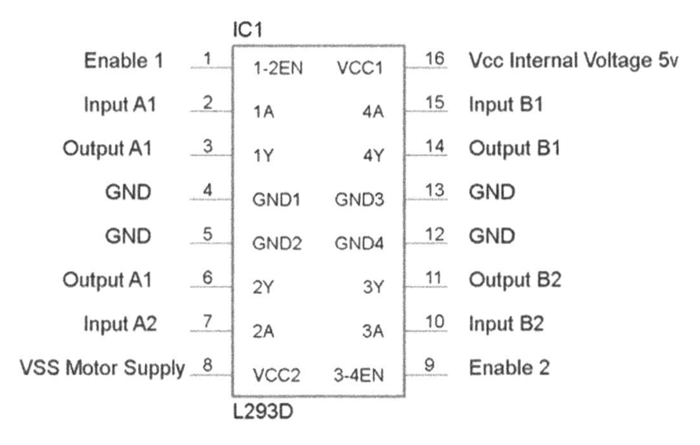

Fig. 6 L293D motor module chip description diagram

Table 1 Pin modes assignments for voice commands

Action	Pin 3	Pin 4	Pin 5	Pin 6
Forward	High	High	Low	Low
Backward	Low	Low	High	High
Left	High	Low	Low	Low
Right	Low	High	Low	Low
Stop	Low	Low	Low	Low

L293D IC works on the concept of H-bridge which allows the voltage to be flown in either clockwise or anticlockwise direction. Since it possesses this capability, H-bridge IC is ideal for driving a DC motor. Inside L293D chip, there are two H-Bridge circuits which can rotate two dc motors independently. This type of ICs is used in a robot for controlling DC motors. The picture of the L293D IC and its pin configuration is given the Figs. 5 and 6, respectively.

L293D IC is a 16 Pin IC which has two Enable pins Pin 1 and pin 9. To active the DC motor, both the pins need to be high. If Pin 1 alone set to high, motor drives in counterclockwise and if pin 9 alone set to high, the motor derives in clockwise direction. If both pin 1 or pin 9 goes low, then the motor in the corresponding section will suspend working. Which is switch off state? Similarly, to make both pins high, simply connect the pin16 VCC (5 V).

3.2 Working Model of the Proposed System

Pin modes description used in the proposed model is shown in Table 1. The action to be forward means, the pin 3 and Pin 4 should be kept high and the other two pins

Step: 1 Establish connection between Bluetooth module and smart phone through voice control bot android application.

Step: 2 Assign pin configuration for the commands (Ref. Table).

Step: 3 Send the command through smart phone.

Step: 4 Micro controller recognizing the voice and control the control the device.

Step: 5 Got to step 3 until the received command is stop.

Fig. 7 Algorithm of proposed system

Fig. 8 Connection establishment between proposed system and android mobile

Fig. 9 Left commands to activate the wheel bed

Fig. 10 Right command to activate the wheel bed

Fig. 11 Commands to end the process

should be low. Similarly, the description for the backward, Left, Right, and Stop operation is defined in the Table 1.

Arduino uno is connected with Bluetooth module H505 and L293D motor module chip through the cable. At the other end, a voice control bot android application is installed in the smart phone. First, the system has been initiated to enable the Bluetooth so as to make communication. Bluetooth should keep on

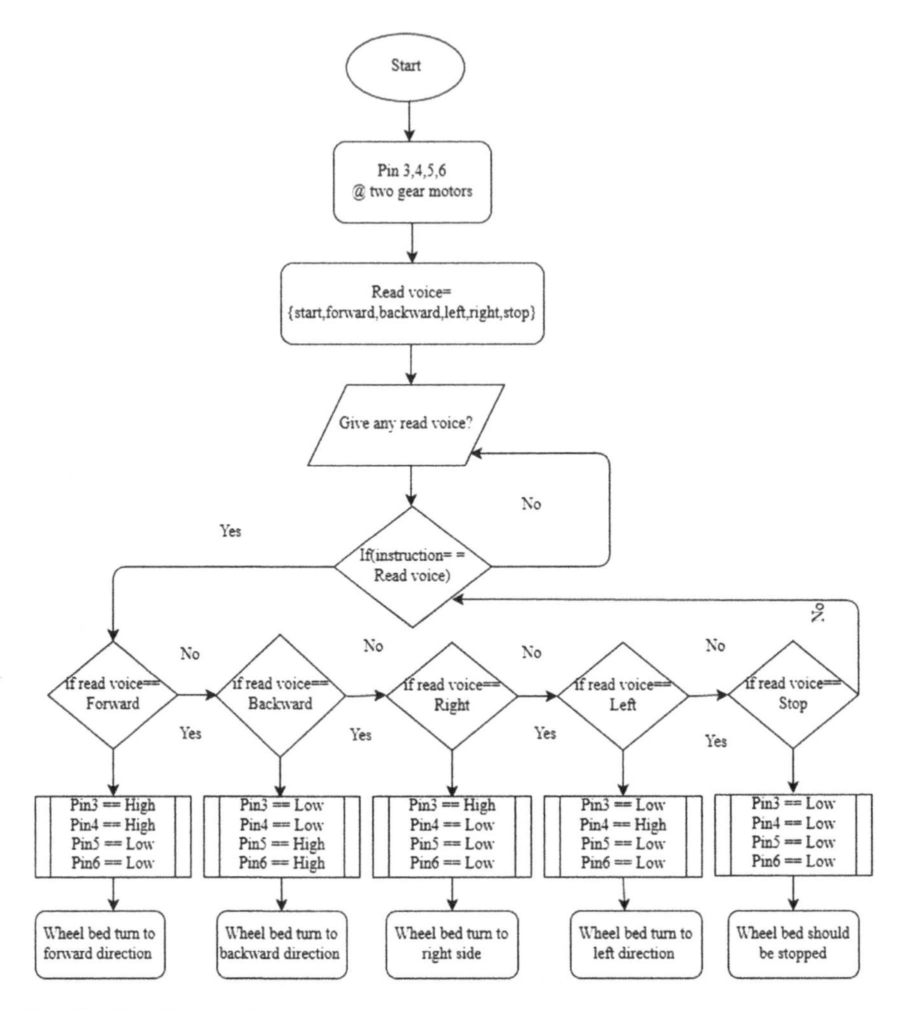

Fig. 12 Flow diagram of proposed system

"ON" position. After connection established, the name of Bluetooth module is displayed on the smart phone screen such as H505 that means the connection has been established successfully. Then user selects "CONNECT" option. Now, the user issue the command such as forward, backward, right, left, stop through a smart phone. At the same time, the Arduino recognizes the voice, interpret the corresponding command, and issue a proper command to the microcontroller for activation of the DC motor. Now, the system has activated according to the command it received. For example, when user said forward the command to the smart phone, the system will move to the forward direction. The Algorithm used in the work is presented in Fig. 7.

Under the chassis board, the system has three numbers of rotating wheel of left side corner, right side corner, and center of the front side, respectively. Two gear motors are fixed in the left and right side wheel. Remaining parts are fixed on the chassis board, i.e., arduino, L293D motor module, batteries are connected on the chassis board. A 9 V battery is used for the experiment conducted. Figures 8, 9, 10, and 11 show the commands to activate the proposed system. The diagram explains how the commands are worked in the voice bot application. The flow diagram of the proposed system is presented in the Fig. 12. The interpretation of the User commands is as follows.

1. Forward: Wheel rotates in the clockwise direction with motor.
2. Backward: Wheel rotates in anticlockwise direction with the motor.
3. Left: both wheels are turned to the left side.
4. Right: both wheels are turned to the right side.
5. Stop: both motors are stopped.

4 Experiment and Results

This experiment is carried out by testing 200 times by five speakers and the system produced an average of 96.5% accuracy. Every speaker checked the system with all the commands by eight times. Whenever the speaker gives command such as

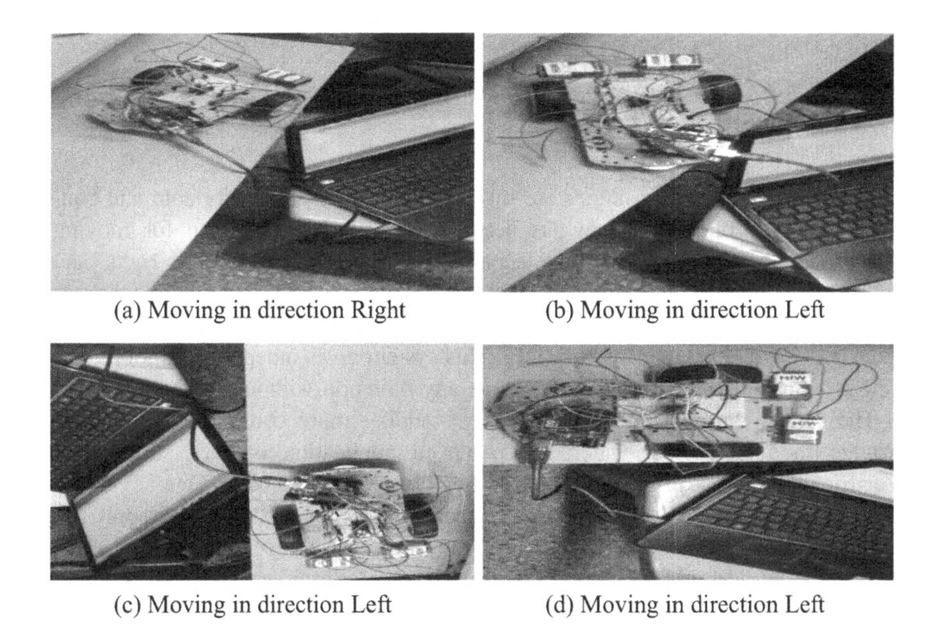

(a) Moving in direction Right

(b) Moving in direction Left

(c) Moving in direction Left

(d) Moving in direction Left

Fig. 13 Working of proposed system with various commands

Table 2 Performance of the proposed system

Speaker	No. of words spoken by the speaker	No. of times system properly worked	Correctness (%)
P1 (Female)	40	36	90
P2 (Male)	40	38	95
P3 (Male)	40	40	100
P4 (Female)	40	39	97.5
P5 (Female)	40	40	100

forward, backward, right, left, and stop the movement of the wheel bed is shown in the Fig. 13. The performance of the proposed system is presented in Table 2.

Advantages of proposed system

- Less hardware requirements needed.
- System is very compact.
- System is very comfortable to the disabled person at hospitals or home.
- It is user-friendly system.
- Handicapped person can use the system with voice independently.
- User can use this system with very low budget.
- The user can use the system without manpower.

5 Conclusion

Nowadays, the new technologies are made human life with comfortable and convenient. This proposed system has been successfully implemented for moving wheel bed at a hospital or home such as forward, backward, left, right, and stop. This work is to be used to disabled persons by providing alternative methods to controls this system through voice. The smart wheel bed using voice recognition is designed and developed successfully. This system gives independent mobility as well as many intelligent facilities to move any direction without any help.

The system further can be enhanced by adding more commands such as to provide food, medicine, water, etc. Android application is very simple and user-friendly and it can be controlled its functions within seconds. One advantage of the proposed system is very flexible and scalable. Hence, the system not only can be used at home or hospital but also in many offices wherever it is needed. This system can be designed in such a way to raise up alarm or send a message to the phone according to the need of the user. In case the patient slipped from the wheel bed, an automatic alarm with beep sound may also be proposed.

References

1. http://www.census.india.gov.in, ArduinoUnoProjects.:http://arduino.cc/en/Main/arduinoBoardUno
2. Jayesh, Kokate, K., Agarkar, A. M:. Voice operated wheel Chair, IJRET: International Journal of Research in Engineering and Technology eISSN: 2319-1163 I pISSN: 2321-7308, Volume: 03 Issue: 02 I Feb-2014
3. Thangadurai., Kartheeka.: Intelligent Control Systems for Physically Disabled and Elderly People for Indoor Navigation, International Journal for Research in Applied science and Engineering Technology, Vol. 2 Issue IX, September 2014, ISSN: 2321-965
4. Kohei Arai., Ronny Mardiyanto.: Eyes Based Eletric Wheel Chair Control System, (IJACSA) International Journal of Advanced Computer Science and Applications, Vol. 2, No. 12, 2011
5. Sudheer, K., Janardhana rao, T.V., Sridevi, C.H., Madhan mohan, M.S.: Voice and Gesture Based Electric-Powered Wheelchair Using ARM, International Journal of Research in Computer and Communication technology, IJRCCT, ISSN 2278-5841, Vol 1, Issue 6, November 2012
6. Bala Krishna, Y., Nagendram, s.: Zigbee Based Voice Control System for Smart Home. IJCTA I JAN-FEB 2012, ISSN:2229-6093
7. Fsaisal Baig., Saira Beg., Muhammad Fahad Khan., Syed Junaid Nawaz.: A Method to Control Home Appliances Based on Writing Commands Over the Air: J Control Autom Electr Syst (2015) 26:421–429, Brazilian Society for Automatics–SBA (2015)
8. Shabana, E.R., Tadvi., Prasad Adarkar., Arvindkumar Yadav and Saboo Siddik. M.H.: Automated Wheelchair using Android Technology, Imperial Journal of Interdisciplinary Research (IJIR) Vol-2, Issue-4, 2016 ISSN: 2454-1362
9. Srishti., Prateeksha Jain., Shalu., Swati Singh.: Design and Development of Smart Wheelchair using Voice Recognition and Head Gesture Control System, International Journal of Advanced Research in Electrical, Electronics and Instrumentation Engineering, ISSN (Print): 2320–3765, ISSN (Online): 2278–8875 Vol. 4, Issue 5, May 2015
10. Sonali Sen., Shamik Chakrabarty., Raghav Toshniwal., Ankita Bhaumik.: Design of an Intelligent Voice Controlled Home Automation System. International Journal of Computer Applications (0975–8887) Volume 121–No.15, July 2015
11. Mohammed Asgar., Mirza Badra., Khan Irshad., Shaikh Aftab.: Automated Innovative Wheelchair, International Journal of Information Technology Convergence and Services (IJITCS) Vol.3, No.6, December 2013
12. Apsana. S., Renjitha G Nair.: Voice Controlled Wheelchair using Arduino, International Advanced Research Journal in Science, Engineering and Technology, ISSN (Online) 2393-8021,ISSN (Print) 2394-1588
13. Prathyusha, M., Roy, K.S., Mahaboob Ali Shaik.: Voice and Touch Screen Based Direction and Speed Control of Wheel Chair for Physically Challenged Using Arduino, International Journal of Engineering Trends and Technology (IJETT), Volume4Issue4- April 2013,ISSN: 2231-5381
14. Nikam, D.S., Joshi Gauri, Shinde Mohini, Tajanpure Mohini, Wani Monika.: Voice Controlled Wheelchair Using AVR, International Journal of Modern Trends in Engineering and Research, Scientific Journal Impact Factor (SJIF): 1.711, e-ISSN: 2349-9745 p-ISSN: 2393-8161
15. Abed, Ali. A.: Design of Voice Controlled Smart Wheelchair, International Journal of Computer Applications (0975–8887) Volume 131 – No.1, December2015

A Secure File Transfer Using the Concept of Dynamic Random Key, Transaction Id and Validation Key with Symmetric Key Encryption Algorithm

Saima Iqbal and Ram Lal Yadav

Abstract As there is enormous increase or rise in the data exchange by the electronic system, the requirements of information security have become a compulsion. The most important concern in the communication system which is between sender and receiver is the security of the information which is to be transmitted. To get rid of the intruders, various cryptographic algorithms are used for example, AES, DES, Triple DES, Blowfish, etc. In this paper, authors have attempted to transfer file securely using randomly generated keys with Blowfish symmetric algorithm for encryption process. A unique transaction id is also generated which is used to fetch the random key which will be used for decryption at the receiver end. A validation key is also proposed which is used to validate that file is decrypted by the intended user only. The combination of random key, validation key, and transaction id as proposed will provide more robustness to secure transfer of file using Blowfish symmetric key encryption.

Keywords Blowfish · Transaction ID · Random key generation

1 Introduction

In the field of the information security, the cryptography algorithms play a very important role. These cryptographic algorithms are categorized into the following two categories, "Symmetric and Asymmetric key cryptography [1]".

In the case of symmetric key encryption [2], a single key is used to encrypt as well as to decrypt the data. In such types of the encryption techniques, the key has the most important role in the process of encryption and decryption of the data. The use of a weak key in the algorithm will let out data to be decrypted very easily. The

S. Iqbal (✉) · R.L. Yadav
Kautilya Institute of Technology and Engineering, Jaipur, India
e-mail: saimaiq1@gmail.com

R.L. Yadav
e-mail: ram.bitspilani@gmail.com

A.K. Somani et al. (eds.), *Proceedings of First International Conference on Smart System, Innovations and Computing*, Smart Innovation, Systems and Technologies 79, https://doi.org/10.1007/978-981-10-5828-8_26

power of the [2] Symmetric key encryption will depend upon the strength of the key, weaker the key, weaker the algorithm, and vice versa. In a similar way, the symmetric algorithms can be classified into two main categories: "block ciphers and stream ciphers [2]". The working of block ciphers algorithms is that it operates on data in forms of groups or blocks. Examples of Symmetric Key Encryption are the Data Encryption Standard (DES), the Advance Encryption Standard (AES), and Blowfish. While in the case of the asymmetric key encryption, we have two types of keys which are used in the process, namely, private keys and public keys. In this, the public key is used in the process of encryption while the private key is used for the decryption process. Example of such an asymmetric encryption is Digital Signatures. Public key is known to all while the private key is known only to the user.

In this paper, we will focus on the creation of the Modified Blowfish algorithm and on the applying the concept of the dynamic key [3] using the concept of the random key generation [4]. Blowfish algorithm was designed in the year 1993 by Bruce Schneider. Blowfish is the symmetric key block cipher, and it makes use of the 64-bit block size and the key is variable in length of 32–448 bits. The Blowfish algorithm has 14 rounds or less. Blowfish algorithm is patents free and there are no copyrights. No attack has been known on this algorithm till date but it has weak keys problem.

2 Dynamic Key Concept

Dynamic keys [3] are one-time symmetric cryptographic keys. Each message in the framework is scrambled by alternate cryptographic key. Therefore, any endeavors to assault the cryptographic framework by re-utilizing a traded off cryptographic key can be easily detected (Fig. 1).

Fig. 1 Difference in session keys [3] and dynamic keys

	Dynamic Key	Session Key
Key Exchange	Once	Every Session
Life time	Within a message	Within a session
Key Reusable	No	Yes
Vulnerable under man in middle attack	No	Yes
From a compromised cryptographic key, adversary can	Decrypt a message	Decrypt all messages in the session
From a compromised pair of public and private keys of the key exchange protocol	Cryptographic system is still safe	Cryptographic system and session are vulnerable

2.1 Synchronization Problem in Dynamic Keys

Dynamic keys [5] must always be same between sender and receiver by use of symmetric cryptography. When these keys differ, communication is hindered because the receiver is unable to decrypt the message received from the sender. This issue is termed as synchronization problem [3].

2.2 Advantages of Using Dynamic Keys

The use of dynamic keys has improved the security of cryptographic system. The probability of breaking cryptographic system has been reduced to 1/2 ms + s [3], from 1/2 s where s is the length of the cryptographic key in bits and m is number of parameters use in creation of dynamic keys.

3 Related Work

Due to congestion in network security is a very important concern. In [6] a dynamic key is made through an LCG [6] for cryptographic algorithm so that for every encryption and decryption a unique key is used and thus it will be hard to break.

In [7] paper the author has discussed Blowfish algorithm, that it is variable-length key block cipher. And in this, he has described in details the working of the Blowfish algorithm and applications where the Blowfish algorithm is used. For this paper, we get the idea regarding the key size which is used for encryption and in the decryption process and rounds which are performed in the data encryption and final output which we get from that.

In [8] the requirement of the cryptographic algorithm in the field of the information security is discussed. In this paper, the author has discussed in details plaintext attack against a variant of Blowfish that is made easier by use of a weak key. Blowfish is more secure and fast processing algorithm. But in this paper the author also identified some problem in the existing Blowfish algorithm, i.e., the Blowfish weak keys produce "bad" S-boxes.

In [9] symmetric key algorithms: DES, the 3DES, the AES, and also the Blowfish are compared on the basis of various parameters [9] that proves Blowfish is much better as compared to other algorithms.

4 Dynamic Key Concept

Phase 1: Encryption of File—In the proposed work, we have first used the random key for the key which is used in the Blowfish encryption and decryption process and the algorithm for the random key generation process is as follows:

Algorithm for Random Key Generation

```
Step 1: Initialize an array KeyChar with the hexadecimal values such that,
KeyChar [1] ='a', KeyChar [2] ='b' and so on KeyChar[16]='9'.

Step 2: Repeat Step 3 to for I: = 1 to 32 do Step 3: Set index: = Rand (1:16).

Step 4: Set s:= s + KeyChar[index]. [End of for loop]
```

Now s will contain the random key generated via the above-mentioned algorithm. In the proposed concept we have created the following framework for the message transferring from the sender and receiver using the modified Blowfish algorithm (Fig. 2).

Phase 2: Sending File—In this, the transaction Id will be generated which is unique for the transfer of the file and it will be used for accessing the random key used for the encryption for the file sharing from the centralized database. In this form in order to further validate the transfer we have used the concept of validation key, which is generated using the following algorithm:

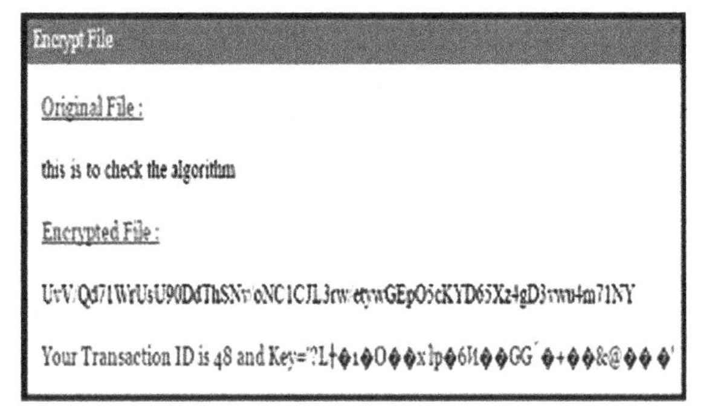

Fig. 2 Encryption form 1

Algorithm for Generating the Validation Key

```
Step 1: Access the Name of Sender and Receiver.
Step 2: Calculate the length of sender username.
Step 3: Calculate the length of receiver username.
Step 4: Encrypt all characters of the sender user name by
        adding up the   length of sender username in the
        characters corresponding to the sender username.
Step 5: Encrypt all characters of the receiver user name
        by adding up the length of receiver username in
        the characters corresponding to the receiver
        Username.
Step 6: Concatenate the string patterns obtained from the
        step 4 and step 5 and the resultant pattern is
        the TRANSFER KEY.
```

We have provided proper validation for unique username while registrations. If any intruder tries to decrypt a file which was not sent to him than validation key will be generated for his name while decrypting and it will not match with the validation key which was made by above algorithm at sending side. File decryption will not be possible and the only intended user will decrypt it.

Phase 3: Receiving and Decrypting File—In this form, the transaction id entered by the user will fetch the key for decrypting the file and another validation key formed from username involved in the transfer is validated. If the validation key matched the file will be decrypted as shown in Fig. 3.

Otherwise, if another user tries to decrypt the file he cannot decrypt the file (Fig. 4).

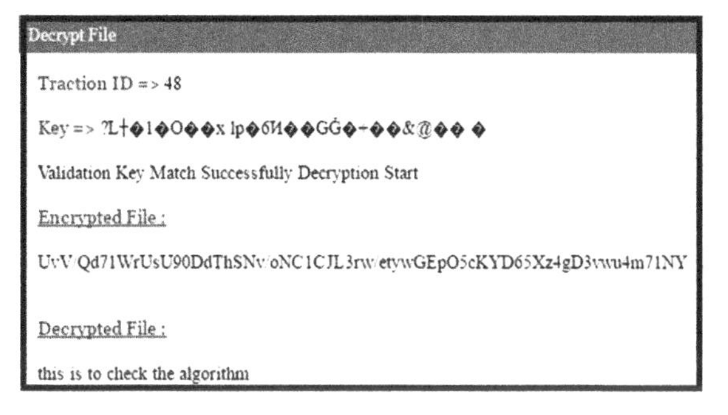

Fig. 3 Decrypted file form

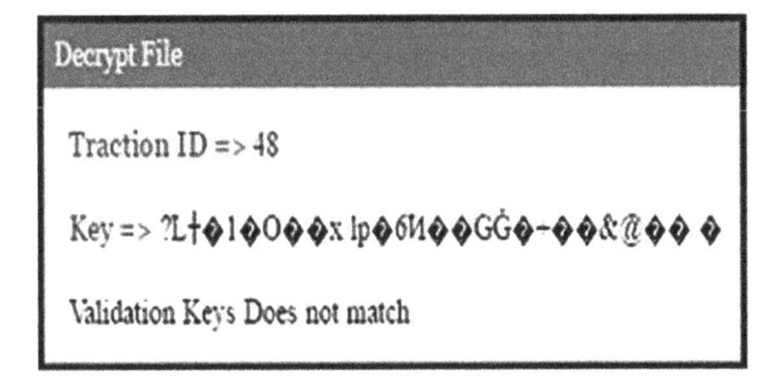

Fig. 4 Error decrypted file form

5 Result Analysis

In the base concept, we have security constraints which we have enhanced using the transaction key and the validation key. Synchronization problem in the dynamic key concept has also been solved with transaction key.

The disadvantages of Blowfish are that it must give the key to the person out of band specifically [10] not through the unsecured transfer channel. So it becomes very mandatory to improve this algorithm by adding new levels of security to make it applicable and can be depended on in any common communication channel. The validation key and random key as proposed by this paper to give more robustness to secure file transfer using Blowfish algorithm and give it strength against intrusion of any kind.

6 Cryptanalysis

Chosen-plain text and chosen-cipher text attacks [11] are performed on private key encryption algorithms. Both attacks are related to selecting a part of the information from a given encrypted text to get an original message or from given plain text to reach encrypted text. In the proposed concept, the key is unique for each block, chosen-plain text, and chosen-cipher text attacks are hard to launch. In differential attack [11] an input is compared with an output value to reach a key. Since the proposed key is made from a strong key, a differential attack is hard to achieve. A distinguishing attack [11] is focused on stream ciphers in which for checking the randomness a given sequence of values are compared. In proposed dynamically generated key there will be no relation between the previous and current keys, so a distinguishing attack is difficult to launch.

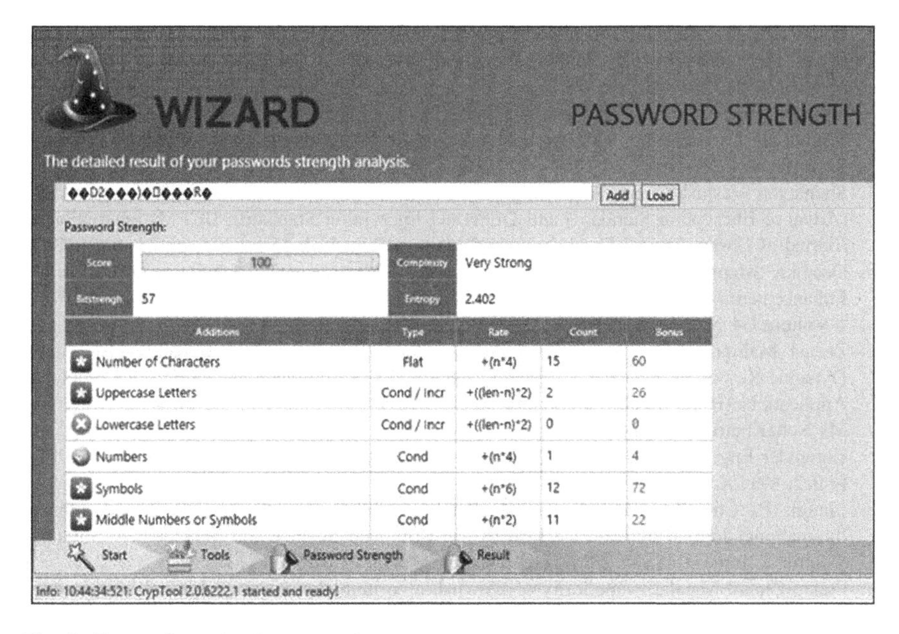

Fig. 5 Test performed using cryptool

7 Key Sensitivity Test

We have used a tool known as Cryptool 2.0 to examine the strength of the keys used in our process.

And the result of the analysis using that tool is shown in the Fig. 5.

8 Conclusion

In last we can say that Modified Blowfish algorithm will further enhance the security aspect for the transfer of data between the sender and receiver. The usage of the random key, transaction id, and the validation key have further boost up the security aspect.

References

1. Himani Agrawal and Monisha Sharma: Implementation and analysis various symmetric cryptosystems, Indian Journal of Science and Technology in Vol. 3 No. 12, ISSN:0974-6846. pp. 1173–1176, (Dec 2010).

2. Pratap Chandra Mandal: Evaluation of performance of the Symmetric Key Algorithms: DES, 3DES, AES and Blowfish, Journal of Global Research in Computer Science, Volume 3, No. 8, (2012).
3. Huy Hoang Ngo, Xianping Wu, Phu Dung Le, Campbell Wilson, and Balasubramaniam Srinivasan: Dynamic Key Cryptography and Applications, International Journal of Network Security, Vol. 10, No. 3, PP. 161–174, (May 2010).
4. Shantayoti Vashishtha on: Evaluating the performance of Symmetric key algorithms: AES (Advance Encryption Standard) and DES(Data Encryption Standard), IJCEM, International Journal of Computational Engineering and Management, Vol. 15 Issue 4, pp 43–49(2012).
5. Deepika, Manpreet: A Review on Various Key Management Techniques for Security Enhancement in WSN, International Journal of Engineering Trends and Technology (IJETT) – Volume 34 Number 4-(2016).
6. Zeenat Mahmood, J. L Rana, Prof. Ashish khare: Symmetric Key Cryptography using Dynamic Key and Linear Congruential Generator (LCG), International Journal of Computer Applications (0975–8887) Volume 50– No.19, (2012).
7. Ms NehaKhatri – Valmik and Prof. V. K Kshirsagar: Blowfish Algorithm, IOSR Journal of Computer Engineering (IOSR- JCE), e-ISSN: 2278-0661, p- ISSN: 2278-8727 Volume 16, Issue2, Ver. X, PP 80–83, (2014).
8. Maulik P. Chaudhari and Sanjay R. Patel: A Survey on Cryptography Algorithms, International Journal of Advance Research in Computer Science and Management Studies, Volume 2, issue 3, ISSN: 2321-7782, (March 2014).
9. Pratap Chand Mandal: Superiority of blowfish algorithm, International Journal of Advance Research in Computer Science and Software Engineering.
10. Blowfish-algorithm-history-and-background-computerscienceessay.phpSmith, T.F., Waterman, M.S.: Identification, https://www.ukessays.com/essays/computer-science.
11. Jayagopal Narayanaswamy, Raghav V. Sampangi and Srinivas Sampalli: HIDE: Hybrid Symmetric Key Algorithm for Integrity Check, Dynamic Key Generation and Encryption, ICISSP 2015-1st International Conference on Information System Security and Privacy (2015).

Smart Control Based Energy Management Setup for Space Heating

Mohammad Zeeshan

Abstract The smart homes are the need of emerging infrastructure in order to reduce the energy needs of society. Room heating is among the major source of energy consumption in such homes. Hence a setup is required in centrally controlled homes to manage the energy consumption. One such arrangement has been discussed in which several heaters in a single room and different rooms can be controlled along with a smart thermostat relay control algorithm so that energy is properly utilized. The controllers have been connected using Zigbee communication for the sake of simplicity and efficiency. The indoor and outdoor temperatures are monitored using appropriate sensors. Alongside, the energy savings and the heating duration in a typical Indian scenario is discussed.

Keywords Smart home · Space heating · Control

1 Introduction

Smart homes are the need of today in the world of smart metering and control. Smart homes can be of great convenience as we can now control the indoor temperature, lightning needs, and door access [1]. The green houses are being promoted by the respective governments to counter the problem of increasing shortage of energy and environmental benefits. Real-time pricing electricity tariff is also introduced as mechanism to encourage consumers to decrease their energy consumption. Energy management system [2] is very essential for implementation in the smart home network. In order to achieve our goals of reduction in energy consumption, the EMS is employed for operation of controllable devices keeping in mind the price of electricity. Room heating is the most primary cause of energy consumption in the homes. According to Planning commission report 2014 [3],

M. Zeeshan (✉)
Faculty of Engineering and Technology, Jamia Millia Islamia, Jamia Nagar, New Delhi 110025, Delhi, India
e-mail: zee123king1@gmail.com

A.K. Somani et al. (eds.), *Proceedings of First International Conference on Smart System, Innovations and Computing*, Smart Innovation, Systems and Technologies 79, https://doi.org/10.1007/978-981-10-5828-8_27

the residential and commercial sectors account for 29% of the total electricity usage. Energy-efficient building designs will lead to reduction in lightning, heating, and other requirements.

The commercial structures in India mostly have an Energy Performance Index in the range of 210–410 kWh/sq/m each year. The design of energy aware building design reduces it to 105–154 kWh/sqm/year. The Indian construction market consists of 75% of residential sector buildings.

The energy consumption in a building is reliant on factors such as open-air climate, erection of the building, modes of operation (i.e., naturally ventilated or air-conditioned), hours of occupancy, occupants comfort, and home appliances. Approximately 55% of the total energy in household buildings is used for heating and cooling purposes and this proportion is supposed to increase constantly in the coming years. Thus, the supervision of the space heating and cooling systems is perilous to the EMS. The various space heaters that are available in the marketplace are furnished with simple controllers [4] which control the temperature in the house and if their mechanism is automatic, it means they can turn on or off whenever the temperature of house rises or falls below a certain level. Some space heaters consist of timers through which people can adjust the time of operation.

In Fig. 1, we can see the space heater circuit diagram. The current passing through the resistance produces heat. The heating power can be calculated using the Joule Effect, and described as $P = I^2R$.

A centrally controlled unit is required to control the several heaters that are present at different portions of the house. The function of the central controller is to turn on/off the heaters in specific time delays through the relay in the heaters.

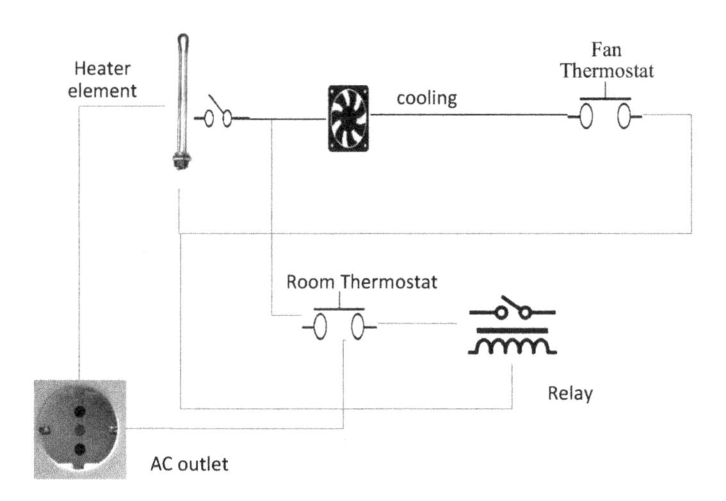

Fig. 1 A general setup of a thermostat controlled room heater

2 Energy Management System

We have proposed a home energy management system (HEMS) which reduces the cost required to run energy intensive equipments, i.e., in our case the heating/cooling system by operation scheduling of space heaters.

Figure 2a shows the schematic diagram for the proposed management system [5]. The system includes two monitoring and control units which will observe the indoor temperature. The outdoor node will monitor the temperature of external environment to the house. We need one portable CPU which runs the EMS program.

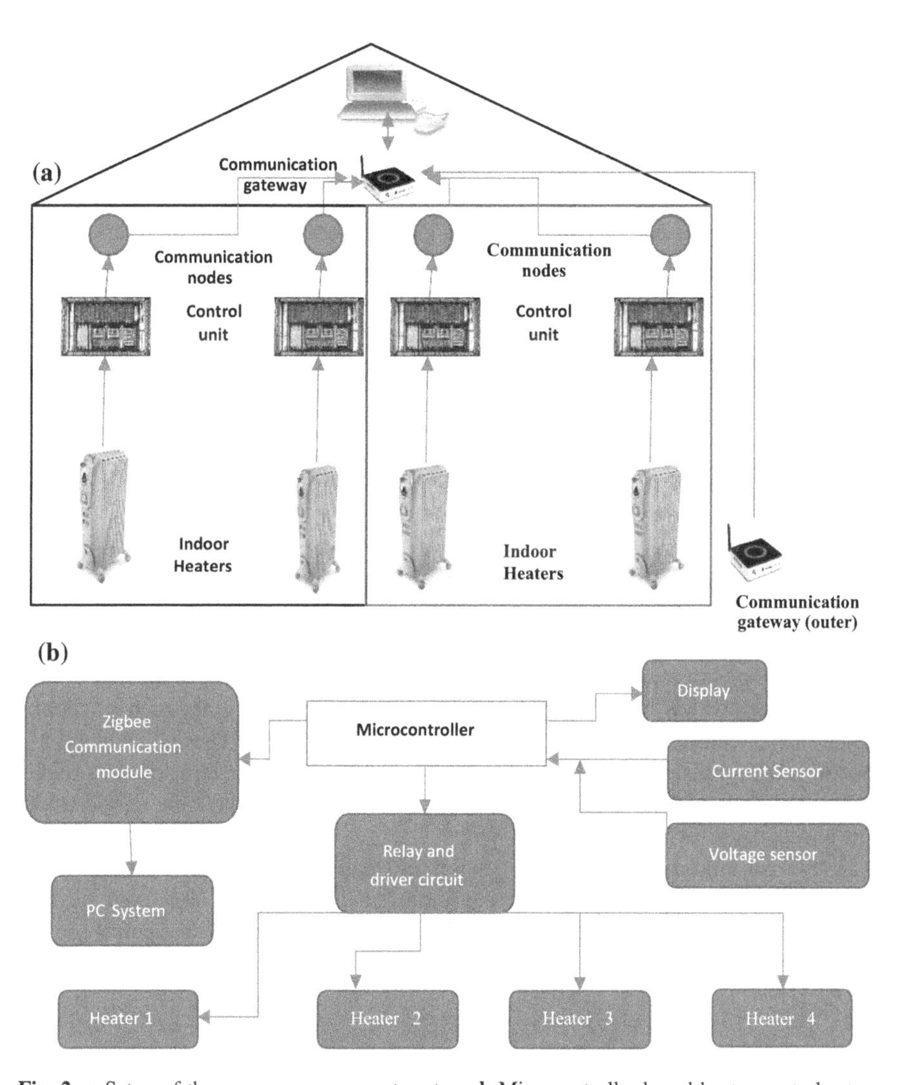

Fig. 2 **a** Setup of the energy management system. **b** Microcontroller based heater control setup

Zigbee wireless transmission technique [6, 7] is used here to transmit data as it is the most efficient and fast technique. Every space heater is connected to a monitoring and control unit and this will be used to monitor the temperature inside the house and this will also record the energy consumption of the heaters and control their operation. The outdoor temperature is also transferred to the EMS circuitry using a communication gateway. The functions of the gateway are to receive data, optimizing it, and then to broadcast control commands.

The graphic user interface for the home energy management system is developed in Visual C++. This GUI is saved in the portable CPU which holds the HEM program [8]. This GUI acts as a console so that the owner of the house can easily monitor the heater's status, the power consumption of all the heaters, all the room's ambient and hot temperatures, and the demand limit. These values are updated every 1 min. Options to change the priority of different loads like, for example, one room is required warmer than the other and other settings can be configured easily. The chosen load controller consists of a microcontroller unit (MCU) with a digital display as shown in Fig. 2b.

It can display measured voltage, current, pf, the real and apparent power in real basis.

The controller device has an embedded ZigBee communication module. It also contains a power relay used for switching ON/OFF the linked load.

There are two Zigbee modules in our setup–one is installed in the home energy management unit and the other is present in the load controller and these are used for communication.

3 Functionality of the Energy Management Setup

We have used four loads at different portions of the laboratory. We will be using one power outlet, and one ZigBee hub. Devices are installed at power outlet for power measurement.

The functionality of these devices is to periodically send data to the Zigbee hub through Zigbee communication gateway. The heaters in itself are connected to the power outlets which have the power measurement function embedded in them; hence power can be directly measured through such outlets. The gathered information is then transferred to the home server which analyses the temperatures of all heaters whether in same room or different rooms. Then it displays the temperature and power usage of each heater.

A user can find out by looking at the server the power each heater consumes in each month by the gathered power consumption data. He can also examine the power usage for each room through the ZigBee hub. He can use the central server and then turn off unnecessarily turned on heaters manually.

4 Heater Smart Control

The timer-based control system can be used to keep the temperature between a certain temperature range [9]. Whenever the temperature rises or falls below this value, the temperature sensor [10] senses this change and brings back the temperature within the given range. If the temperature of the heater is above the onset limit for the required period, the power channel can automatically cut off the AC power and reduce the supplementary power of the heater unit.

A single room is equipped with two heaters of wattages 1000 W. One heater is used to improve the transient condition of the room quickly according to the current room temperature. For instance to provide the initial heat to raise the temperature of the room from a cooler condition to temperature to be reached within the set limits.

It is also used to speedup the operation of heating of the room by a factor of two. After the temperature of the room reaches the prescribed limit, the 1000 W heater maintains the heat within the limit while the other heater is shut down to save useful energy.

The periods of operation of the radiators are configurable through the central control panel, and we can now turn on/off the timers as and when the time signals are triggered. The control system is shown in Fig. 3.

5 Results

5.1 Cost of Energy Consumption (Indian Scenario)

A 1000 W heater running for 1 h equates to 1 unit of electricity and 2000 W heater running for 10 h equates 20 units. The total load = 2 kW and for 1 day equates to 16 units spent. For 30 days the energy consumption is 480 units/month leading to 4.65 Rs/kWh.

5.2 Room Heating Energy Requirement

Volume of air in one room computes to approx 8000 cu ft according to Table 1.

The amount of energy required to raise the temperature of the room by 1 °C equates to the product of specific heat capacity of air, mass of air and temperature difference [11]. The specific heat capacity of air is equal to 1 kJ/kg °C while the density of air is 0.036 kg/cu ft. Hence the mass of air inside the room is calculated to be 288 kg. The temperature differential is 1 °C.

Hence energy needed to reach the desired temperature equates to 289 kJ. A 1000 W heater can transfer 1000 J of energy/sec. The time required to

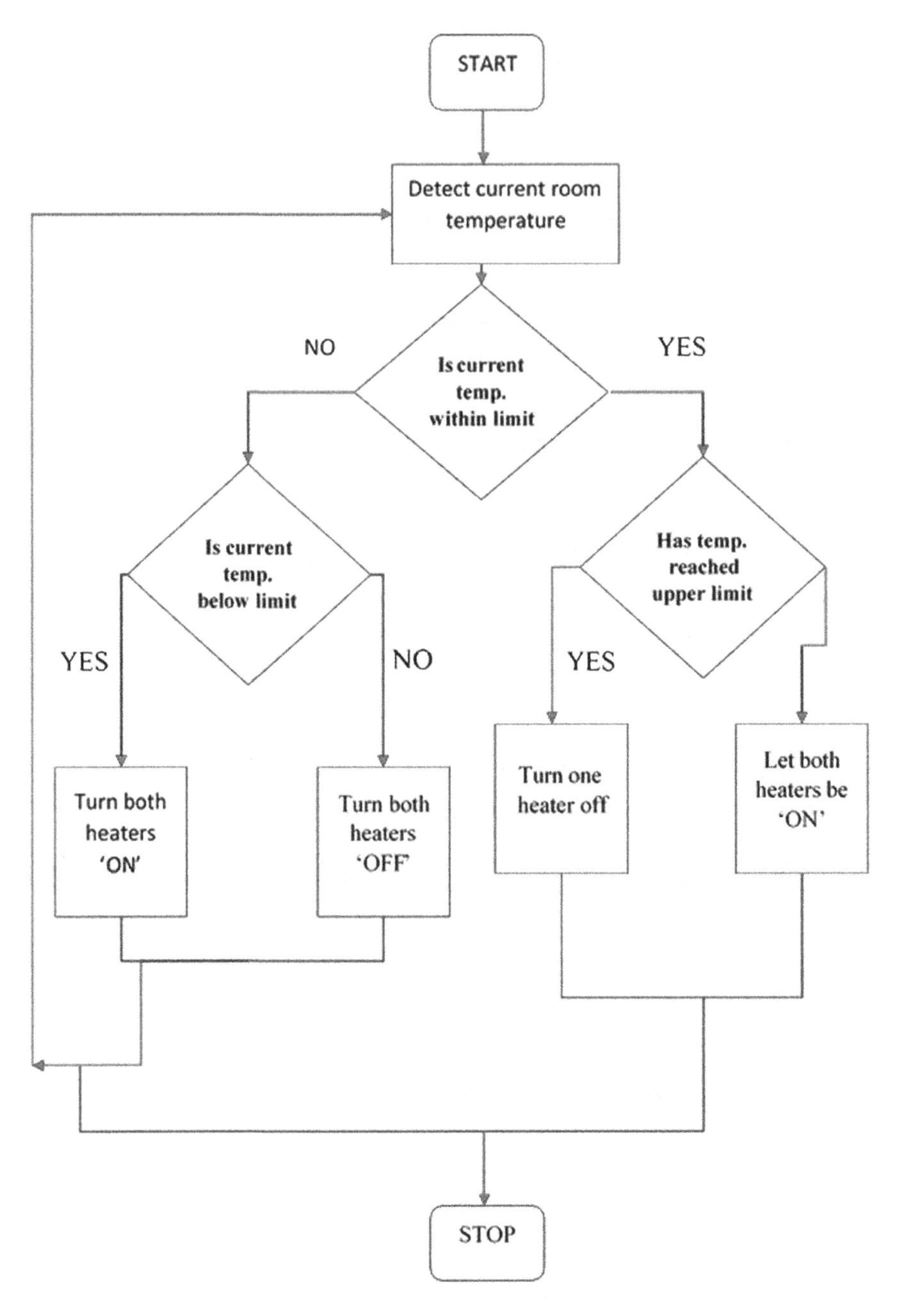

Fig. 3 Smart thermostat based control system

Table 1 Test room parameter assumptions

Parameter	Value	Unit
Single room size (2 heaters)	30 * 21 * 13 = 8190	Cu ft

Table 2 Table depicting the kWh consumption according to variable temperature difference at various intervals of time

Current temperature (°C)	Temp. diff. (°C)	Control condition for both heaters	Energy needed (kJ)	kWh consumption	Time required to reach preset range (min)
7	11	B	3187	0.9	17.7
18	0	B	29	0.01	0.2
19	1	B	289.7	0.08	1.6
21	3	A	869.2	0.24	14.5
19.8	1.8	A	521.5	0.14	8.7
18.5	0.5	A	144.9	0.04	2.4
19.5	1.5	A	434.6	0.12	7.2
22	1	O	0	0	0
21	3	A	869.2	0.24	14.5
19	1	A	289.7	0.08	4.8
20	2	A	579.5	0.16	9.6

raise the temperature of room fulfilling the energy requirements comes out to be 4.8 min and for a 2000 W heater calculates out to be 2.4 min.

For smart control, the time required to reach preset time is shown in Table 2.

5.3 *Energy Savings*

Without smart saving control, a 2 kW heater running for 10 h generates 20 kWh or 20 units/day and monthly 600 units/month. With smart saving control, approx. 1.3 kW of load running for 10 h generates 13 kWh or 13 units/day and monthly 390 units/month. The tariff rate is Rs 4.65/month according to Indian scenario.

The monthly bill for a home without smart saving would be Rs 2790 per month and a home with smart saving would be Rs 1813.5 per month. The total saving for a month would be Rs 976.5 per month. (1 INR equals 0.015 US $ -2017).

Set point: 18 °C–21 °C

Control Condition-A: One 1000 W on B: Both O: OFF

6 Conclusion

In this paper, the HEMS setup has been successfully implemented according to the represented plan according to the environment conditions and tariff rate of India.

The required setup generates a monthly bill of Rs 1813 per month which is a 35% saving in energy consumption from the state when the smart heat control is not installed. All the rooms can be centrally controlled to maintain uniformity in temperature across rooms, hence leading to high efficiency of the setup.

References

1. https://www.smartgrid.gov/the_smart_grid/smart_home.html.
2. Smart Home Monitoring, "www.rogers.com/web/Rogers.portal", 2011.
3. www.moef.nic.in.
4. K. A. Akpado, C. O. Ezeagwu, A. Ejiofor, A. O. Nwokeke: Design, Modeling and Simulation of a Microcontroller Based Temperature Control in a Ventilation System. International Journal of Advanced research in Electrical, Electronics and Instrumentation Engineering vol. 2, 3470–3479 (2013).
5. H. A. Sambo: A microcontroller based automatic room temperature control system, design and implementation. B. Eng. Undergraduate project, Dept. of Elect/Elect. Eng. ATBU, Bauchi (2004).
6. N. Priyantha, A. Kansal, M. Goraczko, F. Zhao: Tiny Web Services: Design and Implementation of Interoperable and Evolvable Sensor Network. Microsoft Research. ACM Sensys, 253–266 (2008).
7. Adnan Afsar Khan and Hussein T. Mouftah: Web Services for Indoor Energy Management in a Smart Grid Environment. IEEE International Symposium on Personal, Indoor, Mobile and Radio Communications *(PIMRC)*, 1036–1040 (2011).
8. A. Kamilaris, V. Trifa, and A. Pitsillides: HomeWeb- An Application Framework for Web-based Smart Homes. International Conference on Telecommunications (ICT), Cyprus, 134–139 (2011).
9. J. S. McDonald: Temperature control using a microcontroller: an interdisciplinary undergraduate engineering design project. Frontiers in Education Conference, Pittsburgh, PA, 1620–1624 (1997).
10. Rami F. Z. Henry and Hussein T. Mouftah: Novel algorithms to attain economical comfort control in residential spaces using Wireless Sensor Networks. IEEE Canadian Conference on Electrical and Computer Engineering (CCECE). 1374–1378 (2011).
11. Jiakang Lu and Tamim Sookoor: The Smart Thermostat: Using Occupancy Sensors to Save Energy in Homes. Proceedings of the 8th ACM Conference on Embedded Networked Sensor Systems, 211–224 (2010).

HAAS: Intelligent Cloud for Smart Health Care Solutions

S. Padmavathi and Sivakumar Sruthi

Abstract The current generation of digitalized lifestyle demands intelligent systems to support all real-time applications. The footpaths of the cloud technology have led to the developments many smart systems which are developed on grounds of machine learning techniques. The health care domain demands smart systems in order to automate functionalities to ease out the patient's problems. The proposed Healthcare As A Service (HAAS) solution describes the smart health care solution which would provide automated functionalities to reach out nearby health care centers to take treatments for their disorders. The predictive analysis of the patient's health information is done based on bagging algorithm in order to alert them regarding their health disorder. In case of emergency treatments, the alerts to the nearby health care centers can be done and the relevant treatments are provided based on the patient's historical information tracked from their unique smart id generated by the HAAS.

Keywords Cloud services • Decision support model emergency alerts
GIS tracking • Smart patient recognizer id

1 Introduction

The developing era of intelligent technological solutions for the real-life problems has paved a way for comfortable lifestyle. Smart systems for the health care domain are designed in such a way that the needs of the end users are satisfied in terms of better connectivity to the health care centers and provide security for maintaining their personal health information and transactions for their bill payments.

S. Padmavathi (✉) · S. Sruthi
Thiagarajar College of Engineering (A Govt. Aided Autonomous Institution), Thiruparankundram, Madurai 625015, Tamil Nadu, India
e-mail: spmcse@tce.edu

S. Sruthi
e-mail: sruthisivakumar22@gmail.com

A.K. Somani et al. (eds.), *Proceedings of First International Conference on Smart System, Innovations and Computing*, Smart Innovation, Systems and Technologies 79, https://doi.org/10.1007/978-981-10-5828-8_28

Smart systems are designed in such a way to predicts the chances of getting disease due to symptomatic health changes in end users [1]. The prediction is based on the Artificial Intelligence Algorithms which would self-learn based on the input provided from the external environment.

Artificial intelligence is the computerized intelligent techniques designed to imitate human brain either by machines or software. Artificial Intelligence (AI) is gaining a popularity in the medical domain as it has enhanced the human life in many life-supporting fields. Research in the area of artificial intelligence has given rise to the many of the developing technology known as expert-based systems, which is widely used to solve the real-life complex problems [2] in various domains such as Medical science, Bioengineering, medical innovations, and disease prediction.

Medical decision support system designed in such a way combines sophisticated representational and computing techniques with an insight of expert opinion to produce tools for improvising health care. The recent trends of incorporation of artificial intelligence and decision support systems in many of the medical applications has offered a good platform for health monitoring capabilities in Ambient Assisted Living (AAL) lifestyle condition [3]. Continuous medical assessment of health indicators for elderly people should be done at regular interval so that they can live on their own without depending on any support [4]. Personalization is an important aspect of remote disease management systems [5]. It incorporates the Markov decision process and also the probabilistic temporal Bayesian-related states do not change over time which can be employed for forecasting any medical decision based on the input. The Hybrid Intelligent Systems provides a solution for the real-world increasingly complex problems which raise the ambiguity issues, uncertainty problems and the high-dimensionality explosion of data. They allow to use both a prior knowledge and compose raw medical data into innovative decision solutions [6]. The hybrid classifier system can easily be implemented in efficient computing environments such as parallel and multithreaded computer architecture which does the implementation of AI techniques in the distributed computing systems like Grid or Cloud computing. Different ideas have been proposed for providing e-health care [7] for the web users. The increasing trend toward social platform oriented web services has resulted in a number of health applications.

Section 2 describes the design and then the functionalities of the HAAS system, Sect. 3 describes the prediction algorithm for the disorder. Section 4 describes the prediction results of the decision support model which is part of the functionality of the HAAS solution and Sect. 5 describes the conclusion of this proposed system.

2 Design of the Smart HAAS: Intelligent Health Care System

2.1 Architecture of the System

The Health care As A Service (HAAS) over the cloud technology is to develop a smart health care application to the end users to monitor their health condition by themselves. The proposed smart health care system is to help out the patients using this application to reach out the nearby health care center to visit and get health guidance for their health disorder. The patient can be recognized by their unique identification number which is generated based on their first registration in any of the health care center using their Aadhaar number to avoid the duplication and fake user intrusion. The main features of this system include predictive analysis of the patient's health information in comparison with their previous health history details to predict the chances of getting the disorder, this feature is mentioned in Sect. 3. The another feature of this system is that Geographical information tracking of the end user of the system, which shows the nearby available primary health care centers to the end users. The GIS option is more helpful in case of Emergency situations to reach the nearby health care centers. The Smart HAAS also has some more functionalities such that it could also share the patient's previous health history to the newly visited health care center, which reduces the time to take medical tests and other medical analysis of the patients. In case of very chronic condition, the alert message is sent to the family members of the patient who is using this application.

The emergency call option is also to be designed with this HAAS services to connect to the nearby ambulance services. The single press call-on button with has auto enabled GIS option to track the location of the patient by the ambulance services. The patient's smart id which is based on the Aadhaar number has the address details of the users and this can be shared to the ambulance services in order to track the patient's location within the shorter time span. The patients could also connect to the nearby health care centers to fix appointments for their treatments. The appointment message notification for visiting hour is sent from the server of health care center to the patient. Hence, this proposed smart HAAS can be deployed in the Amazon EC2 services by enabling private cloud service with advanced security authentication mechanisms for the health information of the end users of the system. The integrity aspects of information which shared between the patients and the health care centers can be enhanced by generating the random one-time passwords notification to the end users so the health data modification issues can be avoided.

2.2 A Case Study for Developing the Prediction Model for Smart HAAS

The predictive analysis is done for the case study which measures people getting affected by heart disease due to various factors like an uncontrolled diabetic, blood pressure, cholesterol levels, stress levels, smoking, and alcohol. Some of the externally observable symptoms such as the excessive sweat, palpitation, dysphonic issues, duration of the chest pain and also other details regarding the family history of having the diabetic disorder, hypertension issues and premature death due to heart disease are considered while developing the prediction model from the obtained dataset. The dataset with nearly 1056 instances was collected in order to develop a decision model to predict the chances getting the heart disorder. The evaluation of the prediction accuracy for the heart disease case study is discussed in the Sect. 4.

2.3 Working of the Smart HAAS

The HAAS has many functionalities as mentioned in the architecture overview of the system. The alert regarding the disorder is based on the prediction done by the machine learning classifier algorithms. The patient's current health data is also obtained from the system and also their past Electronic medical data are analyzed to predict their disorder. If the disorder is predicted, it is intimated to the end users in the form of message alerts and the respective alternative treatments with respect to their disorder can be taken. The patient's condition can be communicated based on their id to the nearby health care center in order to provide medical treatments based on the current health status which is alerted by the AI decision support model of the smart HAAS system.

The predictive analysis of the patient's health information is done by obtaining the real-time health data such as their blood pressure values, sugar values, cholesterol values, etc., are compared with the decision support knowledge model. When the health values are slightly exceeding the threshold level, the disorder of the patient is alerted to consult the doctor in the nearby health care centers (Fig. 1).

The alert notification may vary based on the patient's health state are shown in the Fig. 1, the less critical status is alerted using the alert level-0 [AL = 0] where the health guidance tips could be provided to take care of the health, in case of critic state alert level-1 [AL = 1] is notified by the patient to the doctor at the health care center and then they can consult the doctor by fixing the appointments to take necessary treatments. In case of very chronic disorder state, the notification to their respective nearby ambulance centers and also to their nearby health care centers and also to their relatives are sent to help out the patient by this HAAS system.

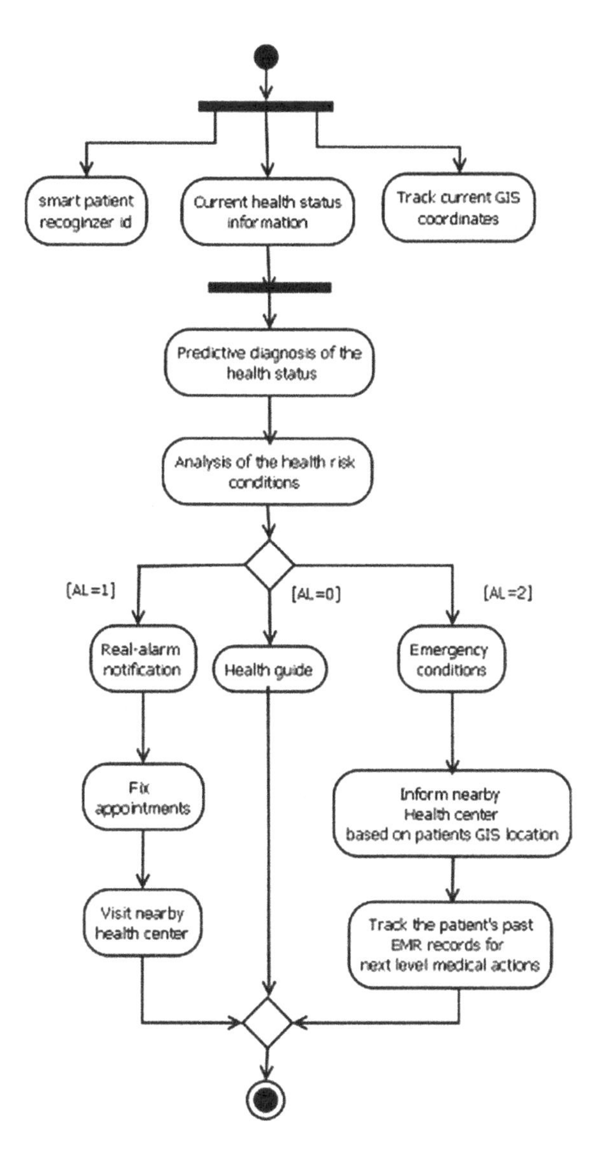

Fig. 1 Activity Diagram which depicts the working of Smart HAAS where the patient health status is analyzed and create necessary alert actions according to the need of the end user

3 Decision Support Model to Predict the Health Disorder

The smart HAAS system which has a decision support model which is developed using Bagging algorithm which ensembles binary classifier algorithms to predict the chances of getting the disorder. The Bayesian network algorithm is used as the base classifier and then some of the other algorithm namely the Naive Bayes classifier, Decision stump, J48, and Linear Regression is used for disorder prediction and then evaluating the accuracy based on votes. The votes are allocated based on the prediction accuracy and also in comparison with the minimized prediction errors.

Algorithm 1: Bagging algorithm to predict the disorder based on the user health data
Training phase:

1. Initialize the parameters
   1.1. $D=\Phi$, the ensemble
   1.2. Set L as the number of binary classifier algorithms to train
2. For i=1,...,L
   2.1. Take a bootstrap data sample S_k from Z.
   2.2. Build a binary classifier D_k from S_k as the training data set
   2.3. Union all the binary classifier to ensemble D, now $D=D \cup D_k$
3. Return D.

Classification Phase:

4. Run $D_1,...,D_L$ on the input data set X which contains the patients health information.
5. The classifier which has the maximum number of votes with minimised error is chosen for disease prediction.

4 Results and Discussion

The proposed HAAS architecture can be simulated by developing health care website with the set of health-oriented questionnaire which collects all the user input value and evaluates it using the Bagging algorithm in order to predict chances of getting the disorder. In case of high severity of disorder predicted, an alert message is sent accordingly by using SMS gateway option enabled along with the health web services. In medical diagnosis prediction, various evaluation metrics are followed in order to check the efficacy of the imparted AI algorithm in terms of accuracy and prediction errors. Various performance evaluation techniques are followed in order validate the decision model are shown in Fig. 2. Test sensitivity measures the correct prediction chance of getting affected by disorder, whereas Test Specificity measures the correct prediction for the chance of not getting affected by the disorder. The precision and recall are based on the test sensitivity and test specificity measure of classification of the dataset while predicting the disorders.

The overall accuracy of the each classifier algorithms is shown in the Fig. 3. The algorithms are ensemble to predict the chances of getting the disease is evaluated by the boosting algorithm. The Bayesian network classifier has good accuracy for

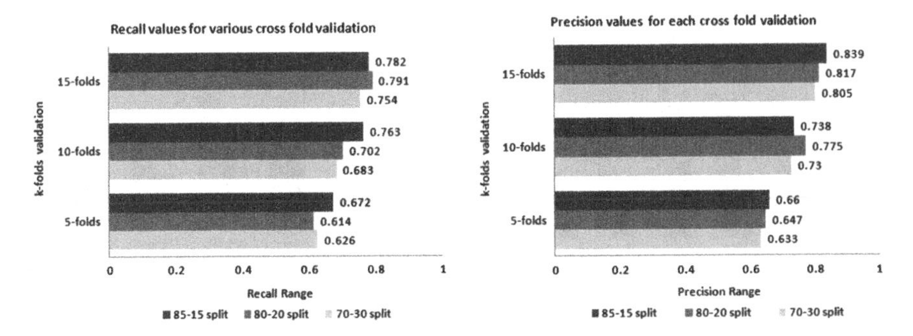

Fig. 2 Precision and Recall values for the base classifier algorithm is evaluated in order to check the accuracy of the prediction level and also to compare with the other classifier based on voting

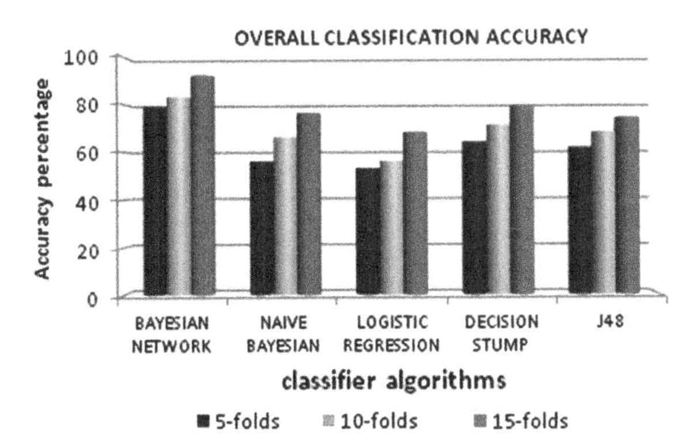

Fig. 3 The overall prediction accuracy of all classifier ensemble algorithms of the proposed Decision support model

predicting the disease since the prediction is computed based on the probabilistic analysis of the values of the health attributes of the dataset.

The cross-folds validations are done in order minimize the prediction errors. The misclassification errors are computed by the Mean absolute error, Root Mean Square error, Relative absolute error, and Root Relative Square errors are shown in the Fig. 4. Overall, the 15-fold validation shows minimized absolute error when compared to other fold validations. Among the various classifiers for validation of prediction accuracy, the Bayesian network has the minimum absolute error of 0.189 along with the minimum RMSE value of 0.381. The RAE value is around 0.357 for the Bayesian network at its 15 cross-fold validation.

Hence from overall misclassification errors of various classifier of the proposed Decision model, the Bayesian network is good in terms of prediction accuracy with the minimized prediction errors.

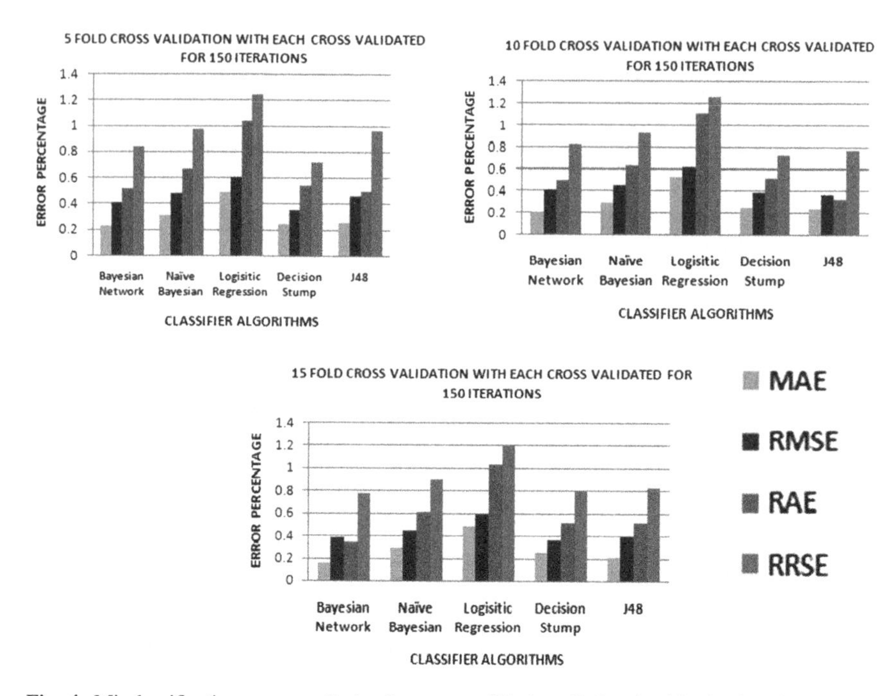

Fig. 4 Misclassification error prediction by cross validating all the classifier by bagging method to minimize the errors while predicting the disorder

5 Conclusion and Future Work

The smart HAAS system proposed in this system is to be implemented using IOT technology by incorporation of mobile and cloud computing technology using the Amazon EC2services. Henceforth, the proposed system helps the patients to make a self analysis of their health care status which alerts them to make necessary treatments. The advancements of mobile applications have also paved a way to incorporate this proposed application as the mobile apps which can easy out the usage of the end users to reach out the health care centers for their consultations. The support for large users scalability and faster network connectivity between the Health care servers and the end users of the application plays a significant role in rendering this HAAS services.

References

1. Jeferey G. Klann Peter Szolovits, Stephen M. Downs, Gunther Schadow: Decision support from local data: Creating adaptive order menus from past clinician behaviour. Journal on Biomedical Informatics. 67, 2016.

2. Anthony Costa Constantinoua, Norman Fentona, William Marsha and Lukasz Radlinskib: From complex questionnaire and interviewing data to intelligent Bayesian Network models for medical decision support. Journal on Artificial Intelligence in Medicine. 67, 75–93, 2016.
3. A.S. Billis, E. I. Papageorgiou, C. A. Frantzidis, M.S. Tsatali, A.C. Tsolaki and P. D. Bamidis..: A Decision-Support Framework for promoting Independent Living and Ageing Well, IEEE Journal of Biomedical and Health Informatics 19, 199–209, 2015.
4. Michal woznaik, Manuel Grana, Emilio corchad.: A survey of multiple classifier systems as hybrid systems. Journal on Information Fusion. 16, 3–17, 2013.
5. Pablo Garcia-Sanchez, Jesus Gonzalez, Antonio M. Mora, Alberto Prieto: Deploying intelligent e-health services in a mobile gateway. Journal on Expert systems with application. 40, 1231–1239, 2013.
6. Marina Velikova, Josien Terwisscha van Scheltinga, Peter J.F. Lucas, Marc Spaanderman: Exploiting causal functional relationships in Bayesian network modeling for personalized healthcare. Journal on Approximate Reasoning. 55, 59–73, 2014.
7. Che-Wei Lina, Shabbir Syed Abdulb, C, Daniel L. Clinciub, D, Jeremiah Scholle, Xiangdong Jinf, Haifei Luf, Steve S. Chenf, Usman Iqbalb, Maxwell J. Heineckg, Yu-Chuan Li.: Empowering village doctors and enhancing rural healthcare using computing in a rural area of mainland China. Journal on Computer Methods and Programs in Bio Medicine. 113, 585–592, 2013.
8. Anthony Costa Constantinou, Mark Freestone, William Marsh, Jeremy Coid: Causal inference for violence risk management and decision support in forensic psychiatry. Journal on Decision support systems. 80, 42–55, 2015.

Compact Slotted Microstrip Patch Antenna with Multiband Characteristics for WLAN/WiMAX

B. Roy, G.A. Raja, I. Vasu, S.K. Chowdhury and A.K. Bhattacharjee

Abstract In this paper, we present a compact slotted microstrip patch antenna whose multiband characteristics have been studied. This antenna is compact in shape and size having a dimension of 24×24 mm^2. This antenna consists of two inverted F-shaped slots in the patch and a defected ground plane for enhancement of impedance bandwidth. This patch is fed by a coaxial probe. The antenna shows multi band characteristics, i.e., from 1.94–1.98 GHz, 2.4–2.52 GHz, 3.2–3.38 GHz, 5.07–5.41 GHz, 5.73–6.09 GHz. After analyzing all necessary characteristics in perspective of gain, bandwidth, polarization, and return loss, the proposed patch is well applicable for WLAN (2.4/3.2/5.2/5.8 GHz) and WiMAX (5.5 GHz) communications.

Keywords Slot antenna · WLAN · WiMAX · F-shaped slot

1 Introduction

The necessity of multiband antenna is increasing due to prevalent diversities in wireless standards of the modern communication system. Modern communication system requires compact, light-weight, low-profile antennas. For this reason,

B. Roy (✉) · G.A. Raja · I. Vasu · A.K. Bhattacharjee
Department of ECE, NIT Durgapur, Durgapur, India
e-mail: bappadittya13@gmail.com

G.A. Raja
e-mail: dittya13@rediffmail.com

I. Vasu
e-mail: bappa13@rediffmail.com

A.K. Bhattacharjee
e-mail: akbece13@gmail.com

S.K. Chowdhury
Department of ET&CE, Jadavpur University, Kolkata, India
e-mail: santoshkumarchowdhury@gmail.com

A.K. Somani et al. (eds.), *Proceedings of First International Conference on Smart System, Innovations and Computing*, Smart Innovation, Systems and Technologies 79, https://doi.org/10.1007/978-981-10-5828-8_29

printed slot antennas find application in various wireless communications ranging from mobile phones to satellites. A number of slotted antennas with different geometry has been proposed for working in WLAN, WiMAX applications [1–5]. In [1], an UWB monopole antenna is studied for the response of different rectangular ground slot for multiband appliances. In [2] a small microstrip antenna consisting of two F-shaped slots radiators and a defected ground has been presented which shows three bands and covers entire WLAN and WiMAX frequency bands. In [3] a planar antenna with folded structure has been proposed which covers dual frequency band for WLAN (2.4/5 GHz). In [4] an upside-down F-shaped antenna with an U-shaped slot has been proposed which covers WLAN frequency bands. In [5], a compact antenna with inverted L-shaped slots has been proposed which covers the WiMAX bands. Several other multiband antennas have been proposed in [6–10]. Though these antennas are efficient in their respective domains, still there exists an issue regarding the compactness of the structure and complexity in feeding technique. Here, we have presented a compact slotted patch antenna with multiband characteristics. The antenna is compact in shape and size having a dimension of $24 \times 24\ mm^2$. The antenna primarily consists of two inverted F-shaped slots in the patch. The ground plane also defected for further enhancement the impedance bandwidth. The patch is fed by a coaxial probe. The antenna shows dual band characteristics, i.e., from 2.70–2.73 GHz and from 5.02–6.02 GHz. After that for getting multiband characteristics, we examine over the ground plane slots (explain in Sect. 4). After examining all necessary characteristics in perspective of gain, bandwidth, polarization, and return loss the proposed patch is well applicable for wireless communications. The antenna is stimulated and studied using Zeland IE3D software whose results have been discussed sequentially.

2 Antenna Design

In this paper, three different antennas have been designed and simulated. Initially, a simple microstrip patch has been designed, which has been subsequently followed by modifications in patch and ground plane. The geometrical structure of Antenna 1 (Reference Antenna) is shown in Fig. 1. The equation used for designing a microstrip patch for the fundamental mode is given as

$$f_r = \frac{c}{2(L+h)\sqrt{\varepsilon_{eff}}} \tag{1}$$

where fr is the resonant frequency, c is the speed of light, ε_{eff} is the effective dielectric constant, which can be calculated using the formula:

$$\varepsilon_{eff} = \left(\frac{\varepsilon r+1}{2} + \frac{\varepsilon r-1}{2}\right)/\sqrt{[1+12(h/W)]} \tag{2}$$

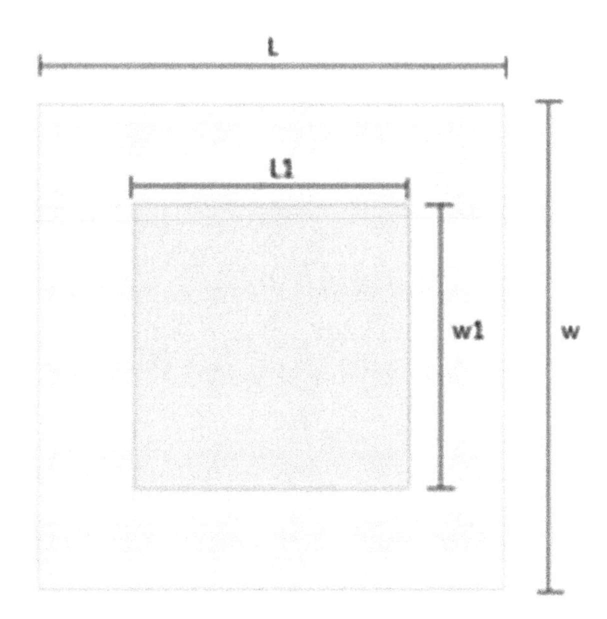

Fig. 1 Reference microstrip patch antenna (antenna 1)

where εr is the relative dielectric constant, L and W are the length and width of the patch, respectively, where h is the thickness of the substrate.

These two formulae are used to predict the substrate thickness and dielectric constant and resonant frequency point(s) of the antenna. Primarily, a Square Microstrip Antenna of patch width W = 20 mm and length L = 20 mm has been considered as Antenna 1 (Reference antenna) whose ground plane dimension is 34×34 mm^2. The antenna has a FR4 epoxy substrate with 1.6 mm thickness and ε_r (relative permittivity) = 4.4, tan δ = 0.02. The feed point is located below the central portion of the proposed antenna patch. SMA connector of 0.4 mm internal radius provides the coaxial feed.

Figure 2 shows the modified patch which consists of a horizontal F-Shaped slot at the lower left arm and an identical inverted F-Shaped slot at the upper right arm. The dimensions have been provided in Table 1. Effect of inclusion of slots in the patch has been discussed in [6]. The patch is the main radiating element in microstrip patch antennas. Presence of slot(s) in patch disturbs the surface current distribution pattern (Fig. 8) of the antenna, thereby creating multiple resonant frequency modes.

Figure 3 shows the modified ground plane which consists of vertical rectangular stripes (dimensions provided in Table 1). This type of etched ground plane is known as a Defected Ground Structure (DGS). The effects of inclusion of DGS in antenna have been discussed in [7]. Defected Ground Structures help in minimizing the return loss of the antenna, thereby improving the realized antenna gain.

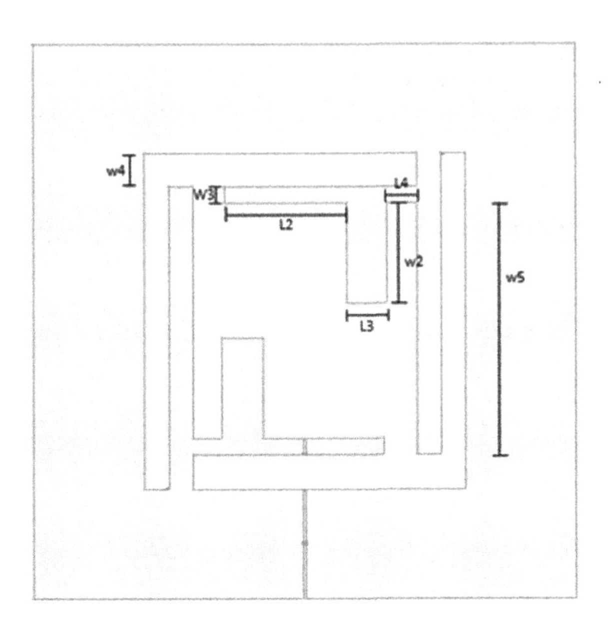

Fig. 2 Modified patch of antenna 1 (antenna 2)

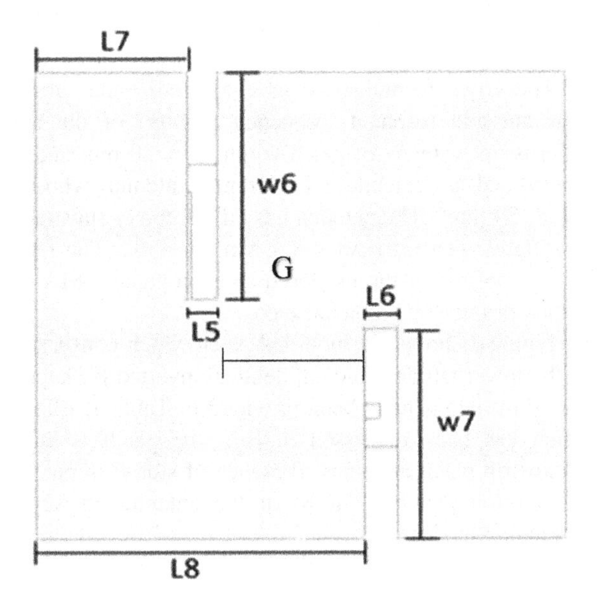

Fig. 3 Modified ground plane of antenna 2 (antenna 3)

Table 1 Parameter dimensions in mm

Parameter	Dimension	Parameter	Dimension
L	24.00	W	24.00
L1	20.00	W1	20.00
L2	07.65	W2	05.98
L3	02.52	W3	01.00
L4	01.85	W4	02.00
L5	01.88	W5	15.00
L6	02.05	W6	16.08
L7	09.75	W7	14.75
G	3.45		

3 Results and Discussion

The Return Loss (S11) plot of Antenna 1, Antenna 2, and Antenna 3 have been depicted in Fig. 4.

On observation, it is seen that the simple microstrip patch (Antenna 1) shows three discrete frequency bands close to 3.5, 6.9, and 8.0 GHz. The antenna with the modified patch (Antenna 2) shows somewhat better return loss characteristics with resonant frequency shifted toward 6.0 GHz. On inclusion of DGS (Antenna 3) better frequency response characteristics have been observed. Two discrete frequency bands have been obtained. One ranging from 2.70 to 2.73 GHz and the other from 5.02 to 6.02 GHz. The latter is applicable for WLAN (5.2/5.8 GHz) and WiMAX (5.5 GHz) communications. The effect of slots in the ground plane, the gain and radiation characteristics of the antenna at these frequency bands have been discussed later along with necessary plots.

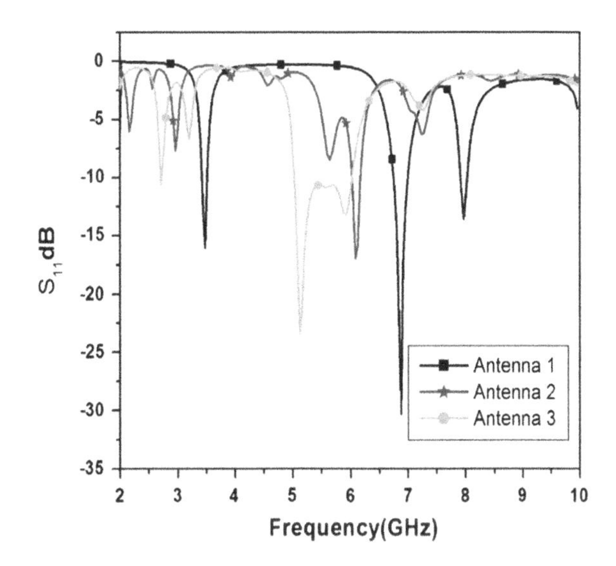

Fig. 4 Return loss plot of antenna 1, antenna 2 and antenna 3

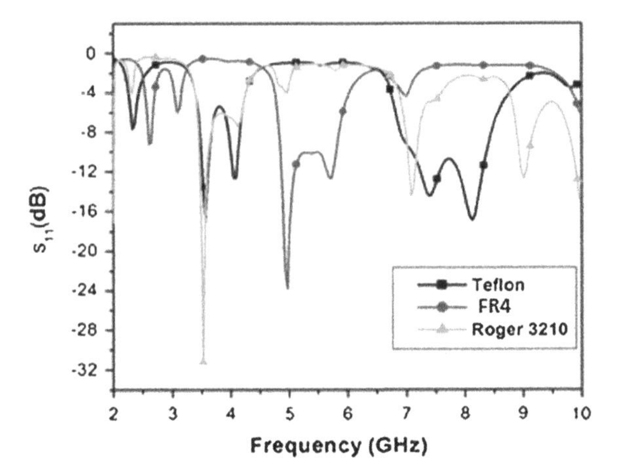

Fig. 5 Effect of various substrates on return loss

Figure 5 shows the substrate effect on return loss. The effect of putting slots in the ground plane has been depicted in Fig. 6. It has been analyzed that there has been an increment in impedance bandwidth as the no of rectangular slots in the ground plane has been increased to two. There has been an increment in impedance bandwidth when two rectangular slots have been introduced in the ground plane of the antenna. Multiple resonant modes are generated due to the inclusion of slot in the ground plane. Due to the presence of multiple slots these generated resonant modes come closer to each other, resulting in a wider bandwidth. Also an improvement in return loss characteristics of the antenna has been observed from Fig. 6.

Figure 7 shows the Gain versus frequency characteristics of Antenna 3. An average gain 1.3 dB has been observed at the frequencies of interest. The Radiation Pattern of the antenna has been depicted in Fig. 8. Well isolation has been observed

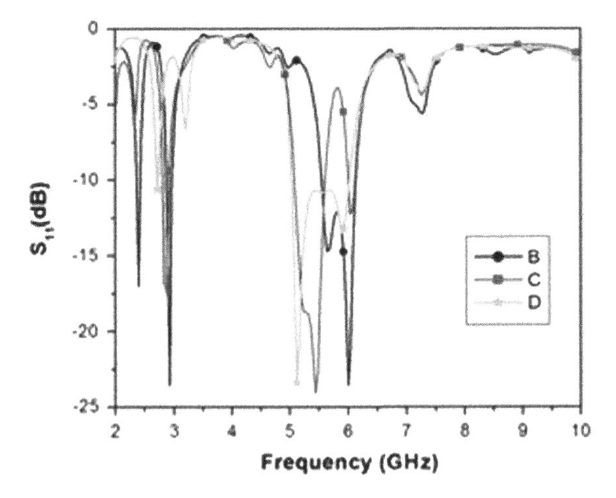

Fig. 6 Effect of ground slots on frequency response (B: Single slot 1, C: Single slot 2, D: Both slots)

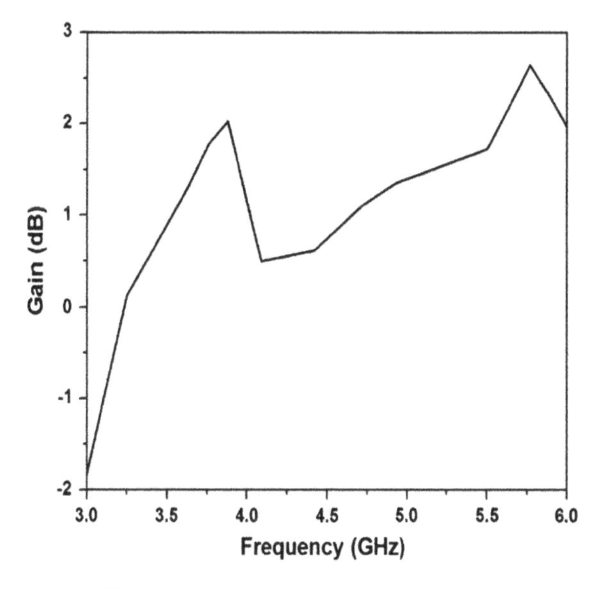

Fig. 7 Antenna gain at different frequencies of interest

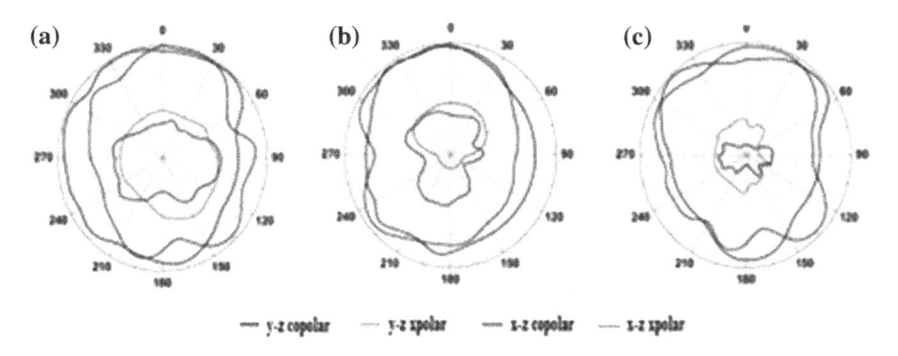

Fig. 8 Radiation pattern **a** 5.2 GHz **b** 5.5 GHz **c** 5.8 GHz

between the co- and cross-polarized components at the frequencies of interest. The y-z copolar/x-polar and x-z copolar/x-polar radiation patterns have been depicted in Fig. 8.

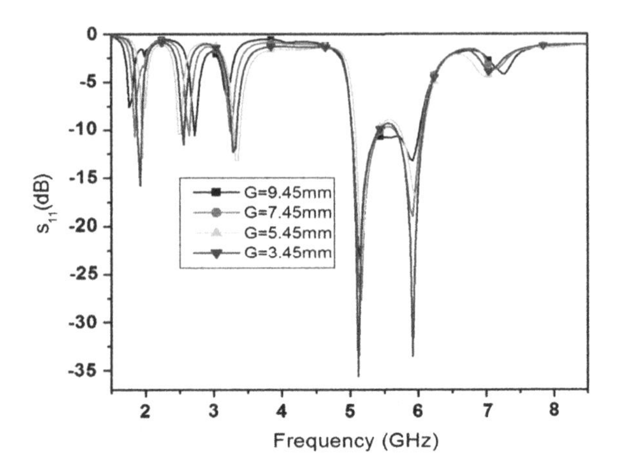

Fig. 9 S_{11}(dB) versus Frequency (GHz) plot for different G

4 Analysis of Ground Slot Gap

After verification of the ground slot gap (G) in Antenna 3 (Shown in Fig. 3), we can get more accurate result and perfect gain. By parametric study, the gap is considered to be 3.45 mm (which is considered as Antenna 4) for getting the desired multiband resonant frequencies which are shown in Fig. 9. It is found that Antenna 4 operate in 2.4, 3.2, 5.2, 5.5, 5.8 GHz (For IEEE 802.11 a WLAN Band).

5 Conclusion

The proposed antenna is compact in shape and size. The compact structure of the antenna makes it a suitable candidate which can be incorporated in wireless devices supporting multiband communication. Moreover, low-cost and the widely available FR4 epoxy substrate are used in this antenna design which makes it even more affordable. The antenna shows multiband band characteristics, i.e., from 1.94–1.98 GHz, 2.4–2.52 GHz, 3.2–3.38 GHz, 5.07–5.41 GHz, 5.73–6.09 GHz. After examining all the necessary characteristics in perspective of gain, bandwidth, polarization, and return loss the proposed antenna is well suitable for WLAN (2.4/3.2/5.2/5.8 GHz) and WiMAX (5.5 GHz) communications.

References

1. N. Kishore, A. Prakash, V.S. Tripathi "A multiband microstrip patch antenna with defected ground structure for its applications" Microwave And Optical Technology Letters, Vol. 58, No. 12, pp. 2787–3015, 2016.

2. A.K. Gautam, L. Kumar, B.K. Kanaujia, And K. Rambabu "Design Of Compact F-Shaped Slot Triple-Band Antenna For Wlan/Wimax Applications", IEEE Transactions on Antennas And Propagation, Vol. 64, No. 3, pp. 1101–1105, 2016.
3. A.G. Alhaddad, R.A. Abd-Alhameed, D. Zhou, C.H. See, I.T.E. Elfergani, P.S. Excell, "Low Profile Dual-Band-Balanced Handset Antenna With Dual-Arm Structure For Wlan Application", IET Microw. Antennas Propag., Vol. 5, Iss. 9, pp. 1045–1053, 2014.
4. P. Salonen, M. Keskilammi, And M. Kivikoski "Single-Feed Dual-Band Planar Inverted-Antenna With U-Shaped Slot", IEEE Transactions On Antennas And Propagation, Vol. 48, No. 8, pp. 1262–1264, 2000.
5. Jui-Han Lu And Bing-Jhang Huang "Planar Compact Slot Antenna With Multi-Band Operation For Ieee 802.16 m Application", IEEE Transactions On Antennas And Propagation, Vol. 61, No. 3, pp. 1411–1414, 2013.
6. B. Roy, A. Bhattacharya, A. K. Bhattacharjee and S. K. Chowdhury, "Effect Of Different Slots In A Design Of Microstrip Antennas," IEEE International Conference On Electronics And Communication Systems, Coimbatore, pp. 386–390, 2015.
7. A. Bhattacharya, B. Roy, S. K. Chowdhury, A. K. Bhattacharjee; "A Compact Fractal Monopole Antenna With Defected Ground Structure For Wideband Communication"; Applied Computational Electromagnetics Society Express Journal, ISSN: 1054–4887, Vol. 1, No. 8, pp. 228–231, 2016.
8. Y. Z. Cai, H. C. Yang and L. Y. Cai, "Wideband Monopole Antenna With Three Band-Notched Characteristics," in IEEE Antennas and Wireless Propagation Letters, vol. 13, no., pp. 607–610, 2014.
9. V. V. Reddy and N. V. S. N. Sarma, "Triband Circularly Polarized Koch Fractal Boundary Microstrip Antenna," in IEEE Antennas and Wireless Propagation Letters, vol. 13, no., pp. 1057–1060, 2014.
10. P. S. Bakariya, S. Dwari, M. Sarkar and M. K. Mandal, "Proximity-Coupled Multiband Microstrip Antenna for Wireless Applications," in IEEE Antennas and Wireless Propagation Letters, vol. 14, no., pp. 646–649, 2015.

Nonlinear Speed Estimator and Fuzzy Control for Sensorless IM Drive

J. Mohana Lakshmi, H.N. Suresh and Varsha K.S. Pai

Abstract The computation of speed in sensorless AC drive has gained much importance over the last few years. The main focus lies in evaluating the accurate value of speed for vector-controlled drives thereby making the use of an encoder at the machine shaft obsolete. Considered to be most simple of its kind, the model-based approach utilizes the linear time invariant and nonlinear time variant systems for speed estimation. In the proposed work two fuzzy controllers are adopted. One fuzzy-based adaptation scheme in the nonlinear feedback is used for obtaining proper speed estimate. In a vector-controlled induction motor, the problem that perseveres even after the evaluation of speed is the harmonics generated due to switching action which decreases the overall efficiency of the drive system. For an improved performance of the drive control system, a second fuzzy-based control is supplemented along with the PI Controller in the outer speed control loop. The significant feature of this work is the implementation of an intuitive fuzzy logic-based learning approach which is fast and effective. The type-1 fuzzy logic-based controller is designed which is less computationally intensive. An artificial intelligence-based machine model scheme which utilizes two fuzzy controllers (Dual Fuzzy) offers most satisfactory performance reducing the Total Harmonic Distortion (THD) to a great extent.

Keywords Adaptive dual fuzzy control • Linear time invariant
Nonlinear time variant • Total Harmonic Distortion

J. Mohana Lakshmi (✉) · H.N. Suresh · V.K.S. Pai
Department of Electrical and Electronics Engineering, Malnad College of Engineering, Hassan 573201, India
e-mail: mohana2024@gmail.com

H.N. Suresh
e-mail: mce.suresh@gmail.com

V.K.S. Pai
e-mail: varsha.cec@gmail.com

A.K. Somani et al. (eds.), *Proceedings of First International Conference on Smart System, Innovations and Computing*, Smart Innovation, Systems and Technologies 79, https://doi.org/10.1007/978-981-10-5828-8_30

1 Introduction

Eliminating the sensor from the machine shaft requires a firm estimation technique. Of all the practices available in the literature, the machine model-based approach is considered as the most simple and effective speed estimator, also known as Model Reference Adaptive Scheme (MRAS), is implemented in electric vehicles, hostile environmental conditions, and emergency operation (i.e., in case of failure of the sensor). A smooth seamless-switching process from the approach with a sensor to a mode without a sensor can keep the drive running until a regular checkup or at least a restoration allows an emergency stop. Implementation of the sensorless scheme in commercial drives considers the following:

- Automatic parameter fine-tuning.
- Compensations on parameter differences on functions of current and temperature.
- Interaction between standard controls.
- Additional ripple in current and torque.

Many applications especially underground oil industries require remote operation of vector-controlled variable speed drives. The use of voltage source inverters introduces over modulated switching signals and harmonic problems. Hence, an advanced control strategy for speed estimation and proper tuning of the VSI is essential.

Various open-loop speed and flux linkage estimators are available in the literature. These estimators use the stator as well as rotor equations of the 3-ϕ motor. But, the accuracy of these estimators strongly depends on the machine parameters. To improve the efficiency, closed-loop estimators are designed. The different parameters of the induction motor can be measured at stand still and even tune the speed and current controllers. However, the factors of the machine differ through normal operation [1]. From the past few years, sensorless speed IFOC is been applied for applications demanding high performance. This is based on rotor speed calculation using the various parameters of the machine [2–7]. The disadvantage posed is the intensive computations involved. Also, the excessive memory space required makes it difficult to tune the parameters. This becomes a dangerous problem during practical implementation. Hence, the use of these methods is limited.

The use of signal flow graphs for speed computation of induction motor model resulted in the satisfactory performance of the drive in steady state but the dynamic performance had to be compromised [8].

MRAS based on a rotor flux technique is simple, stable, and robust against machine parameter variation [9]. However, to achieve the benefits of sensorless control, the speed estimation methods must achieve robustness against model and parameter uncertainties over a wide speed range [10, 11]. Therefore, some kind of parameter adaptation is essential in order to obtain a high-performance sensorless vector control drive, and hence an appropriate control algorithm with improved computational accuracy has to be developed.

Amid all the control patterns, the model-based method delivers acceptable assessments of speed and rotor flux linkages. The work presented in this paper discusses in detail the selection of the adaptation scheme based on Popov's rule of Hyperstability for a fourth-order induction motor. Also, the proposed fuzzy-based adaptation rule gives an exact solution on the selection of gain values accurately. Based on this selection, a proper estimate of speed is obtained. In order to overcome the problems associated with the conventional switching scheme and speed regulation, an adaptive Fuzzy Logic Controller (FLC) is being supplemented in the vector control loop. The fuzzy controller is designed significantly to reduce the harmonics that degrade the machine's performance. The Adaptive Intelligent (AI) technique largely reduces the total harmonic distortion thereby improving the performance of the drive system. Also, the adaptive controller observes the speed error and updates the output such that precise speed monitoring is achieved.

2 Overall Estimation and Control Scheme

The overall structure of the proposed methodology involving a nonlinear estimation system and AI-based adaptive control scheme is shown in Fig. 1. The key units used are speed, rotor flux measuring unit, current calculation unit with i_{ds} and i_{qs} as current reference values, and Current Regulated Pulse Width Modulation (CRPWM) inverter unit. A combination of adaptive load torque estimation with natural observer which uses a PI—Fuzzy-based adaptation scheme is used for speed estimation. The terminal voltages and current from both the estimators are used to measure the speed. The FLC discovers the torque reference of IM using the difference between the reference and estimated speed. The purpose of speed

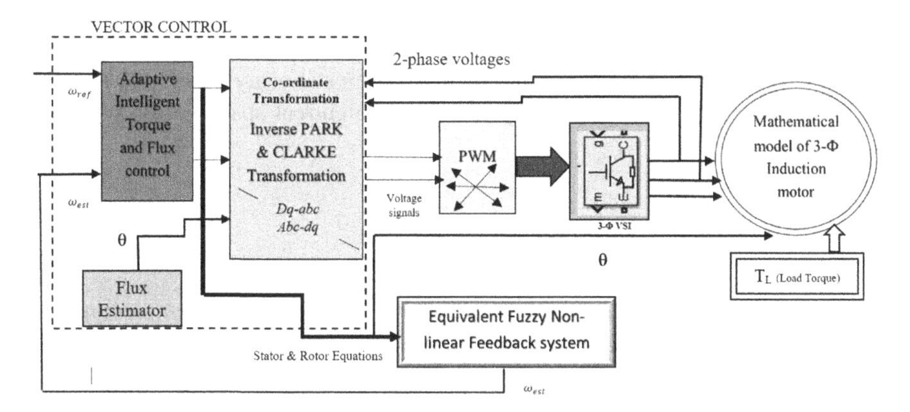

Fig. 1 Overall structure of the adaptive control scheme

estimation is resolved by considering the machine model equations. The voltages and currents in the stator and the rotor flux linkages are used for producing appropriate flux linkage for state vector estimation. Further, an adaptation scheme is involved based on Popov's theorem to evaluate the required state estimate, which in this case is the rotor speed (ω_{est}). The entire control scheme ensures the proper selection of adaptive membership functions and a rule base for fuzzy-based control. The fuzzy rule-based system so designed not only ensures proper speed regulation but also accounts for a large reduction in harmonics in torque and current. This reduces the Total Harmonic Distortion (THD) to a great extent when compared with the conventional AI control methodologies [12]. The control scheme further produces necessary voltage and current signals to the inverse transformation where 2-Φ to 3-Φ transformation is achieved. The control algorithm along with the transformed vectors provides the gating signals to the PWM which in turn generates proper switching states for the 3-Φ inverter. For a wide range of variation in speed and machine parameters the proposed estimation scheme and control algorithm tends to produce a robust solution. It is indeed considered to be the simplest form of the control structure that renders accurate speed estimation, speed regulation and reduction in THD.

2.1 Nonlinear Estimation System

The nonlinear feedback estimation scheme comprises two machine-based subsystems. A feed—forward linear tine invariant subsystem whose inputs will be stator voltages and currents v_{sdq} and i_{sdq} are obtained from stationary reference frame. The space vectors in the generalized form is:

$$\varphi'_{rdq} = x_{dq} = \frac{Lr}{Lm}\left[\int \left(\widehat{v_{sdq} - R_{sdq} i_{sdq}}\right) dt - L_{sdq} i_{sdq}\right] \quad (1)$$

Equation (1) does not contain rotor speed. It only consists of stator equations. Hence, this feedforward system is termed as reference model (Fig. 2).

However, if the rotor equations are represented in stationary frame they contain the term $\widehat{\omega_{r1}}$ which is a dependent parameter. The current space vector takes the form as per Eq. (2):

$$i'_{rdq} = \frac{\varphi'_{rdq} - Lm i'_{sqd}}{L'_{rdq}} \quad (2)$$

The current vector is eliminated from the rotor equations and the expression for the adjustable model is obtained as

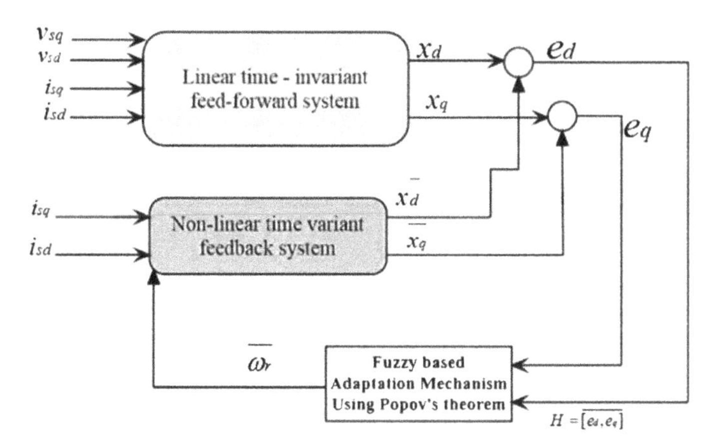

Fig. 2 Structure of Speed Estimator based on Popov's rule

$$\overline{\varphi'_{rdq}} = \overline{x_{dq}} = \frac{1}{\tau_{ra}} \int \left(Lmi'_{sdq} - \varphi'_{rdq} - \widehat{\omega_{r1}} \tau_{ra} \varphi'_{rdq} \right) dt \tag{3}$$

Here τ_{ra} is the time constant of rotor used in adjustable model.

Equation (3) corresponds to a current model which corresponds to rotor speed and rotor fluxes. This model represents a feedback system hence it's coined as an adjustable (adaptive) system.

The feedback and feedforward systems provide the estimated values of rotor flux linkages and angular difference of the output of the two systems. A tuning signal is to be obtained to filter the error between these estimated values. The tuning signal represented by H is governed by Eqs. (4–7):

$$H = \left[ed,\ eq \right]^T \tag{4}$$

$$H = Im\left(\overline{\varphi'_r}, \overline{\varphi^*_r} \right) \tag{5}$$

$$= \varphi_{rq} \widehat{\varphi_{rd}} - \varphi_{rd} \widehat{\varphi_{rq}} \tag{6}$$

$$= x_q \overline{x_d} - x_d \overline{x_q} \tag{7}$$

$$= e_d - e_q \tag{8}$$

The obtained tuning signal is fed to the PI—Fuzzy-based adaptation system. According to Popov's rule of hyperstability, the transfer function of the linear feedforward time invariant system must be strictly positive real and the nonlinear feedback scheme should satisfy the condition:

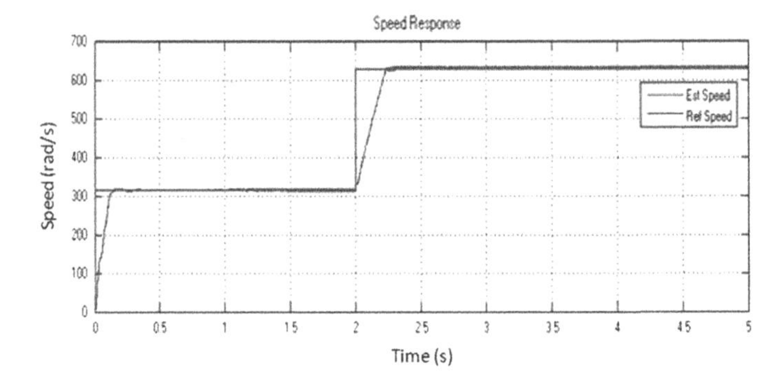

Fig. 3 Speed response characteristics for zero reference load torque condition using PI controller

$$\int H_p^T w dt \geq 0 \forall [0, t_1],\ t_1 \geq 0. \tag{9}$$

Hence, to obtain the adaptation mechanism, the transfer function of feedforward system is obtained first. The values of $\left(e\dot{d}, eq\right)$ obtained by subtracting the state vales of reference system and adaptive system will be purely positive real.

The value of $\omega_e^{'}$ from Eq. (8) is substituted to Popov's integral theorem and the inequality in the rule Eq. (9) is satisfied when:

$$\overline{\omega_r} = \left(K_{p1} + \frac{K_{i1}}{p^{'}}\right)\varepsilon \tag{10}$$

Thus, from Eq. (10), the adaptation scheme is a fuzzy-PI controller that provides a stable nonlinear feedback control. The fuzzy-PI controller adjusts the rotor speed as shown in Fig. 3. If $\overline{\omega_r}$, the estimated rotor speed is varied in the adjustable block such that the difference between the reference block output and adjustable block output is zero, then $\overline{\omega_r}$ will become same as the actual rotor speed $\widehat{\omega_{r1}}$. Any error quantity 'H' present will actuate the algorithm adapted for speed identification and allow the error to converge to zero.

3 Simulation Analysis

For obtaining a precise value of estimated speed without undergoing the tedious PI gain tuning process, a fuzzy-based adaptation scheme is used. This intelligent adaptation scheme renders the accurate value of speed. Also, the principle behind vector control lies in tuning the 3-PI controllers in the speed control loop which there by regulates the current controller to give proper switching signals to the inverter and also regulate the estimated speed to the desired value. The use of PI controllers indeed renders the desired value of speed. But it has no influence on neither reducing the distortions nor reducing the ripple in electromagnetic torque and current.

For the sensorless operation of IM drive and for speed control of IM using PI and FLC simulation studies are carried out using MATLAB. The THD analysis characteristics presented in this work is a unique analysis of the present work. It is clearly observed as a novel contribution of the work, highlighting the relative superiority of the FLC-based operation of IM. The proposed intelligent modules adopt the Gaussian type of membership function rather than the triangular memberships used in [12]. Although the inference process remains the same in both the memberships the Gaussian output membership functions are linear and constant. The method employs Type-I fuzzy controllers for speed computation [13]. The system provides fast response and is less computationally intensive. The two crisp values speed error and change in error are provided as inputs to the fuzzy system. These inputs are fuzzified, following fuzzification is the rule base where various rules are applied, and finally a crisp value is obtained through defuzzification. Here, minimal fuzzy rules are applied and the inference engine is made less intensive computationally. This makes the response faster and exhibits ease of trainability and high generalizability. This fuzzy controller predicts the requisite speed estimate from the machine model-based system and further helps in desired speed control. The overall simulation system of speed-sensorless applied to induction machine (Fig. 1) is simulated and the characteristic variations of speed, current, and torque as a function of time using PI and FLC is observed. A comparative study between the two is performed with the help of responses obtained.

3.1 No-Load Condition

Figures 3 and 4 show the speed versus time response with input reference speed varying from 50 rad/s to 100 rad/s along with zero reference load torque being applied for PI and FLC, respectively. It is observed that estimated speed response characteristics follow reference speed command input and magnitude of distortions are less for fuzzy logic controller compared to PI controller.

Figures 5 and 6 show the torque versus time response with input reference speed varying from 50 rad/s to 100 rad/s along with zero reference load torque being applied for PI and FLC, respectively. It is observed that electromagnetic torque response follows reference torque command input and the magnitude of ripple are less for fuzzy logic controller compared to PI controller.

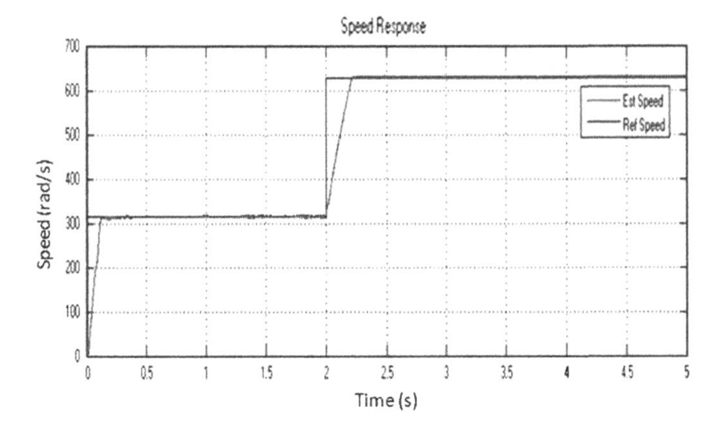

Fig. 4 Speed response characteristics for zero reference load torque condition using fuzzy logic controller

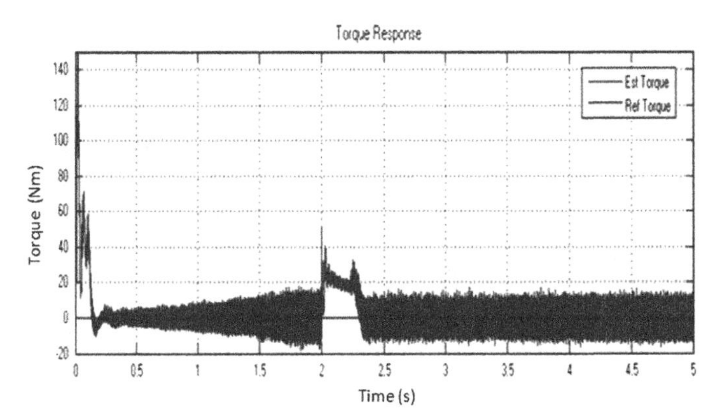

Fig. 5 Torque response characteristics for zero reference load torque condition using PI controller

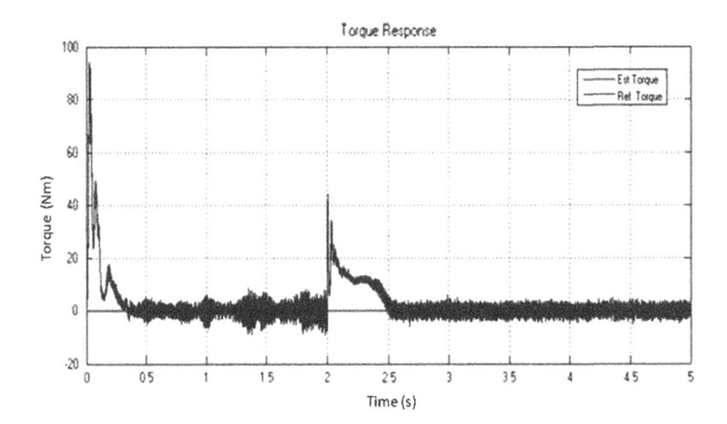

Fig. 6 Torque Response characteristics for zero reference load torque using fuzzy logic controller

3.2 On-Load Condition

The purpose of speed monitoring is that during several transients and dynamics the stability of the machine is ensured. The simulation analysis in the present context is carried out for various loaded conditions between 1 Nm and 10 Nm. The case study for speed response, torque response and the THD analysis for a load of 10 Nm is provided in Figs. 7, 8, 9 and 10. It is observed that estimated speed, electromagnetic

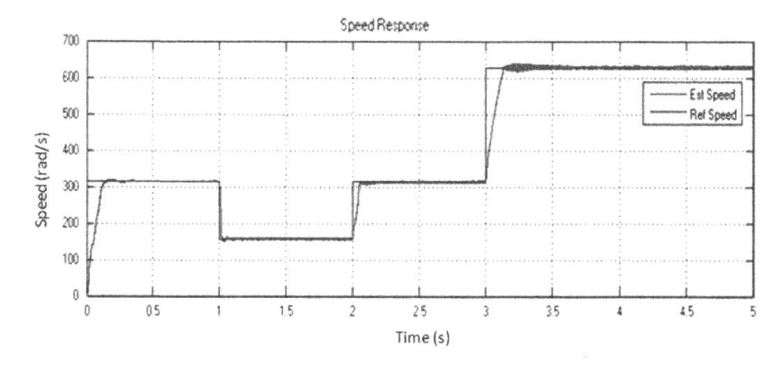

Fig. 7 Speed response characteristics for a reference load torque of 10 Nm Using PI controller

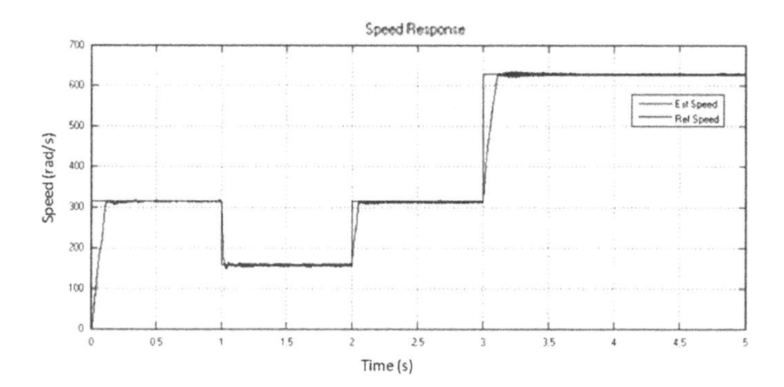

Fig. 8 Speed response characteristics for a reference load torque of 10 Nm using fuzzy controller

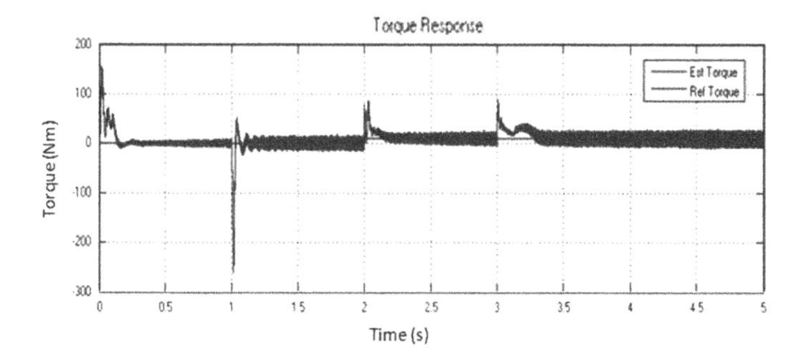

Fig. 9 Torque response characteristics for reference load torque of 10 Nm using PI controller

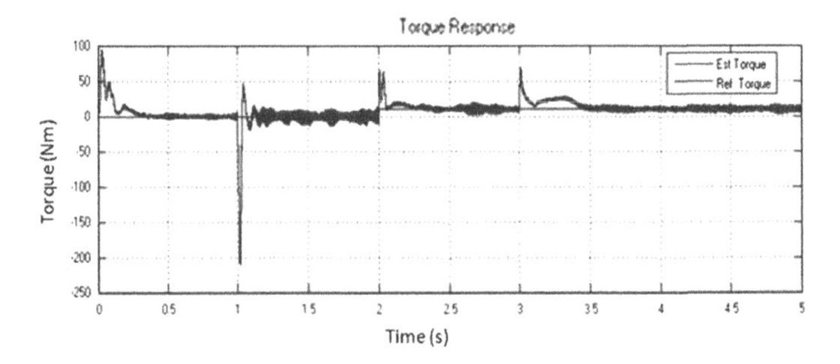

Fig. 10 Torque response characteristics for reference load torque of 10 Nm using Fuzzy controller

torque, and current response characteristics follow their reference command input and the magnitude of distortions are much reduced for fuzzy logic controller compared to PI controller.

3.3 *Total Harmonic Distortion (THD) Analysis*

In AC power system, measurement of power quality must be done at several points in a network. Total Harmonic Distortion (THD) analysis is performed to check whether current or voltage waveforms are deviating from the ideal sinusoidal pattern. In the present work THD analysis is performed for PI and fuzzy logic controller and following Fast Fourier Transform results are obtained as shown in Table 1. The responses obtained are the average value of THD obtained and is performed using Fast Fourier Transforms (FFT). The THD analysis on three-phase stator current characteristics for phase I_a, I_b *and* I_c are analyzed using FFT algorithm for PI and fuzzy logic controller-based Induction motor operations, respectively. THD analysis for PI and the fuzzy logic controller is performed and results obtained are mentioned in Table 1. It clearly shows that total harmonic distortions for the fuzzy logic controller are reduced for speed, torque and three-phase stator current responses in comparisons with PI controller. Hence, the adaptive fuzzy-based intelligent control is preferred for sensorless operation of IM drive than any other AI technique.

Table 1 Comparative THD analysis between PI and fuzzy logic controller

THD analysis performed on waveforms	PI controller (THD in %)	Fuzzy controller (THD in %)
Stator current I_a	96.87	41.24
Stator current I_b	81.20	41.05
Stator current I_c	78.25	40.71

4 Conclusions

With the aim to improve the sensorless speed control phenomenon and hence enhance the drive performance, the adaptive intelligent controller is introduced in the vector control scheme. Fuzzy control is considered as the most familiar technique to reduce the torque and flux distortion of the IM with less current ripples which thereby improves the controller behavior. Significant conclusions obtained from the machine-based model is that the proposed nonlinear estimator involves minimal computations. The processing is less sophisticated and the algorithm is quite simple. Thus, the drive can be operated in hostile environments too. Their feature of low cost with greater reliability without mounting problems have made significant improvements in electric drive technology. Also, modeling of IM makes it possible to operate the drive with different machine parameter which is difficult if a direct asynchronous machine from MATLAB is used. The adaptive fuzzy-based sensorless speed estimation and control strategy are more robust than the conventional one. The use of Type-I fuzzy system enables faster response and lower steady-state error. The inference chosen by the Type-I fuzzy systems ensures a greater stability of the entire drive control system. Due to lesser rise time taken by the intelligent controller, this method gives faster steady-state response and this scheme has better reliability than the conventional PI scheme. The Dual FLC with its well-defined rule base obeys the adaptation mechanism and ensures the stability of the controller in a wide speed range and parameter variation thereby providing a robust control performance.

References

1. Teresa Orlowska-Kowalska and Mateusz Dybkowski: Stator Current based MRAS estimator for a wide range speed Sensorless induction motor drives, IEEE Transactions on Industrial Electronics vol. 51, no. 4, April 2010, pp. 1296–1308.
2. B. K. Bose, Power Electronics and Motor Drives, Pearson Education Inc., Delhi, India, 2003.
3. M. Rodic and K. Jezernik: Speed-sensorless sliding-mode torque control of induction motor, IEEE Transactions on Industrial Electronics, vol. 49, no. 1, pp. 87–95, February 2002.
4. L. Harnefors, M. Jansson, R. Ottersten and K. Pietilainen. Unified sensorless vector control of synchronous and induction motors, IEEE Transactions on Industrial Electronics, vol. 50, no. 1, pp. 153–160, February 2003.
5. M. Comanescu and L. Xu: An improved flux observer based on PLL frequency estimator for sensorless vector control of induction motors, IEEE Transactions on Industrial Electronics, vol. 53, no. 1, pp. 50–56, February 2006.
6. Radu Bojoi, Paolo Guglielmi and Gian-Mario Pellegrino: Sensorless direct field-oriented control of three-phase induction motor drives for low-cost applications, IEEE Transactions on Industrial Applications, vol. 44, no. 2, pp. 475–481, March 2008.
7. Boldea and S. A. Nasar, Electric Drives, Taylor & Francis, New York, 2006.
8. Joachim Holtz: Sensorless control of IM drives. In: Proceedings of IEEE vol. 90 August 2002, pp. 1359–1394.

9. C. Schauder: Adaptive speed identification for vector control of induction motors without rotational transducers, IEEE Transactions on Industrial Electronics, vol. 28, no. 5, pp. 1054–1061, September/October 1992.
10. M. Rashed and A. F. Stronach: A stable back-EMF MRAS-based sensorless low-speed induction motor drive insensitive to stator resistance variation. In: IEEE Proceedings on Electrical Power Applications, vol. 151, no. 6, pp. 685–693, November 2004.
11. F. Z. Peng and T. Fukao: Robust speed identification for speed-sensorless vector control of induction motors, IEEE Transactions on Industrial Applications, vol. 30, no. 5, pp. 1234–1240, September/October 1994.
12. G. Durgasukumar and M. K. Pathak: Torque ripple minimization of vector controlled VSI induction motor drive using neuro-fuzzy controller, International Journal of Advanced Engineering Science, vol. 1, no. 1, pp. 403, 2011.
13. Vijay Bhaskar Semwal, Pavan Chakraborty, G.C. Nandi: Less computationally intensive fuzzy logic (type-1)-based controller for humanoid push recovery. In: Robotics and Autonomous Systems 63 (2015): 122–135.

Steganography Using Bit Plane Embedding and Cryptography

Bharti Rathor and Ravi Saharan

Abstract In the modern era of digital advancement of images, all information are exchanged through Internet, so for exchanging any secret message, either cryptography or steganography is used. Cryptography is a process of converting data in unreadable form. We can embed even the existence of data itself. In this paper, we proposed a new robust and secure method to embed secret message in an image. To increase robustness of message, intermediate significant bits (ISB) are used instead of using LSBs. To increase the security of message, encryption technique is used. The objective of the paper is to embed a message in bit plane such that it is robust against various attack and transformation (like scaling, cropping, filtering, etc.) and also maintaining the perceptual transparency of stego-image.

Keywords LSB · ISB · SIHS · Discrete logarithm · Random numbers

1 Introduction

Nowadays, information exchanging on Internet has rapidly increased. Large amount of information is being exchanged through Internet. As we know, Internet is open to all, information may leak out and can be used for illegal purpose knowingly or unknowingly. So information security plays an important role in Information Technology.

In the growing digital world, multimedia takes an important role in communication. Lots of information are being exchanged through images from one side to other. A user can send secret data/message by hiding in image. Two concepts such as steganography and cryptography are used to exchange information in concealed manner. In cryptography, we can convert our original message in a form that any

B. Rathor (✉) · R. Saharan
Central University of Rajasthan Kishangarh, Kishangarh 305817, Rajasthan, India
e-mail: 2015mtcse003@curaj.ac.in

R. Saharan
e-mail: ravisaharan@curaj.ac.in

A.K. Somani et al. (eds.), *Proceedings of First International Conference on Smart System, Innovations and Computing*, Smart Innovation, Systems and Technologies 79, https://doi.org/10.1007/978-981-10-5828-8_31

unauthorized user cannot read and understand. It is encrypted using a key which is shared between sender and receiver only and without this shared key, user cannot get message back from encrypted message. We encrypt the message in such a way that intruder cannot decrypt the message even know the existence of message. But in steganography, intruder does not have any clear idea even about the existence of message that is hidden in the image being shared. Image steganography helps to provide those methods that embed our secret message without degrading the perceptual quality of carrier image. The carrier image and secret message together produce stego-image by applying algorithm. Then, stego-image is transferred through Internet to the recipient. The recipient gets back the message using extracting algorithm and that key which is provided by the sender from stego-image separately [1, 2, 3].

Architecture of steganography model using the method of cryptography is illustrated in Fig. 1.

In the image processing, various type of steganographic techniques exist which are categorized based on the type of hiding carrier object. Various steganography methods can be used to provide more security and robustness by using good carrier object. Since the growth of digital images in multimedia on Internet, people use images mostly for exchanging information to one end to the other [4, 5]. It can be shown in Fig. 2.

There are several types of methods in image steganography.

(1) LSB (least significant bit) methods
(2) Transform domain methods

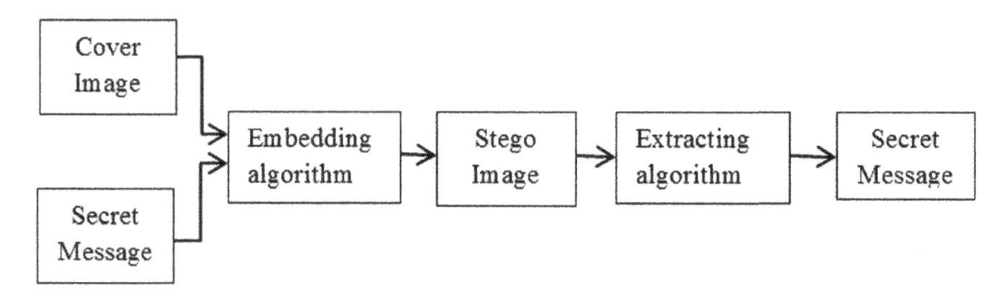

Fig. 1 Architecture of steganographic process using the concept of cryptography

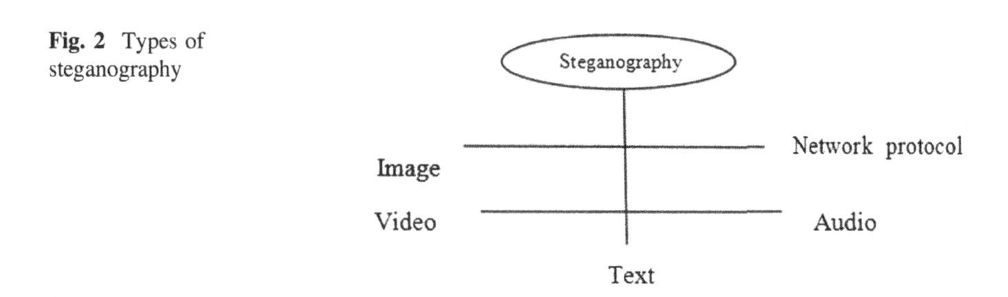

Fig. 2 Types of steganography

(3) Statistical methods
(4) Distortion methods

LSB is a simple and easy method to implement but mostly used to hide the message bits directly into the least significant bits of the cover image. Revamping the least significant bits that does not mean change is perceptible to human because of the less amount of change in intensity value of pixels. But it is not so robust and secure that stego-image can survive against various attacks. In proposed technique, we use ISB method to hide our original message in the intermediate significant bits (2nd–7th LSB's) of cover image depending on its binary coding [6].

In the transform domain technique, we transform the spatial domain method to frequency domain with the help of discrete cosine transform (DCT), fast Fourier transform (FFT), discrete Fourier transform (DFT), and discrete wavelet transform (DWT), etc.

In masking and filtering methods, two signals are embedded together in such a way that only one of the signals is perceptible to the human eye using human visual system (HVS). This method is mostly used in watermarking techniques [7, 8].

2 Related Work

In the area of image processing, many steganography methods exist for hiding message in an image. These methods can be categorized into: (1) spatial domain technique (2) transform domain technique. The spatial domain based steganography directly changes some bits in the image pixel values in hiding data using LSB method.

In [2], Akram and Azizah used ISB method for hiding the message in cover image in such a way to improve the security and robustness of stego-image. The method is based on checking the value of the secret message to the range of 4th bit plane. Author suggests the result after applying technique for every bit plane (1–8) and all possible values of bias (x), best robustness can get with stego-image using the 4th bit plane at the bias value = 6.

The value of 8-bit plane can be represented by 2^{n-1}, where n is the order of the plane starting from 1 to 8. The maximum and minimum value of intensity value of pixel that can fit in 8-bit plane is 255 and 0. Here, n=5 because we are cosidering 4th bit plane for embedding purpose that is 5th LSB of that number. So L = 25–1 = 16 and number of ranges is 256/L = 256/16 = 16 are [1:15] [16–31].... [248–255]. Each range is divided into two ranges that are left-hand group and right-hand group. Author embeds data bits in randomly chosen pixel based on the checking of pixel value lying in the left-hand group and right-hand group and replaced it by new pixel using the bias value.

In [7], Mohammad Ali Shamalizadeh Baei uses SIHS (secure information hiding system) method used to secure the secret message and decode the sequence mapping problem. Author uses discrete logarithm calculation to change the message in encrypted form. The main idea here is to generate a series of random numbers of length equal to the message length that ranging from 3 to 8. These series numbers will be used in random mapping. It will provide more randomness in encrypted form of secret message.

In [9], N.S. Raghav, Ashish Kumar, and Abhilasha Chachal suggest a novel technique to enhance the standard LSB technique by using pseudorandom generation using H'enon chaotic map. This encryption using pseudorandom generator provides more security to the system as the same subset of random numbers cannot be reproduced without knowing the random generator function and thus the secret data cannot be revealed easily. Further, they are used to encrypt to unhide the message in an image by selecting random pixels.

In [3], Anil Kumar and Rohini Sharma use hash-LSB method and encrypt data using RSA algorithm in order to achieve more randomness and hide data in carrier object. Author utilizes a hash function to create a pattern for concealing data bits into LSB of cover image. RSA algorithm gives very secure method of steganography because of using large prime numbers for key that is to be used for encrypting and decrypting message. This technique becomes more security cause of RSA algorithm.

In [4], Bhavana S. and K.L. Sudha explain about a way in which data can be embedded using LSb method along with chaos. Chaos-based techniques provide more security that can be revamped by utilizing multiple chaotic maps for encryption as well as decryption and hiding of message in cover image.

In [10], Y.K. Jain and R.R. Ahirwal have suggested an efficient LSB method which is used for generating a stego-key by dividing the image pixels ranges (0–255). Decide the fixed number of bits insertion into each range and adaptive number of bits insertion into different ranges based on pixel count of cover image in different ranges. K-bits of secret message are substituted into least significant part of pixel value.

The strength of proposed method is its perceptibility and more randomness and high hidden capacity of secret hidden information in stego-image. This method is used to provide high hidden capacity with robustness and more randomness of secret data. RGB image has more capacity to hide the message in cover image because it has 24 bits of each of three channels red, green, and blue [11]. So it provides more randomness and high hidden capacity. Generally, we use grayscale image for simple and efficient technique to hide information.

In this work, we are using SIHS method for providing more security and random selection of pixels, pseudorandom function generator for more randomness in bits of binary form of secret message, watermarking method for hiding message bits in such a manner that message should undamaged as same as possible; detailed procedure is explained in the next section of the paper.

3 Proposed Work

We proposed a new technique to hide secret message in cover image which consists of random cipher technique, embedding, and extracting algorithm.

In order to increase the security and robustness of the method, the secret message is encrypted with random cipher technique with the key which is to be known to sender and recipient only.

3.1 Cipher Technique

This is the process of generating a series of random numbers using key and encrypting a message (series of characters) using randomly generated numbers as, Caesar cipher technique. Random cipher encryption technique is explained in the following example.

Plaintext	A	T	T	A	C	K	A	T	D	A	N
Key	3	9	7	6	2	4	8	5	1	6	3
Cipher text	D	C	A	G	E	O	I	Y	E	G	Q

Security of this cipher technique is depends on the key.

3.2 Discrete Logarithm

It is used to generate a series of random numbers and then added into message characters. Produced series of characters known as encrypted message. These random number values are computed using the following Eq. 1:-

$$x(i) = a^{*}x(i-1)(mod\, p). \quad (1)$$

where i = 1, 2, 3…, m

X (0) = is the sum of k digits.

a = 3 * x (0) and p = K.

These random numbers are also used for selecting pixel position where message is to be hidden [6].

3.3 Embedding Process

For improving the robustness of the method, hide the encrypted message in the cover image using concept of ranges and ISB method. We use the fact of 4th bit

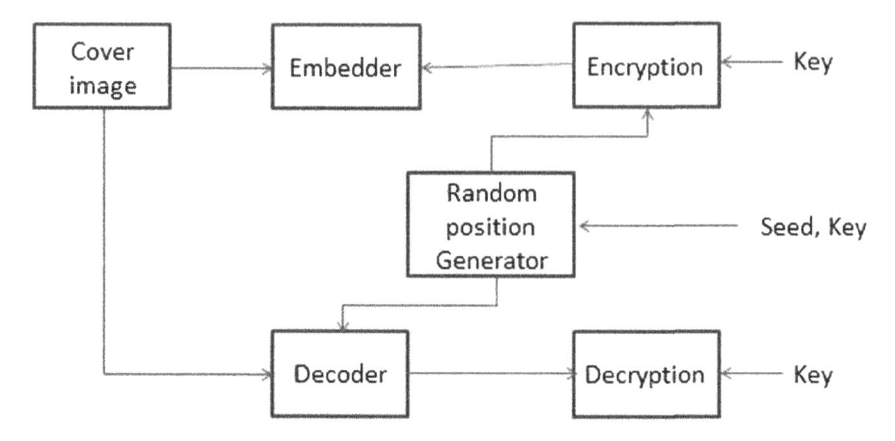

Fig. 3 Architecture of proposed model

plane and bias value is equal to 6 suggested by author in [2]. Proposed work model is given in Fig. 2 (Fig. 3).

4 Proposed Algorithm

In this section, we describe the proposed method to enhance security and robustness of stego-image. That contains embedding extracting algorithm and Vigenere cipher technique. The embedding and extracting algorithms are described below:

4.1 Embedding Algorithm

This algorithm is explained in the following steps.

- Input: An M × N carrier image and a secret message.
- Output: An M × N stego-image.

(1) Read cover image and secret message.
(2) Apply random cipher technique using key to encrypt the secret message for securing the message.
(3) Convert the decimal form of encrypted message to binary form.
(4) Now select random column of cover image matrix pixel using random numbers generated by discrete logarithm.
(5) Calculate range for selecting the 4th bit plane by $l = 2^{n-1}$(n = position of LSB, here n is 5for 4th bit plane) and number of ranges is 256/l.

(6) Every range is broken down into two equal groups L (left hand) and R (right hand); each group length is l/2.
(7) Now start comparing the 4th bit of selected pixel with each bit of binary form of encrypted message, this process is repeated for each row of selected random columns until the message length. If this equation is satisfied then go to Step 8 else go to Step 9.
(8) If

    (a) pixel lies in the left group, calculate difference = original pixel value—minimum value of the range (where d is the distance between the original pixels and closer edge of the ranges).
    Check whether (D >= b) is satisfies or not, if it satisfies then the pixel value will be the same else become (minimum pixel value of the range + b). Here, b = bias value.
    (b) If pixel lies in the right group, d = maximum pixel value of the range—original pixel value. Check whether (D >= b) is satisfies or not, if it satisfies then the pixel value does not changed else become (maximum pixel value of the range –b).

(9) If

    (a) Pixel value lies in the left group or in last group, the intensity value of pixel become (maximum pixel value of the previous range –b).
    (b) Pixel value lies in the Right group or in the first range; the intensity value of pixel is (minimum pixel value of the next range +b).

(10) So hiding of encrypted message is completed and at last, we generate stego-image.

4.2 Extracting Algorithm

This algorithm is explained in the following steps.

- Input: An M × N stego-image
- Output: A secret message.

(1) Read the stego-image.
(2) Store the intensity value of original pixel and changed pixel of cover and stego-image.
(3) Calculate the range of both pixels for each row of selected same random columns as embedding algorithm using seed and random permutation until the length of message.
(4) If both pixel values lie in same range then embedded bit is 4th bit of original pixel else go to step 4.

(5) If 4th bit of pixel is 0 then embedded bit is 1 else 0.
(6) So, we get combination of 1-0 bits and then convert it into decimal form.
(7) Convert decimal form to character form of encrypted message.
(8) Decrypt the encrypted message using reverse of random cipher technique using same key as used before in encryption technique. At last, we get original message which was hidden in cover image.

5 Implementation and Results

5.1 Experimental Setup

The proposed algorithm is implemented on these configurations (Table 1).

5.2 High Capacity

For any steganographic technique, message size to be hidden in an image should be as large as possible. If secret message spread throughout the whole image then conceal the existence of message and revealing is hard to intruder. Proposed method provides high capacity of data which is to be embedded in an image.

5.3 Perceptual Transparency

After embedding, stego perceptual quality will be compromised into stego-image as compared to cover image. Quality of stego-image depends on the data hiding method for embedding message bit in an image. Perceptual transparency should be high. For that, we hide the message bit on the intermediate significant bit instead of least significant bit. Proposed method maintains perceptual transparency of stego-image, which can be shown as in Fig. 4 (Fig. 5).

Table 1 Configuration details

OS	Windows 7 ultimate 64 bit
Processor	Intel core i3 first generation
Memory	4 GB
Software used	MATLAB 2011a
Input image	Standard grayscale images of size 512 * 512 are used

Fig. 4 Cover and stego-images

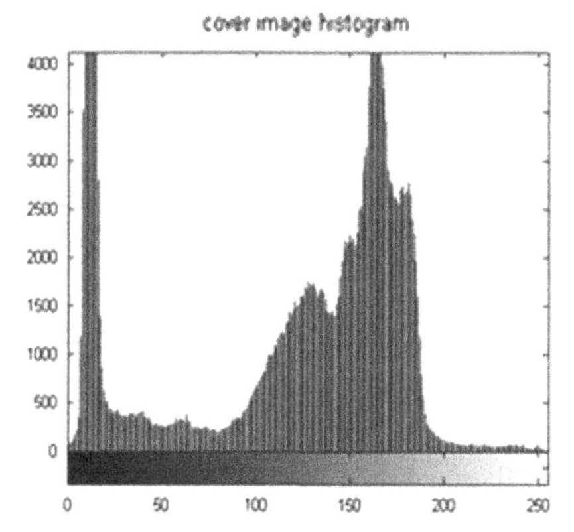

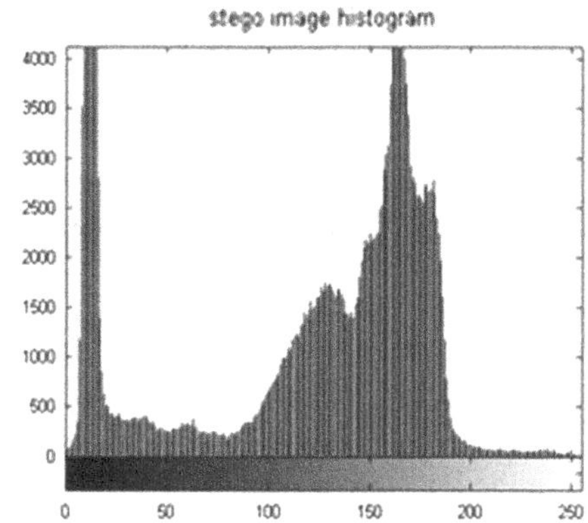

Fig. 5 Histograms of cover and stego-images

5.4 Temper Resistance

If once the message has been hidden into stego-image, it should be complex to modify the message. Any intermediate person cannot alter the secret message after embedding it into carrier object.

5.5 Computational Complexity

Computational cost for embedding and extracting an embedded message using applied algorithm is defined as computational complexity.

The computational complexity of infrequent operations is proportional to nnz, the frequency of nonzero elements in the matrix. Proposed method is less complex, efficient, and takes less time. So, it maintains computational complexity as much as less possible.

5.6 Robustness

After hiding message in the cover image, message should remain same if covered image handles with some operation such as cropping, scaling, filtering, JPEG compression, and addition of noise.

To check the robustness of suggested method, MSE and PSNR will be used. The MSE and PSNR are defined as given by Eq. 2 and 3.

$$MSE = \frac{1}{mn}\sum_{i=0}^{m-1}\sum_{j=0}^{n-1}[I(i,j) - I1(i,j)]^2. \quad (2)$$

MSE = mean square error rate between two objects.

Where I = Cover image, I1 = Stego-image and m. n = Number of pixels.

$$PSNR = 10\,\log_{10}\left(\frac{R^2}{MSE}\right). \quad (3)$$

PSNR = Peak signal-to-noise ratio between two objects.

In the Eq. 3, R is defined as maximum oscillation in the input image data type. For example, if the input images have a double-precision floating-point data type, then R is 1. If input image has an 8-bit unsigned integer data type, R is 255.

The security comparison between images before and after applying attacks is given in Tables 2 and 3, respectively.

Table 2 PSNR and MSE between cover and stego-images

Images	PSNR (dB)	MSE
Cameraman	57.5122	0.1303
Vase	57.7232	0.1107
Woman_blonde	57.7385	0.1103

Table 3 PSNR and MSE of images after applying attacks

Images/PSNR and MSE between images		Gaussian noise attack	Salt and pepper attack	Poison noise attack	Filtering attack	JPEG compression attack
Cameraman	PSNR	56.9814	55.8912	54.8901	53.721	53.8231
	MSE	0.1621	0.1456	0.1483	0.1324	0.1782
Vase	PSNR	56.8761	55.8281	54.8741	53.7665	53.7671
	MSE	0.1456	0.1345	0.1567	0.1324	0.1564
Woman_ blonde	PSNR	56.7981	55.8720	54.8140	53.7634	53.7650
	MSE	0.1345	0.1567	0.1236	0.1546	0.1721

6 Conclusion

The technique is an enhancement over existing methods compared to several methods of steganography while maintaining the method simple enough and keeping it easy to implement according to practical aspects. We have used the random cipher technique using key and random selection of pixel from cover image, which gives extra security to the stego-image. The proposed algorithm gives better security that provides secrecy of the original message and more randomness with robustness. The performance analysis is being done. The values of PSNR and MSE are also being calculated.

References

1. Abbas, Cheddad, Joan Condell, Kevin Curran, PaulMcKevitt. Digital image steganography: Survey and analysis of current methods. In: ELSEVIER (2010).
2. Akram M.zeki and Azihah A. Manaf.: A Novel digital watermarking Technique based on ISB. In: World Academy of Science, Engineering and Technology Vol:3 (2009).
3. Anil Kumar and Rohini Sharma.: A Secure image steganography based on RSA algorithm and Hash—LSB technique. In: International Journal of advance research in Computer Science and software engineering volume 3 Issue, July(2013).
4. Bhavana. S and K.L. Sudha.: text steganography using LSB insertion method along with chaos theory. In: IEEE Explore (2011).
5. Mehdi Hussain and Mureed Hussain.: A Survey of Image steganography Techniques. In: (SZABIST), Islamabad, Pakistan. (2013).
6. R.O.EI Safy and H. H. Zayed.: An Adaptive Steganographic Technique Based on Integer Wavelet Transform. In: IEEE Explore (2009).
7. Mohammad Ali Shamalizadeh Baei and Reza karami.: A new algorithm for embedding message in image steganography.In:International Journal of Engineering Research & Technology (IJERT) Vol. 3 Issue 2, February (2014).
8. Raja Lakshmi C S, Sowjanaya T P and Hemant Kumar C S.: Image Steganography using H-LSb Technique for hiding Image and Text using Dual Encryption Technique. In: SSRG international journal of Electronics and communication Engineering (SSRG-IJECE)-volume 2 Issue 5, May 2015).

9. N S Raghav, Ashish Kumar, Abhilasha Chachal. Improved LSB method for Image Steganography using Henon chaotic Map. In: Open Journal of information Security and applications volume 1, July(2014).
10. Y. K. Jain and R. R. Ahirwal.: A Novel Image Steganography Method With Adaptive Number of Least Significant Bits Modification Based on Private Stego-Key. In: International Journal of Computer Science and Security (IJCSS), vol. 4, (2010).
11. Prashant Manuja and Ravi Saharan.: Statistical Pixel blocks Selection. In: International conference on Advanced communication control and computing technologies(ICACCCT) ISBN No.978-4799-3914-5/14/$31.00@2014 @ (2014).

Minimizing Fuel Cost of Generators Using GA-OPF

Divya Asija, P. Vishnu Astick and Pallavi Choudekar

Abstract In the current scenario of deregulated power system congestion in the network is one of the critical issues, which need to be resolved for better economic and system efficiency. Several measures have been adopted to relieve the system from congestion for maximization of social benefit. Fuel cost minimization is considered as the key factor for social welfare maximization, which provides benefits to both buyer and consumer of electricity. The fuel cost minimization for power system is achieved based on many parameters one of them being generator allocation. This paper uses GA-based optimization algorithm to obtain optimal generator allocation to find the best fit having minimum fuel cost. The proposed approach for fuel cost minimization is tested on IEEE 14 bus standard system.

Keywords Genetic Algorithm (GA) · Linear Programming (LP)
Optimal power flow (OPF) · Quadratic Programming (QP)

1 Introduction

1.1 The Scenario

The electrical power system expanded widely in the past century. The issues related to the load dispatch and the economics behind it has been a prime focus and addressed to since then.

It was in the beginning of 1960s, that the proposed optimal power flow (OPF) solution was taken up. The predecessor of this technique, i.e. economic load

---

D. Asija (✉) · P.V. Astick · P. Choudekar
Department of EEE, Amity University, Noida, Uttar Pradesh, India
e-mail: dasija@amity.edu

P.V. Astick
e-mail: astick.vishnu@gmail.com

P. Choudekar
e-mail: pachoudekar@amity.edu

A.K. Somani et al. (eds.), *Proceedings of First International Conference on Smart System, Innovations and Computing*, Smart Innovation, Systems and Technologies 79, https://doi.org/10.1007/978-981-10-5828-8_32

dispatch considers various controllable variables and constraints for the observed system. The optimal power flow technique determines the optimal setting for such control variables along with considering the defined constraints [1].

Hence as on today, the OPF can solve optimization problems of large, complex and widespread power systems in much less time. Moreover, in OPF technique all the control variables have to be found so as to achieve optimization of the predefined objective.

These techniques minimize the operational and management costs of a power system and consider the operational limitations on various controls and function variables as well as dependent variables.

Various methods and techniques are used to model power dispatch problems. They are based on Langrangian multipliers [2], linear programing methods [2, 3], quadratic programing method [4], Newton's method [5, 6] and steepest descent non-optimized conventional techniques. However, these techniques have unsettled convergence and fail to have a global solution. The reasons could be the nonlinearity of OPF and their dependency on initial guess.

Hence the non-optimized methods based on the Artificial Intelligence like evolutionary programing (EP), genetic algorithm (GA) [7], simulated annealing (SA), particle swarm optimization technique (PSO) [8, 9], tabu search algorithm (TS), ant colony optimization algorithm (ACO) are being adopted to implement and solve the optimal power flow techniques.

1.2 Introduction to Genetic Algorithm

Genetic algorithm is based on the theory of genetic evolution. The formulated problem is used to generate various solutions, which are called as 'population', the individual solutions are called as 'chromosomes'.

The better solutions from the population are selected to form a new set of solutions or 'populations' based on chromosome recombination (crossover and mutation) to get the 'offsprings'. The fitness of the new population is tested and compared to filter out the better optimal solutions. This process is repeated until a defined fitness standard is attained.

Chromosomes are string structures, which are typically concatenated list of binary digits, which are derived from the control parameters of the formulated problem.

The purpose of utilizing this technique is its robustness, fast solution. The conventional OPF techniques were computationally unresponsive, slow and data sensitive leading to corrupted results. Whereas GA overcomes all these issues. Figure 1 represents a flowchart of GA optimal power flow.

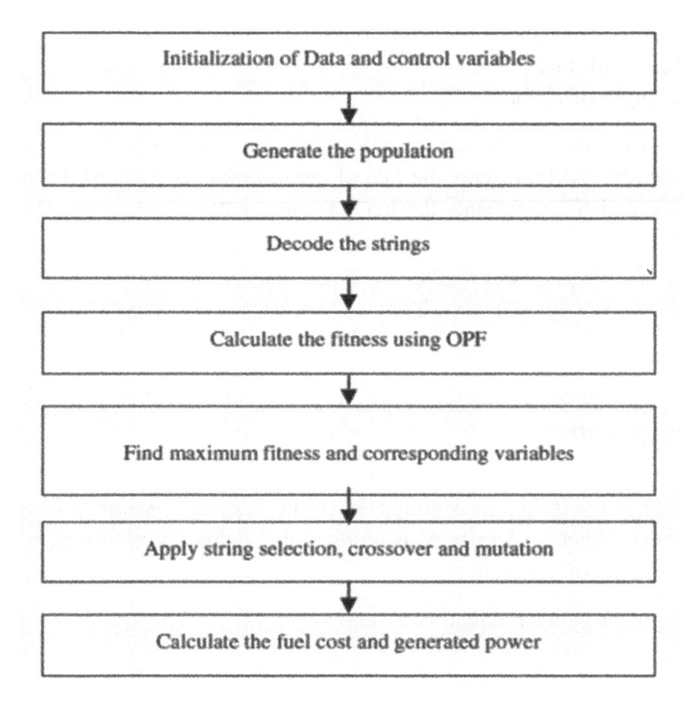

Fig. 1 Flowchart of GA for solving OPF problem

2 Problem Formulation

2.1 Introduction

The problem to be solved is represented as the objective cost function of a generating unit. It is the traditional quadratic function of cost. This function has to be subjected to genetic algorithm-based optimization, which provides the minimal cost subjected to various constraints as defined. Let G represent the objective function, which is the total generation cost.

- N_1 Total No. of generator buses.
- N_2 Total No. of buses.
- N_3 Total No. of voltage-controlled buses.
- P_j Injected Real power.
- Q_j Injected Reactive power.
- P_{lj} Load (Real) on jth bus.
- Q_{lj} Load (Reactive) on jth bus.
- P_{gj} Real power generation on jth bus.

Q_{gj} Reactive power generation on jth bus.
V_j Voltage at jth bus.
d_j Phase angle of jth bus.

The objective function or minimization function is dependent on three parameters specifically, 'A' represents the rate of consumption of fuel, 'B' represents the running cost, and 'C' represents the maintenance costs.

$$G = \sum_{j=1}^{N_1} G_j = \sum_{j=1}^{N_1} \left(A_j Pg_j^2 + B_j Pg_j + C_j \right) \text{ \$/h} \tag{1}$$

2.2 Constraints

The above objective function is subjected to the various constraints of active and reactive power balance, security constraints which define the limits on real power, voltage and angle of generation.

(a) Active power balance equation of network:

$$P_j(V, d) - Pg_j + Pl_j = 0 \quad (j = 1, 2, 3, \ldots N_2) \tag{2}$$

(b) Reactive power balance equation of network:

$$Q_j(V, d) - Qg_j + Ql_j = 0 \quad (j = N_3 + 1,\ N_3 + 2 \ldots N_2) \tag{3}$$

(c) Soft constraints or security-related constraints:

- Limits on real power generation

$$Pg_j^{min} \le Pg_j \le Pg_j^{max} \quad (j = 1, 2, \ldots N_1) \tag{4}$$

- Limits on voltage generation

$$V_j^{min} \le V_j \le V_j^{max} \ (j = N_3 + 1,\ N_3 + 2 \ldots N_2) \tag{5}$$

- Limits on voltage angles

$$d_j^{min} \leq d_j \leq d_j^{max} \quad (\mathrm{j}=2, \ldots \mathrm{N}_2) \tag{6}$$

The real and reactive power flow equation of the bus system defines the stability. It is also expressed or represented as

(d) Real power equation:

$$P_j(V,d) = V_j \sum_{k=1}^{N_2} V_k \left[G_{jk} \cos\left(d_j - d_k\right) + B_{jk} \sin\left(d_j - d_k\right)\right] \tag{7}$$

(e) Reactive power equation:

$$Q_j(V,d) = V_j \sum_{k=1}^{N_2} V_k \left[G_{jk} \sin\left(d_j - d_k\right) + B_{jk} \cos\left(d_j - d_k\right)\right] \tag{8}$$

$Y_{jk} + G_{jk} + jB_{jk}$, are elements of admittance matrix.

2.3 Genetic Algorithm

The chromosomes are generated as binary strings in 1's and 0's format in a random process. These chromosomes represent the various control variables of the system under consideration. In this paper, the various control variables used are as defined in the problem formulation.

For deriving the decimal values of the control variables after the random process of generation, the following equation is used.

For control variables, Let $D_j(x_1, x_2, x_3, \ldots)_2$ be the values of bits in decimal form;

A_j^{min}, the least possible generation value;
A_j^{max}, the largest possible generation value;
reqbits, the total number of bits required, then

$$A_j = A_j^{min} + D_j(x_1, x_2, x_3, \ldots)_2 \times \left(\left(A_j^{max} - A_j^{min}\right) / \left(2^{reqbits} i - 1\right)\right) \tag{9}$$

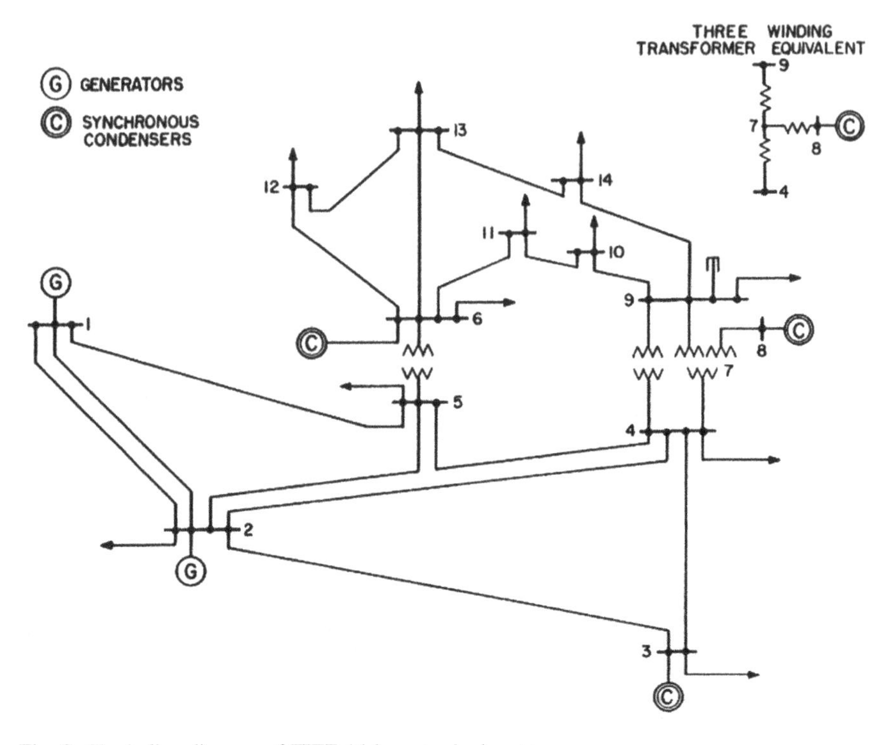

Fig. 2 Single line diagram of IEEE 14 bus standard system

2.4 Fitness Function and Parameters

The fitness function in a GA defines or determines the quality of each chromosome or solution obtained. The output of GA is intended to maximize this fitness function. Maximizing the value of fitness achieves the main objective of the OPF problem is to have the least possible value of the total generation cost while considering the defined constraints as well.

$$\text{The function, } Fitness = \frac{1}{(1 + Function\ value)} \tag{10}$$

The proposed GA algorithm-based OPF is applied and verified on IEEE-based 14 bus standard system. Figure 2 shows the IEEE 14 bus standard system. The 14 bus system consists of 3 generator buses 20 lines, 11 load bus, 3 tap changer transformer and total load capacity of 259.3 MW. The others system parameters are given as below.

2.4.1 IEEE 14 Bus Parameters

The IEEE 14 bus system model developed in MATLAB consists of 14 buses with 20 branches. This system has got three generators connected, which supplies a load of 270.1 MW. In addition, there are three transformer taps also in the system.

Total no. of buses	14
Total no. of branches	20
Total no. of generators	03
System load (MW)	270.1
Total no. of transformer taps	03

2.4.2 GA Parameters

The total population size considered for generation is 50 with a maximum generation limit of the new population up to 100. The selection tool or technique used is a stochastic roulette wheel. The initial crossover and mutation probabilities are considered as 0.9 and 0.001 respectively.

Populace size	50
Maximum generation	100
Operator for selection	Stochastic remainder roulette wheel
Initial crossover probability	0.9
Initial mutation probability	0.001

3 Results and Conclusion

Table 1 shows the line flow of system using Newton Raphson power flow technique with and without constraints.

In Table 2 the superiority of GA-based technique over the classical method in terms of cost optimization and generator allocation at the buses 1, 2 and 8. The GA-OPF method evidently is economic than the classical method providing decrement in cost by 9.575 ($/h).

Figure 3 shows the optimized value of generator fuel cost when GA-OPF is applied. The fuel cost has been reduced by 9.575 $/h with GA-OPF as compared to conventional OPF method which enhances the system efficiency economically. The objective is to have the optimized value of generator fuel cost while sustaining all the system limitations utilizing the genetic algorithm. Obtained results show that the

Table 1 Line flows with and without constraints

Line no	Line flows (MVA) without constraints	Line flows (MVA) with constraints	Line flows limits (MVA)
1	123.63	119.765	120
2	58.602	57.909	65
3	23.547	21.923	36
4	45.089	42.801	65
5	14.467	13.054	50
6	39.72	40.399	63
7	42.621	42.868	45
8	51.120	51.133	55
9	30.024	30.015	32
10	0.737	0.737	45
11	17.862	17.857	18
12	29.828	29.819	32
13	8.646	8.641	32
14	6.371	6.379	32
15	7.937	7.938	32
16	18.245	18.249	32
17	11.324	11.319	18
18	2.476	2.485	12
19	1.527	1.528	12
20	4.809	4.815	12

Table 2 Total fuel cost for generation using GA-OPF

Methods	Generator power (MW)			Total fuel cost ($/h)
	P1	P2	P3	
Classical	153.263	74.643	42.685	1155.502
GA-OPF	153.963	74.953	41.585	1145.927

generation as well as distribution cost function is minimized for the benefit of both the transmitting company as well as the consumer.

Table 3 shows the comparison of proposed technique with the quadratic and conventional techniques. It shows that GA-OPF method gives least cost value, i.e. 1144.9 ($/h) as compared to the classical and quadratic techniques. Thus the proposed method has proved to be a better technique for having minimum fuel cost thereby maximizing the social welfare function providing benefit to generating companies and consumers.

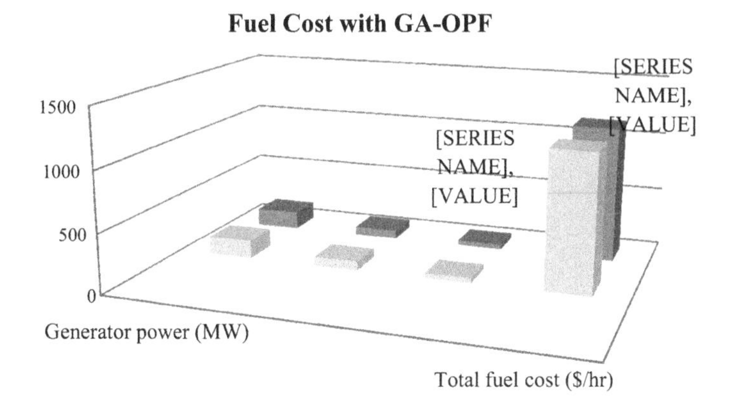

Fig. 3 Fuel cost with GA-OPF

Table 3 Comparison of proposed technique with conventional techniques

Power flow methods	Total fuel cost ($/h)
Classical	1155.5
Quadratic	1146.9
GA-OPF	**1144.9**

References

1. S. Kumar. K. Chaturvedi,: Optimal Power Flow Solution Using GA-Fuzzy and PSO-Fuzzy, In: Journal for the Institution of Engineers (India), 2014.
2. S. P. Shalini; K. Lakshmi, : Solution to Economic Emission Dispatch problem using Lagrangian relaxation method, In: ICGCCEE, 2014.
3. B. Stott, J.L. Marinho, : Linear programming for power system network security applications. In: IEEE Trans. Power Appar. Sys., Volume 98, No.3, pp. 837–848, 1979.
4. Rony Seto Wibowo, Kemas Robby Firmansyah, Ni Ketut Aryani, Adi Soeprijanto, : Dynamic economic dispatch of hybrid microgrid with energy storage using quadratic programming, In: IEEE Region 10 Conference (TENCON), 2016.
5. Enrique Acha, Behzad Kazemtabrizi, Luis M. Castro, : A New VSC-HVDC Model for Power Flows Using the Newton-Raphson Method, In: IEEE Transactions on Power Systems, 2013.
6. Abdullah Umar, Anwar.S.Siddiqui, Naqui Anwer, : Generation Fuel Cost Minimization of Power Grid Using Primal Dual Interior Point OPF (Optimal Power Flow) Method. In: IJIREST, 2014.
7. K. Vijayakumar and R. Jegatheesan, : Optimal Location and Sizing of DG for Congestion Management in Deregulated Power Systems. In: Springer-Verlag Berlin Heidelberg, 2012.
8. J. Rizwana, R. Jeevitha, R. Venkatesh, K. Shiyam Parthiban, : Minimization of fuel cost in solving the power economic dispatch problem including transmission losses by using modified Particle Swarm Optimization. In: IEEE International Conference on Computational Intelligence and Computing Research (ICCIC), 2015.
9. LI Xiang, LIU Yu-sheng, YANG Shu-xia, : Application of improved PSO to power transmission congestion management optimization model. In: Springer J. Cent. South Univ. Technol, 2008.

Stack Automata-Based Framework for Behavior Modeling of Virtual Agents

Saurabh Ranjan Srivastava and Angela Joseph

Abstract This paper presents a framework for modeling the behavior of virtual agents participating in human–computer interaction. For a given input statement, this framework derives the response of the virtual agent from a combination of 2 parameters, context and the emotional state. A stack based automaton is employed to record and track the values of these parameters for constructing the behavior and response of the agent. The approach is later implemented on a quiz based virtual agent response system. The framework can be utilized to enhance the quantum of emotional expressiveness in the response of the virtual agents for more human-like interaction with the machine.

Keywords Automaton · Virtual agent · Transition · Framework
Context · State · Tansduction

1 Introduction

Today automated virtual agents have become a prominent fragment of the human computer interaction domain. Applications such as movie animations, automated voice response systems and games are few examples of this technology [1]. The success of a virtual agent heavily depends upon its quality of realistic and credible response. To generate an output of quality sufficiently close to human response, it is crucial to match its context and event with the specified set of emotional states and vice versa. This paper presents a framework for generation of response by a virtual agent based upon the parameters mentioned above. In later section of this paper, we implement and compare this framework with the responses of ELIZA program [2].

S.R. Srivastava (✉) · A. Joseph
Swami Keshvanand Institute of Technology, Management & Gramothan,
Jaipur, India
e-mail: srs@skit.ac.in

A. Joseph
e-mail: angelsgloryangela@gmail.com

A.K. Somani et al. (eds.), *Proceedings of First International Conference on Smart System, Innovations and Computing*, Smart Innovation, Systems and Technologies 79, https://doi.org/10.1007/978-981-10-5828-8_33

The results of comparison state that responses tagged in accordance to context and relevance are more appropriate compared to responses generated by simple string manipulation chat programs.

2 Components of Proposed Framework

2.1 Response Parameters

The proposed framework constructs the response of a virtual agent on a rule based combination of 2 parameters, namely context and emotional states. These parameters guide the switching of emotional states as well as the generation of output statements for the agent by using the transition rules. These parameters are discussed ahead.

2.2 Context

The environment or the circumstances under which the specific input is provided to the agent can be termed as the context. For constructing a human-like response, it is crucial for the virtual agent to store, recall and analyze the context effectively. In this process, current as well as previous context, both are crucial for switching of emotions and responses. For versatility, it can be also assumed that a virtual agent can process arbitrary inputs linked to different contexts. Hence, a dedicated memory stack will be required for storage of context data. In the proposed framework, the memory stack holds the **relevance tag** relative to the current context. This implies that the context of the current input statement will be compared to the immediate last context stored on top of the stack to decide and record its relevance. This feature can be extended for comparison with short term and long term context values also.

2.3 Emotional State

The proposed framework regulates the emotions of the agent on the basis of Plutchik's model of basic human emotions [3]. On the current stage of development, out of 24 emotions proposed by Plutchik, only 3 emotions have been implemented in the behavior of the basic agent. Limiting the set of emotions derives some pros and cons concurrently. Today the most popular applications of virtual agents are of office assistants, trainers, teachers and personal assistants [4]. In terms of these applications, the emotional states of an agent must be confined in a moderate set to showcase a presentable and professional behavior by the agent [4]. Restricting the set of emotions will also simplify the computational complexity of the agent's behavior model. But in parallel to these advantages, a restricted set of emotions will also confine the possible set of responses of the agent also. Otherwise an unlimited or non-deterministic set of emotions can influence the overall possible output response in an unexpected manner if specified by the transition rules of personality and behavior model.

Hence, with more number of transactions defined over a larger set of emotional states, a more complex model of behavior can be constructed for a virtual agent. Such versatile model of agent's personality can be expected to execute a finer judgement of inputs with more accurate selection of responses.

3 Automata for Behavior Modeling

Formal computational models present a wide range of solutions for applications involving various automata theories. Finite automata (FA), push down automata (PDA) and Turing machines (TM) are examples of automata worth consideration [5–7]. The designs of these formal machines comprise of a complex function called the finite-state controller [8]. This controller is attached to a movable read/write head on an unbound storage tape. The transition conditions specified for the system are implemented by this finite-state-controller. In turn, the inputs accepted and responses generated through the transitions are read and written on the unbound storage tapes [5, 6, 8, 9].

3.1 Finite Automata and Human Behavior

For modeling complex human behavior through the modest finite automata architecture, a skeptic analysis is essential. The advantages of using finite-state automata (FA) [8] for implementing the design of virtual agents can be summarized as follows.

The finite set of states can be modeled as the possible emotional states of the proposed agent. The switching of emotions and moods can be efficiently modeled via these states. The stack of the automata design, specially proposed in push down automata (PDA) [5], can be an effective storage for current and previous values of emotions, contexts and events.

Finally, the finite-state controller function that models the transition rules, can be employed to implement the emotional switching rules regarding the context and events.

But contrary to the gains of using automata discussed above, few limitations must be also considered. A pre-defined finite set of emotional states and transition rules can limit the capacity of the agent to respond in an unfamiliar scenario.

Similarly, combination of 2 or more states may result into an undefined state and/or transition which may be out of the scope of the proposed agent.

Keeping these pros and cons in mind, automata can be effectively employed to model and implement virtual agents with human-like behavior and responses.

3.2 Defining Finite-State Stack Transducer

A finite-state transducer with a memory stack [10–12] can provide an effective transition control function to implement switching of emotional states and response generation. For the proposed virtual agent framework, the finite-state stack transducer (FSST) [10] displayed in Fig. 1 is used. FSST is an assembly of 1 read and 1 write head for accepting and printing of alphabets onto 2 different unbound tapes [10].

Here the unboundedness of tapes presents the infinite possible set of input and output symbols for an automaton. The input symbols are processed into output symbols by a processing element composed of 2 functions. The first is the transition function δ which on accepting an input symbol, maps the transition of the FSST from one state to another state [8]. The second component, known as the transduction function λ, maps an input symbol to an output symbol [7].

The mathematical definition of a finite-state stack transducer (FSST) can be presented as an 8 tuple [10] $< q_0, Q, F, \Sigma, \delta, \lambda, \Lambda, M >$ where:

$q_0 \in Q$ = the initial state
Q = a finite set of states
$F \subseteq Q$ = the final states, a subset of Q
Σ = a finite set of input symbols (the input alphabet)
Λ = a finite set of output symbols (the output alphabet)
M = a finite set of memory stack symbols (the context memory alphabet)
δ = set of transition functions, exactly as for a finite automaton (FA), mapping a pair of a state and an input symbol to a state and a response event

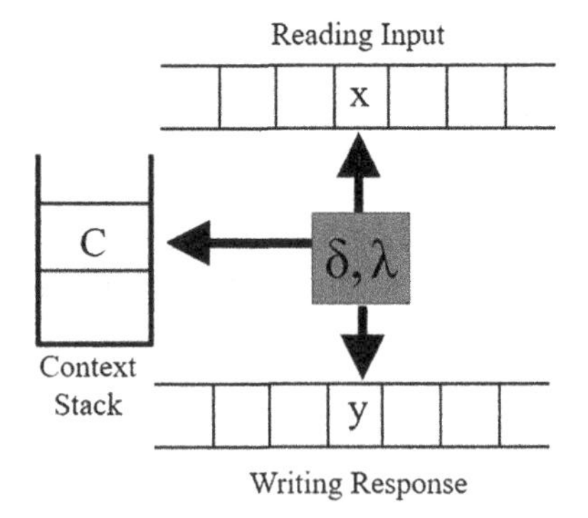

Fig. 1 Conceptual architecture of a finite-state stack transducer [10]

$$\delta: Q \times \Sigma \times M \rightarrow Q \times M$$

λ = the transduction function (emission function), mapping a state and an input symbol to an output symbol a response event.

$$\lambda: Q \times \Sigma \times M \rightarrow \Lambda \times M$$

The context alphabet C as presented in Fig. 1, is added to both the transition as well as the transduction function to improvise the response of the system. This way, now the proposed automaton will be capable of deciding the upcoming response on multiple parameters [9].

4 Framework for Virtual Agent

An example of a virtual agent with a limited set of 3 emotions, Acceptance, Serenity and Annoyance [3] is considered here. The proposed agent accepts input statements of G = Good and B = Bad categories to produce Good or Bad output statements in response. For this scenario, a finite-state stack transducer (FSST) with following specifications can be designed. Here relevance of the good or bad statements will be considered relative to the context of SELF. It means that the virtual agent will evaluate any statement, Good or Bad in context to itself, i.e. whether the context of the input statement belongs to the virtual agent or not. This implies that any statement, Good or Bad may also be R = Relevant or I = Irrelevant to the context of the virtual agent. For this scenario, the finite-state stack transducer (FSST) with given specifications is as follows:

$q_0 \in Q$ = SERENITY
Q = {ACCEPTANCE, SERENITY, ANNOYANCE}
F = {ACCEPTANCE, SERENITY, ANNOYANCE}
Σ = {G, B}
Λ = {G, B}
C = {R, I}
δ = set of transitions mapping the switching of emotions for an input

$$\delta: Q \times \Sigma \times C \rightarrow Q \times C$$

λ = transduction function mapping an emotion and an input statement to an output statement

$$\lambda: Q \times \Sigma \times C \rightarrow \Lambda \times C$$

4.1 State Transition Diagram for the Virtual

As presented in Fig. 2, the switching of emotions and generation of output symbols (Λ) is dependent on the input alphabets GOOD and BAD as well as on context alphabets RELEVANT and IRRELEVANT. Therefore, the possible set of transition rules to define the switching of the emotions and response generation for a given type and relevance of input statement is provided in Table 1.

Similarly, the possible set of transduction functions is also devised. Now the prediction of the possible output statements for a given input statement depends on the type and relevance collectively as follows.

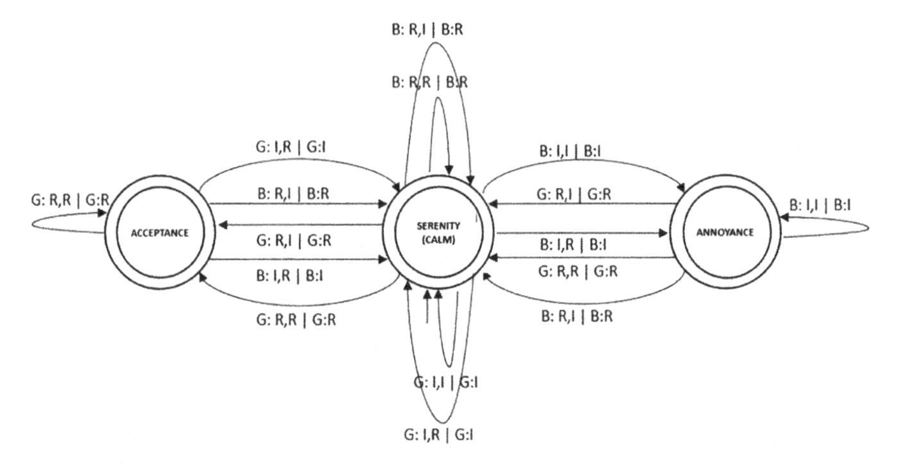

Fig. 2 Transition diagram of finite-state stack transducer for emotions of a virtual agent [10]

Table 1 State transition rules for virtual agent by FSST

Transition Function		State Reached	Transition Function		State Reached
δ: (ACCEPTANCE, G: R)	→	ACCEPTANCE	δ: (SERENITY, G: I)	→	SERENITY
δ: (ACCEPTANCE, G: I)	→	SERENITY	δ: (SERENITY, B: I)	→	ANNOYANCE
δ: (ACCEPTANCE, B: R)	→	SERENITY	δ: (ANNOYANCE, G: R)	→	SERENITY
δ: (ACCEPTANCE, B: I)	→	SERENITY	δ: (ANNOYANCE, B: R)	→	SERENITY
δ: (SERENITY, G: R)	→	ACCEPTANCE	δ: (ANNOYANCE, G: I)	→	SERENITY
δ: (SERENITY, B: R)	→	SERENITY	δ: (ANNOYANCE, B: I)	→	ANNOYANCE

Table 2 Output transduction rules for virtual agent by FSST

Transduction Function		Statement	Transduction Function		Statement
λ: (ACCEPTANCE, G: R)	→	G	λ: (SERENITY, G: I)	→	B
λ: (ACCEPTANCE, G: I)	→	G	λ: (SERENITY, B: I)	→	B
λ: (ACCEPTANCE, B: R)	→	B	λ: (ANNOYANCE, G: R)	→	G
λ: (ACCEPTANCE, B: I)	→	B	λ: (ANNOYANCE, B: R)	→	G
λ: (SERENITY, G: R)	→	G	λ: (ANNOYANCE, G: I)	→	B
λ: (SERENITY, B: R)	→	G	λ: (ANNOYANCE, B: I)	→	B

4.2 Output Transduction of the Virtual Agent

The transduction function given in Table 2 is required to predict the probable response statements by the virtual agent under a specific emotional state for a given context.

5 Implementing the Proposed Framework

The proposed framework for virtual agent was implemented as a C# application[1] on an Intel core i5[2] system in the environment of Windows 7 (See Footnote 1), 64-bit edition (Table 3).

5.1 Input—Response Dataset Corpus

The training and testing of the proposed framework was executed on a corpus of 2060 statements and their corresponding responses developed through a survey of 290 students of undergraduate engineering courses. The participating students were asked to suggest input statements as well as questions. The inputs may belong to a random subject or to themselves which can be grouped under GOOD and BAD categories. They were also instructed to provide relevant answers for these questions under the emotional states of ACCEPTANCE, SERENITY and ANNOYANCE [3]. This way, the collected questions and answers were tagged with context values (SELF, RANDOM) and category values (GOOD, BAD). Here it should be noted that for multiple questions falling under a GOOD or BAD category with a specific context, the response statements can be from a common set as displayed in Table 4.

[1]C# and Windows 7 are registered trademarks of Microsoft Corporation.

[2]Core i5 is the registered trademark of Intel Corporation.

Table 3 Snapshot of quiz session with ELIZA program

Input statements	ELIZA'S response
TYPE: GOOD **CONTEXT**: SELF You are quite good.	Do you sometimes wish you were quite good?
	Does it please you to believe I am quite good?
	Do you sometimes wish you were quite good?
TYPE: GOOD **CONTEXT**: SELF Thank you for help.	We were discussing you—not me.
	What are your feelings now?
	We were discussing you—not me.
TYPE: GOOD **CONTEXT**: RANDOM Great conversation.	What does that suggest to you?
	Does talking about this bother you?
	Tell me more about that.
TYPE: GOOD **CONTEXT**: RANDOM The soup was delicious.	Please go on.
	I'm not sure I understand you fully.
	Please go on.
TYPE: BAD **CONTEXT**: SELF Can I punch you in the face?	Do you punch anyone else?
	Perhaps in your fantasies we punch each other.
	Do you punch anyone else?
TYPE: BAD **CONTEXT**: SELF You are so mean.	Perhaps you would like to be so mean.
	What makes you think I am so mean?
	Perhaps you would like to be so mean.
TYPE: BAD **CONTEXT**: RANDOM Life is a mess.	Does talking about this bother you?
	Please go on.
	That is interesting. Please continue.
TYPE: BAD **CONTEXT**: QUIZ War is a menace.	Do you feel strongly about discussing such things?
	That is interesting. Please continue.
	Does talking about this bother you?

5.2 Response of ELIZA Against Input—Response Dataset Corpus

The ELIZA program [2] was developed in 1967 at MIT. It mimics the behavior of a therapist. ELIZA belongs to a specific class of programs called chat bots, which can interact with the user in simple English language and generate an interactive conversation with a human user. Eliza functions on a repository of static responses, string manipulation and rules.

These responses and rules are repeatedly used to generate the best possible response for an input statement and promote a conversation. Prior to executing the input—response corpus on our proposed framework, we tested our statements on ELIZA program. The responses generated by ELIZA are presented in Table 3. A common observation from these responses is that specific phrases and words are used by the system to generate the output.

Table 4 Snapshot of quiz session with implemented agent

Input statements	Response	Emotion: relevance
TYPE: GOOD **CONTEXT**: SELF You are quite good/ Thank you for help.	Glad to help you.	ACCEPTANCE: R
	Thanks.	SERENITY: R
	Was it a taunt?	ANNOYANCE: R
TYPE: GOOD **CONTEXT**: RANDOM Great conversation/ The soup was delicious.	The pleasure is mine.	ACCEPTANCE: I
	Same here.	SERENITY: I
	The nonsense is finally over.	ANNOYANCE: I
TYPE: BAD **CONTEXT**: SELF Can I punch you in the face?/ You are so mean.	That is a bad idea.	ACCEPTANCE: R
	Are you serious?	SERENITY: R
	Let's hash it out your way.	ANNOYANCE: R
TYPE: BAD **CONTEXT**: RANDOM Life is a mess./ War is a menace.	Simply what you make it.	ACCEPTANCE: I
	What do you think of it?	SERENITY: I
	That's not my subject of interest.	ANNOYANCE: I

For example, "quite good", "punch", and "mean" were captured by the program, and the output was generated by concatenating the responses with these captured words or phrases. As evident from the responses, there is very little scope for emotional and contextual relevance of the output corresponding to the input statement. Besides this, multiple repetitions in output, such as 'Does talking about this bother you?' and 'Please go on' were observed for various input statements. These factors certainly reduce the expressive value of the response.

Later we executed the same set of statements on our proposed framework's agent and mapped its responses in accordance to the 3 emotions and relevance values. The responses of the agent were fetched from a repository of tagged responses for the possible combinations of statement types and contexts. Contrary to responses of ELIZA, the responses of our agent were found to be nonrepetitive and more believable in the conversation. Instead of using string matching and concatenation, our agent mapped the responses with most appropriate tag values by using push and pop mechanism on the context stack of the transducer [10].

5.3 Framework Implementation

As presented in Fig. 3, the implementation of the proposed virtual agent framework proceeds in a prespecified sequence. For example, the input statement "The soup was delicious" is accepted by the speech-to-text interface as an argument. After capturing the input, the processing function derives the type (Σ) as GOOD and context (C) as RANDOM from the statement. Further, the derived statement type (Σ) and the context (C) are forwarded to the next layer. In the second layer, transition function (δ) and transduction function (λ) map the context and response statement for the accepted input. The emotions of the agent will be the result of the

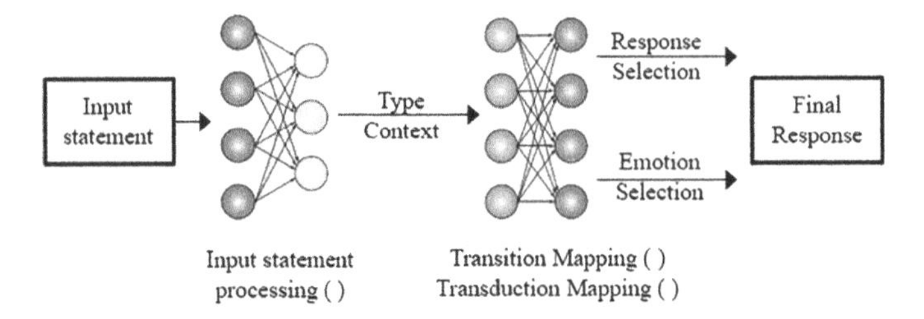

Fig. 3 Framework implementation sequence

previous statements. According to these emotional states, the relevance of the mapped context is recorded into the stack (RELEVANT, IRRELEVANT), while the response statement is printed as output (Λ). If the current emotion of the system is ACCEPTANCE, then the possible response derived from the repository may be "The pleasure is mine". Similarly, for a SERENE emotion, the response can be "Same here". Smaller responses can lead to buildup of dialogue sessions of higher belief. This observation may be useful in design and implementation of virtual agents for automated systems.

5.4 Experiment

After the system was trained on the question—answer corpus, later it was tested for 45 different dialogue sessions. Each session consisted of 100 questions of type GOOD and BAD. In response to a set of common question types, the system responded as a virtual assistant of a limited personality having only 3 emotions by answering these questions based on the emotional transitions and context switches. For every combination of statement type, context and emotion, a set of tagged responses is available in the repository. Responses from this repository are used to construct a conversation session between the user and the proposed agent.

6 Results and Discussion

The observations derived from these test dialogue sessions of ELIZA and our agent led us to following conclusions. Stack coupled finite-state transducers can prove to be potential virtual agent architectures for various automated response applications. Addition of memory stacks can provide efficient recall of parameters such as context, statement type and emotions for the virtual agent leading to transition of states for a wider range of inputs in an automaton [8, 11]. Multiple variables can be recorded in parallel on memory stacks to boost the decision making of the transducer architectures compared to simple string matching response programs like

ELIZA [2]. Memory stack coupled finite-state transducers can provide an effective state and output mapping structure [11, 12].

In future, larger number and variety of questions and answers can be added to the corpus. Techniques such as natural language processing and sentiment analysis, which were excluded from the scope of this work, can be employed for parsing the input statements to improve the response of the framework for a wider range of statements. With increase of statement parameters, more stacks can be added to the stack based transducers to improve their recall capability. Presently used 3 emotional states from Plutchik's model of emotions can be increased to cover a larger set of human emotions such as joy, anger, fear, etc., and enhance the expressiveness of the agent architecture.

Acknowledgements The research material used in this work has been collected from the students of B.Tech final year, seventh semester, Computer Engineering and Information Technology, 2012–2016 batch, at Swami Keshvanand Institute of Technology, (SKIT), Jaipur. The data has been collected through a series of questionnaires, and has been approved by the ethics committee of the respective department. The collected data is intended to be used for the research work 'Behavior Modeling of Virtual Agents' under 'Artificial Intelligence and Machine Learning' stream. It is further certified that this data is collected with duly informed consent of the students for study and usage in the research projects undergoing in the M.Tech program of institute.

References

1. Ali, I. R., Sulong, G., Basori. A. H.: A Natural Conversational Virtual Human with Multimodal Dialog System. JurnalTeknologi, vol. 71(5), pp. 75–78 (2014)
2. Weizenbaum, J.: ELIZA – A Computer Program for the Study of Natural Language Communication Between Man and Machine. Communications of ACM; vol. 9, Issue 1, pp. 36–45 (1966)
3. Plutchik, R.: The Nature of Emotions. American Scientist, vol. 89(4), pp. 348–350 (2001)
4. Oremus, W.: Terrifyingly Convenient. Slate.com., Slate Cover Story, http://www.slate.com/articles/technology/cover_story/2016/04/alexa_cortana_and_siri_aren_t_novelties_anymore_they_re_our_terrifyingly.html (2016)
5. Hopcroft, J.E., Motwani. R, Ullman, J.D.: Introduction to Automata Theory, Languages, and Computation. Pearson Education. ISBN-13: 978–8131720479 (2008)
6. Wright, D.R.: Finite State Machines. CSC215 Class Notes. N. Carolina State Univ. pp. 06–28 (2005)
7. Pieterse, V., Black, P.E.: Algorithms and Theory of Computation Handbook. CRC Press LLC, 1999. Pushdown Automaton, Dictionary of Algorithms and Data Structures, https://www.nist.gov/dads/HTML/pushdownautm.html (2004)
8. Black, P.E.: Finite State Machine, Dictionary of Algorithms and Data Structures, U.S. National Institute of Standards and Technology, https://xlinux.nist.gov/dads//HTML/finiteStateMachine.html (2008)
9. Furia C.A.: A Survey of Multi-Tape Automata, Formal Languages and Automata Theory, arxiv.org, (2012)
10. Srivastava, S. R., Joseph, A.: Finite State Stack Transducer for Functionality in a Multi-Parameter Environment, Proceedings REDSET-2017, 3rd International Conference on Recent Developments in Science, Engineering & Technology, October 21–22, 2016, Gurgaon (Delhi-NCR), India, (2016)

11. Black, P.E.: Pushdown Transducer, Dictionary of Algorithms and Data Structures, U.S. National Institute of Standards and Technology, https://xlinux.nist.gov/dads//HTML/pushdownTransducer.html (2008)
12. Black, P.E.: Finite State Transducer, Dictionary of Algorithms and Data Structures. Black, P. E., Pieterse, V., U.S. National Institute of Standards and Technology, https://www.nist.gov/dads/HTML/finiteStateTransducer.html (2004)

An IoT-Based Innovative Real-Time pH Monitoring and Control of Municipal Wastewater for Agriculture and Gardening

Narendra Khatri, Abhishek Sharma, Kamal Kishore Khatri and Ganesh D. Sharma

Abstract This paper presents an internet of thing (IoT)-based innovative real-time pH monitoring and control of municipal wastewater for agriculture and gardening application. During the past few decades after the green revolution in India, water requirement is increased exponentially in all sectors, viz., agriculture, gardening, industry, etc. The demand and supply relationship is very essential for every country in present time, and is also a big challenge to satisfy this requirement around the world. Regular change in the climate and the urbanization makes the lavish use of the available resources has exhausted the available resources. Water is necessary for the survival of the human being on the earth. So for the survival the conservation and management of the available water resource are also equally important. Moreover, for the healthier society, the access of the clean and safe water resource is also imperative. Nowadays municipal wastewater is recycled and reused for agriculture and gardening application after treatment. This paper describes a smart solution to control the water quality through its pH, thus to treat the municipal wastewater and its reuse in the agriculture and gardening purpose. The idea is to develop a low-cost electronic system and its application with such a quality of maintaining (monitoring and control) the water quality within the prescribed standard.

N. Khatri (✉) · K.K. Khatri
Department of Mechanical-Mechatronics Engineering,
The LNM Institute of Information Technology, Jaipur 302031, India
e-mail: narkhatri@gmail.com

K.K. Khatri
e-mail: kamalkishorekhatri@gmail.com

A. Sharma
Department of Electronics and Communication Engineering,
The LNM Institute of Information Technology, Jaipur 302031, India
e-mail: abhisheksharma@lnmiit.ac.in

G.D. Sharma
Department of Physics, The LNM Institute of Information Technology,
Jaipur 302031, India
e-mail: gdsharma273@gmail.com

A.K. Somani et al. (eds.), *Proceedings of First International Conference on Smart System, Innovations and Computing*, Smart Innovation, Systems and Technologies 79, https://doi.org/10.1007/978-981-10-5828-8_34

Keywords IoT · Wastewater · Wireless · Real-time controlling
Water quality · Smart supervision and control

1 Introduction

The development of the internal of things (IoT)-[1] based innovative real-time pH monitoring [2–4] and control of municipal wastewater for agriculture and gardening application. Real-time measurement of pH and temperature as well as the adjustment are performed through the chemical reactions by the control through microprocessor automatically. This communication deals with the main idea for the development of such product is low cost, easy installation, handling and monitoring of the data online [5], low maintenance and maintenance cost as well as the development of interdisciplinary kind work that includes the sample collection and online raw calculation, analysis, control through the chemistry and computer programming. Thus, the water quality for the agriculture and gardening is to maintain the pH of treated municipal wastewater.

2 The Concept of Real-Time pH Monitoring and Control

Real-time monitoring of the water quality relies on the type of the sensors deployed and measurements of parameters involved in the processes. In general, pH and temperature are the main parameters to be controlled through the phosphate, ammonium, and heavy metal ions in water for agriculture, gardening, and many other applications and play an important role.

The system shown in Fig. 1 consists of a water quality sensor, data processing, and storage and transmission system, connected to automatic system for the control of the parameters to be sensed. Here a wireless communication system [4] is developed for the monitoring and control purpose, that is controlled through the android application or personal computer is used for, the same as well to store the logs.

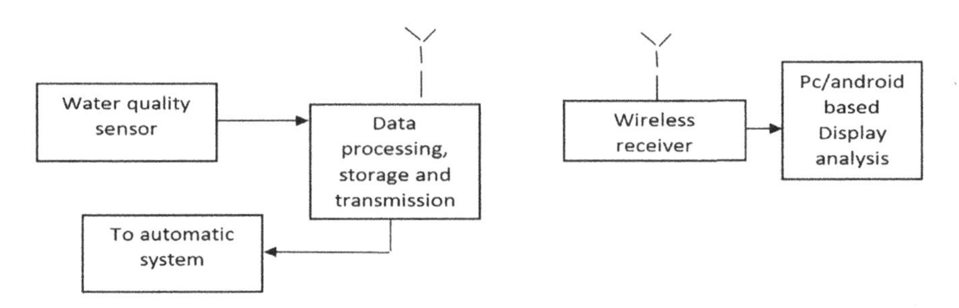

Fig. 1 Block diagram of real-time pH monitoring and control system

2.1 Water Quality Sensing Block

Figure 2 shows the block diagram of the water quality sensor system, consists of sensors to measure pH parameter and an Arduino mega is a microcontroller board based on the ATmega2560 microcontroller with Wi-Fi shield follows 802.11 b/g networks shield. The HDG204 Wireless LAN 802.11 b/g System in-Package is used. In order to measure the pH, pH sensor is used that generates the output in the millivolts. Since this voltage is quite low, therefore a voltage amplifier is employed at the end of sensor probe to amplify the voltage to a sufficient level. The microcontroller will collect the data from the sensor and transmit to the data logger for maintaining the log and also to take proportional action [6, 9] for the control.

The Wi-Fi module with the microcontroller can transmit and receive the data and instruction in the indoor limit of 100 feet and 300 feet for the LOS communication if any. This is specially installed to provide the flexibility of placement for the security of data loggers.

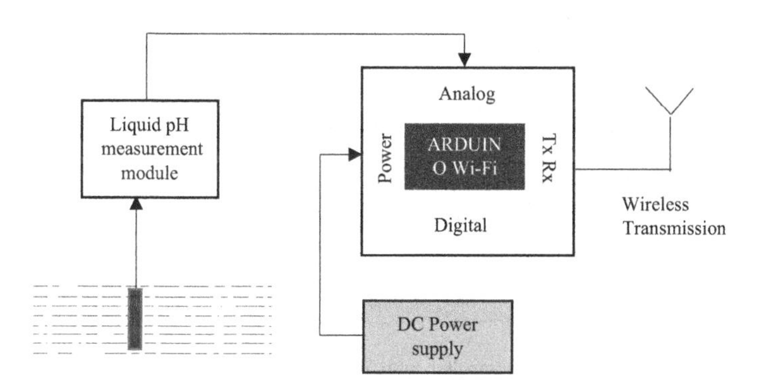

Fig. 2 Water quality sensing block with pH Probes, Arduino and Wi-Fi shield

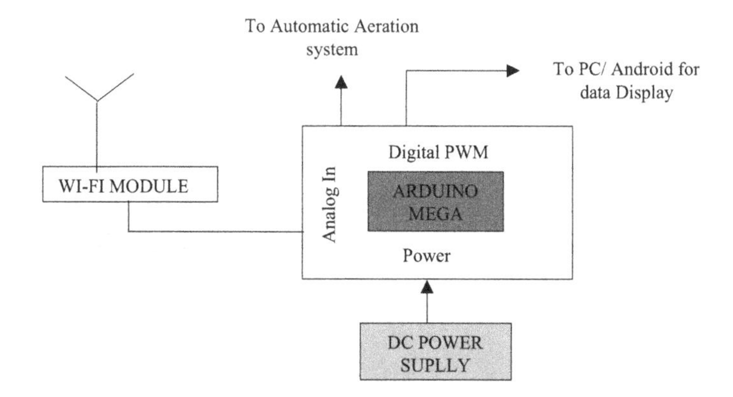

Fig. 3 Data logger with Wi-Fi TX-RX and ARDUINO MEGA microcontroller

2.2 Data Logger

Data logger (as shown in Fig. 3) consists of a Wi-Fi transmitter receiver (Wi-Fi shield) and Arduino mega 2560 (Fig. 4). The microprocessor data is processed in such a manner that it is compared with the predefined threshold.

The real-time-monitored value of the pH falls or rise to the threshold value. The microcontroller will control the value, the valves [7, 8] will operate according to the real-time pH and the predefined threshold's error and the opening of the valve is controlled by the chemical balance equation. The system programmed accordingly and the concern solenoid valve operates automatically and control. The micro-controller sends a signal to operate the respective value of the pH in the prescribed limit. All data (sensed) received stored by the data logger.

Fig. 4 ARDUINO MEGA microcontroller with Wi-Fi shield

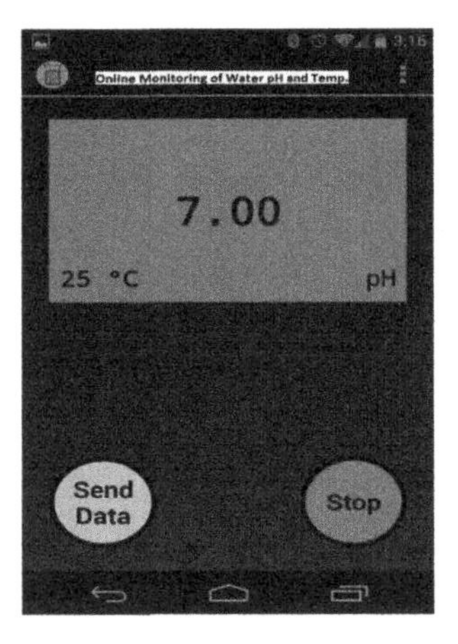

Fig. 5 Android app for online monitoring of water pH and temp

2.3 *Android Display for Water Quality*

Measured parameter is represented on the android display screen. That android interface for the presentation is developed using programming. In Fig. 5, show android app for online monitoring of water [9] pH and Temp.

3 System

3.1 *System Architecture*

Figure 6 represents the system architecture for the real-time monitoring and control of water pH and temperature for the agriculture and gardening. It consist of an Arduino mega with Wi-Fi shield and the three solenoid operated valves operates and control the dosing of the water, acid and base as per the signal generated by the controller after comparing the real-time pH value with the prescribed threshold value, i.e., to neutralize [10–13] the water for the further application. When it archives the prescribed value water will be discharged for agricultural use. This system monitoring and control can be performed by computer or android base application.

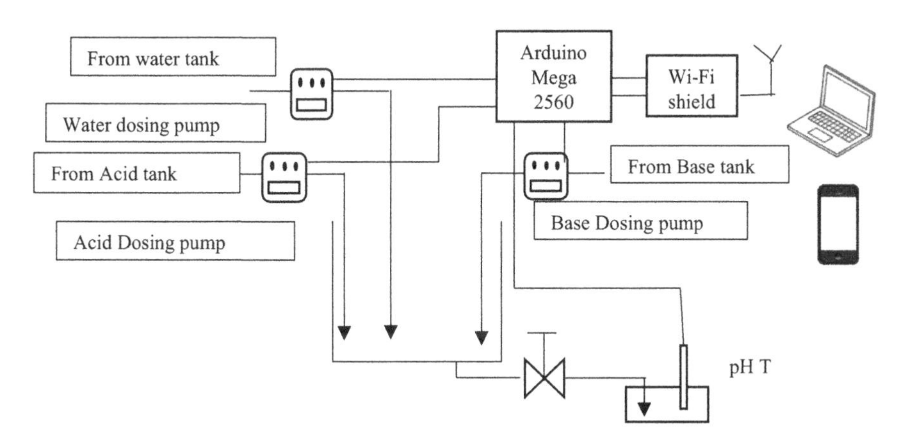

Fig. 6 System architecture RTMC (real-time monitoring and control) of water pH and temperature

Table 1 List of hardware component

S. no.	Name of the component	Make and specification
1	Microcontroller	Arduino mega 2560
2	Temperature sensor	Waterproof LM35
3	pH sensor	pH sensor module v1.1
4	Valve	Solenoid valve
5	Wireless communication module	Wi-Fi shield for Arduino

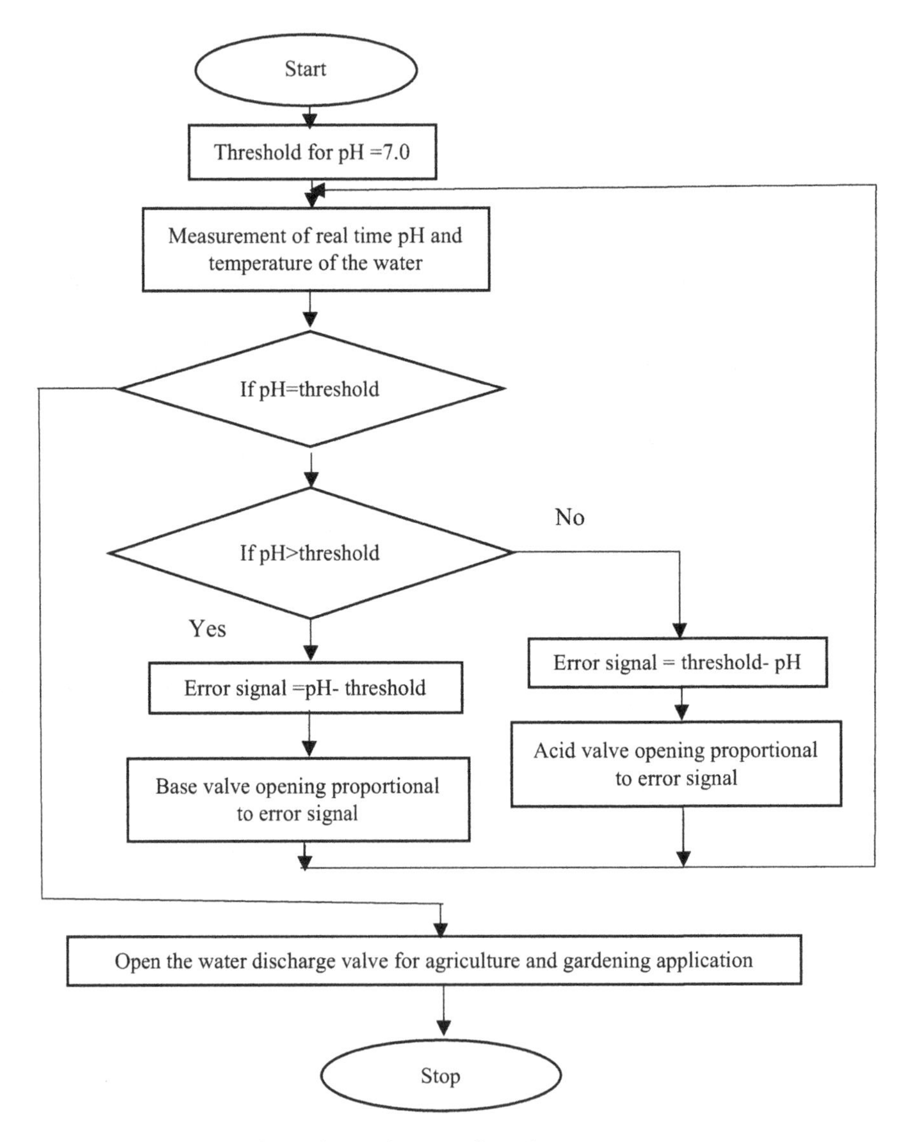

Fig. 7 Real-time monitoring and control process flow chart

3.2 System Hardware Component

Hardware components used for Real-Time pH Monitoring and Control of Municipal Wastewater system is listed in the Table 1.

4 Software Implementation

Figure 7 shows the flowchart for software implementation of Real-Time Monitoring and Control Process.

In the process flow, first the threshold value of pH is defined and the measured value is compared with the threshold. Error signal is calculated between the threshold and the measured value. The dosing acid or base valve opened proportionally.

5 Results and Discussion

Online monitoring and control parameters of pH of wastewater are shown in Tables 2 and 3. Flow rate of the dosing pump is 6l/h.

Online monitoring of the pH and temperature of treated municipal wastewater is completed with the help of pH electrode with the pH module and the temperature is measured with the help of waterproof temperature sensor LM35. The real-time monitoring and control process is shown in Fig. 7. Figures 8 and 9 represent the variation of water pH with time and variation of water temperature with time, respectively.

The sensor data is used for control of the valve opening. That valve opening angle and the opening time are controlled by the Arduino mega 2560.

Table 2 Acid–base Titration, nature of wastewater was base type

S. no.	Volume of wastewater	pH of wastewater	Dosing type	Conc. of acid	Added acid quantity to neutralize	Dosing Pump on time
1	100 l	9	Acid	0.001 mM	100 ml	60 s

Table 3 Acid–base titration, nature of wastewater was acid type

S. no.	Volume of wastewater	pH of wastewater	Dosing type	Conc. of acid	Added base quantity to neutralize	Dosing pump on time
1	100 l	5.3	Base	0.01 M	50.11 ml	30 s

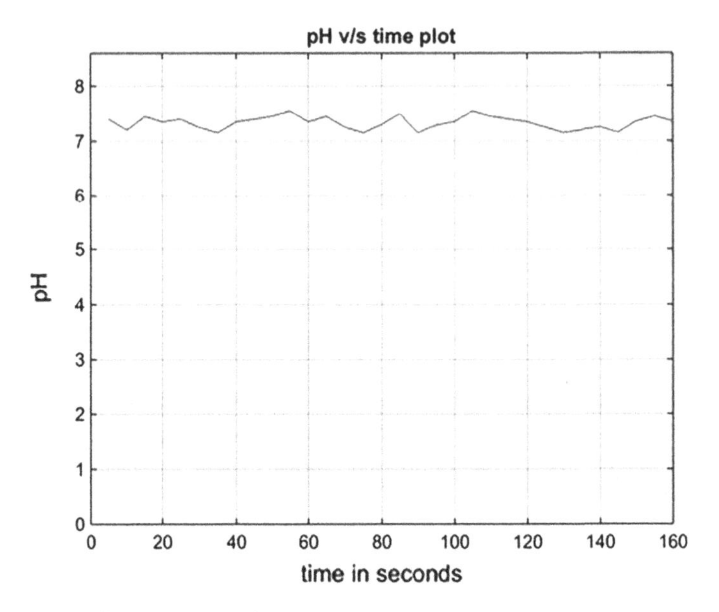

Fig. 8 Variation of water pH with time

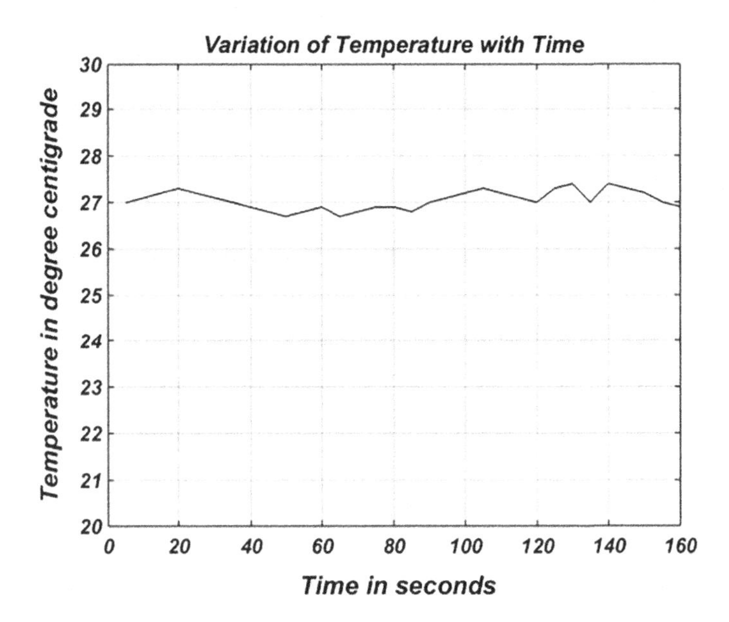

Fig. 9 Variation of water temperature with time

6 Conclusion

This is an IoT-based innovative real-time pH monitoring and control of municipal wastewater for agriculture and gardening application. Wi-Fi shield for wireless communication are applied as the spindle hardware system for building a small wireless sensor network. The energy saving is very significant using this system. The central system is equipped with a computer based programming module that can be used with most popular Android mobile devices connected directly to increase the overall system convenience and timeliness. In the future this system can be widely used in a variety of parameters for water, such as DO (dissolved oxygen), COD (chemical oxygen demand), BOD (biological oxygen demand), etc. Combined with the internet of things, regardless of where the user is located, the user can receive the instant messaging, taking real-time monitoring technology to the next level.

References

1. Anjana S, Sahana M N, Ankith S, K. Natarajan, K. R. Shobha and A. Paventhan, "An IoT based 6LoWPAN enabled experiment for water management," 2015 IEEE International Conference on Advanced Networks and Telecommunications Systems (ANTS), Kolkata, 2015, pp. 1–6. doi:10.1109/ANTS.2015.7413654
2. J. H. Chen, W. T. Sung and G. Y. Lin, "Automated Monitoring System for the Fish Farm Aquaculture Environment," 2015 IEEE International Conference on Systems, Man, and Cybernetics, Kowloon, 2015, pp. 1161–1166. doi:10.1109/SMC.2015.208
3. M. F. Saaid, A. Sanuddin, M. Ali and M. S. A. I. M. Yassin, "Automated pH controller system for hydroponic cultivation," 2015 IEEE Symposium on Computer Applications & Industrial Electronics (ISCAIE), Langkawi, 2015, pp. 186–190. doi:10.1109/ISCAIE.2015.7298353
4. A. M. Telgote, V. Narayanan and N. A. N. Dave, "Design and implementation of water environment monitoring system using GSM technology," 2015 International Conference on Technologies for Sustainable Development (ICTSD), Mumbai, 2015, pp. 1–4. doi:10.1109/ICTSD.2015.7095853
5. V. Aparna, "Development of automated pH monitoring & control system through USB Data Acquisition," 2014 6th IEEE Power India International Conference (PIICON), Delhi, 2014, pp. 1–6. doi:10.1109/POWERI.2014.7117602
6. G. Wiranto, G. A. Mambu, Hiskia, I. D. P. Hermida and S. Widodo, "Design of online data measurement and automatic sampling system for continuous water quality monitoring," 2015 IEEE International Conference on Mechatronics and Automation(ICMA), Beijing, 2015, pp. 2331–2335. doi:10.1109/ICMA.2015.7237850
7. A. H. M. A. Helmi, M. M. Hafiz and M. S. B. S. Rizam, "Mobile buoy for real time monitoring and assessment of water quality," 2014 IEEE Conference on Systems, Process and Control (ICSPC 2014), Kuala Lumpur, 2014, pp. 19–23. doi:10.1109/SPC.2014.7086223
8. P. S. P. Maciel, S. B. da Silva, G. F. B. de Medeiros and T. V. Rodrigues, "Innovative pH control for water: Reusing rainwater," 2013 IEEE Global Humanitarian Technology Conference (GHTC), San Jose, CA, 2013, pp. 288–293. doi:10.1109/GHTC.2013.6713698
9. A. F. O. de Azevedo Dantas, F. A. de Lima, A. L. Maitelli and G. F. Dantas, "Comparative study of strategies for fuzzy control P and PD applied to a pH plant," IECON 2011 – 37th

Annual Conference of the IEEE Industrial Electronics Society, Melbourne, VIC, 2011, pp. 516–521. doi:10.1109/IECON.2011.6119364
10. N. H. S. Abdullah, M. N. Karsiti and R. Ibrahim, "A review of pH neutralization process control," 2012 4th International Conference on Intelligent and Advanced Systems (ICIAS2012), Kuala Lumpur, 2012, pp. 594–598. doi:10.1109/ICIAS.2012.6306084
11. R. Ibrahim and D. J. Murray-Smith, "Design, implementation and performance evaluation of a fuzzy control system for a pH neutralisation process pilot plant," 2007 International Conference on Intelligent and Advanced Systems, Kuala Lumpur, 2007, pp. 1001–1006. doi:10.1109/ICIAS.2007.4658536
12. S. Harivardhagini and A. Raghuram, "Variable structure control of pH neutralization of a prototype waste water treatment plant using LabVIEW," 2015 IEEE Conference on Systems, Process and Control (ICSPC), Bandar Sunway, 2015, pp. 79–84. doi:10.1109/SPC.2015.7473563
13. T. Hong, A. J. Morris, M. N. Karim, J. Zhang and W. Luo, "Nonlinear control of a wastewater pH neutralisation process using adaptive NARX models," 1996 IEEE International Conference on Systems, Man and Cybernetics. Information Intelligence and Systems (Cat. No.96CH35929), Beijing, 1996, pp. 911–916 vol. 2. doi:10.1109/ICSMC.1996.571178

Modular Neural Network for Detection of Diabetic Retinopathy in Retinal Images

Manish Sharma, Praveen Sharma, Ashwini Saini and Kirti Sharma

Abstract Modular feedforward network method is introduced to detect diabetic retinopathy in retinal images. In this paper, the authors present classification method; the Modular Feedforward Neural Network (MNN) to classify retinal images as normal and abnormal. Publically available database such as DIA-RETDB0 including high-quality normal and abnormal retinal images is taken for detection of diabetic retinopathy. Modular Feedforward Neural Network is designed based on the extracted features of retinal images and the train N times method. The classification accuracy by MNN classifier was 100% for normal retinal images and 86.67% for abnormal retinal images. In this paper, the authors have explored such a method using MNN classifier which can detect diabetic retinopathy by classifying retinal images as normal and abnormal.

Keywords Diabetic retinopathy · Modular feedforward neural network
Classification accuracy · Retinal images database DIARETDB0

M. Sharma (✉)
Delhi Technological University, New Delhi, India
e-mail: prof.manishsharma@gmail.com

P. Sharma · A. Saini
Global Institute of Technology, Jaipur, India
e-mail: praveen141sharma@gmail.com

A. Saini
e-mail: aasumi18@gmail.com

K. Sharma
Jagannath University, Jaipur, India
e-mail: kirtisharma259@yahoo.com

A.K. Somani et al. (eds.), *Proceedings of First International Conference on Smart System, Innovations and Computing*, Smart Innovation, Systems and Technologies 79, https://doi.org/10.1007/978-981-10-5828-8_35

1 Introduction

Diabetes is an infection that influences veins all through the body, basically in the kidneys and eyes. Diabetic Retinopathy (DR) is a circumstance when veins in the eyes are influenced. Diabetic retinopathy is a primary general medical issue and this is the fundamental driver of visual impairment in the world so far [1]. It is incited by changes in the veins of the retina prompted this issue.

1.1 Diabetic Retinopathy Has Following Major Symptoms

- Double vision;
- Development of a scotoma or shadow in your field of view;
- Eye floaters and spots;
- Blurry and additionally misshaped vision;
- Fluctuating vision;
- Eye torment;
- Near vision issues disconnected to presbyopia.

Diabetic retinopathy is a microvascular intricacy which may be discovered in patients having diabetes. Diabetic retinopathy's occurrence will bring about the aggravation of visual ability and can eventually lead to visual impairment. The more extended a man has untreated diabetes; there is higher shot of developing diabetic retinopathy as it might be changed over into vision misfortune. When high glucose harms veins in the retina, eyes can spill drain or fluid. This prompts to the retina to swell and shape stores in early phase of diabetic retinopathy. Alongside diabetes, high glucose levels in long stretches can influence little vessels in the retina. Diabetic retinopathy gets to be distinctly symptomatic in its further stage. In the early stage, diabetic patients usually unaware about the infection of the disease [2]. In this way location of diabetic retinopathy at first stage is critical to keep away from vision misfortune. Diagnosing diabetic retinopathy is normally directed by the ophthalmologist by utilizing retinal images of patients [3, 4].

Diabetic retinopathy may grow through mainly four stages:

1. Mild non-proliferative retinopathy—In this small zones of balloon like swelling in the retina's little vein occurs, called smaller scale aneurysms, emerge at this first phase of the illness. These microaneurysms may release fluid/blood into the retina.
2. Moderate non-proliferative retinopathy—At this stage, as the illness turns out to be more terrible; veins that support the retina may swell and mutilate. Their capacity to transport blood may be lost. Both circumstances cause characteristic changes to the presence of the retina and may add to DME.

3. Severe non-proliferative retinopathy—At this stage, numerous more veins are blocked, denying blood supply to ranges of the retina. These influenced areas emit development variables which flag the retina to develop fresh blood vessels.
4. Proliferative diabetic retinopathy (PDR)—During this advanced stage, development variables emitted by the retina trigger the expansion of fresh blood vessels, which develop along within surface of the retina and vitreous gel, the liquid/blood which fills the eye. The fresh blood vessels are delicate, that makes them more inclined to spill and drain. Also scar tissue can contract and further be the reason of retinal separation. The pulling without end of the retina from hidden tissues, similar to backdrop peeling is far from a divider. Retinal separation can lead to permanent vision loss.

In further stages, spillage from veins into the eye's unmistakable, jam like vitreous can bring about genuine vision issues and in the long run results in visual impairment.

In this manner, identification of diabetic retinopathy at introductory level is imperative to maintain a strategic distance from vision misfortune. Finding of diabetic retinopathy is typically directed by the ophthalmologist by utilizing retinal pictures of patients. Ophthalmologist can get retinal pictures from patients to be analyzed by utilizing a fundus camera [5]. From the picture, indications will be recognized physically by an ophthalmologist. In this manner, additional time is required to analyze more number of patients.

A computerized screening framework can be utilized for completely automated mass screening [6]. Such frameworks screen an expansive number of retinal pictures and distinguish anomalous pictures, which are then further inspected by an ophthalmologist. This would spare a lot of workload and time for ophthalmologists, permitting them to focus their assets on surgery and treatment. Ordinary structures of retina are the optic plate, macula, and veins. The trademark elements of diabetic retinopathy are small-scale aneurysms, hemorrhages, and exudates [7]. From the arrangement of parameters like vessel proportion, proportion of exudates range to the aggregate zone of the pictures are appropriated into various gatherings like ordinary, extreme, mellow and irregular diabetic retinopathy, and so forth. Neural systems can be utilized viably in information characterization [8].

2 Database Acquisition

In order to conduct the experiment for detection of retinal images, in public database DIARETDB0 was used. The newest database has 130 colour fundus images out of that 20 are normal and remaining have signs of the diabetic retinopathy [9]. The images were taken in the Kuopio university hospital. The images were dedicatedly elected, however their distribution does not correspond to any typical population, images were captured with few fifty degree field of view digital fundus cameras with unknown camera settings. The images have an unknown quantity of imaging noise and optical aberrations (transverse, dispersion and lateral spherical, chromatic, coma, field curvature, astigmatism, distortion).

Difference over the visual appearance of changed retinopathy discoveries can in this manner be considered as maximal. In any case, the information compared to pragmatic circumstances, and can be utilized to assess the general execution of diagnosis techniques. The general execution relates to the circumstance where no adjustment is played out (no correspondence to this present reality estimations), yet where pictures compare to regularly utilized imaging conditions, i.e., the conditions normally experienced in healing centers. This data set is alluded as "calibration level 0 fundus images" [10]. The database was separated into two sections for training and testing.

A. Feature Extraction

Location of retinopathy includes nearness of exudates, clinical acknowledgment of widening of veins and injuries or whatever other irregularity in the retinal images. Fundus images give the data of these neurotic components with the exception of that anatomical data of the retina in eye. The elements watched are system of veins, macula, and the optic circle in a solid retinal picture. Any change because of diabetic retinopathy or other malady will bring about variety in above talked about components which makes it easy to analyze the sickness. Figure 1 describes about the retinal picture which is influenced with diabetic retinopathy while Fig. 2 describes about the sound dim level fundus picture [11].

In this investigation, diverse elements of retinal images were extricated. The removed elements were: DCT, Entropy, mean, standard deviation, Euler number, normal, relationship, difference, and vitality. Creator's objective was to discover conceivable elements that can really separate retinal picture as typical or anomalous.

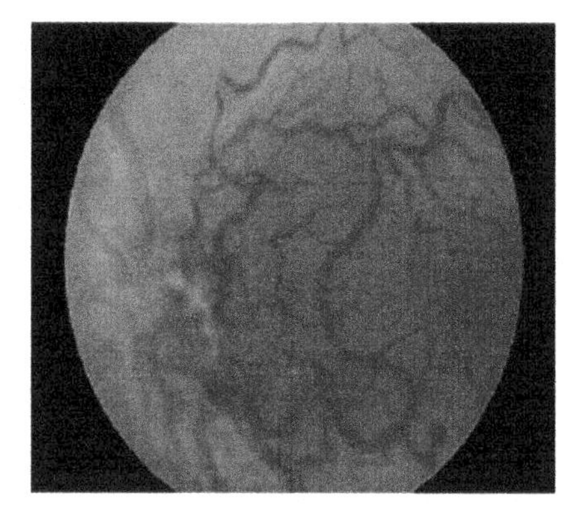

Fig. 1 Affected eye with diabetic retinopathy

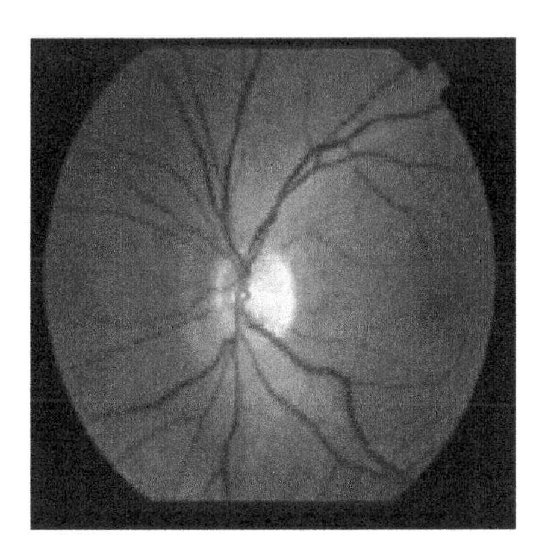

Fig. 2 Non-affected eye

3 Modular Feedforward Neural Networks

As the name shows, the particular bolster forward neural systems are unique instances of Multilayer Perceptron (MLPs) [12], with the end goal that layers are fragmented into modules. This has a tendency to make some structure inside the topology, which will propel specialization of capacity in every sub-module. In science, particular neural systems are exceptionally regular. Rather than the MLP, measured modular feedforward systems do not have full interconnectivity between the layers. In this way, a littler number of weights are required for a similar size system (a similar number of PEs). This tends to speed the preparation and decline the quantity of models required to prepare the system to a similar level of precision. Modular feedforward neural system is best reasonable model to be used [13].

4 Overview of Classifier System

It is proposed to build a choice emotionally supportive network for the identification of diabetic retinopathy utilizing the counterfeit consciousness procedure. The authors expect to utilize neural system for the discovery of diabetic retinopathy utilizing highlights separated from the retinal pictures. The fundus photos will be gathered from information storehouses. These photos were brought with a fundus camera amid mass screening and after that examined by a level bed scanner and spared as picture records. After that, the picture documents were broke down utilizing the calculations discussed further: Fig. 3 demonstrates the square chart of the proposed framework. It has eight modules: Retinal Fundus Image input, Pre-handling, Feature Extraction, Graphical User Interface (GUI), Modular nourish forward Neural Network Classifier, Parameter Acquisition, Graphical User Interface (GUI), and Results (Detection). Here information is the retinal pictures from

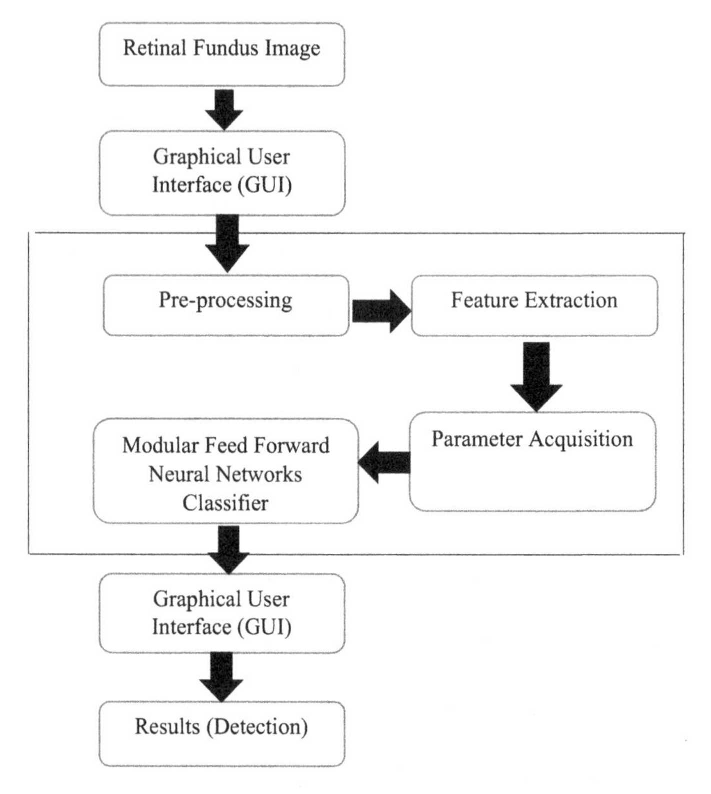

Fig. 3 Block diagram of the proposed system

Table 1 Modular feedforward network's recognition results

Output/desired		Normal (ON)	Abnormal (OA)
Training dataset	ON	19	0
	OA	0	95
Test dataset	ON	1	2
	OA	0	13

Table 2 Modular feedforward networks recognition result based on performance

	Performance	ON	OA
Training dataset	MSE	0.013698564	0.012000019
	Min Abs error	0.0013881	0.000860905
	Percent accuracy	100	100
Test dataset	MSE	0.155588824	0.160918599
	Min Abs error	0.019526124	0.003081384
	Percent	100	86.66666667

DIARETDB0 database [14] and in the wake of removing the picture highlights, modular neural system was prepared to recognize pictures as typical or anomalous yields (Tables 1 and 2).

5 Results

The modular feedforward neural network was used to test the proposed features of retinal images. The neural network was exhaustively designed using one hidden layer with a single neuron. Then progressively numbers of neurons were expanded. The same process was rehashed with two concealed layers. Numerous parameters were changed logically to set ideal neural system with best outcomes and slightest intricacy. The trial results are portrayed in taking after tables.

MSE: The average of the square of the difference of the required response and the actual system output (the error).

6 Conclusion

This paper proposes a system for detection of diabetic retinopathy using neural network. The authors have designed Modular Feedforward neural network for detection of diabetic retinopathy by classifying retinal images as normal or abnormal. Features of retinal images were based on the DCT, Entropy, mean, standard deviation, average, etc., and were given as the input to the neural network. The Train N Times training method was used to train the network. It was observed that with 02 hidden layers and 04 neurons the classification accuracy was 100% for normal retinal images and 86.67% for abnormal retinal images.

References

1. Meindert Niemeijer, "Retinopathy Online Challenge: Automatic Detection of Microaneurysm in Digital Color Fundus Photographs", IEEE Transactions on Medical Imaging vol. 29, no.1, (2010)
2. Gegundez-Arias, Manuel Emilio, et al. "Inter-observer Reliability and Agreement Study on Early Diagnosis of Diabetic Retinopathy and Diabetic Macular Edema Risk." International Conference on Bioinformatics and Biomedical Engineering. Springer International Publishing, (2016)
3. Marwan D. Saleh and C. Eswaran,: "An automated decision support system for non-proliferative diabetic retinopathy disease based on mas and has detection," Computer methods and programs in biomedicine, vol. 108, pp. 186–196, (2012).
4. M. Tamilarasi and K. Duraiswamy, "Genetic based fuzzy seeded region growing segmentation for diabetic retinopathy images," in Computer Communication and Informatics (ICCCI), In: 2013 International Conference on., pp. 1–5, IEEE (2013)

5. Atul Kumar, Abhishek Kumar Gaur, and Manish Srivastava,: "A segment based technique for detecting exudate from retinal fundus image," In: Procedia Technology, vol. 6, pp. 1–9, (2012)
6. Bhatkar, Amol Prataprao, and Govind Kharat. "FFT based detection of Diabetic Retinopathy in Fundus Retinal Images." Proceedings of the Second International Conference on Information and Communication Technology for Competitive Strategies. ACM, (2016)
7. Faust, Oliver, et al. "Algorithms for the automated detection of diabetic retinopathy using digital fundus images: a review." Journal of medical systems 36.1, 145–157. (2012): Zhang, Xiwei, et al. "Exudate detection in color retinal images for mass screening of diabetic retinopathy." Medical image analysis 18.7, 1026–1043 (2014)
8. Akram, Usman M., and Shoab A. Khan. "Automated detection of dark and bright lesions in retinal images for early detection of diabetic retinopathy." Journal of medical systems 36.5, 3151–3162 (2012)
9. K Narasimhan, VC Neha, and K Vijayarekha,: "An efficient automated system for detection of diabetic retinopathy from fundus images using support vector machine and bayesian classifiers," in Computing, Electronics and Electrical Technologies (ICCEET), In: 2012 International Conference on. pp. 964–969, IEEE (2012)
10. Luca Giancardo, Fabrice Meriaudeau, Thomas P Karnowski, Yaquin Li, Kenneth W Tobin, and Edward Chaum, "Automatic retina exudates segmentation without a manually labelled training set," in Biomedical Imaging: From Nano to Macro, In: 2011 IEEE International Symposium on., pp. 1396–1400, IEEE (2011)
11. Li. Tang, M. Niemeijer, Joseph M. Reinhardt, Mona K. Garvin, and Michael D. Abramoff, "Splat feature classification with application to retinal hemorrhage detection in fundus images," Medical Imaging, In: IEEE Transactions on, vol. 32, pp. 364–375, IEEE (2013)
12. Ganesan, Karthikeyan, et al. "Computer-aided diabetic retinopathy detection using trace transforms on digital fundus images." Medical & biological engineering & computing 52.8, 663–672 (2014)
13. Seok Oh, Ching Y. Suen; A class- modular feedforward neural network handwriting recognition. In: Vol. 35, Elsevier, (2002)
14. Valverde, Carmen, et al. "Automated detection of diabetic retinopathy in retinal images." Indian journal of ophthalmology 64.1, 26 (2016)
15. Acharya, U. Rajendra, et al. "An integrated index for the identification of diabetic retinopathy stages using texture parameters." Journal of medical systems 36.3 2011–2020 (2012)

Design and Implementation of Interactive Home Automation System Using LabVIEW

Peeyush Garg, Sankalp Agrawal, Wu Yiyang and Amit Saraswat

Abstract The Interactive automation system provides a user-friendly and secure environment, utilized the virtual instrument based control system. The LabVIEW software offers a software centric approach to define the controls of home appliances and utilities. The DAQ-6009 used for providing an interface between sensors-actuators assembly and software. The overall automated system comprises of independent modules that can be controlled in manual and automated mode. The scheduler system and authenticated log-in scheme add features of optimum utilization with safe and secure environment.

Keywords Interactive automation · LabVIEW · DAQ-6009
Scheduler system

1 Introduction

A home automation system involves advanced mechanized system to provide user with interactive control with home appliance. This system uses 'Interactive features' to provide user with 'Smart' feedback and information by monitoring many aspects of the house. Advanced Software can be used instead of electronics based circuits and subsystem to implement the home automation. The effective way to

P. Garg (✉) · S. Agrawal · A. Saraswat
Department of Electrical Engineering, Manipal University Jaipur, Jaipur, India
e-mail: peeyush01garg@gmail.com

S. Agrawal
e-mail: sankalp.003@gmail.com

A. Saraswat
e-mail: amitsaras@gmail.com

W. Yiyang
School of Electronic Science and Technology, Harbin Institute of Technology, Harbin, China
e-mail: wyy19963@qq.com

A.K. Somani et al. (eds.), *Proceedings of First International Conference on Smart System, Innovations and Computing*, Smart Innovation, Systems and Technologies 79, https://doi.org/10.1007/978-981-10-5828-8_36

approach these demands is to utilize simulated, tested and controlled architectures on software based platform. The design that has been realized here combines both hardware and software technologies, comprises the LabVIEW software and Sensor network with DAQ-6009. The sample house environment and control system has been addressed here consists of various modules. The system can monitor the temperature, internal light ON/OFF control, external light ON/OFF control, water level systems, fire alert alarm and anti-theft alarm, gardening sprinkler system in the house. Hardware implementation of a multifunctional control system for home automation with the help of LabVIEW has been developed.

As advent of electronics based control, opens an era for enhance features for automation. Many circuits and projects were developed to control the house appliances. In 1984, term "Smart home" was introduced by the American Association of Home Builders. Known as the ECHO IV, used to specifically switch home appliances on and off and control home temperatures but it was not commercially accepted. A decade ago, Internet connectivity and development of software based algorithms created a platform to automatize several manual control systems. Most of the home appliances were manually controlled and there was a need to add automatic control with smart and secure features for the home appliances. Many software centric approaches were introduced to control the home appliances [1–3], but the systems were not so interactive and user friendly. The efficient use of water and energy adds the environmental friendly features for smart home [4, 5]. Many programmer have been working to integrate many features for smart home, this tends to increase system complexity [6]. Implementation of these software based home automation algorithms can be embedded on microcontroller and microprocessor based circuits [7], but integrated the large number of home appliances control on processor based hardware platform, become system bulky, slow, and expensive and reduces the reliability and handling features. The LabVIEW software and DAQ Card provide an integrated platform to incorporate the many control schemes in ease manner.

2 LabVIEW

The Graphical system design (GSD) is a modern approach to designing an entire system. For complicated models, high-performance computing with graphical processing units, and multi-core based computers is a key factor. LabVIEW is a multi-core software that provides us space for model solution and simulation. It helps in completing complex project in less time with less human resource in an effective way. It is different because it provides variety of tools on a single platform, ensuring compatibility. It is very user friendly and converts complicated task into easy ones. It is enriched with features like G programming, hardware support, analysis and technical code libraries, UI components and reporting tools, technology abstracts and models of computation, data acquisition, instrumentation control,

Table 1 Specification of NI USB-6009 [9]

Feature	Specifications
AI resolution	14 bits differential, 13 bits single ended
Maximum AI sample rate	48 kS/s
DIO configuration	Each channel individually programmable as open collector or active drive2

and industrial automation on number of platforms including Microsoft Windows paper [8].

Data acquisition card (DAQ): USB-6009 is the way to determine an electrical or physical quantities with a computer. A DAQ system comprises of sensors, DAQ measurement hardware, and a computer with programmable software. As an improvement to old measuring system, the PC-based DAQ system provide a more powerful, flexible, and cost-effective measurement solution. In this project, National Instruments USB-6009 card is used. It has features which provide eight single-ended analog input (AI) channels, two analog output (AO) channels, 12 Digital I/O channels, and a 32-bit counter with a full-speed USB interface (Table 1).

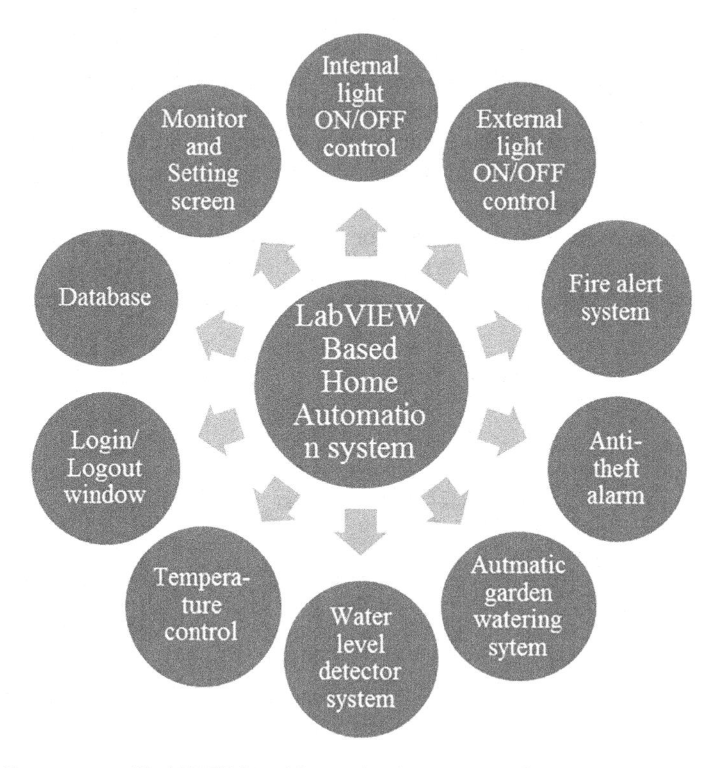

Fig. 1 Components of LabVIEW based interactive home automation system

The designed Interactive Home Automation System (IHAS) comprise the modules and facilities, pointed in the diagram given (Fig. 1).

3 Description of Project

The overall control system comprises the 8 modules, implemented using NI DAQ-6009. An interactive graphical user interface with data logging is facilitated by LabVIEW environment. Each module hardware is designed and implemented on PCBs in laboratory. The detailed explanation of each modules is as follows:

3.1 Internal Light Control

This module has been realized here works on two systems combined, them being the PIR sensor and the time scheduler. The system contains a PIR motion sensor and lights that are connected with the LabVIEW program. The room light automatically switches ON when the motion sensor senses someone's presence in the room. When the PIR Sensor notice that an object is moving, then produce a very small time width pulse, these pulse is hard to detect for further circuits, so a 555-timer based mono-stable multivibrator is used to produce a sufficient large pulse width signal that will be more suitable for lighting inside the home.

The reason we have used this system with a scheduler is because the requirement is to function for a specific period of time only for this purpose, time stamp function is used to start the system.

In this module, HC-SR501 PIR Sensor is used to sense the human availability. Whenever a mammal passes by, it first intercepts one half of the PIR sensor, which makes a positive differential change between the two halves. When the mammal

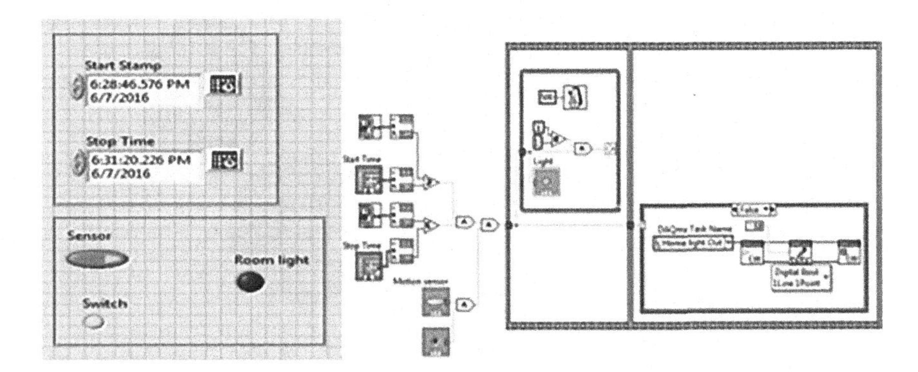

Fig. 2 Front panel and block diagram of internal light system

leaves the sensing area, the reverse happens, then sensor generates a negative differential change. These change pulses are detected. Then the information is provided to DAQ card, future utilize to control the lamp and data logging by LabVIEW. The interactive screen and LabVIEW functional diagram for the internal lighting system is given (Fig. 2).

3.2 External Light Control

The basic idea on which this system works is that as the sun goes down, the lights outside the house must light up automatically. It depends entirely on the data provided by LDR. The hardware submodule, comprises the LDR sensors with conditioning components, generates a digital signal, showing the less availability of sun light. Then the digital signal is transferred to DAQ. Then, user can set the time of morning and night time to control the external light and the hardware signal with the help of LabVIEW program (Fig. 3).

3.3 Fire Alert System

The system consist of three section, first the signal that reaches from the fire alert system sensors and indicator for alerting about the fire, in second section input signal is processed and the output signal is send, and then the controlling system and data processing by LabVIEW. Smoke sensor (MQ7) and heat sensor (thermistor) are used to detect the fire in the house. LabVIEW will send a set signal to

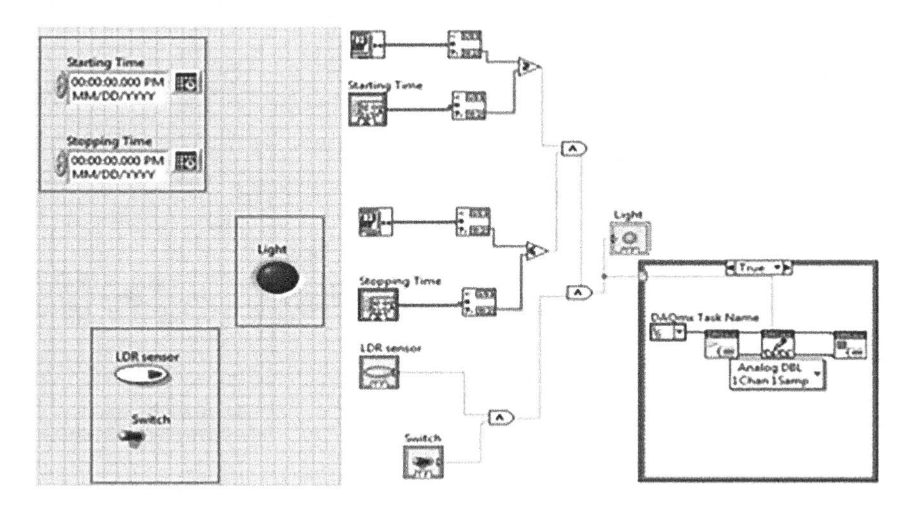

Fig. 3 Front panel and block diagram of external light system

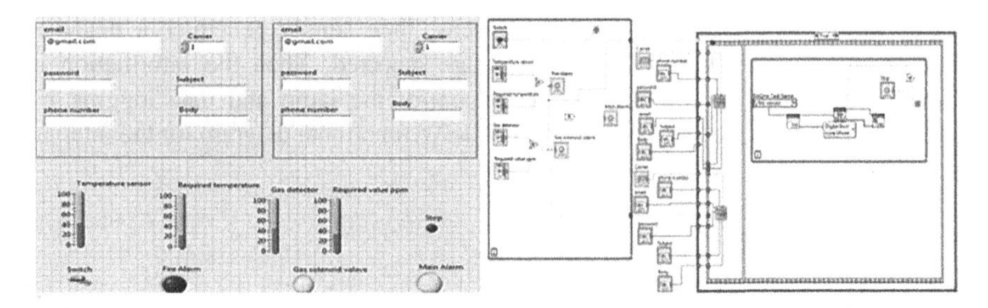

Fig. 4 Front panel and block diagram of fire alarm system

alarm siren to make a loud sound. The interactive screen provides the exact information of smoke concentration level and temperature at real time (Fig. 4).

The fire alarm module also added secure environment in automation system by quickly intimating the house owner, fire safety or security person about fire hazard by e-mail and mobile phone.

3.4 Anti-theft System

It provides us security and a huge sense of safety. The system here is a simple home alarm system that works on a security code. This system works in two parts. One, it compares the code entered by the person entering the house. And, the second part notifies the owners if the codes do not match.

In the VI we have used a very simple logic of comparing two codes, checking status of all the doors and the windows and notifying the owner if wrong code is entered. It will check if the entered code and the set code have same numbers or not.

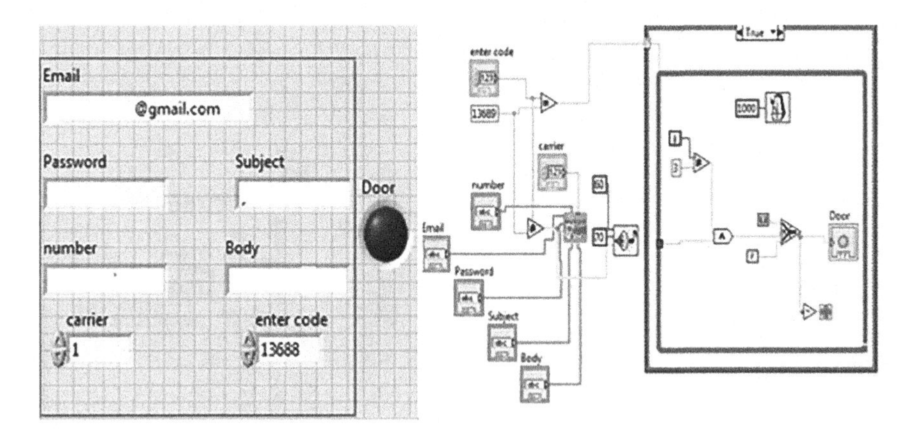

Fig. 5 Front panel and block diagram of anti-theft system

If the numbers match, the front door will open and that will be noted by the software. If the numbers do not match, door will be locked and simultaneously, a text message would be sent to the owner that the passwords don't match and there might be a burglar trying to get into the house and buzzer alarm will ring (Fig. 5).

The Anti-theft module also facilitated with the quick intimation to the owner and security person about fire hazard on e-mail and mobile phone.

3.5 Water Level Detector System

The system realized consists of a calibrated level sensor for depth to water readings. The water level sensor has been connected to the DAQ card so that its measurement can be directly taken as input by the LabVIEW software and indicate us if we need to switch the water pump ON or OFF (Fig. 6).

3.6 Automatic Garden Watering System

This system is used to automatically water the plants so that we don't have to go through the hassle of watering our plants or garden whenever we are busy or not available. This is a very simple control system in which the sprinklers automatically start watering the garden/plants as soon as the scheduler signals, only and if only after checking soil moisture level. For this purpose, sensors are placed in the soil of garden, provide the information of water moisture. The interactive user screen and block diagram for this module is shown (Fig. 7).

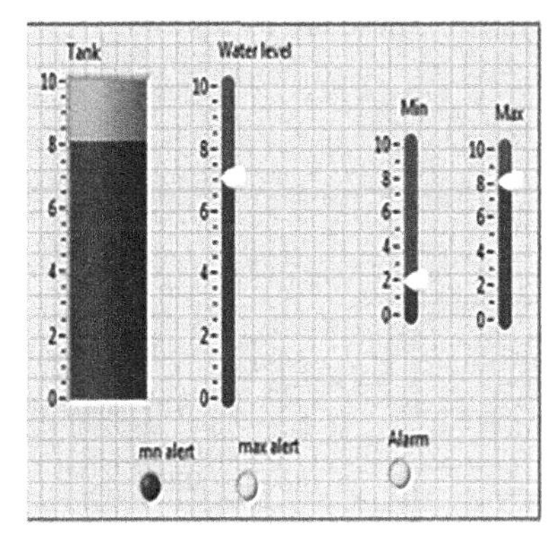

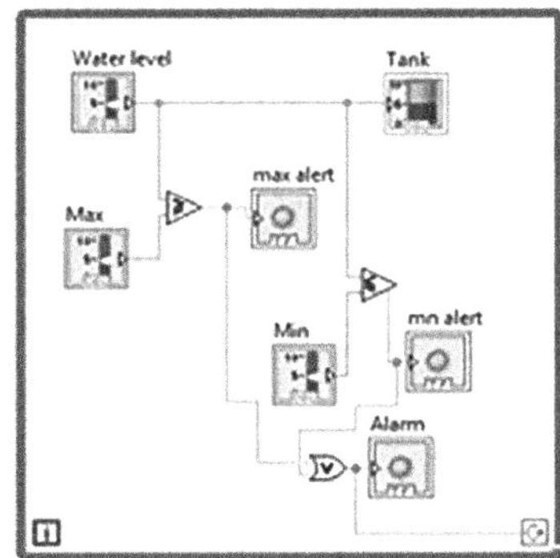

Fig. 6 Front panel and block diagram of water level detector

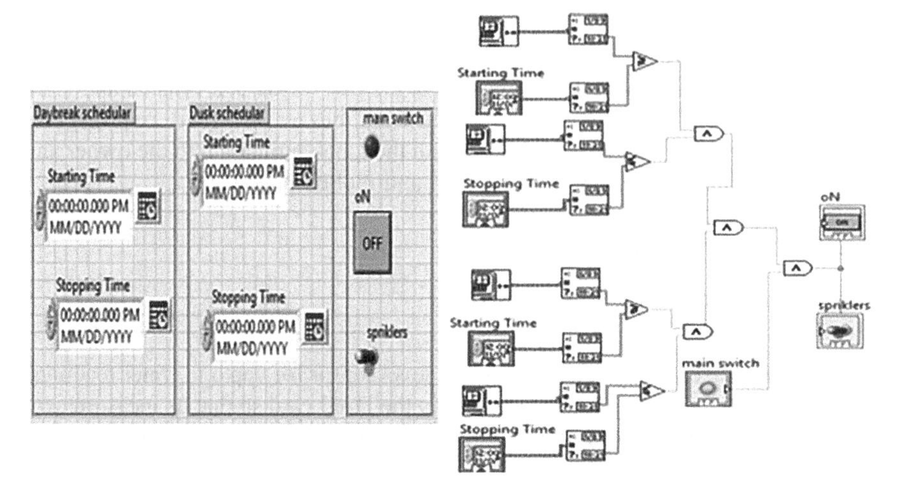

Fig. 7 Front panel and block diagram of automatic garden watering system

3.7 Temperature Control System

Temperature sensor (LM35) based circuit energize/de-energize a relay that will control the current (+5 V supply) to a resistor used as a heating element (Fig. 8).

The LM35 output is connected to a LM393/LM324 op-amp (as a buffer) that helps in switching of the relay. In order to allow efficient thermal contact between the two, we place the resistor and temperature sensor very close to each other. The output of LM35 is also connected to Analog input pin of the DAQ-6009, that is used in VIs to display and log Temperature real values, that can be used for future control action (Fig. 9).

Apart from the modules mentioned above, the system facilitates the user with some advancement of home automation system. These are as follows:

Fig. 8 Switching circuit of relay for temperature control

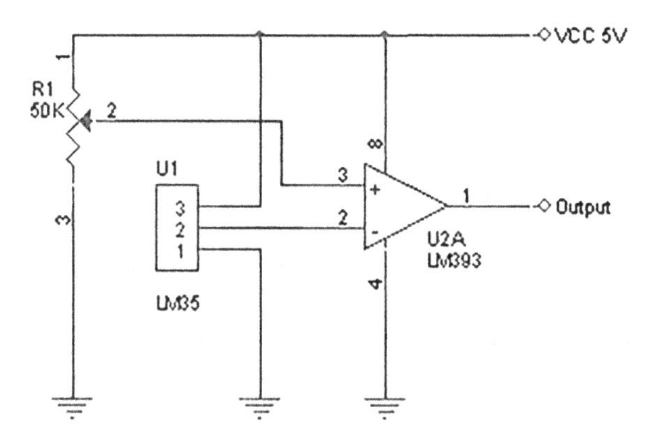

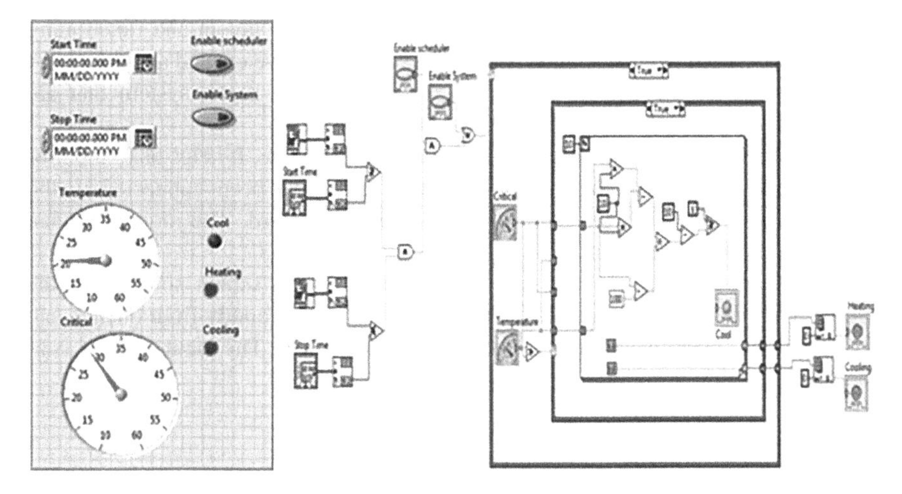

Fig. 9 Front panel and block diagram of temperature control system

Scheduler System: It is simply used to make the control system work in a specified time frame, as per the users demand it to work. We set a starting time point and a stopping time point, and connect it to the sensor and the main switch in a way that their output is only considered if it falls within the specified time frame, so that the room lights are controlled automatically during that time only. The scheduler for the internal lighting system is shown below. The same functional block can be used wherever the scheduler is required (Fig. 10).

DAQ card interface: The action plan of module depends on sensor outputs and the software logics & constraints, well mapped in the functional block diagram of Vis of particular module. LabVIEW VI will acquire the data from sensors through the data acquisition card DAQ-6009. The input from the sensors along with software constraints based Boolean signal are utilized to get proper action plan.

We've created the control system keeping in mind that only the inhabitants of the home can access the automated subsystems and also get a documented

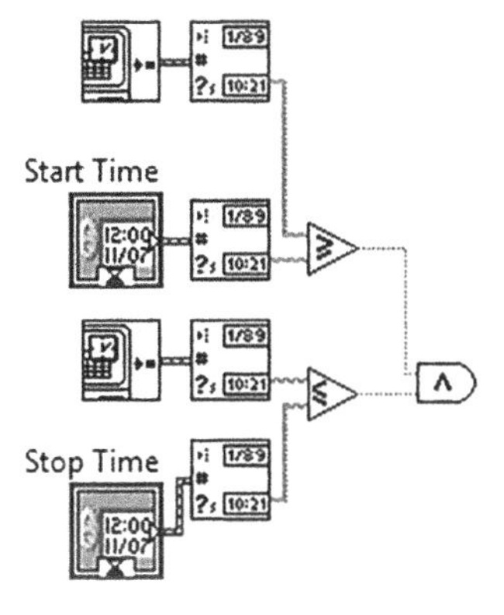

Fig. 10 Scheduler for internal lighting system

information of the previous functioning. In this section the layout and functioning of the entire automation system has been explained. We have created a monitor and setting screen for the user to access all the parts of the house system that are connected with LabVIEW via Data Acquisition Card.

Monitor Screen: It is designed using LabVIEW software. It helps us to monitor all the parts of the Automated House system that have been connected to the LabVIEW software with the help of Data Acquisition Card. The monitor screen looks like the Figure shown (Fig. 11).

Login/Logout Window: It has one login for the monitor screen, so that only the people living in the house can make the system work and nobody from outside can access it. It has been made for the security of the household. The login window is as shown (Fig. 12).

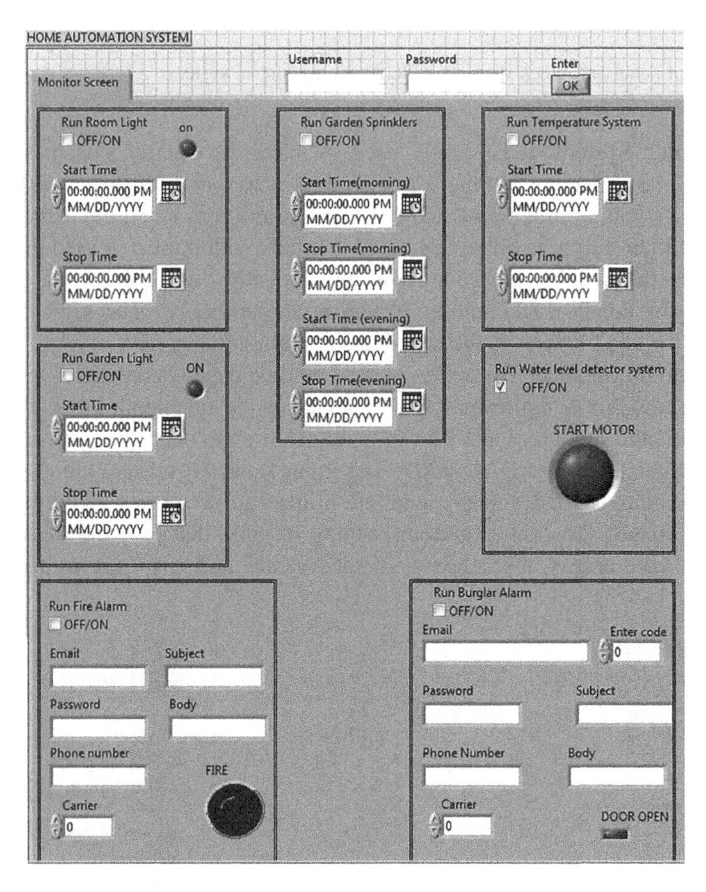

Fig. 11 Monitor screen front panel

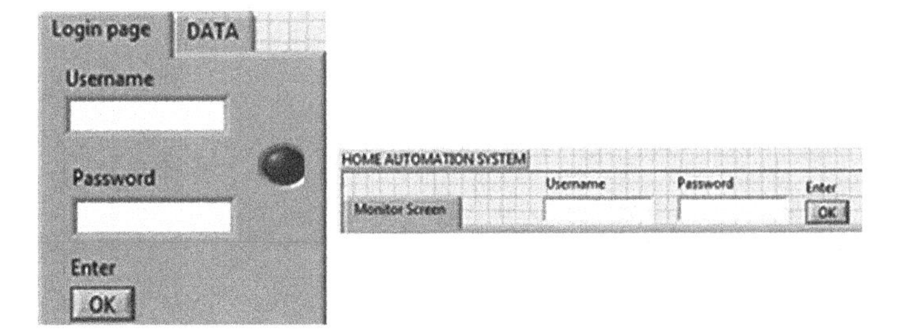

Fig. 12 Login/logout window front panel

4 Conclusion

The interactive home automation system provides the PC based user friendly solutions not only to inspect and log the home events, but also provides a scheduler approach for efficient control of home appliances. The interactive system provides a feature to shift the control from manual to automatic and vice versa. Additionally, the home automation system offers an extensive safe and secure environment with quick mobile and e-mail base informative system for hazards. The authenticated login scheme provides a protective system scheme and prevent unauthorized access of the system. There is a scope of adding more modules to automated the house appliances and equipment.

References

1. Al-Ali, A.R, AL-Rousan. M.: Java-Based Home Automation System. IEEE Transaction on Consumer Electronics, Vol.50, No. 2, pp. 498–504 (2004)
2. Koyuncu, B.: PC Remote Control of Appliances by using Telephone Lines. IEEE Transaction on Consumer Electronics, Vol. 41, Issue 1, pp. 201–209 (1995)
3. Kour G., J.: Lab VIEW Based Alarm Systems In Home: In: UNIASCIT, ISSN 2250-0987, Vol 2, Issue-3, pp. 305–307 (2012)
4. Anil, A. Thampi, AR. John M, P. Shanthi, K J.: Project HARITHA- An automated Irrigation System for Home Gardens. In: India Conference (INDICON), Annual IEEE, pp. 635–639, IEEE (2012)
5. Li-fang, T.: Application of Auto control Technology in Water-saving Garden Irrigation, In: International Conference on Computer Sciences and Information Processing (CSIP), pp. 1311–1314 (2012)
6. Manohar, S. Kumar, D.M.: E-mail Interactive Home Automation System. International Journal of Computer Science and Mobile Computing. Vol.4, Issue.7, pp. 78–87 (2015)
7. Yavuz, E., Hasan, B., Serkan, I., and Duygu. K.: Safe and Secure PIC Based Remote Control Application for Intelligent Home. International Journal of Computer Science and Network Security, Vol. 7, No. 5, pp. 179 (2007)
8. LabVIEW System Design Software - National Instruments, www.ni.com/labview/
9. User Guide of National Instruments (NI) USB-6008/6009, Bus-Powered Multifunction DAQ USB Device www.ni.com/pdf/manuals/371303n.pdf

Power Flow Management in Multiline Transmission System Through Reactive Power Compensation Using IPFC

Divya Asija and Velagapudi Naga Sai

Abstract Interline Power Flow Controller (IPFC) is a series–series type of device belongs to FACTS family. The main purpose of using IPFC instead of remaining FACTS devices is that it can provide series compensation for a required transmission line in system, and also IPFC can control power flow across different lines simultaneously in system. In actual IPFC is a combination of two Voltage Source Converters (VSC's) connected in series by a common dc link in between. By compensating particular line in transmission system, the power flow management becomes effective. It can also possible to exchange reactive power through series compensation within the transmission system. The Voltage Source Converters makes use of snubber capacitance and resistance for reactive power compensation. In this paper a three-phase system is simulated and analyzed by connecting with IPFC Results of the proposed system with output as reactive power are verified for a particular line with and without IPFC.

Keywords FACTS · IPFC · Voltage source converter · Power flow Reactive power compensation

1 Introduction

In Power systems by means of power electronic devices, Flexible AC Transmission Systems, called FACTS, got in the recent years a well-known term for controllability. Different types of fact devices have been introduced worldwide for various applications. In practice a number of new FACTS devices are in stage of being introduced [1].

D. Asija (✉)
Department of EEE, Amity University, Noida, Uttar Pradesh, India
e-mail: dasija@amity.edu

V.N. Sai
Amity University, Noida, Uttar Pradesh, India
e-mail: nagasaiv9@gmail.com

A.K. Somani et al. (eds.), *Proceedings of First International Conference on Smart System, Innovations and Computing*, Smart Innovation, Systems and Technologies 79, https://doi.org/10.1007/978-981-10-5828-8_37

The basic applications of FACTS devices are:

- Power Flow Control
- Increase of transmission capability
- Reactive power compensation
- Improvement in Power quality
- Voltage Control
- Stability improvement
- Power quality improvement
- Power conditioning.

Classification: FACTS devices family is of four types based on the compensation technique and usage of filters in its circuits, they are as follows:

Types of FACTS devices [2]:

1. Series type FACTS devices
2. Shunt type FACTS devices
3. Series–Series type FACTS devices
4. Series–Shunt type FACTS devices (Fig. 1).

The conventional equipment used in FACTS devices circuit has several advantages [3] to enhance the power system controllability of transmission system. They are as follows:

- Series capacitor to control Impendence
- Reactor and Shunt Capacitor to control voltage
- Snubber Capacitance in VSC to control Power Flow.

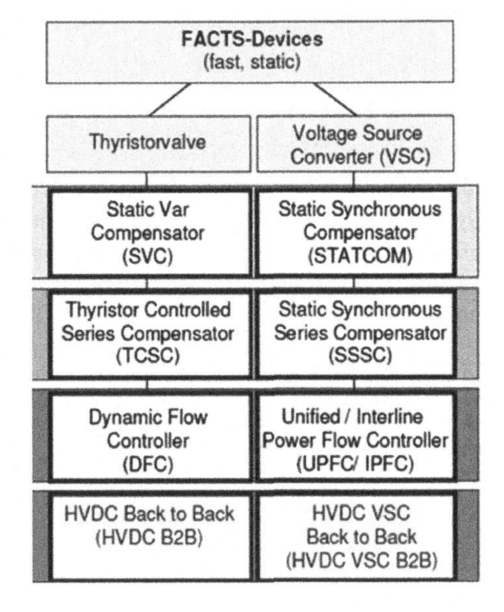

Fig. 1 Different types of FACTS devices

2 Interline Power Flow Controller

Interline power flow controller is a combination of N-number of Voltage Source Converters which are connected by a dc common link [4]. These voltage source converters makes of snubber resistance and capacitance (snubber circuit) for compensating reactive power through series compensation technique. The structure of IPFC used in this paper consists of two voltage source converters connected with a dc link in common.

This paper discuss a new advance technique for management of power flow through reactive power compensation using IPFC [5] controller without significant calculations. The only calculation involved in this approach is basic calculation of real and Volt Ampere Reactive (VAR) powers of a particular line using voltage and current value of that particular line [6]. The best results of the proposed technique is validated with the help of simulated model results at the termination of the paper.

2.1 Structure of IPFC

As discussed earlier IPFC is a series combination of two Voltage Source Converters connected by a dc link in common.

The basic structure of IPFC is as shown in Fig. 2.

From the IPFC basic configuration, the voltage source converters connected in series with a dc link in common. At first level Voltage Source Converter's primary terminal is connected to transmission line between the buses a and b, and the

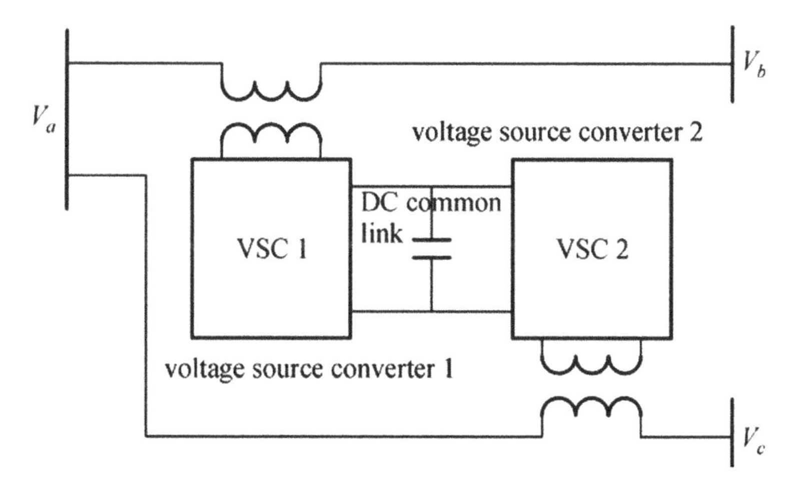

Fig. 2 Basic structure of IPFC

secondary terminal of first VSC is connected in series with primary terminal of second VSC with a dc link in common [7–9].

The secondary terminal of second Voltage Source Converter is connected to a transmission line which is to be compensated. The output variables available at the secondary terminals of VSC-2 are controlled variables, which can control [10, 8] the transmission network. The real and VAR powers of transmission line can be calculated from the obtained controlled voltage and currents of that particular line. It can also possible to exchange reactive power from one line to required transmission line by connecting the secondary terminal of VSC-2 to a point where we need to exchange.

The Real and VAR powers of that particular transmission line which is compensated are given by mathematical formulae:

$$P_n = Real(V_n{*}I_n)KW \tag{1}$$

$$Q_n = Imaginary(V_n{*}I_n)VAR \tag{2}$$

3 Simulation Model of Proposed System

Simulation of system network can be analyzed for two cases:

a. System transmission network without IPFC
b. System transmission network with IPFC

Case (I): System transmission network model without IPFC (Fig. 3).
Case (II): System transmission network with IPFC (Fig. 4).

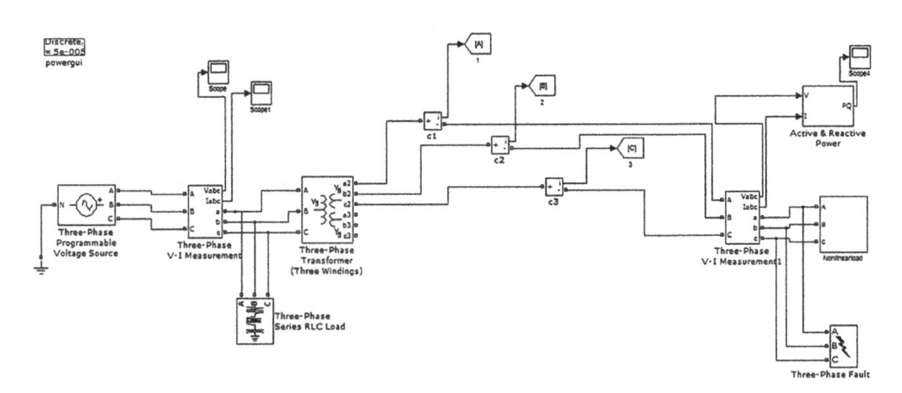

Fig. 3 Simulation model of system network without IPFC

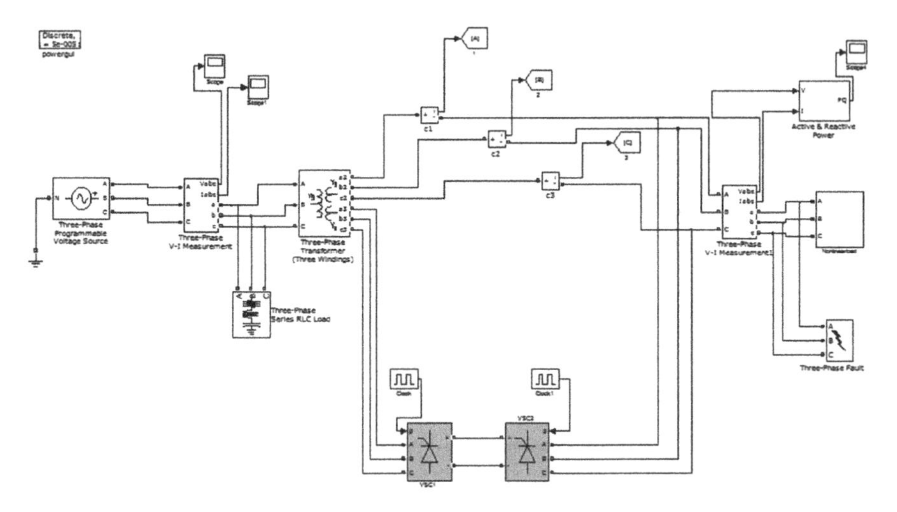

Fig. 4 Simulation model of system network with IPFC

4 Simulation Results

Simulation results show the comparison of reactive powers of the lines which are compensated under series compensation technique. The comparative analysis of reactive power magnitude is expressed in percentage (%). By using the snubber resistance and capacitance in the circuit of Voltage Source Converter, the reactive power of the particular line in transmission network is compensated and the difference in the magnitude of reactive power for considered two cases is shown using graphs. As the input given to the system is low which is about (230 V) and discrete in nature, the difference in the magnitude of reactive powers in both the cases is also least varied. But for a practical transmission system with continuous input at high voltage levels the difference in the magnitudes of reactive powers can be high. The simulation results of Real and VAR powers of both the cases including and excluding IPFC are as shown below.

Waveforms of Real and VAR powers of transmission network excluding IPFC (Figs. 5 and 6).

Waveforms of Real and VAR powers of transmission network including IPFC (Figs. 7 and 8).

- The reactive power magnitude of system without IPFC is = 0.0506 VAR
- The reactive power magnitude of system with IPFC is = 0.1097 VAR

Percentage change in magnitude = 51%.

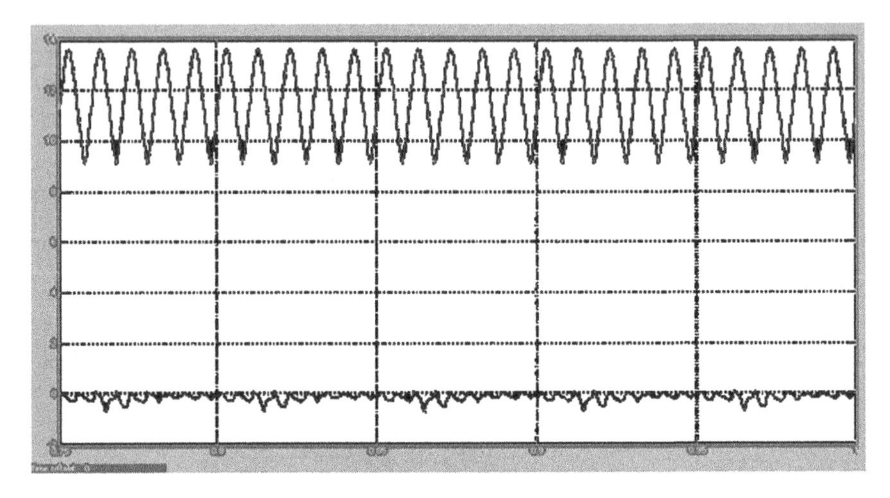

Fig. 5 Real and VAR powers waveforms of system excluding IPFC

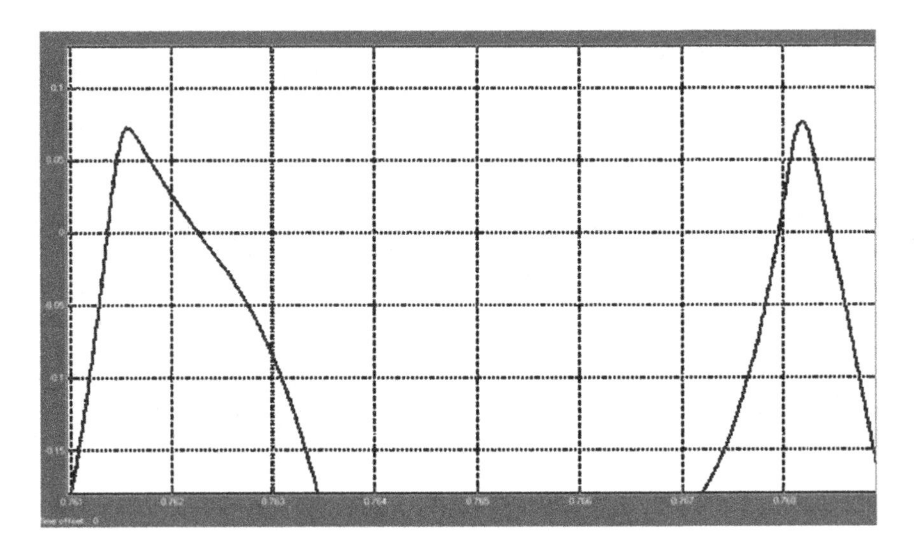

Fig. 6 VAR power magnitude of system excluding IPFC

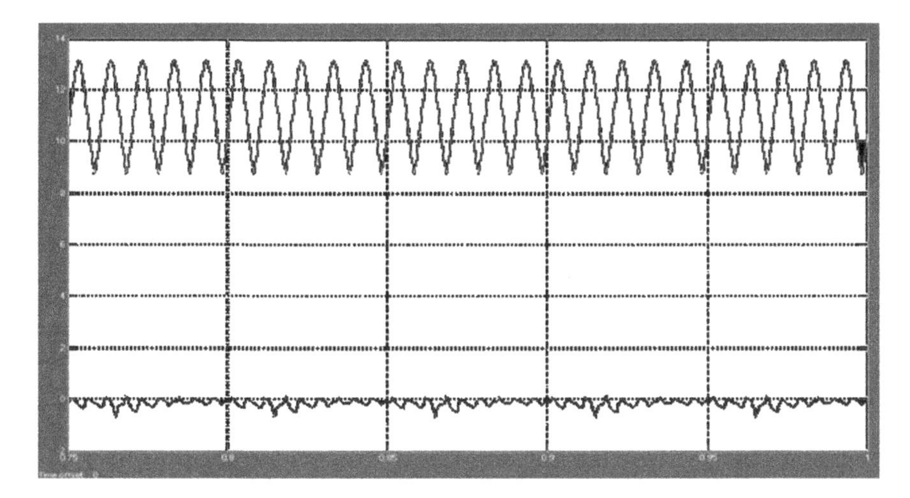

Fig. 7 Real and VAR powers waveforms of system including IPFC

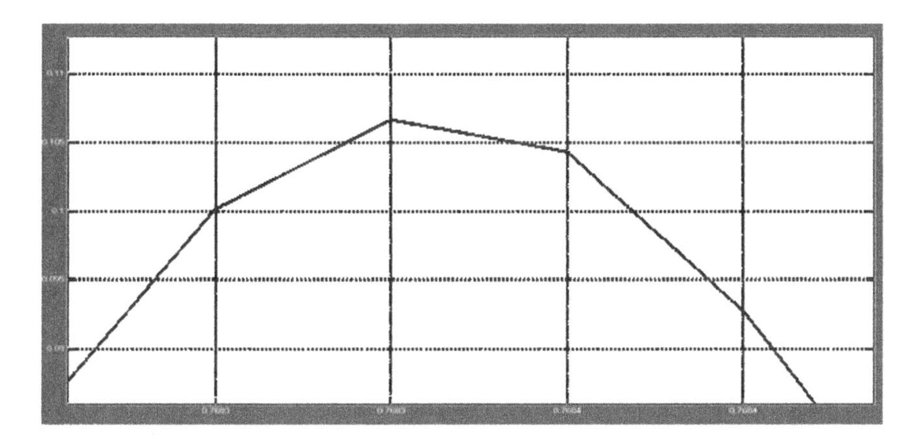

Fig. 8 VAR power magnitude of system including IPFC

5 Conclusion

Using IPFC in series compensation of reactive power the power transfer capacity of transmission line network can be increased. In addition to that it is also verified that the compensation of reactive power of the particular line is done effectively by making use of snubber resistance and capacitance in the Voltage Source Converter circuit. The percentage change in magnitude of VAR power for the considered two cases that is without and with IPFC have significant value that is 51% which is worth noticeable. Thus it has been concluded that series reactive power compensation using IPFC reduces the overall system power losses and can be considered as viable compensating element. The unique characteristics of IPFC enables compensation of transmission lines.

References

1. N.G. Hingorani, L. Gyugyi, "Understanding FACTS: concepts and technology of flexible AC transmission system, IEEE PRESS, Standard Publishers Distributors, Delhi", 2000.
2. N. G. Hingorani and L. Gyugyi, "Understanding FACTS", Nework: IEEE Press, 2000.
3. Nadarajah Mithulananthan, Claudio A. Canizares, John Reeve, and Graham J. Rogers, "Comparison of PSS, SVC, and STATCOM controllers for damping power system oscillations," *IEEE Trans. Power Systems*, vol. 18, no. 2, pp. 786–792, May 2003.
4. L. Gyugyi, K. K. Sen and C. D. Schauder, "The interline power flow controller concept: a new approach to power flow management in transmission systems," *IEEE Trans. Power Del.*, vol. 14, no. 3, pp. 1115–1123, Jul. 1999.
5. R. Strzelecki, P. Smereczynski and G. Benysek, "Interline power flow controller- properties and control strategy in dynamic states".
6. Indra Prakash Mishra, Sanjiv Kumar, "Control of Active and Reactive Power Flow in Multiple Lines through Interline Power Flow Controller (IPFC)", International Journal of Emerging Technology and Advanced Engineering, Vol. 2, No. 11, 2012.
7. Jianhong Chen, Tjing T. Lie, D.M. Vilathgamuwa "Basic Control of Interline Power Flow Controller" IEEE, Int.Confon Power electronics, India, pp 1–7, 2011.
8. Jianhong Chen, Tjing T. Lie, D.M. Vilathgamuwa, "Design of An Interline Power Flow Controller", 14th PSCC, 2002.
9. Laszlo Gyugyi, Kalyan k. Sen, Colin D. Schauder, "The Interline Power Flow Controller concept: A New Approach to Power Flow Management in Transmission System", IEEE Transactions on Power Delivery, Vol.14, No.3, 1999.
10. RL Vasquez-Arnez "Main Advantages and Limitations of Interline Power Flow Controller: A Steady State Analysis".

A Machine Learning Approach for User Authentication Using Touchstroke Dynamics

Dinesh Soni, M. Hanmandlu and Hukam Chand Saini

Abstract Touchstroke dynamics is an essential component of computer security. In recent years, we are heavily dependent on computers for communication, banking, security applications, and many other areas. This dependency has increased the chances of malicious attacks, so there is a need for high security to protect user's secured data from unauthorized access. Currently, we are using PINs and passwords for access in computers, but these methods are not sufficient as the computer systems are accessed globally. So we propose a method for touchstroke dynamics in touchscreen mobile devices to improve security. The behavioral biometric gives a confidence measurement instead of accept/reject measurements. We have used an android mobile device for assessing the security using the touchstroke behavior of users. This provides us with confidence measurements for security purpose as compared to physiological biometric in which FRR/FAR cannot be changed by varying threshold at individual level.

Keywords Keystroke · Touchstroke · FRR/FAR

D. Soni (✉) · H.C. Saini
Department of Computer Science and Engineering, Rajasthan Technical University, Kota, Rajasthan, India
e-mail: dsoni@rtu.ac.in

H.C. Saini
e-mail: hukams.saini@gmail.com

M. Hanmandlu
Department of Electrical Engineering, Indian Institute of Technology Delhi, New Delhi, India
e-mail: mhmandlu@ee.iitd.ac.in

A.K. Somani et al. (eds.), *Proceedings of First International Conference on Smart System, Innovations and Computing*, Smart Innovation, Systems and Technologies 79, https://doi.org/10.1007/978-981-10-5828-8_38

1 Introduction

Keystroke dynamics is also known as keystroke biometric or typing dynamics. It is a behavioral biometric like: voiceprints, signatures. Keystroke dynamics is a science of typing rhythm and manner on keyboard. It tells about how the individual users type on the keyboard. It gives the timing and pressure information about individual users. When one user types on the keyboard at 30 words per minute and another user types 50 words per minute then if user one suddenly starts typing at a speed 55 words per minute, so we can say clearly that user one is not real user of the system because it is not possible for a user to get about double the typing speed.

Keystroke dynamics gives the signatures of individuals about typing behavior. If a user is typing slow or fast, the combinations of keys a user employs to type capital letters make a difference to the stroke dynamics. For this, one can use (shift+keys) combinations or Caps Lock on keyboard. Which hand whether left hand or right hand the user is accustomed with, how frequently the backspace key is used; these observations tell us about the possible mistakes he/she commits [1]. In the recent years, the popularity of keystroke dynamics has been increasing leaps and bounds as can be seen in its applications in online education sites like Coursera to authenticate users, Psylock, KeyTrac, BioTracker, TypeWATCH, etc. On the other side, touchstroke dynamics have rich set of features than keystroke dynamics. Touchstroke dynamics has become popular to enhance the security on touchscreen mobile devices. With the increased touch behavior on mobile devices, the possibility of attacks on mobile devices is increased. It is necessary to develop some method to prevent the loss of secure information. We present a secure method to enhance user's authentication on touchscreen mobile devices with the increased number of features [2]. We apply support vector machine, TreeBagger (random forest), distance measure, and neural networks for the classification of users.

The organization of this paper is as follows. Section 2 gives a literature survey of the current work on touchstroke dynamics. Section 3 presents the feature extraction from touchstrokes. Section 4 discuss about JAVA language and Android technologies which are used for the development of software. Section 5 presents the simulations and experimental results and discusses the accuracy rates achieved from different classifiers, and Sect. 6 concludes the paper.

2 Literature Survey

Touchstroke dynamics is a behavioral biometric authentication method. In the current mobile market, sensor-rich smartphones are growing in popularity. Smartphones have many sensors like

Gyroscope: It detects the orientation details of a mobile using up or down, left or right and rotation around X, Y, Z-axes.

Accelerometer: It detects the movement and orientation information using acceleration, vibration, and tilt.
Magnetometer: It detects the user's location using direction of magnetic north and GPS.
Proximity Sensor: It detects the face of mobile when making a call and disable touchscreen display.
Light Sensor: It detects the lighting level around and display accordingly. **Fingerprint sensor**: It is used to secure the device using biometric verification and website authentication.
Infrared Sensor: It is used to recognize the gestures on mobile and identify the user's movement.

The current research work on touchstroke dynamics authentication is now presented. In [3] neural network approach is used for the classification purpose in touchstroke dynamics. This approach achieves the mean EER 7.8% for completely different users. If the number of users and touch gestures are increased then better performance is achieved. In [4] touchscreen behavior of only six users is used for 16 min interaction with device. This approach identifies 90 s sample correctly with 83% of time. In [5] touchstroke behavior of users is employed by using their hand dominance features which the users apply either right or left hand while interaction with device. This approach achieves better results and finds the gesture-size and gesture-time touch characteristics. In [6] multi-touch gestures of device are used to study about the hand and muscle behavior to identify users. The score-based classifier is applied on time features and EER of 4.46% is achieved. By using pinch, rotation, and zoom gestures EER of 58% is achieved. In [2] experiments with different age groups are performed. The experiments are conducted on 14 adults and 16 children by using features like X-cord, Y-cord, touch time, touch size, and touch pressure for each move, up and down events. It is observed that adults are identified more than children because of their outside screen touches. In [7] 3D-gestures with mobile having 3-axis accelerometer are captured while user holds a mobile. The gestures are collected on 25 users for analysis of 3D-gestures and EER of 2.01% is achieved in zero effort and EER of 4.82% is achieved for active attack. In [8] multi-touch gestures and mixed gestures inputs are used. Users asked to be comfortable with touch gestures by practicing on mobile. The system has recorded ten trials of touch gestures for each user. In [1] lock pattern is used on mobile. The finger-move times are employed for data collection. They applied the machine learning random forest classifier that combines different lock patterns three times. An EER of 10.39% is achieved for 32 users.

3 Feature Extraction

3.1 Touchstroke Dynamics System

The block diagram in Fig. 1 explains about touchstroke dynamics system. Touchstroke dynamics system has been implemented by the following steps:
Step1: In the enrollment step, each user types username and performs touch gestures (4 different touch locations, repeat 20 times).
Step2: In the training step, when user performs touch gestures features are extracted and then they are normalized. By using the extracted features, a user template is created and stored in a database.
Step3: In the verification step, distance between the test sample and train sample is used for getting score.
Step4: In this step each users score is compared with the threshold for making a decision.

The following features are extracted from touch gestures as shown in Fig. 2.
X-coordinate: Touch position on X-axis.
Y-coordinate: Touch position on Y-axis.
Pressure: Pressure applied by finger on touch.
Size: Approximate size of screen area on a touch.
Hold time: Time between press and release on a touch.
Inter time: Time between one touch to another touch.
Pitch: Rotation angle on X-axis of mobile.

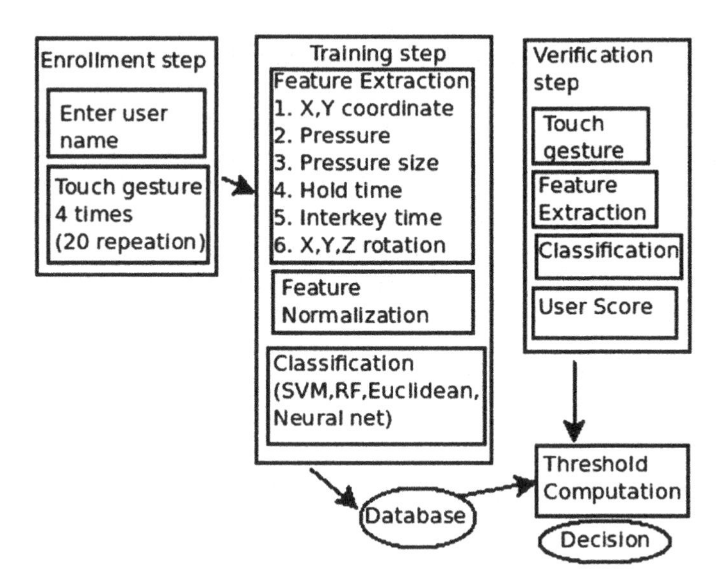

Fig. 1 Touchstroke dynamics system

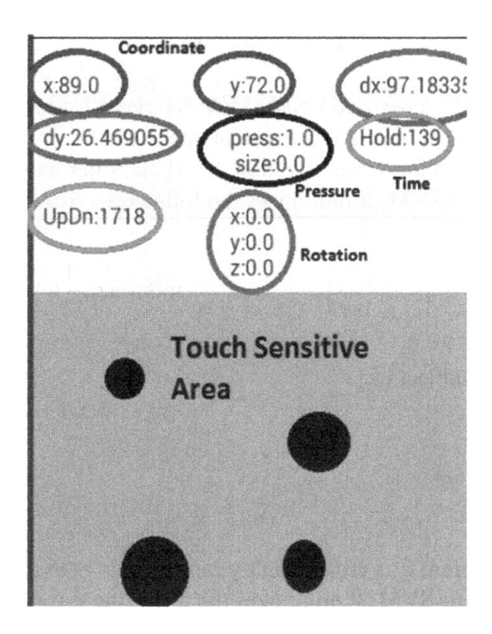

Fig. 2 Features in touchstroke dynamics

Roll: Rotation angle on Y-axis of mobile.
Yaw: Rotation angle on Z-axis of mobile. We have used the users scores and threshold for the classification by different machine learning classifiers such as Euclidean distance, SVM, TreeBagger, and ANNs and compared their accuracy.

3.2 *Classification Methods*

3.2.1 Euclidean Distance

It is defined as the distance between the test vector and the mean vector. It is computed from

$$E = \sqrt{\sum_{i=1}^{n} (X_i - Y_i)^2} \tag{1}$$

here

$$X = (X_1, X_2, X_3 ... X_n) \tag{2}$$

represent **test vector** and

$$Y = (Y_1, Y_2, Y_3 ... Y_n) \tag{3}$$

represent **mean vector** for training sample.

3.2.2 SVM Classification

We have used binary SVM for classification between two users. It constructs hyperplane in high-dimensioal space to classify between two classes. It uses kernel for nonlinear classification. SVM classifier has the following types:
C-SVM: It minimizes the following error function.

$$\text{minimize } \frac{1}{2}w^T w + C\sum_{j=1}^{N} \xi j \tag{4}$$

subject to,

$$y_j(w^T \phi(x_j) + b) \geq 1 - \xi j$$

and

$$\xi j \geq 0$$

Here C is error-penality ranges from zero to infinity.
nu-SVM: It minimizes the following error function.

$$\text{minimize } \frac{1}{2}w^T w + \frac{1}{N}\sum_{j=1}^{N} \xi j - \upsilon\rho \tag{5}$$

subject to,

$$y_j(w^T \phi(x_j) + b) \geq \rho - \xi j$$

and

$$\xi j \geq 0$$

and

$$\rho \geq 0$$

Here nu is error-penality ranges from zero to one.

We have used RBF kernel. It is defined as

$$exp(\gamma \left\| x - x' \right\|^2) \tag{6}$$

for two samples x and x'. Here

$$\gamma = -\frac{1}{2\sigma^2}$$

3.2.3 TreeBagger Classification

TreeBagger is a type of ensemble learning algorithm. It uses bagging (bootstrap aggregation) for the classification. Random forest is invoked by setting "NVarToSample" parameter to "all" in the treebagger. The decision tree is constructed from randomly selected samples. Many decision trees are formed in RF. Votes are given for each class and the class belonging to the unseen data gets the highest number of votes. Data samples are divided into 2/3 for training and 1/3 for testing purpose. Then out-of-bag samples are used to calculate classification error. It measures variable importance by permutation. OOB errors give measure of generalization error. Then proximity matrices are used for getting outliers information, missing data values, and data views in lower dimension space.

3.2.4 Artificial Neural Network (ANN) Classification

We have used the FITNET (function fitting neural network) machine learning tool for classification. It is a two layer feed forward neural network. In hidden layer, it uses tan-sigmoid transfer function. In output layer, it uses linear transfer function. For training purpose, we have used **TRAINLM** algorithm. It is the fastest backpropagation and supervised algorithm. It uses Jacobian for all calculations and it updates the weights and bias values by using Levenberg–Marquardt optimization algorithm.

The LM algorithm applies the following approximation to the Hessian matrix.

$$X_{i+1} = X_i - [G^T G + \gamma I]^{-1} G^T e \tag{7}$$

Here: G is the 'Jacobian matrix', e is the 'error vector', and γ is a scalar parameter. When γ is zero, it becomes Newton's method, which uses approximate Hessian matrix. For large value, it becomes gradient descent learning law having lower step size. Performance function is measured by using MSE. So γ is decreased after each successful step, it reduces performance function and when γ is increased, it increases the performance function. So, at each iteration step the performance function is decreased.

4 Technology

4.0.1 JAVA JAVA is most popular known language now-a-days. It is designed by Sun Microsystems. It has write once, run anywhere feature. JAVA program is compiled into bytecode that can be run on any JVM, so it is platform independent. JAVA supports object oriented, structured, class-based, concurrent, imperative, and multi platform features. JAVA is simple, robust, portable, secure, threaded, high performance, architecture independent, interpreted, and dynamic language with automatic garbage collector. All these features make JAVA a powerful programming language.

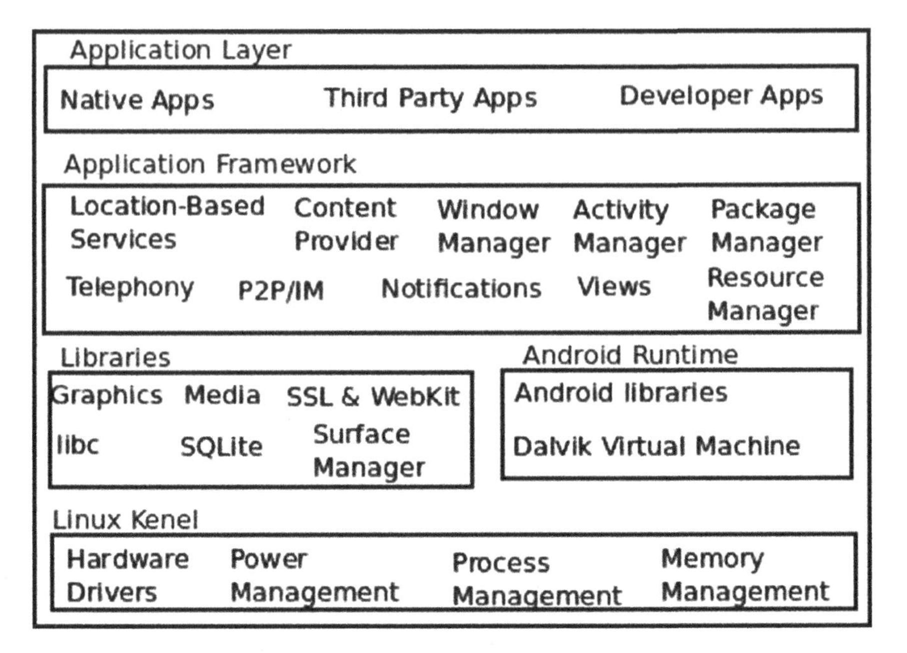

Fig. 3 Android architecture

4.0.2 ANDROID It is an open source operating system in many mobile devices developed by Google. It is also used in smartphones. Android has an ARM architecture shown in Fig. 3 with Monolithic linux kernel. Note that in old versions, Dalvik is used as a process virtual machine. JAVA classes are compiled into dex-code during Android runtime.

5 Simulations and Experimental Results

The experiment is performed for 10 users. Each user enters user name and gives four touchstrokes on different locations on the Touch Sensitive Area. User repeats this 20 times. For each touchstroke, nine features are extracted and stored in the database as shown in Fig. 4.

5.0.1 Euclidean Distance Classifier For each user, we compute the Euclidean distance as a threshold between two users after which classification is done as shown in Fig. 5.

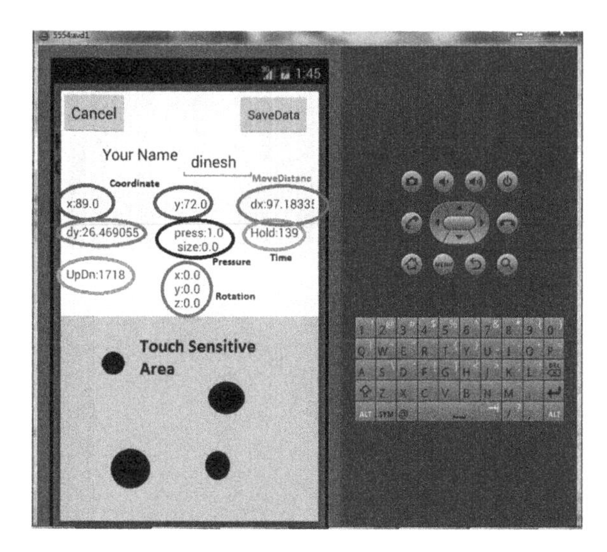

Fig. 4 Android GUI for touchstroke

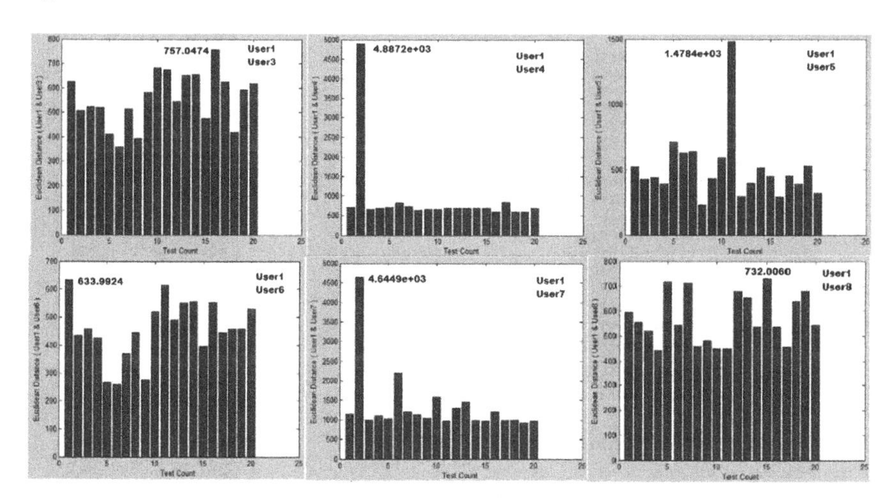

Fig. 5 Euclidean distance between user1 to user8

5.0.2 SVM Classifier We have taken five users with five test samples per user and then applied SVM for binary classification. As shown in Fig. 6 the test samples are classified correctly.

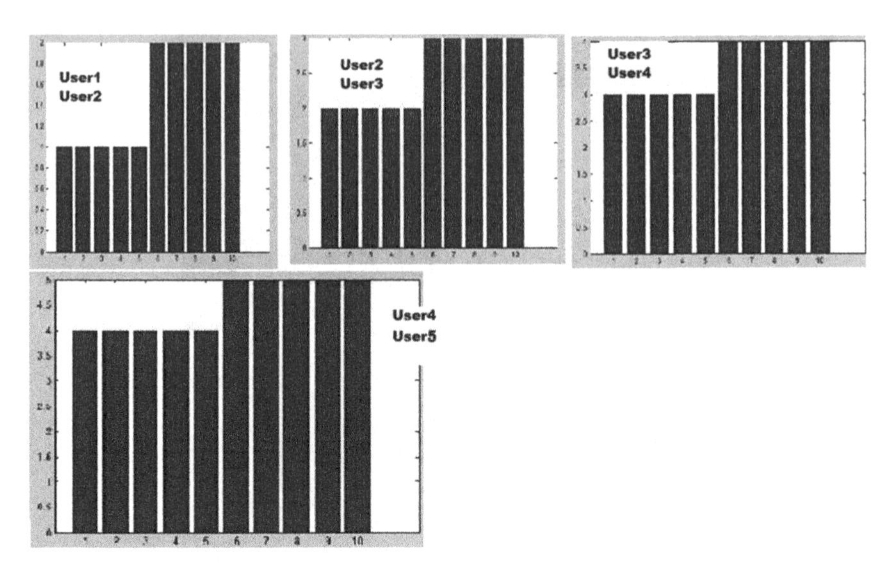

Fig. 6 SVM output for user1 to user5

```
User1BagpredictedClass =1
User1BagpredictedClass =1
User2BagpredictedClass =2
User2BagpredictedClass =2
User3BagpredictedClass =7
User3BagpredictedClass =7
User4BagpredictedClass =7
User4BagpredictedClass =7
User5BagpredictedClass =7          (18 rows train)
User5BagpredictedClass =7         (2 row test)
User6BagpredictedClass =7
User6BagpredictedClass =7             8/20 =40% wrong   12/20 = 60% correct
User7BagpredictedClass =7
User7BagpredictedClass =7
User8BagpredictedClass =8
User8BagpredictedClass =8
User9BagpredictedClass =9
User9BagpredictedClass =9
User10BagpredictedClass =10
User10BagpredictedClass =10
```

Fig. 7 TreeBagger output (18 rows train, 2 rows test)

5.0.3 TreeBagger (Random Forest) Classifier TreeBagger classifier makes use of 18 rows of training and 2 rows of testing samples which yield 12/20 (60%) correct predictions and 8/20 (40%) wrong predictions; but with 19 rows of training and 1 row of test sample yield 10/10 (100%) correct predictions as shown in Figs. 7 and 8. TreeBagger class observations are shown in Fig. 9. All 10 user-classes are shown in different colors. TreeBagger Out-Of-bag classification error that decreases as the number of grown trees increases is shown in Fig. 10.

```
User1BagpredictedClass =1
User2BagpredictedClass =2          (19 row train)
User3BagpredictedClass =3         (1 row test)
User4BagpredictedClass =4            10/10 =100%
User5BagpredictedClass =5
User6BagpredictedClass =6
User7BagpredictedClass =7
User8BagpredictedClass =8
User9BagpredictedClass =9
User10BagpredictedClass =10
```

Fig. 8 TreeBagger output (19 rows train, 1 rows test)

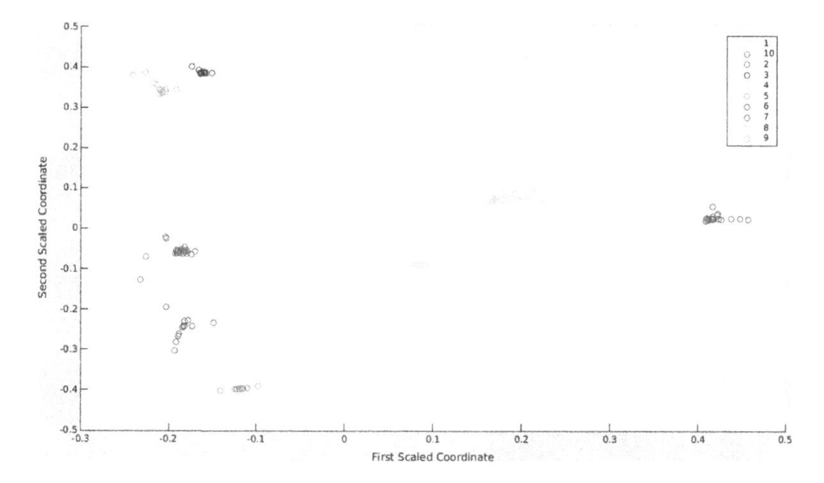

Fig. 9 TreeBagger: second-scaled coordinate versus first-scaled coordinate

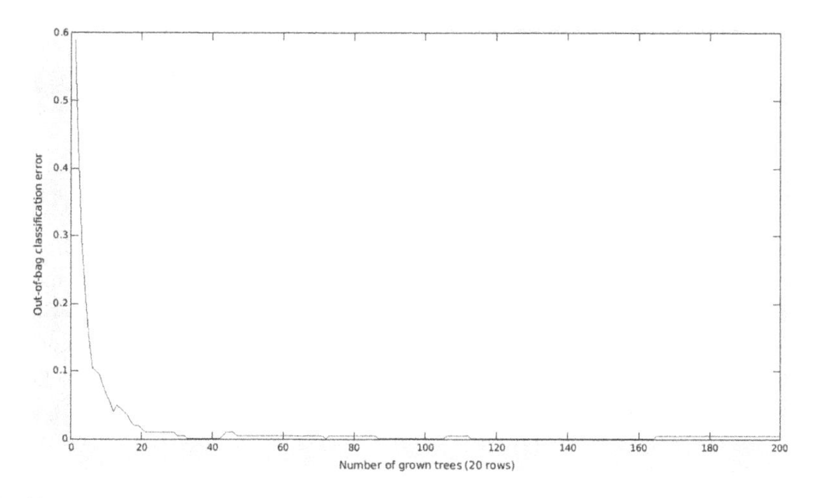

Fig. 10 Out-of-bag error versus number of grown trees

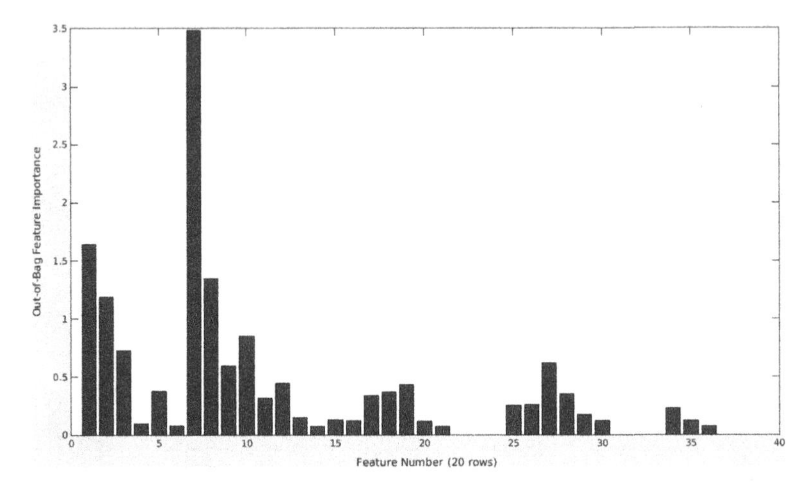

Fig. 11 Out-of-bag feature importance versus feature number

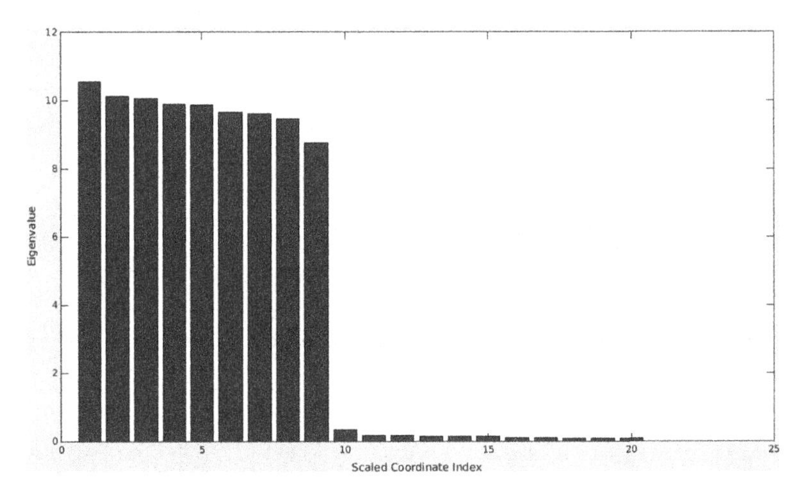

Fig. 12 Eigen values for 20 training samples

TreeBagger Out-Of-bag Feature Importance against feature number is shown in Fig. 11. This shows the feature importance of Rotation angle on X-axis of mobile. TreeBagger Eigen values for 20 training samples are shown in Fig. 12. TreeBagger outlier measure and number of observations are shown in Fig. 13. TreeBagger Out-of-Bag mean classification margin accuracy increases as the number of trees grown increases as shown in Fig. 14.

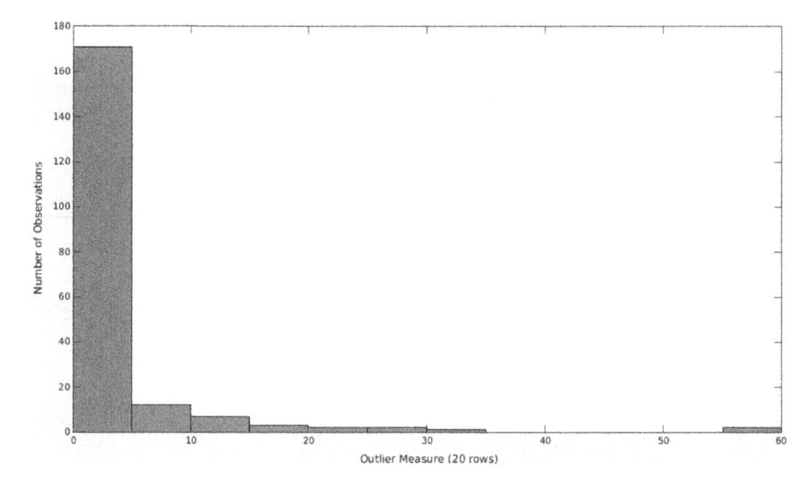

Fig. 13 Number of observations versus outlier measure

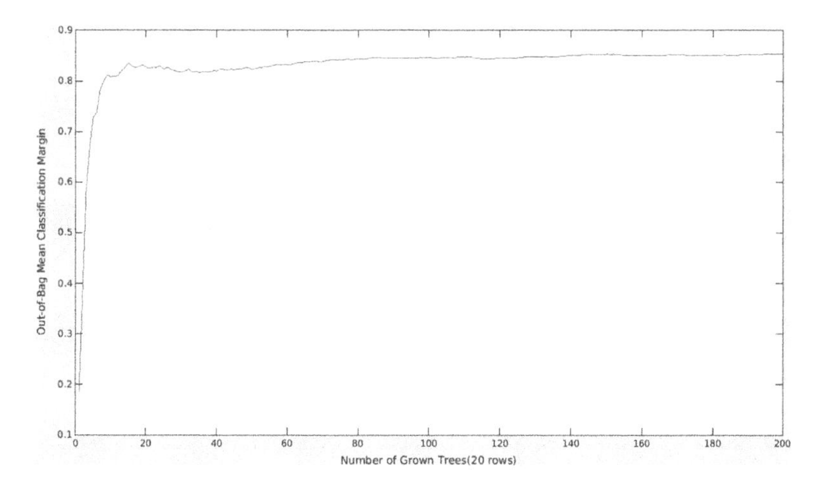

Fig. 14 Out-of-bag mean classification margin versus number of grown trees

5.0.4 Artificial Neural Network Classifier The neural network architecture for 20 rows of training samples is shown in Fig. 15. Data samples are divided into 80% training, 10% testing, and 10% validation samples. The accuracy achieved is 89.3770%. Performance plots for training, testing, and validation samples are shown in Fig. 16.

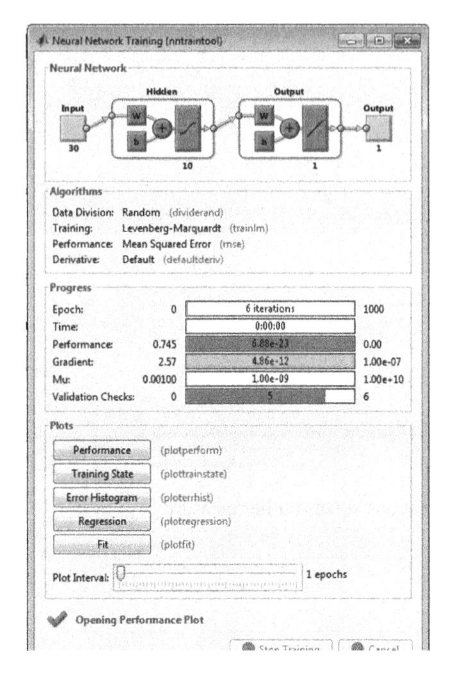

Fig. 15 Neural network structure

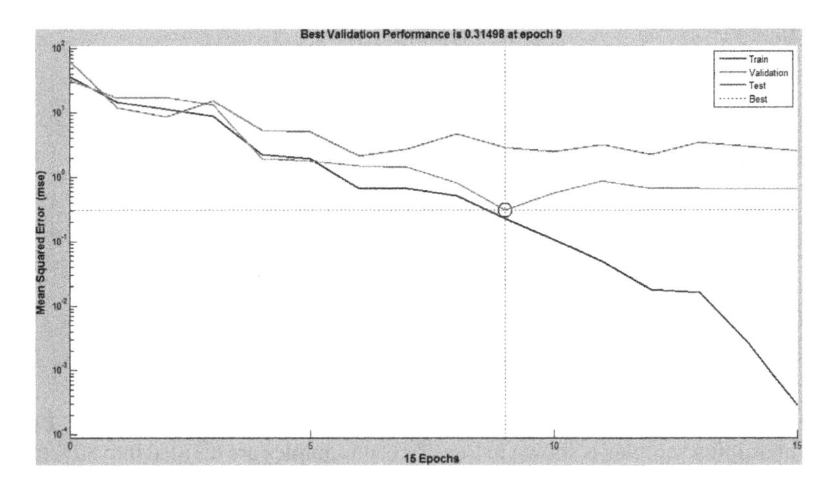

Fig. 16 Performance curve for training, testing, and validation

Error histogram plots for training, testing, and validation samples are shown in Fig. 17. Regression plots for training, testing, and validation samples are shown in Fig. 18. Training state plots for training, testing, and validation samples are shown in Fig. 19.

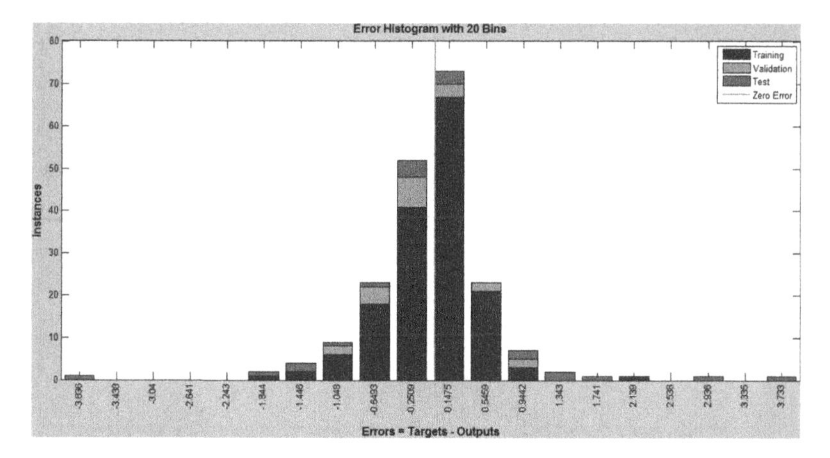

Fig. 17 Error histogram curve

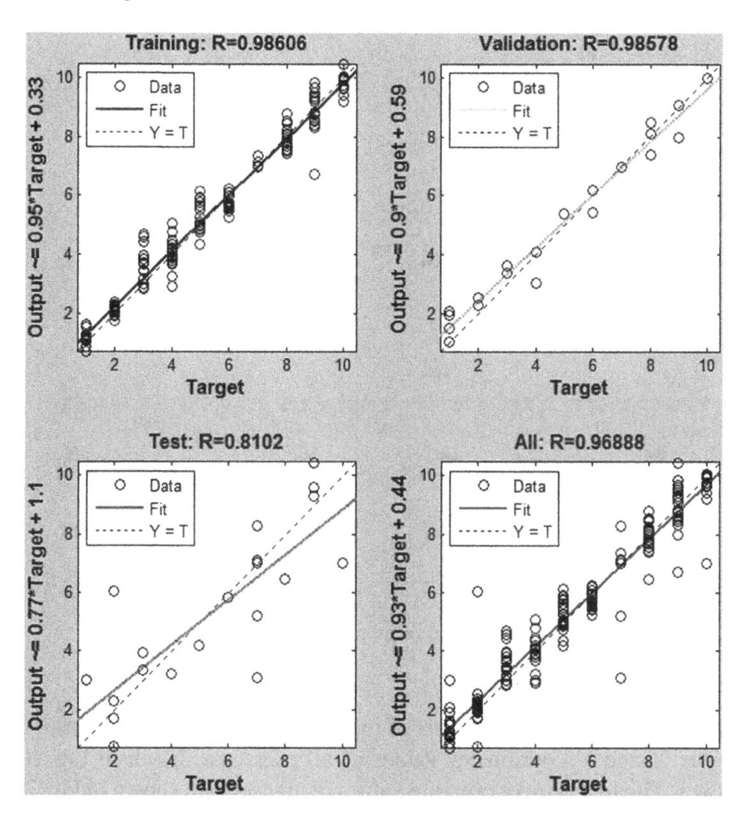

Fig. 18 Regression curve

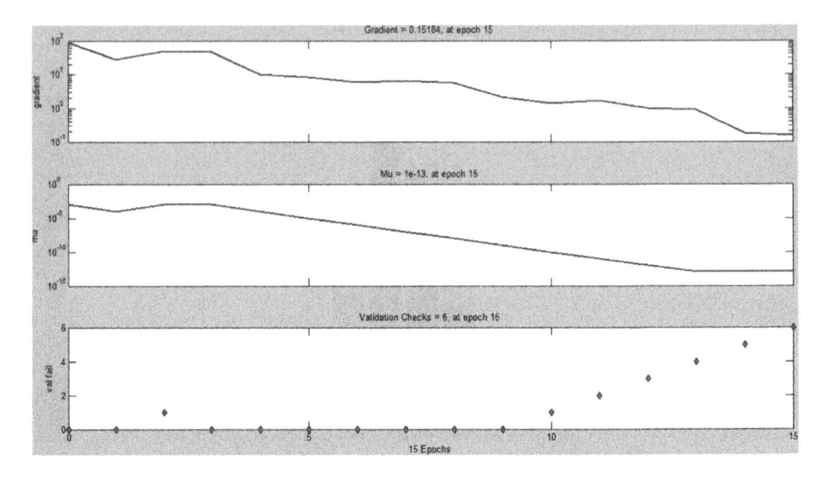

Fig. 19 Training state curve

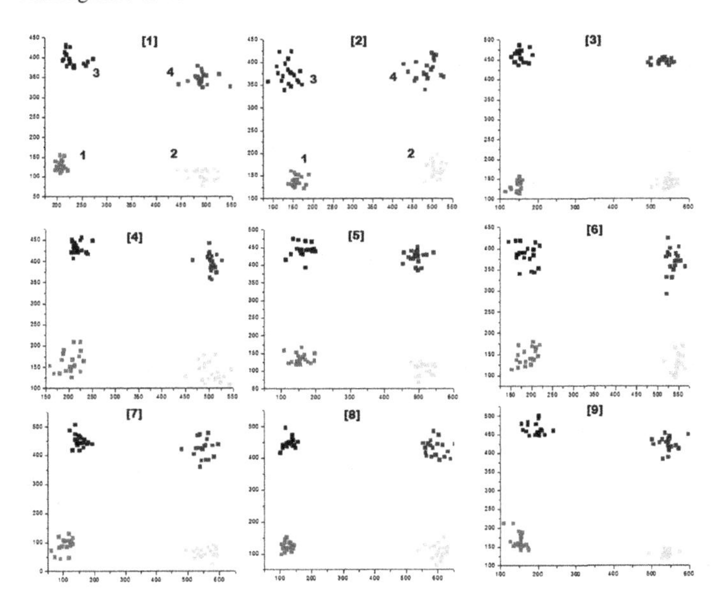

Fig. 20 X-Coordinates and Y-Coordinates on all touchstroke

We will now resort to the analysis of touchstroke behavior of different users on mobile. The X and Y coordinates values of all users are shown in Fig. 20 on all touchstrokes. The touchstroke pressure values of all users are shown in Fig. 21 on all touchstrokes.

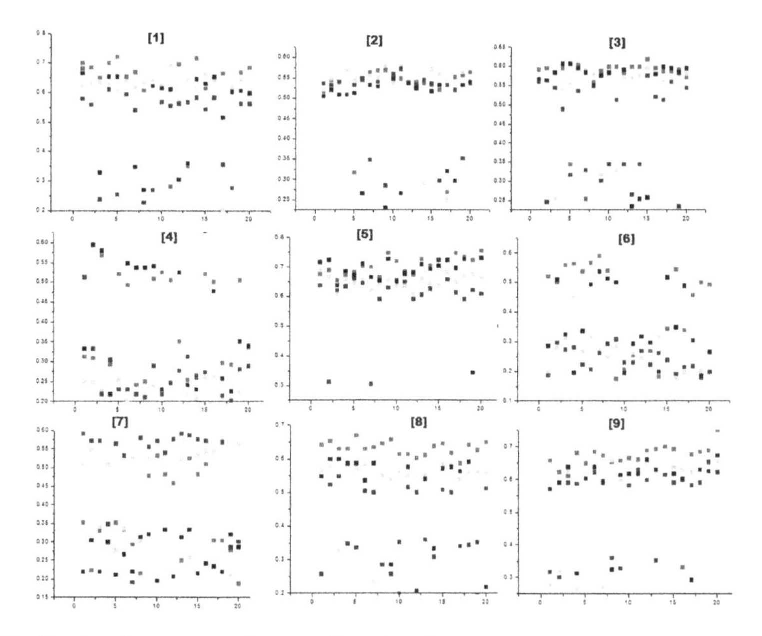

Fig. 21 Touchstroke pressure values on all touchstrokes

The touchstroke size values of all users are shown in Fig. 22 on all touchstrokes. The hold and latency time values of all users are shown in Fig. 23 on all touchstrokes. The rotation angles about X, Y, and Z axes of all users are shown in Fig. 24 on all touchstrokes.

As a result, we have experimented on 10 users with 20 training samples per user, out of these 18 samples are used for the training and the rest two are used for testing the samples. We get Euclidean distance of 757 for user1 and user3, 753 for user1 and user9, and 1267 for user1 and user10 which is used to set a threshold for classification. The SVM classifier output accuracy is 100% for five user samples from user1 to user5 with five testing samples per user. In the TreeBagger classifier output, we have 18 training samples and two testing samples. We get 60% correct and 40% wrong predictions. But with 15 training samples and five test samples, we get 100% correct predictions. The ANN classifier gives performance 89.377% accuracy for the randomly selected 80% training samples, 10% testing samples, and 10% validation samples.

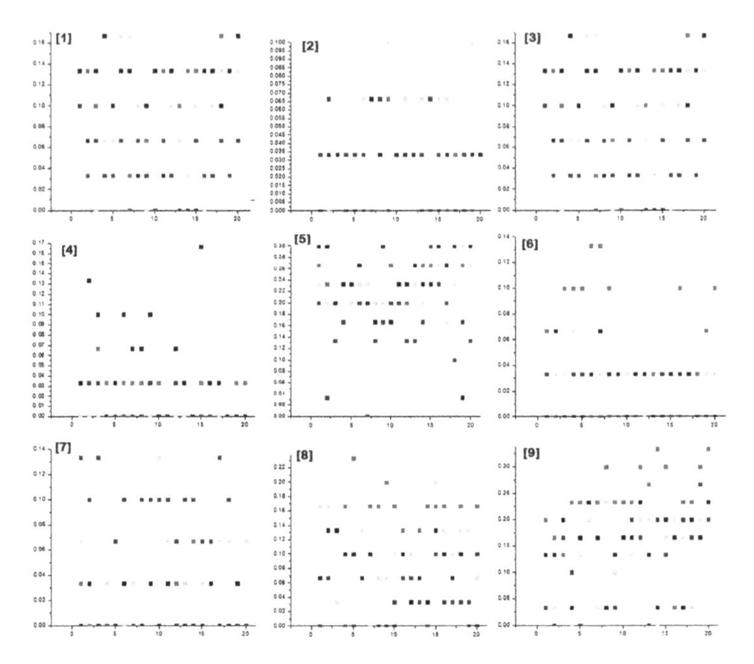

Fig. 22 Touchstroke size values on all touchstroke

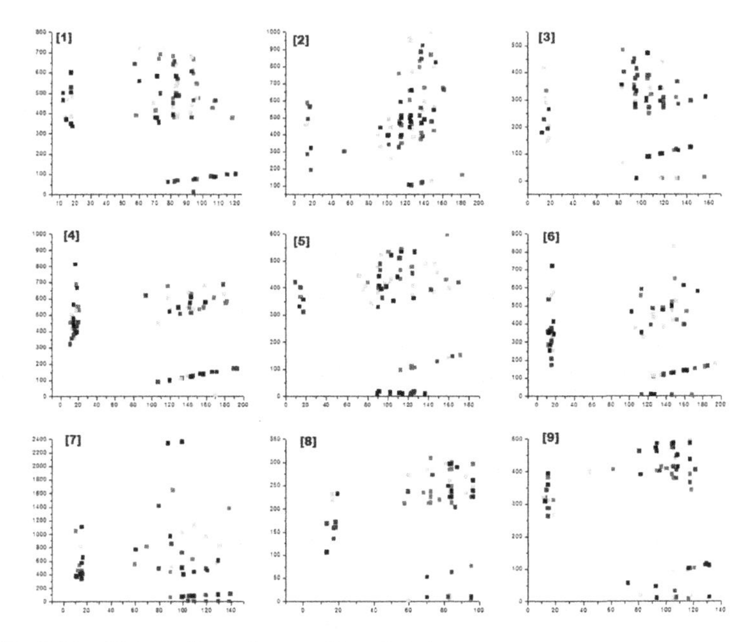

Fig. 23 Hold and latency time values on all touchstroke

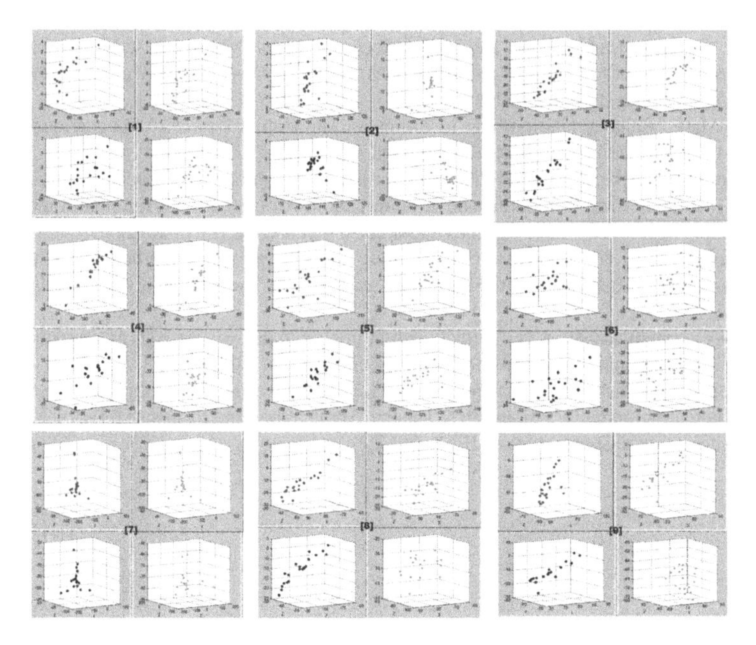

Fig. 24 Rotation angle about X, Y, and Z axes on all touchstrokes

6 Conclusion

Behavioral biometric mechanisms for authenticating users are surveyed on the basis of their interaction with the android mobile devices. Several classifiers provide varying accuracy with different training and testing samples. As regards the touchstroke, dynamics of the single-touchstroke behavior on mobile is promising but the use of multi-touch gestures on mobile would increase the classifier accuracy at the cost of the increased computation for a large number of features in the Android App. Presently, most devices offer cloud-based services, which are useful for future development. As there is no single biometric mechanism which can provide better authentication, it is desirable to create a multimodal biometrics system by combining other biometric modalities.

References

1. Angulo, Julio, and Erik Wstlund. "Exploring touch-screen biometrics for user identification on smart phones." Privacy and Identity Management for Life. Springer Berlin Heidelberg, 2012. 130–143.
2. Anthony, Lisa, et al. "Interaction and recognition challenges in interpreting children's touch and gesture input on mobile devices." Proceedings of the 2012 ACM international conference on Interactive tabletops and surfaces. ACM, 2012.

3. Meng, Yuxin, Duncan S. Wong, and Roman Schlegel. "Touch gestures based biometric authentication scheme for touchscreen mobile phones." In Information Security and Cryptology, pp. 331–350. Springer Berlin Heidelberg, 2013.
4. Wolff, Matt. "Behavioral Biometric Identification on Mobile Devices." Foundations of Augmented Cognition. Springer Berlin Heidelberg, 2013. 783–791.
5. Sandnes, Frode Eika, and Xiaoli Zhang. "User identification based on touch dynamics." Ubiquitous Intelligence and Computing and 9th International Conference on Autonomic and Trusted Computing (UIC/ATC), 2012 9th International Conference on. IEEE, 2012.
6. Sae-Bae, Napa, et al. "Biometric-rich gestures: a novel approach to authentication on multi-touch devices." Proceedings of the SIGCHI Conference on Human Factors in Computing Systems. ACM, 2012.
7. Guerra-Casanova, Javier, et al. "Authentication in mobile devices through hand gesture recognition." International Journal of Information Security 11.2 (2012): 65–83.
8. Sae-Bae, Napa, Nasir Memon, and Katherine Isbister. "Investigating multi-touch gestures as a novel biometric modality." Biometrics: Theory, Applications and Systems (BTAS), 2012 IEEE Fifth International Conference on. IEEE, 2012.

GPU-Based Parallelization of Topological Sorting

Rahul Saxena, Monika Jain and D.P. Sharma

Abstract Topological sort referred to as topo sort or topological ordering is defined as constraint-based ordering of nodes (vertices) of graph G or DAG (Directed Acyclic Graph). In other words, it gives a linearized order of graph nodes describing the relationship between the graph vertices. Many applications of various fields in computer science require a constraint-based ordering of tasks and, thus, topological sorting holds a big place of importance for many applications like semantic analysis in compiler design, Gantt chart generation in software project management and many more. In this paper, a parallel version of this ordering algorithm over CUDA (Compute Unified Device Architecture) has been discussed by identifying an approach to process-independent portions of the graph simultaneously for load flow analysis over radial distribution networks. The serial implementation of topological sort has been first discussed followed by its implementation on thread-block architecture of CUDA modifying the serial algorithm. Finally, the efficiency of this parallel version of topo sort has been investigated on various structures of graph modeled from radial distribution networks and has been reported.

Keywords Topological sort · Directed acyclic graph · Radial distribution network · CUDA · Parallel

R. Saxena (✉) · M. Jain · D.P. Sharma
School of Computing and Information Technology,
Manipal University Jaipur, Jaipur, India
e-mail: rahulsaxena0812@gmail.com

M. Jain
e-mail: monikalnct@gmail.com

D.P. Sharma
e-mail: dps158@gmail.com

A.K. Somani et al. (eds.), *Proceedings of First International Conference on Smart System, Innovations and Computing*, Smart Innovation, Systems and Technologies 79, https://doi.org/10.1007/978-981-10-5828-8_39

1 Introduction

Topological sorting is a classical graph algorithm, to solve problems of sequential nature with ordering restriction. It provides linear ordering of the vertices in a DAG (directed acyclic graph). The rule of the linear ordering is given as: every edge (u, v) from node u to v, means that u must come before v in the linear ordering of the graph [1, 2].

Topological sorting plays an important role in the scheduling of jobs such as instruction scheduling, determining the order of compilation tasks to perform in make files, resolving symbol dependencies in linkers, and deciding in which order to load tables with foreign keys. Invocation of the apt-get command uses topological sorting in order to determine the best way to install or remove packages [3]. Topological sort is also being used to find the best ways to schedule a large amount of jobs in distributed networks [4]. This paper aims to investigate whether it is possible to gain any performance by implementing a naive parallel implementation of topological sorting on a GPU rather than a CPU. The parallel implementation is constructed using CUDA, a parallel programming API developed by NVIDIA to use with their graphics cards. Starting from the previous work done in this regard, several papers have documented the use of parallel algorithms to solve the problem, but they are old and theoretical in nature [5–7]. The parallel algorithms here need an exponential amount of processors, making it impractical to implement on available GPU hardware. The most common approach to perform a topological sort is an algorithm based on either Breadth First Search (BFS) or Depth First Search (DFS). Tarjan [8] was the first one to describe the DFS approach in 1976. However, it has been shown that parallelization of DFS algorithms is extremely difficult and performance gains are limited and hard to achieve [9]. Kahn [2] described the BFS approach first. Parallelizing a BFS implementation has proven to be a suitable solution for similar problems when using CUDA [10]. Memory access in graph algorithms are often highly irregular, resulting in badly coalesced accesses. This is often due to graph traversal in algorithms, where nodes connected by an edge might reside far from each other in the device memory [11]. Despite this, there are studies that have shown that running graph algorithms on a GPU can be beneficial [12].

This paper presents a new GPU implementation of topological sorting over special class or types of graph called radial network topology graphs (RDN). It uses a reverse traversing of graph structure from leaf to junction. The organization of the paper is as follows—Sect. 2 describes the storage of graph G in memory as a 2-D array. Section 3 describes the identification of junction and leaf nodes in the graph. Then, in Sect. 4 an algorithmic implementation of topological sort is presented over CPU. Section 5 shows how the algorithm in Sect. 4 is modified to present a parallel implementation over GPU for independent parts of graph (from leaf to junction). Finally, Sects. 6 and 7 sums up the paper with experimental results and conclusions respectively. The experimental results in Sect. 7 shows that the time complexity of algorithm from O(|V|), where |V| corresponds to number of vertices in graph G for sequential implementation, drops to O(|V|’) where |V|’ corresponds to the number of nodes traversed in the longest path from sink node to source (root) node and, it is quite obvious that |V|' < |V|.

2 Modeling of RDN as Graph and Its Memory Representation

The relationship between graph nodes can be represented in many ways. The node-to-node connectivity can be explained either through adjacency matrix or through adjacency list or likewise in any other way. Here a 2-D matrix (array) has been used to depict this node to node connectivity of graphs defining which node is connected to which other node. The graph G (V, E), where V corresponds to set of vertices (nodes) and where E corresponds to set of edges, is stored as a 2-D array 'A' of order N*K where 'N' corresponds to the number of nodes in graph and 'K' corresponds to maximum degree of branching for graph G at any given node. Figure 2 shows the memory representation of graph G shown in Fig. 1.

The base array index locations like 1, 2, 3……9 for which the node connectivity is defined, column locations for a row defines the nodes to which the connections are made. If the node is a leaf node, i.e., no connections, then the entry in the column location corresponding to that row will be some null value or zero (see Fig. 2). For example, node 1 has connectivity with node 2, so A(i, 1) = 2, and since it has no more node connections, the other two index locations has zero stored at other two index locations.

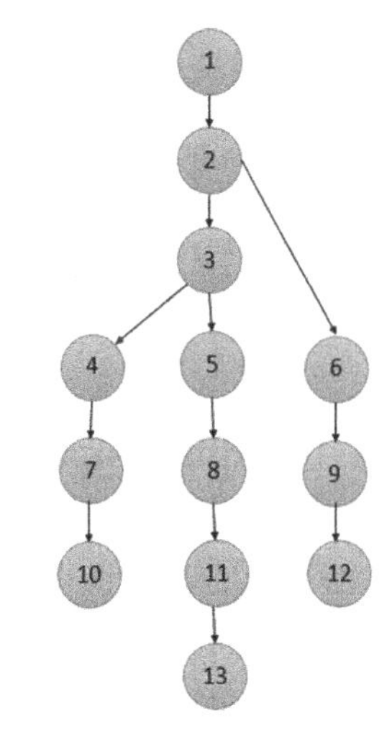

Fig. 1 Diagrammatic representation of graph G

Fig. 2 Memory representation of Graph G

1	2	0	0
2	3	6	0
3	4	5	0
4	7	0	0
5	8	0	0
6	9	0	0
7	10	0	0
8	11	0	0
9	12	0	0
10	0	0	0
11	13	0	0
12	0	0	0
13	0	0	0

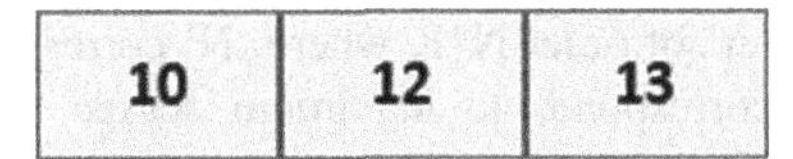

Fig. 3 Leaf node array

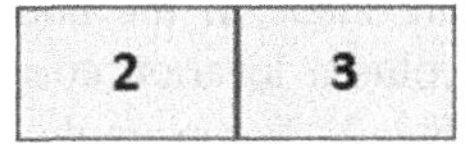

Fig. 4 Junction node array

3 Identifying Leaf and Junction Nodes of Graph

As per the 2-D array defined in Sect. 2, leaf node corresponds to those nodes for which the entry at first index location is zero, i.e., A (i, 1) = 0. This implies that they are the end points of DAG (Directed Acyclic Graph) and has no connection with any other node. Similarly, for detection of junction nodes, if the entry at A (i, 2) does not corresponds to zero, then, this implies that node has more than two connections and, thus, it is named as junction node of the graph. For example, as per the array shown in Fig. 2, nodes 10, 12, and 13 refer to leaf nodes (see Fig. 3) while nodes 2 and 6 form the junction nodes (see Fig. 4).

4 Finding Topological Order of Nodes of Graph

Topological sorting of a graph is a mechanism to find out a linearized order of graphs. Based on this ordering for a graph, one can find out which node appears prior to which other node in the graph. Figure 5 shows the topological order of graph G shown in Fig. 1. This shows that no node of graph G moving from left to right can appear prior to its node on left.

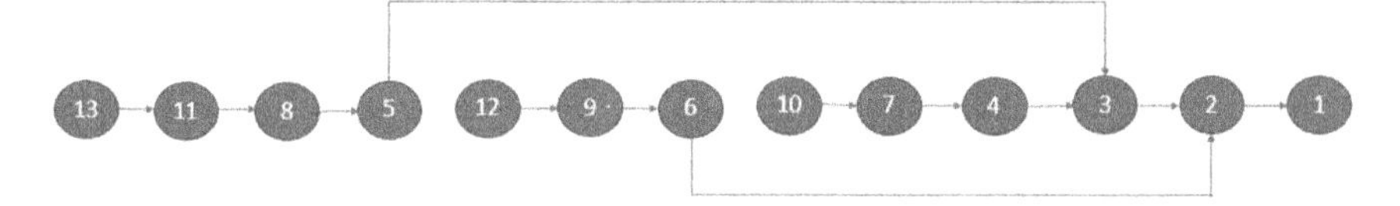

Fig. 5 Topological order of nodes

4.1 Procedure to Find Topological Order of Graph G from Leaf to Junction Nodes

Step1:—Identify the leaf and junction nodes in the graph G as per the process discussed in sect. 3.
Step2:—Move from leaf node towards the junction node in the path finding parent of each node.
Step3:—Stop the process as soon as a junction node is encountered. Start from next leaf node.
Step4:—Repeat steps 2 and 3 for all the leaf nodes.

Following the above process will yield us the topological order of nodes in reverse manner starting from leaf to junction nodes. Pseudo-code for this is given as

ALGORITHM 1: Topological_order_leaf_to_junction

```
for each leaf
      item = leaf
      add item to TOPO array and mark visited

L1:      for j=1 to size of array A
               for k=1 to maximum degree of branching
                        if (item=A (j, k) and item is not a junction)
                                 item = j
                                 add j to TOPO and mark visited
                                 goto L1
                        end
               end
      end
end
```

Figure 6 shows an array named TOPO which stores the topological order of nodes till the junction nodes for graph G.

Fig. 6 TOPO array till junction nodes

4.2 *Finding Order of Remaining Nodes*

Once the order of nodes has been determined for the graph up to all the junction nodes starting from leaf nodes, we propagate from the junction node encountered first in reverse manner of graph to next junction node in the path (if exists). This procedure is repeated for all the junction nodes till root node is reached. The check that has to be made over here is to identify whether all the children of the junction node has been added in the list or not. Algorithmic procedure for this is as follows:

ALGORITHM 2: Topological_order_junction_to_root

```
While (root node is not reached)
    for each junction node
        for i=1 to maximum degree of branching
            if ( A(item, j) is not visited )
                Add A(item, j) in TOPO and mark visited
            end if
        end for
    end for

    while (another junction node in the path is not reached)
        for i=1 to size of A
            for j=1 to maximum degree of branching
                if (A (i, j) equals to item)
                    get i and break loop i
                end if
            end for
        end for

        if (i is not a junction node)
            add i to TOPO and mark visited
        end if
end while
```

Adding this order of nodes to TOPO array completes the topological order of nodes of graph G.

5 Parallel Implementation

5.1 Compute Unified Device Architecture (CUDA)

The NVIDIA graphics processor (Quadro-600) based on the GeForce 400 series Fermi-architecture implements a four-stage pipeline. All the 128 processors (cores) are of the same type and memory access speeds are almost analogous, thus making it to be a massively parallel processor. CUDA is a programming interface to use this parallel architecture. It is a set of library functions which is an extension of standard C language. CPU views CUDA as a multicore co-processor. All available memory on the device is accessible through CUDA with no restriction on the representation for different kind of memories. This allows the programmer to take advantage of power of parallel processors for general purpose computing in a more comprehensive way.

CUDA Hardware Model: Quadro-600 [13] is a collection of 16 multiprocessors with eight processors on each. Each multiprocessor has its own shared memory which is shared by all eight processors within it. It also contains 32-bit registers, texture, and constant memory caches. Every processor executes the same instruction on different data set which makes each processor a SIMD or rather SPMD processor. Multiprocessors communicate through the device memory available to all the processors.

CUDA Programming Model: CUDA [14–16] is a collection of working units called threads. The kernel launch with a given number of threads is a programmer decision. Block unit refers to the collection of such threads. There can be multiple blocks assigned to a single multiprocessor and they share time execution resources. Moving up in this hierarchy, collection of blocks refers to grid. Each thread and block has a unique ID [13] which enables each thread to access their respective data. CUDA is an SPMD (Single Program Multiple Data) Model where each thread executes the same kernel code or function.

5.2 CUDA Implementation

Section 4 described a serial approach for determining topological order of nodes based upon the leaf and junction node identification of graph G. The basic idea used here is to propagate first from all the leaf nodes to junction nodes in the graph. By doing so, it can be realized that each path originating from a leaf node to the junction node is independent of each other and, hence, can be traversed simultaneously together.

Thus, CUDA thread-block architecture is employed where each thread is responsible to traverse down each independent branch from leaf to junction node simultaneously and independently.

The basic idea for this is to first identify a parent child relationship, i.e., which node is parent of which node or which node is the predecessor of which node. Pseudo-code for determination of this array is as follows:

ALGORITHM 3: *Parent_child_rel (A, Par) // kernel function*

```
Begin
        id ← get Thread id for each thread
        for j ← 1 to maximum degree of branching
                if (A(id) is not equal to zero)
                                Par(id)← id
                          //Storing parent of id node handled by thread at that
                          location into array Par

                end if
        end for
end
```

Once we have evaluated this information, next step is to identify how the parallel code should be initialized for path determination. Pseudo-code for determination of topological order of nodes of graph G from leaf to junction nodes is as follows:

ALGORITHM 4: Topology (Par, list) // kernel function to evaluate order from leaf to junction

```
Begin
        id ← get Thread id for each thread
        if(Par(id) is not junction)

                list(id)=par(id)
        else
                list(id)=-1
        end if
end
```

Here we have initialized as much threads in a block as the number of leaf nodes. Now, each thread will fetch up its parent node based upon the p_c array evaluated above and place it into a 2-D array list. This process named as order (kernel function) is called from host till all the threads have not reached their respective junctions. When all the nodes have reached their respective junctions, the code terminates. The terminating condition here has been identified as -1, i.e., when all the threads have a write-up of -1 value at their respective index locations in list_array, then, code will terminate or kernel execution stops. So by the end, list_array will have an entry at all its index locations filled with -1. Now each column of list_array (up till -1) will give the path from leaf to junction node as required (see Fig. 7).

10	12	13
7	9	11
4	6	8
-1	-1	5
-1	-1	-1

Fig. 7 Leaf to junction traversing of graph through each thread and as soon as junction is encountered thread start writing −1 at their respective index location

The order of remaining nodes will be evaluated serially discussed in Sect. 4.2, as now there are no more independent paths which can be explored simultaneously.

6 Experimental Results

The following results have been worked out on NVIDIA Quadro-600 [13] having 16 multiprocessors with eight processors each with CUDA 7.0 runtime environment on different topologies of the network (Table 1).

7 Conclusion

As per the experimental results and approach followed, we can infer the performance of a naive parallel implementation of Kahn's topological sorting algorithm on a GPU compared to the sequential CPU implementation. The study has been performed only on a special kind of DAGs with a limitation that no node can have more than one child in the distributed region. The algorithm works well for very large size graphs with more branching involved into it of such nature. As this kind of distributed regions in the graph increases, the speed up gained by GPU implementation will proportionally increase with respect to the serial implementation over CPU as per Gustafson's law of parallel programming. The analysis in Sect. 6 shows the complexity of the algorithm comes to linear order, i.e., the number of

Table 1 Timing comparison for CPU and GPU implementation of topological sort algorithm

Network size	CPU execution time	GPU execution time	Speed up factor (CPU/GPU)
33 nodes	650.380 μs	64.195 μs	10.131
69 nodes	1125.771 μs	86.222 μs	13.056
500 nodes	10.004 ms	748.6 μs	13.363
1000 nodes	27.248 ms	840.056 μs	32.435
5000 nodes	192.107 ms	3.492 ms	55.013
10000 nodes	405.996 ms	5.443 ms	74.592

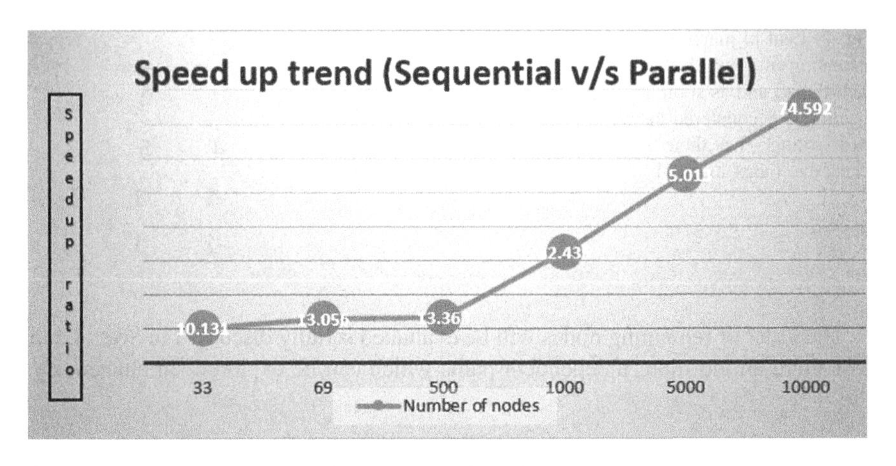

Fig. 8 Graphical representations of the results in Sect. 6

nodes in the longest path from root to leaf. A graphical analysis of the experimental results is shown in Fig. 8 supports the inference discussed. Further enhancement in the efficiency of the algorithm can be attained by using shared programming model of CUDA and making changes in the storage pattern of the node information matrix which eases out the accessibility.

References

1. D. J. Pearce and P. H. J. Kelly, "A dynamic topological sort algorithm for directed acyclic graphs", Journal of Experimental Algorithmics (JEA), Vol. 11, pp. 1–24, 2006.
2. A. B. Kahn, "Topological sorting of large networks", Communications of the ACM, Vol. 5, pp. 558–562, 1962.
3. R. Fox, "Linux with operating system concepts" CRC Press, Pp. 544, 2014.
4. A. Jarry, H. Casanova and F. Berman. DAGsim, "A simulator for DAG scheduling Algorithms", 2000.
5. E. Dekel, D Nassimi, S. Sahni. "Parallel Matrix and Graph Algorithms", SIAM J. Computing, 10(4). pp 657–675, 1981.
6. M. C. Er.," A Parallel Computation Approach to Topological Sorting", The Computer Journal, 26(4). pp. 293–295, 1983.
7. J. Ma, K. Iwama, T. Takaoka, Q. Gu., "Efficient Parallel and Distributed Topological Sort Algorithms. Parallel Algorithms/Architecture Synthesis",Proceedings, Second Aizu International Symposium. Pp. 378–383, 1997.
8. R.E. Tarjan, "Edge-disjoint spanning trees and depth-first search", Acta Informatica. 6(2), pp. 171–185, 1976.
9. Harish P, Narayanan PJ, "Accelerating large graph algorithms on the GPU using CUDA In High performance computing"–HiPC 2007, Dec 18. pp. 197–208, 2007.
10. F. Dehne, K. Yogaratnam., "Exploring the Limits of GPUs with Parallel Graph Algorithms", 2010.

11. V. W. Lee, C. Kim, J Chhugani, et al., "Debunking the 100X GPU vs. CPU Myth: An Evaluation of Throughput Computing on CPU and GPU", ACM SIGARCH Computer Architecture News-ISCA '10, 38(3), pp. 451–460, 2010.
12. P. Harish, P.J Narayanan, "Accelerating large graph algorithms on GPU using CUDA. High Performance Computing", HiPC 2007, Vol. 4873, pp 197–208, 2007.
13. Wen-mei W. Hwu, "GPU COMPUTING GEMS", Emrald Edition, Published by Elsevier Inc, ISBN 978–0-12-384988-5, 2011.
14. David B. Kirk and Wen-mei W. Hwu, "Programming Massively Parallel Processors", Published by Elsevier Inc, ISBN: 978–0-12-415992-1, 2013.
15. NVIDIA Corporation, CUDA Programming Guide 1.0, http://www.nvidia.com, 2007.
16. Rahul Saxena, Jaya Krishna, D.P. Sharma, "Faster Load Flow Analysis", Proceedings of International Conference on ICT for Sustainable Development, Vol. 408 of series Advances in Intelligent Systems and Computing, pp. 45–54, 2016.

An Approach for Adapting Component-Based Software Engineering

Nitin Arora, Devesh Kumar Srivastava and Roheet Bhatnagar

Abstract Traditionally, software was developed by writing a main method which invoked many subroutines. Each subroutine was programmed as a specific part of the program based on the given requirements and function partitions. Software engineers called for enhanced software quality, timely, at reduced costs and hence adopted the use of reusable components. This work intends at designing and augmenting generic software components for admission management system domain using OOPs methods. The analysis of major admission management system functions, data and behaviors has been taken herewith. Also, pattern-based domain engineering was conducted so as to identify the structure points thereby factoring out generically reusable components.

Keywords Feature model · Component-based software engineering
Object-oriented paradigm

1 Introduction

The paradigm shift of constructing software from scratch to the engineering of grouping components to create software embroils numerous problems. In order to curtail the existing software crisis, it demands a lot of precision for each component to make this paradigm as the first-rate solution. Component-Based Development (CBD) has been derived from object-oriented paradigm [1]. However, earlier OOPs supported reusability only through inheritance. Single instance classes being

---

N. Arora (✉) · D.K. Srivastava · R. Bhatnagar
SCIT, Manipal University Jaipur, Jaipur, India
e-mail: nitin9407@gmail.com

D.K. Srivastava
e-mail: devesh988@yahoo.com

R. Bhatnagar
e-mail: roheetbhatnagar@yahoo.com

A.K. Somani et al. (eds.), *Proceedings of First International Conference on Smart System, Innovations and Computing*, Smart Innovation, Systems and Technologies 79, https://doi.org/10.1007/978-981-10-5828-8_40

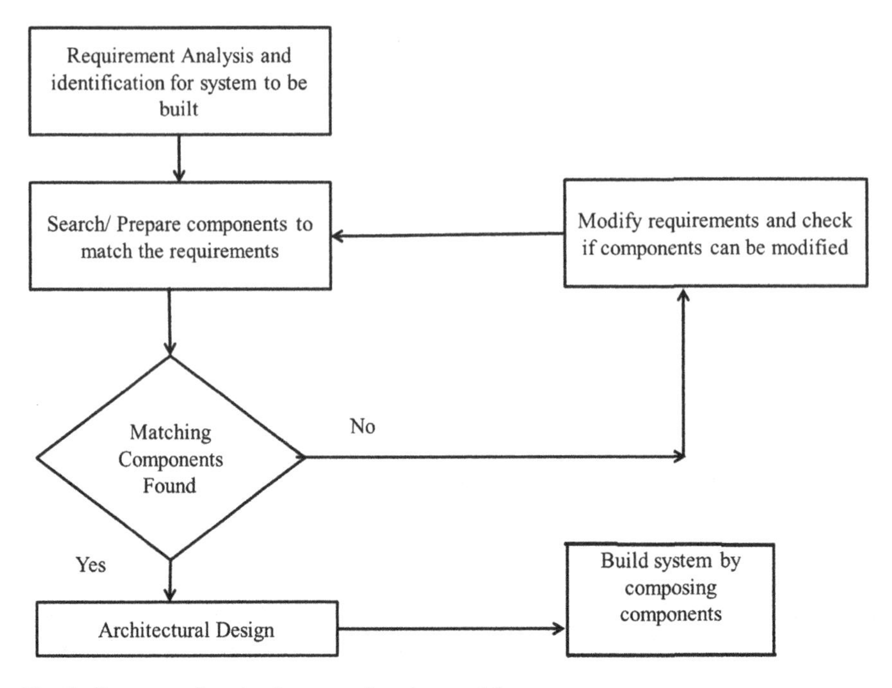

Fig. 1 Component-based software engineering workflow

descriptive and precise and abstraction being the characteristic of components allows them to be stand-alone service providers (Fig. 1).

The concept of creating software by acquiring available components which could fulfill the user requirements gave rise to the field of Component-Based Software Engineering (CBSE). Software companies are following CBD methodology for developing cost and time-optimized software and meeting the requirements. More efforts would be required for handling pre-built components, finding them, choosing the best possible component and testing components [2].

2 Component Design Principles

Different experts have defined the term "software component" in different perspectives. According to Szyperski, a software package element may be a unit of composition with contractually fixed interfaces and context dependencies solely [3]. Briefly, components provide a key abstraction for the design and development of flexible solutions. They enable designers to focus on business level concerns which are independent of the lower level component implementation details. It is important to recognize that components can be composed at runtime without compilation [4]. Component renders service irrespective of its execution and language. The survey conducted by the Manchester University, highlighted the important characteristics of a component—a component constitutes a name, interface, and code [5]. The component interface is published and every interaction is through the published interface. A component lifecycle has three phases—design, deployment, and runtime. The three phases heavily affect each other. They need to be coordinated.

3 Literature Review

Component modeling of software was urged by McIlory (1969) as the way of confronting the software crisis. However, the concept of component-based software development has been launched only within the last decade. Component-based development is based on software reuse. The CBD obligates the reduction in development prices. Component structures are versatile and maintainable due to the plug-and-play nature of components. As said by Brown, CBD has several benefits incorporating practical management of complexity, reduced time to market, enhanced productivity, improved excellence, an extra degree of consistency, and a broader variety of usability [6]. Heineman and Council briefed that the key goals of Component-Based Software Development (CBSD) are the availability of help for the event of structures as assemblies of components, the improvement of parts as reusable entities, and also the maintenance and upgradation of systems by using customization and alteration of parts [7]. Crnkovic and Larrson stated that each customer and supplier have had splendid expectations from CBSD but their expectations were not met [8].Considering the project execution, as soon as the requirement are defined, the development phase begins from scratch which leads to budget and time anomaly. Reusability is considered to be the revolution, cutting the time and cost of software development.

4 Methodology

The methodology embodies analyzing major admission management system functions, information and behaviors, and conducting a pattern-based domain engineering to spot structure points and repeating patterns. The generic structure points known within the chosen application domain were accustomed to design reusable components by applying a general distributed Component-Based Software Development technology model.

5 Admission Management System (AMS): Empirical Study

5.1 Admission Business Functions

The below listed business functions and processes were identified where component modeling can be applied.

1. Admission
2. Registration
3. Result Management

5.1.1 Admission

The primary actors pertaining to admission are Student and Registrar.

Figure 2 depicts the use cases for Student and Registrar.

5.1.2 Registration

Actors in registration are Student, Faculty Coordinator (F.C), Finance, Medical Officer (M.O), and Hall Administrator (H.A). Figure 3 depicts the use cases for Registration actors.

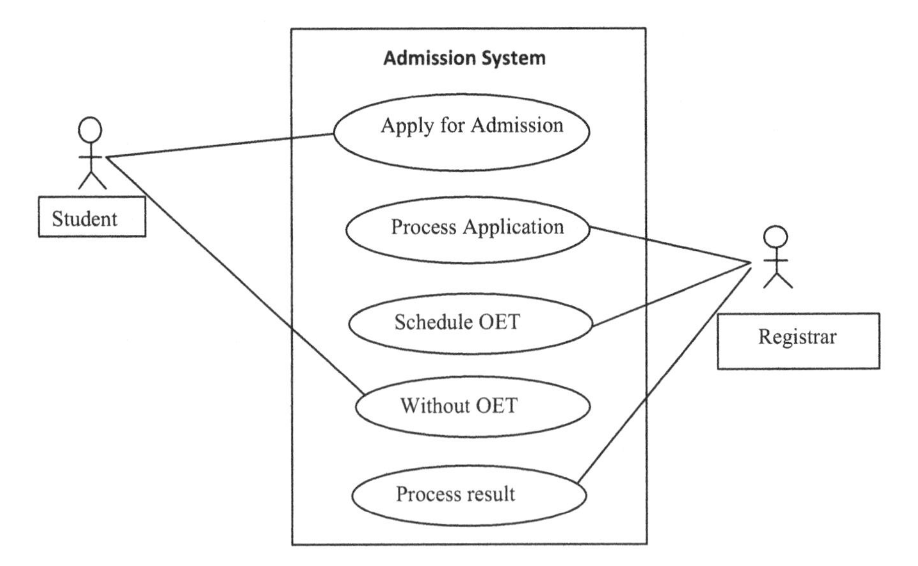

Fig. 2 Use case for admission system

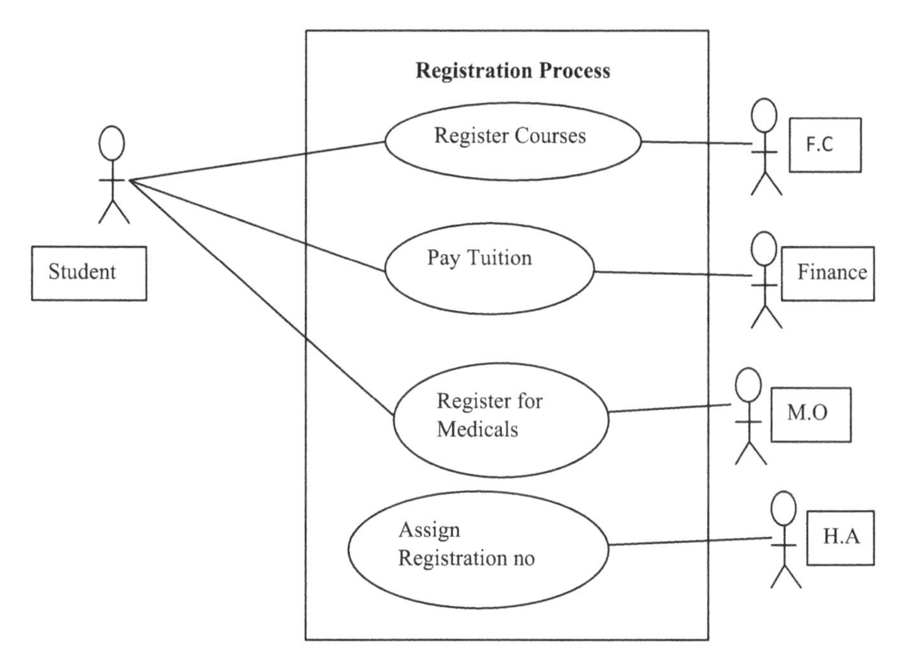

Fig. 3 Use case diagram for registration process

5.1.3 Result Management

Actors in result management are Professor, Head of Department (HoD), Dean, and Student. The corresponding use case is presented in Fig. 4.

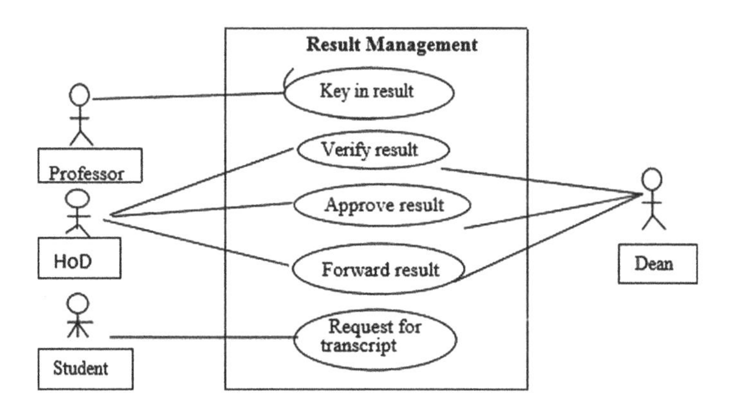

Fig. 4 Use case diagram for result management

5.2 *Fully Dressed Text Based Use Case for the Registration System*

Text-based used case is paramount to analysis and design of successful enterprise solution. Test-based use case might be casual or fully dressed version [9]. The fully dressed used here exhaustively presents the detailed interaction in order to avoid missing requirements.

Use Case 1: Login to the System Primary Actor: Student Scope: Registration System Level: User goal Stakeholders and Interests: *Student*- to gain access into the system *Registry*—expects a bonafide student to gain access to the system Precondition: Account already exist or created for student Guarantee: admission management system (AMS) logs in a valid student Main Success Scenario: (1)Student enters username and password into AMS (2)AMS validates username and password (3) AMS re-directs student into the system (3)Extension: 1a. Student does not have username and password 1a1. AMS prompts student to get a username and password 2a. Wrong username/password entered 2a1. AMS denies access and request for correct username and password *Use Case 3: Add Course* Primary Actor: Student Scope: Registration System Level: User goal Stakeholders and Interests: *Students*- add course to be taken for the semester *Course Advisor*—ensures the right courses and the right unit is taken Precondition: None Guarantee: None Rule: Total credit unit not exceeded	*Use Case 2: Browse Course List* Primary Actor: Student Scope: Registration System Level: User goal Stakeholders and Interests: *Student*- wants to see a list of courses available to be taken Precondition: All courses pre-loaded Guarantee: courses are available online Main Success Scenario: (1) Students select department and semester (2) Student scroll through course list up and down (3) AMS display course listing based on level and department Extension: 1a. Course not on the list of displayed or available courses 1a1. Administrator adds the omitted courses *Use Case 4: Remove Course* Primary Actor: Student Scope: Registration System Level: User goal Stakeholders and Interests: *Student*- remove added course from the selected Precondition: At least one course exist in the course list Guarantee: course is successfully removed
Main Success Scenario: (1)Browse Course List (2)Select applicable courses (3)Add selected course to student course item list Extension:	Main Success Scenario: (1)Students view selected courses (2)Students select course to remove (3)AMS removes course successfully (4)AMS updates and display list

(continued)

(continued)

2a. Course not added because total unit exceeded 2a1. Remove (Delete) Course	
Use Case 5: Fee payment Primary Actor: Student Secondary Actor: Finance Scope: Registration System Level: Summary Stakeholders and Interests: *Finance*—wants to ensure student does not owe institution *Institution*—wants to be sure no student owes Precondition: Admitted/bonafide student Guarantee: payment successfully approved Main Success Scenario: (1)Students pay fees to university account (2)Finance converts bank slip to university fees receipt (3)Finance acknowledges and approves payment (4)AMS grants student permission to register	*Use Case 6: Approve Registration* Primary Actor: Faculty Coordinator, Head of Department(HoD) Scope: Registration System Level: Summary Stakeholders and Interests: Faculty Coordinator—wants to ensure students register correctly Precondition: Add course Guarantee: Registration successfully approved Main Success Scenario: (1)Faculty coordinator views student registration detail for correctness (2)Faculty coordinator approves student registration (3)AMS sends approval and notification email message to student
Extension: 1a. Student pays insufficient fee 1a1. Bursar acknowledge and disapprove and advise students to pay up outstanding balance	Extension: 2a. Open issues—Any reason for rejection of approval

5.3 *Sequence Diagram*

Sequence diagrams emphasize the sequence of the messages instead of the relationships between the objects [10]. Objects or roles of the system are lined up in a row. A dashed line, called lifeline, which flows from each object downward, defines the time axis. The arrows between the lifelines represent messages being sent between the objects. The following sections, Sects. 5.3.1–5.3.3, present certain illustrations of some sequence diagrams involved in the design.

5.3.1 Sequence Diagram for Login to the System

The candidate enters a login name and password, and clicks the submit button on the login page. The browser sends HTTP GET request to the Web server. The servlet controller receives the client request and delegates responsibility to the UserManagerFacade enterprise bean to invoke is ValidUser() method which validates entry (Fig. 5):

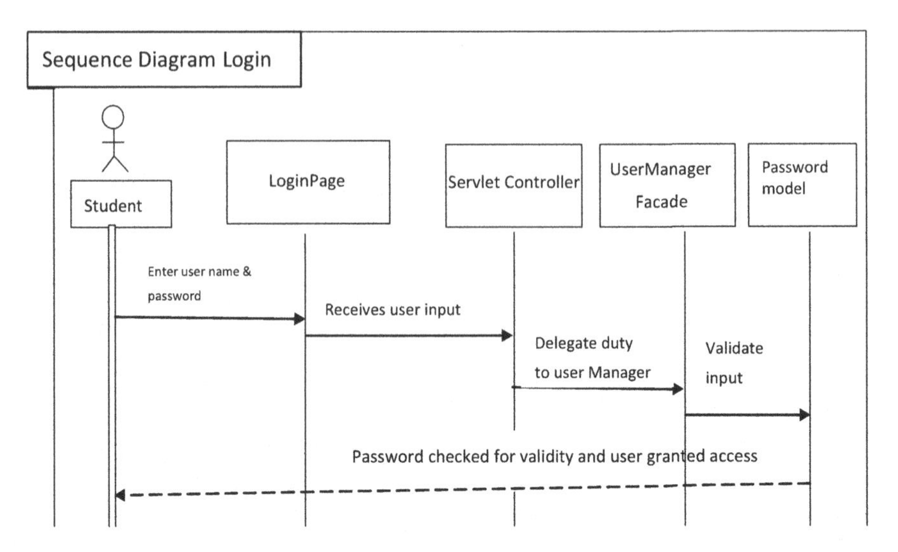

Fig. 5 Sequence diagram for student login to the system

5.3.2 Sequence Diagram for Add Course

Figure 6 shows the flow of what happens at Add Course use case as the operation is carried out by student. The course façade is the duty of the Enterprise Java Beans (EJB) component helping out in the interaction.

5.3.3 Sequence Diagram for Pay Tuition

Figure 7 illustrates the process of registration and the systems expected behavior. Various delegations of duties are shown:

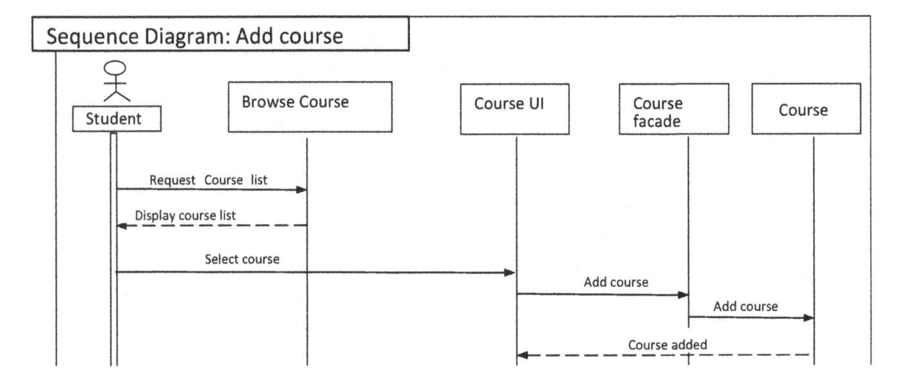

Fig. 6 Sequence diagram for add course

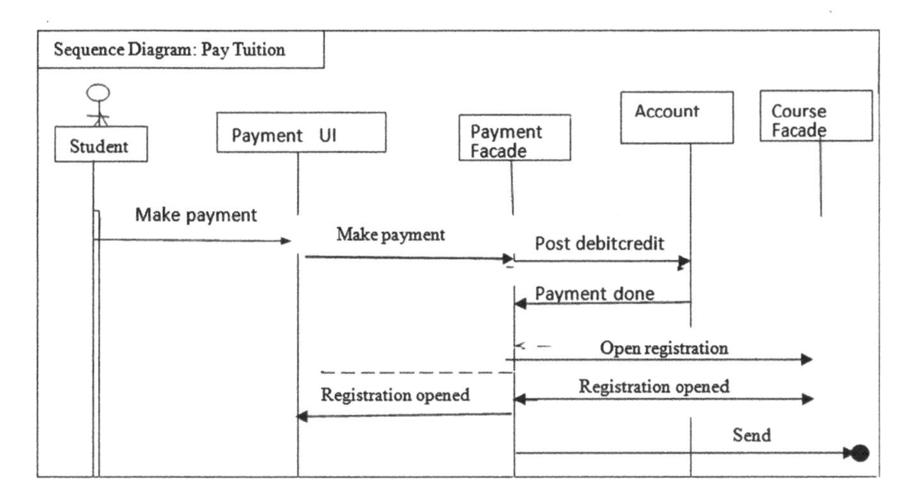

Fig. 7 Sequence diagram for tuition

6 Conclusion and Future Work

The traditional software processes does not consider the software reuse wholly and often the risks associated with component reuse is overlooked. Component-based discipline is a systematic way to succeed the software reuse. A substantial benefit of component modeling beyond building libraries or Application Programming Interface (API) is that it is expected to expedite the development process and curtail failure. While reusing libraries or API, the developer gets more work to do either at the business logic level or at the enterprise information system level. A component is designed to be self-contained, offering an end-to-end service to the user.

Over and above, if there are vast numbers of reusable components that are easily accessible in a common repository, software development endeavor would be a more attractive task. Given more attention, component-based development could be one of the potent and promising approaches to confront unremitting software crisis.

References

1. International Journal of Advanced Research in Computer Science and Software Engineering, Volume 4, Issue 9, September 2014.
2. Morisio, M., et al., "Investigating and Improving a COTS-Based Software Development Process," Proc. 22nd Int. Conf. on Software Engineering, Limerick, Ireland, ACM Press, 2000.
3. Szyperski C, Gruntz D and Murer S - Component Software: Beyond Object Oriented Programming||, Addison Wesley, Second Edition 2002.

4. Aoyama, M., "New Age of Software Development: How Component-Based Software Engineering Changes the Way of Software Development," Proc. 1st Workshop on Component Based Software Engineering, 1998.
5. Lau K K, Wang Z -A survey of Software Component Models, second ed., School of Computer Science, University of Manchester, May 2006.
6. Brown, A.W. Large-Scale, Component-Based Development, Englewood Cliffs, NJ: Prentice Hall, 2000.
7. Heineman G T, Councill W T, -Component Based Software Engineering: Putting the pieces together.
8. Ivica Crnkovic, Magnus Peter Henry Larrson Building Reliable Component Based Software Systems, 2002.
9. Alistair C., "Writing Effective Use Cases", Michigan: Addison-Wesley (2001).
10. M. Fowler, "UML Distilled", A brief Guide to the Standard Object Modeling Language, Forewords by Cris Kobryn, Grandy Boosh, Ivar Jacobson and Jim Rumbaugh. Addison Wesley (2003).
11. An approach based on feature models and quality criteria for adapting component-based systems L. Emiliano Sanchez1, J. Andres Diaz-Pace1, Alejandro Zunino1, Sabine Moisan2 and Jean-Paul Rigault2, 2015.

Mitigating Primary User Emulation Attack in Cognitive Radio Network Using Localization and Variance Detection

Rukhsar Sultana and Muzzammil Hussain

Abstract Cognitive Radio (CR) mechanism brings the solution to the difficulty of spectrum inadequacy by distributing the unexploited spectrum to unlicensed users (Secondary Users) when the licensed users (Primary Users) are not using them so that transmission by primary user would not be distorted. To attain this, the CR dynamically regulates their physical and transmission parameters. The key characteristics of Cognitive radio, Cognitive Capability and Reconfigurability make the radio network vulnerable to security threats. One of the foremost intimidations is Primary User Emulation Attack (PUEA) performed by secondary user (or an attacker) by imitating the primary user characteristics at physical layer. In this paper, an algorithm is projected to distinguish primary user signal from others, through an energy-efficient localization mechanism and channel parameter variance, which is simulated in MATLAB environment and it is found to be an effective mechanism to authenticate primary user and mitigate the primary user emulation attack.

Keywords Cognitive Radio Network (CRN) · Authentication Key (AuK) Authentication Center (AuC) · Levelling and Sectoring · Variance

1 Introduction

As stated in the report of measurement campaigns coordinated by Federal Communications Commission (FCC) in the United States, many primary users (PUs) do not use their allotted spectrum bands continuously and ubiquitously. A measurement results report shows that in the GSM (Global System for Mobile Communication) the uplink (UL) channel in 850 and 900 MHz bands has on average less

R. Sultana (✉) · M. Hussain
Central University of Rajasthan, Kishangarh 305817, Rajasthan, India
e-mail: emannaaz@gmail.com

M. Hussain
e-mail: mhussain@curaj.ac.in

A.K. Somani et al. (eds.), *Proceedings of First International Conference on Smart System, Innovations and Computing*, Smart Innovation, Systems and Technologies 79, https://doi.org/10.1007/978-981-10-5828-8_41

than 10% duty cycle over a 24 hour period measurement time. This indicates that in GSM most of the bands are unutilized and they can be allotted to secondary users (SUs) for their transmission in opportunistic mode without creating interference to primary users. Cognitive Radio technology makes possible to use licensed spectrum by the secondary users in an effective and dynamic manner. This concept of cognitive radio is initiated by Mitola (2000), based on Software Defined Radio (SDR) which is driven by software and is field programmable. They can sense the radio environment and can change their communication parameters depending on the current situation of the environment [1].

Due to spectrum scarcity, now FCC is also allowing for opening of licensed bands for the unlicensed users with the strategy of non-intervention. Thus FCC is permitting for opportunistic spectrum sharing. To be aware about the unoccupied or occupied spectrums bands cognitive radio makes use of spectrum sensing. If there are no more vacant channels in the environment, then cognitive radio have a duty to differentiate between primary and secondary user signal, otherwise some greedy secondary or malicious users can impersonate the primary user signal properties to get access to more spectrum or to interrupt the transmission of primary users respectively [2–4].

An algorithm is suggested to authenticate primary user and to mitigate the primary user emulation attack (PUEA) in cognitive radio which takes advantage of Levelling and Sectoring and Variance Detection mechanism.

Section 2 defines the existing mechanisms for avoiding PUEA. In Sect. 3, the proposed algorithm is narrated. Simulation of algorithm is traced in Sect. 4. Section 5 outlines the obtained results and Sect. 6 concludes the whole work.

2 Related Work

Here, we summarize existing two mechanisms for avoiding PUEA in Cognitive Radio Network (CRN).

2.1 Primary User Authentication Through Signal Properties [5]

In this mechanism to authenticate the primary user signal its signal properties as Distance (d) and Angle of Arrival (AoA) are deployed. At the time of registration, these values are approximated for each primary user and stored in a table. Whenever a request for spectrum is detected, the calculated distance and AoA is compared with the stored values. If they are equal, the requesting user is authenticated as primary user.

The PUs are mobile so at every instant when they move, their signal properties need to be recalculated and updated.

2.2 Defense Against Primary User Emulation Attack [6]

In this mechanism, the defender models an attack based on the strategy used by attacker to imitate the primary user signal and designs a defense strategy using variance detection. Variance is a channel parameter which is invariant. This channel parameter can be used as a signature of spectrum users and to detect advance primary user emulation attack, this invariant is approximated. To find received power from transmitted power here, the path loss and the log-normal shadowing of a channel is considered. When the signals with different energies are received from primary user, the received energy is determined as

$$y_i = P_t * r^{-\alpha} * G_p. \quad (1)$$

Here P_t is the transmitted power from PU, r is distance of PU from receiver, α is path loss exponent and G_p is shadowing random variable. Here $G_p = 10^{\beta/10} = e^{a\beta}$, where $a = \frac{ln10}{10}$ and β follows a normal distribution $\beta \sim N(0, \sigma_p^2)$.

Estimation. For the PU, variance is estimated and used further for variance detection.

If for n number of observations, $y_1, y_2, y_3, \ldots, y_n$ are the received power, then the variance $(\sigma_p^2)_D$ using unbiased estimator is

$$\left(\sigma_p^2\right)_D = \frac{1}{(n-1)a^2}\sum_{j=1}^{n}(lny_j - \frac{1}{n}\sum_{i=1}^{n} lny_i)^2. \quad (2)$$

Determination. Whenever the sequence of signals is received (from PU or the attacker), for m number of received signal variance is estimated using formula in (2) as

$$\left(\sigma^2\right)_D = \frac{1}{(m-1)a^2}\sum_{j=1}^{m}(lny_j - \frac{1}{m}\sum_{i=1}^{m} lny_i)^2. \quad (3)$$

$$\begin{cases} \text{If } \left|(\sigma^2)_D - \left(\sigma_p^2\right)_D\right| \le k\left(\sigma_p^2\right)_D \\ \text{then the signal is from primary user.} \\ \text{if } \left|(\sigma^2)_D - \left(\sigma_p^2\right)_D\right| > k\left(\sigma_p^2\right)_D \\ \text{then the signal is from the attacker.} \end{cases}$$

where k is a threshold factor, $k > 0$.

This technique has enhanced performance than simple energy mean detection approach but in case when $\sigma_s^2 = \sigma_p^2$, it cannot distinguish among primary and secondary user signals.

3 Proposed Algorithm

The algorithm is designed for authentication of primary user and detection of primary user emulation attack when cognitive radio is incorporated into GSM. PUs are the licensed users of GSM bands and SUs are the unlicensed users who utilize the licensed spectrum band when PUs are not using them.

The PUs and SUs are randomly deployed in cognitive radio field and GSM base station works as the Cognitive Radio Base Station (CBS).

The algorithm comprises of four phases as

1. Initialize
2. Register
3. Update
4. Authenticate.

3.1 *Initialize*

Levelling and sectoring mechanism is used for localization of cognitive users (CUs) without estimation of their actual location. The location of each cognitive user (secondary user) is identified with level ID and sector ID, so the entire cognitive network area is divided into various levels depending on signal strength [7–9].

Levelling. Cognitive base station (CBS) sends packets containing level ID for level 1 with minimum power level. All the user nodes that receive the signal set their levels as 1. Next the CBS increases its signal power level to reach the next level and sends packet containing next level ID. All the nodes that receive this signal, if have not already fixed their level ID, set their levels to 2. This process is continued until the CBS has sent signals corresponding to all levels. The number of levels into which the network gets divided is equal to the number of different power levels at which the CBS has transmitted the signal.

Sectoring. After levelling, the field is divided into sectors. Using the directional antenna, the CBS will send signals with maximum power in one direction with sector ID 1, then the directional antenna is rotated in anticlockwise direction and send signal with sector ID 2. Each node receiving this signal will set its sector ID as one received through signal from directional antenna. This process is continued until the whole field is divided into sectors.

Now, the cognitive users know their location in terms of (L_i, θ_j). CBS broadcasts a REQ packet into the network requesting the each of the nodes for their location information. Upon receiving this, all the nodes in the network respond back by sending their location as level ID, sector ID, node ID to CBS. CBS store this information in a location table.

3.2 Register

Whenever a new node enters into the radio cell, then through control channels it requests for registration by sending its ID and authentication key (AuK) to CBS.

1. CBS forwards these credentials to authentication center (AuC). If AuC verifies this, then the node is identified as PU and registered to CBS.
2. At CBS the distance is calculated with PU by using Received Signal Strength Indicator (RSSI) mechanism and phase is calculated with the help of Array of Antenna [5].
3. Based on estimated distance and phase, CBS determines Level ID and Sector ID for PU.
4. CBS receiver observes different signal energies from PU as y_1, y_2, y_3,…, y_n. To calculate these energies here path loss and log-normal shadowing of a communication channel is considered as (1).

$$y_i = P_t * r^{-\alpha} * G_p.$$

Here P_t is the transmitted power from PU, r is distance of PU from receiver, α is path loss exponent, G_p is shadowing random variable. Here $G_p = 10^{\beta/10} = e^{a\beta}$, where $a = \frac{ln10}{10}$ and β follows a normal distribution $\beta \sim N(0, \sigma_p^2)$.

5. Mean and variance for PU is calculated by using formula based on unbiased estimator as (2).

$$\mu_p = \frac{1}{n} \sum_{i=1}^{n} lny_i. \tag{4}$$

$$\left(\sigma_p^2\right)_D = \frac{1}{(n-1)a^2} \sum_{j=1}^{n} \left(lny_j - \frac{1}{n} \sum_{i=1}^{n} lny_i\right)^2. \tag{5}$$

6. Node ID, Level ID, Sector ID, mean and Variance for PU are stored in location table as Table 1 for PU at the CBS (Fig. 1).

```
BEGIN
INPUT the ID and AuK to AuC
IF these credentials are verified by AuC THEN
        COMPUTE distance & phase
        PROCESS distance & phase to determine Li & θi for PU
        COMPUTE μp and σp^2 for PU
        SET location table with ID, Li, θi, μp and σp^2 for PU
END IF
END
```

Fig. 1 Register phase

Table 1 Primary user location table

Node ID	Level ID (L_i)	Sector ID (θ_j)	Mean (μ_p)	Variance (σ_p^2)

3.3 Update

Whenever a PU moves from its current location, its movements are reported to CBS.

1. If the PU is changing its location within the cell then new L_i, θ_j, μ_p and σ_p^2 for PU are determined for the new location and information is updated in the location table
2. If the PU moving out from the cell then its entry from the location table is deleted (Fig. 2).

```
BEGIN
INPUT Request Signal
IF requesting node is new THEN
        CALL Register Phase
ELSE
        IF channels are required THEN
                CALL Authenticate Phase
        ELSE
                IF node is moving out of cell THEN
                        CLEAR entry of node from location table
                ELSE
                        COMPUTE new Li, θi, μp and σp^2 and RESET location table
                END IF
        END IF
END IF
END
```

Fig. 2 Update phase

3.4 Authenticate

Whenever the request for spectrum is detected and if free channels are scarcely available for allocation. Then authentication phase is applied.

1. When a user requests for channels by sending its ID. ID is searched in the table and if found then L_i, θ_j for the requesting user are extracted for further process. If L_i, θ_j are not found then it is identified as malicious user (MU) and authentication ends.
2. If L_i, θ_j are found then μ and σ^2 as (3), are calculated and a condition is checked (Fig. 3).

$$(\sigma^2)_D = \frac{1}{(m-1)a^2}\sum_{j=1}^{m}(lny_j - \frac{1}{m}\sum_{i=1}^{m} lny_i)^2.$$

$$\begin{cases} \text{If } \left|(\sigma^2)_D - \left(\sigma_p^2\right)_D\right| \leq k\left(\sigma_p^2\right)_D \\ \text{then the signal is from primary user.} \\ \text{if } \left|(\sigma^2)_D - \left(\sigma_p^2\right)_D\right| > k\left(\sigma_p^2\right)_D \\ \text{then the signal is from the attacker.} \end{cases}$$

where k is threshold factor, $k > 0$.

```
BEGIN
INPUT Request Signal
IF spectrum is free THEN
        Allocate channels to requesting node
ELSE
        IF requesting node is new THEN
                CALL Register Phase
        ELSE
                GET Li, θj for node
                IF not found in table THEN
                        PRINT "SU"
                ELSE
                        COMPUTE μ and σ²
                        IF |(σ²)D - (σp²)D| ≤ k(σp²)D THEN
                                PRINT "PU"
                        ELSE
                                PRINT "SU"
                        END IF
                END IF
        END IF
END IF
END
```

Fig. 3 Authenticate phase

3.5 Brief Explanation

The algorithm requires the locations of cognitive users so initially, through levelling and sectoring mechanism Level ID and Sector ID is assigned to each cognitive users.

Registration phase registers the PU only when they are verified from AuC. Distance is calculated according to RSSI mechanism and to calculate phase, array of antenna technique is used. By using unbiased estimator formula for mean and variance as (3) and (4), mean and variance are calculated for PU. After these estimations, an entry for PU is made into PU location table.

Update occurs when a new node arrives into the vicinity of radio or PU is moving from its current position. Authenticate phase is applied only when there are no more available channels and demand for spectrums is increasing enormously, then to distinguish SU from PU authentication is performed.

4 Security Analysis

The proposed algorithm is secure and efficiently authenticates primary user at physical layer in CRN. Thereby it also mitigates following security attack which is performed at physical layer.

4.1 Primary User Emulation Attack

CBS registers the requesting users as primary user if they have valid authentication key. A malicious node with an invalid authentication key cannot register itself as PU. Thus only verified users are registered as PU and it makes impossible to occur PUEA at the registration stage. This perception is already approved.

In the proposed algorithm, CBS asks to cognitive users for location information and they reply back to CBS by sending their location (Node ID, Level ID and Sector ID). Malicious node can emulate this information and can send spectrum requesting signal. In authentication phase, variance is detected for the requesting signal, when the location information is found in the location table. But for the malicious user, it is almost impossible to emulate the variance of the received signal even the location is emulated, because variance is invariant of a communication channel and it is unlike for each user. Thus localization (through levelling and sectoring) and variance detection are used together here to authenticate primary user and PUEA is avoided successfully.

5 Implementation

The proposed algorithm was simulated in MATLAB environment. The simulation is based on some assumptions as

- The primary and secondary users are deployed randomly in the network.
- CBS is able to transmit power signal up to five levels.
- The directional antenna has a sector angle of 30°.

The network field was divided into levels and sectors and cognitive users know their location in terms of (L_i, θ_j). It is presumed that PUs are registered to CBS and there is an entry in the location table corresponding to each PU.

The mean and variance for the PU are estimated by taking channel parameter (variance) equal to 8. The simulation is performed for the three values of variance for SU as 4, 8 and 12.

Figure 4 shows the cognitive field after levelling and sectoring. PUs are shown with star and SUs are shown with filled circle. CBS is at the center of the circular field. Parameters used for simulation are shown in Table 2.

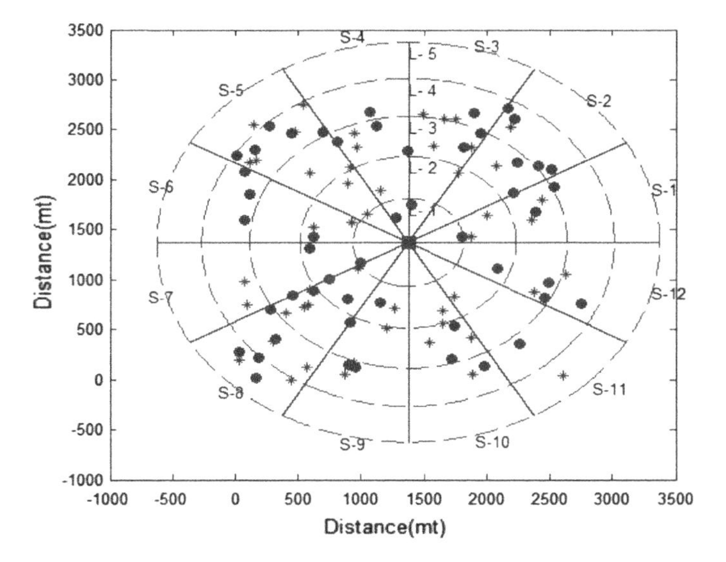

Fig. 4 Cognitive field after levelling and sectoring

Table 2 Simulation parameter

Parameter	Value
Area of cognitive field	2000 m × 2000 m
Number of PUs	50
Number of SUs	50
PU transmission power (dBm)	28
SU transmission power (dBm)	25
GSM band (MHz)	900
PU variance (σ_p^2)	8
CU variance (σ_s^2)	4, 8, 12

6 Results and Analysis

Based on the observations from the simulation of proposed algorithm, we plotted three graphs for three scenarios of variance being 4, 8 and 12, respectively. The graphs are plotted based on numbers of PU's and SU's allocated spectrum with time in all three scenarios. X-axis is time and Y-axis is number of users.

Figure 5 shows the number of SUs and PUs to which spectrum are allocated for $\sigma_p^2 = 8$ and $\sigma_s^2 = 4$.

Figure 6 shows the number of SUs and PUs to which spectrum are allocated for $\sigma_p^2 = 8$ and $\sigma_s^2 = 12$ over period of time.

Figure 7 shows the numbers of SUs and PUs to which spectrum are allocated, when the variance of PU is equal to variance of SU, before authentication almost all the SUs emulates the PUs.

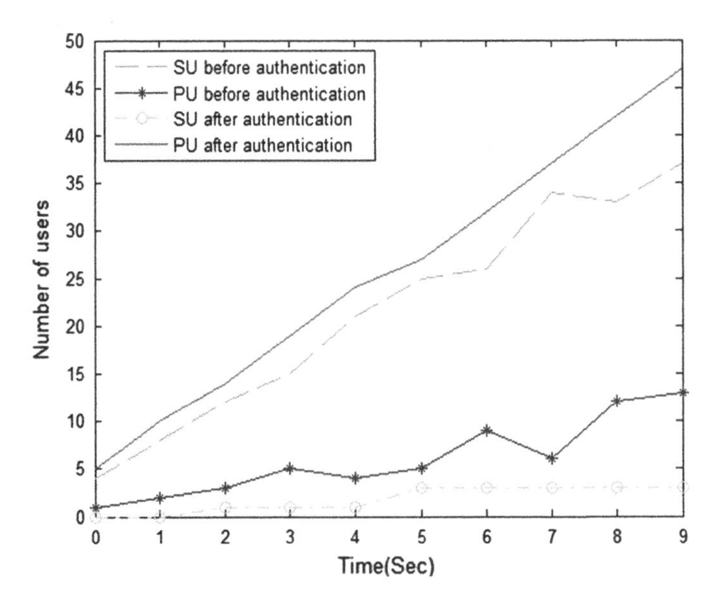

Fig. 5 Allocation of spectrum to PU v/s SU, when $\sigma_p^2 = 8$ and $\sigma_s^2 = 4$

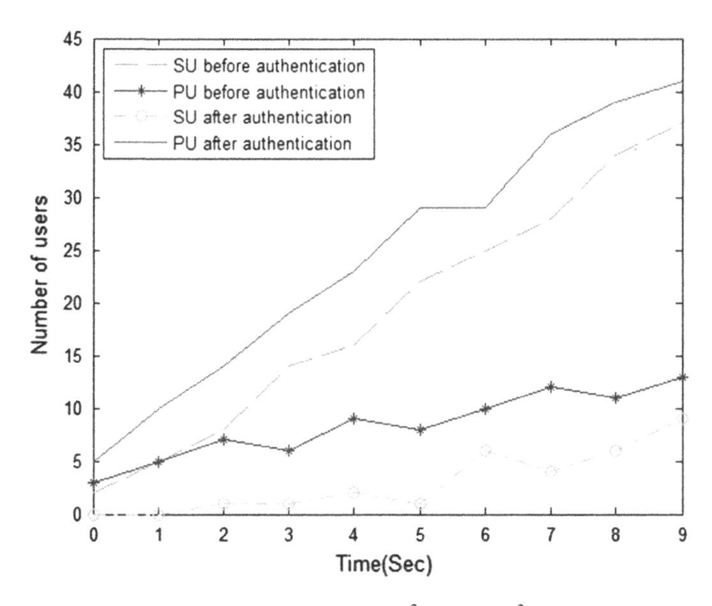

Fig. 6 Allocation of spectrum to PU v/s SU, when $\sigma_p^2 = 8$ and $\sigma_s^2 = 12$

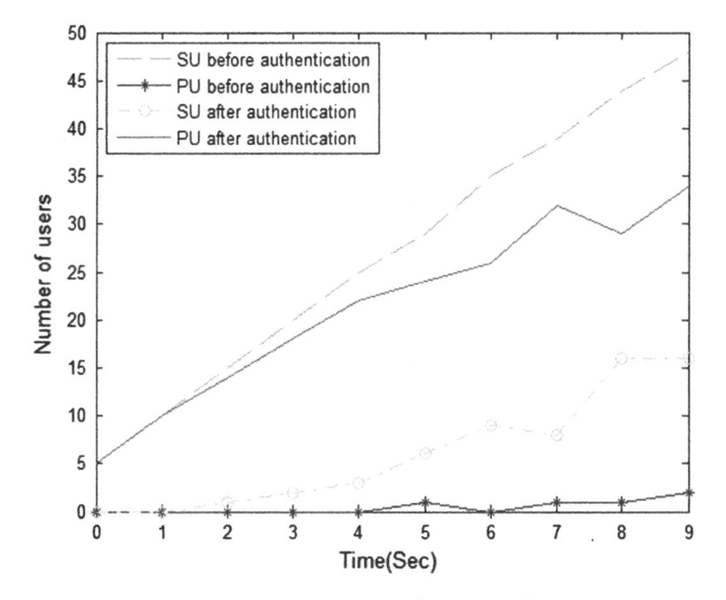

Fig. 7 Allocation of spectrum to PU v/s SU, when $\sigma_p^2 = 8$ and $\sigma_s^2 = 8$

Each of the graphs plotted clearly shows that before applying the proposed algorithm most of the SUs are emulating the PUs and are allocated spectrum. But after the proposed mechanism is applied, SUs are unable to emulate PUs and spectrum allocation to PUs has increased drastically.

7 Conclusion

CRN improves the spectrum usage optimally but security is a major concern and any SU or illegitimate user may emulate the PU and can access the spectrum illegitimately. Hence, authentication of a PU is very much essential because it avoids many security issues and enhances the usage of spectrum. Most of the PU authentication mechanisms are based on signal properties and SU or attacker node can emulate these properties. Here, we have proposed an algorithm for authentication of PUs in CRN. We have used the signal properties as distance, angle of arrival, variance of received signal and also the field is divided into levels and sectors. The authentication is based on these properties and it is impossible to emulate all these properties simultaneously. The proposed algorithm is simulated in MATLAB and its performance is analyzed. It has been found that the proposed algorithm is efficiently authenticating PUs, thereby the PUEA is mitigated and spectrum is allocated to PUs than SU.

References

1. Gao, Jason: Channel Capacity of a Cognitive Radio Network in GSM Uplink Band. In: Communications and Information Technologies, 2007. ISCIT'07, International Symposium on, pp. 1511–1515. IEEE (2007).
2. Sharma, Himanshu, and Kuldip Kumar: Primary User Emulation Attack Analysis on Cognitive Radio. Indian Journal of Science and Technology 9.14 (2016).
3. Akyildiz, Ian F.: NeXt Generation/Dynamic Spectrum Access/Cognitive Radio Wireless Networks: A Survey. Computer networks 50.13, 2127–2159 (2006).
4. Parvin, S., Hussain, F.K., Hussain, O.K., Han, S., Tian, B. and Chang, E.: Cognitive Radio Network Security: A Survey. Journal of Network and Computer Applications, 35(6), pp. 1691–1708.
5. Suditi Choudhary and Muzzammil Hussain: Primary User Authentication in Cognitive Radio Network using Signal Properties. In: Proceeding of the Third International Symposium on Women in Computing and Informatics, pp. 290–296. ACM (2015).
6. Chen, Zesheng: Modeling Primary User Emulation Attacks and Defenses in Cognitive Radio Networks. In: 2009 IEEE 28th International Performance Computing and Communications Conference, pp. 208–215. IEEE (2009).
7. Lakshmi Phani G, Venkat Sayeesh K, Vinod Kumar K, Rammurthy: ERFLA: Energy Efficient Combined Routing, Fusion, Localization Algorithm in Cognitive Wsn. In: 2010 7th International Conference on Wireless and Optical Communication Networks, Colombo - (WOCN), pp. 1–5. IEEE (2010).
8. Md. Aquil Mirza, Rama Murthy Garimella: PASCAL: Power Aware Sectoring based Clustering Algorithm for Wireless Sensor Networks. In: 2009 International Conference on Information Networking, pp. 1–6. IEEE (2009).
9. Bore Gowda S B, Puttamadappa C, Mruthyunjaya H S, Babu N V: Sector based Multi-hop Clustering Protocol for Wireless Sensor Networks. In: International Journal of Computer Applications (0975–8887) vol 43– No.13 (2012).

A Study on Shape Detection: An Unexplored Parameter in the Gallstones Identification

Sakshi Garg, Angadpreet Walia, Abhilasha Singh and Anju Mishra

Abstract Elimination of the gallstones is gaining popularity due to significant increase in the count of the people suffering from Cholelithiasis. Cholelithiasis is one of the most reported diseases in India. This paper presents a comparative study of various approaches of gallstones detection and further analyzing the ultrasound images of the patients suffering from Cholelithiasis. Our research mainly focuses on examining the fissures in the identification of the shapes of the Gall Bladder stones automatically by applying various techniques such as image processing, segmentation, and a combination of other preprocessing morphological techniques to scrutinize gallstones with a motive of aiding medical science with more competent techniques for the proficient, effortless and cost-effective removal of gallstones.

Keywords Cholelithiasis · Image processing · Segmentation
Morphological techniques · Shape detection · Techniques · Preprocessing

1 Introduction

Gallstones are hard deposits in Gall Bladder. Gall Bladder is a small organ that stores bile, which is a digestive fluid and is made in the liver [1]. Lithiasis is the stone formation and Cholelithiasis [2, 3] is the nearness of gallstones, which are solidifications that frame in the biliary tract, more often than not in the gallbladder.

S. Garg (✉) · A. Walia · A. Singh · A. Mishra
Amity School of Engineering and Technology, Amity University,
Noida, Uttar Pradesh, India
e-mail: sakshijyotigarg@gmail.com

A. Walia
e-mail: waliaangad08@gmail.com

A. Singh
e-mail: asingh19@amity.edu

A. Mishra
e-mail: amishra1@amity.edu

A.K. Somani et al. (eds.), *Proceedings of First International Conference on Smart System, Innovations and Computing*, Smart Innovation, Systems and Technologies 79, https://doi.org/10.1007/978-981-10-5828-8_42

Hence, automated shape detection of stones will serve doctors of the world in painless, economical and timeliness removal of the gallstones. According to the latest surveys, it is concluded that this disease is more prevalent from the age of 40 onwards and primarily found in women. This disease is spread worldwide and a major concern in India [2, 3].

The preceding years' background and researchers makes it easier to classify gallstones [4]. Japanese classify the stones broadly into two types: the Cholesterol stone and the Pigment stone [5] each of which are further categorized as the Cholesterol Stone into Pure Cholesterol stone, the Combination stone, Mixed stone and the Pigment stone into the Black stone and Calcium Bilirubinate stone. The proportion of Cholesterol is the decisive factor in differentiating between the Cholesterol and the Pigment stones. More often than not, to be called as Cholesterol stones they should be no less than 80% cholesterol by weight (or 70% as indicated by the Japanese classification system). Something else, the stone is resolved as a Pigment stone with Calcium Bilirubinate as its prime essential. The concentration of this review lies in investigating the states of the gallstones consequently utilizing morphological image processing and image detection strategies. Additionally, on the premise of site of presentation, gallstones can likewise be named Primary and Secondary stones [4, 6]. Cholelithiasis is one of the most spread diseases affecting the gastrointestinal tract [6–8]. Henceforth, it is of foremost significance to know the attributes of the stones for applying the real measurement level for the treatment of a specific kind of the stones [8, 9]. The purpose of this analysis is to examine the gaps and propose some effective algorithms and techniques for the detection of the shapes of the gallstones. We are also working on the implementation of automated shape detection algorithm.

The main contribution of this paper is to identify and examine the fissures analyzed in the study of the size and shape of gallstones for a speedy and automated detection of gallstones which will assist in easing the removal of stones [10–14]. The aspiration of this study is to lend a helping hand to the doctors and physicians in the society offering them with the detailed investigation of gallstones for their uncomplicated and easy elimination. So the later sections, Sect. 2 gives the Literature Review. Section 3 discusses about the comparative analysis of various approaches available of gallstones segmentation [15–20], detection and selection of one of the most apt approach for further analysis. Section 4 states the Significance and Justification of this research. In Sect. 5, describes about the General Procedure to be followed for shape detection of stones. Section 6 specifies the Results and Discussions. Section 7 contains the Conclusions on the comprehensive study of the gallstones in Gall Bladder and finally, Sect. 8 gives the Future Scope and directions.

2 Literature Review

Cholelithiasis is one of the most spread diseases in India as well as outside world. Hence, the identification of gallstones becomes a mandatory part of it.

Webb [21], described in his study that women are most frequently affected, and by 50 years of age, very nearly 20% of women and 5% of men have gallstones. A huge segment of the stones are made of cholesterol and are formed due to cholesterol settling out of the bile as bile is being secured in the gallbladder between suppers. The best danger segments for making gallstones join family history of gallstones, ethnicity, being female, pregnancy, taking oral contraceptives, weight, diabetes, and age more than 40. Diverse diseases, for instance, Crohn's ailment, sickle cell illness, and thalassemia are for the most part associated with gallstones too. An association between specific dietary admission and gallstone advancement is not clear, yet rather there may be an extended threat of gallstones for those eating diets high in direct sugars and doused fat and low in fiber.

Frencesco Cotta [22] explained gallstones should never again be considered as a remarkable component, yet as a heterogeneous disease, which consolidates no under three unmistakable subgroups: cholesterol stones, blended stones with cholesterol as the essential section and shade stones, which are perceived as dull or chestnut shading. Despite these three major sorts of gallstone, there are furthermore blend stones and composite gallstones. The past consolidate stones with a central hubs of one compose (cholesterol or dim shade) and an outer piece of another sort (cocoa or calcified periphery); the last happen when unadulterated cholesterol stones are found inside a comparative gallbladder or bile channel together with perfect shading stones, i.e., there are no under two various stone peoples in a comparable subject.

Various techniques and algorithms are available through the study of the literature of the gallstones detection [23] mainly Preprocessing, Image Enhancement, Histogram Equalization techniques and a host of other techniques. Out of which Segmentation and Edge Detection grabs the prominent role in Gallstones Identification and its Size detection. History is full of literature on Segmentation. It is an efficient technique for size detection of gallstones. But segmentation in terms of shape detection requires new school of thought. Medical community has ample number of technologies to detect the size of the gallstones but lacks in techniques and algorithms for Shape Detection. In the early years of 1990–95, segmentation techniques were described for image processing with restricted scope. Then after an automated segmentation technique for gallstones detection came into existence.

For much more than most recent three decades, a few reviews about medical image segmentation have been put out. Pal and Pal [17] portrayed limit strategies from iterative pixel characterization, surface-based division procedures, edge location techniques and fuzzy set strategies. Hu et al. [18] grouped among limit, edge and locale based strategies. Pham et al. [19] recognized division strategies into four gatherings: area-based, boundary-based, hybrid and atlas-based. Zuva et al. [20] utilize eight classifications: thresholding approaches, area developing strategies, classifiers, bunching techniques, Markov arbitrary field models, artificial neural networks, deformable models, and atlas-guided methods. Shivi et al. [24] proposed a method to consequently distinguish gallstones in ultrasound pictures, dedicated as, Automated Gallstone Segmentation (AGS) Technique. He smothered Speckle Noise in the ultrasound picture utilizing Anisotropic Diffusion Technique. The edges were improved utilizing Unsharp Filtering. NCUT Segmentation Technique sectioned the picture. Later, edges were recognized utilizing Sobel Edge Detection. From there on, Edge Thickening Process was utilized to smoothen the edges and after that, the picture was jotted utilizing Automatic Scribbling Technique. At last, the fragmented gallstone inside the gallbladder utilizing the Closed Form Matting Technique was gotten. Erdt et al. [15] gave even a more non-specific view on the arrangement of division methodologies. He proposed a scientific categorization in view of a continuum between two extremes: simply picture based calculations and solid shape-based techniques. All division calculations were ordered inside this continuum as indicated by the measure of shape data utilized by the strategy. His article inspected the writing on image division techniques and presented a novel scientific classification in view of the measure of shape information being fused in the segmentation procedure.

Narang et al. [25] researched with an objective of the audit to see whether there is any association of number, size and sort of irk stones and patient's lipid profile with the occasion of annoy bladder carcinoma (GBC) as closeness of gallstones is thought to be the most basic possibility component for rankle bladder sickness. Subsequently, for this 200 cases of post-cholecystectomy gallbladder were mulled over. The number, size, and kind of stones and lipid profile were considered in each one of these cases. Net and histo-neurotic examination of irk bladders example was considered. Along these lines it was examined that as the number, size, and cholesterol irritate stone augmentation; the peril of rankle bladder tumor furthermore increases with no association with lipid profile. Outlined in writing is not completely unswerving and two principle classes of calculations can be recognized: image-based calculations and shape-based calculations.

Hence, all these techniques and algorithms gave ways of detecting size of the gallstones by applying image segmentation but none of them gave a vibrant technique or algorithm for the Shape Detection of the gallstones. Therefore, this paper highlights on analyzing the gaps analyzed during the literature reviews in the medical domain, which can be further explored for the easy and economical removal of gallstones [5].

3 Approaches to Gallstones Detection

There are multiple techniques for gallstones detection like Ultrasound [4, 5, 13, 26], Abdominal Computed Tomography Scan, Gall Bladder Radionuclide Scan, and Blood tests.

3.1 Ultrasound

Ultrasound tests [13, 14, 26] produce images of abdomen. This technique is preferred to initially confirm the presence of stones in Gall Bladder.

3.2 Abdominal Computed Tomography Scan

Abdominal CT Scan is an imaging that captures pictures of abdominal region and liver.

3.3 Gall Bladder Radionuclide Scan

Gall Bladder Radionuclide Scan completes in an hour. A radionuclide substance is injected into the veins by a specialist. The substance travels through the blood to the Gall Bladder and the liver. This technique reflects the infection or blockages in these organs (Fig. 1).

3.4 Blood Tests

Blood tests are done to measure the amount of bilirubin in blood that is used to check the signs of infection or obstruction, and to rule out other conditions [3, 27].

Considering the current trend in biomedical and industrial applications, ultrasonography is one of the most widely used techniques for real-time imaging [7, 12].

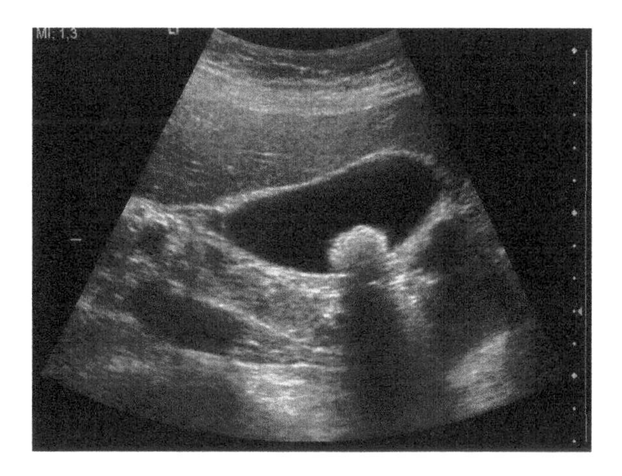

Fig. 1 Gallstones in Gall Bladder

It is a non-responsive and non-obtrusive procedure where the framework contains a pillar shaping part known as transducer. It is utilized to create the ultrasonic soundwaves at the time of reception. The gallstones are seen as opacities having a distal shadow with high particular gravity [1, 26]. Numerous other factors also imply in opting for ultrasound as one of the most efficient techniques for distinguishing the shape of gallstones. Sometimes, ultrasound exam might be incidentally uncomfortable, however it is not excruciating. Ultrasound is generally accessible, intelligible and prudent than other imaging strategies. Ultrasound imaging is greatly secure and does not utilize any ionizing radiation. Ultrasound filtering gives an evident picture of delicate tissues that do not appear well on x-beam pictures [5, 13]. Ultrasound gives constant imaging, making it a decent instrument for controlling negligibly obtrusive systems, for example, needle biopsies and liquid goal.

4 Justification of the Study

The Gall Bladder (GB) stones [1] are also seen as layered crystalline composites which consist of complex chemical composition. Factors like diet, age and other climatic and environmental conditions bring a large variation in the chemical composition of the stones. The gallstones happen in the gall bladder, and fluctuate in size and shape. In this manner, they turn into an issue of distress and torment to the patients. Hence, it is important to remove the gallstones identifying the shape and size of the stones [8, 9]. Many researchers have progressed with plentiful techniques and algorithms to examine the medical images. We emphasis on scrutinizing the gaps analyzed during the literature reviews in the medical domain, which can be further explored for trouble-free, painless, and economical removal of gallstones as an application of the work [5].

5 General Procedure

Detection of the shape and size of the gallstones provides a comprehensive view of the stones in the Gall Bladder which makes the removal of the gallstones comparatively more rigorous. Size detection is basically a five step process as shown in Fig. 2.

Step 1: Load the image from the computer and then convert them into gray level image. The general syntax in MATLAB for this is

$$I = \text{rgb2gray}(\text{RGB}), \quad (1)$$

where; I = Image, RGB is the notation for colored image.

Step 2: Adjust the brightness of the image. The general syntax for this in MATLAB is

Fig. 2 General size detection procedure [Source: 5]

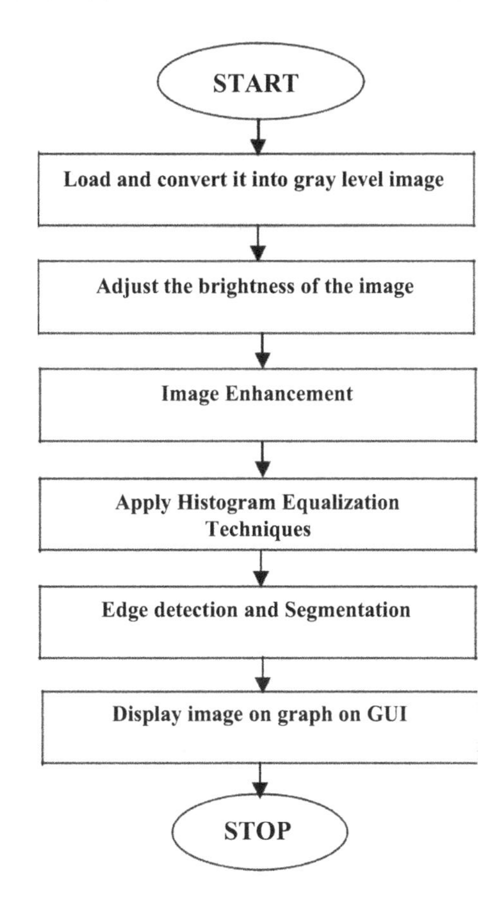

$$\text{OutputImage} = \text{imadjust}(\text{InputImage}, [\text{lowin}; \text{highin}], [\text{lowout}; \text{highout}]), \quad (2)$$

where; low_in,high_in,low_out and high_out are between 0 and 1.

Step 3: Perform image enhancement.

Step 4: Apply histogram equalization techniques.

Step 5: Perform edge detection and segmentation.

Step 6: And, finally display the image and graph on GUI. The general syntax for this in MATLAB is:

$$\text{imshow}(\text{I}) \quad (3)$$

Edge Detection [28] is an essential preprocessing step for image segmentation [15, 16, 17–20]. It changes the information picture to a binary picture, which indicates either the nearness or the nonappearance of an edge. Finally after all processes, gallstones can be detected along with their size [23].

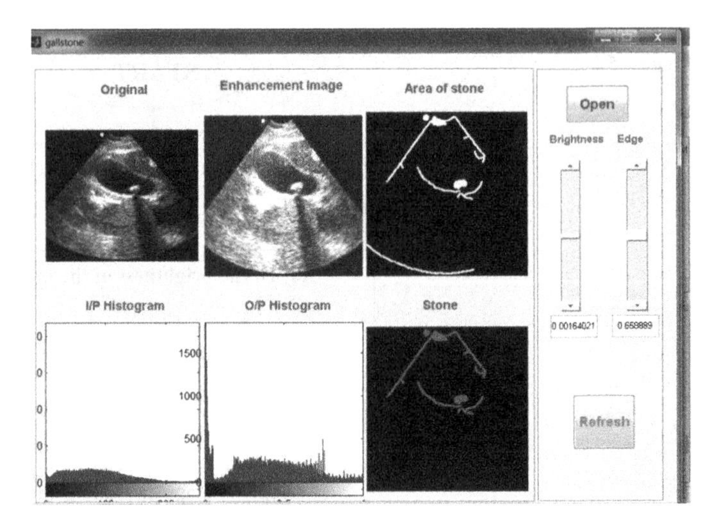

Fig. 3 Result of the general procedure [Source: 5]

6 Results and Discussions

The general procedure follows various steps that results in the extraction of the size of the gallstones, the area of the Gall Bladder, in removing shadowing effect and further refinement, enhancement and automated segmentation of the gallstones from the ultrasound images. But, we could not find neither a technique nor an algorithm that gives the automated implementation of the shape of the gallstones irrespective of the fact that size and shape are of utmost importance in case of Cholelithiasis and its treatment (Fig. 3).

7 Conclusion

Shape identification and extraction calculations are required for naturally distinguishing the states of the gallstones. This paper gives exhaustive study on different ways to deal with the gallstones recognition strategies. In this paper we have experienced the review and characterized the distinctive strategies with their necessities, advantages, and confinements. Additionally, we reasoned that to dispose of gallstones, shape and size assumes a key part out of which calculations and strategies so far given and composed do the robotized identification of gallstones giving just size of the gallstones. For the shape of the stones the physician has to manually look for it in the ultrasound images. Hence, it can be concluded that there is no algorithm or technique till date for automated shape detection of the gallstones and we are working on the implementation model in future to fill the gaps identified.

8 Future Scope

This paper presents a thorough study of the techniques and algorithms for gallstones detection identifying the gaps in the study till date. The future scope of this study is to work on the gaps identified during this literature survey and propose a technique or algorithm for Automated Gallstones shape detection in form of an implemented model. Further, capable applications might be created and expanded utilizing that method for the finding of various diseases other than Cholelithiasis.

References

1. I. Sook Kim, Seung Jae Myung, Sang Soo Lee, Myung Hwan Kim, "Classification and Nomenclature of Gallstones Revisited", Yonsei Medical Journal, Vol. 44, No. 4, pp. 561–570, 2003.
2. National Center for Biotechnology Information, https://www.ncbi.nlm.nih.gov/pmc/aticles/PMC3899548.
3. Dray X, Joly F, Reijasse D, Attar A, Alves A, Panis Y, et al. "Incidence, risk factors, and complications of cholelithiasis in patients with home parenteral nutrition". J Am Coll Surg. 2007;204:13–21. [PubMed].
4. http://www.healthline.com/health/gallstones#Diagnosis6.
5. Neha Mehta, SVAV Prasad, Leena Arya, Milie Pant, "A novel approach for the analysis of US images using morphological image processing techniques", Computing for Sustainable Global Development (INDIACom), 2015 2nd International Conference, 4, May 2015.
6. M. Tseng, J.E Everhart and R.S. Sandler, "Dietary intake and Gallbladder disease: a review", Public Health Nutrition, Vol. 2(2), pp. 161–172, January 1999.
7. P. Gupta, V. Malik, M. Gandhi, "Implementation of Multilevel Threshold Method for Digital Images Used In Medical Image Processing", International Journal of Advanced Research in Computer Science and Software Engineering, Vol. 2, No. 2, February 2012.
8. V.K. Kapoor, A.J.Mc Michael, "Gallbladder cancer: An, Indian Disease", The National Medical Journal of India, Vol.16, No. 4, 2003.
9. V.R. Sigh, Suresh Singh & Urmil Dhawan "Structural Analysis Of Gallbladder Stones", Conference Sewing Humanity, Advancing Technology Od 12–15, 99, Amb. GA, USA.
10. V.R. Singh, "Study Of Gall Bladder Stones As A New Piezoelectric Sensor Material", IEEE International Workshop on Medical measurements and Applications Ottawa, Ontario, Canada - May 9–10, 2008.
11. A. Mittal, S.K. Dubey, "Analysis of MRI Images of Rheumatoid Arthritis through Morphological Image Processing Techniques", IJCS, Vol. 10, Issue 2, No. 3, pp. 118–122, March 2013.
12. M.M. Foghi, et.al "Application of three-dimensional digital image processing for dynamic pore structure characterization", Global Journal of Science, Engineering and Technology, Issue 5, pp. 203–208, 2013, (ISSN: 2322-2441).
13. J. Jago, A.C. Billon, C. Chenal, J. M. Jon and S.M. Ebeid, "XRES: adaptive enhancement of ultrasound images", MEDICAMUNDI, Vol. 46, No. 3, pp. 36–41, November 2002.
14. Wan M. Hafizah and Eko Supriyanto, "Automatic Generation of Region of Interest for Kidney Ultrasound images using Texture Analysis", International Journal of Biology and Biomedical Engineering, Vol. 6, No. 1, p. 634, 2012.

15. M. Erdtl, S. Stegerl, G. Sakasl, 2 1 Fraunhofer IGD, Darmstadt, Germany 2 Technical University Darmstadt, Darmstadt, Germany, "Regmentation: A New View of Image Segmentation and Registration" ISSN: 1663-618X, J Radiat Oncol Inform 2012:4:1:1–23.
16. Varshali Jaiswal1, Aruna Tiwari2 1 Department of Computer Science and Engineering SGSITS, Indore, MP, India 2 Department of Computer Science and Engineering IIT Indore, Indore, MP, India "A Survey of Image Segmentation based on Artificial Intelligence and Evolutionary Approach" IOSR Journal of Computer Engineering (IOSR-JCE) e-ISSN: 2278-0661, p- ISSN: 2278-8727Volume 15, Issue 3 (Nov–Dec. 2013), PP 71–78.
17. Pal NR, Pal SK A review on image segmentation techniques. Pattern Recognition. 1993;26 (9):1277–1294. http://www.sciencedirect.com/science/article/pii/003132039390135J.
18. Hu A, Grossberg B, Mageras C. Survey of recent volumetric medical image segmentation techniques. Biomedical Engineering. 2009;321–346. http://www.intechopen.com/source/pdfs/8807/InTechSurvey_of_recent_volumetric_medical_image_segmentation_techniques.pdf.
19. Pham DL, Xu C, Prince JL. A survey of current methods in medical image segmentation. Annual Review of Biomedical Engineering, 2000;2:315–338. http://www.tecn.upf.es/~afrangi/ibi/CurrentMethodsInImageSegmentation_Phan2000.pdf.
20. Zuva T, Olugbara OO, Ojo SO et al. Image segmentation, available techniques, developments and open issues. Canadian Journal on Image Processing and Computer Vision 2011;2(3):20–29. http://www.ampublisher.com/Mar%202011/IPCV-1103-011-Image-Segmentation-Available-TechniquesDevelopments-Open-Issues.pdf.
21. Travis P. Webb, "Gallstone Disease", Pancreas and Biliary Disease pp 221–229, 10 August 2016. doi:10.1007/978-3-319-28089-9_10.
22. Francesco Cotta, "Classification, Composition and Structure of Gallstones. Relevance of these Parameters for Clinical Presentation and Treatment", Biliary Lithiasis pp 51–65. doi:10.1007/978-88-470-0763-5_4.
23. Gupta, A., Gosain, B. & Kaushal, "A comparison of two algorithms for automated stone detection in clinical B-mode ultrasound images of the abdomen" S. J Clin Monit Comput (2010) 24: 341. doi:10.1007/s10877-010-9254-0.
24. Shivi Agnihotri, Harsh Loomba, Abhinav Gupta, VineetKhandelwal, "Automated segmentation of gallstones in ultrasound images", 2nd IEEE International Conference on Computer Science and Information Technology, pp. 56–59, 2009.
25. Shveta Narang et al, "Gall bladder cancer analysis", International Journal of Cancer Therapy and Oncology, June 26, 2014. ISSN: 2330-4049.
26. D. Kouame, "Ultrasound Imaging: Signal Acquisition, new advanced processing for Biomedical and Industrial Application", ICASSP, Vol. 993, 2005.
27. S. Jagadeesh, et.al "Image Processing Based Approach to Cancer Cell Prediction In Blood Samples", ME&HWDS, 2015 2nd International Conference on Computing for Sustainable Global Development (INDIA Com) 871_International Journal of Technology and Engineering Sciences, Vol. 1(1), 2013, (ISSN: 2320-8007).
28. Bertan Karahoda1, Gülden Köktürk2, "Gall Bladder Ultrasonic Image Analysis by Using Discrete Wavelet Transform", University Mechanical Engineering Department, Izmir, Turkey, University Electrical-Electronics Engineering Department, Izmir, Turkey.

Assistive Dementia Care System Through Smart Home

K.S. Gayathri, Susan Elias and K.S. Easwarakumar

Abstract As per the statistics of World Health Organization, a major percentage of elderly society across the globe is affected with dementia (age related memory loss). Dementia care entails prolonged effort in terms of money, time and manpower. Assistive health care system is therefore essential and is probable through a smart home, that offers Ambient Assisted Living (AAL) to its occupants. The objective of this work is to model an Intelligent Decision Support System (IDSS) for dementia care through the smart home. The innovation in the design of IDSS is to offer two levels of decision-making (1) Short-Term Decision-Making (STDM)—to raise suitable alerts for the abnormality detected in ADL (2) Long-Term Decision-Making (LTDM)—to decide on the progress in occupant's developmental stage of dementia. The novelty in the design of STDM is to assimilate Random Forest (data driven) decision-making and Rule-based (knowledge-driven) decision-making within a single framework. Random Forest modeling provides better predictive accuracy through ensemble learning which is later combined with the domain specific knowledge to offer context-based decision-making. On the other hand, LTDM decides on occupants developmental stage of dementia through automation of Barthel score. Barthel score is a clinical measure to assess the stage of dementia through the level of dependency required by the occupant to complete his activities. The experimental analysis confirms the proficiency of the proposed IDSS in decision-making is better than existing approaches.

K.S. Gayathri (✉)
Department of Computer Science and Engineering,
Sri Venkateswara College of Engineering, Sriperumbudur, Chennai, India
e-mail: gayasuku@svce.ac.in

S. Elias
School of Electronics Engineering, VIT University, Chennai, India
e-mail: susanelias70@gmail.com

K.S. Easwarakumar
Department of Computer Science and Engineering, Anna University, Chennai, India
e-mail: easwarakumarks@gmail.com

A.K. Somani et al. (eds.), *Proceedings of First International Conference on Smart System, Innovations and Computing*, Smart Innovation, Systems and Technologies 79, https://doi.org/10.1007/978-981-10-5828-8_43

Keywords Dementia care · Smart home · Decision-making
Machine learning · Random forest · Progressive assessment

1 Introduction

Smart home, an ambient intelligent environment comprises of a network of sensors, computing systems, actuators to proficiently compute and rationally respond to the context of environment [1]. Smart home proactively assists its occupant in their Activities of Daily Living (ADL) in an attempt to enhance their safety and comfort within the living environment. Autonomy is its primary objective that enables and empowers independence to the occupant to meet their social, emotional and rational needs [2, 3]. Ambient Assisted Living (AAL) offered by smart home facilitate its integration in modeling health care applications [3, 4]. Dementia care is one such application that has gained focus with 35 million citizens across the world affected by dementia and is likely to double by 2030 [5, 6]. Excessive care is required for the dement occupant to carry out their daily living which engrosses enormous amount of resources in terms of time, money and manpower.

A survey on dementia care shows that the care taker contributed 12.5 billion hours of chargeless care which represents an average of 21.9 h per caretaker per week and 1,139 h per annum. It's estimated that more than $148 billion was spent for dementia care in 2005 and $604 billion in 2010 for the same which is around 1% of world's GDP. From these statistics, it is noticeable that the overhead incurred in dementia care is immense and hence assistive health care systems are indispensable [7]. Smart home aid dementia care modeling through the design of activity recognition and decision-making system. Abnormal detection ought to be incorporated within the activity recognition system to recognize the abnormal occupant behavior [8]. The rationale of decision-making system is to decide on a suitable action for the abnormal situation so as to increase the well being and comfort of the occupant [9].

Timely diagnosis and recommendations to handle the drifted situations are required and provided through short-term decision-making. Various machine learning algorithms, artificial intelligence strategies are employed in STDM modeling that discovers the association between abnormal factors and recommended alert. Furthermore, dementia is distinguished by persistent and dynamic deterioration in cognitive functionality including the symptoms progression over time. According to the degree of risk estimated the treatment varies all over the course, the dementia patients require disease severity assessment and treatment according to their current stage. Patients with early stage just require proper care planning but at later stage there is a need for the improvement of patients and caretakers lifestyle to promote optimal functionality to maximize comfort [10]. Functional status of an occupant's ability is described in terms of his/her own daily activities and these assessments provide a baseline for assessing the effect of deviation or any intervention from a standard lifestyle to handle the drifted situations [11]. These assessments are performed and results are reviewed after a period of time. The results lead to the development of a crucial decision-making process for the ability to adapt to the changing needs of patients and caretakers and later render them with solutions for the treatment and care of their betterment. Therefore, IDSS require making a decision on the developmental stage of dementia and is dealt with through Long-Term Decision-Making system. Therefore, this paper proposes an Intelligent Decision Support System framework for dementia care through smart home to offer both short-term and long-term decision-making. Quick alerts to handle abnormality situation is provided through short-term decision-making and assessment of the developmental stage of dementia is provided through long-term decision-making system.

2 Related Work

Machine learning and artificial intelligence strategies are employed to model intelligent decision support system and this section presents details on associated IDSS modeling approaches. Current research in the design of STDM focuses on data driven and knowledge-driven approaches [12, 2]. The use of probabilistic and statistical methods for IDSS modeling is the crux of the data driven approach. Application of knowledge engineering and management techniques to model the IDSS system is viewed as the knowledge-driven paradigm. Various data mining, machine learning, and artificial intelligence strategies are employed in modeling both data driven and knowledge-driven decision-making system. Frequent pattern mining and clustering are the commonly used data mining approaches in modeling smart home [12]. Machine learning approach that is used to model data driven approach is either generative or discriminative in nature [12]. Generative approach uses a probabilistic model which builds a complete description of the input sensor data space whereas discriminative approach builds the system as a classification model rather than representation. The most commonly used generative approach includes Naive Bayes classifier (NBC), Markov Models and Dynamic Bayesian Networks. The discriminative approaches represent the decision-making system as a classification model rather than representation model. Support Vector Machines (SVM), Decision Trees, Artificial Neural Network (ANN) are some of the commonly used discriminative approaches.

Domain knowledge representation of IDSS enables context-based decision-making and is feasible through knowledge engineering techniques [12, 13]. Propositional, first order, and description logic are the knowledge representational approaches. Domain knowledge is represented in the form of rules through various logical formalism. Inference mechanisms such resolution, subsumption, abduction, and deduction are executed over the represented knowledge to recognize appropriate decision given the situation. Though symbolic approaches enable context-based reasoning it has a limitation in modeling uncertain data. Thus, the proposed system employs a data driven approach to make initial decisions and later enhances the decision through knowledge-driven approach. The existing systems have either each of these approaches individually. Moreover, the proposed system enables both short-term and long-term decision-making within a single framework. The proposed data driven STDM employs Random Forest machine learning algorithm [14] for decision-making as it makes a collective decision on different decision tree modeled. Whereas, first-order rule is used for modeling knowledge-driven STDM.

Long-Term Decision-Making is required to assess the developmental stage of dementia. A range of assessment score is available in the medical literature to evaluate the progress of dementia [7, 10]. Barthel score is the best measure to appraise the developmental progress of dementia through ADLs. The score is estimated in terms of dependency of the occupant to execute his ADLs. The existing approaches assess only the quality of activities [15] but none of the approaches assess the developmental stage of dementia through automated calculation of Barthel score. Thus, proposed LTDM formulate its decision on the progress of dementia by automated evaluation of Barthel score with the detected abnormality factors. Moreover, the existing systems have modeled either STDM or LTDM separately. This research innovatively integrates STDM and LTDM within a single framework for the design of dementia care system through the smart home.

3 Proposed Decision-Making System for Dementia Care

As shown in Fig. 1, the novelty of the proposed design is to model two different levels of decision-making namely short-term decision-making and long-term decision-making within a single framework. As the name suggests, short-term decision-making decides on the action immediately for the recognized abnormality while long-term decision-making records the abnormality factors for a period of time (say for about two months) to evaluate the developmental stage of dement occupant. The activity recognition and abnormality detection approach of dementia care [16] is a Markov Logic Network based hierarchical framework. Each layer of this framework is functionally different based on their associated abnormality factors. The lower layer is designed to collectively recognize activity and detect abnormality based on objects used. Whereas, higher layers considers attributes of the activity such as time, location and duration for abnormality detection. The MLNs at each layer are designed to utilize the associated factor and detect abnormality independently. The hierarchy among the MLNs is defined based on the priority that is associated with each factor. This priority is assigned based on the ability of the factor to contribute to efficient context-based abnormality detection in a hierarchical manner. Hence the hierarchical approach and MLN gives a quick response and better recognition results. The proposed system of dementia care extends this activity recognition framework [16] to model decision-making system that effectively utilizes the recognized abnormality factors for decision-making.

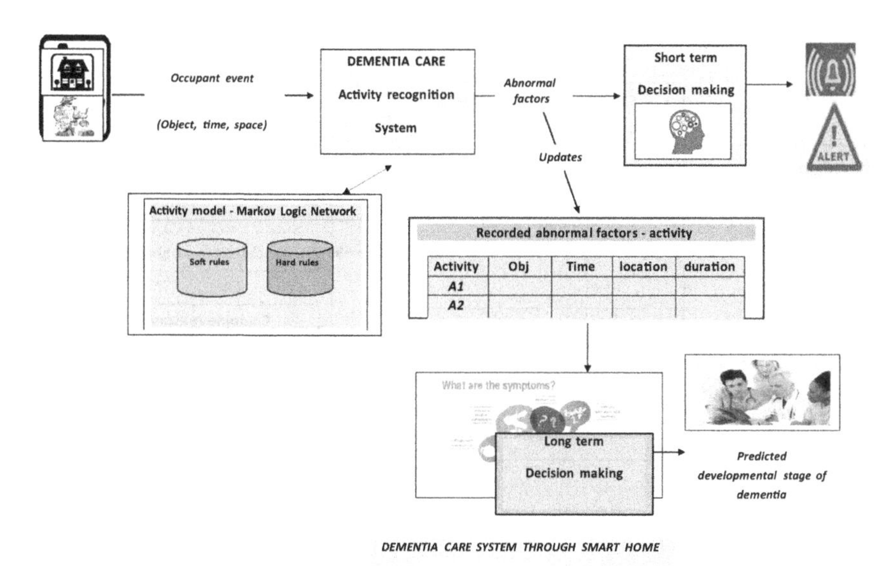

Fig. 1 Overall framework of dementia care system using smart home

3.1 Short-Term Decision-Making (STDM) System

Short-Term Decision-Making system (STDM) raises alerts instantaneously for any abnormality detected by the activity recognition system. The inputs to STDM are the abnormality factors (object, time, space, concurrency, and duration) together with its corresponding ADL. The output is generally an alert (low, high or emergency) that handles the current abnormal situation. The proposed STDM as shown in Fig. 2 employs both data driven and knowledge-driven to decide on an appropriate alert given the abnormal factors. Random Forest machine learning algorithm is engaged in data driven STDM to learn the association of abnormal factors and alerts at varied perspective. Random Forest constructs multitude decision trees with abnormality factors as its attributes for modeling. This is feasible through the random selection of features and samples for modeling the decision trees. Randomization is done by selecting a random subset of the predictors for each split. Number of predictors to try at each split is known as mtry. Typically mtry = $\sqrt{k}$ for classification, where 'k' is the total number of attributes available. Random Forest is an ensemble learning approach that employs Bootstrap aggregation to make a collective decision [14]. Majority voting aggregation of multiple decision trees affords enhanced recommendation on an action that augments the overall accurateness in decision-making. The strategy of majority voting scheme at all times gives high preference to a decision that is outputted by most of the classifier. Thus, Random Forest improves the overall accuracy of the proposed STDM that coalesce the choice of various experts (decision trees).

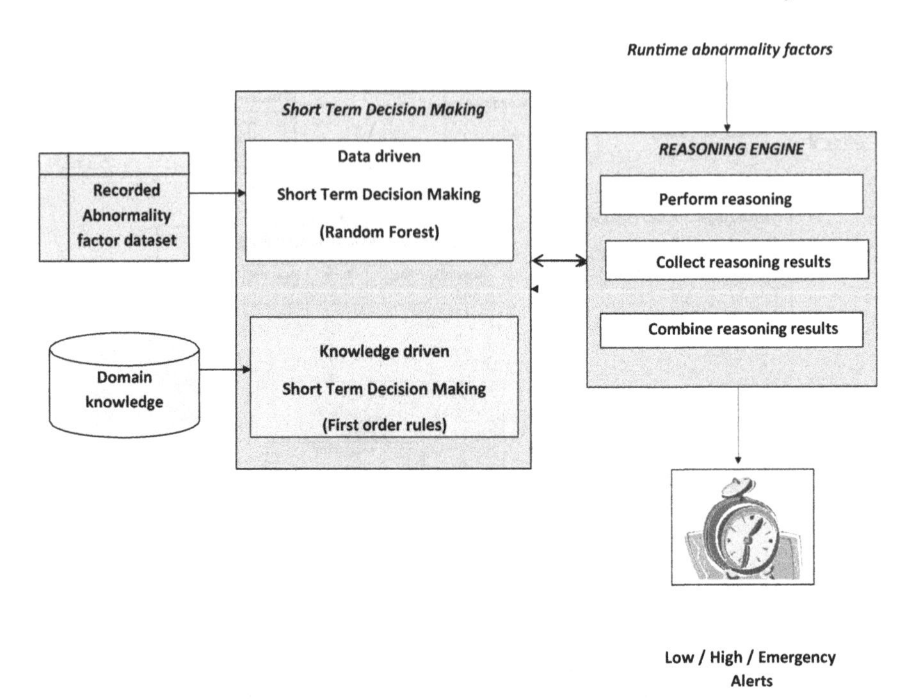

Fig. 2 Short-term decision-making system for dementia care

The knowledge-driven STDM focuses to provide context-based decision-making. Health care application requires modeling human vital signals like glucose monitoring, oral, rectal, skin temperature, heart rate and blood pressure for decision-making. The proposed system models the vital parameters through knowledge-driven STDM. There is no restriction on the number of rules to be implemented and it varies according to the requirement of doctors/care taker. This could be extended to model the other parameters by just including additional rules. The explicit specification of rule aids the reasoning system to become a scalable model to model complex health care applications. Other than vital parameter modeling, rules are framed to provide context-based alerts. For, e.g., consider a situation in which the occupant does not react to the previous alert raised by data driven STDM for a specific duration, in such circumstances knowledge-driven STDM raises context prcised alerts to tackle the situation. Therefore, the context specific decisions are made by the first-order rules modeled within knowledge-driven STDM.

3.2 Long-Term Decision-Making (LTDM) System

The role of Long-Term Decision-Making system (LTDM) is to provide a decision on the developmental symptom of dementia in an attempt to present progressive feedback to the care taker/doctors. Onset and developmental stage of dementia can be described through the level of dependency the occupant requires from the care taker to complete his ADL and Barthel Index is one such clinical measure [11, 10]. The general procedure to measure the Barthel index is to put the occupant under observation for a period of time and manually record his dependency. Such, labor-intensive examination is an overhead to the care taker and is mostly prone to errors. As a result, LTDM system ably automates the process of Barthel index estimation in order to conclude on the developmental stage of dement occupant as Fig. 3.

Consequently, abnormality factors recognized through activity recognition are utilized for long-term decision-making. Dependencies of the occupant to complete ADL can be estimated through the factors that stimulate abnormality. The level of dependency in executing ADL is proportional to a number of abnormality factors acknowledged. The use case of dementia showcase the priority among abnormality factors varies in dementia care and for this reason, different weights are assigned to each of these abnormal factors. The hierarchical framework of MLN-based activity recognition [16] is exploited to characterize weights of the abnormal factors on the basis of the priority assigned in modeling the abnormal detection system. Any ADL is exemplified in terms of the physical objects involved in performing an activity

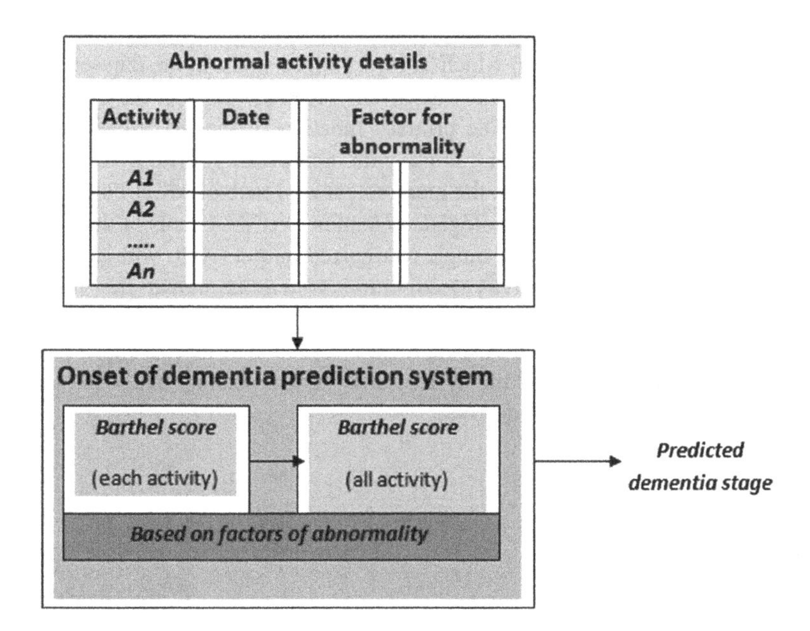

Fig. 3 Long-term decision-making system for dementia care

rather than time, location and duration. Thus high weight is assigned to object abnormality factor and subsequent low weights for time, location and duration.

$$BarthelScore = \sum \frac{wt_i * ab_i}{N} \quad (1)$$

The weighted average is the statistical measure used in the proposed LTDM system to evaluate the Barthel score. The expression to assess Barthel score of an ADL is provided in the Eq. 1. Where, 'wti' represents the weight of abnormal factor i, 'abi' specifies the presence of abnormality i and 'N' represents the frequency of ADL and the outputted score is a value between 0 and 10. The Barthel score of individual ADLs are summed to assess the dependency of the occupant. The LTDM concludes the developmental stage of dementia as early, middle or later based on the calculated Barthel score. Early stage is identified with Barthel score above 60, middle stage with score between 60 and 30, later stage with score less than 30. Thus Barthel score furnishes the dependency of the occupant to execute his ADL with which stage of dementia is predicted to be early, middle or later.

4 Experimental Section

The proposed Intelligent Decision Support System is experimented to demonstrate the decision-making ability within the context of dementia care. IDSS is modeled both from data as well as domain knowledge. Abnormality dataset required for modeling was obtained through several runs of experiments with MLN-based activity recognition system [16] which in turn utilizes smart home dataset [17] for its modeling. Domain specific knowledge of occupant preferences is provided through the care taker/doctors. The primary function of the MLN-based activity recognition is to recognize ADL along with abnormality factors (if any). As described in the proposed Sect. 3, the hierarchical approach of MLN-based activity recognition system perform chronological detection over the events of the occupant to recognize abnormal factors. A sample recognized output (ADL with abnormality factors) from MLN-based activity recognition system is shown in Fig. 4 The extracted dataset contains abnormality of an ADL in terms of object, location, time, duration, concurrency followed with the recommended alert labels. These recorded data were labeled manually by the system designer (annotator) that was later utilized for modeling the proposed IDSS.

4.1 Experimental Setup

Short-term decision-making system comprises of data driven STDM and knowledge-driven STDM. The data driven STDM is evaluated with different machine learning algorithms, which is subjected to tenfold cross validation. The hyper parameters engrossed in machine learning algorithms were fine tuned so as to make better accuracy in decision-making. Accuracy and F-measure are the evaluation metrics used to conclude the efficiency of the proposed system. STDM is developed using "R" [18] and "Prolog" [19]. "R" is a statistical language that contains built-in libraries to effectively model various machine learning algorithms. Random Forest, the approach preferred to model STDM is compared with other base line decision-making algorithms. Decision Trees (DT), Support Vector Machine (SVM) and Artificial Neural Network (ANN) are the machine learning algorithms considered for the comparative study. Figure 5 shows the modeling of Random Forest using R.

```
> Dataset
  object location time duration concurrency ADL RECOMMENDATIONS
1      0        1    1        1           0   0        NO_ALERT
2      0        1    1        0           0   0       LOW_ALERT
3      1        1    1        1           0   0        NO_ALERT
4      1        1    1        0           0   0        NO_ALERT
5      0        2    1        1           0   0       LOW_ALERT
6      0        2    1        0           0   0       LOW_ALERT
```

Fig. 4 Sample dataset for ADL 1

```
Call:
 randomForest(x = x, y = y, mtry = param$mtry)
               Type of random forest: classification
                     Number of trees: 500
No. of variables tried at each split: 2

        OOB estimate of  error rate: 4.49%
Confusion matrix:
           EMG_ALERT HIGH_ALERT LOW_ALERT NO_ALERT class.error
EMG_ALERT         34          0         0        0  0.00000000
HIGH_ALERT         0         67         0        1  0.01470588
LOW_ALERT          0          0        20        4  0.16666667
NO_ALERT           0          2         1       49  0.05769231
> print(rf_model)
Random Forest

178 samples
  6 predictor
  4 classes: 'EMG_ALERT', 'HIGH_ALERT', 'LOW_ALERT', 'NO_ALERT'

No pre-processing
Resampling: Cross-Validated (10 fold)
Summary of sample sizes: 160, 160, 158, 159, 160, 161, ...
Resampling results across tuning parameters:

  mtry  Accuracy   Kappa
  2     0.9609443  0.9444442
  4     0.9432564  0.9200986
  6     0.9377008  0.9123400

Accuracy was used to select the optimal model using  the largest value.
The final value used for the model was mtry = 2.
```

Fig. 5 Random Forest approach to model STDM

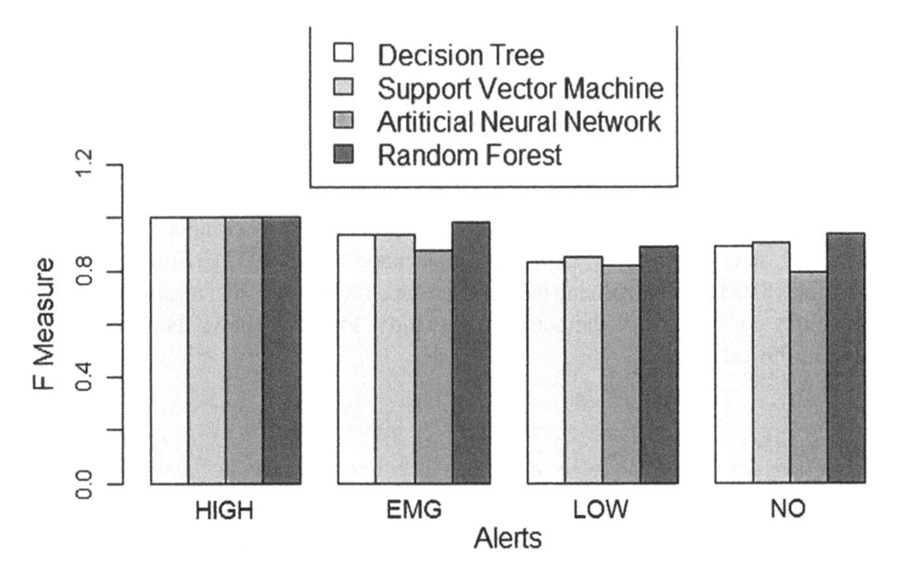

Fig. 6 F-measure—Random Forest-based STDM

The comparative study to analyze F-measure of various STDM system is visually presented in Fig. 6. It is observed from the experimental study, F-measure is high for Random Forest approach of decision-making. The reason for high F-measure is that Random Forest makes a collective decision on different decision tree modeled through random selection of features and sub samples. This ensemble learning approach uses majority voting bagging scheme to conclude on an alert given the abnormal factors. Moreover, Random Forest avoids over fitting and has low generalization error. Whereas, Decision Tree approach over fits the data and has high generalization error. SVM though models the decision-making system with better accuracy compared to the decision tree. Figure 7 compares each of these models in terms of accuracy and shows that Random Forest approach of modeling decision support system has an accuracy of 0.961. The reason for better accuracy compared with other models is that Random Forest constructs a large number of decision trees from the random subsamples of dataset and outputs an alert based on majority voting scheme. While the other learning methods do not exhibit any kind of ensemble learning due to which accuracy is low compared to Random Forest. The reason for SVM to have less accuracy when compared to the random forest is that it does not perform a random selection of subsamples for modeling the decision-making system, no bagging to make a collective decision and does not produce certainty score as its output. Artificial Neural Network has low accuracy than Random Forest because neural network learning overfits and is more of black box approach which hides the details of how the decision-making system is modeled.

During run time data driven STDM maps the recognized abnormal factors onto the learned Random Forest model to decide on an action to be executed. Whereas,

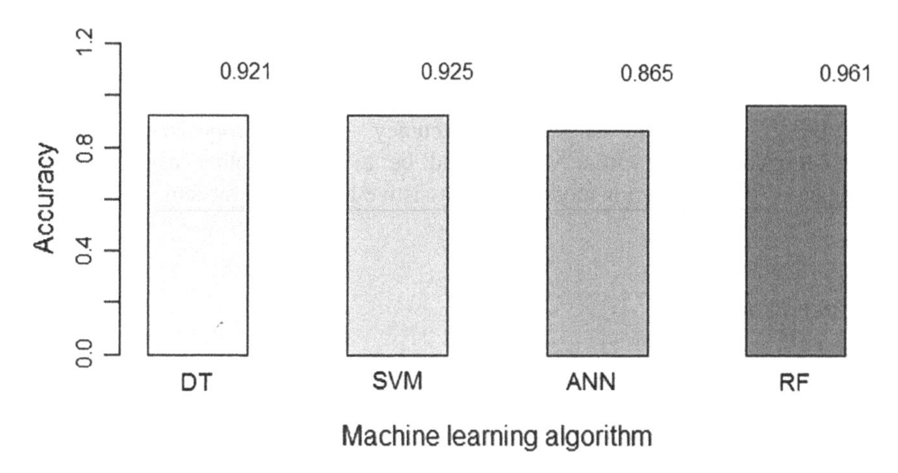

Fig. 7 Accuracy comparision

knowledge-driven STDM models human vital parameters such as blood pressure level and sugar level. The system raises appropriate alerts as soon as the occupant deviates from the expected level of vital signs. Different rules can be modeled to enable context-based system; here a rule is framed that raises the emergency alarm at a situation where there is no response from the occupant for a specific duration for the previous alert raised. The outputs of both data driven and knowledge driven are combined and the privileged alerts that augment the well being of the occupant are given back to the environments.

Long-Term Decision-making, calculates Barthel score for each ADL using weighted average statistical measure from the recorded dataset. Barthel score is later premeditated for all ADLs and the summarized score is given for assessing the developmental stage of dementia. The experimental results from various runs of analysis demonstrate the weighted average approach to model the proposed LTDM has high F-measure of above 0.9 and hence the LTDM is efficient to assess the developmental stage of dementia.

5 Conclusion

Dementia care system modeled through smart home offers ambient assisted living that aid the occupant to lead an independent everyday life. The proposed system innovatively integrates short-term and long-term decision-making to provide intelligent support for dementia care. Quick alerts to handle the abnormal situation are presented by short-term decision-making while the assessment of the progressive stage of dementia is presented by long-term decision-making. Random Forest, an ensemble learning raises appropriate through majority voting scheme that makes a collective decision on the various decision trees model. Further, more human vital signs and context-based reasoning are enabled through knowledge-driven STDM. LTDM system automates the evaluation of Barthel scores through a statistical measure that aids the assessment of the developmental stage of dementia. The experimental evaluation shows the accuracy of the proposed system in decision-making. The future work would be to include other psychosomatic parameters in dementia for modeling the assistive health care system.

References

1. Muhammad Raisul Alam, Mamun Bin Ibne Reaz, and Mohd. Alauddin Mohd. Ali. A review of smart homes - past, present, and future. IEEE Transactions on Systems, Man, and Cybernetics, Part C, 42(6):1190–1203, 2012.
2. JuanCarlos Augusto, Hideyuki Nakashima, and Hamid Aghajan. Ambient intelli- gence and smart environments: A state of the art. Handbook of Ambient Intelligence and Smart Environments, Springer US, pages 3–31, 2010.
3. Diane J. Cook and Sajal K. Das. How smart are our environments? an updated look at the state of the art. Pervasive Mob. Comput., 3(2):53–73, March 2007.
4. Ulises Cort´es, Cristina Urdiales, and Roberta Annicchiaric. Intelligent healthcare managing: An assistive technology approach. pages 1045–1051, 2007.
5. Michele Amoretti, Sergio Copelli, Folker Wientapper, Francesco Furfari, Stefano Lenzi, and Stefano Chessa. Sensor data fusion for activity monitoring in the persona ambient assisted living project. J. Ambient Intelligence and Humanized Computing, 4(1):67–84, 2013.
6. Anders Wimo and Martin Prince. World Alzheimer Report 2010: The Global Economic Impact of Dementia, September 2010.
7. Dementia performance measurement set-american academy of neurology-american geriatrics society-american medical directors association-american psychiatric association-physician consortium for performance improvement.
8. Ahmad Lotfi, Caroline S. Langensiepen, Sawsan M. Mahmoud, and M. Javad Akhlaghinia. Smart homes for the elderly dementia sufferers: identification and prediction of abnormal behaviour. J. Ambient Intelligence and Humanized Com- puting, 3(3):205–218, 2012.
9. Claire A.G. Wolfs, Marjolein E. de Vugt, Mike Verkaaik, Marc Haufe, Paul-Jeroen Verkade, Frans R.J. Verhey, and Fred Stevens. Rational decision-making about treatment and care in dementia: A contradiction in terms? Patient Education and Counseling, 87(1):43–48, 2012.
10. Ladislav Volicer. Response to sheehan, b.(2012) assessment scales in dementia. Ther adv neurol disord 5: 349–358. Therapeutic advances in neurological disorders, page 1756285613489764, 2013.
11. https://www.healthcare.uiowa.edu/igec/tools/function/bartheladls.pdf.
12. Liming Chen, Jesse Hoey, Chris D. Nugent, Diane J. Cook, and Zhiwen Yu. Sensor- based activity recognition. IEEE Transactions on Systems, Man, and Cybernetics, Part C, 42 (6):790–808, 2012.
13. Liming Chen, Chris D. Nugent, and Hui Wang. A knowledge-driven approach to activity recognition in smart homes. IEEE Trans. Knowl. Data Eng., 24(6):961–974, 2012.
14. Leo Breiman. Random forests. Machine Learning, 45(1):5–32, 2001.
15. Prafulla N Dawadi, Diane J Cook, Maureen Schmitter-Edgecombe, and Carolyn Parsey. Automated assessment of cognitive health using smart home technologies. Technology and health care, 21(4):323–343, 2013.

16. K.S. Gayathri, Susan Elias, and Balaraman Ravindran. Hierarchical activity recognition for dementia care using markov logic network. Personal and Ubiquitous Computing, pages 1–15, 2014.
17. F.J Ordonez, de Toledo, and Sanchis P. Activity recognition using hybrid genera- tive/ discriminative models on home environments using binary sensors. Sensors 2013, pages 5460–5477, 2013.
18. https://en.wikipedia.org/wiki/r.
19. http://www.cpp.edu/jrfisher/www/prologtutorial/contents.html.

Fault Aware Trust Determination Algorithm for Wireless Body Sensor Network (WBSN)

A. Chitra and G.R. Kanagachidambaresan

Abstract The design of a fault tolerant and eminent Wireless Body Sensor Network (WBSN) has become very necessary since Wireless Sensor Network (WSN) serves as a leading solution to all sorts of monitoring and surveillance problems. The Body Sensor Network (BSN) has reduced the occupancy of patients in the hospital. The sophistication and remote monitoring of BSN make the system more vulnerable to attacks and faults. Designing a trusted fault tolerant system is mandatory to meet these issues. A Fault Aware Trust Determination (FATD) algorithm for WBSN is proposed in this paper. The trust of the node is identified based on the battery terminal voltage, receiver signal strength, and speed of movement of the node. A value is assigned for each node in the range of −1 to 1 called trust value, to rate the trustworthiness. This trust value of the node is assigned as per the trustworthy algorithm. The Hierarchical Hidden Markov Model (HHMM) predicts the transition between states and the packet classifier classifies the packet and communicates it to the sink through a high trusted path. The simulation results show that the proposed Fail Aware Trust determination algorithm outperforms the existing LEACH and QPRR protocols in terms of lifetime and throughput.

Keywords Wireless Body Sensor Network (WBSN) · Mobility
Reliability · Energy

A. Chitra (✉)
Department of Computer Application, PSG College of Technology,
Coimbatore 641004, India
e-mail: ctr.psg@gmail.com

G.R. Kanagachidambaresan
Department of EEE, Sri Krishna College of Technology, Coimbatore 641042, India
e-mail: kanagachidambaresan@gmail.com

A.K. Somani et al. (eds.), *Proceedings of First International Conference on Smart System, Innovations and Computing*, Smart Innovation, Systems and Technologies 79, https://doi.org/10.1007/978-981-10-5828-8_44

1 Introduction

Wireless Body Sensor Network (WBSN) is the collection of sensor nodes which monitors the physiological signals from the human body [1]. The sensor nodes may be wearable or stuck to the body surface or may also be implanted in nature. The implanted sensor nodes monitor physiological or biological changes of the body [2]. The data from the implanted nodes and other non-implanted node reaches the Central Mote (CM) a data collection center, wearable over the human body. The CM collects the data from all the nodes, both implanted and non-implanted nodes and relays the data to the Health Monitoring System (HMS) [3]. Figure 1 illustrates the WBSN with implanted and non-implanted node. The data from the sensor nodes share the information with the CM wirelessly. The implanted node is more vulnerable to attack, McAfee demonstrated hacking of an insulin pump inside the human body and making it excrete more insulin causing mortality [4, 5]. Thus, attack to the implanted node causes worse conditions to the human life than attack to the wearable non-implanted nodes. In the case of communication failure, the data is re requested from the implanted node causing additional burden to the implanted node. The implanted node communicates the same packet several times due to

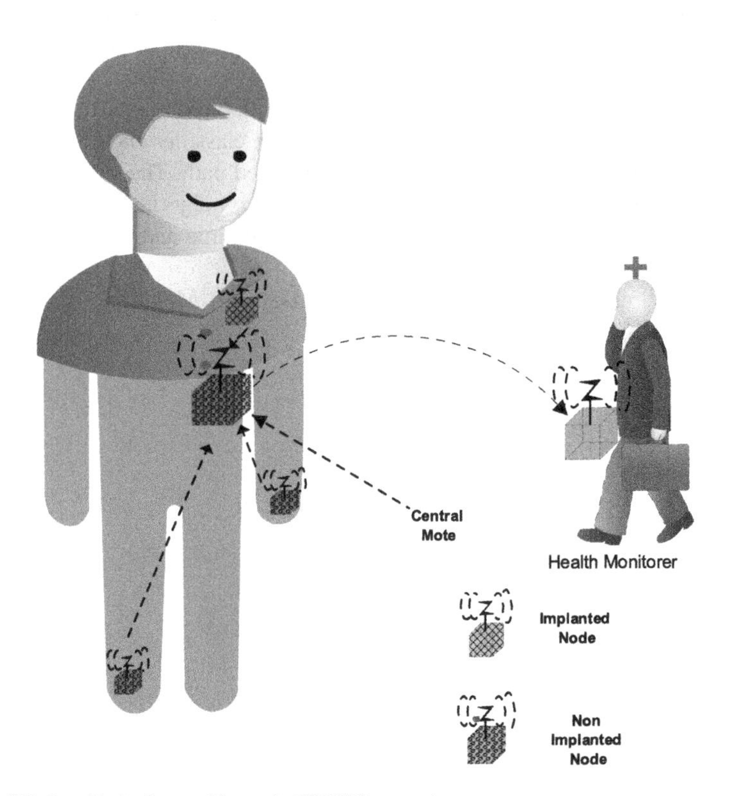

Fig. 1 Wireless Body Sensor Network (WBSN) scenario

insufficient trust in the communication. The implanted node either sensing-based or actuating-based the data communication should be fault tolerant in nature in order to make the node energy efficient for usage during critical time and to avoid tissue damage due to high power consumption. Thus, the data from implanted nodes should be taken special care [6–9]. The major challenges faced by the real time sensor network are addressed in [3, 5, 6, 9]. The WBSN serves to be the key for out of hospital monitoring environment. The WBSN serves to be the solution for avoiding the feeling of getting monitored to the patients. The communication of node by WBSN should be uninterrupted so that it is available during critical conditions. Creating the fault tolerant communication system is the major challenge in the real time Body Sensor Network.

2 Related Work

The Low Energy Adaptive Clustering Hierarchical (LEACH) algorithm satisfies the Energy trust of the nodes. LEACH protocol has two phases namely setup phase and routing phase. The Cluster Head is selected in the setup phase and routing data to the sink is proceeded in the routing phase [1]. Group-Based Trust Management Scheme (GTMS) [7], Agent-Based Trust and Reputation Management (ATRM) [8] propose a trust-based WSN which deals about trust of the node based on the authentication and encryption. However, this scheme only discuses about the authentication mechanism. ATRM and GTMS lack in dealing about other trust parameters such as mobility, reliability, and energy on the whole. Energy-Aware Peering Routing (EPR) aimed at reducing network traffic with energy efficient routing thereby creating an energy-aware WBSN [10]. QoS—Aware Peering Routing for Delay-Sensitive Data (QPRD) [11] intends to improve EPR, the data packet is classified into an ordinary packet and Delay Sensitive Packets. The QPRD focuses on delay sensitive packets making the packet to reach sink in delay sensitive manner. The QoS-Aware Peering Routing for Reliability (QPRR) [12]—sensitive data algorithm intends to increase the throughput with reduced energy. QPRR focuses on signal strength and energy cost with respect to distance in creating an energy-aware routing with increased throughput [11, 13, 14]. The proposed algorithm considers mobility, energy, and reliability as a parameter for designing trust in the WBSN. The factor avoids loss of data due to lack of signal strength and lack of energy faults.

3 Fault Aware Trust Determination Algorithm

The Overall trust depends on the Initial trust, Energy, Reliability, and Mobility trusts status of the nodes. When the residual energy is high, the potential difference across the battery is high showing that the node is not with lack of energy. Battery

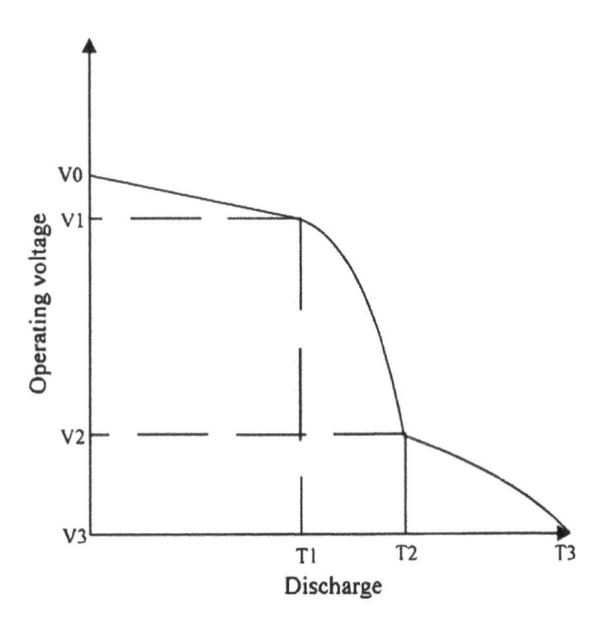

Fig. 2 Battery curve of the sensor node

terminal voltage disintegrates on the decrease in residual energy. The battery is rated in Ampere-Hour (Ah), increased loading of a node causes increased current intake from the battery. This results in decreased voltage, which in turn increases the current intake making it a chain effect. Figure 2 illustrates the Voltage curve of the battery

Battery curve depicts the voltage (vs) discharge time, the terminal voltage of the battery decays on increasing discharge rate. Equation 1 describes the battery model.

$$E_t = a_1 \sin(b_1 x + c_1) + a_2 \sin(b_2 x + c_2) \tag{1}$$

where E_t: Energy trust, a_1, b_1, c_1, a_2, b_2, c_2: Curve constants.

Equation 1 describes the voltage (vs.) residual energy of the battery in the CM from Fig. 2. The equation is used for calculating the energy trust of the sensor node. When the point of E_t is within the region $V_0 \rightarrow V_1$ the Trust of the node is $0.5 < E_t < 1$, when it is within the region $V_1 \rightarrow V_2$ the Trust of the node is $0 < E_t < 0.5$, when it is within the region $V_2 \rightarrow V_3$ the Trust of the node is $-0.5 < E_t < 0$, when it is above the region V_3 the Trust of the node is $-1 < E_t < -0.5$. This trust distribution determines whether the node is suffering from lack of power. The unnecessary failure of the link due to lack of power is avoided by E_t. High Trust in E_t makes the node to extract less current from the node thereby increasing the lifetime of the network. This trust avoids the increased current intake, decrease the tissue damage to the subject in case of the implanted node. The Reliability trust is related to the data reliability and link strength between the sender and receiver. The Receiver Signal Strength Indicator (RSSI) signifies the strength of the connection between the sender and receiver. Figure 3 shows the

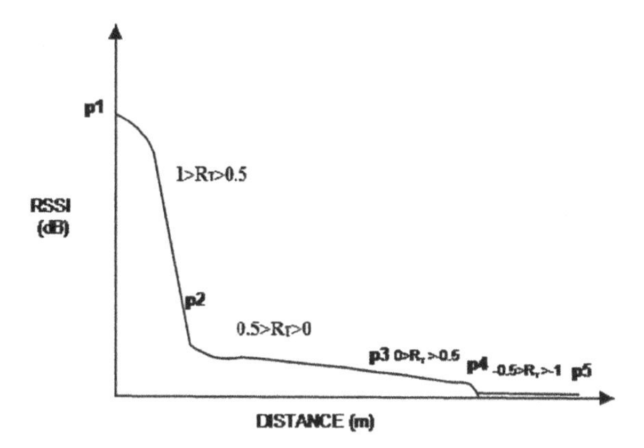

Fig. 3 Receiver Signal Strength Indicator (RSSI)

RSSI, the quality of signal between the receiver and sender. Higher the signal strength lesser the probability of failure of packet delivery [11].

The RSSI value decreases on increase in distance between the sender and receiver nodes. Equation 2 describes the signal strength of the sensor node with respect to distance.

$$R_t = e^{-\lambda d} \tag{2}$$

where R_t: Reliability trust, λ: Slope of the curve, d: distance between the sender and receiver.

When Rt is distributed in the curve as $1 > R_t > 0.5, 0.5 > R_t > 0, 0 > R_t > -0.5, -0.5 > R_t > -1$. Unnecessary data failure due to low link quality is avoided using this trust. Mobility trust is denoted by the movement of the node within the ROI. The data hopped through the moving node has the higher probability of loss of data. The immobile nodes have lesser probability of losing data. The packet containing the data of the implanted nodes are being hopped mainly through the immobile nodes. Figure 4 represents the walking speed of the patients with respect to their age. The fast moving nodes may lose its data because of its mobility.

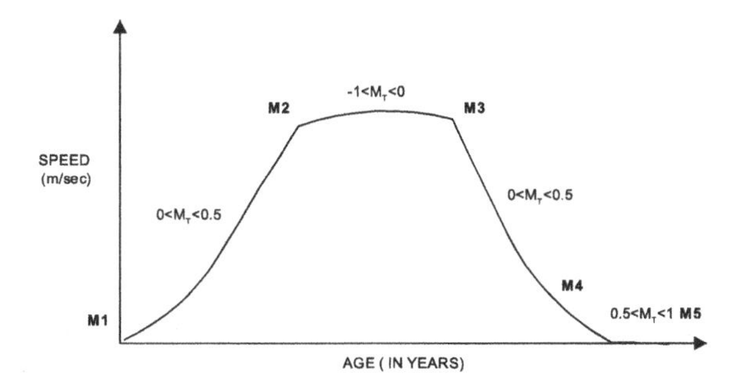

Fig. 4 Mobility curve of the patient

Given O_t=Σ(I_t, M_t, R_t,E_t)/4;O_t, I_t, M_t, R_t, E_t are trust values of the node;
The Initial Trust I_t of the node is Ratio of successful Interaction to Total Interaction;
While data is requested

1) Compute Initial Trust **I_t** of the neighbor nodes (Successive Interaction (I_s)/Total Interaction (I_T))
2) Compute Mobility Trust **M_t** of the neighbor nodes
3) Compute Reliability Trust **R_t** of the neighbor nodes
4) Compute Energy Trust **E_t** of the neighbor nodes and compute the Overall Trust **O_t**

for(P€NP)
Check for Overall trust;
if(Overall Trust >0.5) Communicate NP;
for(P€CP)*//CP is Critical Packet*
Check for Overall trust and Individual trust;
if(Overall Trust >0.5 && Individual Trust >0.5)
Communicate CP;
else
deny communication;
end all;

Fig. 5 FATD algorithm

Equation 3 represents the mobility trust Mt.

$$Mt = \alpha\beta^{-\alpha} x^{\alpha-1} e^{-(x/\beta)^{\wedge\alpha}} \tag{3}$$

where M_t: Mobility M_t of the patients inside the ROI is modeled.α, β: Shape parameter and scale parameter of the curve.

The trust values are exchanged between the Cluster Head to Cluster Head by hello packet communication. The sender sends the hello packet to the neighboring nodes and enquiries about the trust values of the nodes. The nodes reply its individual trust to the sender node. The sender collectively gathers the information and communicates the packets to the node as per the FATD algorithm in Fig. 5. The trust values are calculated as per the curves and CP are transmitted through the high trust path as given in FATD algorithm. When the trust value of the sensor node is more than 0.5 it is allowed to hop the NP and when both individual and overall trust is more than 0.5 the node is allowed to hop the CP.

4 Results and Discussions

The scenario is simulated using Matlab 7.0 with 100 nodes. The proposed FATD algorithm is compared with LEACH and QPRR algorithms. The one out of four packets is termed as CP. Figure 6 illustrates the lifetime comparison of the

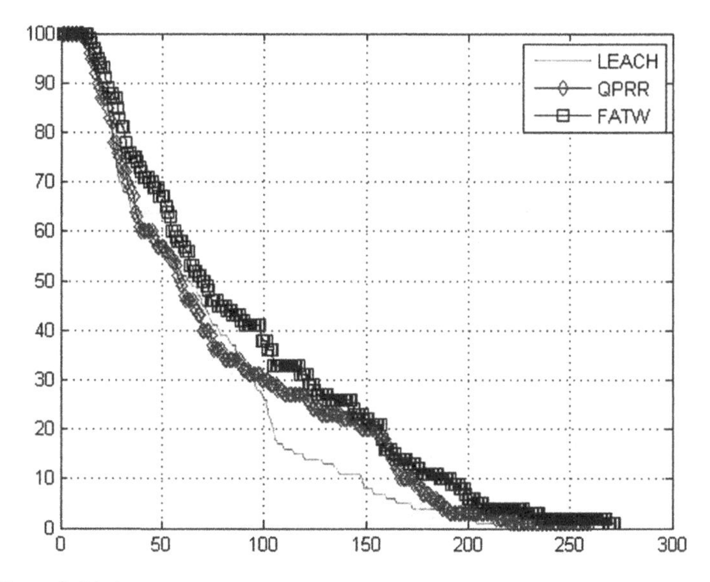

Fig. 6 Network lifetime

Fig. 7 Network throughput

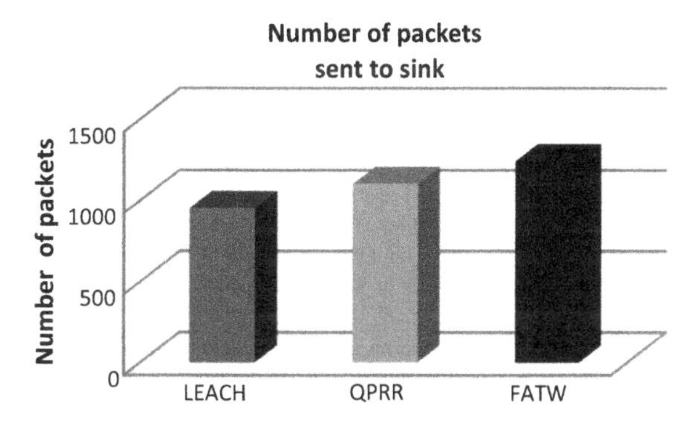

proposed algorithm with LEACH and QPRR algorithms. The proposed algorithm outperforms the existing algorithm by approximately 1.23 times.

Figure 7 illustrates the throughput comparison of the proposed algorithm with LEACH and QPRR algorithms. The proposed algorithm outperforms the LEACH and QPRR algorithms by 1.26 times.

5 Conclusion

The proposed algorithm serves to be the better solution for trust issues. The algorithm enhances the lifetime of the network and provides increased throughput to the network. The algorithm provides data transmission to the node with low mobility, thereby ensuring low drop packet and reduced energy utilization. The future work of this paper includes real time testing of the proposed algorithm.

References

1. Akyildiz, I.F., Sankarasubrararnanian, Su. W., Cayirci, W.E.: Wireless Sensor Networks:A Survey. Computer Networks, Elsevier Science, 393–422 (2002)
2. Kanagachidambaresan, G.R., SarmaDhulipala, V.R., Vanusha, D., Udhaya, M.S.: Matlab Based Modeling of Body Sensor Network Using ZigBee Protocol. CIIT 2011, CCIS 250, 773–776 (2011)
3. Kanagachidambaresan, G.R., Chitra, A.: Fail Safe Fault Tolerant mechanism for Wireless Body Sensor Network. Wireless Personal Communication. 78, 247–260 (2014)
4. Huawei Zhao., JingQin., JiankunHu.:An Energy Efficient Key Management Scheme For Body Sensor Networks. IEEE Transaction Parallel and Distributed systems. 24, 11 (Nov 2013)
5. Judge, P.: Hack Attacks Warning on Medical Implants. TechWeekEurope, http://www.techweekeurope.co.uk/news/hack-warning-on-medical-implants-72025, (2012)
6. Kanagachidambaresan, G.R., SarmaDhulipala, V.R., Udhaya, M.S.: Markovian Model Based Trustworthy Architecture. ICCTSD, ProcediaEngineering, Elseiver (2011)
7. Riaz Ahmed Shaikh., HassanJameel., Brian, J.d., Auriol, Heejo., SungyoungLee., Young-Jae Song.: Group-Based Trust Management Scheme for Clustered Wireless sensor network. IEEE Transactions on Parallel and Distributed Systems, 20, 11 (Nov 2009)
8. Wei Zhang., Sajal, K. Das., Younghe Liu.: A Trust Based Framework for Secure Data Aggregation in Wireless Sensor Networks. Proceedings of IEEE SECON (2006)
9. Momani, M., Challa, S., Alhmouz, R.: Can we Trust Trusted Nodes in Wireless Sensor Networks?. The International conference on Computer and Communication Engineering ICCCE -08, Kuala Lumpur, Malaysia (2008)
10. Khan, Z., Aslam, N., Sivakumar, S., Phillips, W.: Energy-aware peering routing protocol for indoor hospital body area network communication. Procedia Comput. Sci., 10, 188–196 (2012)
11. Khan, Z., Sivakumar, S., Phillips, W., Robertson, B.: QPRD: QoS-Aware Peering Routing Protocol for Delay Sensitive Data in Hospital Body Area Network Communication. In Proceedings of 7th International IEEE Conference on Broadband, Wireless Computing, Communication and Applications (BWCCA), 178–185 (November 2012)
12. Khan, Z.A., Sivakumar, S., Phillips, W., Robertson, B.: A QoS-aware routing protocol for reliability sensitive data in hospital body area networks. Procedia Comput. Sci, 19, 171–179 (2013)
13. Razzaque, M.A., Hong, C.S., Lee, S.: Data-centric multiobjective QoS-aware routing protocol for body sensor networks. Sensors, 11, 917–937 (2011)
14. Kanagachidambaresan, G.R., Chitra, A.: TA-FSFT Thermal Aware Fail Safe Fault Tolerant algorithm for Wireless Body Sensor Network (WBSN), Wireless Personal Communication, 89, 1–16 (2016)

Classification of ECG Signals Related to Paroxysmal Atrial Fibrillation

Shipra Saraswat, Geetika Srivastava and Sachidanand Shukla

Abstract Paroxysmal atrial fibrillation is a life threatening arrhythmia which leads to sudden cardiac death. Cardiac professionals are always looking to obtain a maximum accuracy in identifying and treating heart disorders. The new method of automatic feature extraction and classification of paroxysmal atrial fibrillation is proposed in this paper. The first step toward classifying paroxysmal disorder is to decompose the ECG signals (healthy and unhealthy) using wavelet transformation techniques. Corresponding to these decomposed levels, the values of ECG signals are computed on the basis of entropy by using the method of cross recurrence quantification analysis. The classification was implemented by probabilistic neural network (PNN) concept. Overall gained accuracy by using PNN classifier is 86.6%. The purpose of this work is to develop a smart method for the proper classification of paroxysmal AF arrhythmias. Long-Term AF Database (ltafdb) and MIT-BIH Fantasia Database (fantasia) have been chosen from Physio Bank ATM for carrying out this work.

Keywords Paroxysmal atrial fibrillation · Wavelet transformation
Cross recurrence quantification analysis · Probabilistic neural network classifier

S. Saraswat (✉) · G. Srivastava
Amity University, Noida, Uttar Pradesh, India
e-mail: sshipra1510@gmail.com

G. Srivastava
e-mail: geetika_gkp@rediffmail.com

S. Shukla
RML Avadh University, Faizabad, India
e-mail: sachida.shukla@gmail.com

A.K. Somani et al. (eds.), *Proceedings of First International Conference on Smart System, Innovations and Computing*, Smart Innovation, Systems and Technologies 79, https://doi.org/10.1007/978-981-10-5828-8_45

1 Introduction

For identifying any heart disorder, electrocardiograph (ECG) is a necessary and popular tool in order to achieve this goal. Paroxysmal atrial fibrillation (AF) is an abnormal heart rate described by speedy and asymmetrical beating. Sometimes, there may be heart palpitations, fainting and chest pain is realized by the patients of this disorder [1]. It can be stated as one type of supraventricular tachycardia. A typical normal ECG cycle is showing in Fig. 1. An ECG report having Paroxysmal AF consists no P wave. If there is an absence of P wave while extracting the features of ECG signals, it may provide a warning of paroxysmal AF. The aim of this paper is to classify ECG signals and to make a proper separation between healthy and unhealthy signals. In order to achieve this, feature extraction is performed and by making the use of extracted features, signals are classified on the basis of decomposed levels. For carrying out the purpose of this work, three lead ECG systems were used.

2 Related Work

The study based on the incremental benefit of vagal denervation by radiofrequency for preventing AF. This study was based on 297 patients underwent CPVA test and recommends adjunctive CVD during CPVA in order to reduce recurrence of AF at 12 months [2]. Research based on 80 patients having Paroxysmal AF aged around 10–52 years by using the pulmonary vein isolation has been shown in the paper [3].

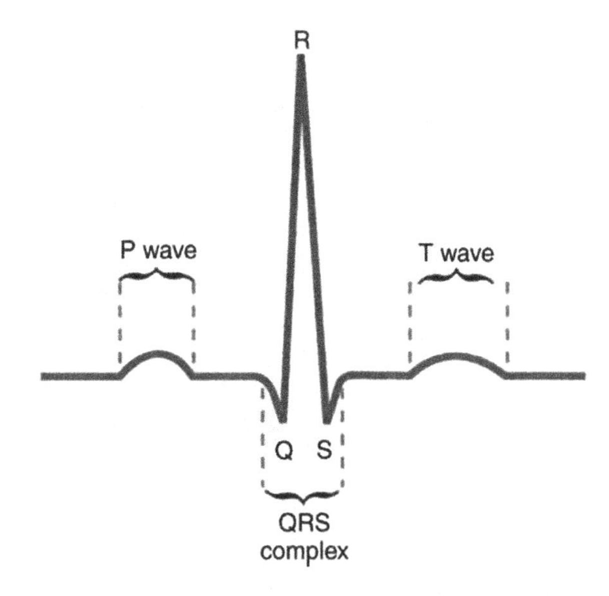

Fig. 1 Typical ECG cycle

Classification of PAF signals has been done with the help of 3D mapping and mean radiofrequency duration in order to achieve a fast automatic method. The results showed that pulmonary ablation could be helpful in patients with PAF if used by 3D mapping and double lasso technique [4]. Paroxysmal AF is characterized as the various schemes of classification in which various arrhythmic groups are not defined clearly. In paper [5], new methods and information about classification of AF have been discussed. These are capable in classifying atrial fibrillation. AF arrhythmias increases the risk factor of mortality, due to this, authors of paper [6] focuses the pathophysiology of paroxysmal arrhythmias and discusses the pharmacological treatment procedure. The malignant ventricular ectopic beats have been classified in the paper [7] by using the concept of wavelet transformation and probabilistic neural network method. The detection of QRS complex is an important step in recognition of any cardiac disorder [8].

The authors of this paper propose the DWT with RQA method for feature extraction of experimental ECG signals in order to classify paroxysmal atrial fibrillation. For experimental analysis, authors have taken the ECG signals from MIT-BIH database. Section 3 contains the research methodology used in this work, Sect. 4 deals with results and discussion and finally conclusion and future work are given in Sect. 5.

3 Methodology Used

3.1 Discrete Wavelet Transformation

The discrete wavelet technique is very popular and multipurpose tool in the field of signal processing for solving the complex scientific and engineering problems and it is gaining popularity in the field of biomedical applications. It has a giant number of applications in almost every area like engineering, science, mathematics, artificial intelligence, and many more. In this work, DWT is used for decomposing the ECG signals (unhealthy and healthy). The decomposition process is performed till the fifth level of resolution for extracting the signals. The output of wavelet filter analysis method is based on two discrete words approximations and details [9]. The Paroxysmal AF and normal signals are analyzed through a series of filters used for decomposition.

$$Y[n] = (X * g)[n] = \sum_{k=-\infty}^{\infty} X[k]g[n-k] \tag{1}$$

After this, the filter bank theory is used for generating the filter outputs of PAF and normal signals and perform the down sampling by 2. The formulas used for this analysis is given below where g and h represent low and high pass filters, respectively.

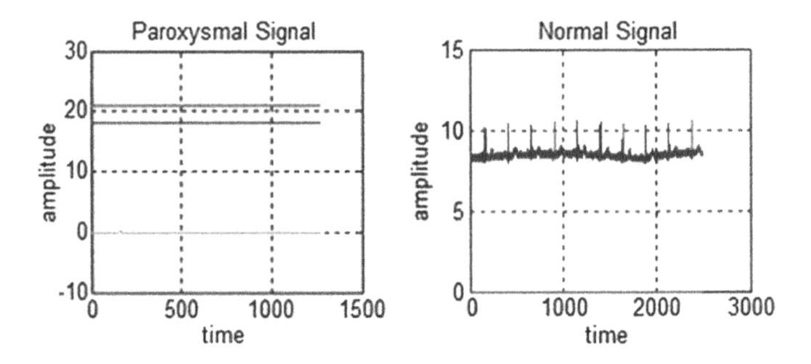

Fig. 2 Decomposed experimental ECG signals (Paroxysmal and normal)

$$Y_{low}[n] = \sum_{k=-\infty}^{\infty} x[k]\, g[2n - k] \quad (2)$$

$$Y_{high}[n] = \sum_{k=-\infty}^{\infty} x[k]\, h[2n - k] \quad (3)$$

The decomposition tree produces outputs in terms of approximate coefficients, i.e., (AC1–AC5) and detailed coefficients (DC1–DC5) after passing the signal through high pass and low pass filters and then down sampled. Following Fig. 2 represents the decomposed experimental ECG signals (Paroxysmal and normal) and in Figs. 3 and 4 authors showing the approximate and detailed levels of decomposition of Paroxysmal AF (unhealthy) and normal signals, respectively.

3.2 *Cross Recurrence Quantification Analysis*

RQA is a versatile tool involved in the observation of nonlinear data and identifies vibrant changes. Data analysis can be performed on Cross Recurrence Plot (CRP) toolbox which is especially designed to work only on Matlab. This toolbox is used for evaluating the parameters of quantification analysis in order to determine any anomalies in the ECG signal [10]. Out of various parameters, authors chose the entropy as a basis parameter for carrying out this work. The entropy of decomposed Paroxysmal and normal signals can be estimated from a probability value which depends on a diagonal line length L. The formula for calculating the entropy values of decomposed signals at fifth level are as follows and p (l) shows the distribution of frequency at length L.

$$E = -\sum_{l=l_{min}}^{N} p(l) \ln p(l) \quad (4)$$

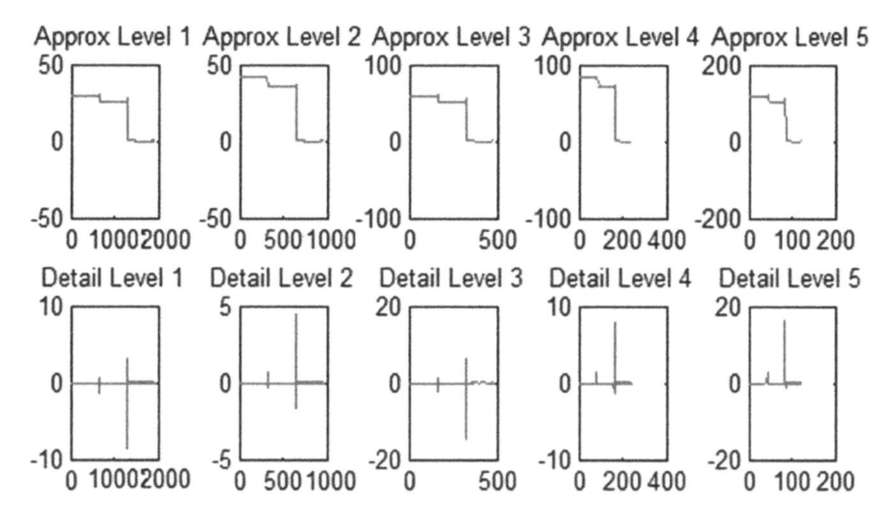

Fig. 3 Approximate and detailed coefficients of Paroxysmal AF signal (unhealthy)

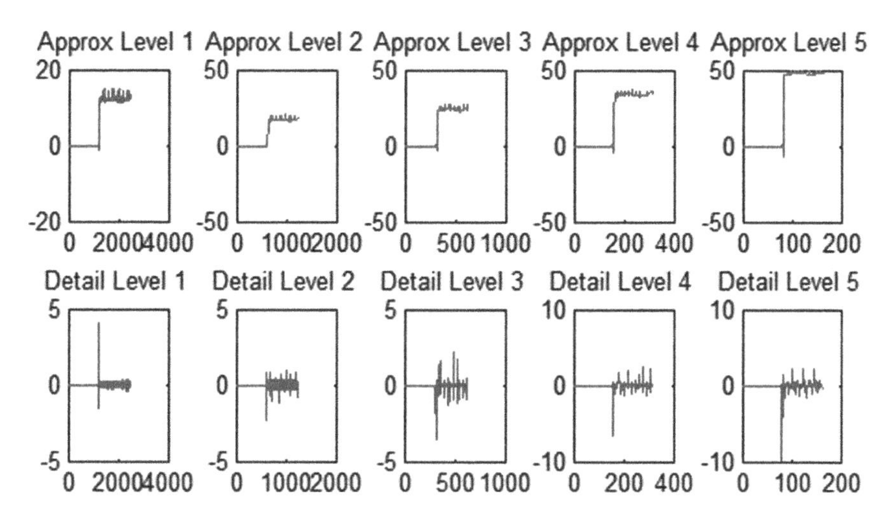

Fig. 4 Approximate and detailed coefficients of normal signal (healthy)

Following tables represents the entropy values by using the CRQA method. Table 1 showing the values of 10 ECG signals corresponding to its approximate coefficients and Table 2 producing the entropy values of 10 ECG signals related to detailed coefficients, where unhealthy signals represent paroxysmal AF and healthy signals represent normal persons.

Table 1 Entropy values of ten ECG signals on five approximate levels of decomposition

ECG signals	AC1	AC2	AC3	AC4	AC5
Unhealthy signal 1	6.4603	5.7708	4.4008	3.7255	4.6501
Unhealthy signal 2	6.4614	5.7708	5.0853	4.4008	3.7311
Unhealthy signal 3	7.0971	6.4077	5.7146	5.0361	4.1635
Unhealthy signal 4	6.4603	5.7708	5.0824	4.4008	3.7255
Unhealthy signal 5	7.0971	6.4077	5.7146	5.0361	4.3485
Healthy signal 1	0.3077	0.3471	0.3637	0.2856	0.3544
Healthy signal 2	0.3883	0.4321	0.3970	0.4025	0.2819
Healthy signal 3	0.3152	0.2852	0.3330	0.3392	0.2925
Healthy signal 4	0.4369	0.4237	0.3932	0.3965	0.3326
Healthy signal 5	0.3167	0.2589	0.2457	0.4568	0.2456

Table 2 Entropy values of ten ECG signals on five detailed levels of decomposition

ECG signals	DC1	DC2	DC3	DC4	DC5
Unhealthy signal 1	4.6501	4.5024	5.0664	4.3517	3.6376
Unhealthy signal 2	5.3263	5.3536	5.0661	4.3547	3.6544
Unhealthy signal 3	4.3485	4.2467	5.0808	4.4806	3.8747
Unhealthy signal 4	4.0707	4.0266	5.0663	4.3517	3.6376
Unhealthy signal 5	3.2261	3.2273	5.0805	4.4086	3.8747
Healthy signal 1	0.4073	0.2833	0.2590	0.3484	0.2775
Healthy signal 2	0.3216	0.2453	0.3144	0.2733	0.4609
Healthy signal 3	0.2871	0.2575	0.2467	0.2141	0.1870
Healthy signal 4	0.2639	0.2734	0.2638	0.2562	0.2420
Healthy Signal 5	0.3456	0.5688	0.4678	0.3567	0.4667

3.3 *Probabilistic Neural Network*

PNN was announced by a famous researcher named D.F. Specht planned for the classification process using Bayesian network. A Probabilistic Neural Network (PNN) concept includes four layers. They are input layer, pattern layer, summation layer, and output layer for performing the operations of feedforward neural network [11]. This method is used basically for classifications and is very famous for its fast learning capability. The authors used PNN classifier for evaluating the performance of proposed method. The training and testing of ten ECG signals have been performed on the basis of calculated entropy values by using the concept of PNN.

4 Results and Discussion

For realizing the efficiency of this proposed method, the output of PNN classifier has been evaluated on the basis of following formulas [12, 13]. They are defined as:

$$\text{Hit rate (HR)} = \text{TP}/(\text{TP} + \text{FN})$$
$$\text{False Alarm Rate (FAR)} = \text{FP}/(\text{FP} + \text{TN})$$
$$\text{TN Ratio (TNR)} = \text{TN}/(\text{TN} + \text{FP})$$
$$\text{Missed Hypothesis (MH)} = \text{FN}/(\text{FN} + \text{TP})$$
$$\text{Accuracy (AC)} = \text{TP} + \text{TN}/(\text{TP} + \text{TN} + \text{FP} + \text{FN})$$

where TP, TN, FP, FN represents true positive, true negative, false positive, and false negative, respectively. Hit rate describes the number of paroxysmal signals correctly identified, false alarm rate is called as healthy signals recognized as unhealthy, TN ratio represents the normal signals classified correctly and missed hypothesis described the unhealthy signals identified as healthy ones and accuracy is defined as the sum of true positive and true negative by the sum of total number of cases. The authors of this paper have performed the experimental analysis on total 30 ECG signals including 15 signals each. In Table 3, authors showing the results of ten ECG signals generated with the help of Matlab programming while training and testing process of PNN classifier and verified their results on the basis of confusion matrix which is shown in Table 4.

Out of 30 ECG signals, authors have taken 15 signals of diseased persons and 15 signals of non-diseased persons. The accuracy of the proposed method in this paper can be calculated by using the concept of confusion matrix. The results generated by using Probabilistic neural network method has been shown with the help of confusion matrix in Table 4.

Table 3 Classification results based on PNN classifier

ECG signals	Training					Testing				
	HR (%)	FAR (%)	TNR (%)	MH (%)	AC (%)	HR (%)	FAR (%)	TNR (%)	MH (%)	AC (%)
Unhealthy signal 1	100	3.53	89.26	0.00	89.92	73.85	24.29	74.29	24.75	71.52
Unhealthy signal 2	98.46	10.90	89.10	1.54	93.78	90.46	26.67	73.33	9.54	87.50
Unhealthy signal 3	94.86	3.50	90.33	4.80	88.30	71.00	29.00	66.00	26.67	66.00
Unhealthy signal 4	100	4.86	89.33	0.00	93.17	65.86	27.00	68.67	30.17	62.00
Unhealthy signal 5	100	0.00	86.67	0.00	86.67	80.60	28.00	58.00	18.60	73.28
Healthy signal 1	100	0.00	100	0.00	100	92.00	60.00	40.00	8.00	89.96
Healthy signal 2	92.05	9.36	90.64	7.95	91.35	86.52	34.67	65.33	13.48	82.92
Healthy signal 3	86.99	20.15	79.85	13.01	83.42	72.58	37.14	62.86	27.42	69.62
Healthy signal 4	100	0.00	100	0.00	100	92.00	60.00	40.00	8.00	89.96
Healthy signal 5	97.78	22.22	77.78	2.22	87.78	90.68	61.33	38.67	9.32	83.44

Table 4 Confusion matrix with results

Unhealthy	Healthy	Total ECG signals
True positive = 13	False positive = 2	Total number of positive results = 15
False negative = 2	True negative = 13	Total number of negative results = 15
Total number of diseased persons = 15	Total number of non-diseased persons = 15	Total cases = 30

5 Conclusion and Future Work

This paper proposes a unique combination of method for classifying paroxysmal atrial fibrillation arrhythmias. Authors decompose the ECG signals for feature extraction process and then calculated the entropy values at each decomposed level by using the concept of recurrence quantification. The obtained entropy values are fed into PNN classifier for training and testing the signals. The accuracies of PNN classifier while training and testing are 93.60 and 79.60%, and the overall accuracy achieved by using this method is 86.6%. The findings of this paper may be useful while treating the patients of paroxysmal atrial fibrillation arrhythmias. The future work of this paper can be extended for developing another model for the identification of various cardiac disorders.

References

1. Lip, G. Y., Hee, F. L. S.: Paroxysmal atrial fibrillation. QJM. Vol. 94, pp. 665–678. (2001).
2. Pappone, C., Santinelli, V., Manguso, F., Vicedomini, G., Gugliotta, F., Augello, G., Mazzone, P., Tortoriello, V., Landoni, G., Zangrillo, A., Lang, C.: Pulmonary vein denervation enhances long-term benefit after circumferential ablation for paroxysmal atrial fibrillation. Circulation American Heart Association. Vol. 109, pp. 327–334. (2004).
3. Oral, H., Scharf, C., Chugh, A., Hall, B., Cheung, P., Good, E., Veerareddy, S., Pelosi, F., Morady, F.: Catheter ablation for paroxysmal atrial fibrillation segmental pulmonary vein ostial ablation versus left atrial ablation. Circulation, vol.108, pp. 2355–2360. (2003).
4. Ouyang, F., Bänsch, D., Ernst, S., Schaumann, A., Hachiya, H., Chen, M., Chun, J., Falk, P., Khanedani, A., Antz, M, Kuck, K.H.: Complete isolation of left atrium surrounding the pulmonary veins new insights from the double-Lasso technique in paroxysmal atrial fibrillation. Circulation, Vol. 110, pp. 2090–2096. (2004).
5. Gallagher, M. M., Camm, J.: Classification of atrial fibrillation: The American journal of cardiology, Vol. 82, pp. 18 N-28 N. (1998).
6. Markides, V. Schilling, R. J.: Atrial fibrillation: classification, pathophysiology, mechanisms and drug treatment. Heart, Vol. 89, pp. 939–943. (2003).
7. Saraswat, S., G. Srivastava, S.N. Shukla.: Malignant Ventricular Ectopy Classification using Wavelet Transformation and Probabilistic Neural Network Classifier. Indian Journal of Science and Technology, Vol. 9, pp. 1–5. (2016).
8. Saraswat, S., G. Srivastava, S.N. Shukla.: Review: Comparison of QRS detection algorithms. IEEE International Conference on Computing, Communication & Automation (ICCCA), pp. 354–359. (2015).
9. Khorrami, H., & Moavenian, M. A.: comparative study of DWT, CWT and DCT transformations in ECG arrhythmias classification. Expert systems with Applications, Vol. 37, pp. 5751–5757. (2010).
10. Marwan, N., & Kurths, J.: Cross recurrence plots and their applications. Mathematical Physics Research at the Cutting Edge, CV Benton Editor, pp. 101–139. (2004).
11. Specht, D. F.: Probabilistic neural networks. Neural networks, vol. *3*, pp. 109–118. (1990).
12. Ge, D., Srinivasan, N., & Krishnan, S. M.: Cardiac arrhythmia classification using autoregressive modeling. Biomedical engineering online, *1*(1), 1. (2002).
13. Yu, S. N., & Chou, K. T.: Integration of independent component analysis and neural networks for ECG beat classification. Expert Systems with Applications, Vol. 34, pp. 2841–2846. (2008).
14. Addison, P. S.: Wavelet transforms and the ECG: a review. Physiological measurement, Vol. *26*, R155. (2005).
15. Prasad, G. K., & Sahambi, J. S.: Classification of ECG arrhythmias using multi-resolution analysis and neural networks. In TENCON. Conference on Convergent Technologies for the Asia-Pacific Region. Vol. 1, pp. 227–231. IEEE. (2003).
16. Al-Fahoum, A. S., & Howitt, I.: Combined wavelet transformation and radial basis neural networks for classifying life-threatening cardiac arrhythmias. Medical & biological engineering & computing, Vol. 37, 566–573. (1999).
17. Marwan, N., Romano, M. C., Thiel, M., & Kurths, J.: Recurrence plots for the analysis of complex systems. Physics reports, Vol. 438, pp. 237–329. (2007).

An Image Encryption Scheme Using Chaotic Sequence for Pixel Scrambling and DFrFT

Harish Sharma and Narendra Khatri

Abstract Now a day digitalization is everywhere in the world, and there is a huge amount of data transfer takes place every second on the internet. In this digital world major contribution of data as Images. The entire data transfer will take place on the internet, i.e., a public network, so the security of these Image data is very important. Image encryption can be done through different methods, these are cryptographic algorithms and various transforms were used for Image encryption. In this paper, an innovative Image encryption algorithm is proposed using Image Pixel Scrambling and DFrFT. The Image pixel scrambling operation is performed using the sequence generated by the chaotic function. The scrambled Image is encrypted through DFrFT. The developed algorithm improves the information security and the robustness of the system. The effectiveness and the efficiency of the proposed algorithm are verified from the simulation results.

Keywords Image scrambling · Chaotic sequence · Image decryption · MSE

1 Introduction

Importance of encryption comes in the existence with in the development of computer network and internet. Data theft and hacking incidences continuously increasing on the internet. So security of the information over the network is very important. Encryption is one of approach through we can secure information. Huge amount of data transfer on the internet in the form of Images. Image encryption process is used to ensure the security of the Image during the transmission over the computer network/internet.

H. Sharma (✉)
Department of Computer Science and Engineering, Manipal University, Jaipur, India
e-mail: harish27j@gmail.com

N. Khatri
Department of MME, The LNM Institute of Information Technology, Jaipur, India
e-mail: narkhatri@gmail.com

A.K. Somani et al. (eds.), *Proceedings of First International Conference on Smart System, Innovations and Computing*, Smart Innovation, Systems and Technologies 79, https://doi.org/10.1007/978-981-10-5828-8_46

Various Image encryption algorithm based on the DFrFT [1], wavelet packet decomposition [2, 3] with DFrFT has used for the Image encryption. DFrFT has used for Image encryption. This paper proposes the combination of the Image pixel scrambling with the DFrFT. The combination of the Image pixel scrambling using the chaotic sequence and DFrFT enhance the security in terms of key Size. The application of DFrFT increased recently in the research due to digitalization internationally.

In this paper the combination of the Image scrambling and DFrFT. The Combination raises the key dimensions making the Image encryption more robust and significant. Simulation results are presented.

Organization of the paper is as follows. In the Sect. 2, we briefly review the details of Image pixel scrambling through the chaotic sequence. In Sect. 3, DFrFT and its application on Digital Image. In Sect. 4, proposed algorithm for the Image encryption. In Sect. 5 Simulation results of the proposed algorithm and Sect. 6 paper conclusion.

2 Image Scrambling

Image pixel scrambling technique is used for making the digital Image into the undetectable Image; it is also used in the Image watermarking. Image scrambling is the transformation of the digital Image into a vain Image to improve the security of the Image from the various attacks [4]. Image scrambling can be accomplished in three domains, i.e., space, frequency, and color or gray domain.

Different kind of Image scrambling algorithms were proposed, one of the important type is chaos-based Image scrambling algorithm. The key space available with chaos-based Image scrambling algorithm is large. More number of key enhances the security of the image.

2.1 Image Scrambling Algorithm

Image scrambling performed is based on the chaos theory. Image is scrambled using the chaotic iteration and then the sorting.

Digital Image I can be scrambled using an uninformed chaotic iteration

$$y_{n+1} = f(y_n), y_i \in A \tag{1}$$

The chaotic function adopted is 1-dimension Logistic mapping used [5, 6]

$$y_{n+1} = 1 - 2y_n^2 \quad y \in \{-1, 1\} \tag{2}$$

is used for generation of the chaotic sequence of real numbers. Statistical analysis is developed by selecting the size of Image I is 256 × 256 pixels. First, the value of x_1 is selected and it acts as a secret key. Then the following procedure is adopted for scrambling and unscrambling of digital Image I.

2.1.1 Scrambling Algorithm

Step 1. Initialize the scrambling key y_1 adopt $k = 1$.
Step 2. The chaotic iteration (1), Iterate it N − 1 times, it generate real number sequence $\{y_1, y_2, \ldots, y_N\}$.
Step 3. Real numbers Sequence $\{y_1, y_2, \ldots, y_N\}$ is sorted ascending order, sorted sequence is as follows $\{y_{¢1}, y_{¢2,} \ldots, y_{¢N}\}$.
Step 4. Scrambling address codes set calculated of $\{s_1, s_2, \ldots, s_N\}$, where $s_i \widehat{I}\{1, 2, \ldots, N\}$. s_i is the new subscript of y_i in the sorted sequence $\{y_{¢1}, y_{¢2,} \ldots, y_{¢N}\}$.
Step 5. Permute the kth row of the Image I with permuting address code $\{s_1, s_2, \ldots, s_N\}$, namely, replace the s_ith column pixel with the ith column pixel for i from 1 to N.
Step 6. If $k = M$, terminate. Else, let $y_1 = y_N$, and $k = k + 1$. Repeat the step 2–5.

2.1.2 Unscrambling Algorithm

The procedure for unscrambling the scrambled Image is same as scrambling, using the key element y_1 with it's initial value. Except step 5 changed: the ith column pixel with the s_ith column pixel for i from 1 to N.

3 DFrFT

DFrFT defined using the close form of the discrete fractional Fourier transform. The kernel matrix of the DFrFT is calculated for the image encryption is in [7].

Proposed algorithm implements the fast algorithm for digital computation of an approximation of continuous transform. Using the digital computation technique for computation enabled the DFrFT's implementation in discrete domain of image encryption.

Conventional Fourier transform's one of the generalization is the fractional Fourier transform. A Distinct property of the DFrFT is time-frequency analysis, and it depends on order of transform p. It is taken as the rotation in the Wigner distribution of signals, mathematically the angle for rotation is ($\alpha = p\pi/2$).

Mathematical definition of kernel matrix of DFrFT is as follows [8–12]:

$$F_{N,\alpha} = F^{2\alpha/\pi} = \widehat{V}_N D_N^{2\alpha/\pi} \widehat{V}_N^T \tag{3}$$

$$= \begin{cases} \sum_{k=0}^{N-1} e^{-jk\alpha} \widehat{V}_k \widehat{V}_k^T \\ \sum_{k=0}^{N-2} e^{-jk\alpha} \widehat{V}_k \widehat{V}_k^T + e^{-jN\alpha} \widehat{V}_N \widehat{V}_N^T \end{cases} \tag{4}$$

where

$$\widehat{V} = \left[\widehat{V}_0 \middle| \widehat{V}_1 \middle| \ldots \middle| \widehat{V}_{N-3} \middle| \widehat{V}_{N-1} \right], \text{ for odd N} \tag{5}$$

$$\widehat{V} = \left[\widehat{V}_0 \middle| \widehat{V}_1 \middle| \ldots \middle| \widehat{V}_{N-2} \middle| \widehat{V}_N \right], \text{for even N} \tag{6}$$

Here normalized eigenvector V_k is equivalent to Hermite function's *k*th-order [13].

4 Proposed Algorithm for the Image Encryption

Image encryption algorithm using the Image pixel scrambling and DFrFT is as follows:

Step 1: image scrambling is performed as in the scrambling algorithm

Step 2: generate the DFrFT matrix using order of the transform p.

Step 3: apply DFrFT matrix to the Scrambled Image to encrypt the Image.

5 Results and Analysis

5.1 Parameters for Analysis

Parametric analysis of the proposed algorithm using Image pixel scrambling and DFrFT is developed using two parameters, i.e., histogram and mean square error.

5.1.1 Histogram

Histogram is a demonstration of the rate of change in the gray scale pixel values. Histogram of an Image represents the intensity variation. The histogram for the encrypted image is random, the reason for the randomness is the non-correlation in the image pixel after the scrambling operation.

5.1.2 Mean Square Error

Mean square error is calculated to analyze the sensitivity of the order of transform with the quality of Image encryption. MSE is calculated between the input image and the decrypted image after image encryption.

Analysis of sensitivity is developed by varying the order of transform at the decryption side by order of 0.01, and the limit for that variation of order is –0.5 to +0.5.

Mathematically the MSE [14, 15] is calculated as follows:

$$\mathrm{MSE} = \frac{1}{M \times N} \sum_{j=1}^{N} \sum_{i=1}^{M} |I_D(i, j) - I_E(i, j)|^2 \tag{7}$$

where the value of M is the no. of rows, N is no of column comes from the size of the Image. $I_D(i, j)$ and $I_E(i, j)$ are the representation of the decrypted Image pixel and the input Image pixel values at particular position defined by (i, j), respectively.

Input Image used for the simulation is gray scale Lena Image of 256 × 256, as shown in Fig. 1a. Input Image is scrambled using the chaotic sequence generated then after the scrambled Image is encrypted using the kernel matrix of the DFrFT [16]. Figure 1b represents the scrambled Image and Fig. 1c represents the encrypted Image.

Figure 2a represents histogram of input Image and Fig. 2b represents histogram of the scrambled Image and Fig. 2c represents histogram of the encrypted Image

Mathematical Analysis of the encryption quality can be done by calculating the mean square error. Mean square error is calculated between the input Image and the decrypted Image after the encryption of the input image.

Figure 3 represents the relative variation in the order of transform with mean square error (MSE).

Max mean square error with the change in order of transform is 9237.3 with the minimum change of order is 0.001 and the range of change selected is –0.5 to 0.5.

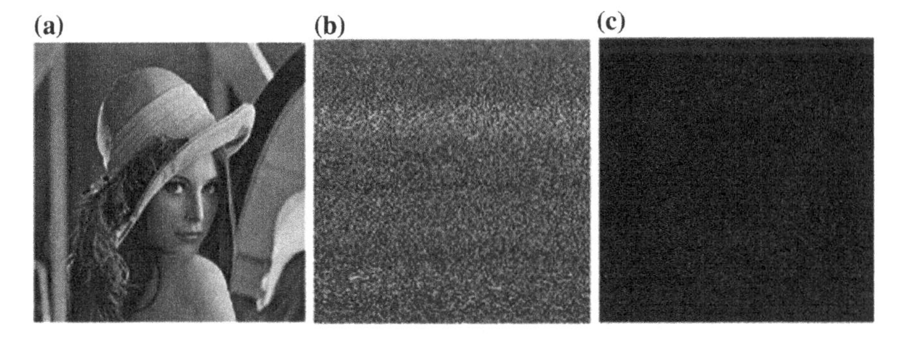

Fig. 1 Images **a** Input **b** Scrambled **c** Encrypted

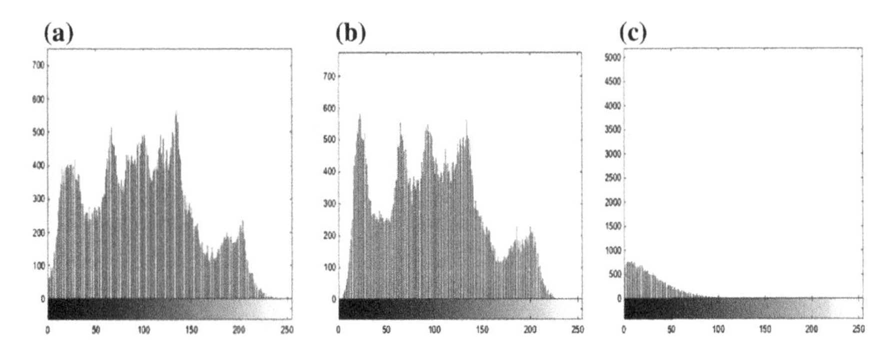

Fig. 2 Histogram of images **a** Input **b** Scrambled **c** Encrypted

Fig. 3 Change in the order of transform with mean square error (MSE)

6 Conclusion

Image encryption is done for achieving high security. Using Image pixel scrambling and DFrFT enhanced the key space for the security of the Image and the quality of encryption is good. This can be verified from the results.

References

1. X. Feng, X. Tian and S. Xia, "A novel Image encryption algorithm based on fractional fourier transform and magic cube rotation," *2011 4th International Congress on Image and Signal Processing*, Shanghai, 2011, pp. 1008–1011. doi:10.1109/CISP.2011.6100319
2. W. Zheng, Z. G. Cheng and Y. l. Cui, "Image Data Encryption and Hiding Based on Wavelet Packet Transform and Bit Planes Decomposition," *2008 4th International Conference on Wireless Communications, Networking and Mobile Computing*, Dalian, 2008, pp. 1–4. doi:10.1109/WiCom.2008.773
3. K. K. Sharmaand Heena Jain, "Image encryption algorithm based on wavelet packet decomposition and discrete linear canonical transform", Proc. SPIE8760, International Conference on Communication and Electronics System Design, 87601P (January 28, 2013); doi:10.1117/12.2012323
4. H. Y. Zhang, "A New Image Scrambling Algorithm Based on Queue Transformation," *2007 International Conference on Machine Learning and Cybernetics*, Hong Kong, 2007, pp. 1526–1530. doi:10.1109/ICMLC.2007.4370387
5. L. Xiang dong, Z. Junxing, Z. Jinhai, H. Xiqin, "Image Scrambling Algorithm Based on Chaos Theory and Sorting Transformation", in International Journal of Computer Science and Network Security, VOL.8 No.1, 2008, pp. 64–68
6. Hai-Yan Zhang, "A new Image scrambling algorithm," *2008 International Conference on Machine Learning and Cybernetics*, Kunming, 2008, pp. 1088–1092. doi:10.1109/ICMLC.2008.4620566
7. A. Koc, H. M. Ozaktas, C. Candan and M. A. Kutay, "Digital Computation of Linear Canonical Transforms," in *IEEE Transactions on Signal Processing*, vol. 56, no. 6, pp. 2383–2394, June 2008. doi:10.1109/TSP.2007.912890
8. S. Oraintara, "The generalized DFrFTs," *2002 IEEE International Conference on Acoustics, Speech, and Signal Processing*, Orlando, FL, USA, 2002, pp. II-1185-II-1188. doi:10.1109/ICASSP.2002.5744012
9. X. Kang, F. Zhang and R. Tao, "Multichannel Random DFrFT," in *IEEE Signal Processing Letters*, vol. 22, no. 9, pp. 1340–1344, Sept. 2015. doi:10.1109/LSP.2015.2402395
10. S. C. Pei and W. L. Hsue, "Random DFrFT," in*IEEE Signal Processing Letters*, vol. 16, no. 12, pp. 1015–1018, Dec. 2009. doi:10.1109/LSP.2009.2027646
11. N. Jindal and K. Singh, "Image Encryption Using Discrete Fractional Transforms," *2010 International Conference on Advances in Recent Technologies in Communication and Computing*, Kottayam, 2010, pp. 165–167. doi:10.1109/ARTCom.2010.9
12. Soo-Chang Pei and Wen-Liang Hsue, "The multiple-parameter DFrFT", in *IEEE Signal Processing Letters*, vol. 13, no. 6, pp. 329–332, June 2006. doi:10.1109/LSP.2006.871721
13. Jae Wan Shim and Ren´ee Gatignol, "How to obtain higher-order multivariate Hermite expansion of Maxwell–Boltzmann distribution by using Taylor expansion?" in R. Z. Angew. Math. Phys. (2013) 64: 473. doi:10.1007/s00033-012-0265-1
14. H. Agrawal, D. Kalot, A. Jain and N. Kahtri, "Image encryption using various transforms-a brief comparative analysis," *2014 Annual International Conference on Emerging Research Areas: Magnetics, Machines and Drives (AICERA/iCMMD)*, Kottayam, 2014, pp. 1–4. doi:10.1109/AICERA.2014.6908280
15. N. Khatri and N. Agrawal, "Double Image encryption using double pixel scrambling and linear canonical transform," *2013 Fourth International Conference on Computing, Communications and Networking Technologies (ICCCNT)*, Tiruchengode, 2013, pp. 1–5. doi:10.1109/ICCCNT.2013.6726546
16. X. Luo, J. Fan and J. Wu, "Single-channel color Image encryption based on the multiple-order DFrFT and chaotic scrambling," *2012 IEEE International Conference on Information Science and Technology*, Hubei, 2012, pp. 780–784. doi:10.1109/ICIST.2012.6221754

Dynamic Route Optimization Using Nature-Inspired Algorithms in IoV

Nitika Chowdhary and Pankaj Deep Kaur

Abstract Internet of Vehicles (IoV) has gained immense popularity with increasing research on fully integrated smart cities. The live data collectively gathered by various types of sensors or cameras installed on running vehicles, traffic lights, etc., forms an integral part of smart city ecosystem. Various emergency situations require vehicles to divert from their respective default routes to some more feasible and optimal routes. In this work, we implement two well-known algorithms Ant Colony Optimization and Particle Swarm Optimization to allow a coordinated dynamic route customization among the vehicles for an overall optimal traffic management. A centralized decision-making module is implemented by applying ACO and PSO on the data that continuously gathered from the vehicles in the live environment. The experimental setup consists of two well-known simulator software tools SUMO and NS2 that are accompanied by TraNS to provide mutual interactions between them. The results confirm an increase in the overall performance of the system with application of the optimization approaches.

Keywords IoV · Optimizing route · Ant colony · Particle swarm

1 Introduction

The increase in the popularity of self-sustainable smart cities along with Internet of Things (IoT) has led to the advancement in the existing technological infrastructure. IoT defines a seamlessly integrated environment where virtually everything communicates with each other and maintains a coordinated environment [1]. Various applications supported by IoT include smart industrial control smart city, smart

N. Chowdhary (✉) · P.D. Kaur
Department of Computer Science and Engineering,
GNDU RC Jalandhar, Jalandhar, India
e-mail: nitikachowdhary@ieee.org

P.D. Kaur
e-mail: pankajdeep.csejal@gndu.ac.in

A.K. Somani et al. (eds.), *Proceedings of First International Conference on Smart System, Innovations and Computing*, Smart Innovation, Systems and Technologies 79, https://doi.org/10.1007/978-981-10-5828-8_47

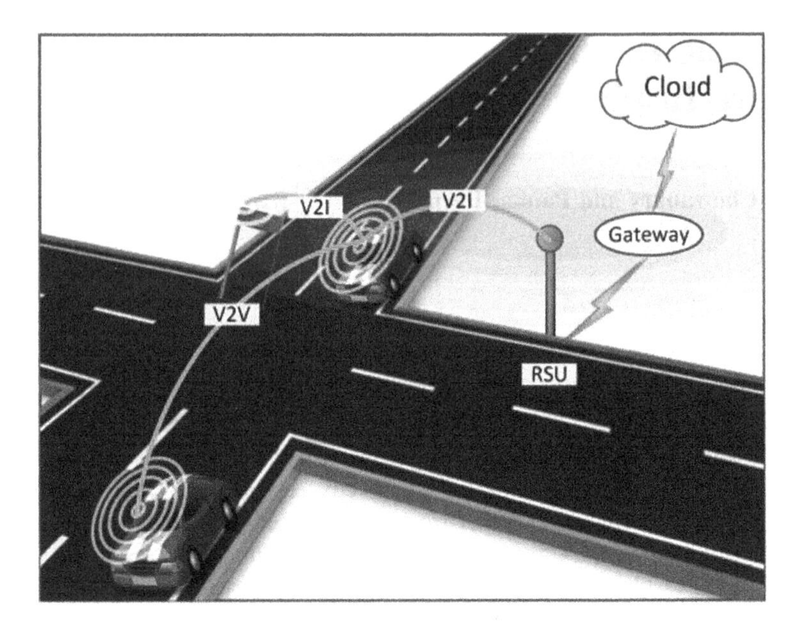

Fig. 1 Architecture of Internet of Vehicles

metering smart water and smart logistics. A part of IoT, where all the vehicles are interconnected to each other as well as other entities like traffic lights and pedestrians, (as shown in Fig. 1) is popularly known as Internet of Vehicles (IoV).

The continuous movement of vehicles generates the humungous amount of data that needs to be processed by high-end infrastructure. Keeping this in mind, the whole ecosystem is well connected to the cloud through gateways. This setup may assist in the centralized decision-making process that is applicable throughout the system under consideration. Many vital decisions like emergency vehicle passing, disaster management, dynamic route management, etc. This may also aid in providing the vehicles an optimal route toward their destinations. This requires a constant process of data gathering, route computation, and data dissemination. This paper presents a unified route optimization approach based on Ant Colony Optimization [2] and Particle Swarm Optimization [3] algorithm to enhance the overall delay in the network traffic. The routes are dynamically computed to provide the vehicles with updated and optimal routes toward their respective destinations. Three different road attributes are taken into consideration while evaluating the cost a vehicle inherits while traveling through that road.

The organization of the paper is given below. A brief background and related work are presented in Sect. 2. The optimization approach and solution methodology are detailed in Sect. 3. Sections 4 and 5 present the experimental setup used in this study and its corresponding results, respectively. The paper is concluded in Sect. 6.

2 Background and Literature Review

There are a number of elements that form a complete system to comprise a Smart City. It could provide a sustainable growth as well as economic benefits to the inhabitant in various aspects of their day-to-day routines. Sensors and camera play an important role when collecting the data from the environment. The modern day vehicles are also now equipped with various types of smart sensors to provide the capability of advanced sensing like temperature, object closeness, etc. Vehicles in IoV are themselves capable of processing and storing the information instead of acting as an intermediate node. The cameras installed on a vehicle could provide a 360° view of the current status of the traffic around it. However, the major advantage of IoV is to provide various kinds of services by taking into consideration the information pertaining to the whole ecosystem. The subsequent sub-sections discuss the need of dynamic routing to improve the state of traffic congestion in a network.

2.1 Background

Routing issues have always prevailed in a system that is filled with moving nodes. A unanimous decision-making is required so as to provide an optimal path taking into consideration not only s single node but also the whole network. This becomes challenging with an increase in the size (region of deployment) and density (number of moving nodes) of the system. There exist various complex parameters when dealing with such situation in a real world environment. This work aims at computing a set of feasible and optimal routes so as to provide an overall congestion free environment. Below we discuss some of the state-of-art literature available on this issue.

2.2 Literature Review

Decades ago, the problem of route optimization considered only *distance* as the input parameter to compute path toward the destination [4]. However, it was clear after further research that considering only this parameter is not yielding suitable results. Earlier the distribution of road networks was sparse. There were only few connection points available between two places. But with the rise in population and the road connectivity, it became complex to only rely on few parameters. McGinty [5] introduced two new metrics (path quality and user satisfaction) in order to provide with a more comprehensive solution to the roaring problem. Moving further, Choi et al. [6] designed routing mechanisms keeping driver the prime focus. However, a complete view of the network was necessary to provide with a suitable route. This led to the use of various traditional algorithms like A-star, Dijkstra to solve the above-said problem. The study by Wang [7] on the other hand depicted the ineffectiveness of these traditional algorithms in finding out optimal routes under complex and variable parameters.

To support logistical operation, Dethloff [8, 9] designed an equation that is supposed to provide a minimal total distance being traveled in relation to maximum vehicle capacity. A meta-heuristic-based approach for VRP was in the beginning presented by Crispim and Brandao [10] that combined Tabu Search and Variable Neighborhood Search algorithms. Other meta-heuristics-based methodologies include Chen and Wu [11], and Bianchessi and Righini [12].

Sahoo et al. [13] used Ant Colony Optimization technique for optimizing the vehicle routes in VANETs. They combined ACO with zone-based clustering to further enhance the routing results. AbdAllah et al. [14] solve dynamic routing problem by applying multiple modifications to the genetic algorithm. Their modified genetic algorithm outperformed a number of previously published literature works available on dynamic route management. Oranj et al. [15] designed an algorithm for efficient routing based on ant colony optimization and DYMO (Dynamic MANET On-demand) protocol. Their algorithm can effectively monitor changes in the environment to perform re-routing of vehicles. Pan et al. [16] solved the problem issues pertaining to the privacy of a vehicle location and intensive computation resulting in increasing vehicles. They proposed a scheme named DIVERT, which can effectively avoid forming any type of congestion in the network by performing dynamic re-routing of the selected vehicles.

3 Solution Methodology

Various bio-inspired optimization algorithms exist in the literature that has successfully been implemented under various domains. This section discusses the different road factors that were taken into consideration while evaluating the optimality of possible routes. This is followed by a brief presentation on the functioning of Ant Colony Optimization method.

3.1 Road Factors

To assess the elements of a road network, we define three metrics namely quality, length, and congestion based on which the optimal route is computed. These three metrics are defined below.

- Quality, Ω: The road quality depends on various factors like potholes, no. of lanes, etc. This makes it an important metric that should definitely be considered while computing an optimal route.

- Length, ꓒ: Multiple length road elements could coexist between two junctions. Selection purely depends on the shorter but high-quality road.
- Congestion, ℂ: Some road elements are busier than the other due to which the preference of an optimal route might vary. Longer but less congested routes are sometimes preferred more than a short and congested route.

We define a tuple with three metrics ⟨ꓒ, Ω, ℂ⟩, which is assigned to every road element in the network. The following fitness function is used for the optimization for the route between a specific starting and ending points.

$$\mathrm{F} = \sum_{i=1}^{n} \pm \begin{pmatrix} C_1^i & \cdots & C_m^i \\ \vdots & \ddots & \vdots \\ C_m^i & \cdots & C_m^i \end{pmatrix} \tag{1}$$

Here n is the number of metrics (which is three), $C(x, y)$ defines the ith metric value of the road element starting from point x to point y.

3.2 Optimization Algorithms

ACO. It is based on a meta-heuristic approach to solve combinatorial NP-hard optimization problems. Dynamic traffic variations aid in finding the shortest path between two junctions. The algorithm for ACO is shown in Algorithm 1. The optimization process follows the given process. n ants are initialized at the starting. Each ant is assigned random starting point in the applicable domain. Ants iteratively construct routes by constantly traversing every a new graph vertex using the probabilistic rule of nearest neighbor transition. When the route has been constructed completely, pheromone quantity is changed to enhance the best route.

PSO. This optimization algorithm is based on the behavior of fish schooling or flocking birds. Each particle represents a point carrying two values, velocity and location. Every particle gains knowledge from their neighbors. This algorithm initializes with random particle count. This is followed by varying values of position and velocity to search optimum solution. With each update in location, objective function sampling is performed. After a few iterations, all the particles converge to an optimal location in the search space that defined the optimal solution.

4 Experimental Setup

The experimental portion of this study is divided into two phases. In the first phase, a specific section of the map is selected followed by the extraction of the road network. Figure 2 represents the experimentation flow diagram followed by this study.

The second phase comprises of the actual implementation where the data gathered is used to compute optimized routes for vehicles.

Algorithm 1: Pseudocode for Ant Colony System

```
01: Input: Prob_Size, Populationsize, n, α, γ, Θ, g0
02: Output: O
03: O ← Solution (Prob_Size);
04: Ocost ← CostCalculation (Ph);
05: Init ← 1.0 / (Prob_size X O_cost);
06: Pheromone ← Initialize(init);
07: while Stop() do
08:   for x = 1 to n do
09:     Pi ← Solution(Pheromone, Prob_Size, γ, g0);
10:     Pxcost ← CostCalculation (Px);
11:     if Pxcost ≤ Ocost then
12:       Ocost ← Pxcost;
13:       O← Px;
14:     end
15:     LocalDecay (Pheromone, Px, Pxcost, Θ);
16:   end
17:   GlobalDecay (Pheromone, Pbest,Ocost, α);
18: end
19: return O;
```

The complete experimental setup comprised the integration of two open source tools NS2[1] and SUMO[2] using TraNS.[3] In order to provide support for cloud, GreenCloud is utilized where all the processing of data takes place. We took two time windows, long (T_L) and short (T_S) time windows, with 30 and 120 s, respectively. The data collection is initialized after every short time interval and optimal route computation is performed on that data after every long time window. Therefore, any re-routes, if desired are decided only after the completion of the long time window. Various communication parterres and other related simulation parameters are listed in Table 1.

The map selection contains a very busy portion of *New York* city. This selection is converted to a road network defined by a graph with various nodes and edges. As shown in Fig. 3, each edge corresponds to any one of the road elements with nodes as the end points. Visibly, there are a number of different options available for a driver to follow till the destination is reached.

[1]www.isi.edu/nsnam/ns/.

[2]www.sumo.dlr.de/.

[3]http://lca.epfl.ch/projects/trans/.

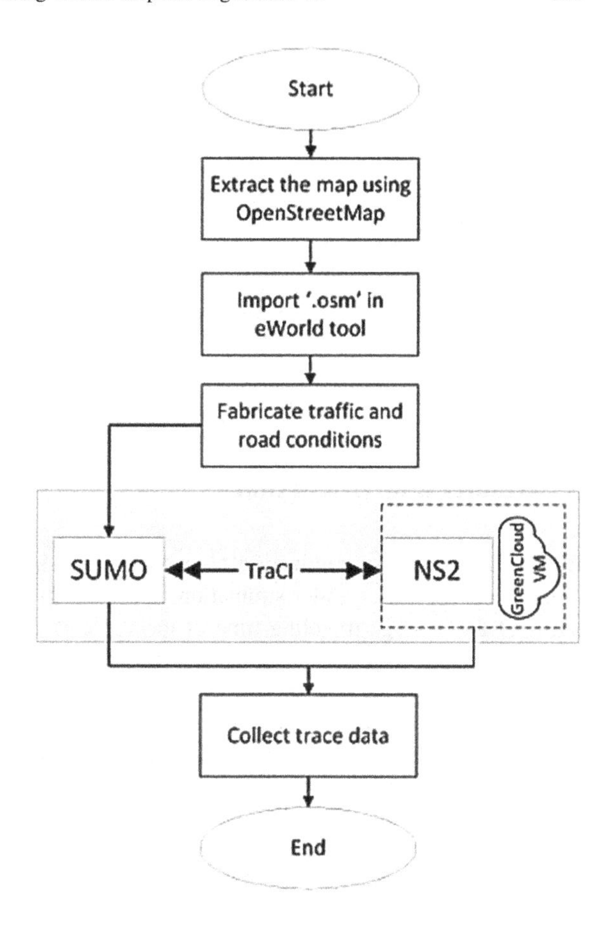

Fig. 2 Flow diagram of experimental setup

Table 1 Simulation parameters

Parameter	Value
Technology	ITS-G5/IEEE 80211.p
Frequency band	10 MHz with 5.9 GHz
Fading	Log-distance
Tx power	12 dBm
T_L length	120 s
T_S length	30 s
TSD, TSM size	100 bytes
Vehicle average speed	20–40 m/s
Simulation time	2000 s

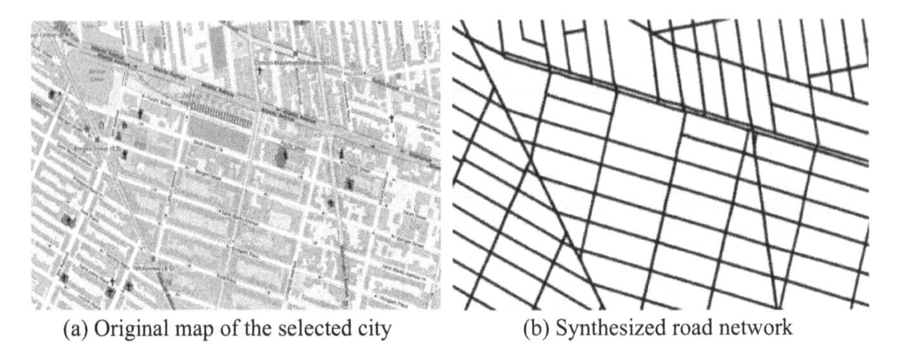

(a) Original map of the selected city (b) Synthesized road network

Fig. 3 Extraction of road network from the city map

5 Results and Discussion

Both of the applied optimization algorithms provided results better than the randomly selected routes under simulation. ACO has given better results than PSO in terms of the average traveling time of the vehicles. The randomly selected routes will eventually result in congestion with time. However, when congestion is detected on a specific or a set of road elements, the dynamic route estimated using the considered algorithms provides an updated path toward the destination. With the new routes, vehicles are more likely to avoid congestion and reach their destinations earlier than the usual timings.

Figure 4 shows the travel times of vehicles for the system under the application of ACO and PSO for different iterations in the simulation. ACO comprehensively outperforms PSO as shown in Fig. 4a. As the number of vehicles grows in the city, average time gradually increases due to possible congestions and effect on the average speed, as shown in Fig. 4b. ACO performs better in maintaining the

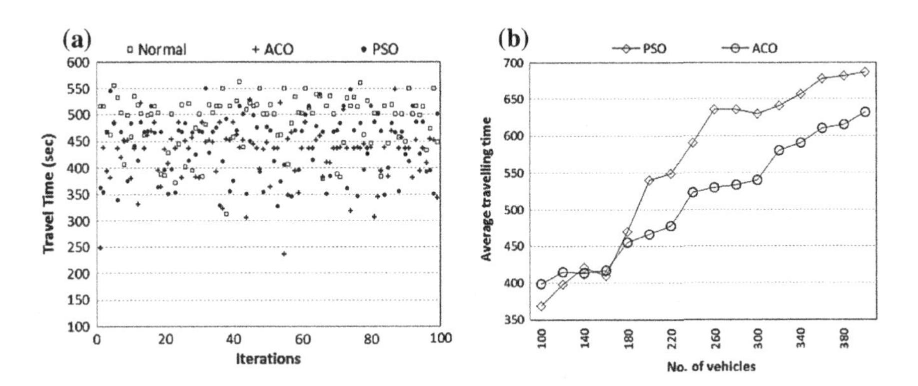

Fig. 4 **a** Travel time of vehicles during different simulation iterations. **b** Average travel time with an increase in the number of vehicles

average traveling time with the increase in the number of vehicles. ACO also converges fast in contrast to PSO to provide with the optimal routes. The vehicles, therefore, will receive quicker updates about the route in case of the former as compared to the former.

6 Conclusion

Various domains have adopted bio-inspired algorithms to solve optimizations problems. Dynamic route management becomes crucial in various situations like congestions, earthquakes, floods, etc. This work comprehensively evaluates the performance of two well-known bio-inspired algorithms ACO and PSO to estimate possibilities of better available routes toward different destinations. The simulation is primarily executed with fabricated traffic conditions using an integration of SUMO and NS2. The results depicted superior performance of Ant colony optimization over the Particle Swarm Optimization in decreasing the average traveling time of a vehicle by providing continuously updated routes. In future, we will examine a few other bio-inspired algorithms and extend its applicability to a bigger section of a map.

References

1. Medagliani, P., Leguay, J., Duda, A., Rousseau, F., Duquennoy, S., Raza, S., Ferrari, G., Gonizzi, P., Cirani, S., Veltri, L. and Monton, M.,: Internet of Things Applications-From Research and Innovation to Market Deployment. The River Publishers, pp. 287–313 (2014)
2. Jun, M., Xingzhi, T. and Wenxia, X.: Study on VRP based on improved ant colony optimization and internet of vehicles. In: IEEE Conference and Expo on Transportation Electrification Asia-Pacific, pp. 1–6 (2014)
3. Kaiwartya, O., Kumar, S., Lobiyal, D.K., Tiwari, P.K., Abdullah, A.H. and Hassan, A.N.: Multiobjective Dynamic Vehicle Routing Problem and Time Seed Based Solution Using Particle Swarm Optimization. Journal of Sensors, vol. 2015 (2015)
4. Christofides, N. and Eilon, S.: An Algorithm for the Vehicle-dispatching Problem. Journal of Operational Research Society, vol. 20, no. 3, pp. 309–318 (1969)
5. McGinty, L. and Smyth, B.: Personalised Route Planning: A Case-Based Approach. In: Advances in Case-Based Reasoning, E. Blanzieri and L. Portinale, Eds. Springer Berlin Heidelberg, pp. 431–443 (2000)
6. Choi, W.K., Kim, S.J., Kang, T.G. and Jeon, H.T.: Study on Method of Route Choice Problem Based on User Preference. In: Italian Workshop on Neural Networks Conference on Knowledge-based Intelligent Information and Engineering Systems, Berlin, Heidelberg, pp. 645–652 (2007).
7. Wang, Z. and Crowcroft, J.: Quality-of-service routing for supporting multimedia applications. IEEE Journal on Selected Areas in Communications, vol. 14, pp. 1228–1234 (1996)
8. Dethloff, J.: Vehicle routing and reverse logistics: The vehicle routing problem with simultaneous delivery and pick-up. OR Spektrum, vol. 23, no. 1, pp. 79–96 (2001)
9. Dethloff, J.: Relation between Vehicle Routing Problems: An Insertion Heuristic for the Vehicle Routing Problem with Simultaneous Delivery and Pick-Up Applied to the Vehicle

Routing Problem with Backhauls. The Journal of the Operational Research Society, vol. 53, no. 1, pp. 115–118 (2002)

10. Crispim, J. and Brandão, J.: Metaheuristics applied to mixed and simultaneous extensions of vehicle routing problems with backhauls. Journal of Operational Research Society, vol. 56, no. 11, pp. 1296–1302 (2005)
11. Chen, J.-F. and Wu, T.-H.: Vehicle Routing Problem with Simultaneous Deliveries and Pickups. The Journal of the Operational Research Society, vol. 57, pp. 579–587 (2006)
12. Bianchessi, N. and Righini, G.: Heuristic algorithms for the vehicle routing problem with simultaneous pick-up and delivery. Computers & Operations Research, vol. 34, no. 2, pp. 578–594 (2007)
13. Sahoo, A., Swain, S.K., Pattanayak, B.K. and Mohanty, M.N.: An optimized cluster based routing technique in VANET for next generation network. In: Information Systems Design and Intelligent Applications, Springer India, pp. 667–675 (2016)
14. AbdAllah, A.M.F., Essam, D.L. and Sarker, R.A..: On Solving Periodic Re-Optimization Dynamic Vehicle Routing Problems. Applied Soft Computing (2017)
15. Oranj, A.M., Alguliev, R.M., Yusifov, F. and Jamali, S.: Routing algorithm for vehicular ad hoc network based on dynamic ant colony optimization. Int. J. Electron. Elect. Eng., vol. 4, no. 1, pp. 79–83 (2016)
16. Pan, J.S., Popa, I.S. and Borcea, C.: Divert: A distributed vehicular traffic re-routing system for congestion avoidance. IEEE Transactions on Mobile Computing, vol. 16, no. 1, pp. 58–72 (2017)

A Novel Cross Correlation-Based Approach for Handwritten Gujarati Character Recognition

Ankit K. Sharma, Dipak M. Adhyaru and Tanish H. Zaveri

Abstract One of the major reasons for poor recognition rate in handwritten character recognition is the lack of unique features to represent handwritten characters. In this paper, an attempt is made to utilize the similarity already exist in different parts of the Gujarati characters. A novel feature extraction technique based on normalized cross correlation is proposed for handwritten Gujarati character recognition. An overall accuracy of 53.12%, 68.53%, and 66.43% is obtained using Naive Bayes classifier, linear and polynomial Support Vector Machine (SVM) classifiers, respectively, with the proposed feature extraction algorithm. Experimental results show significant contribution by proposed technique and improvement in recognition rate may be obtained by combining these features with some other significant features. One of the significant contributions of proposed work is the development of large and representative dataset of 20,500 isolated handwritten Gujarati characters.

Keywords Normalized cross correlation · Gujarati · Naive Bayes classification
Support vector machine

A.K. Sharma (✉) · D.M. Adhyaru
Instrumentation and Control Engineering Department, Institute of Technology,
Nirma University, Ahmedabad, Gujarat, India
e-mail: ankit.sharma@nirmauni.ac.in

D.M. Adhyaru
e-mail: dipak.adhyaru@nirmauni.ac.in

T.H. Zaveri
Electronics and Communication Engineering Department, Institute of Technology,
Nirma University, Ahmedabad, Gujarat, India
e-mail: ztanish@nirmauni.ac.in

A.K. Somani et al. (eds.), *Proceedings of First International Conference on Smart System, Innovations and Computing*, Smart Innovation, Systems and Technologies 79, https://doi.org/10.1007/978-981-10-5828-8_48

1 Introduction

The Optical Character Recognition (OCR) system is mainly used to classify optical patterns, which are contained in the digital image, corresponding to alphabets, numbers, and other symbols. OCR system can read the information from paper documents and convert it to a standardized digital format. The main application of character recognition systems for handwritten Gujarati script will be to convert a scanned handwritten document in an editable format. This can also be used for digitizing historical documents.

Gujarati language is mainly evolved from Sanskrit and a widely spoken native language. In spite of 65 million speakers of this language across the world, we are still not completely successful in recognition of handwritten Gujarati characters. This might be because of its complex shape and different style of writing which makes it difficult for recognition. Few of the Gujarati characters have much similar shape, which also create difficulty in recognition [1].

The development of OCR system for Gujarati script has recently been a serious area of research because of its large market potential. This paper aims at the utilization of a novel feature extraction technique based on normalized cross correlation for handwritten Gujarati character recognition.

Handwritten Gujarati characters considered for proposed work are shown in Fig. 1. The rest of the paper is organized as follows. Section 2 provides some details regarding previous attempts in handwritten Gujarati numerals and character recognition. Section 5 describes classifiers adapted for the classification process. Experiment results are discussed in Sect. 6 and conclusion is provided in Sect. 7.

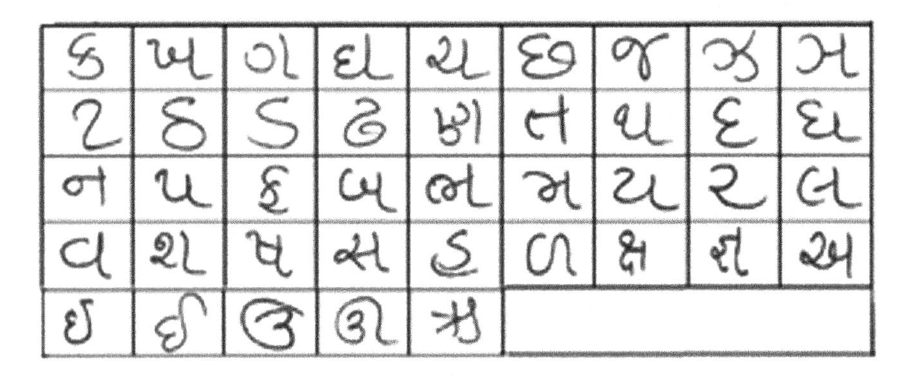

Fig. 1 Gujarati characters considered for proposed work

2 Review of Related Work

In other Indian languages like Hindi, Kannada, Bangla, Tamil, Telugu there has been tremendous progress in the area of handwritten character recognition as compared to Gujarati language [2–6]. Research work in the area of handwritten Gujarati character recognition is very limited.

First attempt in the area of handwritten Gujarati numerals recognition is made by Desai [7]. Recognition of handwritten Gujarati numerals is attempted through horizontal, vertical, and two diagonal projection profiles and Artificial Neural Network. Maloo and Kale [8] made an effort for handwritten Gujarati character recognition by using affine invariant moments-based features with Support Vector Machine classifier.

Desai [9] utilized image subdivision-based feature extraction approach for handwritten Gujarati characters. An accuracy of 86.66% is achieved with support vector machine classifier. This one is the only noticeable work in the area of Gujarati handwritten character recognition. Authors presented a comparative analysis of zoning-based methods for handwritten Gujarati numerals [10]. Gujarati numeral recognition is attempted using $16 \times 16, 8 \times 8, 4 \times 4$, and 2×2 zoning with neural network classifier. Goswami and Mitra [11] proposed a low-level stroke features extraction technique for the recognition of handwritten numerals from Gujarati and Devanagari scripts.

3 Dataset Generation and Preprocessing

Generation of a good database that represents the wide variation of handwriting styles is one of the most challenging aspects of handwritten character recognition. One of the biggest obstructions for the research in the area of handwritten characters recognition for Gujarati script is the unavailability of benchmark database. One of the significant contributions of proposed work is the development of large and representative dataset for handwritten characters. Gujarati handwritten character dataset of 20,500 images (500 images of each character) is generated. As shown in Fig. 1, total 41 Gujarati characters are considered for proposed work.

A special data set collection form was provided to the writers for dataset generation. Gujarati handwritten data collection form is shown in Fig. 2. For dataset collection, writers having different education levels and professions are considered. The purpose of data collection was not disclosed to writers in order to let them write in their natural handwriting style. Equal proportions of male and female writers are maintained while form filling.

Handwritten forms are digitized using a flatbed scanner at 300 dpi resolution in color format. The scanned images of forms are stored in JPEG format for future use. Segmentation algorithm is applied on all the form images in order to obtain isolated characters. Total 21,500 isolated character images are generated from these forms,

Fig. 2 Special form designed for handwritten database collection

which includes 500 images of each character from 500 writers. These images are used for the training and testing purpose.

Preprocessing of the isolated handwritten character images is required before going for the feature extraction and classification. Preprocessing steps are adapted in order to convert all the handwritten character images into uniform form. Preprocessing includes binarization, median filtering, resizing, and thinning operations. Otsu's thresholding algorithm [12] is used to convert isolated character images into binary images. In order to remove any salt and pepper noise from the binary image, Median filtering is applied. All the character images are resized to the size of 48×48 pixels with bilinear interpolation algorithm. Morphological thinning operation is applied on resized images, in order to obtain one pixel wide thinned images.

4 Feature Extraction

After dataset generation and preprocessing, next step is to extract unique features from all dataset images and represent the character images in form of unique feature vectors. Many techniques are developed for extracting features from images for pattern recognition purpose, but very few feature extraction techniques are feasible for handwritten characters recognition of Indian scripts. The proposed feature extraction methodology is based on generating feature vector based on a calculation of normalized cross correlation coefficient between the different half parts of the character image. Different half parts of the character image considered for feature

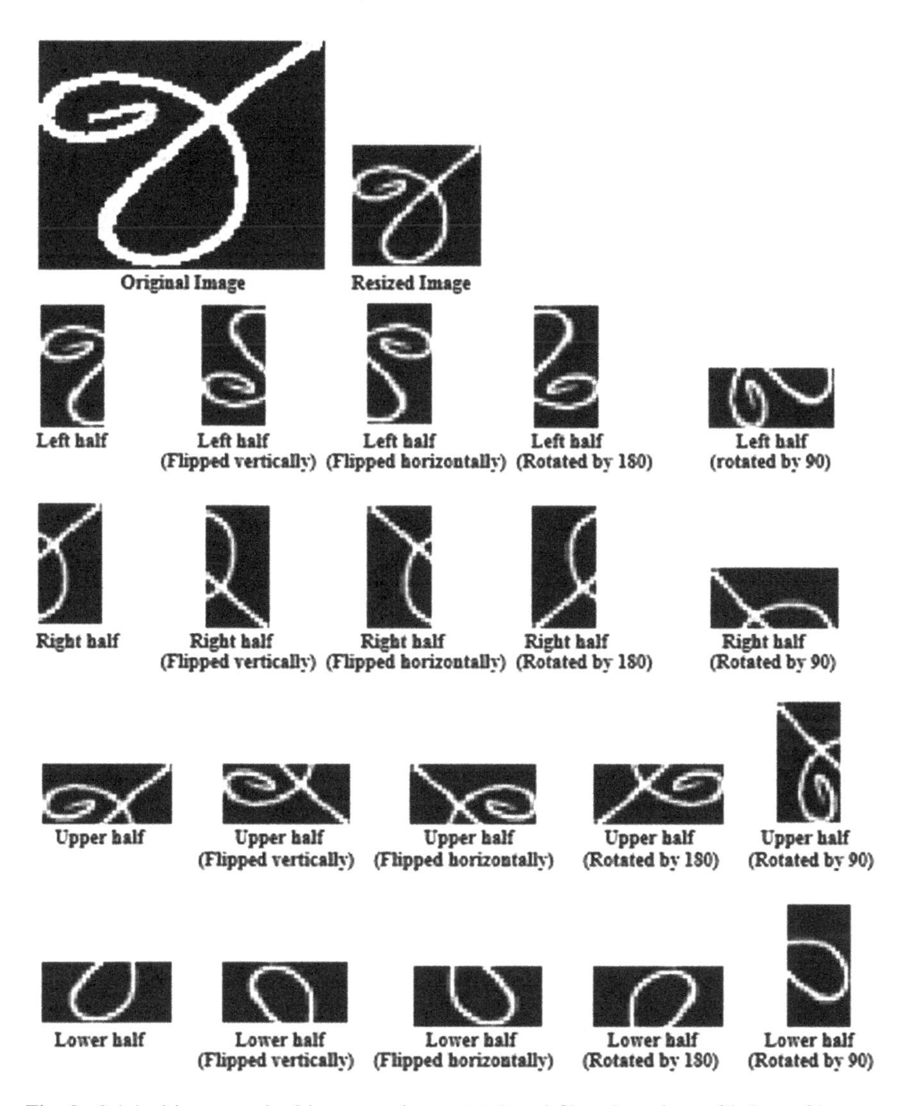

Fig. 3 Original image, resized image, various rotated, and flipped versions of halves of image

vector generation are shown in Fig. 3. In order to maintain uniformity, all the character images were resized to 48 × 48.

Cross correlation between two images is a well-known tool for matching two images giving the degree up to which they are similar [13]. For extracting feature, using normalized cross correlation, two components: source image and template image are required. Here, source and template image will be the two half portions of the character image (Table 1).

Table 1 Pairs of half parts of the characters images considered for calculation of normalized cross correlation coefficient in order to generate feature vector

Sr. no.	First half of character image	Second half of character image
1	Upper half	Lower half
2	Upper half	Upper half (vertically flipped)
3	Upper half	Upper half (horizontally flipped)
4	Upper half	Upper half (rotated by 180°)
5	Upper half	Lower half (vertically flipped)
6	Upper half	Lower half (horizontally flipped)
7	Upper half	Lower half (rotated by 180°)
8	Lower half	Lower half (vertically flipped)
9	Lower half	Lower half (horizontally flipped)
10	Lower half	Lower half (rotated by 180°)
11	Lower half	Upper half (vertically flipped)
12	Lower half	Upper half (horizontally flipped)
13	Lower half	Upper half (rotated by 180°)
14	Left Half	Right half
15	Left Half	Left half (vertically flipped)
16	Left Half	Left half (horizontally flipped)
17	Left half	Left half (rotated by 180°)
18	Left half	Right half (vertically flipped)
19	Left half	Right half (horizontally flipped)
20	Left half	Right half (rotated by 180°)
21	Right half	Right half (vertically flipped)
22	Right half	Right half (horizontally flipped)
23	Right half	Right half (rotated by 180°)
24	Right half	Left half (vertically flipped)
25	Right half	Left half (horizontally flipped)
26	Right half	Left half (rotated by 180°)
27	Left half (rotated by 90°)	Upper half
28	Left half (rotated by 90°)	Lower half
29	Right half (rotated by 90°)	Upper half
30	Right half (rotated by 90°)	Lower half
31	Upper half (rotated by 90°)	Left half
32	Upper half (rotated by 90°)	Right half
33	Lower half (rotated by 90°)	Left half
34	Lower half (rotated by 90°)	Right half

Let A and B are two images with size $m \times n$ and $p \times q$, respectively. Here, $p \leq m$ and $q \leq n$, then the normalized cross correlation between A and B are defined as follows:

Table 2 SVM design parameters and their values tested in cross validation of training set

Parameters	Values
Degree of Kernel function (d)	2, 3, 4
Regularization parameter (c)	0.01, 0.1, 1, 10, 100

$$\lambda(x,y) = \frac{\sum_s \sum_t \delta_{A(x+s,y+t)}\delta_{B(s,t)}}{\sum_s \sum_t \delta^2_{A(x+s,y+t)} \sum_s \sum_t \delta^2_{B(s,t)}}$$

where

$$\delta_{A(x+s,y+t)} = A(x+s,y+t) - \bar{A}(x,y),\ \delta_{B(s,t)} = B(s,t) - \bar{B} \quad (1)$$

$$s \in \{1,2,3,\ldots,p\},\ \text{and } t \in \{1,2,3,\ldots,q\}.$$

$$x \in \{1,2,3,\ldots,m-p+1\},\ y \in \{1,2,3,\ldots,n-q+1\}$$

$$\bar{A}(x,y) = \frac{1}{pq}\sum_s \sum_t A(x+s,y+t)$$

$$\bar{B} = \frac{1}{pq}\sum_s \sum_t B(s,t)$$

For matching purpose, the image B slides over A and value of cross correlation coefficient, γ is calculated for each coordinate (x, y). Range of γ is [−1, +1]. After completing this calculation, the point which exhibits maximum γ is considered as γ_{max}. This maximum value γ_{max} was calculated between all the flipped and rotated versions of the preprocessed image, as per the pairs of images shown in Table 2. These set of all 34 values of γ_{max} formed the feature vector. Hence, for every image, a feature vector having 34 elements was obtained.

5 Classification

Naive Bayes and Support Vector Machines classifier was used for classification of handwritten Gujarati characters. SVM was employed with linear and polynomial kernels. For experimentation purpose, regularization constant (c) was considered as the design parameter for linear SVM. For polynomial SVM both the degree (d) of a kernel function and regularization constant (c) was considered as design parameters. Values of the parameters which were tested are summarized in Table 2. In order to decide the best values of design parameters, 5-fold cross validation of training set was performed. After identification of best values of design parameters, prediction models were configured according to these values.

Table 3 Performance of proposed feature extraction algorithm with Naive Bayes classifier, linear SVM, and polynomial SVM

Classifier	Best value of design parameters	Accuracy (%)
Naive Bayes	–	53.12
Linear SVM	c = 10	68.53
Polynomial SVM	c = 100, d = 2	66.43

6 Experiments and Results

Data set consists of 21,500 handwritten Gujarati characters were used for experimentation purpose. For Experimentation purpose, 5-fold cross validation was used, in which dataset was divided into 4/5 training and 1/5 test samples using stratified random sampling in every fold. Hence, 17,200 images were used for training purpose and 4300 images were used for testing purpose. Table 3 shows the performance of linear SVM, polynomial SVM and NB classifier with the normalized cross correlation-based feature. Best value of design parameters is also shown in the same table.

7 Conclusion

In this proposed work, a novel approach based on normalized cross correlation is attempted for recognition of handwritten Gujarati characters. The proposed technique is based on the similarity exist in different half portions of the Gujarati characters. Dataset consist of 21,500 handwritten Gujarati characters is generated for experimentation purpose. Naive Bayes and Support Vector Machines classifier is used for classification purpose. Highest accuracy of 68.53% is obtained by Linear SVM with proposed feature extraction algorithm. Experimental results show significant potential in proposed feature extraction algorithm and accuracy can be improved by combining proposed feature with other feature extraction techniques.

Acknowledgements The authors are thankful to Institute of Technology, Nirma University for their support to carry out this research.

References

1. M. Chaudhary, G. Shikkenawis, S. K. Mitra, and M. Goswami, "Similar looking Gujarati printed character recognition using Locality Preserving Projection and artificial neural networks," in International Conference on Emerging Applications of Information Technology, EAIT, 2012, pp. 153–156.
2. Pal, U., and B. B. Chaudhuri, "Indian script character recognition: a survey," Pattern Recogniion., pp. 1887–1899, 2004.

3. N. Sharma, U. Pal, F. Kimura and S. Pal, "Recognition of off-line handwritten Devnagari characters using quadratic classifier", *Computer Vision, Graphics and Image Processing*. Springer Berlin Heidelberg, 2006. 805–816.
4. U. Pal, N. Sharma, T.Wakabayashi, and F. Kimura, "Off-line handwritten character recognition of Devnagari script," in Proc. 9th Conference on Document Analysis and Recognition, 2007, pp. 496–500.
5. U. Pal, S. Chanda, T. Wakabayashi, and F. Kimura, "Accuracy improvement of Devnagari character recognition combining SVM and MQDF," in Proc. 11th Int. Conf. Frontiers Handwrit. Recognit., 2008, pp. 367–372. Dr. P. S. Deshpande, Latesh Malik, Sandhya Arora, "Fine classification recognition of handwritten devnagari characters with regular expressions minimum edit distance method", JOURNAL OF COMPUTERS (2008). VOL. 3, NO. 5, MAY 2008.
6. U. Pal, T. Wakabayashi, and F. Kimura, "Comparative study of Devanagari handwritten character recognition using different features and classifiers," in Proc. 10th Conf. Document Anal. Recognit., 2009, pp. 1111–1115.
7. Apurva A. Desai, "Gujarati handwritten numeral optical character reorganization through neural network", Pattern Recognition 43 (2010) 2582–2589.
8. Mamta maloo, K.V. Kale, "Support vector machine based Gujarati numeral recognition", International Journal on Computer Science and Engineering (IJCSE), ISSN: 0975-3397 Vol. 3 No. 7 July 2011.
9. Desai, Apurva A. "Support vector machine for identification of handwritten Gujarati alphabets using hybrid feature space." CSI Transactions on ICT, pp. 1–7, 2015.
10. Ankit K. Sharma, Dipak M. Adhyaru, Tanish H. Zaveri, and Priyank B. Thakkar. "Comparative analysis of zoning based methods for Gujarati handwritten numeral recognition.", 5th Nirma University International Conference on Engineering (NUiCONE), pp. 1–5. IEEE, 2015.
11. M. Goswami and S. Mitra, "Offline handwritten Gujarati numeral recognition using low-level strokes," *Int. J. Appl. Pattern Recognit.*, 2015.
12. N. Otsu, A threshold selection method from gray-level histograms, Automatica 11 (1975) 23–27.
13. Lewis, J. P. "Fast normalized cross-correlation." Vision interface. Vol. 10. No. 1. 1995.

Realization of Junctionless TFET-Based Power Efficient 6T SRAM Memory Cell for Internet of Things Applications

Anju, Sunil Pandey, Shivendra Yadav, Kaushal Nigam, Dheeraj Sharma and P.N. Kondekar

Abstract The Internet of Things (IoTs) applications have garnered its interest to realize low-power memory circuit based on emerging nanoscale transistors for its data processing unit. Therefore, in this work, we focussed on tunneling mechanism-based tunnel field-effect transistor (TFET) which can be a suitable option beyond-CMOS devices for designing reliable and efficient memory circuits for its key sensing and data processing unit. However, this work is further extended toward low-power design strategy to meet the essential requirements of IoT applications. For this purpose, a junctionless (JL) TFET based on work-function engineering is reported in this work, where a high-k material (HfO_2) adjacent to the SiO_2 toward source side is considered underneath the gate region to improve the ON-current of the proposed device. The main benefits of junctionless architecture is that it reduces the fabrication complexity, high thermal budget, and is free from random dopant fluctuations (RDFs). The significant benefits in terms of hold, read, and write static noise margin (SNM) of JLTFET-based six-transistor (6T) memory cell enables its potential application for IoT memory unit.

Keywords Band-to-band tunneling · Internet of Things (IoTs) · Memory TFET

Anju (✉) · S. Pandey · S. Yadav · K. Nigam · D. Sharma · P.N. Kondekar
Indian Institute of Information Technology, Design and Manufacturing,
Jabalpur 482005, MP, India
e-mail: anju@iiitdmj.ac.in

S. Pandey
e-mail: sunilpandey@iiitdmj.ac.in

S. Yadav
e-mail: shivendra1307@gmail.com

K. Nigam
e-mail: kaushal.nigam@iiitdmj.ac.in

D. Sharma
e-mail: dheeraj@iiitdmj.ac.in

P.N. Kondekar
e-mail: pnkondekar@iiitdmj.ac.in

A.K. Somani et al. (eds.), *Proceedings of First International Conference on Smart System, Innovations and Computing*, Smart Innovation, Systems and Technologies 79, https://doi.org/10.1007/978-981-10-5828-8_49

1 Introduction

In recent years, Internet of Things (IoTs) have found much attention in human life where the world community are closely connected with each other through intelligent devices. This exaggerated the demand of its core unit such as memory unit which is responsible for controlling power, area, and speed [1, 2]. Several researchers in recently reported works have contributed toward memory design based on conventional CMOS-based technology [3, 4]. However, CMOS-enabled memory circuits are suffering from its fundamental KT/q limit and subthreshold swing (SS) greater than 60 mV/dec. Moreover, computation time and communication methods are facing enormous challenges with CMOS technology for further power reduction [5, 6]. Its main reason is that voltage scaling after a certain limit in CMOS process reduces the dynamic computation, but, at the same time it is creating severe problem related to exponentially increasing leakage power which can be disadvantageous for speed of the device. It restricts the integration of functionality and scenarios where IoT devices are efficiently operated by batteries.

To address the aforementioned issues, tunnel field-effect transistor (TFET) based on tunneling phenomenon has been studied as an appropriate option over conventional MOSFET due to its various features such as SS < 60 mV/dec, low power consumption, scaling with supply, and free from random dopant fluctuations (RDFs) [6–9]. However, it overcomes another limitation of fabrication complexity of the conventional FET-based architectures. This challenge makes FET- based memory circuits inferior for IoT sensing and signal processing unit. Therefore, we report a JLTFET-based 6T static random access memory (SRAM) cell which is efficient and reliable for IoT applications. We analyzed the design-level issues in JLTFET-based memory circuit and demonstrate its memory capability. However, work-function engineering concept is introduced at the source electrode to induce p^+, so that, it becomes like a p^+-i-n^+ gated structure of conventional TFET. Further, a high-k dielectric HfO_2 adjacent to the SiO_2 is used beneath the gate region to enhance the ON-state current of the device. It also provides substantially additional features that could be captured for new functionality.

Remaining part of this work is organized as follows: Sect. 2 includes the IoT systems block diagram, while, Sect. 3 covers the device description and simulation setup for JLTFET. The detailed simulation results of JLTFET-based memory circuit are elaborated in Sect. 4. Finally, in Sect. 5, conclusion with some important key findings are highlighted.

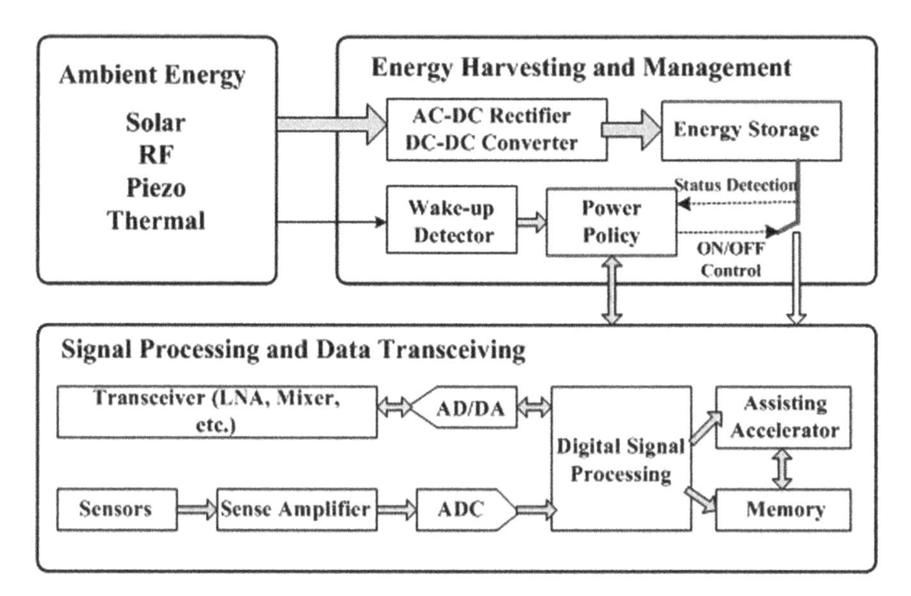

Fig. 1 A general IoT systems diagram with their bottleneck

2 General IoT Systems Diagram with Their Bottleneck

IoT systems diagram with their bottleneck are shown in Fig. 1. It includes ambient energy unit, energy harvesting and management unit, sensors and signal processing unit [10]. The first block includes ambient source which can be of RF, solar, Piezo-electric and thermal gradient type [11, 12]. The energy density of different ambient sources varies depending upon gradients. However, they can also differ in particular range of voltage, generating principle, AC/DC conversion rate. The main task of harvesting system is to convert the total power received from antenna to the rectifier circuit. It can be done by selecting appropriate type of matching networks. But, due to nonlinear behavior of its impedance with power and frequency, suitable broadband matching networks are desired for complete transfer of maximum power. If RF circuit in the complete network is not properly matched, then we get reflected power which can generate standing waves between the source and load. A rectenna circuit can convert AC signal into DC output efficiently. In this concern, a DC–DC boost converter is needed to achieve lower or higher DC voltages. On the other hand, the design optimization strategy such as tracking and adaptive technique can also be adopted to improve the energy density [13].

The memory unit is the most critical and important building block in IoT systems. Therefore, realization of efficient memory architecture based on JLTFET is required to improve both speed and power. In recent literature, many authors have investigated the possibility of MOSFET-based memory circuits. But, there is no significant improvement in its SNM characteristics due to which MOSFET- based memory cells are not efficient for the IoT memory unit. On the other hand, JLTFET-based 6T SRAM memory cell can provide significant advantages to increase the speed of IoT systems. Therefore, an investigation is required to check its potential for IoT memory applications.

3 Device Description and Simulation Setup For JLTFET

Figure 2a, b show the cross-sectional views of proposed p-type and n-type JLTFET. In this, p-channel and n-channel is formed over heavily doped silicon wafer with concentration of (1×10^{19} cm^{-3}) by doping of p-type and n-type dopants. Further, by work-function engineering, source and channel region for both p-type and n-type JLTFET is created in such a manner that it forms a structure of n^+-i-p^+ for p-type JLTFET and p^+-i-n^+ for n-type JLTFET. However, a high-k material is used underneath the gate region toward the source side to improve coupling between gate and

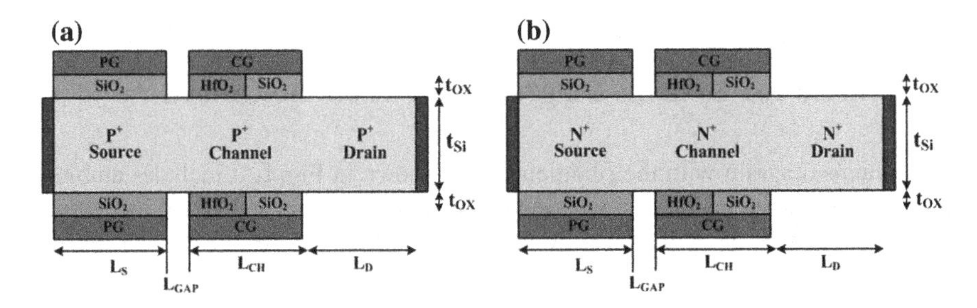

Fig. 2 Cross-sectional views of **a** p-type, and **b** n-type junctionless TFET

Table 1 Device design parameters used in the simulation

Parameter name	Symbol	Value	Unit
Source length	L_S	30	nm
Channel length	L_{CH}	30	nm
Drain length	L_D	30	nm
Oxide thickness	t_{OX}	1	nm
Silicon film thickness	t_{Si}	5	nm
Control gate length	L_{M2}	30	nm
Source/Channel gap length	L_{GAP}	3	nm
Work-function of polarity gate(PG) for p-type	ϕ_{PG}	3.9	eV
Work-function of control gate(CG) for p-type	ϕ_{CG}	5.0	eV
Work-function of polarity gate(PG) for n-type	ϕ_{PG}	5.93	eV
Work-function of control gate(CG) for n-type	ϕ_{CG}	4.67	eV

channel region which would result in an improved ON-state current. The structural parameters of the presented device are listed in Table 1. All the simulations of both devices are performed using Silvaco ATLAS TCAD simulator [14]. However, some key models such as band-to-band tunneling model, Band Gap Narrowing (BGN), Shockley Read Hall (SRH), Quantum Confinement (QC) model is incorporated in the simulation [14]. In this, band-to-band tunneling model is used for considering tunneling effects at the source-channel junction. However, BGN is included for the reason of heavy concentration in channel region. SRH model is used to account recombination of minority carriers [14] and since the substrate thickness is less than 10 nm, therefore, QC model is also incorporated in the simulation [7]. Moreover, WKB approximation method is adopted for TFETs simulation.

Figure 3a, b show the energy band of p-type and n-type JLTFET in OFF and ON-state. From the first figure, it can be seen that the barrier width is large in OFF condition. However, in ON-state, it shifts in upward direction at the source/channel interface for the case of p-type JLTFET when $V_{gs} = -0.6$ V is applied. Similarly, when a positive gate voltage is applied in case of n-type JLTFET, more band bending causes reduction in tunneling barrier width which increases the number of electrons to travel from source to channel region. Thus, it results a significant improvement in ON-current of the device. Figure 4a, b show the transfer characteristics of p-type and n-type JLTFET, where both devices achieves same order of current for different V_{ds}. Due to same order of current with respect to V_{ds}, we have considered drain voltage as 0.6 V for further low-power circuit analysis as it is the minimum possible voltage at which both devices are working appropriately. Figure 4c, d show the output characteristics of p-type and n-type JLTFET, where an increment in drain current is achieved in triode region and then it reaches in saturation region.

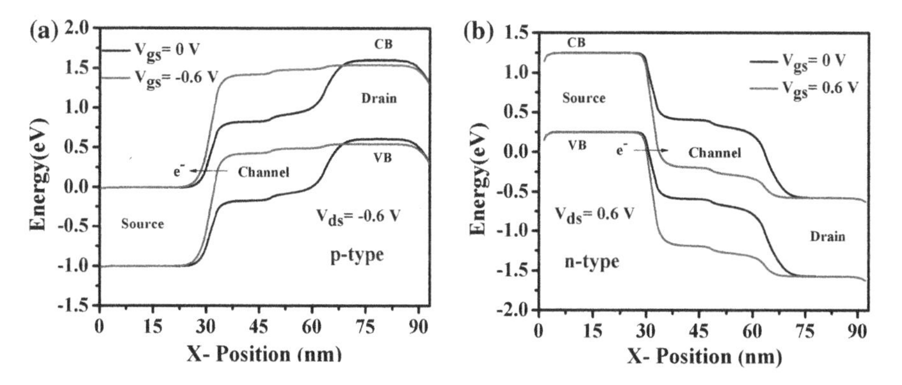

Fig. 3 Energy band diagram under OFF and ON-state of **a** p-type, and **b** n-type JLTFET

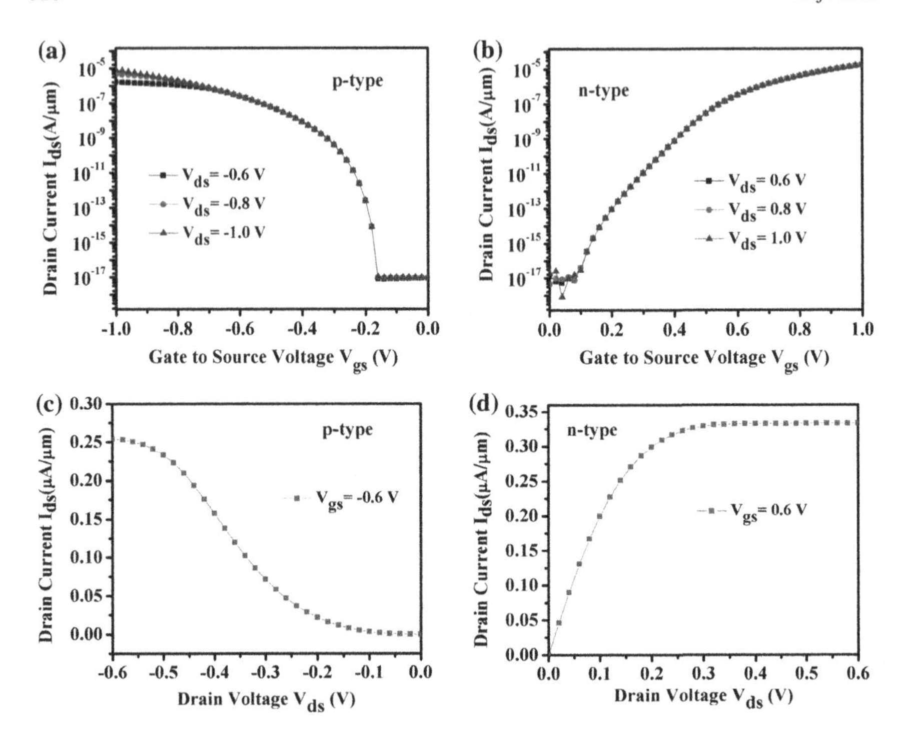

Fig. 4 Transfer characteristics of **a** p-type, and **b** n-type JLTFET. Output characteristics of **c** p-type, and **d** n-type JLTFET

4 Design of 6T SRAM Cell

Figure 5 shows the JLTFET-based standard 6T SRAM memory cell, where the basic circuit of SRAM cell is formed by connecting two inverter back-to-back and access transistor is used to store one bit information. The output Q of the first inverter is input $\overline{Q}$ of the second inverter and vice-versa for second inverter. For SRAM analysis, read and write operations of SRAM cell are performed by using bit line (BL) and

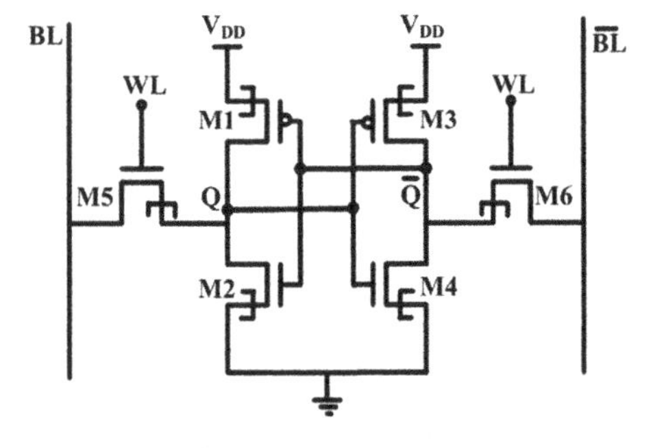

Fig. 5 Junctionless TFET-based power-efficient 6T SRAM memory cell

word line (WL). The transistors M5 and M6 which is used to access the storage bit are turned ON when the WL is high. Similarly, when the WL is low, the transistors M5 and M6 are in OFF condition and the SRAM cell turns into hold state. In hold state, the information remains stored as it was stored previously. Read-Write operation stability depends upon cell ratio and pull-up ratio, where cell ratio is defined as the ratio between width of pull-down transistor (M2 or M4) to access transistor (M5 or M6). To overcome read failure, pull-down transistor driving strength is made greater than the access transistor, so cell ratio is always considered as greater than 1. Similarly, pull-up ratio is defined as the ratio between pull-up transistor (M1 or M3) to access transistor (M5 or M6). Normally, pull-up ratio is assumed as one for better write operation. In this work, cell ratio is considered as two while pull-up ratio is taken as one.

Initially, data stored at node Q is considered as 0 which results M1 and M4 in OFF condition and M2 and M3 in ON condition. For the read operation to be executed, both BL and $\overline{BL}$ are pre-charged to V_{DD} which makes WL high (WL = 1). In this condition, $\overline{BL}$ will not show any variation, therefore, no current will flow through M5 and M6 and M2 will conduct since small amount of current flow which results the voltage of the BL starts slightly decreasing. Thus, the node voltage starts increasing from its initial voltage 0. For the write operation to be executed, initially data stored at node Q is considered as 1 which results M2 and M3 in OFF mode and M1 and M4 in ON mode. To write 0, BL voltage is considered as 0 and $\overline{BL}$ voltage is pre-charged to V_{DD} depending upon the data stored at node Q (if Q = 0, BL = 1 and $\overline{BL}$ = 0) and select wordline. The voltage of BL is forced to change the information at node Q, so that, write 0 operation can be executed.

The implementation of JLTFET-based 6T SRAM memory cell is performed by using the device electrical characteristics, where a lookup table-based Verilog-A model is built to examine its circuit behavior [15]. In this concern, SRAM cell is a standard circuit for analyzing the device circuit interaction and the success of its achievement depends upon proper realization of SRAM memory cell. For this purpose, stability of SRAM cell is realized by static noise margin (SNM) which is defined as the highest DC noise voltage for which the cell does not flip its state during its access operation [15, 16]. SNM can be calculated by superimposing both the voltage transfer characteristics (VTCs) and its inverse VTCs which results into a butterfly curve. Further, a square fitting method is used in the butterfly curve to determine the SNM value [15, 16]. Figure 6a–c show the hold SNM, read SNM and write SNM of 6T JLTFET-based SRAM memory cell, where hold, read, and write SNM are achieved as 250, 130, and 200mV, respectively. These results indicate that JLTFET can be utilized for IoT applications.

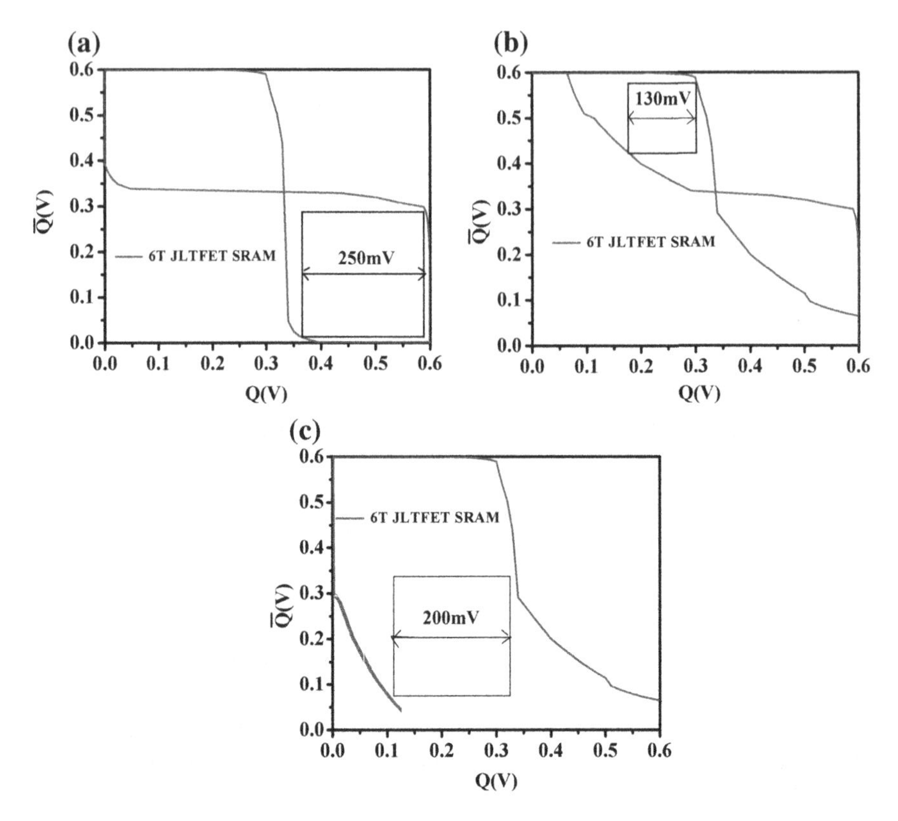

Fig. 6 **a** Hold SNM **b** read SNM, and **c** write SNM of 6T SRAM memory cell

5 Conclusion

In this work, we analyzed JLTFET-based 6T SRAM memory cell for low-power IoT applications which reduces the fabrication complexity and is free from RDFs. The electrical characteristic shows that the proposed device can be used beyond-CMOS devices for realizing efficient memory circuits. The designed SRAM cell based on JLTFET works on a of $V_{DD} = 0.6$ V with a power consumption of 39 μW in hold state and 1.02 μW in read mode. The significant SNM characteristics in hold, read, and write mode enables its potential for low-power IoT applications.

References

1. L. Atzori, A. Iera and G. Morabito, The internet of things: A survey, Comput. Netw., vol. 54, no. 15, pp. 2787–2805, 2010.

2. H. Scott et al., A low-voltage processor for sensing applications with picowatt standby mode, JSSCC 2009.
3. K. Roy and S. Prasad, Low-Power CMOS VLSI Circuit Design. 1st ed. New York, NY, USA: Wiley, 2000.
4. B. Zimmer et al., SRAM Assist Techniques for Operation in a Wide Voltage Range in 28-nm CMOS, IEEE Transactions on Circuits and Systems II: Express Briefs, vol. 59, no. 12, pp. 853–857, 2012.
5. A. C. Seabaugh and Q. Zhang, Low-voltage tunnel transistors for beyond CMOS logic. Proc. IEEE, vol. 98, no. 12, pp. 2095–2110, Dec. 2010.
6. M. A. Ionescu and H. Riel, Tunnel field-effect transistors as energy efficient electronic switches, Nature, vol. 479, pp. 329–337, 2011.
7. K. Boucart and A. M. Ionescu, Double gate tunnel FET with high-k gate dielectric, IEEE Trans. Electron Devices, vol. 54, no. 7, pp. 1725–1733, Jul. 2007.
8. M. J. Kumar and S. Janardhanan, Doping-Less Tunnel Field Effect Transistor: Design and Investigation, IEEE Trans. Electron Devices, vol. 16, no. 10, pp. 3285-3290, Oct. 2013.
9. B. Ghosh, M. W. Akram, Junctionless Tunnel Field Effect Transistor, IEEE Electron Device Letter, vol. 34, no. 5, pp. 584–586, May. 2013.
10. K. Swaminathan, H. Liu, X. Li et al., Steep slope devices: Enabling new architectural paradigms, 2014 51st ACM/EDAC/IEEE Design Automation Conference (DAC), San Francisco, CA, pp. 1–6, 2014.
11. X. Li et al., RF-powered systems using steep-slope devices, New Circuits and Systems Conference (NEWCAS), 2014.
12. S. Kim, et al., Ambient RF Energy-Harvesting Technologies for Self-Sustainable Standalone Wireless Sensor Platforms, Proc. IEEE, vol. 102, no. 11, pp. 1649–1666, 2014.
13. M. A. G. de Brito et al., Evaluation of the Main MPPT Techniques for Photovoltaic Applications, IEEE Transactions on Industrial Electronics, vol. 60, no. 3, pp. 1156–1167, March 2013.
14. Silvaco Interantional, ATLAS Users Manual, A 2D-3D Numerical Device Simulator, (http://www.silvaco.com).
15. S. Ahish, D. Sharma, M. H. Vasantha, Y. B. N. Kumar, Device and circuit level performance analysis of novel InAs/Si heterojunction double gate tunnel field-effect transistor, Superlattices and Microstructures, vol. 94, pp. 119–130, Jun. 2016.
16. S. Strangio, P. Palestri, M. Lanuzza, F. Crupi, D. Esseni and L. Selmi, Assessment of InAs/AlGaSb Tunnel-FET Virtual Technology Platform for Low-Power Digital Circuits, IEEE Trans. Electron Devices, vol. 63, no. 7, pp. 2749–2756, July 2016.

A Charge Plasma Based Dielectric Modulated Heterojunction TFET Based Biosensor for Health-IoT Applications

Deepika Singh, Sunil Pandey, Dheeraj Sharma and P.N. Kondekar

Abstract The health Internet-of-Things (IoTs) applications have motivated to design nanoscale transistor based biosensors. Therefore, in this work, we mainly focus on how emerging transistors such as tunnel field-effect transistor (TFET) can be an alternative beyond-CMOS features for biosensor design and to further increase the low power design strategy to enable it for IoT applications. For this purpose, a charge plasma based dielectrically modulated TFET is proposed in this work, where source region is having SiGe material (with Si and Ge composition equal to 0.5) to improve the ON-state current. The charge plasma concept is employed to make the fabrication process simpler and to avoid the random dopant fluctuations (RDFs) and higher thermal budget. The significant improvement in the sensitivity of the proposed dielectrically modulated junctionless (DMJL) TFET biosensor with different dielectric constant and charge density, and a comparison with conventional MOSFET-based sensor enables its potential application for the Health-IoT applications.

Keywords TFET · Charge plasma · Biosensor · Internet-of-things application

D. Singh (✉) · S. Pandey · D. Sharma · P.N. Kondekar
Indian Institute of Information Technology, Design and Manufacturing,
Jabalpur 482005, MP, India
e-mail: deepikasingh@iiitdmj.ac.in

S. Pandey
e-mail: sunilpandey@iiitdmj.ac.in

D. Sharma
e-mail: dheeraj@iiitdmj.ac.in

P.N. Kondekar
e-mail: pnkondekar@iiitdmj.ac.in

A.K. Somani et al. (eds.), *Proceedings of First International Conference on Smart System, Innovations and Computing*, Smart Innovation, Systems and Technologies 79, https://doi.org/10.1007/978-981-10-5828-8_50

1 Introduction

In recent years, Internet-of-Things (IoTs) have attracted huge attention in the field of information and communication technology (ICT) due to rapid growth in making intelligent devices required for better human life [1]. In many places, poor medication becomes a major problem for patient as well as a healthcare providers. As a earlier report, it is find that about 25 % of the aging population do not follow their given medication which leads to poor health and an increment in death. Therefore, improvement in technology for facilities and services for health care are primarily important of this population. So, we have been considered a technology based on IOT and proposed a home-based biosensor for health care [2]. This technology can be used to connect sensors, healthcare systems and other devices to the Internet. It also plays an important role in our daily life by connecting different types of smart devices through wireless medium, and improves significantly in our living standard [3]. In this concern, biosensing element is an essential component for Health-IoT systems which is used for real-time monitoring, early stage detection of diseases and monitoring of life-threatening adverse events [1–4]. However, it can also be used for checking the prescribed treatment given for a patient. For this purpose, many biosensors based on CMOS technology have been explored [5, 6]. However, CMOS-enabled devices are suffering from better sensing capability due to its fundamental KT/q limit and subthreshold swing (SS) greater than 60 mV/dec [6]. To overcome this issue, tunneling mechanism based tunnel field-effect transistor (TFET) has been explored as an alternative option over conventional MOSFET due to its salient features such as low power consumption, SS < 60 mV/dec, scaling with supply, and free from short channel effects (SCEs) [7–10]. Moreover, with the exploration of emerging nanoscale devices, tunneling transistor gives the opportunity to save the power consumption. Another limitation with the conventional FET-based sensor devices are fabrication complexity [6]. This challenge makes FET-based sensors inferior for Health-IoT applications. In this respect, TFET-based sensor design based on biomolecule conjugation effect can play an important role to meet the essential requirement of Health-IoT systems. Further, TFET-based biosensor devices with low SS can be preferred to increase the speed of complete IoT systems [7, 8]. However, carrier diffusion from S/D to the channel also makes a problem for the creation of an abrupt junction profile, and hence suffers from RDFs. To address this issue, junctionless configuration based TFET has been studied in [9] which reduces the fabrication complexity. Still, none of the authors have investigated the sensing capability of dielectrically modulated junctionless tunnel field-effect transistor (DM-JLTFET) based biosensor for IoT applications.

Therefore, in this work, a DM-JLTFET based biosensor is introduced which works on phenomenon of biomolecule conjugation effect and addresses the requirements of Health-IoT systems. We analyzed the design-level issues in proposed DM-

JLTFET biosensor and demonstrate its sensing capability. However, instead of conventional doping, a charge plasma concept [10], is employed in the proposed structure for the formation of p^+ source and n^+ drain regions which avoids RDFs and thermal budget, and also reduces the fabrication complexity desired for Health-IoT applications. It also provides substantially new features that could be captured for new functionality.

2 General Building Blocks for the Health-IoT Systems

The application scenario and building blocks for the Health-IoT system are shown in Fig. 1. It includes base station, biochip, RFID chip, weight sensor, signal processing unit and Health-IoT cloud. From base station, ambient source such as RF, solar, piezoelectric and thermal gradient can be considered. The energy density of different ambient sources varies depending upon gradients. However, they can also differ in particular range of voltage, generating principle, AC/DC conversion rate. The main task of RF to DC conversion system is to convert the total power received from antenna to the rectifier circuit. It can be done by selecting appropriate type of matching networks. But, due to nonlinear behavior of its impedance with power and frequency, suitable broadband matching networks can be employed for complete transfer of maximum power. If RF circuit in the complete network is not properly matched then we get reflected power which can generate standing waves between the source and load. In this concern, a rectifier circuit converts AC signal into DC output efficiently [11]. Apart from this, power conversion efficiency generated from conventional CMOS technology enabled devices is very small and have high resistive loss in low operating voltages. For this purpose, a super capacitor based energy

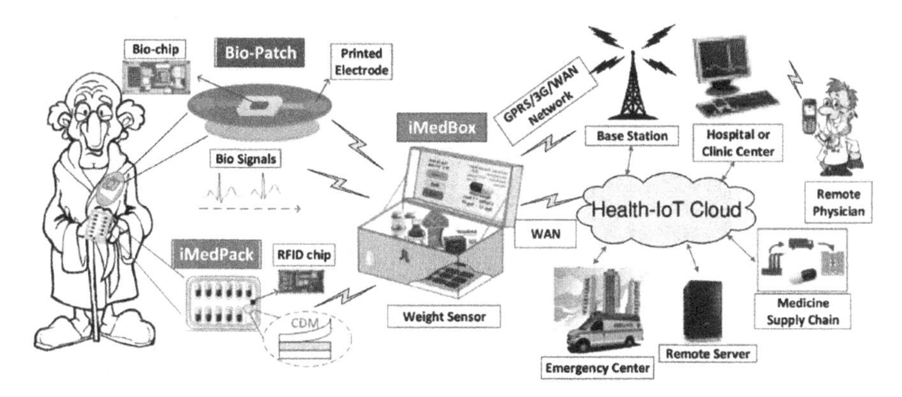

Fig. 1 Application scenario and building blocks for the Health-IoT systems [1]

scavenging system can be employed to further improve the efficiency [11]. In various building blocks of Health-IoT system, the biosensor design (inside biochip unit) is really a challenging task. However, its realization with emerging transistor is also not an easy task in terms of size and power. To address these issues, notable work has been done in past few years. But, the sensing capability based on FET biosensors is not enough for the Health-IoT systems. However, passive sensing has also been studied to reduce the power consumption. But, it still suffers with the limitation of the sensed type's signals or the speed. Apart from above mentioned bottlenecks, some other concerns for IoT systems are RFID chip, memory unit, security, privacy etc., but these are not the main focus of this work.

3 Device Structure and Simulation-Setup

The device structure of a double gate DM-JLTFET based biosensor is shown in Fig. 2. In this structure, source (p^+) and drain (n^+) regions are induced with the charge plasma concept by using suitable metal electrodes with an appropriate work functions of Platinum (Pt) = 5.93 eV and Hafnium (Hf) = 3.9 eV, respectively. To increase the tunneling current, Silicon-Germanium (SiGe) material have been considered in the source region with Ge composition of 0.5 in the device structure. The length spacer between source and gate electrode is 2 nm, whereas, the spacer between gate and drain electrode is taken as 5 nm. However, other parameters of the device are: Silicon body thickness = 10 nm, gate length (L_g) = 53 nm, length of nanogap cavity (L_{cavity}) = 23 nm, thickness of the cavity (t_{cavity}) = 5.5 nm, thickness of silicon oxide (t_{ox}) = 0.5 nm, gate length (ϕ_g) = 4.1 eV, length of source (L_s) and drain regions (L_d) are considered as 45 nm. The 2-D Atlas device simulator [12], is used

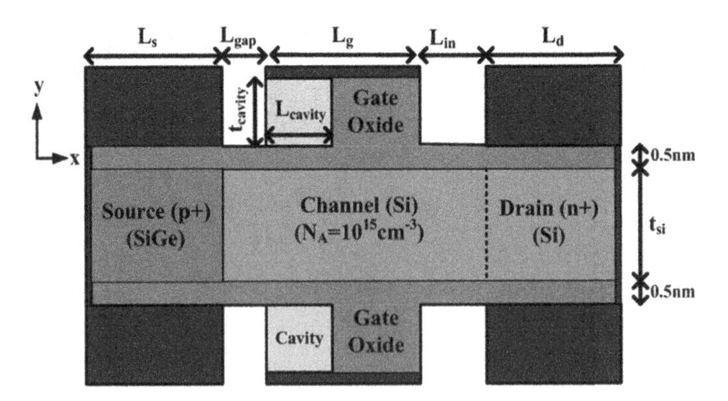

Fig. 2 Schematic view of a charge plasma based hetero DM-JLTFET biosensor

for simulating the proposed device. Moreover, for the simulations of TFET device, a non-local band to band tunneling (BTBT) model is preferred because it calculates the tunneling rate across energy band profile [7]. Thus, a quantum tunneling is defined at the source/channel region to enable the reverse tunneling. Whereas, for recombination SRH and Auger for minority carriers is used. Further, Fermi-dirac Statistics for carrier statistics and field-dependant mobility models are also considered in the simulation. In DM-JLTFET biosensor, the nanogap cavity is formed by etching the gate dielectric material at the source side because effect of gate-drain underlap is not significant on the electrical characteristics of the proposed device. To detect the presence and absence of biomolecules in the cavity region, the dielectric constant (k) and charge density (ρ) are considered as the controlling parameters to analyze the gate modulation [7]. For this purpose, as the dielectric constant increases from k = 1 (air) to k > 1 (biomolecules) in the cavity, a significant improvement in the tunneling current is observed. However, to analyze the behavior of the charge density for DNA analysis, the charge density in the range of -10^{11} to -10^{12} cm^{-2} is considered at the Si/SiO_2 interface [6].

4 Results and Discussion

First, we focus on tunneling mechanism through BTBT process from source to channel region with different dielectric constant and charge density of biomolecules.

For this purpose, Fig. 3a, b show the energy band profile along X-position under ON-state with different dielectric constant and charge density. In Fig. 3a, the tunneling barrier width reduces at the source/channel interface with an increment in dielectric constant and thus resulting an improvement in tunneling current. On the other hand, barrier width increases with an increment in the negative charge density and hence resulting in a reduced drain current, as shown in Fig. 3b. Figure 4a, b show the drain current along V_{gs} with different dielectric constant and charge density. As the dielectric constant increases from k = 1 (air) to k > 1 (biomolecules) in the cavity, an increment in drain current is observed. Whereas, with an increment in the negative charge density in the range of -10^{11} to -10^{12} cm^{-2}, a decrement in the drain current is noticed, which can be analyzed from Fig. 4b. In the same way, Fig. 4c, d show the similar phenomenon for output characteristics with different k and ρ, respectively.

To detect the presence of charged and non-charged biomolecules in the cavity, it is important to measure the drain current sensitivity, where its expression with detailed description are reported in [8]. To analyze the sensitivity, the drain current sensitivity along gate bias for different dielectric constant and charge density are shown in Fig. 5a, b. As we observed from Fig. 5a that the drain current sensitivity increases for a higher value of dielectric constant and decreases for a negative value of charge density as noticed from Fig. 5b. Whereas, Fig. 5c, d shows the drain cur-

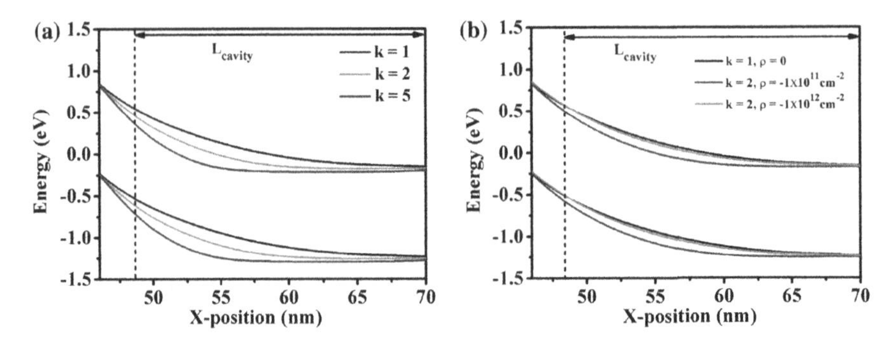

Fig. 3 Energy band diagram of DM-JLTFET biosensor along X-position with different **a** dielectric constant ($\rho = 0$), and **b** charge density (k = 2) in ON-state ($V_{gs} = 1.2$ V and $V_{ds} = 0.7$ V)

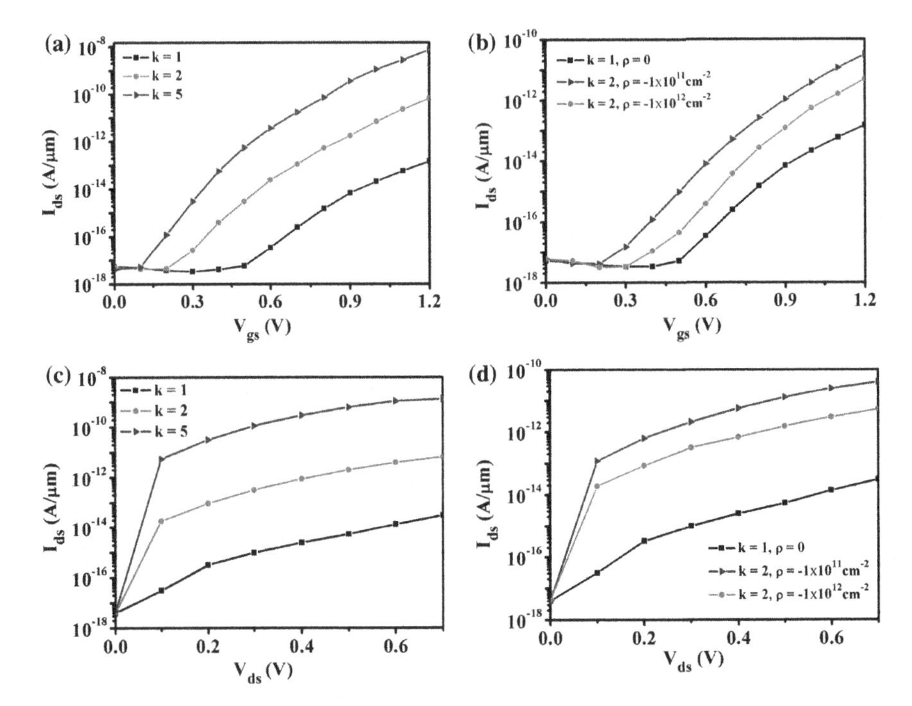

Fig. 4 I_{ds}–V_{gs} characteristics of DM-JLTFET biosensor with different **a** dielectric constant ($\rho = 0$), and **b** charge density (k = 2) is in the μC range from -10^{11} to -10^{12} cm^{-2}) at $V_{ds} = 0.7$ V. I_{ds}–V_{ds} characteristics of DM-JLTFET biosensor with different **c** dielectric constant ($\rho = 0$), and **d** charge density (k = 2) is in the μC range from -10^{11} to -10^{12} cm^{-2}) at $V_{gs} = 1.2$ V

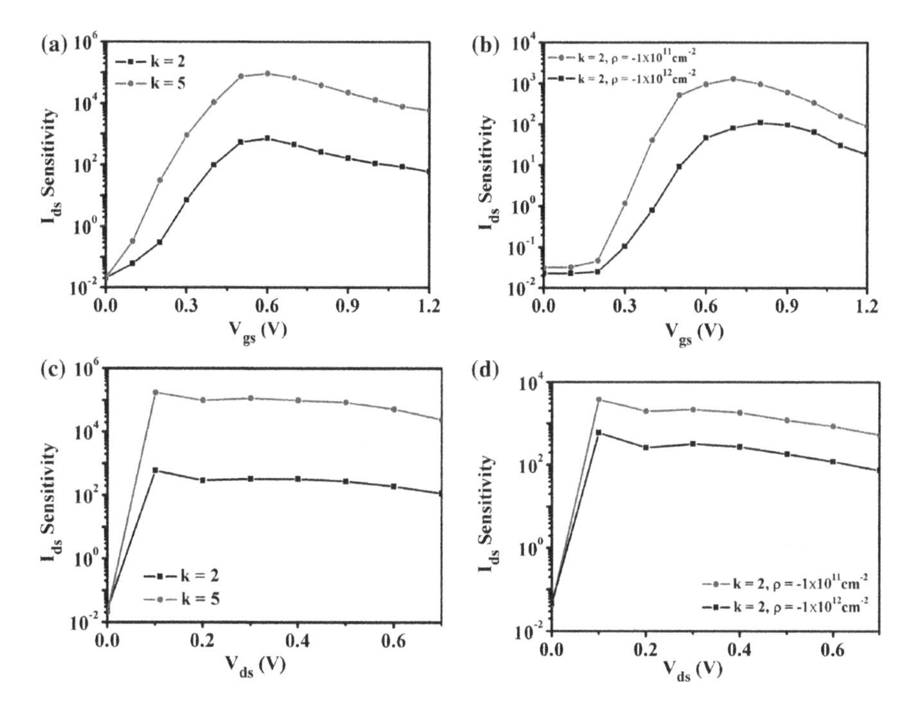

Fig. 5 Characteristics of I_{ds} sensitivity of the DM-JLTFET biosensor along V_{gs} with different **a** dielectric constant ($\rho = 0$), and **b** charge density (k = 2). Characteristics of I_{ds} sensitivity of the DM-JLTFET along V_{ds} with different **c** dielectric constant ($\rho = 0$), and **d** charge density (k = 2)

rent sensitivity along drain bias for different dielectric constant and charge density, respectively and it follows the same trend as shown in Fig. 5a, b. It can be noticed from these plots that the higher sensitivity is achieved at moderate gate bias and at lower drain bias. It confirms that the selected bias conditions can play an important role in significant improvement of sensitivity of the device. Further, it is necessary to compare the DM-JLTFET with DM-JLMOSFET biosensor to prove its potential for IoT applications. For this purpose, the sensing parameters such as I_{ON} and I_{ON}/I_{OFF} ratio have been considered as a measuring parameter, where a large variation in its characteristics denotes its better sensing capability. Figure 6a, b show I_{ON} along ρ for different dielectric constant and with an increment in dielectric constant the I_{ON} increases of DM-JLTFET and MOSFET, as we observed from given plots that the variation in I_{ON} of DM-JLTFET is more in comparison to MOSFET, whereas, Fig. 6c, d show I_{ON}/I_{OFF} ratio along k for different length of cavity (L_{cavity})of DM-JLTFET and MOSFET and with a decrement in length of cavity the value of I_{ON}/I_{OFF} ratio increases. From these plots, it is apparent that the relative variation in the sensitivity is higher for DM-JLTFET biosensor which results in better sensitivity of

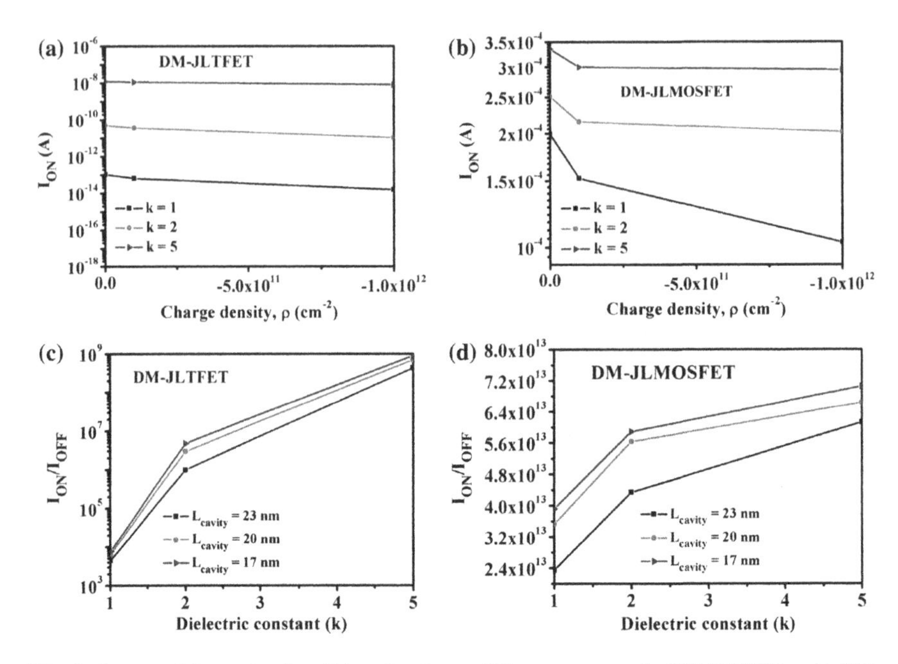

Fig. 6 Impact of charge density of biomolecules on ON-state current of **a** DM-JLTFET, and **b** DM-JLMOSFET biosensor for various dielectric constant values at V_{gs} = 1.2 V and V_{ds} = 0.7 V. Plots of I_{ON}/I_{OFF} ratio of **c** DM-JLTFET, and **d** DM-JLMOSFET biosensor as a function of dielectric constant at different length of cavity (L_{cavity}) at V_{gs} = 1.2 V and V_{ds} = 0.7 V

the device as compared to DM-JLMOSFET biosensor despite the fact, that current order is lower in DM-JLTFET biosensor than the DM-JLMOSFET based biosensor. Therefore, DM-JLTFET biosensor can be chosen as a better sensitive device for biomedical diagnostic and Health-IoT applications with greater sensitivity and improved response time.

5 Conclusion

In this work, we investigated the performance of DM-JLTFET biosensor for Health-IoT applications. The SiGe material has been considered in the source region with Ge composition of 0.5 to improve the tunneling current. However, the charge plasma concept has been employed to induce p^+ source and n^+ drain regions in the silicon body. It makes the proposed device immune from random dopant fluctuations (RDFs) and also it reduces fabrication steps with the lower cost. Further, the comparison of DM-JLTFET with DM-JLMOSFET biosensor has been performed. From comparison, it is noticed that the DM-JLTFET biosensor has higher sensitivity with better response time. Thus, the investigated results of DM-JLTFET biosensor makes it a suitable candidate for the Health-IoT applications.

References

1. G. Yang et al., A Health-IoT Platform Based on the Integration of Intelligent Packaging, Unobtrusive Bio-Sensor, and Intelligent Medicine Box, in IEEE Transactions on Industrial Informatics, vol. 10, no. 4, pp. 2180–2191, Nov. 2014.
2. B. Schuz et al., Medication beliefs predictmedication adherence in older adults with multiple illnesses, J.Psychosom. Res., vol. 70, no. 2, pp. 179–187, 2011. medication adherence in older adults with multiple illnesses, J. Psychosom. Res., vol. 70, no. 2, pp. 179–187, 2011.
3. K. Yelmarthi, A. Abdelgawad and A. Khattab, An architectural framework for low-power IoT applications, 2016 28th International Conference on Microelectronics (ICM), Giza, Egypt, 2016, pp. 373–376.
4. G. Kortuem, F. Kawsar, D. Fitton, and V. Sundramoorthy, Smart objects as building blocks for the Internet of things, IEEE Internet Comput., vol. 14, no. 1, pp. 44–51, Feb. 2010.
5. M. Barbaro, A. Bonfiglio, and L. Raffo, A charge-modulated FET for detection of biomolecular processes: Conception, modeling, and simulation, IEEE Trans. Electron Devices, vol. 53, no. 1, pp. 158–166, Jan. 2006.
6. C. H. Kim, C. Jung, K. B Lee, H. G. Park, and Y. K. Choi, Label-free DNA detection with a nanogap embedded complementary metal oxide semiconductor, Nanotechnology, vol. 22, no. 13, pp. 1–5, Apr. 2011
7. R. Narang, K. V. S. Reddy, M. Saxena, R. S. Gupta, and M. Gupta, A dielectric modulated tunnel FET based biosensor for label free detection: Analytical modeling study and sensitivity analysis, IEEE Trans. Electron Devices, vol. 59, no. 10, pp. 2809–2817, Oct. 2012.
8. S. Kanungo, S. Chattopadhyay, P. S. Gupta, and H. Rahaman, Comparative performance analysis of the dielectrically modulated full-gate and short-gate tunnel FET-based biosensors, IEEE Trans. Electron Devices, vol. 62, no. 3, pp. 994–1001, Mar. 2015.
9. B. Ghosh and M. W. Akram, Junctionless tunnel field-effect transistor, IEEE Electron Device Lett., vol. 34, no. 5, pp. 584–586, May 2013.
10. M. J. Kumar and S. Janardhanan, Dopingless tunnel field-effect transistor: Design and investigation, IEEE Trans. Electron Devices, vol. 60, no. 10, pp. 3285–3290, Oct. 2013.
11. P. Nintanavongsa, U. Muncuk, D. R. Lewis, and K. R. Chowdhury, "Design Optimization and Implementation for RF Energy Harvesting Circuits. IEEE Journal on Emerging and Selected Topics in Circuits and Systems," vol. 2, no. 1, pp. 24–33, Mar. 2012.
12. ATLAS Device Simulation Software, "Silvaco Int.," Santa. Clara, CA, Version 5.19.20, 2016

FP-Tree and Its Variants: Towards Solving the Pattern Mining Challenges

Anindita Borah and Bhabesh Nath

Abstract Mining patterns from databases is like searching for precious gems which is a gruesome task but still a rewarding one. The frequent patterns are believed to be valuable assets for the researchers that provide them useful information. The frequent and rare pattern mining paradigm is broadly divided into Apriori and FP-Tree-based approaches. Experimental results and performance evaluation available in the literature have established the fact that FP-Tree-based approaches are superior to the Apriori ones on various grounds. This paper explores the various modifications of FP-Tree that were developed to tackle the major pattern mining research challenges. Through this paper, an attempt has been made to review the usefulness and applicability of the most eminent data structure in the domain of pattern mining, the FP-Tree.

Keywords Frequent patterns · FP-Tree · Pattern mining · Challenges

1 Introduction

Pattern mining has established itself as a significant field of data mining over the years. The notion of frequent pattern mining emphasizes that useful information may be hidden among the frequently occurring patterns in a database. Since its inception, there were several attempts from the frequent pattern mining researchers for extracting such interesting patterns. The significance of frequent patterns was

A. Borah (✉) · B. Nath
Department of Computer Science & Engineering, Tezpur University,
Napaam, Sonitpur 784028, Assam, India
e-mail: anindita01.borah@gmail.com

B. Nath
e-mail: bnath@tezu.ernet.in

A.K. Somani et al. (eds.), *Proceedings of First International Conference on Smart System, Innovations and Computing*, Smart Innovation, Systems and Technologies 79, https://doi.org/10.1007/978-981-10-5828-8_51

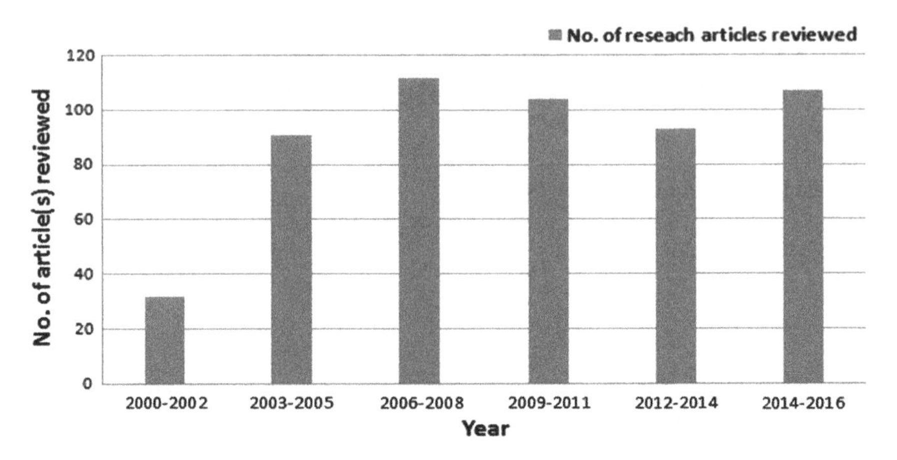

Fig. 1 Number of articles published based on FP-Tree

first identified in [1]. They developed a level-wise approach named Apriori for mining the frequent patterns by generating candidates. This primer approach proved to be inefficient as it generates an enormous quantity of candidate patterns as well as performs multiple scanning of the database. With a view to overcome the shortcomings of Apriori method, Han et al. [2] proposed a data structure called Frequent Pattern Tree (FP-Tree) that retains the complete database information. Initially, it collects the item frequencies present in the database and then generates the conditional bases and conditional FP-Trees from which the frequent patterns are extracted.

Performance evaluation illustrates that FP-Growth is the most efficient frequent pattern mining method and performs better than Apriori on various grounds. A year-wise distribution of the FP-Tree-based articles published is shown in Fig. 1. Only relevant and recent articles have been considered, excluding review articles or general surveys. The graph in the figure indicates that a commendable amount of research based on FP-Tree has been carried out over the years. Considering the significance and popularity of the FP-Tree data structure, many of its variants were proposed and various pattern mining algorithms employed the same for handling the research issues. An overview of all those attempts is therefore necessary to let the researchers get introduced with the usefulness of FP-Tree.

The remaining paper, followed by the introduction is systematized as follows: Sect. 2 discusses the various research issues that were resolved using FP-Tree based approaches. To illustrate the efficiency of FP-Tree-based approaches, a comparative analysis of the same with Apriori and its variants is provided in Sect. 3. Finally some future prospects for FP-Tree-based approaches and conclusion is discussed in Sect. 4.

2 Research Issues Handled by FP-Tree-Based Approaches

Since its outset, FP-Tree-based approaches have attempted to solve various research challenges confronted by the pattern mining community. Due to its importance and significance, the usability of FP-Tree is widespread. This section discusses the different endeavors made by FP-Tree-based approaches to handle the major pattern mining research challenges.

2.1 *Improvement in Efficiency*

In spite of being one of the widely accepted pattern mining technique, there are many instances where the efficiency of FP-Growth demands improvement. Several variations of FP-Tree have been proposed over the years that attempted to enhance the effectiveness of FP-Growth algorithm. Liu et al. [3] tried to lessen the search space of FP-Growth by employing ascending frequency order to construct the prefix trees as well as for search exploration. With a view to upgrade the effectiveness of FP-Growth algorithm on grounds of space and execution time, Racz [4] developed an alternative data structure that is more condensed than FP-Tree and also allows faster traversal and allocation.

One of the major drawbacks of FP-Growth is its imperative demand for memory. TD-FP-Growth developed by [5], avoids the generation of conditional pattern bases as well as conditional FP-Trees to reduce the amount of memory consumption. Sucahyo and Gopalan [6] symbolize the transaction items in main memory using Compressed FP-Tree (CFP-Tree) in order to lessen the memory usage. FP-Growth algorithm is also inefficient in handling sparse datasets. FPGrowth algorithm developed by Grahne and Zhu [7] attempts to solve this issue using an array based implementation.

2.2 *Mining Frequent Closed Itemsets*

Setting up a low support threshold might generate an enormous number of frequent itemsets in the database. It has been found that instead of looking for each and every frequent itemset in the database, it is better to obtain a significant class of frequent itemset called *frequent closed itemset*. A *frequent closed itemset* is one, all of whose proper subsets are frequent. Pei et al. [8] developed the CLOSET algorithm with the purpose of generating the *frequent closed itemsets*. For fast exploration of *frequent closed itemsets*, a modified compression technique called CLOSET+ [9] is used based on single-prefix path. Another algorithm called FPClose [10], employs a tree data structure called CFI-Tree to check how close the frequent itemsets are.

2.3 Secondary Memory Based

Limitation of main memory is one of the serious bottlenecks of pattern mining techniques while mining large databases. Adnan and Alhajj [11] recognized this major issue and developed a secondary memory-based approach for mining the frequent patterns. Their proposed data structure behaves similar to FP-Tree upon construction in main memory but gets modified to a disk-resident structure only when the FP-Tree can no longer be accommodated in physical memory. A similar approach was adopted by Bonchi and Goethals [12] to cater to the needs of main memory.

2.4 Handling Incremental Datasets

A major challenge encountered by the pattern mining techniques is to deal with dynamic or incremental datasets. It is worthless starting the entire mining process from scratch if any update occurs in the database. Incremental Frequent Pattern Growth (IFP-Growth) [13] is an enhanced version of the traditional FP-Growth algorithm that handles the addition or deletion of data in dynamic databases. A similar strategy is adopted by the AFPIM algorithm [14]. Cheung and Zaiane [15] developed CATS Tree that executes only a single scan of the database and gives better results in terms of storage compression. CanTree [16] and CP-Tree [17] are other single-pass algorithms for incremental mining of the database.

2.5 Mining Frequent Patterns from Uncertain Data

Mining patterns from uncertain data have established itself as an emerging field of research over the years. Leung et al. [18] proposed a new technique called UF-Growth in order to generate patterns from uncertain data. Calders et al. [19] proposed a variation of FP-Growth called UFP-Growth that focuses on extracting frequent patterns from uncertain data. Their UFP-Tree data structure stores the probabilistic information of the items in each node and computes the expected support of every item during the first scan.

2.6 Mining High Utility Itemsets

Identifying high utility or profitable itemsets from databases is an indispensable task of data mining. High utility itemsets signify those itemsets that have high importance or are more profitable for the users. To generate the high utility itemsets,

Tseng et al. [20] developed a FP-Growth-based approach called Utility Pattern Growth (UP-Growth) where the information about the high utility itemsets is stored in a tree structure called Utility Pattern Tree (UP-Tree) which is further extended in [21]. Lin et al. [22] exploited the property of FP-Tree in their proposed data structure called High Utility Pattern Tree (HUP-Tree). Their approach integrates the strategies of two-phase algorithm.

2.7 *Mining Sequential Patterns*

Sequential pattern mining has established itself as an important data mining application over the years. Lin et al. [23] proposed a sequential pattern mining data structure called Fast Updated Sequential Pattern Tree (FUSP-Tree). In order to represent the sequence relation between two connecting nodes, the link between them is imprinted with the symbol *s* whereas to represent the relation between items, the link is imprinted with the symbol *i*. Pei et al. [24] proposed another FP-Tree-based data structure called Web Access Pattern Tree (WAP-Tree) for mining frequent sequential patterns from web logs. The data structure holds information about access patterns and the mining algorithm then generates the access patterns from log.

2.8 *Mining Maximal Frequent Itemsets*

Tremendous number of itemset generation has always been the major issue encountered by frequent pattern mining techniques. To lessen the number of candidate subsets formed, some of the existing pattern mining techniques emphasize the generation of *maximal frequent itemsets*. In case of *maximal frequent itemsets*, none of the superset is frequent. Grahne and Zhu [25] developed a recursive algorithm called FPMax to mine the MFI's where a linked list is used to store the items of the conditional pattern bases during the current call. Yan et al. [26] developed the Frequent Pattern Tree for Maximal Frequent Itemsets (FPMFI) that employs a projection based superset checking technique to find out the MFI's.

2.9 *Handling Data Streams*

Frequent pattern mining over data streams comes up with diverse challenges. The rapid and continuous flow of data stream makes the mining and update of frequent patterns a difficult task. Giannella et al. [27] proposed a variation of FP-Growth algorithm called FP-stream with a view to generate the frequent patterns from data streams. The algorithm is based on a *tilted time window* framework and employs two tree data structures. DS-Tree data structure developed by Leung et al. [28], stores the information of the data streams in a canonical order. Tanbeer et al. [29] proposed a data structure CPS-Tree that employs a sliding window strategy and partitions the window into a series of transactions called *pane*.

2.10 Mining Rare Patterns

Recent studies show that in several domains, the rare items are of greater interest as compared to the frequent ones. This insisted on mining and retaining rare items that are removed during the frequent itemset generation phase. Hu and Chen [30] modified the FP-Growth algorithm by incorporating multiple minimum supports during itemset generation. Another variant of FP-Growth was proposed by Tsang et al. [31] called the Rare Pattern Tree (RP-Tree) algorithm that generates only the rare itemsets, discarding the frequent ones. Bhatt and Patel [32] further extended the RP-tree algorithm using the maximum constraint model to improve its efficiency.

2.11 Handling Big Data

Extracting frequent patterns from big data using traditional pattern mining techniques is a difficult and problematic task. To handle big data, Chen et al. [33] developed the Parallel FP-Growth (PFP-Growth) algorithm to allow parallel processing of big data. Leung and Hayduk [34] developed the MR-Growth algorithm that generates frequent patterns from big uncertain data. The concepts of FP-Growth and MapReduce are combined in this novel method to extract frequent patterns from large quantities of uncertain data. To incrementally update the frequent itemsets in big data, Chang et al. [35] proposed a method that employs a heap data structure and outperforms existing algorithms in terms of complexity as well as execution time.

3 Comparison with Apriori and Its Variants

The FP-Tree and Apriori-based approaches have been extensively used for handling the pattern mining challenges. This section presents a general comparison of the research work carried out, employing these two standard strategies. The graphical analysis given in Fig. 2, illustrates a comparison between the number of approaches developed under Apriori and FP-Growth. The blue bar in the graph represents FP-Tree-based approaches while the red bar represents Apriori-based approaches respectively. From the graph it can be observed that except two issues, that is extracting rare patterns and frequent pattern mining from

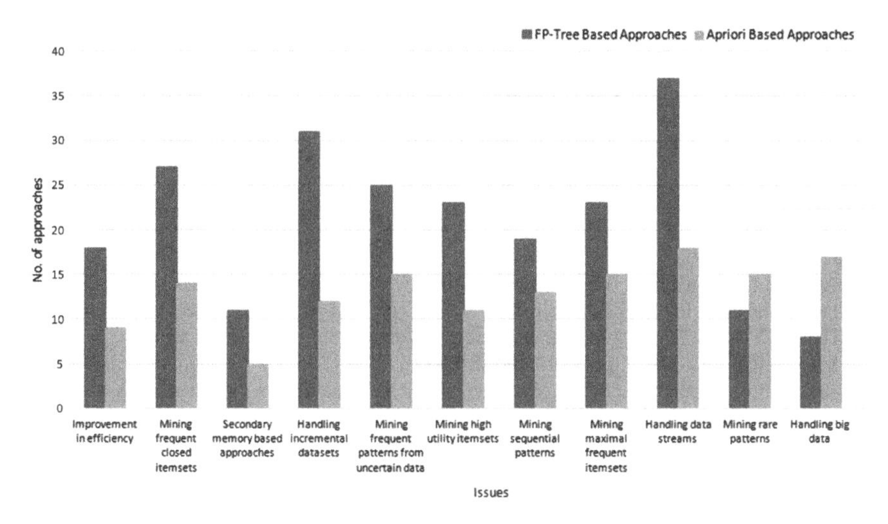

Fig. 2 Comparison between Apriori and FP-Tree-based approaches

big data, FP-Tree-based approaches are extensively explored over Apriori-based approaches. This clearly establishes the superiority of FP-Tree-based approaches over the Apriori ones.

4 Conclusion and Future Prospects

The pattern growth approaches have been found to be more efficient among the pattern mining techniques. Among the pattern growth approaches, the most prominent and favorable one is the FP-Growth algorithm. From Sect. 2, the relevance of FP-Tree based approaches can be recognized. The comparative study given in Sect. 3, establishes the superiority of FP-Tree-based approaches over Apriori ones. However, there are still some issues that are not pervasively addressed by the FP-Tree-based approaches.

FP-Tree based approaches are inefficient in handling sparse datasets. In spite of this fact, only one attempt has been made to improve its efficiency in this regard. Rare pattern mining being a new and emerging area, have not explored the FP-Tree based approaches to much extent. Only a limited number of approaches based on FP-Tree can be found in the literature. Another less explored issue by FP-Tree-based approaches is mining patterns from big data. The approaches based on FP-Tree are quite less than Apriori approaches for handling this issue. Considering the popularity and effectiveness of FP-Tree-based approaches, the researchers need to work on the less explored issues to contribute some fruitful pattern mining techniques to the research community. Even though, literature has endowed plentiful FP-Tree-based approaches to the pattern mining community, there is still much room for expansion.

References

1. Agrawal, R., Imielinski, T., Swami, A.: Mining association rules between sets of items in large databases. In: Acm sigmod record. vol. 22, pp. 207–216. ACM (1993).
2. Han, J., Pei, J., Yin, Y.: Mining frequent patterns without candidate generation. In: ACM Sigmod Record. vol. 29, pp. 1–12. ACM (2000).
3. Liu, G., Lu, H., Yu, J.X., Wang, W., Xiao, X.: Afopt: An efficient implementation of pattern growth approach. In: FIMI (2003).
4. Racz, B.: nonordfp: An fp-growth variation without rebuilding the fp-tree. In: FIMI (2004).
5. Wang, K., Tang, L., Han, J., Liu, J.: Top down fp-growth for association rule mining. In: Pacific-Asia Conference on Knowledge Discovery and Data Mining. pp. 334–340. Springer (2002).
6. Sucahyo, Y.G., Gopalan, R.P.: Ct-pro: A bottom-up non recursive frequent Itemset mining algorithm using compressed fp-tree data structure. In: FIMI. vol. 4, pp. 212–223 (2004).
7. Grahne, G., Zhu, J.: Efficiently using prefix-trees in mining frequent itemsets. In: FIMI. vol. 90 (2003).
8. Pei, J., Han, J., Mao, R., et al.: Closet: An efficient algorithm for mining frequent closed itemsets. In: ACM SIGMOD workshop on research issues in data mining and knowledge discovery. vol. 4, pp. 21–30 (2000).
9. Wang, J., Han, J., Pei, J.: Closet+: Searching for the best strategies for mining frequent closed itemsets. In: Proceedings of the ninth ACM SIGKDD international conference on Knowledge discovery and data mining. pp. 236–245. ACM (2003).
10. Grahne, G., Zhu, J.: Fast algorithms for frequent itemset mining using fp-trees. IEEE transactions on knowledge and data engineering 17(10), 1347–1362 (2005).
11. Adnan, M., Alhajj, R.: Drfp-tree: disk-resident frequent pattern tree. Applied Intelligence 30 (2), 84–97 (2009).
12. Bonchi, F., Goethals, B.: Fp-bonsai: the art of growing and pruning small fp-trees. In: Pacific-Asia Conference on Knowledge Discovery and Data Mining. pp. 155–160. Springer (2004).
13. Xu, B., Yi, T., Wu, F., Chen, Z.: An incremental updating algorithm for mining association rules. Journal of Electronics (China) 19(4), 403–407 (2002).
14. Koh, J.L., Shieh, S.F.: An efficient approach for maintaining association rules based on adjusting fp-tree structures. In: International Conference on Database Systems for Advanced Applications. pp. 417–424. Springer (2004).
15. Cheung, W., Zaiane, O.R.: Incremental mining of frequent patterns without candidate generation or support constraint. In: Database Engineering and Applications Symposium, 2003. Proceedings. Seventh International. pp. 111–116. IEEE (2003).
16. Leung, C.K.S., Khan, Q.I., Li, Z., Hoque, T.: Cantree: a canonical-order tree for incremental frequent-pattern mining. Knowledge and Information Systems 11(3), 287–311 (2007).
17. Tanbeer, S.K., Ahmed, C.F., Jeong, B.S., Lee, Y.K.: Cp-tree: a tree structure for single-pass frequent pattern mining. In: Pacific-Asia Conference on Knowledge Discovery and Data Mining. pp. 1022–1027. Springer (2008).
18. Leung, C.K.S., Carmichael, C.L., Hao, B.: Efficient mining of frequent patterns from uncertain data. In: Seventh IEEE International Conference on Data Mining Workshops (ICDMW 2007). pp. 489–494. IEEE (2007).

19. Calders, T., Garboni, C., Goethals, B.: Efficient pattern mining of uncertain data with sampling. In: Pacific-Asia Conference on Knowledge Discovery and Data Mining. pp. 480–487. Springer (2010).
20. Tseng, V.S., Wu, C.W., Shie, B.E., Yu, P.S.: Up-growth: an efficient algorithm for high utility itemset mining. In: Proceedings of the 16th ACM SIGKDD international conference on Knowledge discovery and data mining. pp. 253–262. ACM (2010).
21. Tseng, V.S., Shie, B.E., Wu, C.W., Philip, S.Y.: Efficient algorithms for mining high utility itemsets from transactional databases. IEEE transactions on knowledge and data engineering 25(8), 1772–1786 (2013).
22. Lin, C.W., Hong, T.P., Lu, W.H.: An effective tree structure for mining high utility itemsets. Expert Systems with Applications 38(6), 7419–7424 (2011).
23. Lin, C.W., Hong, T.P., Lu, W.H., Lin, W.Y.: An incremental fusp-tree maintenance algorithm. In: 2008 Eighth International Conference on Intelligent Systems Design and Applications. vol. 1, pp. 445–449. IEEE (2008).
24. Pei, J., Han, J., Mortazavi-Asl, B., Zhu, H.: Mining access patterns efficiently from web logs. In: Pacific-Asia Conference on Knowledge Discovery and Data Mining. pp. 396–407. Springer (2000).
25. Grahne, G., Zhu, J.: High performance mining of maximal frequent itemsets. In:6th International Workshop on High Performance Data Mining (2003).
26. Yan, Y.J., Li, Z.J., Chen, H.W.: Efficiently mining of maximal frequent item sets based on fp-tree. Ruan Jian Xue Bao (J. Softw.) 16(2), 215–222 (2005).
27. Giannella, C., Han, J., Pei, J., Yan, X., Yu, P.S.: Mining frequent patterns in data streams at multiple time granularities. Next generation data mining 212, 191–212 (2003).
28. Leung, C.K.S., Khan, Q.I.: Dstree: a tree structure for the mining of frequent sets from data streams. In: Sixth International Conference on Data Mining (ICDM'06). pp. 928–932. IEEE (2006).
29. Tanbeer, S.K., Ahmed, C.F., Jeong, B.S., Lee, Y.K.: Efficient frequent pattern mining over data streams. In: Proceedings of the 17th ACM conference on Information and knowledge management. pp. 1447–1448. ACM (2008).
30. Hu, Y.H., Chen, Y.L.: Mining association rules with multiple minimum supports: anew mining algorithm and a support tuning mechanism. Decision Support Systems 42(1), 1–24 (2006).
31. Tsang, S., Koh, Y.S., Dobbie, G.: Rp-tree: rare pattern tree mining. In: Data Warehousing and Knowledge Discovery, pp. 277–288. Springer (2011).
32. Bhatt, U., Patel, P.: A novel approach for finding rare items based on multiple minimum support framework. Procedia Computer Science 57, 1088–1095 (2015).
33. Chen, M., Gao, X., Li, H.: An efficient parallel fp-growth algorithm. In: Cyber-Enabled Distributed Computing and Knowledge Discovery, 2009. CyberC'09. International Conference on. pp. 283–286. IEEE (2009).
34. Leung, C.K.S., Hayduk, Y.: Mining frequent patterns from uncertain data with map reduce for big data analytics. In: International Conference on Database Systems for Advanced Applications. pp. 440–455. Springer (2013).
35. Chang, H.Y., Lin, J.C., Cheng, M.L., Huang, S.C.: A novel incremental data mining algorithm based on fp-growth for big data. In: Networking and Network Applications (NaNA), 2016 International Conference on. pp. 375–378. IEEE (2016).

Game-Theoretic Method for Selection of Trustworthy Cloud Service Providers

Monoj Kumar Muchahari and Smriti Kumar Sinha

Abstract In spite of lucrative features, customers are still afraid of deploying their business in the Cloud. The main hindrance in the acceptance of cloud computing is lack of trust on Cloud service providers by promising customers. This paper introduces a game-theoretic technique for selection of a trustworthy Cloud service provider in strategic and extensive-form games. Potential Cloud service providers as well as consumers are assumed as the two rational players, whose actions are measured in terms of ordinal payoffs. We then solved those games based on game theory solution concepts. The solutions are demonstrated in this paper through experiment.

Keywords Game theory · Cloud computing · Trust · Nash

1 Introduction

Cloud computing is very large, open and dynamically scalable where trustors may deceit to boost their benefits [1]. The main hindrance in the acceptance of cloud computing is lack of trust on Cloud service providers by promising customers [2, 3]. Trust has several uses for the autonomous and multi-agent systems and Game Theory (GT) can be one of the significant perspectives in related research [4]. Cloud Service Consumer (CSC) will decide certain strategies to consume some services provided by different Cloud Service Providers (CSPs) depending upon the strategies of those CSPs. Similar scenarios are well analyzed in GT, where choices of others determine the achievement of an individual in making choices. In this paper through GT, we attempt to explain and predict how CSP and CSC behave. We

---

M.K. Muchahari (✉) · S.K. Sinha
Department of Computer Science & Engineering, Tezpur University, Napaam, Sonitpur 784028, Assam, India
e-mail: memonoj01@yahoo.com

S.K. Sinha
e-mail: smriti@tezu.ernet.in

A.K. Somani et al. (eds.), *Proceedings of First International Conference on Smart System, Innovations and Computing*, Smart Innovation, Systems and Technologies 79, https://doi.org/10.1007/978-981-10-5828-8_52

provide three most commonly used solution concepts namely Dominance, Nash Equilibrium (NE) and Subgame Perfect Nash Equilibrium (SPNE) to solve the game. The remaining paper proceeds by modeling the trustworthy CSP selection game in different types of games according to the possible market scenario, solving those models, elementary proof of existence of a NE and experimental results are provided. Finally, a conclusion section and suggestions for future research are given.

2 Related Works

According to Niyato et al. [5], to offer infrastructure-as-a-service (IaaS) in cloud, game models may help in analyzing the behavior and decision-making process of the self-interested Cloud providers. A broad approach describing the notion of institution in terms of game's equilibrium may be considered to examine the factors related to privacy and security issues of the Cloud [6]. Today various solutions relating to scheduling [7], security [8], resource allocation [9] and more, in Cloud make use of GT. Zeng et al. [10] suggests that GT can improve the performance and efficiency of operational mechanism of Cloud storage. For many years, GT is the most prevailing paradigm employed for the design of reputation models and computational trust [11]. But to the best of our knowledge till date GT have not been used for choosing trustworthy CSP.

3 Modeling the Game

To characterize the key aspects and for simplicity, we confine our analysis to two-person games namely CSP and CSC. For modeling a service provider and service consumer game, following suppositions are assumed: CSP and CSC are rational, independent and their actions are measured in terms of *ordinal payoffs*; CSP and CSC may change their requirements and preferences that will affect the service behavior; CSP has advertised a Cloud service presenting an intention to satisfy requests that conform to the service; CSC has its own service requirements and will opt for the best service.

As *cardinal payoffs* are difficult to calculate, we take *ordinal payoffs* into consideration for modeling the game. For the calculation of payoff, we assume outright information of strategies of CSP and CSC. Further to convert the information into payoffs we need to evaluate the following cost of services: providers' gain in providing complied services (G_{cp}), consumers' gain in consuming complied services (G_{cc}), providers' loss when intended complied services are not consumed (L_{icnc}), consumers' loss when complied services are not consumed (L_{cnc}), providers' gain when intended non-complied services are consumed (G_{inc}), consumers' loss when non-complied services are consumed (L_{nc}), providers' loss when intended non-complied services are not consumed (L_{innc}) and consumers' gain when non-complied services are not consumed (G_{ncnc}). GT models being abstract, it is a common practice to ignore certain costs or combine all costs or vary costs accordingly. For example, if we consider natural numbers for some general scenario of cloud service providing and consuming, the payoff calculation can have general form as Eq. 1:

$$G_{inc} > G_{cp} = G_{cc} > L_{cnc} = L_{innc} = G_{ncnc} > L_{icnc} = L_{nc} \quad (1)$$

3.1 Dominance

CSP and CSC are considered as *Player I* and *Player II*, respectively, for the game of providing and consuming Cloud services. *Player I* has two strategies to either *Comply* or *Cheat*, labeled as *C* and *D*. *Player II* has two strategies to either *Consume* or *Don't Consume*, labeled as *c* and *d* respectively. It is difficult to verify the quality of service of *Player I* by *Player II*. Assuming that *Player I* will always try to cheat in order to minimize their cost or not to take-up the burden of eradicating the complexities while providing services. Complying of *Player I* is more preferable for *Player II* to cheating.

To describe the situation, Fig. 1 gives the possible payoffs. CSP and CSC gets payoff of 2 each for complying and consuming respectively as both of them are equally happy. If CSP complies with the service qualities and if CSC does not consumes that service, gives payoff of 0 and 1. CSC earns 0 payoff or nothing while CSP cheating and still having consumer earns them 3 payoff as they have customers even if their service is bad. CSP not complying and CSC not consuming that very service bring in payoff of 1 each as both have chances to opt for other options. *Player I* plays the row, either *C* or *D*, and simultaneously *Player II* plays one of the columns *c* or *d*. Combination (*C*, *c*) gives payoff 2 for each player, and combination (*D*, *d*) gives each player payoff 1. Payoff is 0 when combination is (*C*, *d*), for *Player I* and 1 for *Player II*, and when (*D*, *c*) is played, *Player I* gets 3 and *Player II* gets 0. *Player II* is aware that it is difficult for *Player I* to comply and keeping in mind the rationality of *Player I*, *Player II* will choose not to consume (payoff 1) than to consume (payoff 0) the service. Thus, both the players being rational provider will cheat of what is promised and consequently service will not be consumed. If *Player I* chooses strategy *D*, *Player I* dominates *C* because if *Player II* chooses *c*, then payoff of *Player I* is 3 when choosing *D* and 2 when choosing *C;* if *Player II* chooses *d*, then *Player I* gets 1 for *D* and 0 for *C*. We can conclude that strategy *d* dominates *c* and also *D* is always better and dominates *C*. For strategic combination (*C*, *c*) that is CSP complies and so CSC consumes, is not possible as CSP would always tend to cheat. As, the individually rational outcome is worse for both the players than another outcome so, the outcome cannot be deemed as determinative.

Fig. 1 Game of providing and consuming cloud services

Player I (CSP) \ Player II (CSC)	Consume (c)	Don't Consume (d)
Comply (C)	2, 2	0, 1
Cheat (D)	3, 0	1, 1

The indecisive outcome of the game can be said to portray the real scenario in cloud computing today. It is indeed due to the unethical CSPs and their inability to cope with the issues, makes the task of CSCs difficult [12].

3.2 Extensive-Form Game with Imperfect Information

CSC might not have information about CSP, if they are new. Also at times, CSC do not always have relevant information about the CSP. This situation can be modeled as extensive game with imperfect information. Let us speculate the situation, when CSC comes across a new CSP with perfect advertisements about their service. The CSP can only know where they stand among the other Cloud competitors. The CSC can believe 50% of whatever the CSP claims. CSP has two choices; they can strive to deliver better service or not. Alternatively, CSC can either consume the service or only repent for consuming the service keeping in mind the non-availability of interoperability of service. CSC does not have adequate information about the CSP's quality of service. If the CSP does not comply with the agreement or advertisement, CSC will discontinue consuming the service.

Information set indicates a state where a *player* cannot distinguish between two or more states. In extensive-form games with imperfect information, backward induction can be complex because a player's favorable action depends on which node the player is at in that *information set.* So, we will introduce another solution concept called *subgame perfect equilibrium* or *subgame perfect Nash equilibrium.*

In Fig. 2, CSP or *Player I,* if choose not to provide service, yields payoff of 1. On the contrary, if *Player I* decides to provide service then the player has the option to either *comply* (*C*) or *cheat* (*D*). Dotted line represents information set for CSC or *Player II.* In this position, *Player II* is not aware of the decision of *Player I.* There are strategies for *Player II* namely *consume* (*c*) and *don't consume* (*d*) giving earns of 1, 0, 1 and 2 as payoffs. We know that a *subgame* begins with a singleton information set, contains all decision nodes following this information set and does not split any information sets. Figure 3a and b cannot be *subgames.* Only the entire game itself is a *subgame* as shown in Fig. 4a. From normal form game Fig. 4b, we

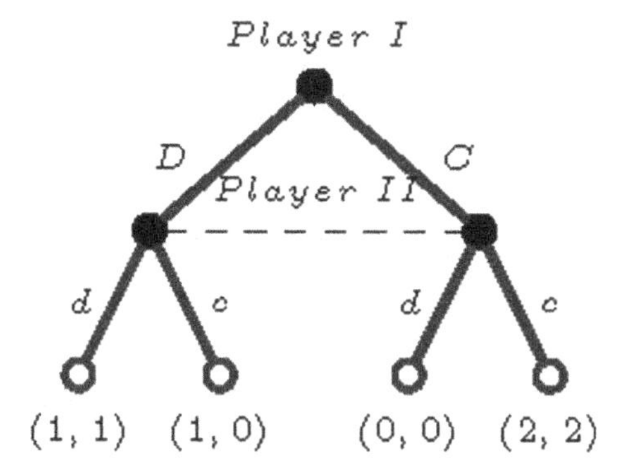

Fig. 2 Game tree with imperfect information

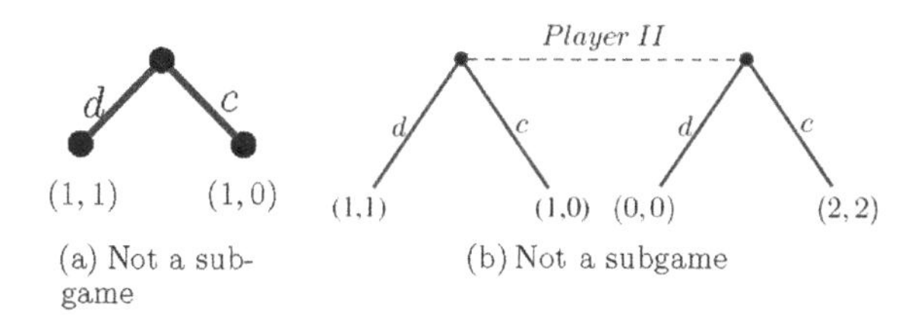

(a) Not a subgame (b) Not a subgame

Fig. 3 Not a subgame

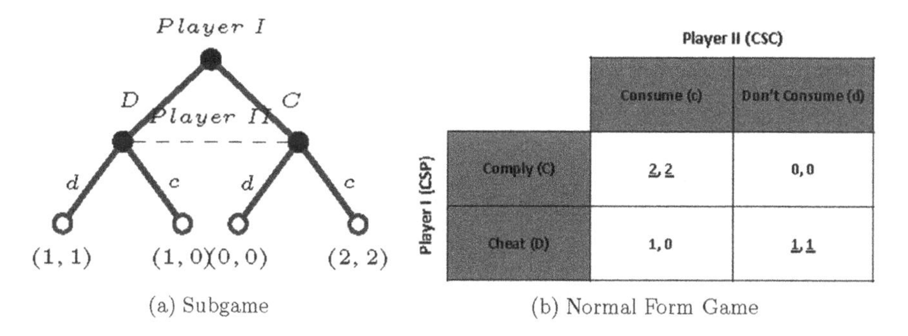

	Player II (CSC)	
Player I (CSP)	Consume (c)	Don't Consume (d)
Comply (C)	2,2	0,0
Cheat (D)	1,0	1,1

(a) Subgame (b) Normal Form Game

Fig. 4 Subgame Perfect Nash Equilibria (SPNE)

have two NE, (*C*, *c*) with payoff (2, 2) and (*D*, *d*) with payoff (1, 1). The strategy (*D*, *d*) is not a credible solution. Optimal play for *Player I* in the root is to *provide* giving payoff of 2 rather than 1. Hence, the *SPE* for this game is (*comply, consume*).

Past history of CSPs can be beneficial for building trust but what if the CSP is new and have no history. This kind of situation can be a daunting task for CSCs. Better is to truncate the game instead of the overall game. In that truncated game or subgame, the action of both CSP and CSC is sequentially rational at information set which is a listing of all the possible distinguishable circumstances in which CSC might be called upon to move. In solution concept of SPNE, the strategy profile of both CSP and CSC is optimal not only at the start of the game but also after every history. So, definitely optimal solution for the whole game will lead to trustworthiness.

3.3 Existence of Nash Equilibrium

NE is the most influential solution concept in GT. Every finite game has an equilibrium point [13]. So, we try to prove the existence of NE in this game using elementary technique.

Let, *Player I*'s strategy be represented by α such that $\sigma 1 = \alpha C + (1 - \alpha)D$ while *Player II*'s strategy be $\sigma 2 = \alpha c + (1 - \alpha)d$ as in Fig. 5a.

So, the best response of *Player II* to *Player I* playing strategy α:

$$u2(c, \alpha C + (1-\alpha)D) = 2\alpha; \quad u2(d, \alpha C + (1-\alpha)D) = 1 - \alpha$$

So, *Player II* will strictly prefer strategy c if $2\alpha > 1 - \alpha$ which implies $\alpha > 1/3$. Therefore, the best response correspondence of *Player II* is

$$BR_2\alpha = \begin{cases} 1, \ if \ \alpha < 1/3 \\ [0,1], \ if \ \alpha = 1/3 \\ 0, \ if \ \alpha > 1/3 \end{cases}$$

Similarly, best response of *Player I* to *Player II* playing strategy β:

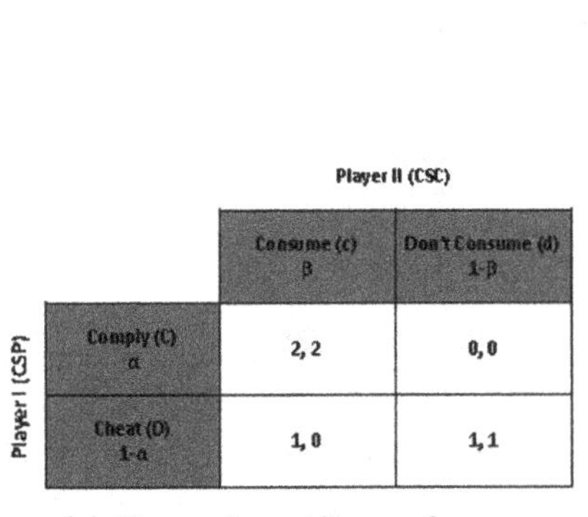
Player II (CSC)

Player I (CSP)	Consume (c) β	Don't Consume (d) 1-β
Comply (C) α	2, 2	0, 0
Cheat (D) 1-α	1, 0	1, 1

(a) Game of providing and consuming Cloud services for existence of NE

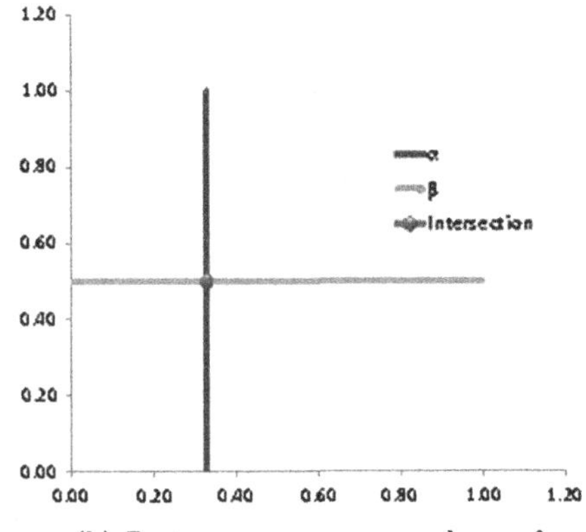

(b) Best-response correspondences for the game

Fig. 5 Existence of NE

$$u1(C, \beta c + (1-\beta)d) = 2\beta; \quad u1(D, \beta c + (1-\beta)d) = 1$$

So, best response correspondence of *Player I* is

$$BR_2\alpha = \begin{cases} 1, \; if \; \alpha < 1/2 \\ [0,1], \; if \; \alpha = 1/2 \\ 0, \; if \; \alpha > 1/2 \end{cases}$$

If we draw both best response correspondences in a single graph, we see in Fig. 5b that both correspondences intersect in the single point $\alpha = 1/3$ and $\beta = 1/2$. If each player plays a best response to the strategies of all other players, we have a NE, which is the unique (mixed) NE of the game. Existence of NE in this game means both CSPs and CSCs can be in a state where they can make best decisions taking into account the decisions of the others. NE provides the way of predicting the best decisions of CSPs to how to be trustworthy and of CSCs to how much they can trust the CSPs.

4 Experimental Results

Experiment is carried out using publicly available trust feedback dataset called as "Armor" by Noor et al. [14] and two synthetic dataset named as "Synthetic 1" (No. of CSCs—10000, No. of CSPs—150, No. of QoS attributes—14 and Rating Range—3–5) and "Synthetic 2" (No. of CSCs—9668, No. of CSPs—100, No. of QoS attributes—9 and Rating Range—1–5), generated using random integer number function in Matlab.

To determine the NE, we evaluated the payoffs using Eq. 1 on the basis of average QoS rating values. The NE payoffs of "Armor" dataset is shown in Fig. 6a. Along with positive payoffs that indicate trustworthy, there are some negative payoffs indicating untrustworthy possibilities. In "Armor", the payoffs of CSP 9, 10, 11, 41, 95, 96, and 97 have much better payoffs but NE payoffs is negative. CSPs are cheating and CSCs are not consuming those services imply that no player has anything to gain by changing only their own strategy and thus the CSPs can be deemed as untrusthworthy. The result on "Synthetic 1" as seen in Fig. 6b, that have rating range of 3–5 shows all positive payoffs and depending on the payoffs we can be aware of the relevant strategies. Figure 6c, depicting the NE for "Synthetic 2", shows variation according to the dataset. General conclusion that can be drawn is that with an appropriate payoff calculation equation and dataset, NE can be calculated to interpret the possible strategies to judge trustworthiness. Varying the payoff equation based on cost of services results in different NE.

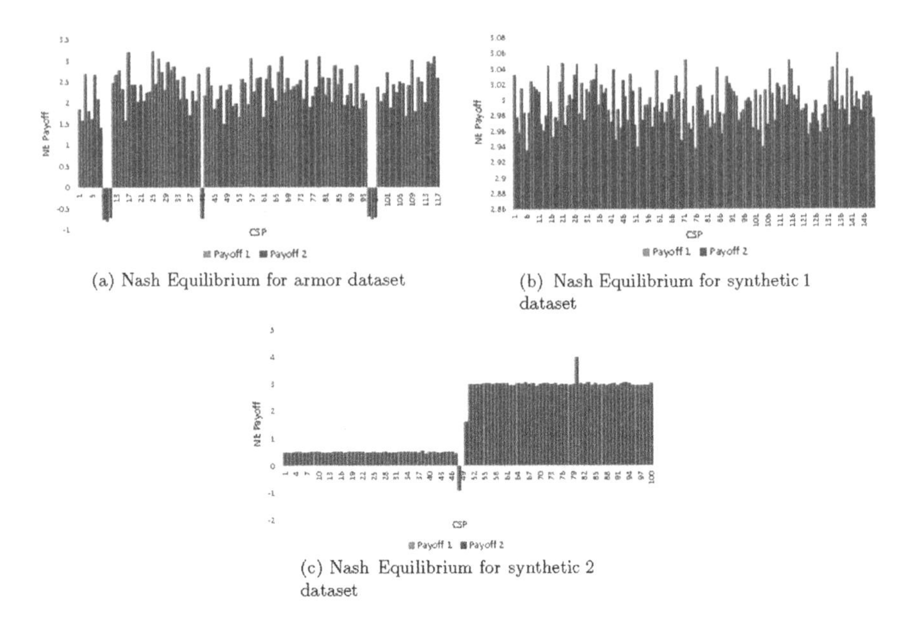

(a) Nash Equilibrium for armor dataset

(b) Nash Equilibrium for synthetic 1 dataset

(c) Nash Equilibrium for synthetic 2 dataset

Fig. 6 Experimental results

5 Conclusion

This paper presents a game-theoretic method for selection of trustworthy CSP. Different strategies for both CSP and CSC are analyzed based on ordinal payoffs. We propose a way to calculate payoffs based on cost of services, assuming it as the outright information of strategies. We have tried to model the Cloud service provision and consumption so as to mimic real-life scenario and provided solution to those game models. Those solutions can help both CSPs and CSCs to decide their strategies according to their needs to increase the trustworthiness level. Using elementary technique, existence of NE is proved in the game which can be very useful benchmark and a starting point for the game analysis. Experiments using real world cloud dataset and synthetic dataset were carried out. Possibility of cooperation between CSPs and CSCs indicates scope for increase in trust level of CSPs by CSCs. We will be extending our work concentrating more on repeated and multi-player game that interpret real cloud service scenario. Besides, substantive cost of services for payoff calculation is an issue that also need to be addressed in future.

References

1. Sun, D., Chang, G., Sun, L., Li, F., Wang, X.: A dynamic multi-dimensional trust evaluation model to enhance security of cloud computing environments. International Journal of Innovative Computing and Applications 3(4), 200–212 (2011).
2. Muchahari, M.K., Sinha, S.K.: A survey on web services and trust in cloud computing environment. In: National Workshop on Network Security 2013, Tezpur University. pp. 91–102.
3. Muchahari, M.K., Sinha, S.K.: Mmh: An effective clustering algorithm for trustworthy cloud service provider selection. International Journal of Computer Science and Information Security 14(5), 281 (2016).
4. Artz, D., Gil, Y.: A survey of trust in computer science and the semantic web. Web Semantics: Science, Services and Agents on the World Wide Web 5(2), 58–71 (2007).
5. Niyato, D., Vasilakos, A.V., Kun, Z.: Resource and revenue sharing with coalition formation of cloud providers: Game theoretic approach. In: Cluster, Cloud and Grid Computing (CCGrid), 2011 11th IEEE/ACM International Symposium on. pp. 215–224. IEEE (2011).
6. Kshetri, N.: Privacy and security issues in cloud computing: The role of institutions and institutional evolution. Telecommunications Policy (2012).
7. Mohammadi Fard, H., Prodan, R., Fahringer, T.: A truthful dynamic workflow scheduling mechanism for commercial multi-cloud environments (2013).
8. Kırlar, B.B., Ergün, S., Gök, S.Z.A., Weber, G.W.: A game-theoretical and cryptographical approach to crypto-cloud computing and its economical and financial aspects. Annals of Operations Research pp. 1–15 (2016).
9. Wei, G., Vasilakos, A.V., Zheng, Y., Xiong, N.: A game-theoretic method of fair resource allocation for cloud computing services. The Journal of Supercomputing 54(2), 252–269 (2010).
10. Zeng, W., Zhao, Y., Ou, K., Song, W.: Research on cloud storage architecture and key technologies. In: Proceedings of the 2nd International Conference on Interaction Sciences: Information Technology, Culture and Human. pp. 1044–1048. ACM (2009).
11. Sabater, J., Sierra, C.: Review on computational trust and reputation models. Artificial intelligence review 24(1), 33–60 (2005).
12. Muchahari, M.K., Sinha, S.K.: A new trust management architecture for cloud computing environment. In: Cloud and Services Computing (ISCOS), 2012 Inter- national Symposium on. pp. 136–140. IEEE (2012).
13. Nash, J.: Non-cooperative games. The Annals of Mathematics 54(2), 286–295 (1951).
14. Noor, T.H., Sheng, Q.Z., Ngu, A.H., Alfazi, A., Law, J.: Cloud armor: a platform for credibility-based trust management of cloud services. In: Proceedings of the 22nd ACM international conference on Conference on information & knowledge management. pp. 2509–2512. ACM (2013).

Design of Low-Power Full Adder Using Two-Phase Clocked Adiabatic Static CMOS Logic

Dinesh Kumar and Manoj Kumar

Abstract In this paper, a full adder using two-phase clocked adiabatic static CMOS logic (2PASCL) has been presented. A six-transistor X-OR gate has been used with transmission gate multiplexer. The simulations of proposed and other designs have been performed in 0.18 μm CMOS technology. The proposed design shows improved power delay product (PDP) in the range of 0.34×10^{-21} J to 1.12×10^{-21} J as compared to 4.53×10^{-21} J to $6.7\ 8 \times 10^{-21}$J (static energy recovery full adder), $3.4\ 1 \times 10^{-21}$ J to 7.36×10^{-21} J (10 transistor), 6.40×10^{-21} J to 19.17×10^{-21} J (transmission gate) with a supply voltage variation of 1.2 V–2.8 V respectively. The proposed design also performs better at varying temperature conditions as compared to other existing designs. Simulation results of proposed design have been compared with existing designs reported in the literature and proposed design shows better performance in terms of PDP.

Keywords Adiabatic logic · CMOS · Low-power full adder
Power delay product

1 Introduction

Ultra low-power portable electronic devices have become the backbone of today's electronic world. Portable electronic devices used for mundane activities, health monitoring instrument, advanced biomedical implantable devices (hearing aids, pacemakers, etc.), and modern warfare instruments needs to be power efficient for prolonged battery life. These devices composed of ultra large-scale integrated circuits (ULSICs) which consist of complementary metal–oxide semiconductor (CMOS) field effect transistors. Fundamentally there are three main components of power dissipation in an MOS device, dynamic power dissipation, short circuit

Dinesh Kumar (✉) · Manoj Kumar
University School of Information, Communication and Technology,
Guru Gobind Singh Indraprastha University, Delhi, India
e-mail: dinesh4saini@gmail.com

A.K. Somani et al. (eds.), *Proceedings of First International Conference on Smart System, Innovations and Computing*, Smart Innovation, Systems and Technologies 79, https://doi.org/10.1007/978-981-10-5828-8_53

power dissipation, and static or leakage power dissipation [1]. As the feature size decreases with the amelioration of technology, according to Moore's law [2] the component of power dissipation due to leakages increases. As the significant improvement in scaling has been achieved already, it becomes very difficult for researchers to improve scaling further as it imposes many limitations. Therefore, researchers need to explore new techniques of power reduction for longer battery life. Energy recovery is such an energy saving techniques which are based on adiabatic logic. In electronics devices and systems, high-speed power-efficient processors play a vital role for computations. The processers of electronic devices which are used for implementations of dedicated algorithms like filtering and convolution [3] primarily depends on arithmetic and floating point units. These units consist of multipliers which require more than one number of adders or compressors [4] stages for complex computations. Therefore, adders contribute a major part of power dissipation in electronic systems. There are several designs of adders based on different techniques, which are reported in literature such as energy efficient single bit full adder [5], a new X-OR, X-NOR-based 4-2 compressor [6]. Body biasing technique has been used for implementation of the full adder in [7]. Sub-threshold adiabatic logic has been reported for ultra low-power applications in [8]. In this paper, a full adder with 2PASCL [9] based six transistors X-OR gate has been presented. Full adder based on different techniques which are reported in literature such as static energy recovery full adder (SERF), 10 transistor (10T) and transmission gate (TG) [10] have been simulated and compared with the proposed design.

1.1 Adder Architecture and Functionality

Adder fundamentally does the addition of bits; a full adder adds three bits and gives output as the sum and carry. X-OR gates primarily used for addition with AND gate or multiplexer to calculate carry. In this work, X-OR with multiplexer has been used for the realization of a full adder. The block-level structure of a full adder has been shown in Fig. 1. Four transistors-based X-OR gate [10] modified with 2PASCL, and transmission gate (TG) MUX has been used for the implementation

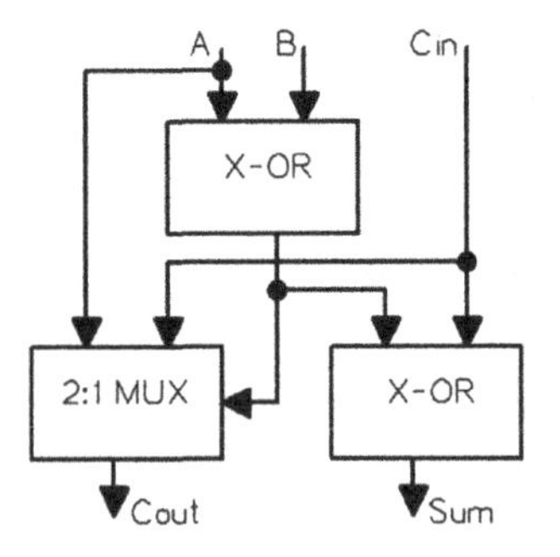

Fig. 1 Full adder

of adder. The logical expression of a full adder can be obtained with the help of K-map by using a standard truth table. If the three input bits are *A, B, C* then the logical expressions for sum and carry can be given as follows.

$$Sum = A \oplus B \oplus C$$

$$Carry = AB + BC + CA$$

1.2 Adiabatic Logic

The term adiabatic generally used in thermodynamic which means, when a system goes from one state to another there will be no loss or gain of heat ideally. During charging and discharging of a conventional CMOS inverter CV_{dd}^2 energy is required from the supply. The node capacitance stores $1/2CV_{dd}^2$ energy and rest $1/2CV_{dd}^2$ is dissipated by the PMOS ON state resistance during charging of node capacitance. The energy stored in node capacitance will go to the ground through NMOS and wasted away. In the case of adiabatic logic circuits, the stored energy goes back to the power supply which reduces the power dissipation. Adiabatic charging and discharging can be modeled by an *RC* circuit as shown in Fig. 2.

The capacitor *C* represents the node capacitance which is charged through an MOS transistor. The ON state resistance of MOS represented by *R* and a constant current *I*(*t*) source has been used to charge the *C* as compared to a fixed voltage V_{dd} in traditional CMOS inverter. The capacitor voltage can be represented by *Vc*(*t*) if the initial charge at the capacitor is zero and is given by the following expression:

$$V_c(t) = \frac{1}{C} i(t) * t \tag{1}$$

And

$$i(t) = C \frac{V_c(t)}{t} \tag{2}$$

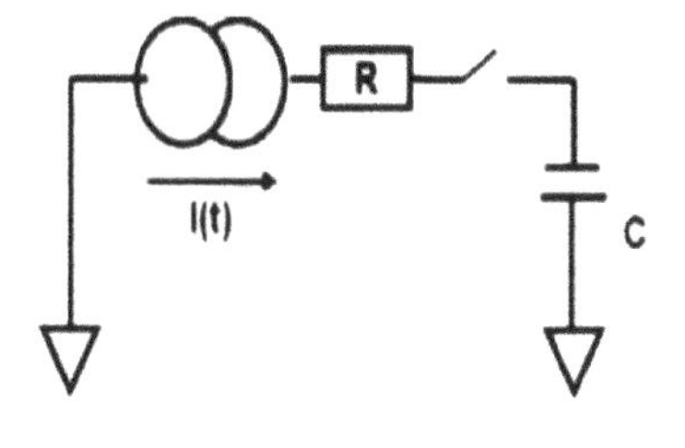

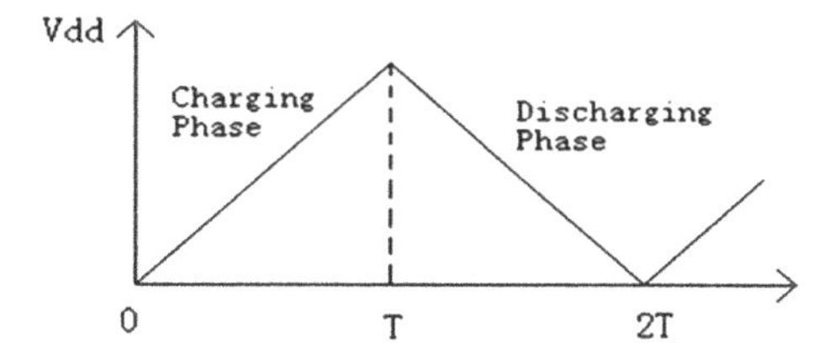

Fig. 2 Adiabatic charging

A constant current source has been used therefore the energy dissipation during the process of charging and discharging can be approximated by the following equation:

$$E_{diss} = R\int_{0}^{T} i^2(t)dt = Ri^2(t) * T = \frac{RC}{T}CV_c^2(T) \quad (3)$$

where $V_c(T)$ is the voltage across the capacitor at time T. It is evident from (3) that the energy loss decreases as compared to conventional charging at V_{dd} supply voltage, when $T > 2RC$. Further, the energy loss can be reduced by increasing T, i.e., by using a slowly time varying voltage supply. The energy loss can be reduced further by reducing R as it depends on R.

2 Proposed Work

2.1 2PASCL X-OR

In this paper, a new six-transistor X-OR gate using 2PASCL has been used for implementation of full adder design. The functioning of this X-OR gate can be explained with the help of Fig. 3. In this two extra MOS transistors function as diodes D_1, D_2 are used with complementary power supply clocks PC and $\overline{PC}$. These diodes like MOS transistors help to recycle the charge from load capacitance. One power clock supply is 180° out of phase with the other one and the voltage magnitude of both power clock supplies is governed by (4) and (5) [9].

$$V_{PC} = \frac{V_{DD}}{4}\sin(\omega t + \theta) + \frac{3}{4}V_{DD} \quad (4)$$

$$V_{\overline{PC}} = \frac{V_{DD}}{4}\sin(\omega t + \theta) + \frac{1}{4}V_{DD} \quad (5)$$

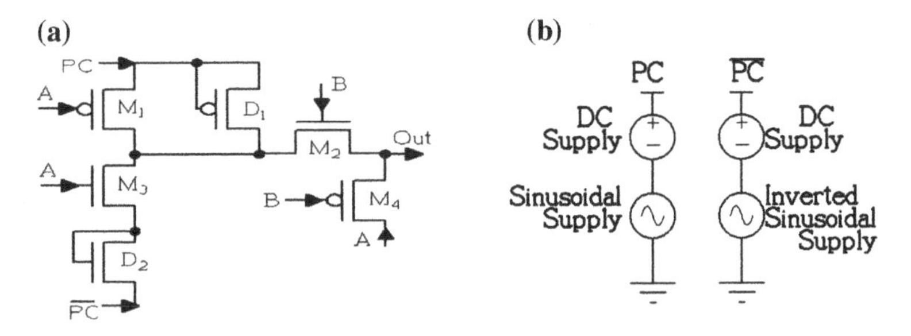

Fig. 3 **a** 2PASCL based X-OR gate **b** Power clock supplies

It is depicted from (4) and (5) that power dissipation decreased due to the reduced voltage difference between the electrodes as the magnitude of V_{PC} is twice of $V_{\overline{PC}}$. Reduction in power dissipation further achieved as the proposed circuit uses split-level clock supply for charging/discharging the load capacitance. The functioning of this circuit can be explained as follows. When both input signals are low, M_1, M_4 turns ON and M_2, M_3 remains OFF, and output remains low. When both input signals are high, M_1, M_4 remains OFF, M_2, M_3 turns ON and whatever charge has present at node out will be discharged through M_3, D_2 and output will be low. For complementary input signals, the output will be high when input signal A is high and B is low M_4 turns ON and pass the high input of A at node out. In the case of the input signal A is low and B is high M_1, M_2 turns ON and output node gets charged through M_1, M_2. During hold step energy loss can be decreased by reducing the switching activity, if no transition takes place at input side then the output will also be held by the load capacitance.

2.2 Performance Analysis and Results

In this paper, all the circuits have been simulated in 0.18 μm CMOS technology. The performance of all circuits has been analyzed and compared with varying voltage supply and temperature. The proposed (2PASCL) design of full adder containing 16 transistors is illustrated in Fig. 4a. The SER full adder has been simulated and the circuit diagram is shown in Fig. 4b. The design of 10T full adder as shown in Fig. 4c also simulated for comparison. Transmission gate-based full adder design which consists of 20 transistors is shown in Fig. 4d.

The performance of all designs in terms of power efficiency (PDP) has been calculated and analyzed with the help of Tables 1 and 2 at varying temperature and supply voltage respectively. The graph in Fig. 5 shows the comparison of power efficiency with existing designs at varying temperature of 15–50 °C, whereas Fig. 6 shows at varying supply voltage of 1.2–2.8 V.

2.3 Comparison with Existing Designs

The proposed design using 2PASCL has been compared with existing designs reported in the literature in terms of various parameters as given in Table 3.

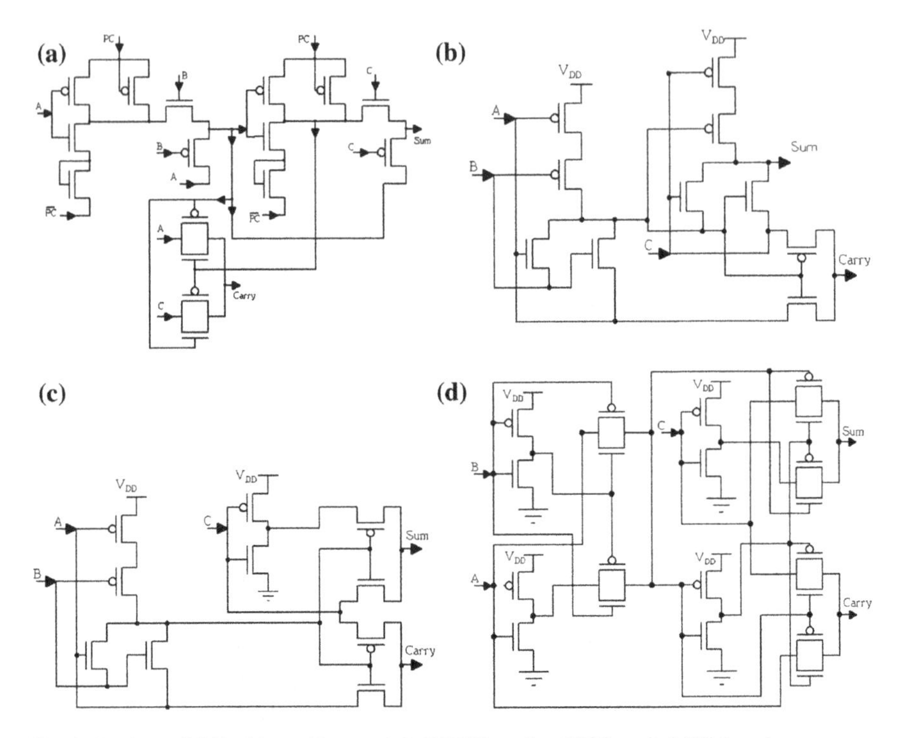

Fig. 4 Designs of full adder, **a** Proposed, **b** SERF based, **c** 10T based, **d** TG based

Table 1 PDP comparison at varying temperatures

Temperature (°C)	2PASCL ($J \times 10^{-21}$)	SERF ($J \times 10^{-21}$)	10TFA ($J \times 10^{-21}$)	TG FA ($J \times 10^{-21}$)
15	0.68	4.03	3.28	8.76
20	0.72	4.46	3.57	9.51
25	0.76	4.89	3.87	10.31
30	0.81	5.35	4.20	11.14
35	0.87	5.81	4.55	12.04
40	0.95	6.30	4.96	13.05
45	1.06	6.88	5.43	14.29
50	1.25	7.63	6.06	15.90

2.4 Waveforms

Figure 7 shows the all possible three-bit input combinations and output waveforms of simulated circuits for sum and carry. Therefore, the functioning of the proposed design has been verified from the waveforms with output voltage level, above 1.1 V

Table 2 PDP comparison at varying voltages

Supply voltage (V)	2PASCL FA ($J \times 10^{-21}$)	SERFFA ($J \times 10^{-21}$)	10TFA ($J \times 10^{-21}$)	TGFA ($J \times 10^{-21}$)
1.2	0.34	4.53	3.41	6.40
1.3	0.41	4.24	3.57	6.96
1.4	0.46	4.35	3.66	7.58
1.5	0.54	4.52	3.75	8.36
1.6	0.62	4.59	3.89	9.08
1.7	0.70	4.89	4.10	9.85
1.8	0.78	5.07	4.32	10.63
1.9	0.86	5.32	4.54	11.40
2.0	0.91	5.49	4.73	12.27
2.1	0.97	5.82	5.03	13.11
2.2	1.02	6.09	5.32	13.94
2.3	1.04	6.15	5.62	14.80
2.4	1.03	6.38	5.90	15.69
2.5	1.08	6.35	6.29	16.65
2.6	1.12	6.39	6.61	17.46
2.7	1.11	6.62	6.98	18.45
2.8	1.12	6.78	7.36	19.17

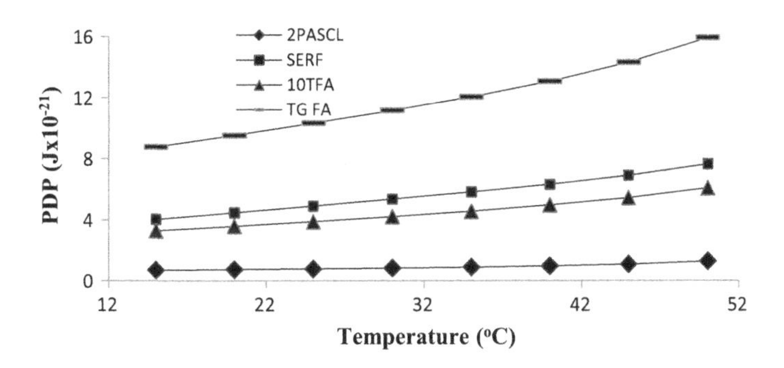

Fig. 5 PDP comparison at varying temperatures

for high logic and below 0.7 V for low logic, which is sufficient for adiabatic logic families for digital switching circuits as reported in the literature.

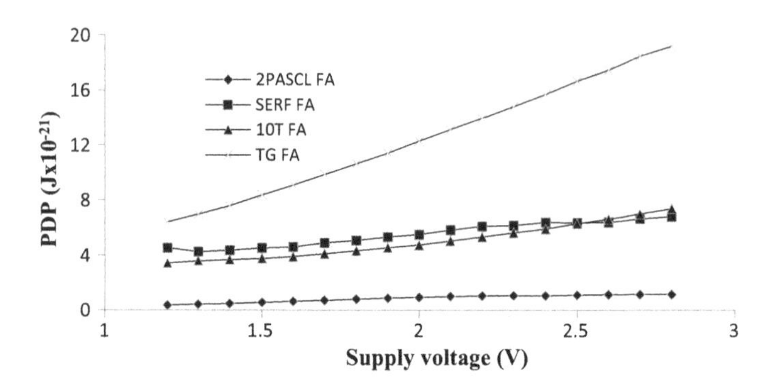

Fig. 6 PDP comparison at varying voltages

Table 3 Comparison with existing designs

Technology (μm)	PDP (J)	Design
0.15	8.61×10^{-16}	[11] Dual threshold node design FA
0.13	4.11×10^{-17}	[12] Hybrid CMOS logic
0.13	3.81×10^{-17}	[12] GDI-MUX
0.18	7.84×10^{-22}	Proposed

3 Conclusion

A new 2PASCL-based full adder has been reported in this paper. The simulation results show the PDP of the proposed design is 0.78×10^{-21} J as compared to 5.07×10^{-21} J (SERF), 4.32×10^{-21} J (10T), and 10.63×10^{-21} J (TG) at a supply voltage of 1.8 V. The proposed design also performs better in terms of PDP for varying temperatures conditions. The proposed full adder has been compared with existing designs and show improvement in terms of PDP. As the limitations in ICs increasing with scaling the proposed design is a better candidate for low-power applications.

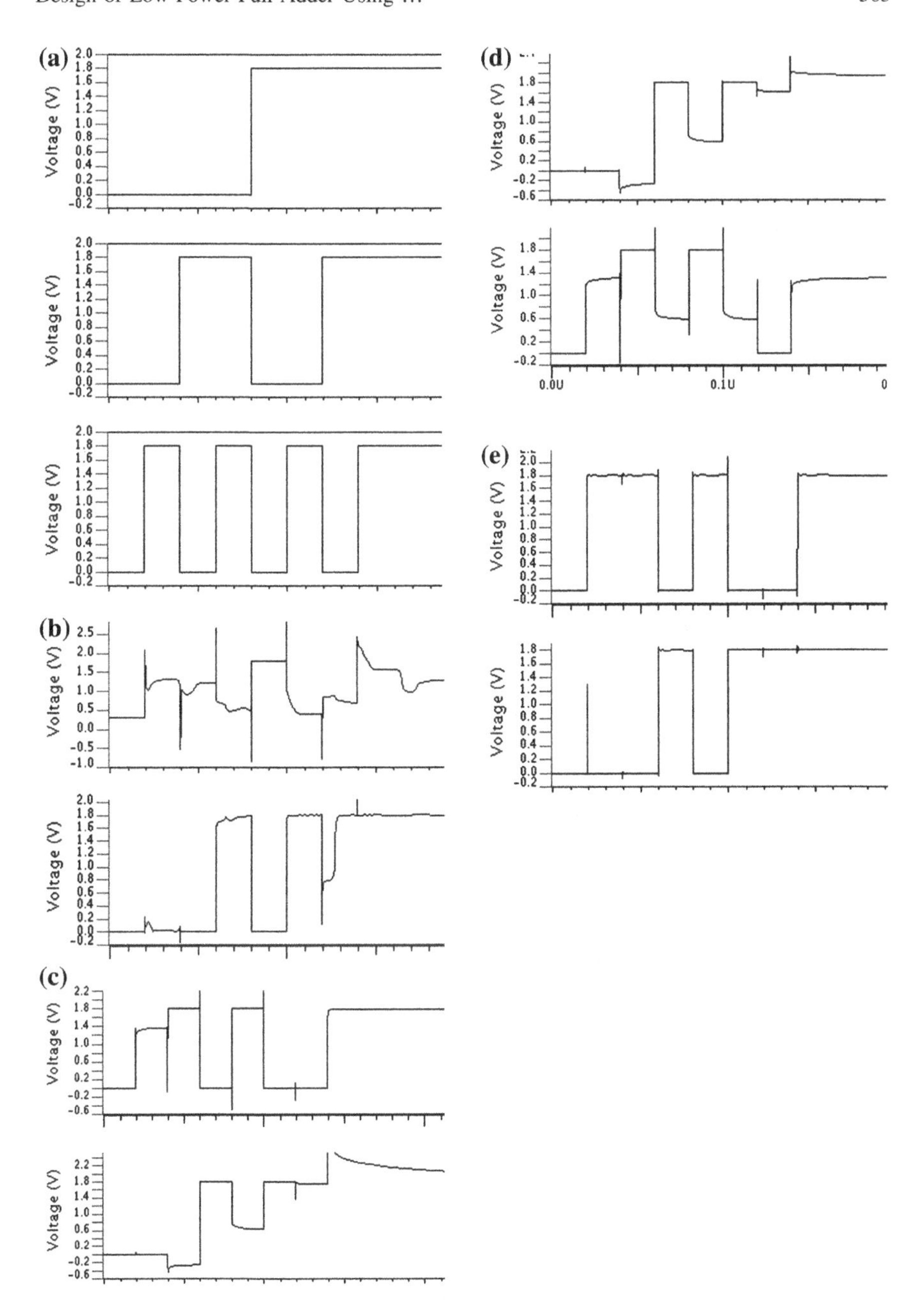

Fig. 7 Waveforms of **a** Three-bit input combinations, sum and carry of full adder with **b** 2PASCL, **c** SERF, **d** 10T, **e** TG

References

1. Roy, K., Prasad, S.: Low Power CMOS VLSI Design In: Low power `CMOS VLSI Circuit Design, 1st Ed. USA, Wiley-Interscience, ch. 1, sec. 1.2, pp. 2–3, (2000).
2. Moore, G.E.: Cramming More Components on to Integrated Circuits: Electronics Magazine, 38 (8) (1965).
3. Chang, C. H., Gu, J. M., Zhang, M.,: Ultra Low-voltage Low-power CMOS 4-2 and 5-2 Compressors for Fast Arithmetic Circuits. IEEE Transactions on Circuits and Systems I, Regular Papers, vol. 51, no.10, pp. 1985–1997, (2004).
4. Kumar, D., Kumar, M.: Modified 4-2 Compressor using Improved Multiplexer for Low Power Applications. In: 2016 International Conference on Advances in Computing, Communications, and Informatics (ICACCI), pp. 236–242, IEEE, Jaipur (2016).
5. Kumar, M., Pandey, S., Arya, S.K.: Design of CMOS Energy Efficient Single Bit Full Adders. In: High Performance Architecture and Grid Computing, pp. 159–168, Springer Berlin Heidelberg (2011).
6. Kumar, S., Kumar, M.: 4-2 Compressor Design with New XOR-XNOR Module. In: 2014 Fourth International Conference on Advanced Computing & Communication Technologies, pp. 106–111, IEEE, (2014).
7. Kumar, M., Pandey, S., Arya, S.K.: Low Power CMOS Full Adder Design with Body Biasing Approach. J. of Integrated Circuits and Systems, 6, pp. 75–80, Brazil (2011).
8. Chanda, M., Jain, S., De, S., Sarkar, C. K.: Implementation of Sub threshold Adiabatic Logic for Ultra low-Power Application. In: IEEE Transactions on Very Large Scale Integration (VLSI) Systems, vol. 23, no. 12, pp. 2782–2790, (2015).
9. Anuar, N., et.. al Two Phase Clocked Adiabatic Static CMOS Logic and its Logic Family, J. of Semiconductor Tech. and Science, vol. 10, pp. 1–10, Japan (2010).
10. Lin, J. F., Hwang Y. T., Sheu, M. H., Ho, C. C.: A Novel High-Speed and Energy Efficient 10-Transistor Full Adder Design. In: IEEE Transactions on Circuits and Systems—I: Regular Papers, vol. 54, No. 5, (2007).
11. Panda, S., Kumar, N. M., Sarkar, C. K.: Transistor Count Optimization of Conventional CMOS Full Adder and Optimization of Power and Delay of New Implementation of 18 Transistors Full Adder by Dual Threshold Node Design with Submicron Channel Length. In: 2009 4th International Conference on Computers and Devices for Communication (Codec), pp. 1–4, Kolkata, (2009).
12. Foroutan, V., Taheri, M.R., Navi, K., Mazreah, A. A.: Design of two Low-Power Full-adder Cells using GDI Structure and Hybrid CMOS Logic Style: Integration the VLSI Journal, vol. 47, no. 1, pp. 48–61, (2014).

Design and Implementation of Montgomery Multipliers in RSA Cryptography for Wireless Sensor Networks

G. Leelavathi, K. Shaila and K.R. Venugopal

Abstract In this work, the architecture and modeling of two different RSA encryption and decryption public key systems are presented, for a key size of 128 bits. The systems that require different levels of security can be utilized easily by changing the key size. Two different architectures are proposed with and without MMM42 multiplier to check the suitability for implementation in Wireless Sensor Nodes to utilize the same in Wireless Sensor Networks. Synthesis and simulation of VHDL code is performed using Xilinx-ISE for both the architectures. Architectures are related in terms of area and time. The RSA encryption and decryption algorithm implemented on FPGA with data and key size of 128 bits, without modified MMM42 multiplier gives good result with 50% less utilization of hardware. As device utilization is less in the second architecture, the key size can be increased to have more security with good speed for Wireless Sensor Networks.

Keywords FPGA · Modular multiplication · Modified montgomery multiplication (MMM) · RSA cryptosystem · VHDL

1 Introduction

A Wireless Sensor Network (WSN) refers to a heterogeneous system, composed of autonomous devices called sensor nodes. It has limited data transmission and low computational power that clues to a demanding environment to afford security [1]. Security mechanisms necessitate certain quantity of resources for the implemen-

G. Leelavathi (✉) · K. Shaila
VTU-Research Centre, Vivekananda Institute of Technology, Bengaluru, India
e-mail: nisargamodini@gmail.com

K. Shaila
e-mail: shailak17@gmail.com

K.R. Venugopal
University Visvesvaraya College of Engineering, Bengaluru, India
e-mail: venugopalkr@gmail.com

A.K. Somani et al. (eds.), *Proceedings of First International Conference on Smart System, Innovations and Computing*, Smart Innovation, Systems and Technologies 79, https://doi.org/10.1007/978-981-10-5828-8_54

tation. For WSNs memory, area and energy to control the sensor are very limited in tiny wireless sensor nodes. For a long time, it was understood that the public key cryptography was not suitable for WSNs because it requires elevated processing power. However, in the course of studies of encryption algorithms it proved the feasibility of that technique in WSN [2]. Rivest–Shamir–Adleman (RSA) is the extensively used public key cryptosystem. RSA crypto-accelerators and their hardware implementations are important for high security [3–5].

Motivation: The computation costs associated with public key cryptosystems is restricted due to the limited computing resources. Crypto-accelerators are capable by way of they classically attain better power efficiency and performance over software implementation with a generic processor. The proposed work aims to design arithmetic architectures for RSA Cryptosystems that are optimized for contemporary FPGAs. Major contribution of this work is the FPGA implementation of the proposed architectures using Montgomery algorithm to increase the speed of RSA cryptosystem [3–6].

Organization: Various surveys connected to security mechanisms, public key cryptography (PKC) and RSA, dissimilar multiplication algorithms are enlightened in Sect. 2. Problem definition is specified in Sect. 3 and RSA processor model is described in Sect. 4. In Sect. 5 results with simulations are given. Conclusions discussed in Sect. 6.

2 Literature Survey

Alan et al. [5] exploits the feature maximum carry chain length of FPGA that is utilized to implement operation of modular exponentiation. Speed and area results with different multiplier sizes for the pipelined modular multiplier and RSA encryptor/decryptor are provided. Shian-Rong et al. [6, 7] proposed architecture with less energy consumption and higher throughput. The whole design manipulates the MM52 and MM42 algorithm for MM, in turn modifying modular exponentiation algorithm. This MMM42 multiplier analysis w.r.t area and delay in comparison with MM52 and MM42. These features attracted to select this MMM42 architecture to implement on 128 bit with Xilinx FPGA. The synthesize of MMM42 multiplier is performed using Synopsys Design Compiler.

Abdullah et al. [8] highlighted the essentiality of strong cryptography is WSNs, with improvement over time, efficiency and reduction in power consumption. Vincent et al. [9] defines RSA cryptosystem in which, to reduce critical path a Kogge–Stone Adder, i.e., fast parallel prefix adder is used. For RSA cryptosystems scalable systolic hardware architecture is described by Gudu et al. [10]. Their design enables making area–time tradeoff with different precision of inputs. The throughput and performance of the Radix-216 design is increased with the DSP48E slices.

In our work, input is split into 64 bits for Modified Montgomery Multiplication 4 to 2 Carry Save Adder architecture [MMM42]. Input of 128 bits is used without

MMM42. Modified Modular Multiplication algorithm MMM42 is implemented with the modification in handling the input data.

3 Problem Definition

RSA algorithm's security is with its incapability to competently factorize large integers. Application-specific integrated circuit (ASIC) solutions have compact flexibility and high nonrecurring (NRE) cost. Cryptographic algorithms implementation on FPGA gives the solution with speed and physical security of hardware along with flexibility. This proposed work investigates the performance of RSA encryption algorithm using Montgomery Modular Multiplication in software.

4 RSA Cryptosystem Model

In Fig. 1 private and public keys are generated and assumed that they are shared between the sender and receiver. The plaintext size considered is 64 and 128 bits. The sender must first discover public key (*M*, *E*) i.e., the modulus and exponent. He computes the cipher-text (*C*) by:

$$C = P^E (mod\, M) \tag{1}$$

To decipher the message, Receiver practices private key (*D*) to recuperate the plain text using:

$$P = C^D (mod\, M) \tag{2}$$

From Eq. (1) plaintext *P*, Exponent (*public key*) *E* and Modulus *M* are represented as indata, inExp, and inMod respectively.

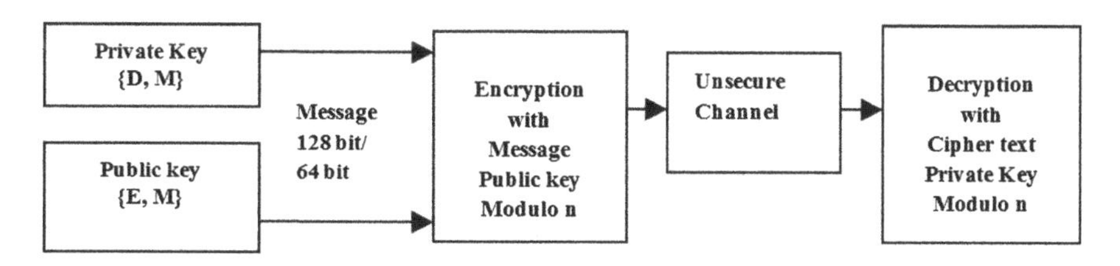

Fig. 1 Block diagram of RSA cryptosystem model

4.1 Montgomery Modular Multiplication

Montgomery multiplication algorithm is employed in the design and implementation of modular multiplication. RSA encryption and decryption with and without MMM42 algorithm is executed and performance is compared to check the suitability for WSNs in terms of execution speed and hardware utilization. Montgomery's algorithm for modular multiplication is shown in Algorithm 1 that calculates the product of two integers modulo and third one without carrying out division by M. This produces the compact product with sequence of additions in which C, S represents Carry and Sum respectively.

Algorithm 1: Basic Montgomery Algorithm

```
MontProd(A,B,M)
{
S-1=0
for i=0 to n-1 do
Ci=(Si-1+BiA)Mod 2
Si=(Si-1+qiM+BiA)/2
end for
Return Sn-1
}
```

Algorithm 1 is modified with carry save representation of the data A, B, and M in [6]. To perform modular multiplication, the same algorithm [6] is utilized and modified. In the modified algorithm, the data is represented individually as 64 bits to perform RSA-128 bit operations. The pseudocode without modified MMM42 multiplier to perform RSA encryption and decryption is given in Table 1.

Portmap is like a function call statement through which the message, public key, private key, and modulus value is passed to encryption and decryption modules are shown in Fig. 2. Schematic diagram of the entity for MMM42 multiplier is given in Fig. 2. The 64 bit inputs include $(A1, A2) = A$, $(B1, B2) = B$ and N and output $(S-1, S-2) = S$. *Clk*, *lda rst* and *done* are control signals for input and output respectively.

4.2 Modular Exponentiation

In this work ***L*** algorithm or Right to Left algorithm is selected because independent square and multiply operations can be executed in parallel. To accomplish the modular exponentiation 50% of clock cycles are essential. However, to accomplish the speeding up of the algorithm the two physical multipliers are required. As

Table 1 Pseudocode without modified MMM42

```
Entity Main
{ Inputs:clk,rst, Publickey, Privatekey, modin, data_in,
Outputs:cyphertext, original_msg

Component RSACipher
{ Generic(KEYSIZE:integer :=32);
      Inputs: indata, inExp, inMod, clk, ds, reset;
      Outputs: Cypher, ready;
Encryption: RSACypher generic map(keysize=>128)
Portmap(data_in, public key, modin, enc_msg, clk, ds1,
rst, rdy1);
Decryption: RSACypher generic map(keysize=>128)
Portmap(enc_msg, private key, modin, original_msg, clk,
ds2, rst, rdy2);}}
```

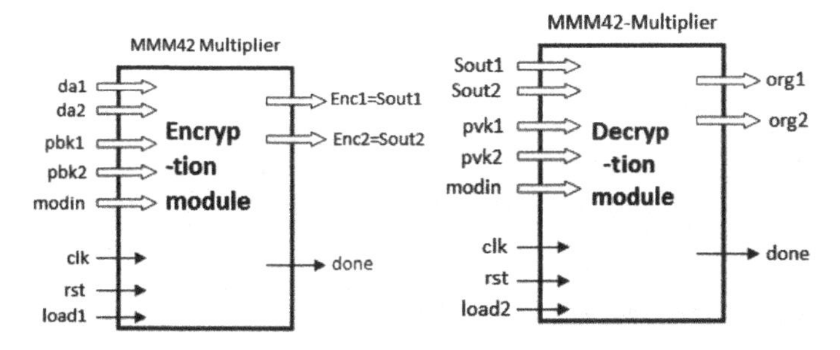

Fig. 2 Schematic of MMM42 multiplier for encryption and decryption

shown in Eqs. (1) and (2) implementation of Modular exponentiation is illustrated in Fig. 3.

To implement RSA cryptosystem the MMM42 algorithm is utilized and modified [9]. The structural VHDL coding of the RSA includes two MMM42, one for encryption and the other for decryption. The second architecture RSA-128 is carried out to compare the performance.

Figure 4 shows Schematic Diagram of Main Unit which consists of both encryption and decryption of implemented VHDL code using Xilinx on FPGA. The input data size for the system is 128 bit. The inputs to the module are $(da1, da2) = da$, da is data which a plaintext input to the system is represented in individual 64 bits. Similarly $(pbk1, pbk2) = pbk$, $(pvk1, pvk2) = pvk$ and *modin* are the 64 bit data given to cryptosystem. The $(enc1, enc2) = enc$, $(org1, org2) = org$ are the output data.

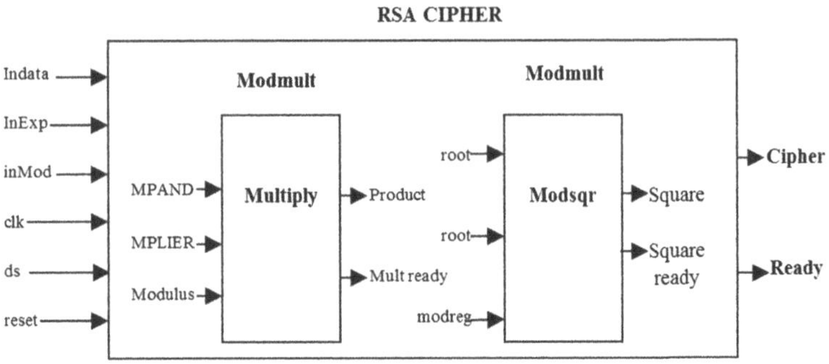

RSA CIPHER
Indata
InExp
inMod
clk
ds
reset
Modmult
MPAND
MPLIER
Modulus
Multiply
Product
Mult ready
Modmult
root
root
modreg
Modsqr
Square
Square ready
Cipher
Ready

Fig. 3 Implementation of modular exponentiation

The RSA module consists of two Modified MMM42 multiplier as shown in Fig. 4. The (*da*1, *da*2) = *da*, (*pbk*1, *pbk*2) = *pbk* and *modin* are the inputs to the encryption unit. The encrypted data output (*enc*1, *enc*2) = *enc* is obtained. This data is fed to the Decryption unit as (*Sout*1, *Sout*2) = *Sout* along with (*pvk*1, *pvk*2) = *pvk* and *modin*. The decrypted original plaintext (*org*1, *org*2) = *org* is obtained from decryption unit. The separated multiplier diagrams are shown in Fig. 2.

MMM42_Multiplier
Decryption
MMM42_Multiplier
Encryption
TOP

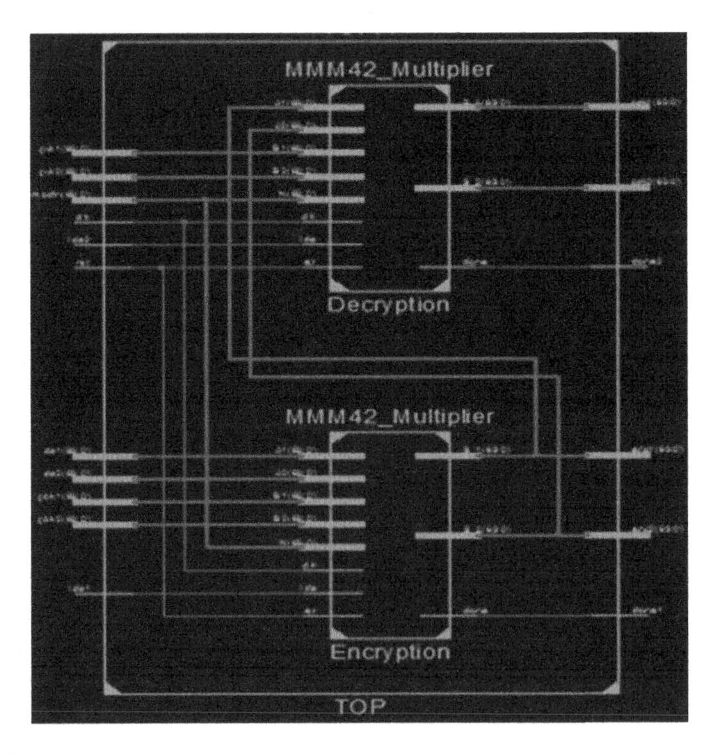

Fig. 4 Complete RSA module

5 Results and Discussion

5.1 Simulation and Results with Modified MMM42 Multiplier

In this work, FPGA devices utilized are Spartan 3 XC3S400-5pq208 chosen for software and hardware implementation. The VHDL code is simulated with ISE simulator and waveforms are observed with ModelSim. The corresponding RTL schematic and timing diagrams are obtained. The data input ($A1$, $A2$), ($B1$, $B2$) exponent (*public key*) and modulus (N) of 64 bits are applied and output (S_1, S_2) is obtained when done signal goes high. The timing diagram for the encryption and decryption unit with modified MMM42 is shown in Fig. 5. The inputs considered for encryption are ((*0011223344556677*), (*8899AABBCCDDEEFF*)) = ($a1$, $a2$) = A, (*0000000000000000*), (*0000000000010001*) = ($b1$, $b2$) = *public key*. It is observed from the waveform that transmitter uses the public key for the encryption process which is shared between the sender and receiver. *The following output is obtained* ((*125DB969FF426764*), (*FD832F8B30971598*)) = (s_1, S_2) = *cipher output. Clk*, *rst*, *lda,* and *done* are control signals.

The inputs considered for decryption are ((*125DB969FF426764*), (*FD832F8B30971598*)) = ($a1$, $a2$) = A, ((*40F76A229EE3C0763*), (*F598241C-5E3DCC01*) = ($b1$, $b2$) = *private key*. It is observed from the waveform that transmitter uses the private key for the encryption process, which is not shared between the sender, and receiver, *only receiver knows about private key.*

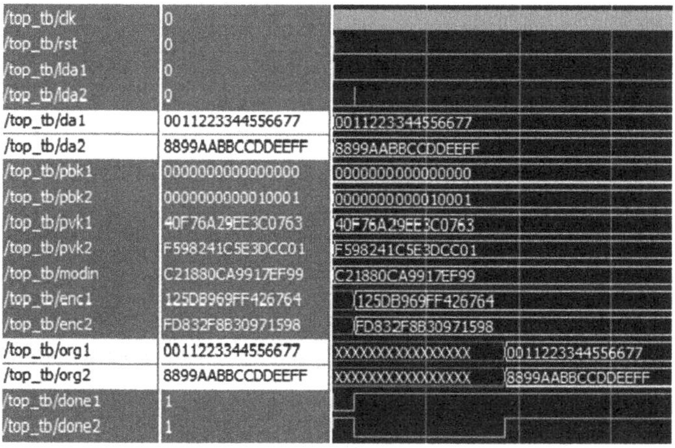

Fig. 5 Timing diagram of encryption and decryption

5.2 Simulation and Results Without Modified MMM42 Multiplier

The RTL Schematic of the RSACipher_128 is shown in Fig. 6. The inputs to the RSA CIPHER-128 encryption includes *Public key, (modulus M) modin and data_in (plain text message)* as inputs of 128 bits and Cipher as output. Decryption unit takes Cipher, Private key, and Modulus (*M*) as inputs to produce the plaintext message output, i.e., original_msg. Figure 7 gives the corresponding timing diagram of RSA CIPHER_128 shown in Fig. 6. The device utilization for the complete RSA system is very less, i.e., below 50% compared to the earlier architecture discussed in Sect. 5.1. It is also observed from the execution and timing diagrams, for encryption and decryption the time required is also around 50%. Through comparison, the second architecture is appropriate for WSNs as it satisfies the constraints and limitations in terms of hardware usage.

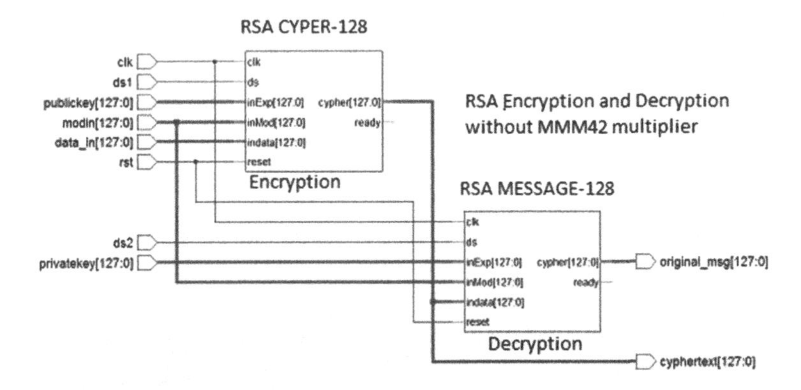

Fig. 6 RTL schematic of RSACipher_128

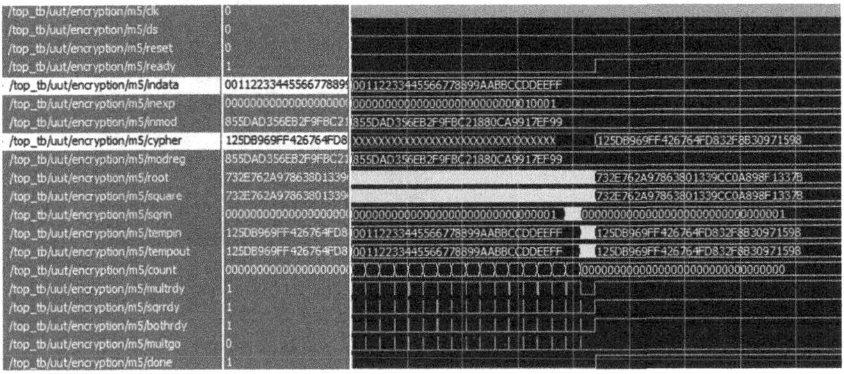

Fig. 7 Timing waveform RSA system without modified MMM42 multiplier

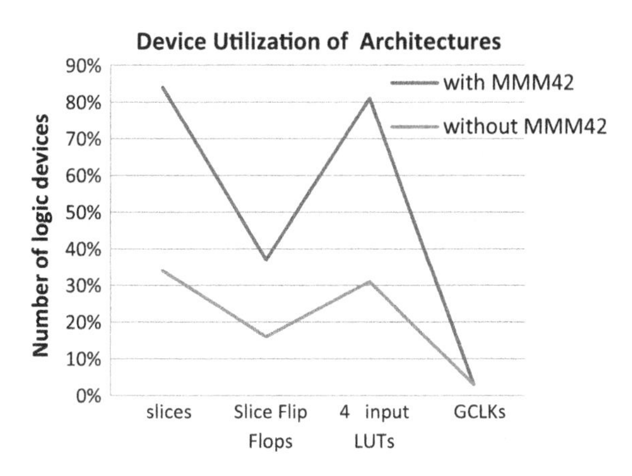

Fig. 8 Comparison of device utilization

6 Performance Analysis

Two different architectures are designed and implemented to compare the performance. RSA cryptosystem without MMM42 multiplier gives the good performance with respect to speed and area. It is observed from the results shown in Fig. 8, Tables 2 and 3 with modified MMM42 multiplier is not suitable for WSN nodes as it consumes 50% more hardware in FPGA.

Table 2 Device utilization of two different architectures

Logic utilization	Available	With MMM42 (%)	Without MMM42 (%)
Slices	5472	84	34
Slice FFs	10944	37	16
LUTs	10944	81	31
Bonded IOBs	320	221	161
GCLKs	32	3	3

Table 3 Timing analysis of two different architectures

Timing criterion	With MMM42	Without MMM42
Min. period (ns)	9.847	8.981
Max. frequency (MHz)	101.554	111.343
Min. arrival time of input (before clock) (ns)	7.003	2.366
Max. required time for output (after clock) (ns)	7.545	0.917
Max. combination path delay (ns)	9.555	No path found

7 Conclusions

To achieve better security and improve the speed constraints, flexibility pertaining to the cryptographic processes is necessary in WSNs. RSA cryptosystem without modified MMM42 multiplier gives the good performance with respect to speed and area. It is observed from the results that the implementation with modified MMM42 multiplier is not suitable for WSN nodes as it consumes 50% more hardware in FPGA. The design without modified MMM42 can be selected for data encryption and decryption in WSN Nodes.

References

1. Ian, F., Akylidiz, Weilian Su, YogeshSankarasubramaniam and E., Cayirci.: Wireless Sensor Network: Survey on Sensor Networks. In: IEEE Communication Magazine, ISSN: 0163-6804, vol. 40, no. 8, pp. 102–114, 2002.
2. Shaila, K., S. H., Manjula, Thriveni J., Venugopal K. R., L. M., Patnaik: Resilience Against Node Capture Attack using Asymmetric Matrices in Key Predistribution Scheme in Wireless Sensor Networks. In: International Journal on Computer Science and Engineering, ISSN: 0975-3397, vol. 11, no. 3, pp. 31–41, 2011.
3. P. L., Montgomery: Modular Multiplication Without Trial Division: Mathematical Computations, vol. 44, no. 170, pp. 519–521, 1985.
4. C. Mclvor, M. McLoone, J. V., McCanny: Fast Montgomery Modular Multiplication and RSA Cryptographic Processor Architectures. In: proceedings of 37th Asilomar Conference on Signals, Systems, Computations, vol. 1, pp. 379–384, 2003.
5. Alan Daly, William Marnane: Efficient Architectures for Implementing Montgomery Modular Multiplication and RSA Modular Exponentiation on Reconfigurable Logic. In: Proceedings of the 2002 ACM/SIGDA tenth International Symposium on FPGAs, pp. 40–49, ACM 1-58113-452-5/02/2002, Monterey, California, USA.
6. Shiann-Rong Kuang, Jiun-Ping Wang, Kai-Cheng Chang, Huan-Wei Hsu: Energy-Efficient High-Throughput Montgomery Modular Multipliers for RSA Cryptosystems, In: IEEE Transactions on Very Large Scale Integration (VLSI) Systems, vol. 21, no. 11, pp. 1999–2009, 2013.
7. Shiann-Rong Kuang, Kun-Yi Wu, Ren-Yao Lu: Low Cost High Performance VLSI Architectures for Montgomery Modular Multiplication, In: IEEE Transactions on Very Large Scale Integration (VLSI) Systems, vol. 24, no. 2, pp. 434–443, 2016.
8. Abdullah Said Alkalbani, Teddy Mantoro, Abu Osman Md Tap: Comparison between RSA Hardware and Software Implementation for WSNs Security Schemes, In: Proceedings of Third International Conference on ICT4M, pp. E84–E89, 2010.
9. Desiree Juby Vincent: Fast and Area Efficient RSA Cryptosystem Design Using Modified Montgomery Multiplication for FPGA Applications, In: International Journal of Scientific Engineering Research, vol. 4, no. 7, pp. 2221–2228, 2013.
10. Tamer Gudu: A New Scalable Hardware Architecture for RSA Algorithm, In: Proceedings of International conference on Field Programmable Logic and Applications, pp. 670–674, 2007.

Load Optimization in Femtocell Using Iterative Local Search Heuristics in an Enterprise Environment

Rajalakshmi Krishnamurthi and Mukta Goyal

Abstract Femtocells are required to connect the end users and also provide better services to access various applications such as web 2.0 services, mobile multimedia, etc. The major challenge for mobile network operators is, to provide such services with better coverage and capacity to its end users. This requires efficient load balancing of femtocells corresponding to number of users, such that, the cellular network can provide services to the end users in smooth manner. This paper presents a heuristic search technique to optimize the load on femtocells to provide better services to the end users.

Keywords Femtocells · Load balancing · Pilot power · Enterprise Self-optimization

1 Introduction

In today's era, the mobile network operator has to provide various services to their customers such as mobile multimedia, video on demand, E-learning services, social networking services, etc. The mobile network operators need to develop efficient mechanisms to provide these challenging services. In this aspect, the two immediate issues to be handled are coverage and capacity in the current 3G generation and also in future 4G generations. For the reason that, impact of providing such highly demanding services, is sensed to be severe in these potential mobile networks, in comparison to conventional mobile network technique of 2G generation [1].

To address better coverage and capacity in potential mobile networks, the femtocells provide vital solution to connect the end users. Femto cells are low cost, low-power cellular base station installed by end users. It provides macrocellular

R. Krishnamurthi (✉) · M. Goyal
Jaypee Institute of Information Technology, Noida, India
e-mail: k.rajalakshmi@jiit.ac.in

M. Goyal
e-mail: mukta.goyal@jiit.ac.in

A.K. Somani et al. (eds.), *Proceedings of First International Conference on Smart System, Innovations and Computing*, Smart Innovation, Systems and Technologies 79, https://doi.org/10.1007/978-981-10-5828-8_55

coverage, high data rates in residential or enterprise environment [6, 7]. It operates in an unstructured manner to connect mobile terminals to mobile operator's network. Mohjazi et al., uses a genetic algorithm to optimize the network coverage in terms of handled load, coverage gaps, and overlaps, which provides a dynamic update of the downlink pilot powers of the deployed femtocells [1].

Femtocells require broadband connection such as digital subscriber line (DSL), cable modem, or a separated radio frequency. To make the successful communication with the cellular network, the femtocells parameters are essential. One of the essential parameter is Common pilot channel (CPICH) as in universal mobile telecommunication system which is used by mobile terminals for larger cell coverage area. CPICH requires between 10 and 20% of maximum transmit power in conventional macrocell network [8, 9]. Figure 1 depicts the architecture of mobile network with femtocells.

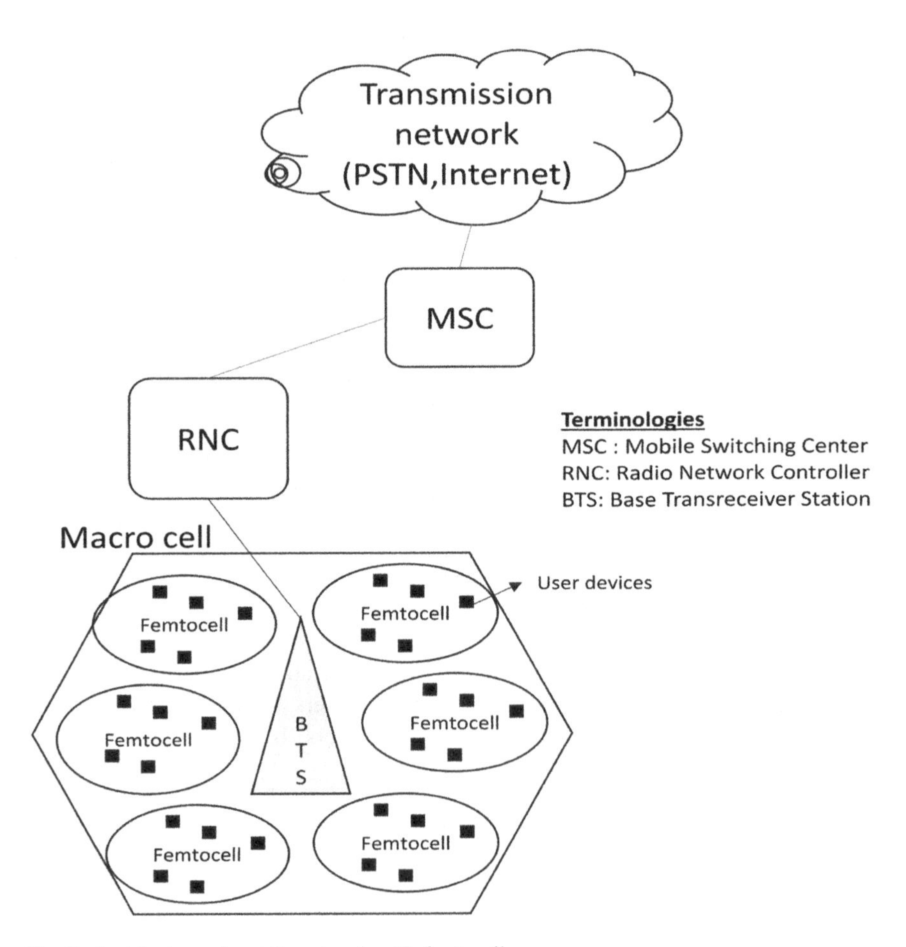

Fig. 1 Architecture of mobile network with femtocells

Another important feature of femtocell is, to deliver challenging services in a smooth manner, both at indoor and outdoor, using efficient load balancing mechanisms [5]. Due to unstructured organization of femtocells, the cellular network may suffer overload situations, which may degrade the network performance. Some authors have proposed automatic and intelligent mechanisms to tune the network parameters using fuzzy approach to improve the overall cellular network performance. Load balancing mechanisms are used to adapt the number of users at a time on available bandwidth of the femtocell. Some of the authors have proposed the fuzzy approach, to distribute the traffic demand between Long-Term Evolution (LTE) femtocells in an enterprise scenario [2]. Other authors have used Particle Swarm Optimization (PSO) to provide the maximum cellular coverage for large indoor areas [3]. Different self-optimization techniques for residential femtocell deployments have been used. The leakage of the pilot signal to the outside of a house causes highly increased signaling load to the core network. Further, this leads into higher number of mobility events caused by passing users [4].

2 Preliminaries

To optimize load in femtocells different notations are used as given below.

2.1 *Notations*

Symbol	Meaning
I	Number of femtocells
J	Number of users
L_i	Load of a femtocell i
L_j	Load required by a user j
L_{th}	Threshold load of femtocells
Pt_{si}	Pilot power of a femtocell i in individual s
S	Population size
PL_{ij}	Path loss between user j and femtocell i

2.2 *Objective Function*

The optimization of load handled by all Femtocells in enterprise environment by maximizing the mean load handled by each femtocell while satisfying the power and load constraint.

$$FL(s) = \max(L_i) \tag{1}$$

Subject to

$$L_{ij} \leq L_{th} \tag{2}$$

$$Pt_{si} - PL_{ij} \geq \varepsilon \tag{3}$$

2.3 Load Constraint

It is to be noted that the load L_i of femtocell i should be within the threshold load L_{th}. This provides two advantages. First, as given in Eq. (2), the femtocell can accommodate load of further users until it reaches the threshold value L_{th}. Second, the load of femtocell is optimized to its maximum as given in Eq. (1).

2.4 Pilot Power Constraint

The primary objective pilot power constraint is to maintain the minimal amount of pilot power by femtocell, without compromising the load requirement by each user. For this purpose, difference between the pilot power and path loss of signal should be greater than or equal to parameter ε, where ε is maintained as −104 db. This ε value is the minimum signal requirement for voice traffic at user equipment.

2.5 Femtocell Load (L_i)

The load handled (L_i) by a femtocell i for its J users is given by

$$L_i = \sum_{j=0}^{J} L_j \tag{4}$$

2.6 Pilot Power (Pt_{si})

The pilot power (Pt_{si}) is the CPICH transmitted from femotcell i in an individual of population s.

2.7 *Path Loss (PL_{ij})*

The path loss between femotcell i and the user device j is given by

$$PL = 3.85 * \eta * 10Log10(d) + \sum_{j=1}^{F} a_f w_f \quad (5)$$

where

- η the distance power decay factor
- d the distance between transmitter and receiver in meters
- a_f the number of penetrated walls of type f
- F the different types of walls
- W_f the attenuation due to wall type of f.

3 Iterative Local Search Algorithm

The iterative local search algorithm is simple and powerful metaheuristic algorithm. The ILS algorithm is unique as it incorporates effective mechanisms within it, namely, local search, perturbation, acceptance criteria, and shuffle tolerance method. These features exhibit ILS algorithm different from other existing heuristic algorithms like Tabu search, Simulated Annealing and Genetic algorithm. Figure 2 exhibits the block diagram of ILS algorithm.

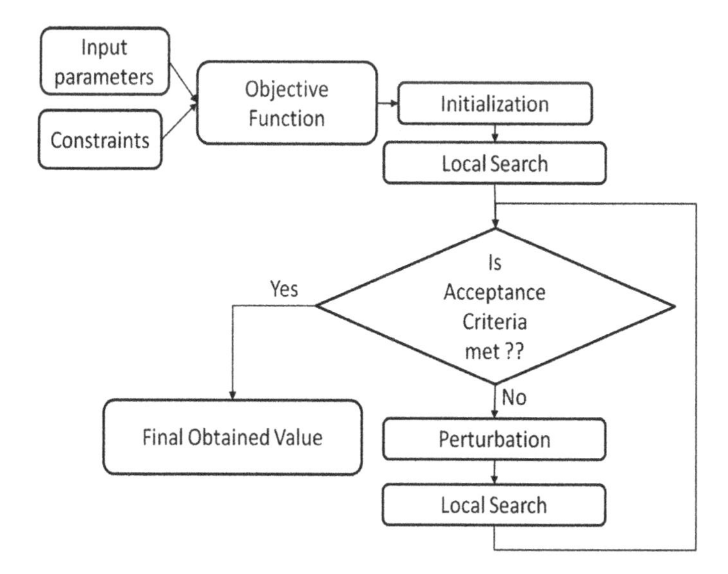

Fig. 2 Block diagram of iterative local search heuristics algorithm

To begin with, the input parameters and constraints of the femtocell network is provided as input for the objective function. Next, an initial assignment of user to femtocell is performed using uniform distribution. The initial objective function is estimated and updated as current known solution for the defined objective function.

3.1 Local Search Method

Next, the local search mechanism in ILS algorithm is used to find the best local minimum within the search space of current know solution. In this paper, 2-opt local search mechanism is used. In 2-opt local search mechanism, two random positions are selected at any given time for shuffling. 2-opt shuffle enhances further to explore the current local maximum. For example, in Fig. 3 given, the two locations at 3 and 6 are selected for local exchange. This way, the local maximum is explored for further optimization of objective function.

3.2 Perturbation

The perturbation is a very powerful technique, which is to avoid the local search method to stuck into local solution. Sometimes, the local search algorithm fails to explore all the possible search spaces within the objective function. This can be avoided by randomly shuffling the values within the current known best solution.

3.3 Acceptance Criteria

The acceptance criteria decides whether the local solution obtained so far is better than the current known best solution. If there is any improvement, then the

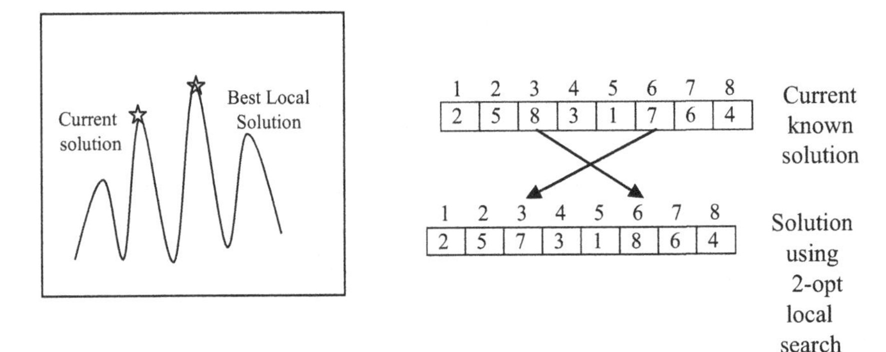

Fig. 3 2-opt local search method

objective value is updated with the best solution. The acceptance criteria also track of the number of iteration to be carried out by the algorithm. The ILS algorithm terminates its operation whenever the limit for number of iterations is reached.

4 Experimental and Results

The experiment was conducted on Intel Core CPU @2.00 GHz processor with 64-bit Windows 7 operating system. The ILS algorithm was implemented using Visual C++. The load matrix of IxJ is given as input matrix to the ILS algorithm. Let, I be the number of femtocell, J be the total number of users. Figure 4 depicts, the simulated area is an enterprise environment. In this enterprise environment, 8 femtocells with total configuration for 21 Erlangs is considered. Further, the simulated area has four types of wall, which is depicted as different thickness in the Fig. 4. Let, the total of J users within this simulated area has to share 8 femtocells. Thus, the objective function is optimal assignment of J users to 8 femtocells.

Inorder to obtain the objective function of optimal assignment of J users to each femtocell, the constraints like load and pilot power has to be satisfied. Now, pilot power of femtocell is nothing but CICPH signal transmitted between femtocell i and user j. In this experiment, the CICPH signal of femtocell i is considered between 0 and 7.8 db. The user pilot is considered between 0.2 and 0.5 db. However, the pilot power transmitted by each user j will have effect of path loss of signal. The path loss of signal experienced by each user j is estimated using Eq. 5. For path loss calculation, in this experiment, four different types of walls are considered. The attenuation of signals due to heave wall is considered as 5.7 db and for light wall as 3.8 db. Based on the reference paper [1], the various other parameter are listed in Table 1.

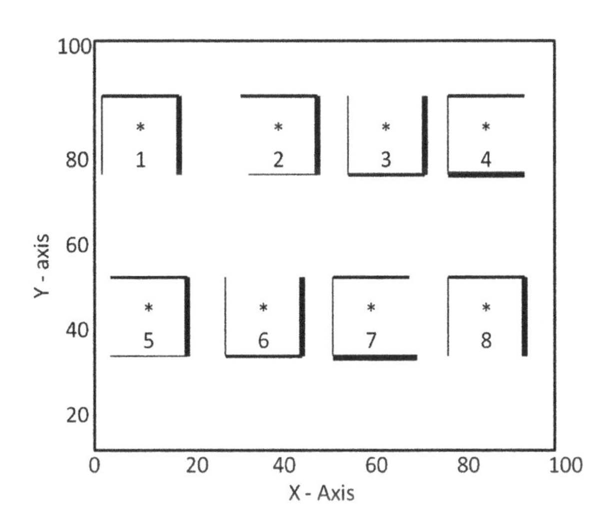

Fig. 4 Simulated enterprise area layout with 8 femtocell

Table 1 Parameters consider for experimentation

L_{th}	8 Erlangs
L_j	0.2 db to 0.5 Erlangs
ε	−104.2 db
Pt_{si}	0–7.78 db
W_f	3.8 db (light wall)
	5.7 db (heavy wall)
η	2

Now, for iterative local search algorithm, the inputs are the load required by each user, load offered by each femtocell, the total load of all 8 femtocells. Similarly, the objective function is obtained by maximizing the load assigned to each femtocell, based on finding the best possible assignment of user to each femtocell. While, solving for objective function, the load constraint as given in Eq. 2 and the pilot power constraint as given in Eq. 3 are verified.

The ILS algorithm was examined for different iterations as given in Table 2. It is observed that, the computation time taken by ILS algorithm to converge to the objective function value is almost same for different iterations. Thus, the proposed ILS algorithm is effective in terms of time consumed. This is possible, due to the simplicity and effective local search methods, perturbation, and acceptance criteria techniques incorporated in this ILS algorithm Table 2.

Figure 5 depicts the pilot power settings of each femtocells. The ILS algorithm has performed optimization and provided the best feasible solution of pilot power. It is to be noted that, the pilot power of each femtocell is maintained at desired level. Also, the allocation of pilot power is maintained to be evenly distributed among femtocells. This shows, the pilot power constraint as given in Eq. (3) is maintained throughout the optimization procedure by ILS algorithm.

Figure 6 illustrates the optimized load allocation obtained by ILS algorithm. It is to be noted that the load to be offered by femtocells has been distributed almost evenly among these femtocells. Also, it is observed that, the proposed ILS

Table 2 Execution time for ILS algorithm

ILS algorithm	
No. of iterations	CPU (s)
20	0.075
50	0.079
70	0.08
100	0.081
150	0.083
200	0.085
250	0.087
300	0.094
400	0.096

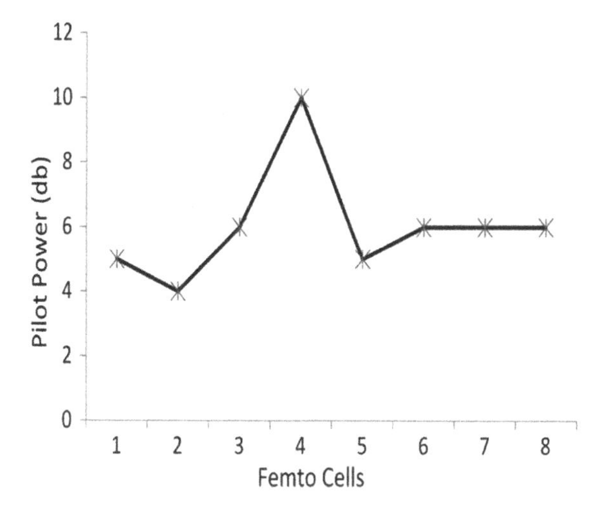

Fig. 5 Pilot power planning of femtocells

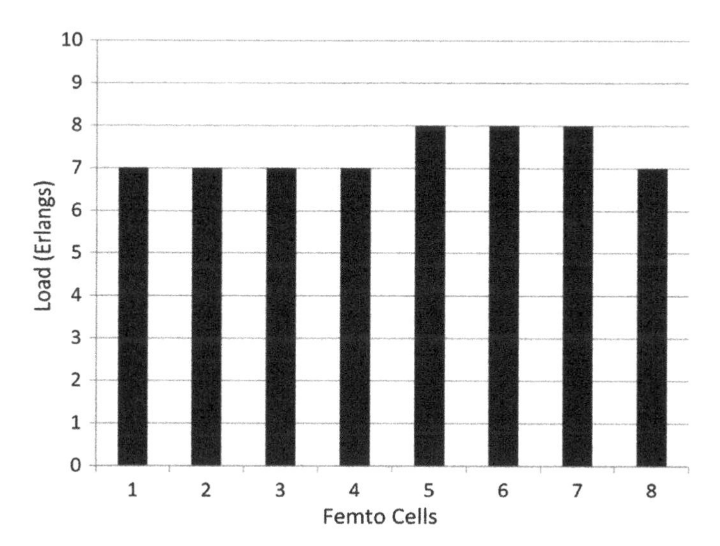

Fig. 6 Optimized load handling of femtocells

algorithm has obtained the optimized load balancing effectively for all the femtocells, by maintaining the load balancing constraint at each level of femtocells. Further, the ILS algorithm has converged to the optimal or best feasible solution within a few computation times. This property of ILS algorithm computation time is best suitable for solving sensitive enterprise environments.

5 Conclusion

The load balancing and self-optimization of femtocells plays critical role in performance and economical perspectives in future mobile network. The femtocell proves to be an effective solution for enhanced wireless radio coverage. In this paper, an Iterative local search heuristic algorithm is proposed and proved to be effective approach for optimization of femtocells in terms of load balancing. The results of proposed ILS algorithm demonstrated its performance to be outperforming for dynamic load balancing in femtocells network within minimized pilot power of femtocells along with distributed traffic requirement of users. Future work will incorporate comparative study of different optimization techniques for the load balancing in femtocells network.

References

1. Mohjazi, L. S., Al-Qutayri, M. A., Barada, H. R., Poon, K. F., & Shubair, R. M.,: Self-optimization of pilot power in enterprise femtocells using multi objective heuristic. Journal of Computer Networks and Communications, (2012), Article ID 303465 pp 130–143.
2. Ruiz-Aviles, J. M., Luna-Ramirez, S., Toril, M., & Ruiz, F.,: Fuzzy logic controllers for traffic sharing in enterprise LTE femtocells. In Vehicular Technology Conference (VTC Spring), IEEE 75th (2012) pp. 1–5.
3. Huang, L., Zhou, Y., Hu, J., Han, X., & Shi, J.,: Coverage optimization for femtocell clusters using modified particle swarm optimization. In 2012 IEEE International Conference on Communications (ICC), IEEE (2012) pp. 611–615.
4. Claussen, H., Pivit, F., & Ho, L. T.,: Self-optimization of femtocell coverage to minimize the increase in core network mobility signalling. Bell Labs Technical Journal, Volume 14, issue 2, (2009) 155–183.
5. Aguilar-Garcia, A., Fortes, S., Molina-García, M., Calle-Sánchez, J., Alonso, J. I., Garrido, A., & Barco, R.,: Location-aware self-organizing methods in femtocell networks. Computer Networks, 93, (2015) 125–140.
6. Jover, R. P.,: Interference in Femtocell Networks, (2009).
7. Hamalainen, J.,: Femtocells: Technology and Developments. Wireless Information Theory Summer School, Center for Wireless Communications, Aalto University (2011).
8. Chandrasekhar, V., Andrews, J. G., & Gatherer, A.,: Femtocell networks: a survey. IEEE Communications magazine, volume 46, issue 9, (2008) 59–67.
9. Akinlabi, O. A., Paul, B., Joseph, M., & Ferreira, H.,: A review of femtocell. In Proc. Int. MultiConf. Eng. Comput. Sci. (IMECS) Vol. 2, (2014).

Cuckoo Search Strategies for Solving Combinatorial Problems (Solving Substitution Cipher: An Investigation)

Ashish Jain, Jyoti Grover and Tarun Jain

Abstract Approximate algorithms have been well studied in order to solve combinatorial problems. This paper addresses cryptanalysis of the substitution cipher which is an interesting combinatorial problem. For this purpose, we utilize one of the latest approximate algorithms which is referred to as cuckoo search. Here, we point out that the proposed cuckoo search algorithm is not only an effective and efficient approach for solving the considered cryptanalysis problem, rather it can be a true and efficient choice for solving similar combinatorial problems.

Keywords Evolutionary computing · Cuckoo search · Scatter search · Genetic algorithm · Tabu search · Automated cryptanalysis

1 Introduction

Cryptology is very important in day-to-day life of common people, because cryptology is one set of techniques that provides information security. Cryptology field is commonly divided into two subfields cryptanalysis and cryptography.

Cryptography is the study of art and science for designing cryptosystems (or ciphers). A crucial component of the cipher is known as encryption algorithm. The encryption algorithm uses a secret key to transform a given plaintext into a ciphertext. Historically, cryptography is commonly connected with surveillance, warfare, and the like. However, with the advent of the information civilization and the digital

A. Jain (✉)
Discipline of Information Technology, Manipal University Jaipur, Jaipur, India
e-mail: ashish.jain@jaipur.manipal.edu

J. Grover · T. Jain
Discipline of Computer Science and Engineering, Manipal University Jaipur, Jaipur, India
e-mail: jyoti.grover@jaipur.manipal.edu

T. Jain
e-mail: tarun.jain@jaipur.manipal.edu

A.K. Somani et al. (eds.), *Proceedings of First International Conference on Smart System, Innovations and Computing*, Smart Innovation, Systems and Technologies 79, https://doi.org/10.1007/978-981-10-5828-8_56

revolution, cryptography is more and more important also in the peaceful lives of common people, e.g., when buying something over the Internet through credit card, withdrawing money from the ATM machines using smart-cards, and locking and unlocking luxury cars.

Cryptanalysis is related to finding flaws or oversights in the design of cryptosystems. The person who performs cryptanalysis is known as the cryptanalyst. The aim of cryptanalyst is to systematically recover the plaintext and/or secret key by attacking the cipher [1]. The attack involves the use of intelligent computer algorithms, just the ciphertext (i.e., encrypted plaintext) and/or some plaintexts. In the case of "ciphertext-only" attack, the attacker recover the plaintext and/or determine secret key by utilizing only the given ciphertexts. This type of attack is most challenging because we have just the knowledge of some ciphertexts and using that ciphertexts we need to determine secret key. This paper presents such a challenging cryptanalytic attack on a classical substitution cipher.

Cryptanalysis problems are considered as combinatorial search problems where the search space is consists of all possible combinations of key elements. The cryptanalytic attack via exhaustive (brute-force) searching, in theory, can be used against any encrypted text. However, in the worst case, this would involve traversing the entire search space. For instance, if the period of a substitution cipher is 26, then there are $26! = 4.03 \times 10^{26}$ permutation keys. If a supercomputer is available that could verify one billion million, i.e., 10^{15} "26-tuple" keys per second, would, in theory, require about 12788.28 years. Hence, these ciphers are secure from brute-force attack because the keyspace size is so large the resources and time are not available for searching the key exhaustively. On the other hand, search heuristics are capable of efficiently reducing the search space to a considerable extent. In the literature, several search heuristics have been applied that presents successful automated attacks on various classical ciphers. The advantage of automated attacks is that they run without time-consuming interaction of humans with a search process and finish when the key is determined. Thus, many cryptanalyst are interested in developing automated attacks of cryptographic algorithms. In this paper, we consider the automated cryptanalysis of the simple substitution cipher which is a most general form of classical ciphers.

Substitution cipher: A key of the substitution cipher is denoted as a permutation of plaintext letters (e.g., if 26 English alphabets represent plaintext letters, then a permutation of these 26 plaintext letters usually forms a substitution cipher key). Upon encryption, each letter of the plaintext message is replaced by the corresponding key element results in a ciphertext of length equal to the length of the plaintext. The original plaintext from the ciphertext is then recovered by intended recipient using decryption process. For a detailed description on the substitution cipher the interested reader can refer [2].

1.1 Related Work and Our Contributions

In literature, various search techniques have been applied for successfully attacking classical ciphers (e.g., substitution and transposition ciphers). Here, we mention only some of the previous research work where the main aim was to develop efficient automated attacks using search heuristics. Simulated annealing (SA) has been utilized by Forsyth and Safavi-Naini [3] to demonstrate attacks on the substitution ciphers. Another attack on these ciphers has been reported by Spillman et al. [4] using a genetic algorithm (GA). Clark [2, 5] has presented tabu search (TS) attack on the classical ciphers along with enhancement in previously proposed SA and GA attacks. Garici and Drias [6] have investigated the efficiency of scatter search (SS) for breaking the substitution cipher. In 2014, Boryczka and Dworak [7, 8] extended the study of attacks on the transposition ciphers using evolutionary algorithms to speed up the cryptanalysis process. The main deficiency in the above-mentioned attacks (i.e., in [6–8]) is that they fail to balance effectiveness and efficiency property of the GA attack that was proposed by Clark [2]. In other words, no cryptanalytic attacks of substitution cipher have been proposed in the literature that satisfies all the following objectives:

1. Number of ciphertext characters required by the proposed attack should be lesser among other attacks. In addition, the success rate should not be decreased. That is a more effective attack than other reported attacks.
2. Time required by the proposed attack should be lesser among other attacks. That is a more efficient attack that other reported attacks.
3. Number of keys examined by the proposed attack in finding the best solution should be lesser than other reported attacks.

This paper achieves the above-stated objectives by presenting two novel attacks based on the cuckoo search (CS). The authors of this paper are first researchers determining the efficiency of CS technique in developing automated attacks of the substitution cipher. In order to optimize the CS technique, some of its parameters are fine-tuned. Some of th efficient attacks of the substitution cipher that are based on the GA and TS techniques are also implemented. It should be noted that the SS attack presented by Garici and Drias [6] is not efficient with respect to the time complexity, however, this attack is considered in order to obtain a significant comparison between various attacks. Note that the CS algorithm is being designed in such a way that can be easily modified and can be used to solve similar combinatorial problems.

1.2 Fitness Function for Assessing Candidate Substitution Key

For determining the secret key of the substitution cipher, we start with a pool of candidate keys. Afterwards, we find suitability of each candidate key using following strategy. First of all, the known ciphertext is decrypted using candidate key. Afterward comparison is obtained between n-gram statistics of the decrypted text and language statistics which are considered to be known. Generally, these statistics are compared using the following equation:

$$Cost_k = \alpha\left(\sum_{i\in\mathcal{A}} |\mathcal{K}_i^u - \mathcal{D}_i^u|\right) + \beta\left(\sum_{i\in\mathcal{A}} |\mathcal{K}_i^b - \mathcal{D}_i^b|\right) + \gamma\left(\sum_{i\in\mathcal{A}} |\mathcal{K}_i^t - \mathcal{D}_i^t|\right), \quad (1)$$

In Eq. 1, $\mathcal{A}$ represents the alphabet of the language (for example, in English language: A, B, ..., Z), and $\mathcal{K}$ and $\mathcal{D}$ represent known language statistics and decrypted text statistics, respectively. The symbols u, b and t represent the unigram statistics of the language, bigram statistics of the language and trigram statistics of the language, respectively. Different value in 0.0, 0.1,..., 1.0 can be assigned to α, β and γ. However, $\alpha + \beta + \gamma = 1.0$ condition should be satisfied to keep the number of combinations of α, β and γ workable. Note that the substitution cipher attack is more effective by utilizing the cost functions that are designed using only bigrams than one which utilizes trigrams only [2]. Due to this fact, the cost function given below is used in this work which is purely based on the bigrams.

$$Cost_k = \sum_{i\in\mathcal{A}} |\mathcal{K}_i^b - \mathcal{D}_i^b| \quad (2)$$

1.3 Implementation of GA and TS Attacks of the Substitution Cipher

It is surprising that in last two decades no cryptanalytic attacks of the substitution cipher (except CS attack proposed by Jain and Chaudhari in [9]) have been proposed in the literature that can give results better than GA and TS algorithms of Clark [2]. Therefore, our task is to first implement GA and TS attacks and then compare two different CS attacks (one which is developed in this paper and another that was reported in [9]).

GA Attack: GA technique is initialized by a pool of n solutions (i.e., by n candidate keys of the substitution cipher). The main components of the GA attack are: fitness function which is already defined in the previous section, selection method which is a tournament selection method that selects a best candidate key from five randomly chosen candidate keys, crossover operator is defined in [2], and mutation operator is simply based on swapping two random elements of the investigated candidate key. Lack of space prevents us to present the pseudocode of GA attack in this paper. However, for a detailed description of GA attack, the interested reader can refer [2, 9].

First of all, a pool of n candidate keys called tabu is initialized. The TS attack maintains this tabu list as a short-term memory list. In each iteration, the investigated key is appended in the tabu list, and the key remains in tabu for a fixed number of iterations. Lack of space prevents us to present the pseudocode of TS attack in this paper. However, for a detailed description of TS attack, the interested reader can refer [2, 9].

2 Proposed CS Attacks: CS Attack1 and CS Attack2

The "standard" CS technique has been proposed in 2009 by Yang and Deb [10, 11] which is a new nature-inspired population-based search heuristic. In this paper and in [9], we have utilized the standard CS technique and incorporated our attack idea in the standard template. The "standard" template has already been presented in [9, 10]. For pseudocode of CS attack1, we refer the reader to [9]. For pseudocode of CS attack2, the reader can refer Algorithm 1 of this paper. The main difference between CS attack1 and CS attack2 are as follows: (1) we repeat Steps (6) to (8) of Algorithm 1 n times in the case of CS attack2. (2) We abandon the 0.02 fraction of worst nests in the case of CS attack2, while in the case of CS attack1 we abandon 0.01 fraction of worst nests. These changes have been introduced to identify the better CS attack. The set of equations that are employed in Algorithm 1 are as follows:

$$\mathbf{x}_j(t+1) = \mathbf{x}_j(t) + \mu\, l \tag{3}$$

$$l = \frac{u}{|v|^{1/\lambda}} \tag{4}$$

$$\sigma_u(\lambda) = \left[\frac{\Gamma(1+\lambda)\,\sin(\pi\lambda/2)}{\Gamma((1+\lambda)/2)\lambda 2^{(\lambda-1)/2}}\right]^{1/\lambda} = 0.696575 \text{ and } \sigma_v(\lambda) = 1(\text{if: } \lambda = 1.5), \tag{5}$$

Due to lack of space we cannot give details here, therefore, for a detailed description of terms used in the above set of equations, the reader can refer [9]. Here we describe the main idea of the attack. First of all, we map major parts of the CS technique that is nest, egg and Lévy flights to the considered problem. In almost all problems where CS technique is applied, usually it is assumed that each nest has one egg. But, in the case of cryptanalysis of the substitution cipher, we consider each nest has N distinct eggs/elements (i.e., N distinct characters of a key: $n_1, n_2, ..., n_n$, where, $N = 26$ (i.e., A–Z character). For the sake of simplification, we consider a unique number ($\in [1, N]$) is associated with each of the eggs. It means there must be a unique identity of each element in the nest, i.e., a substitution cipher key cannot have two similar characters. The difficulty is how to preserve distinctness property of the key elements, because it can be observed that the updating Eq. (3) will disturb the distinctness property of the key elements during update. It should be noted that the importance of Eq. (3) is that it build the new solution from an existing solution via Lévy flights which is an efficient approach, because the step-size is heavy-tailed and any large step is possible. That is, the candidate key has more chance to get

Algorithm 1 : CS Attack2 [proposed in this paper]

1: **Initialization:** Generate a pool of n host nests randomly, where host nests are actually the candidate keys of the substitution cipher. Call this pool $Existing_{keys}$, where each key is a vector.
2: **repeat**
3: Initialize a counter COUNT=1.
4: Compute the cost of each of the keys of the $Existing_{keys}$ using Eq. (2).
5: **repeat**
6: Select a lowest cost key from the $Existing_{keys}$, call this key $BEST_{key}$.
7: Choose a key randomly from $Existing_{keys}$, e.g., $Existing^i_{key}$. Construct a new key called NEW_{key}) via Lévy flights as: NEW_{key}=$Existing^i_{key}$+μ l, where l is evaluated using Eq. (4) with λ=1.5 and μ=0.01.
8: Compute the cost of NEW_{key} using Eq. (2). If the cost of NEW_{key} is lesser than the cost of $BEST_{key}$ then it becomes the new $BEST_{key}$,
9: **until** (COUNT>n)
10: Update the $Existing^i_{key}$ using $BEST_{key}$ and NEW_{key} as follows.
11: Generate a random number in the range [1, 26], call this number m.
12: **repeat**
13: Find the "next" element in NEW_{key} that has the identity less than m. Assuming that the element is found at P_1 position. In the case of first iteration, read "first" instead of "next".
14: Note the identity of the element which is located at position P_1 in the $BEST_{key}$. Call this identity n_1.
15: Assuming that P_2 is the position in the $Existing^i_{keys}$ where the element of identity n_1 is located. Swap elements of the $Existing^i_{nests}$ that are located at positions P_1 and P_2. This operation will store the element of identity n_1 at position P_1 because P_1 is the best position of the same identity element in the $BEST_{key}$. It is important to note that this swapping operation is carried out in order to preserve the element that have already placed in $Existing^i_{keys}$ to its best position.
16: **until** (NEW_{key} list has been traversed completely)
17: If $BEST_{key}$ and updated $Existing^i_{keys}$ are identical, then swap the elements that are located at two different positions in $Existing^i_{keys}$, where positions are determined randomly.
18: Abandon a fraction (p_a) of worst nests and create new nests, i.e., create new keys as replacement of abandon keys. The new keys are created again from the best key by swapping two randomly chosen elements of the best key. In the experiment, we fixed p_a=0.02.
19: **until** (Maximum-Iterations (or) All key characters are determined correctly)
20: Output the best fitness key from the $Existing_{keys}$.

converted in the exact key in less computations. Hence, we do not change this equation, reasonably we apply its efficiency in improving existing keys using the best key and new keys generated by Eq. (3) (see Algorithm 1, Steps (8) to (13)).

3 Results

The attack algorithms GA, TS, and CS have been implemented in Java 2.0 on an Intel Quad-Core processor i3 CPU (@2.20 GHz). The input to each of the algorithms is bigram statistics of the English language, known ciphertext, and ciphertext length. As an output, each algorithm determines full or partial key that has been used in the substitution cipher for encrypting plaintexts. Parameters such as population size in the case of GA attack, tabu list size in the case of TS attack and a pool of host nests in the case of CS attack have been fine-tuned by performing extensive experiments. The fine-tuning has been carried out separately for each of the techniques in order to optimize the cryptanalysis process. In some scenario, guidelines helps us, e.g., in the case of CS attack, we choose $\mu = 0.01$ and $\lambda = 1.5$ that has been published in [11] as a generic choice. For obtaining a fair comparison between presented attacks, the guidelines suggested by Clark [2] is followed, i.e., three criteria are used: number of ciphertext characters are available for attack, number of keys examined for finding the best solution, and time needed to find the best key.

We tested each of the attacks on 200 different known ciphertexts. For each ciphertext, each attack has been tested three times, i.e., a total of 600 times, and the best of the three has been recorded. In this way, 200 best-recorded results corresponding to each of the algorithms is then averaged. This recording and averaging process is repeated for known ciphertext of size 100, 200, 300, ..., 800 characters. The average results of each algorithm are shown in Figs. 1 and 2. From Fig. 1, we can observe that each algorithm perform well and approximate equally. However, the approximate algorithms should be compared based on the following two criteria for determining the accurate efficiency: state space searched (i.e., the number of keys examined before evolving the best result) and complexity of the attack (i.e., time taken to determine the best key). For these two criteria, we have tested each attack algorithm on 100 different known ciphertexts of 1000 characters length and recorded the number of keys examined and the time taken by the attack. The average results are shown in Figs. 3 and 4. Note that the SS method of Garici and Drias [6] is not implemented and therefore not considered in the Figs. 3 and 4, because Garici and Drias have concluded that the SS attack takes 75% more time than the GA attack, while the quality of the results is only 15% better. From Figs. 3 and 4, it can be observed that the average number of keys examined and average performance time, respectively in the case of TS attack are lesser than the GA attack. It is also clear from Figs. 3 and 4 that the CS attacks outperform TS and GA attacks in both respects.

Finally, we conclude that the CS attack1 is superior in performance than CS attack2.

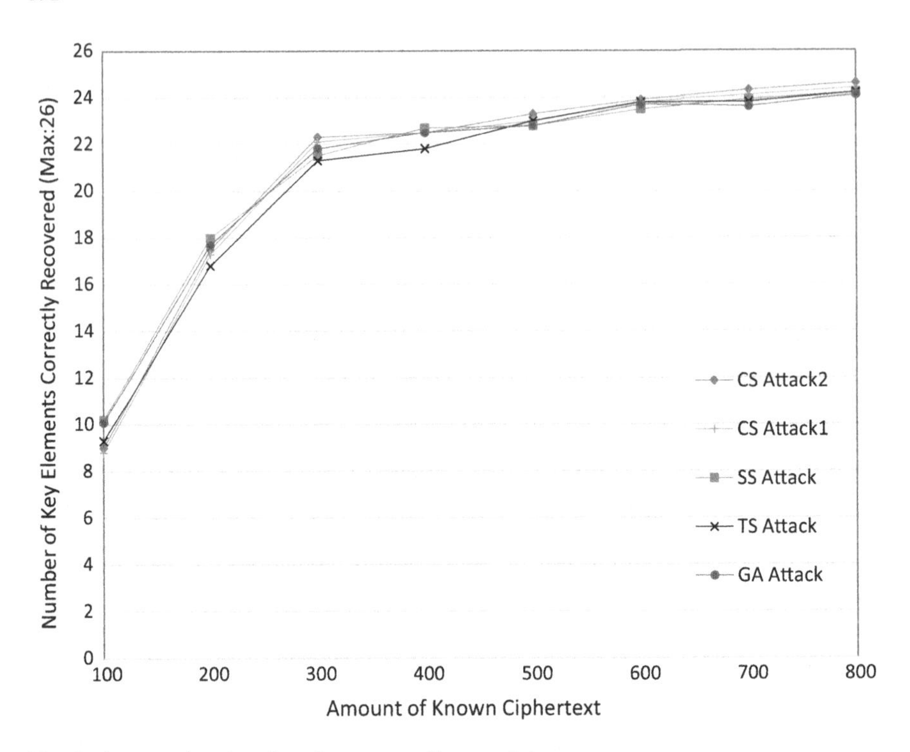

Fig. 1 A comparison based on the amount of known ciphertext

Number of ciphertexts	GA Attack		TS Attack		SS Attack		CS Attack1		CS Attack2	
	X	SD	X	SD	X	SD	X	SD	X	SD
100	10.1	4.52	9.3	4.87	10.2	4.63	8.8	5.54	9.0	5.43
200	17.7	5.58	16.8	6.84	18.0	5.32	17.3	4.76	17.5	5.12
300	21.8	4.62	21.3	5.74	21.5	5.65	22.1	4.69	22.3	4.57
400	22.5	4.22	21.8	5.37	22.7	4.13	22.5	4.24	22.5	4.19
500	22.8	3.54	23.0	4.51	22.8	3.38	23.0	3.98	23.3	4.04
600	23.7	2.43	23.8	4.21	23.5	2.79	23.8	4.11	23.9	3.98
700	23.6	2.57	23.8	4.15	23.9	2.13	24.1	3.21	24.3	3.06
800	24.1	2.27	24.2	4.12	24.2	2.07	24.4	2.98	24.6	2.77

Fig. 2 Mean (X) and standard deviation (SD) corresponding to Fig. 1

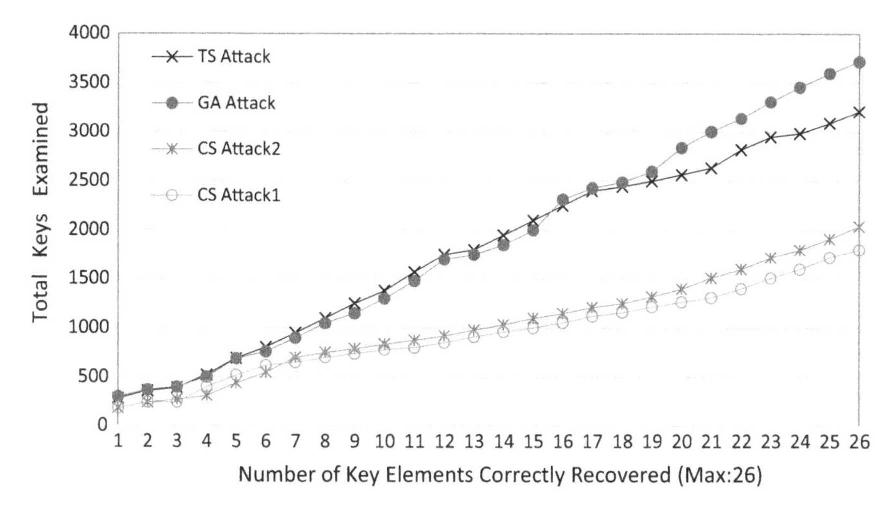

Fig. 3 A comparison based on the number of keys examined

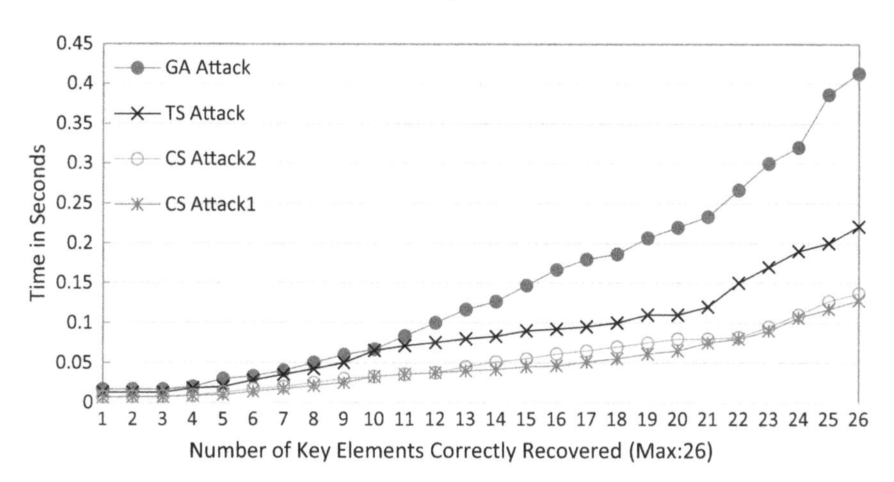

Fig. 4 A comparison based on time

4 Conclusion and Open Problems

This paper has demonstrated various attacks on the substitution cipher. The attacks have been developed by utilizing search heuristics, namely, GA, TS, SS, and CS. Results indicate the performance of CS attacks are superior than GA, TS, and SS attacks. It is important to note that in this work, we have to fine-tuned lesser number of parameters for both type of CS attacks than GA, TS, and SS attacks. This study indicates that the CS strategy is capable to produce results that are much better than the previously proposed GA, TS, and SS techniques. Therefore, the proposed CS strategies are true and efficient alternatives for solving this kind of combinatorial problems, for example, for optimal placement of phaser measurement units in a power system, for graph coloring, for solving Boolean satisfiability problem, etc., but, at what extent the proposed CS strategy can be utilized we left as an open problem.

References

1. Stinson, D.R.: Cryptography: theory and practice. CRC press (2005)
2. Clark, A.J.: Optimisation heuristics for cryptology. PhD thesis (1998)
3. Forsyth, W.S., Safavi-Naini, R.: Automated cryptanalysis of substitution ciphers. Cryptologia **17**(4) (1993) 407–418
4. Spillman, R., Janssen, M., Nelson, B., Kepner, M.: Use of a genetic algorithm in the cryptanalysis of simple substitution ciphers. Cryptologia **17**(1) (1993) 31–44
5. Clark, A.: Modern optimisation algorithms for cryptanalysis. In: Intelligent Information Systems, 1994. Proceedings of the 1994 Second Australian and New Zealand Conference on, IEEE (1994) 258–262
6. Garici, M.A., Drias, H.: Cryptanalysis of substitution ciphers using scatter search. In: Artificial Intelligence and Knowledge Engineering Applications: A Bioinspired Approach. Springer (2005) 31–40
7. Boryczka, U., Dworak, K.: Genetic transformation techniques in cryptanalysis. In: Intelligent Information and Database Systems. Springer (2014) 147–156
8. Boryczka, U., Dworak, K.: Cryptanalysis of transposition cipher using evolutionary algorithms. In: Computational Collective Intelligence. Technologies and Applications. Springer (2014) 623–632
9. Jain, A., Chaudhari, N.S.: A new heuristic based on the cuckoo search for cryptanalysis of substitution ciphers. In: Neural Information Processing, Springer (2015) 206–215
10. Yang, X.S., Deb, S.: Cuckoo search via lévy flights. In: Nature & Biologically Inspired Computing, 2009. NaBIC 2009. World Congress on, IEEE (2009) 210–214
11. Yang, X.S., Deb, S.: Engineering optimisation by cuckoo search. Internl. Journal of Mathematical Modelling and Numerical Optimisation **1**(4) (2010) 330–343

A WSN-Based Landslide Prediction Model Using Fuzzy Logic Inference System

Prabhleen Singh, Ashok Kumar and Gaurav Sharma

Abstract This paper proposes a new WSN-based Landslide Prediction Algorithm, developed using Fuzzy Logic Inference System. Three factors conditioning landslides are considered, namely: slope angle, soil moisture and topographical elevation. These conditioning factors are sensed using WSN and analysed using proposed algorithm at sink node. A Mamdani-type Fuzzy Inference System (FIS) is used to develop the algorithm. Triangular membership functions are considered for all FIS parameters. A total of 45 rules have been developed in this FIS, which holds capability to generate a three-level alarm to warn residents of area about any impeding danger due to landslide. For results, surface plots are generated which show us the variation of landslide susceptibility with the change in three parameters considered.

Keywords Wireless sensor networks · Fuzzy logic · Mamdani · Landslide prediction

1 Introduction

The advancement in design of sensor technology, low-power electronics, and low-power radio frequency (RF) have enabled the development of low-power sensors, called micro-sensors, which connect themselves to form a network called Wireless Sensor Network (WSN) [1]. WSNs have numerous applications that include event monitoring, disaster prediction, and so on [2]. In this paper we focus on single application: Landslide Prediction.

P. Singh (✉) · A. Kumar · G. Sharma
Department of Electronics and Communication Engineering, National Institute of Technology, Hamirpur, Himachal Pradesh, India
e-mail: singh.prabhleen123@gmail.com

A. Kumar
e-mail: ashok@nith.ac.in

G. Sharma
e-mail: ergaurav209@yahoo.co.in

A.K. Somani et al. (eds.), *Proceedings of First International Conference on Smart System, Innovations and Computing*, Smart Innovation, Systems and Technologies 79, https://doi.org/10.1007/978-981-10-5828-8_57

Landslide is a natural disaster responsible for great losses. Landslide or slope failure phenomenon is mostly unavoidable but not unpredictable. The destruction it causes motivates us for Landslide Prediction, often referred to as Landslide Susceptibility. Since last few decades, the assessment of landslide susceptibility has became an important subject for engineers [3]. Real-time prediction of landslide can help avoid loss of life and property. In the literature of this field, numerous techniques were observed assessing landslide susceptibility with the absence of any inference system [4, 5]. Recent advancements in this field include calculation of rainfall thresholds, landslide modelling and hazard zonation [6–8]. These use data-driven analyses, which has numerous disadvantages, like, the methods used are black box techniques, from which user gets complete isolation [5].

Recently, Mamdani FIS is being used extensively to compute complex problems of engineering and geology. Its advantage is its transparent problem description. Although several studies using similar type of system have been published, a Mamdani FIS has been used for a constricted number of landslide prediction assessments till date. There are some studies on the use of WSNs for landslide monitoring [9, 10]. Reference [11] proposed a WSN measuring strain in the rocks using strain gauges and tested it in a laboratory test bed. Algorithm proposed makes use of Mamdani FIS and is very easy to use in terms of simplicity and understandability. It holds capability to generate three levels of warning depending upon the susceptibility of landslide in area and works at the sink node of a WSN. The tiny nodes of WSN sense different geological parameters and this geological data is aggregated at the sink node. The sink node analyses the real-time data using proposed algorithm and generates a warning if need be. Geological parameters sensed in real time are angle of slope, soil moisture and topographical elevation.

2 Proposed Landslide Prediction Model

2.1 Landslide Causative Factors

Slope Angle. One of the most important factors that control slope stability is slope angle. Slope Angle is measured using Inclinometers and measure slope angle in degrees (°). A Micro-Electro-Mechanical Systems (MEMS) single-axis inclinometer can be used [12]. Landslide frequency is highest between 26–35°. However, it was also found that when slope angle exceeds 45° the frequency of landslides decreased.

Soil Moisture. The second input parameter, Soil Moisture, simply tells the wetness conditions of the soil in real time. Most studies have estimated the soil moisture based on rainfall [13] while they do not account for variable soil moisture prior to and during rainfall events. Garich [14] has made use of soil probes (ECH20^{TM} EC-5) for measurement of soil moisture.

Topographical Elevation. Research has shown that relief at high elevation show greater susceptibility to slope failure. However, some of the research has shown that

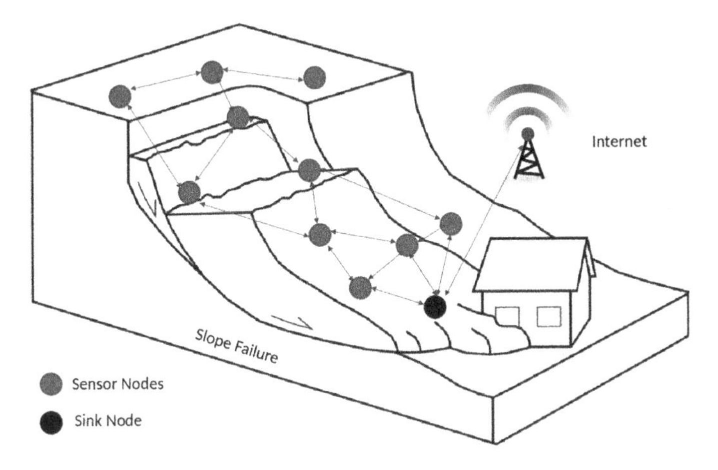

Fig. 1 WSN deployed at landslide area

higher topographical elevations constitute of lithological units resistant to sliding [3]. Thereby we need the knowledge of soil moisture and slope angle to find landslide susceptibility at different topographical elevations.

2.2 Wireless Sensor Network Scenario

An application of WSN for which the algorithm is being proposed has been explained briefly in this section. A WSN as shown in Fig. 1 can be deployed for real-time remote monitoring of an area susceptible to landslides. Deployment area can be hill-side or some similar topology. The deployed nodes construct a network by communicating with each other through radio communication. Sensor nodes have the capability to sense real-time data using sensors onboard. One or more sensor nodes can possess the capability to communicate the information collected to the user, often referred as sink node. To achieve this a sink node can transmit the information to a Base Station (BS), which further communicates it to the user via Internet.

2.3 Algorithm Design

FIS Structure. The proposed algorithm is designed using Fuzzy logic. Tool used for design process is MATLAB, which inherently supports development of Fuzzy Models with in-built Fuzzy Logic Designer interface. Our proposed work is, technically, a Fuzzy Inference System (FIS). We have chosen Mamdani-type FIS [15] for design. One of the advantages of Mamdani FIS is it can combine numerical data with

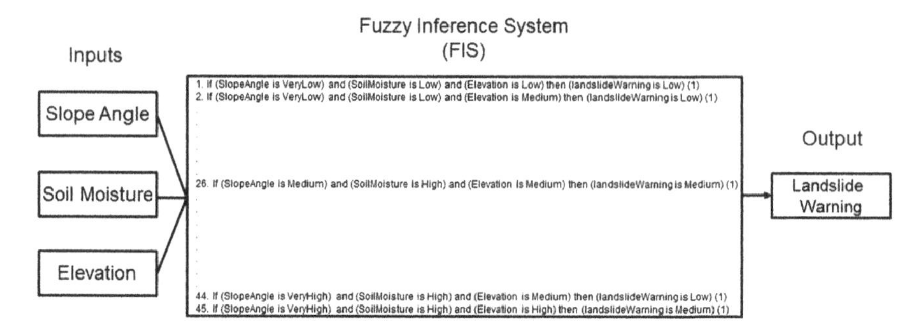

Fig. 2 Structure of a Mamdani FIS

linguistic data. A Mamdani FIS consists of three main parts as shown in Fig. 2: the membership functions of input(s), fuzzy if-then rules, and output membership functions. The Mamdani FIS for the assessment of landslide prediction includes a total of three inputs: slope angle, soil moisture and topographical elevation. The number of outputs is one, namely Landslide Warning. The if-then linguistic rule-set contains a total of 45 rules.

Membership Functions. A fuzzy set can be described by the below formula as:

$$A = \{x, f_A(x)\}, \quad x \in R. \tag{1}$$

where A is a fuzzy set, x is an element of universal set R, and f(x) is the fuzzy membership function. A fuzzy set inherits continuous membership in the range (0, 1). Landslide prediction requires determination of fuzzy membership functions of causative functions. To minimize the uncertainty, a 50% overlap is applied between the fuzzy sets for each input parameter. Triangular membership functions are used for each input parameter. The output Landslide Warning consists of three fuzzy sets in shape of triangular membership functions. Slope angle contains five fuzzy sets in the form of triangular membership functions as shown in Fig. 3a. The five fuzzy sets are named as very low, low, medium, high and very high. Whereas, soil moisture, topographical elevation and landslide warning have three triangular membership functions, namely: low, medium and high as shown in Fig. 3b–d.

Linguistic Rules. The general structure of if-then rules in Mamdani FIS [15] is given as

$$R_i : \textit{if } x_i \textit{ is } A_{il} \textit{ and} \ldots \textit{then } y \textit{ is } B_i \; (\textit{for } i = 1, 2, \ldots k) \tag{2}$$

where x_i is the input (antecedent) variable; y is the output (consequent) variable (Landslide Warning); and k is the numbers of rules.

Since the number of input and output parameters totals to four, the number of if-then rules came out to be 45 in our algorithm. While describing if-then rules following considerations deduced from knowledge of landslides are used:

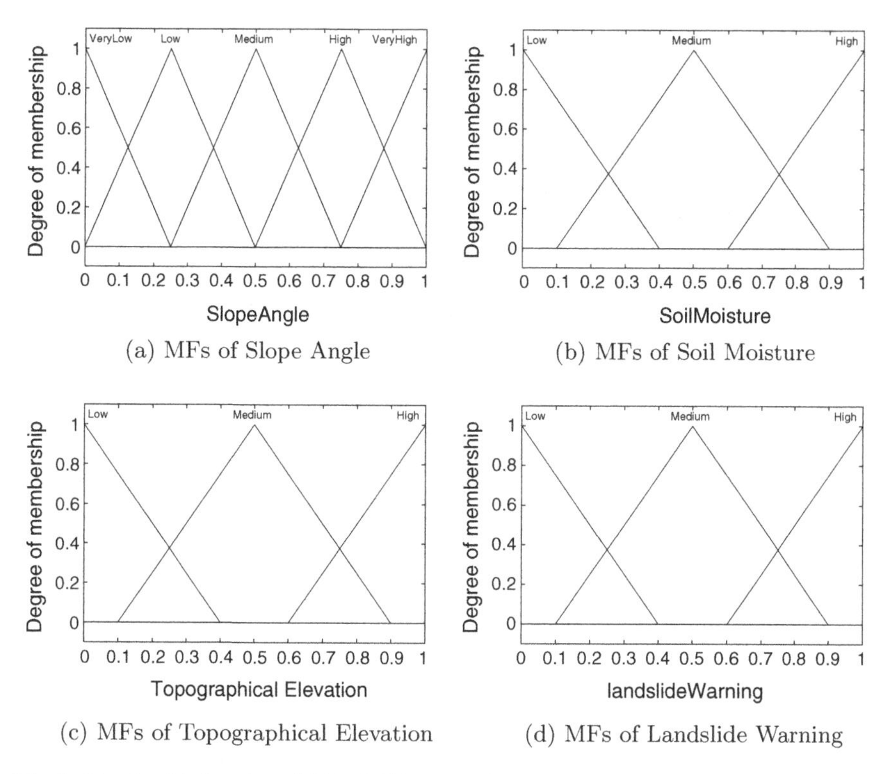

(a) MFs of Slope Angle

(b) MFs of Soil Moisture

(c) MFs of Topographical Elevation

(d) MFs of Landslide Warning

Fig. 3 Membership functions (MFs) of all inputs and outputs

(1) Landslide susceptibility is medium to high, if the rule includes "high" in soil moisture input, i.e. output landslide warning is "medium" or "high".
(2) If the rule includes "very high" for slope angle input then output is accepted as "low", as at very steep slopes the number of landslides was observed to be less.
(3) If two inputs are positive, i.e. include "high" or "very high" and one is negative, i.e. includes "low" or "very low", the output is considered as "medium".
(4) If two inputs are positive and one is moderate, i.e. includes "medium", the output is accepted as "medium" or "high".
(5) If all three inputs are positive, then output is always considered as "high".
(6) If the slope angle is "very low" then output is always accepted as "low".
(7) If all three inputs include "medium" then output is always "medium".

Proposed work contains one output which is the three-level warning for probability of landslide to occur. The FIS analyses input data according to the linguistic rules and gives one of the three warnings as output. The output depicts the susceptibility of a landslide to occur, i.e. "low" at output depicts there are no chances for a landslide to occur, whereas "high" at output depicts it is very likely for landslide to occur.

3 Results and Discussion

We have proposed an algorithm which is installed on the sink node of a WSN. The sink node aggregates all the data from the network and analyses it using the proposed algorithm. The output of algorithm consists of three different levels of warning, i.e. “low”, “medium” and “high”. The algorithm was designed using Fuzzy Logic Inference (FLI) System. In an FLI, all input and output parameters have different membership functions as shown in Fig. 3. The numerical values of the different range in which the membership functions of input parameters are defined is given in Table 1. The range of values in Table 1 were observed in Kullu Area of Himachal Pradesh, India [16]. The division of numerical values into different sets of ranges and then assignment to membership functions is according to their impact observed on landslide phenomenon. For example, there is a huge variation in the impact of slope angle on landslide, therefore, it was divided into five sets of numerical values and further assigned to membership functions. Similarly, numerical values for different parameters were chosen for corresponding membership functions. If the output warning is “low”, it depicts there no chances of landslide in the area. A “medium” output means there are enough chances of landslide to alert the residents of area to be careful. A “high” output depicts the area is very dangerous and needs to be evacuated immediately.

Surface plots help us visualize the result of proposed algorithm with variation in different outputs. Surface plots for input and output parameters have been generated and shown in Fig. 4a. Since we have three input parameters three surface plots were generated. Surface plot in Fig. 4a shows output with change in elevation and slope angle. Similarly Fig. 4b, c show the change in output with change in soil moisture versus elevation and slope angle versus soil moisture. For example, if value of Slope Angle is 33°, Soil Moisture is 55% and Elevation is 8000 ft, then our algorithm generates a “medium” warning which alerts the residents to be careful as their are chances for a landslide to occur.

Table 1 Numerical values of membership functions

Parameter	Membership function	Value
Slope angle	Very low	≤15°
	Low	16–25°
	Medium	26–35°
	High	36–45°
	Very high	≥45°
Soil moisture	Low	≤40%
	Medium	40–60%
	High	≥60%
Elevation	Low	≤4000 ft
	Medium	4000–7000 ft
	High	≥7000 ft

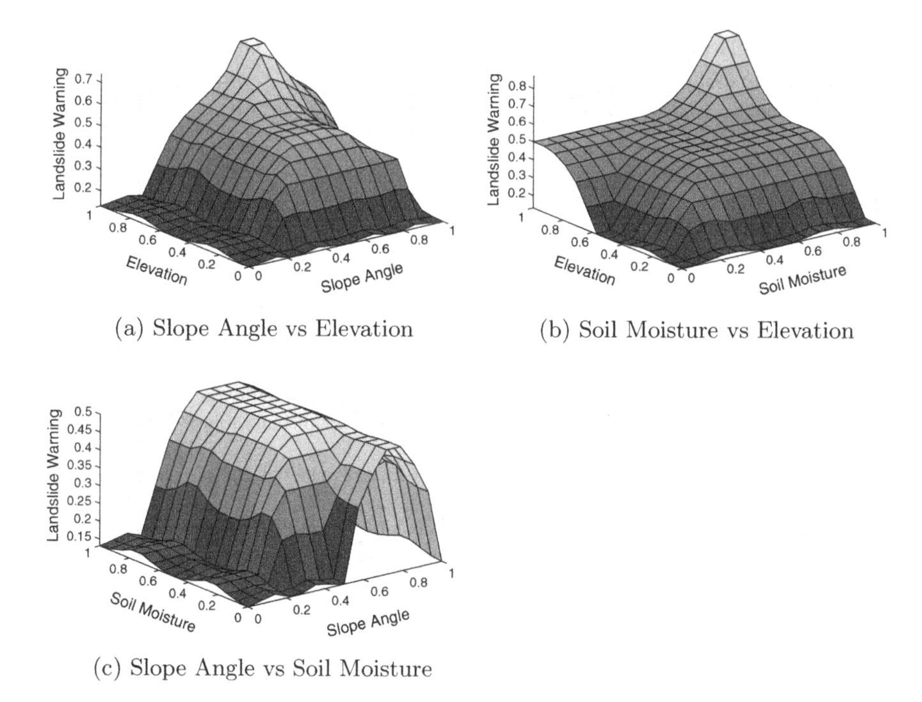

(a) Slope Angle vs Elevation

(b) Soil Moisture vs Elevation

(c) Slope Angle vs Soil Moisture

Fig. 4 Surface plots for various input and output parameters

4 Conclusion

The WSN-based system model consists of sensor nodes deployed in an area prone to landslides. These nodes sense three causative factors of landslides, namely: slope angle, soil moisture and topographical elevation. Data collected by these sensor nodes is aggregated at sink node. Sensed data is processed using the proposed WSN-based Landslide Prediction Algorithm using Mamdani FIS at sink node, which issues a warning to residents of area via Internet, by transmitting warning to a nearby base station. Output of algorithm can be one of three levels of warning, i.e. low, medium and high. To check the output, surface plots have been generated which depict the variation of output with change in input parameters. From the surface plots it is clearly evident that

(1) When slope angle changes from low to medium, the output goes from low to medium to high. But, when it goes from medium to high, output goes from high to medium to low.
(2) When soil moisture changes from low to medium and if slope angle is between medium and high then output is high, otherwise output is low.
(3) When elevation changes from low to medium to high, output changes from low to medium to high.

The output levels depict the chances for landslide to occur. Thus loss of life and property can be avoided.

References

1. Heinzelman, W.B., Chandrakasan, A.P. and Balakrishnan, H.: An application-specific protocol architecture for wireless microsensor networks. IEEE Transactions on wireless communications 1(4), pp. 660–670 (2002)
2. Wang, J., Ghosh, R.K. and Das, S.K.: A survey on sensor localization. Journal of Control Theory and Applications 8(1), pp. 2–11 (2010)
3. Ercanoglu, M. and Gokceoglu, C.: Assessment of landslide susceptibility for a landslide-prone area (north of Yenice, NW Turkey) by fuzzy approach. Environmental geology 41(6), pp. 720–730 (2002)
4. Juang, C.H., Lee, D.H. and Sheu, C.: Mapping slope failure potential using fuzzy sets. Journal of geotechnical engineering 118(3), pp. 475–494 (1992)
5. Akgun, A., Sezer, E.A., Nefeslioglu, H.A., Gokceoglu, C. and Pradhan, B.: An easy-to-use MATLAB program (MamLand) for the assessment of landslide susceptibility using a Mamdani fuzzy algorithm. Computers & Geosciences 38(1), pp. 23–34 (2012)
6. Rosi, A., Peternel, T., Jemec-Aufli, M., Komac, M., Segoni, S. and Casagli, N.: Rainfall thresholds for rainfall-induced landslides in Slovenia. Landslides 13(6), pp. 1571–1577 (2016)
7. De Luca, D.L. and Versace, P.: A comprehensive framework for empirical modeling of landslides induced by rainfall: the Generalized FLaIR Model (GFM). Landslides, pp. 1–22 (2016)
8. Hung, L.Q., Van, N.T.H., Van Son, P., Khanh, N.H. and Binh, L.T.: Landslide susceptibility mapping by combining the analytical hierarchy process and weighted linear combination methods: a case study in the upper Lo River catchment (Vietnam). Landslides, pp. 1–17 (2016)
9. Rosi, A., Mamei, M., Zambonelli, F. and Manzalini, A.: Landslide monitoring with sensor networks: a case for autonomic communication services. In International conference on wireless technology for rural and emergency scenarios, IEEE CS, Roma (2007)
10. Terzis, A., Anandarajah, A., Moore, K. and Wang, I.: Slip surface localization in wireless sensor networks for landslide prediction. In Proceedings of the 5th international conference on Information processing in sensor networks (pp. 109–116). ACM (2006)
11. Sheth, A., Thekkath, C.A., Mehta, P., Tejaswi, K., Parekh, C., Singh, T.N. and Desai, U.B.: Senslide: a distributed landslide prediction system. ACM SIGOPS Operating Systems Review, 41(2), pp. 75–87 (2007)
12. Lu, P., Wu, H., Qiao, G., Li, W., Scaioni, M., Feng, T., Liu, S., Chen, W., Li, N., Liu, C. and Tong, X.: Model test study on monitoring dynamic process of slope failure through spatial sensor network. Environmental Earth Sciences 74(4), pp. 3315–3332 (2015)
13. Rosso, R., Rulli, M.C. and Vannucchi, G.: A physically based model for the hydrologic control on shallow landsliding. Water Resources Research 42(6), (2006)
14. Garich, E.A.: Wireless, automated monitoring for potential landslide hazards (Doctoral dissertation, Texas A&M University) (2007)
15. Grima, M.A.: Neuro-fuzzy modeling in engineering geology. AA Balkema, Rotterdam, 244 (2000)
16. Data Source : Field Visits & Google Earth Image(c) 2016 INEGI, Imagery Date 2002–2016

Task-Enabled Instruction Cache Partitioning Scheme for Embedded System

Bhargavi R. Upadhyay and T.S.B. Sudarshan

Abstract Energy reduction is an important factor and a challenge in the design of the embedded system. In this work, we propose TEST, a process-aware partitioning scheme to study the impact of partitioning scheme based on process aware in instruction cache for multitasking embedded system. Process-aware partitioning will partition instruction cache which is based on the mapping between process and cache memory. This technique results in 70–80% reduction in dynamic energy and 50–70% in static leakage energy as compared to the base set-associative cache architecture. Results of the TEST are evaluated using the simple scalar 3.0 simulator using the Mi-bench-embedded benchmarks.

Keywords Cache partitioning scheme • Process-aware cache
Embedded system

1 Introduction

A major requirement for embedded devices design is high performance with the low power consumption. Recent embedded processors execute multiple processes on a single core as it competes for shared resources like CPU, bus, memory, etc. Memory subsystem consumes a significant amount of power in the embedded system. Cache memory being a main segment can consume up to 50% of a microprocessors total power [1]. Cache memory is a major contributor to total dynamic and static leakage energy. The conventional N-way set-associative instruction cache enables all the N tag-ways to perform the parallel comparison for

B.R. Upadhyay (✉)
Department of Computer Science and Engineering, Amrita School of Engineering, Bengaluru, India
e-mail: u_bhargavi@blr.amrita.edu

T.S.B. Sudarshan
Amrita Vishwa Vidyapeetham, Amrita University, Coimbatore, India
e-mail: sudarshan.tsb@gmail.com

A.K. Somani et al. (eds.), *Proceedings of First International Conference on Smart System, Innovations and Computing*, Smart Innovation, Systems and Technologies 79, https://doi.org/10.1007/978-981-10-5828-8_58

finding the requested instruction in the cache. This results in high dynamic power consumption in the system. In an ideal situation, a direct-mapped cache for a cache hit must have only one comparison of a tag entry and access of one data entry. Based on this observation, providing access only to a subsection of the cache memory results in a reduction of the dynamic energy with negligible performance impact. Enabling only the required instruction way for access will reduce the cache energy consumption. In the modern embedded system, many applications are partitioned into several tasks known as processes. The cache is being shared between multiple processes which will incur additional inter-task cache architecture. Here data belonging to one process evict the data of another process when the context switch initiated by the operating system. This can be the cause of the performance degradation for the task performance. In a Cache partitioning scheme, different processes are mapped to different cache-ways of the cache memory. This, in turn, eliminates the task interference to a larger extent. We study the impact of basic static partitioning based on mapping of the task and particular partition of cache. We proposed a new Task-Enabled Selective Trimming cache mechanism, which partitions the instruction cache to enable subsection of cache memory saves a significant amount of dynamic energy consumption and static leakage power in multitasking embedded system. The system becomes more predictable as it helps in the task interference problem. The proposed Task-Enabled Selective Trimming Instruction Cache Scheme (TEST) partitions the cache-ways between the processes. This scheme mapped the processes to set of cache-ways in N-way set-associative cache memory architecture. This paper is organized as follows. Section 2 describes the TEST scheme with the architecture. The experimental setup is mentioned in the Sect. 3. Section 4 evaluates the proposed approach and analyzes the results. Section 5 describes the related work and finally concludes in Sect. 6.

2 The Test Architecture

The proposed work is based on partitioning the cache such that several tasks coexist in the instruction cache. The TEST scheme is applied to instruction cache where several tasks which are driven to wait state by OS still can reside in the cache and wait for their turn to execute. Without being replaced by a different set of instructions belonging to a task which enters the run state. Here in a N-way set-associative cache, each cache-way or group of cache-ways are dedicated to the task. The instruction pages corresponding to the currently executing process are available in a reserved way. This results in enabling only the assigned cache-way for tag comparison, as shown in Fig. 1. The cache-ways which do not belong to the currently executing process are shut down to reduce dynamic power consumption. A significant amount of dynamic power reduction is possible with this scheme. Power consumption can be reduced as only the cache-way reserved for an executing task will be activated. This scheme has an N-way set-associative cache, one K bit register per cache-way to store the process-related information in pid, log2N bits

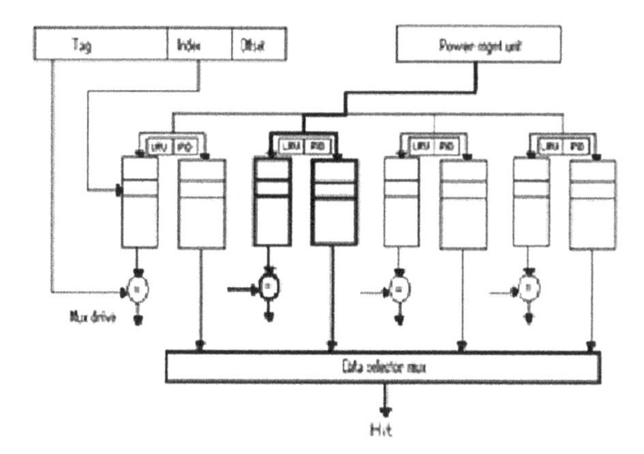

Fig. 1 The TEST architecture

per cache-way to store the LRU cache-way information as shown in Fig. 1 which shows the TEST architecture for the 4-way set-associative cache.

2.1 *TEST Algorithm*

Begin Divide the address into three filed of Tag, Index and Offset.

```
Way_Sel = DedicatedCachewayof the currentlyRunning Process
Way Select way enabled;
all other N–1cache-ways in deep sleep;
If(CACHE HIT in Way_Sel)then Read/Write data
from/to offset location of I Th block in Way
Select way
If(Reference is CACHE MISS in iTh set)
Then Transfer data from Main memory to the Way select
End
```

The task-enabled selective trimming (TEST) set-associative cache works as follows. The cache obtains the physical address of the requested reference similar to the conventional cache. For every cache access, only the dedicated cache-way of that process is enabled with the help of the power management unit, process-id and cache controller. This is similar to operating a set-associative cache virtually as a direct-mapped cache. The corresponding line index of the dedicated cache way is searched for a tag match, as shown in Fig. 1. On a cache hit, the data from the corresponding offset location is accessed. This results in 75% dynamic power reduction as compared to a conventional 4-way set-associative cache. In case of a miss, a cache miss here is equivalent to a way-miss. In this scenario, the data will be transferred from main memory to the cache-way memory, which is active. If the number of processes is more than the number of available ways, than a replacement policy is enabled to decide upon mapping.

2.2 Test Replacement Policy

When the number of tasks is less than the number of cache-ways, we can achieve more reduction in static leakage power as more number of cache-ways can be put into low power mode. When the number of tasks is more than the number of cache-ways in the cache memory, it is difficult to figure out the optimum victim for replacement. LRU replacement policy is widely used as the replacement policy for improving the hit rate. LRU replacement policy is used to assign the way for the process, when the number of processes is more than the number of cache-ways. The non-executing cache-ways are kept in low power—drowsy mode [2]. This, in turn, reduces static power consumption. The dynamic power consumption is reduced by reduction in the number of active cache cache-ways.

3 Experimental Setup and Simulation

This section explains the simulator and simulator flow to evaluate the cache architecture.

3.1 Simulator

A two-part simulation environment was created for the purpose of evaluating the proposed TEST cache architectures. Figure 2 illustrates the simulation flow discussed in the following sections.

3.2 Cache Trace Extraction

We used Simple Scalar tool set for the ARM processor architecture [3]. Cache traces for various benchmark programs were extracted. Simple Scalars cache.c file was modified to dump all level-1cache accesses to a file. The simple scalar tool set with the modifications needs to be recompiled. We used benchmark program binaries in the simulator. We simulated those binaries for extracting their unique cache traces.

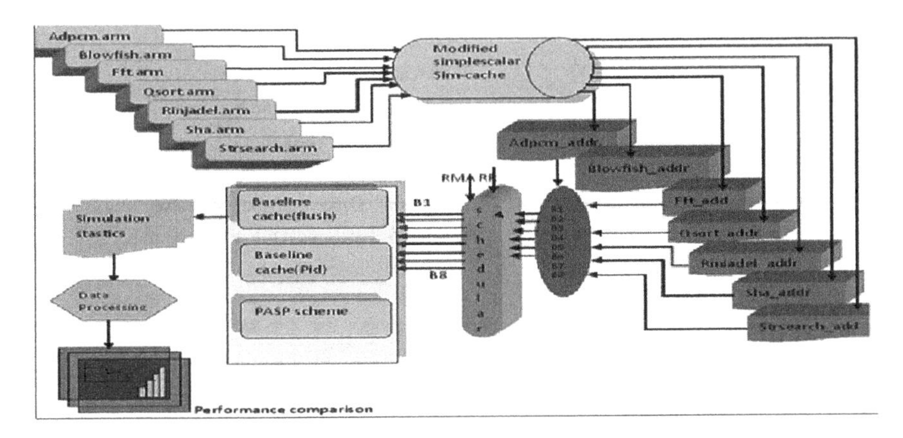

Fig. 2 Cache simulation flow for TEST scheme

Table 1 Different cache configurations

Cache parameters	Range
Cache size	16 K, 32 (in Bytes)
Block size	32 (in Bytes)
Cache associativity	4, 8

3.3 *Simulation Parameters*

Table 1 shows the configuration of cache memory for the system. A configuration similar to high-end embedded processors such as the Intel XScale and ARM9 were used for experimentation.

3.4 *Workloads*

Benchmark programs from the Mi-Bench suite have been selected as multitasking workloads [4]. We covered a wide range of applications from the several embedded application domains. Table 2 shows the selected benchmarks from the benchmark suite from different domains of the embedded system. We have used the Adpcm, Blowfish, FFT, Qsort, Rinjadel, Sha ans string search benchmarks for the experiments.

We grouped the randomly four tasks for multitasking workload. Table 2 shows the combinations of tasks used for each group.

Table 2 Multitask benchmark sets

B1	Qsort, Adpcm
B2	Strsearch, Sha
B3	Adpcm, Qsort, Strsearch
B4	Dijkstra, Sha, Adpcm
B5	Qsort, Adpcm, Strsearch, Sha
B6	Bitcount, Dijksstra, Sha, Strsearch
B7	Adpcm, Bitcount, Dijkstra, Strsearch
B8	Crc, Fft, Strsearch, Sha, Qsort

Table 3 RMA configurations

Bench1	Ei	Pi
Qsort	1780417	8092804
Adpcm	1117207	4468828
Strsearch	8297	103712
Sha	396493	23323311
Dijkstra	976766	5156663
Bitcount	133809	955778
Crc	379	3790
Fft	337	2592

3.5 Comparison Approaches

We compared the proposed TEST cache with the conventional set-associative cache using flush method and a conventional cache using process id without partitioning. Each of the three cache structures was simulated with the benchmark cache traces. In the conventional cache, flush scheme, flush the entire cache, in case of a context switch. In process id scheme, multiple tasks sharing the cache by using unique process identifier PID for each process. Round-robin scheduling policy and RMA static priority scheduling policy used to model multitasking. We used context switch frequency of 33 K instructions in round-robin to 7TDMI using 12 MHz frequency to find the simulation results for RMA scheduling. RMA assigns priority to task based on the occurrence of a task. Table 3 shows the RMA configuration.

4 Evaluation

4.1 Performance

The obvious advantage of the cache partitioning scheme is faster hit time and reduction in energy consumption. The drawback is increasing miss-rate because of less cache space for the task. The miss-rate of baseline cache without partitioning are obtained using flush and processed id scheme. Baseline cache is compared with

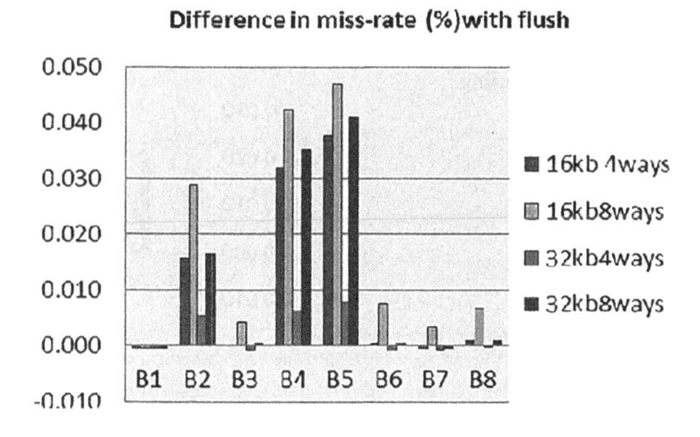

Fig. 3 Difference in miss-rate with flush for RR scheduling

the proposed TEST partitioning scheme for round-robin scheduling with respect to the difference of miss-rate as shown in Figs. 3 and 4. The same observation is made for RMA scheduling which is shown in Fig. 5 and Fig. 6 respectively. In TEST scheme 16 kb 4-way occupy 4 k size of the entire cache, 16 kb 8-way cache occupy 2 k size of the entire cache, 32 kb 4-way occupy 8 k size of the entire cache and 32 kb 8-way occupy the 4 k size of the entire cache. We can find an increase in miss-rate because of the capacity misses in 16 kb 8-way cache. We can observe in benchmark B1 miss-rate has not increased and in some cases, there is a slight improvement in hit rate.

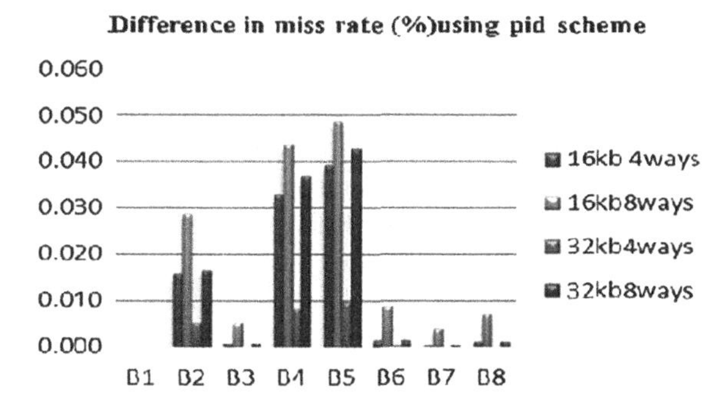

Fig. 4 Difference in miss-rate using pid scheme for RR scheduling

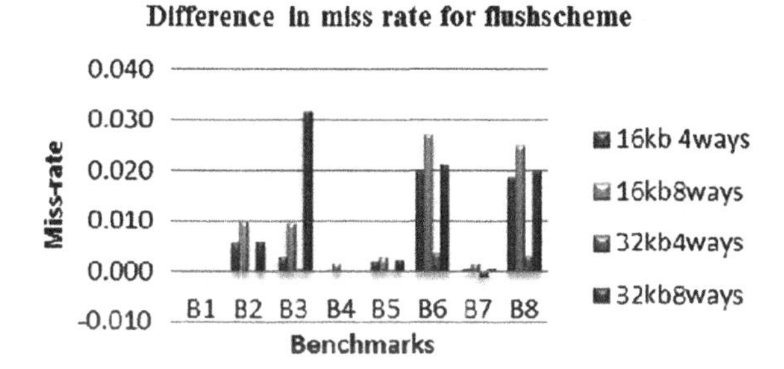

Fig. 5 Difference in miss-rate using flush scheme for RMA scheduling

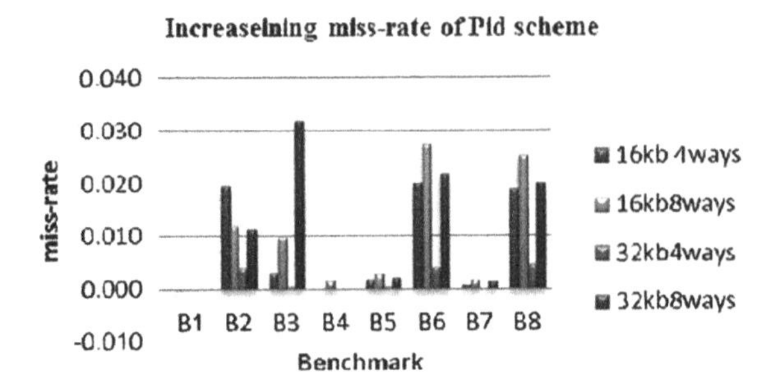

Fig. 6 Differences in miss-rate using pid scheme for RMA scheduling

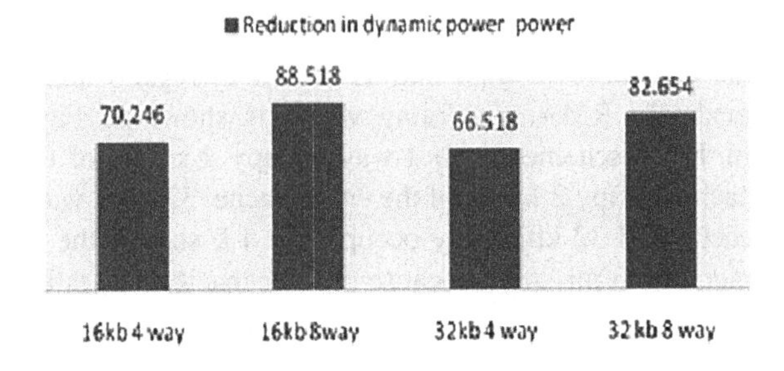

Fig. 7 Reduction in dynamic energy

4.2 Impact on Dynamic and Static Leakage Power

The impact of the TEST scheme on dynamic and leakage power is shown in Fig. 7 and Fig. 8 respectively. Based on CACTI results, total dynamic energy consumption for conventional and proposed cache scheme is obtained. Furthermore, in the proposed cache scheme inactive cache region is kept in low power drowsy mode. In this mode, static leakage power consumption is around 28% of that of conventional scheme. Dynamic and leakage power were modeled using CACTI [5] with 70 nm CMOS technology. In conventional cache configuration, the tasks share the entire cache. Dynamic energy reductions relative to the conventional set-associative scheme is shown in the Fig. 7. The reduction in dynamic energy observed as 88–70% for the 16 K cache and 66–82% for the 32 K cache. The reduction in the fraction of cache which is not used results in leakage power. So it is placed in low-leakage power mode. As per TEST scheme results, the usage of cache resource is below 50%. This shows that at any time about 50% of the instruction cache resources are kept in drowsy mode. So there is a scope for reduction in leakage power of TEST cache. Figure 8 shows the actual leakage power reduction relative to set-associative cache scheme obtained by our methodology. For the 16 K cache, reduction in leakage energy is in the range of 50–70%. Also, reduction in leakage energy of 54–62% is achieved for the larger cache.

Fig. 8 Reduction in leakage power

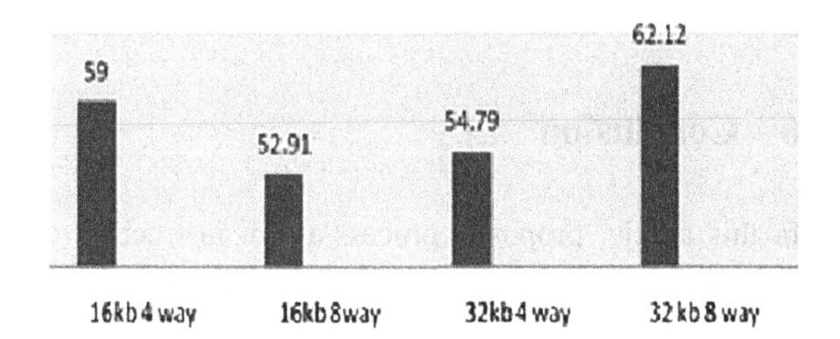

5 Related Work

Several partitioning schemes have been proposed to reduce the energy consumption of set-associative caches for the embedded scheme. In general purpose processors, interference problem can be mitigated by increasing cache size. The same solution cannot work for an embedded system because of resource constrained. In [6], for different cache configurations, cache interference was analyzed in terms of its impact on performance over an extended period of time. The primary focus of this paper was the interference between user and kernel space. The effect of cache flushing on context switch or using PIDs was evaluated. TEST compared with the both the schemes. In WCET analysis, the unpredictability of caches with multiple tasks remains an important problem. Since there are often real time constraints to be met, WCET is a critical component in embedded systems. Unlike general purpose systems, operating in a timely manner is a necessity rather than a convenience. A more accurate knowledge of WCET is needed to have better processor utilization. Cache partitioning can be done to have an accurate WCET analysis. This ensures that tasks are restricted to a subset of the cache. This, in turn, improves predictability [7]. The most recently used way for each set is recorded in way-prediction. So it predicts that this way is going to be accessed again next. For tags and data, it is needed to access only one way. The predicted way contains the requested data if a successful tag comparison occurs, otherwise a standard, parallel N-way lookup is performed next [8]. By enabling or disabling one or more sets respectively, the cache size is incremented or decremented in a set-only cache [9]. Albonesi [10] proposed a cache design which reduces dynamic power where size and associativity are varied by enabling or disabling cache-ways, while using less resource. TEST set-associative cache is interpreted as a direct-mapped cache by dedicating one way to a process. The TEST scheme reduces the number of tag comparisons by enabling only the selected tag and data way based on process information given by the operating system in the first cycle. This implementation does not require replacement circuitry as it is converted to direct mapping. TEST is the scheme which partitions the cache based on the knowledge of the process to way mapping. It creates the process-aware cache architecture with the help of compiler and operating system support.

6 Conclusion

In this article, proposed process-aware instruction cache partitioning scheme for dynamic and leakage power reduction. The scheme divides the instruction cache for multitasking embedded system by mapping the tasks to separate portions of the cache. Only executing tasks partition is kept active, thus achieving significant power savings in the cache memory. This work is a basic study of how the mapping of process task to partition improves the cache performance. Further improvement can be done by making it dynamic and also can be compared with data cache partitioning.

References

1. Malik, A., Moyer, B., and Cermak, D. A Low-Power Unified Cache Architecture Providing Power and Performance Flexibility. In ISLPED'00: Proceedings of the 2000 International Symposium on Low Power Electronics and Design, page 241–243, August 2000.
2. N. Kim, K. Flaunter, D. Blaauw and T. Mudge, "Single-vDD and single-vT super-drowsy techniques for low-leakage high performance instruction caches", Proceedings of ISLPED, pp. 54–57, 2004.
3. T. Austin, E. Larson, and D. Ernst. Simplescalar: An infrastructure for computer system modeling. IEEE Computer, 35(2):59–67, February 2002.
4. M. Guthaus, J. S. Ringenberg, D. Ernst, T. Austin, T. Mudge, and R.Brown. Mibench: A free, commercially representative embedded benchmark suite. In WWC-4: Workshop on Workload Characterization, pages 3–14, December 2001.
5. Ahn S. Thoziyoor, N. Muralimanohar and N. Jouppi, "CACTI 5.1", Technical report, Technical report, HP Laboratories Palo Alto, 2008.
6. A. Agarwal, J. Hennessy, and M. Horowitz. Cache performance of operating system and multiprogramming workloads. ACM Transactions on Computer Systems, 6(4):393–431, 1988.
7. Rakesh Reddy, Peter Petrov "Eliminating Inter-Process Cache Interference through Cache Reconfigurability for Real-Time and Low-Power Embedded Multi-Tasking Systems" CASES'07, September 30–October 3, 2007, Salzburg, Austria.
8. Bhargavi R. Upadhyay, TSB sudarshan," Low power predictive placement scheme for embedded system", In proceedings of international conference on embedded system, 4–5 July, 2014, Coimbatore, India.
9. Se-hyun Yang, Michael Powell, Babak Falsafi, Kaushik Roy, and T. N. Vijaykumar. Dynamically resizable instruction cache: An energy-efficient and highperformance deep-submicron instruction cache. Purdue University, 2000.
10. David H. Albonesi. Selective cache cache-ways: on-demand cache resource allocation.In Proceedings of the 32nd annual ACM/IEEE international symposium on Micro architecture, MICRO 32, pages 248–259, Washington, DC, USA, 1999. IEEE Computer Society.

Patient Tracking Using IoT and Big Data

A. Jameer Basha, M. Malathi, S. Balaganesh and R. Maheshwari

Abstract In bucolic area, the majority of the people become extinct owed to unacquainted of their health conditions and inadequacy of hospital facilities. In view of this, patient monitoring scheme plays a most imperative role. This article mainly includes remote monitoring progression using WBAN with ZigBee protocol, abet for tracking the human's health care in and out of the hospital, which facilitates to monitor precisely. Patient's details are stored in IoT cloud and data analytics technique is exploiting to analyze that details. All the time, the patient's records can be stored and retrieved in IoT cloud. Any variation occurs in the patient's wellbeing stipulation means, it afford a message to the specified doctor and the patient protectorate.

Keywords Wireless Body Area Network · IoT cloud · ZigBee protocol
Temperature sensor · Health care

1 Introduction

In recent technology, IoT renovate life, business and economy mostly lie of people. IoT has advancement in sensors, which will make a world to wireless communication. IoT have three major technology overviews hardware (chips, sensors), communication and software (data storage, analytics and frontend applications) [1].

A. Jameer Basha (✉) · M. Malathi · S. Balaganesh · R. Maheshwari
Sri Krishna College of Technology, Coimbatore, India
e-mail: jameer@skct.edu.in

M. Malathi
e-mail: m.malathi@skct.edu.in

S. Balaganesh
e-mail: 14n107@skct.edu.in

R. Maheshwari
e-mail: 14n137@skct.edu.in

A.K. Somani et al. (eds.), *Proceedings of First International Conference on Smart System, Innovations and Computing*, Smart Innovation, Systems and Technologies 79, https://doi.org/10.1007/978-981-10-5828-8_59

Wireless sensor with the communication has invaded the part of health care largely. In general, sensor can sense the human health either by injecting the sensor inside them or without injecting the sensors on human body. There are six sensors, which support to monitor human health without injecting sensors to the body. The performances of IoT have a benefit of monitoring and gathering the information about the patients [2]. In current approach, an increasing fast growth in the pasture of sensor. It has increased the premonition about the health activity, Health care and fitness about particular human [3]. It refers the patient not only who is sick and who is pretentious by disease. This system refers the patient who is healthy and wants to maintain fitness regarding their health. The researchers are now going on based on whether these wearable sensors improve the patient health or not. Besides the niche recreational fitness arena catered to by modern devices, researches have also considered its usage in the remote health monitoring systems for long research footage, management and medical access to patient's physiological information [2]. Now and then on the technological world, we can assume that the future generation will make use of the wearable sensor for all health related events. The sensors continuously record the data, store them in the database of the particular patient, and record their health from time to time.

This detail about the health of patient can be retrieved as required by patients or doctor. When the doctor observes physical information, they provide lab and clinic test and then could consult the doctor. Now as the technology, increases work of human become lesser at the same situation each people in this world are the busy and they don't care about their own health. Suppose health care is also provides via some technology, it provides guidelines to them, to increase their view over their health. The exploit of sensor not only provides the information about physiological and metabolic measurement of the patient but also provide the rich longitudinal information [2]. The wireless technology that we make use is the better solution in case of critical situation as natural adversity or human emergency and so on. Wireless technology will be the good solution in the crises like natural or adversity happened due to human and sometimes related to military where patients' records such as previously held medical details, patient identification and other significant information are necessary. Health care colleague now need ingress to promising new knowledge. This detail is in the appearance of "BIG DATA" due to its intricacy variety and appropriateness. Big data analyzation is to obtain intuition. Although these efforts are still in their starting stages [4]. Many innovative private sectors are building an application and analysis the tools which help patients and health care stakeholders to identify the worth and opportunities [4].

2 Related Work

Wireless sensor consisting some spatially devices, which may assist to monitoring the physical conditions and ecological conditions and these sensors are trouble free to install and preserve because of wireless technology [5]. To analyze and enrich the overall system performance characteristics in wireless state [6]. Wearable sensors are to take out clinically applicable details from the physiological data. Sensor transfers the data to the receiver through wireless communication. Physical information of patient can be recognized by using wearable sensor [7]. Sensors are mainly broadcast through gateway protocols and supply more security compared with the other protocols [5]. It is trouble-free to use, apply and reapply of sensor on patient's body is very simple [7]. Data are processed based on data analyzing applications and data management methods [8]. To accumulate the enormous volume information of the patient's health condition the cloud database is used [3]. Recent technology in health care system which is Big data which handles extremely large capacity of data such as search, store, capture and reduce complexity in Big data. By using 'V', declare Big data: volume, velocity, variability, identifies the patient's future condition takes long time for the doctor. However, in Big data it takes a single day to achieve the result [9]. Hadoop framework can be used to analyze the health sensor data [10]. In health care, organization needs the support of cloud computing to hold the PHI and to enhance the critical situations by means of cloud based virtual server. IoT cloud is used to storage and computation purpose [11]. Wearable device collect the ECG details and transfer the data to think speak IoT analyzer through ZigBee. To provide visual and timely ECG, IoT cloud use HTTP and MQTT protocol [12]. The context-based system sends information through Bluetooth audio device and also able to adjust the work routines, measurement and interfaces based upon the disease it should be monitored. The system will be presented visually and earshot on the mobile phones using the audio device configured in home [13]. Embedded system is involved in monitoring and track of a patients' position. To elevate the local database from the central database wireless sensor Xbee is used and in simultaneously, uploaded on an online database. Active RFID cards are used to map out the position of the patient [14].

3 IoT Cloud Based Patient Monitoring

In proposed work, as shown in Fig. 1, the sensor devices are used to get the information of the patient and are monitored. Sensor acquires more sufficient process in their particular working comportment and it can be processed among it. The wearable sensors like watch are placed in patient hands and it is sufficient for us to hold. Patient health details are always updated in the database as soon as the prescribed doctor watches over the patient health. The patients need not inject the sensors used here but it is wearable as digital watches. Sensing device provides the information about their health, for example, if patient is attacked by stroke, health of normal person and him differs. The patient have to maintain appropriate temperature, weight, food and also control their emotions in sometimes. The sensor could detect the sentimental analyze of the patients. They transfer and store the information to the cloud from where the details can be retrieved at any site, so that patient need not bring their medical files at the time of emergency or at critical

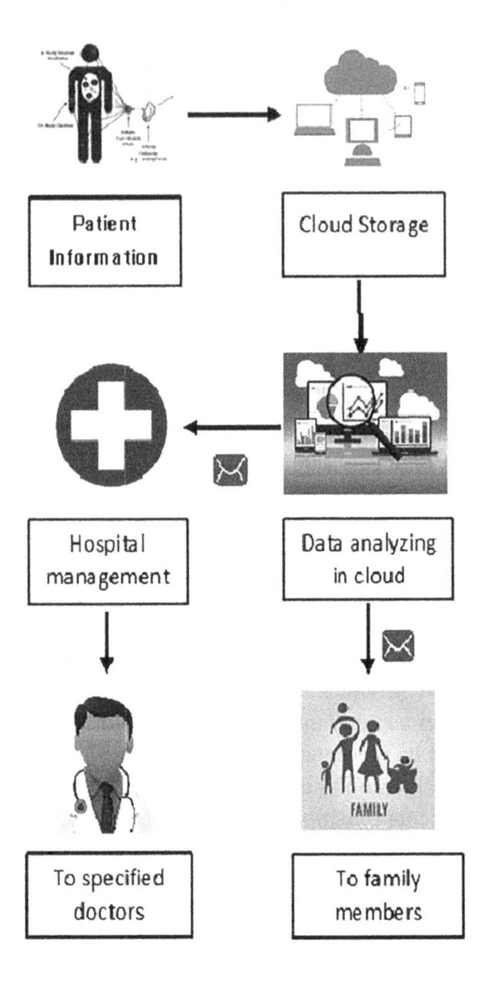

Fig. 1 Health care monitoring with WBAN, ZigBee and IoT cloud

situation. It makes the job of doctors easier and also provides accurate information of the patient. If the patient is within the hospital, they can use RFID tags or WBAN sensors to analyze the current situation. If the patients are available at any other location away from hospital, they can use the WBAN with ZIGBEE or UWB standard.

From this sensor, moreover the signals are send out with the help of ZigBee protocol. The transmitted information is send to the IoT cloud and from that cloud process; the details can be stored and accessed through a device namely REST and SDK APIs. For security purpose, the cloud provides a private and public key to the users. IoT cloud uses BSN to monitor the patient health conditions. It act as a virtual server for storing the patient's details in third party server. Data is analysed in the cloud using Big data concepts, which handles a huge volume of information. While analyzing the data the information is compare with previous details of the patient. After analyzation, if any disparity in the health condition they provide the message to the respective hospital management system and to the patient's caretaker.

IoT sensor detects the information in the arrangement of electrical signals or electrical waveform. These waveforms are basically a optical symbol of graphs and it is of two types which are unidirectional and by their needs are constantly growing. IoT sensors are used for monitoring the heartbeat rate and the data can be transmitted through the IoT sensors, complex algorithm analyze the data and the information can be retained in the cloud. In our application, the oxymetry sensor is used to perceive the oxygen saturation level in human blood and also calculate the percentage of haemoglobin. In patient's body we fit this sensor in thin part like fingertip, earlobe. Compare to all other IoT sensors it is inexpensive and it provide secure information about the health conditions of the patient. Skin conductance sensor is used for measuring the galvanic skin response, it can able to sense the sweating level contained in-patient's hand and it also detect the quantity of calories burn. Its range is from 0.1 to 1000 micro-Siemens. Skin temperature IoT sensor is also used for calculate the body temperature.

The IoT cloud aid to store the bulk quantity of data and also reduce the data complexity. It assists to retrieve the data at any time as required. By the way, the consultant doctor must retrieve at the data of human all time. This system must also provide secure access because sometimes the information may be stolen or misused by the attackers. To reduce the activity of attackers many protocols are present in IoT field. Data analytics abet to analyze the data, the current details of patient are transmit by using the sensor and it will be accumulate in the cloud. In every 10 minutes, the information is composed and stored in the cloud. The divergence between the current details and the previous records of the patient are to be observed. Normal temperature of the patient is 37 °C (98.6 °F), suppose any digression in-patient's temperature level means, the variant result is send to the specified doctor and patient protectorate. The early records are retained and maintained, and the upcoming records are accumulated. Allows viewing the patient details and their updates from anywhere as required, also issues an alert message about the health condition to the patient protectorate.

3.1 Tenable Data Transmit Using ZigBee

ZigBee is IEEE 802.15.4 standard, and low cost, stumpy data rate and low energy consumption these are the advantages of ZigBee. It is helpful in health care and health monitor. ZigBee network make use of CSMA/CA method [15]. ZigBee make use of star topology and mesh topology. There are totally three devices in ZigBee, one is Full Function Device, the other is Reduced Function Device and the third one is Network or ZigBee Coordinator. Controlling the communication in networking coordinator is used. One coordinator is facilitated in a ZigBee network. The wireless standard protocols that defined for wireless sensor network can be defined by FFD device. It also acts similar to the coordinator. The Reduced Function Device provides the limited functionality. Hence, either the coordinator or FFD can be used in ZigBee network in case of rich functionality Fig. 2.

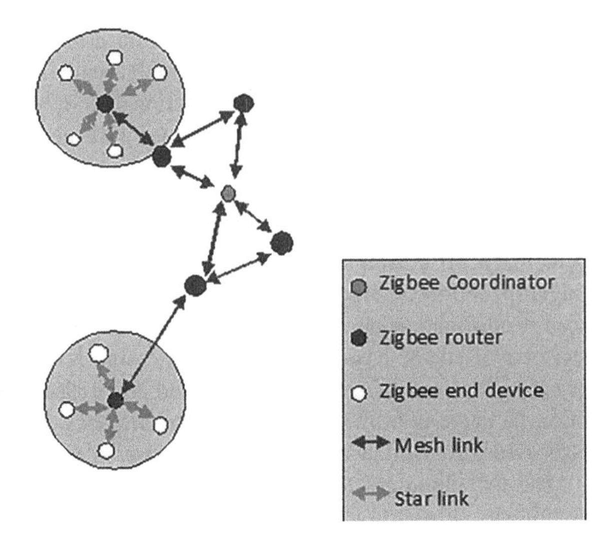

Fig. 2 ZigBee functionality [16]

3.2 Stalk Patient's Health Stipulation Using Wearable Technology

The WBAN is used in low power sensor and integration purpose in and around the person. It helps to alarm when condition changes. It is also used in auto medication in case any emergency. It is RF based network technology. It is of two types as Wearable WBAN and Implantable WBAN. It utilizes the wireless sensors [17]. WBAN increases the excellence of life by continuous health monitor and ubiquitous monitoring [18]. Wearable WBAN is runs on the vicinity body system where in outcome plantable WBAN is run inside the patient's body [19]. It has reliable data rate. The patients may move around anywhere but the mobility pattern remains the same when compared with other wireless sensor network. Replaceable of cells is very easy one and the extension of the battery life is moreover avoided [20]. WBAN is mainly developed from WPAN and it could provide more information about the patient.

In WBAN there are three nodes namely, master node, coordinator and distribution network. Sensors are present in master node. Coordinate node is intermediate between master node and Distribution system, it act as a hub. Distribution system supply the information about the patient and can retrieve at all time.

Fig. 3 Graphical outlook of Our Result

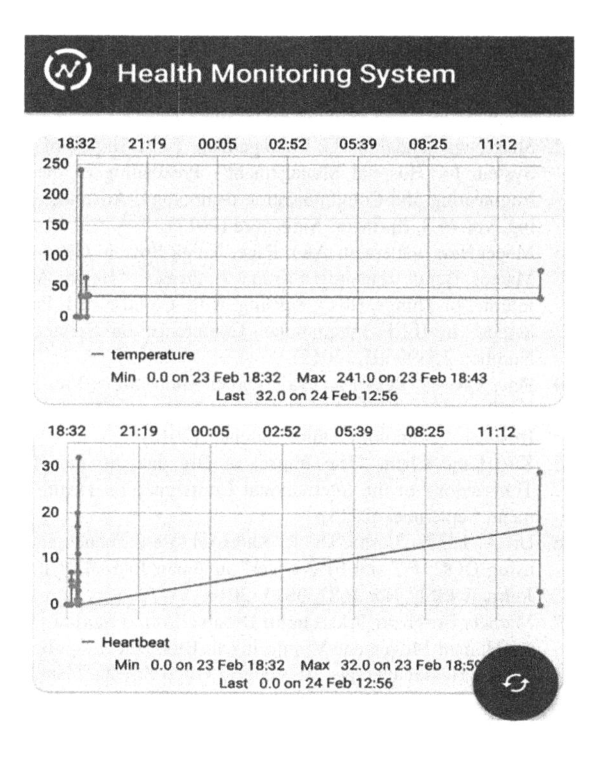

4 Temperature and Heartbeat Sensor Result

This article designate the functioning process with the sensing device WBAN with ZigBee and Fig. 3 represent the outcome of the system. It indicates both the temperature and heartbeat condition of the patient. The co-ordinate X axis denotes the temperature and Y axis denotes the time. In that graph it represents the health care information details which includes an accurate time and date as follow.

5 Conclusion

WBAN with ZigBee is a secure health monitoring system to monitor patient with specified doctor and maintain the individual information in the database. At all time the information can be retrieved form the database. In the system the sensor placed in patient's body, it detects the information about the patient and the information is send to IoT cloud through ZigBee technology. The temperature and heartbeat pulsing waves detected in the patient's body and these pulses are easily able to recognize patient health conditions. If there is any modification in the patient's health condition it notifies a message, which can be send to the corresponding doctor.

References

1. Knud Lasse Lueth "IoT basics: Getting started with the Internet of Things", March (2015).
2. Shanmuga sundaram. G, Thiyagarajan. P "A Survey of Cloud based Healthcare monitoring System for Hospital Management", Proceeding of the International Conference on Data Engineering and Communication technology. Advances in Intelligent Systems and Computing, vol 45.8. Springer, Singapore (2017).
3. MoeenHassanalieragh, Alex Page, Tolga Soyata, Gaurav Sharma, Mehmet Aktas, Gonzalo Mateos Burak Kantarci, Silvana Andreescu "Health Monitoring and Management Using Internet-of-Things (IoT) Sensing with Cloud-based Processing: Opportunities and Challenges" in IEEE International Conference on Services Computing, INSPEC Accession Number: 15399605, (2015).
4. Peter Groves, Bassel Kayyali, David Knott, Steve Van Kuiken. "The Big data revolution in health care: Accelerating Value and Innovation" In: Center of US Health System Reform Business Technology Office, January (2013).
5. Kuo Lane Chen, "The Impact of Big data on the Healthcare Information System" In: Transactions of the International Conference on Health Information Technology Advancement, September (2013).
6. Dr. A. Jameer Basha, Dr. R. Kanmani "Performance analysis of Wireless OCDMA System using OOC, PC and EPC codes" in Asian Journal of Information Technology In: Medwell Journals ISSN No: 1682–3915, 2016.
7. Michela Borghetti, Aleaaandro Dionisi, Emilio Sardini, Mauro Serpelloni, Wearable Sensors for Human Movement Monitoring in Biomedical Applications: Case Studies (2015).
8. Moeen Hassanalieragh, Alex Page, Tolgo Soyata, Gaurav Sharma, Mehmet Aktas, Dept of ECE, University of Rochester, "Healthcare Monitoring and Management using Internet-of-Things Sensing with Cloud-based Processing: Opportunities and Challenges" In: IEEE International Conference on Services Computing (2015).
9. Sricasthan M, Yogesh Arjun K, "Healthcare Monitoring System by Prognotive Computting using Bigdata Analytics" Published by Elseiver BV.
10. P. Dineshkumar, R. Senthilkumar, K. Sujatha, "Big data analytics of IoT based Health care monitoring system" In: IEEE International Conference on Computer and Electronics Engineering. (UPCON), December (2016).
11. Botts N, Thoms B, Noamani A, Horan TA: "Cloud computing architectures for the underserved: public health cyberinfrastructures through a network of health ATMs". *System Sciences (HICSS), 2010 43rd Hawaii International Conference on; 5–8 Jan. 2010* 2010, 1–10.
12. Zhe Yang, Qihao Zhou, Lei Lei, Kan Zheng, Wei Xiang, An IoT-cloud Based Wearable ECG Monitoring System for Smart Healthcare, Springer US, ISSN 1573–689X, doi:10.1007/s10916-016-0644-9 (2016).
13. Jorge Gomez, Byron Oviedo, Emilio Zhuma, "Patient Monitoring System based on Internet of Things" In: the 7th International Conference on Ambient System, network and Technologies, Proceeding Computer Science 83, (2016) 90–97 Elsevier.
14. Amitabh Yadav, Vivek Kaundal, Abhishek Sharma. Wireless Sensor Network based Patient Health Monitoring and Tracking System and Proceeding of International Conference On Intelligent Communication, Control and Devices. Advances in Intelligent Systems and Computing, vol 479. Springer, Singapore (2017).
15. P. Rohitha, P. Ranjeet Kumar Prof. N. Adinarayana, Prof. T. Venkat Narayana Rao, "Wireless Networking through ZigBee Technology" International Journal of Advanced Research in Computer Science and Software Engineering Volume 2, Issue 7, July (2012).
16. Aleksandar Milenkovic, Chris Otto, Emil Jovanov, "Wireless Sensor Networks for Personal Healthcare Monitoring": Issues and an Implementation, Published by Elsevier BV.
17. Chris Otto, "An implementation of a wireless body area network for ambulatory health monitoring" In: The University of Alabama in Huntsville (2006).

18. Bo yu, Prof. Liuqing Yang, Ph.D. "Wireless Body Area Networks for Healthcare: A Feasibility Study" Published in: IEEE, March (2009).
19. Ashwini Singh, Ajeet Kumar, Pankaj Kumar "Body Sensor Network: A Modern Survey & Performance Study in Medical Perspect" Selected from International Conference on Recent Trends in Applied Sciences with Engineering Applications, vol.3, No.1, (2013).
20. Ms. Dharmistha D. Vishwakarma Research Scholar, India, "IEEE 802.15.4 and ZigBee: A Conceptual Study" In: International Journal of Advanced Research in Computer and communication Engineering, vol. 1, Issue 7, September (2012).
21. Shaltis PA, Reisner A, Asada HH: "Wearable, cuff-less PPG-based blood pressure monitor with novel height Sensor". ConfProc IEEE EngMed Biol Soc 2006, 1: 908–911.

An Enhanced Approach to Fuzzy C-means Clustering for Anomaly Detection

Ruby Sharma and Sandeep Chaurasia

Abstract In the recent years, the improvement in the security is a challenging task in the Internet environment The Intrusion Detection System (IDS) is one of the significant tools used to detect the attacks. Various IDS techniques have been proposed to identify the attacks and alert the user or administrator about the attacks. However, they are unable to manage new attacks. This paper proposes an Intrusion Detection System based on the density maximization-based fuzzy c-means clustering (DM-FCC). In this approach, cluster efficiency is improved through a membership matrix generation (MMG) algorithm. Dissimilarity Distance Function (DDF) has been used to compute the distance metric while creating a cluster in proposing an IDS. The proposed enhanced fuzzy c-means algorithm has been tested upon ADFA Dataset and the model performs highly appreciable in terms of accuracy, precision, detection rates, and false alarms.

Keywords Intrusion detection · Anomaly detection · Clustering
Fuzzy C-means

1 Introduction

Intrusions are the activities that violate the security policy of the computer systems [1]. But unfortunately, intrusion free network is hard to survive. This leads to the major field of study named as Intrusion Detection System (IDS) [2–4]. The intrusion detection is the process of detecting the actions that compromise the integrity and security of a computer resource based on the assumption about the difference in the behavior of the intruder and legitimate users [5]. The IDS can be classified into several categories [6]: Network IDS (NIDS) [7–9] analyzes the network traffic.,

R. Sharma (✉) · S. Chaurasia
Department of Computer Science Engineering, Manipal University, Jaipur, India
e-mail: study.ruby@gmail.com

S. Chaurasia
e-mail: sandeep.chaurasia@jaipur.manipal.edu

A.K. Somani et al. (eds.), *Proceedings of First International Conference on Smart System, Innovations and Computing*, Smart Innovation, Systems and Technologies 79, https://doi.org/10.1007/978-981-10-5828-8_60

Host-Based IDS (HIDS) [10] monitors an individual host device, Signature detection system (SDS) [11, 12] recognizes mismatching patterns and Anomaly detection system (ADS) [11, 13, 14] identifies the anomalous events that differ from the normal behavior. IDS widely follow data mining procedures to detect and identify the intruder in networks. The data mining techniques have been proved advantageous in term of handling large different types of databases and in data analysis, summarization, and visualization [15, 16]. Initially, the work was carried on misuse-based detection techniques which use signatures of the known attacks. But later it shifted to anomaly based detection techniques which can be easily pointed out through Table 1 highlighting the related work in the same field. Anomaly detection techniques have overridden misuse-based techniques as they are unlikely able to detect the known attacks as well. Anomaly detection technique is based on detection the abnormality. Most of the researchers are working to improve the performance of the anomaly based techniques in terms of reducing high false alarm.

The paper is structured as follows: Sect. 2 reviews the literature related to data mining-based IDSs. Moreover, it involves the comparative analysis the techniques used in this paper. Section 3 introduces the proposed anomaly detection-based intrusion detection approach. Section 4 presents the experimental setup and results Sect. 5 discusses the conclusion and future scope.

2 Literature Review

A number of research work has been done in this field of intrusion detection; some of the works are taken into account as proceeds for the proposed work. Various data mining techniques are described as follows [16, 17]: Classification, Association rule, Frequent Episode rules, and Clustering. From the data mining techniques, clustering is widely used for intrusion detection [18] due to the numerous reasons. Table 1 illustrates the comparison of various IDSs. From the survey, it is evident that the clustering-based techniques achieve better performance when compared to the existing data mining-based intrusion detection techniques. Hence, the clustering-based techniques are used for intrusion detection.

From the above discussion, some points get highlighted. First, Fuzziness-based classification method is proposed to identify the intrusion by the divide—and—conquer strategy where the unlabeled samples are classified with the predicted labels based on the magnitude of the fuzziness value, but this technique can classify only two classes namely normal and attack data. But it was not able to identify the type of the attack. Second, Cluster center and feature selection approach is proposed which is the distance-based feature representation method; this technique also suffers to detect the type of the attack that is imposed. Lastly, Q-learning-based intrusion Detection system has been proposed where the Rough set Theory is used, even though the feature selection and the accuracy parameter is calculated

Table 1 Comparison of related work

Work	Year	Methodology	Evaluation parameters	Pros	Cons
Ashfaq et al. [19]	2016	SSL algorithm based on fuzzy has been used	Accuracy and response time	The model has been proved very accurate	It cannot detect multiple types of attacks
Lin et al. [5]	2015	A distanced based feature representation model with cluster has been used	Accuracy, DR[a], FAR[b]	1. It is a very efficient and accurate model with a good detection rate and very low false alarm rate	U2R[c] and R2L[d] attacks cannot be detected by this model
Elhag et al. [20]	2015	The model of the combination of genetic Fuzzy systems and pair wise learning has been used	Accuracy, recall, precision, MFM[e], DR, FAR	The model effectively detect the attacks and produce less false alarms	The number of classifiers to be learned increases with the number of classes and requires high training time
Sengupta et al. [21]	2013	The rough set theory and Q-learning algorithm are integrated for detecting intrusions using the network traffic data	Accuracy	The model is a good classifier with low computational cost	It is not suitable for the data having variations with time and space
Xie et al. [22]	2016	The provenance is used to identify the dependency between the objects in a structured manner for detecting and analyzing the intrusions	DR, FP[f], run-time	It is an efficient model which has less frequency of wrong identification of attacks	The model is not suitable for the intrusions using the covert channel and the intruders can also leverage the files
Pandeeswari and Kumar [23]	2015	A hybrid approach of FCM with ANN has been applied	Accuracy and FAR	It is good at detecting the abnormal behavior with less rate of false detection	The cost of computation is high

(continued)

Table 1 (continued)

Work	Year	Methodology	Evaluation parameters	Pros	Cons
Amini et al. [24]	2016	The combination of radial basis function (RBF), neural networks, and fuzzy clustering is used to build ensemble classifier	Accuracy and DR	It is a stable model which can produce less complex training datasets	The results are less accurate
Hosseini et al. [25]	2015	A multiple criteria linear programming with particle swarm optimization	DR, FP, run-time	It can easily detect the attacks with less wrong predictions	It has high execution time complexity with respect to its running time complexity

[a]DR: Detection Rate
[b]FAR: False Alarm Rate
[c]U2R: User to Remote
[d]R2L: Remote to local
[e]MFM: Mean F-Measure
[f]FP: False Positive

simultaneously, but the anomaly detection and attack type classification by existing knowledge are not implemented.

3 Proposed Approach of Anomaly Detection

The intrusion detection system continuously uses additional resources in the system as it monitors even when there are no intrusions occurring. This is the *resource usage problem.* An intruder can potentially disable or modify the programs running on a system, rendering the intrusion detection system useless or unreliable. This is the *reliability problem.* To overcome all the issues in the existing clustering techniques [5] which are also highlighted in Table 1, an enhanced technique of Fuzzy C-Means Clustering algorithm is proposed to clusters the data by classifying data based on density and clustering. Figure 1 depicts the proposed model of enhanced Fuzzy C-means-based IDS. Initially, the dataset is cleaned and preprocessed to separate data into the labels and other features. Then the membership matrix is generated which is applied for feature extraction. The training dataset is trained with features, labels, and ground truth. Then, testing dataset features are extracted and the group features on the basis of computed density of each cluster. Lastly, the cluster can be categorized and anomaly gets reported.

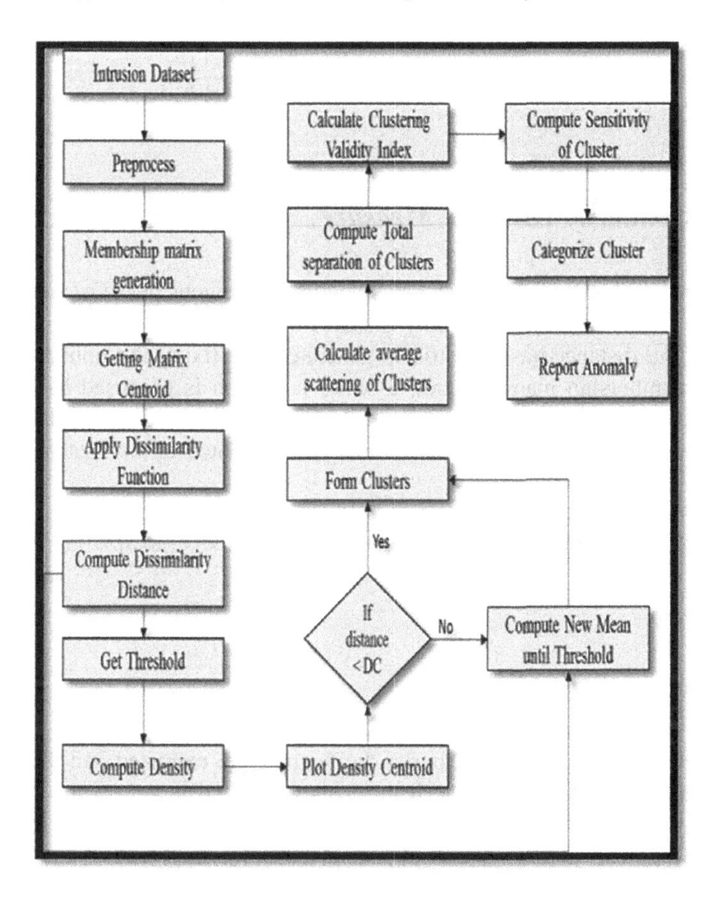

Fig. 1 The proposed model

3.1 Membership Matrix Generation (MMG)

In the initial preprocessing phase, the raw dataset will be preprocessed. The data values and the labels that refer to the type of attacks are extracted along which the ground truth values are also extracted that are needed for the final verification. The Membership matrix composed of the total number of attributes be *An* is generated in reference to each type of attack referred as *Lb* from the computation of the minimum, maximum, mean and standard deviation values obtained by the following equations. Let $A_n(j)$ be the list of values in A_n where the selected attribute is j.

$$\text{Compute}\, Memb\_mat_{xy} = \text{Min}[A_n(j)] \tag{1}$$

$$\text{Compute}\, Memb\_mat_{xy} = \text{Max}[A_n(j)] \tag{2}$$

$$\text{Compute}\, Memb\_mat_{xy} = \frac{\text{Sum } [A_n(j)]}{\text{Size } [A_n(j)]} \tag{3}$$

3.2 Dissimilarity Distance Measure

The Generated Membership Matrix generated in the previous phase is considered as one of the inputs of the dissimilarity distance calculation. Based on this distance algorithm, the distance-based matrix is generated. By fixing the boundary values from the membership matrix values, a set of condition is proposed based on the number of attributes for each type of attack. From this condition, the data records are processed to compute the distance values and updated in the Distance matrix.

$$\text{Compute}\, T_{val} = \sqrt{\frac{\sum [A_n(j)_k] - Memb_{mat_{xy}}}{\text{Size } [A_n(j)]}} \quad \text{Where k} = 1 \text{ to Size } [A_n(j)] \tag{4}$$

3.3 Feature Extraction

From the dissimilarity distance Matrix, the features are extracted and the Features are trained to the machine. These features are extracted based on the conditions generated with the reference to the dissimilarity matrix, based on these conditions the features are extracted for the available attributes for each type of attack to train the machine.

3.4 Density-Based Clustering

The proposed clustering algorithm is implemented in this phase. Based on the size of the data, the cluster validity index is set. The density of each cluster is computed based on this index value. Based on the number of available attributes, the data is clustered into groups by the conditions that are generated by the cluster validity index. From the generated clusters, the features of the test data are extracted. One of the major problems is the fixation of fuzziness index problem. It will be overcome by using this density maximization technique that will help to compute new mean values and check if that value will be added to membership range until that value converges inside the membership range.

3.5 Anomaly Detection

The training features that are extracted are compared with the extracted testing feature. From the comparison, the anomaly is detected effectively from the test data.

4 Experimental Results

4.1 Experimental Setup

Even though there were several bench mark data sets available to test an anomaly detector, but the advancements in operating systems made these datasets outdated. So, for examining the performance of our model in intrusion detection, we used Australian Defense Force Academy (ADFA) dataset. It has a wide collection of system call traces representing modern vulnerability exploits and attacks. The tool used for the experiment is Mat lab R2009b. There are ten types of attacks in the dataset that are analysis, backdoor, DOS, exploits, fuzzers, generic, normal, reconnaissance, shell code, and worms that are having multi-class label from 1 to 10 and binary label 1 for attack and 0 for normal.

4.2 Performance Metrics

We have used the following common performance metrics to evaluate the performance of the proposed model.

True Positive (TP): These are cases in which predicted attacks are actual attack's traces.
False Positive (FP): These are cases in which predicted attacks are actual normal traces.
True Negative (TN): These are cases in which predicted normal are actual normal traces.
False Negative (FN): These are cases in which predicted normal are actual attack traces.

Precision: It reflects reproducible predictions of attack even if they are far from accepted value out of total predictions of attack traces.

$$\text{Precision} = \text{TP}/\text{Predicted Attack Traces}$$

Recall: It is a measure of how often the predicted attack is an actual attack.

$$\text{Recall} = \text{TP}/\text{Actual Attack Traces}$$

Accuracy: It measures the closeness of attacks identified to the actual traces. Overall, it measures how often the classifier is correct.

$$\text{Accuracy} = \text{TP} + \text{TN}/\text{Total traces}$$

Receiver Operating Characteristics (ROC) Curve: It is a graph that summarizes the performance of a classifier to overall possible thresholds.

4.3 Results

Three approaches are applied to evaluate the proposed anomaly detection model. The proposed AD scheme was used to form clusters for the different categories of attacks mentioned in Table 2. Three scenarios of the ADFA dataset had been tested for ADFA dataset having total number of instances 1000, 10000, and 20000 with 234, 3031, 5999 test instances, respectively. The accuracy of the proposed scheme is above 92% in all three dataset size results as shown in Tables 2, 3 and 4. With the increase of the size of the dataset, it is highly appreciable that the precision of the detection scheme goes to 99.4567. Figure 5 depicts the variation in accuracy with a change in data size. Figure 2 shows the ROC characteristics of the enhanced fuzzy

Table 2 ADFA dataset with 1000 instances

TP	223
TN	53
FP	20
FN	4
Precision	91.7695
Recall	98.2379
Accuracy	92%

Table 3 ADFA dataset with 10000 instances

TP	2804
TN	52
FP	21
FN	154
Precision	99.2888
Recall	94.7938
Accuracy	94.2263%

Table 4 ADFA dataset with 20000 instances

TP	5675
TN	42
FP	31
FN	251
Precision	99.4567
Recall	95.7644
Accuracy	95.2992%

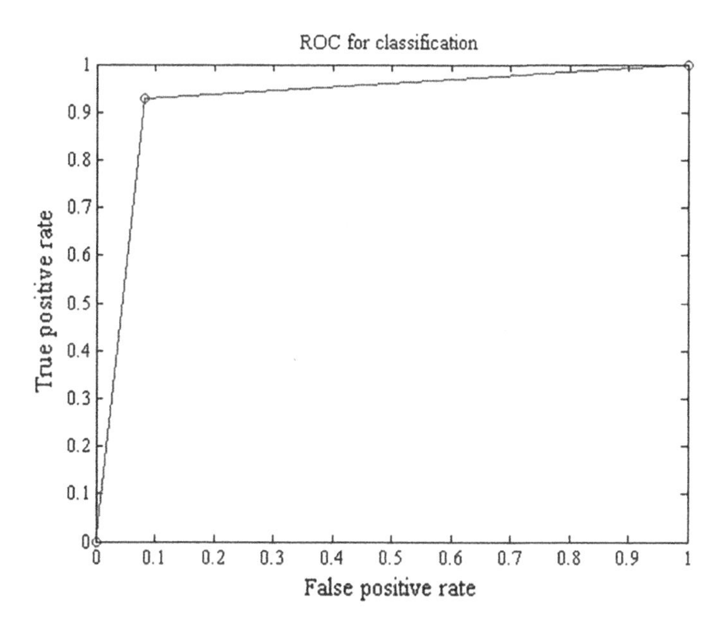

Fig. 2 Receiver operating characteristics for ADFA dataset with 1000 instances

c-means scheme for the ADFA dataset of 1000 instances. Figure 3 shows the ROC characteristics of 10000 instances and Fig. 4 of 20000 instances. The proposed model has outperformed in terms of higher detection rate, false alarm rate and in accuracy and precision.

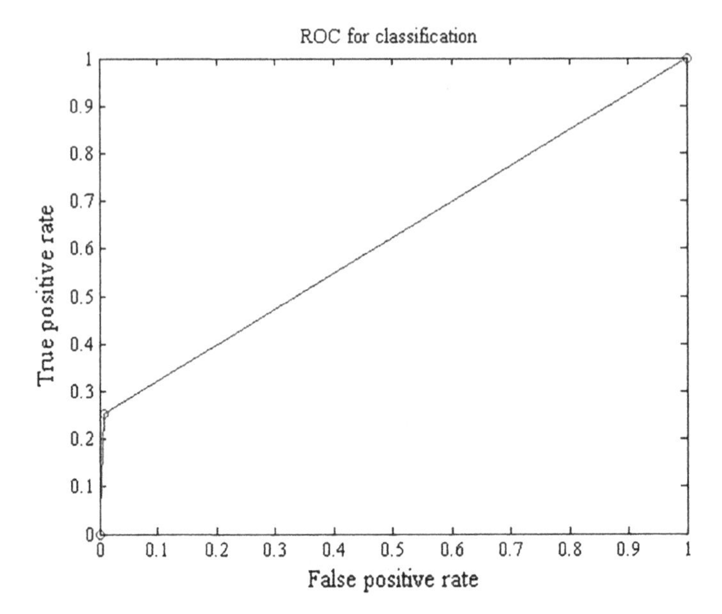

Fig. 3 Receiver operating characteristics for ADFA dataset with 10000 instances

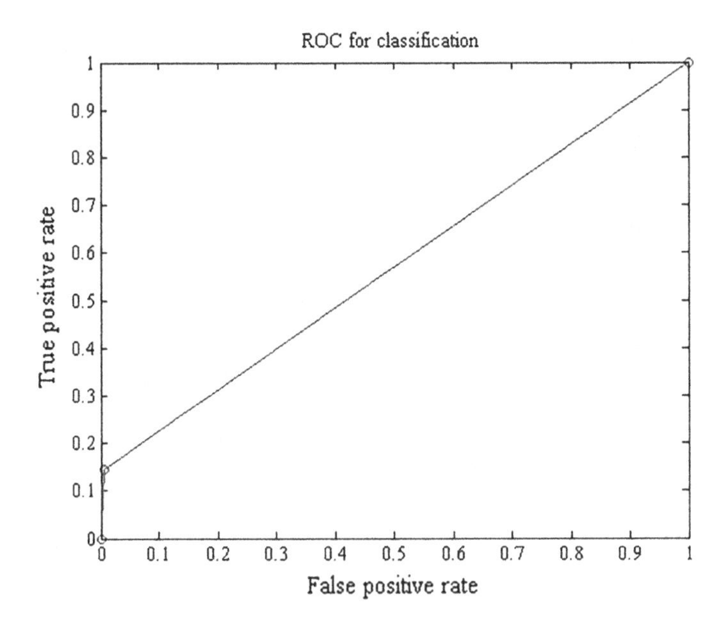

Fig. 4 Receiver operating characteristics for ADFA dataset with 20000 instances

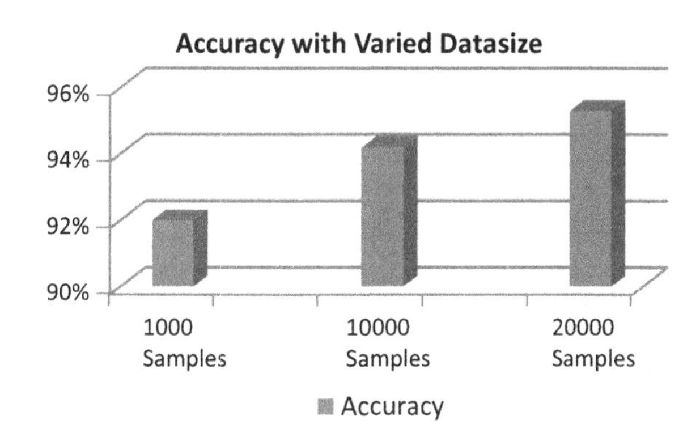

Fig. 5 The variation in accuracy of the proposed model with varied sample sizes

5 Conclusion

In the recent years, many researchers have been conducted to develop effective data mining-based anomaly detection system which can perform the task with both high accuracy and low false alarm rates. In this paper, we proposed a model based on clustering with density maximization. The performance of the proposed model has been examined by ADFA dataset with three different sets of a varied number of instances. The experimental study proved that the proposed model get better performance based on accuracy and precision. As future work, the model will be applied on KDD Cup 99 to examine the performance of our model on detecting the different attacks simultaneously. Also, we will test the performance of the model on other parameters—Fault Acceptance Rate (FAR), Fault Rejection Rate (FRR), Global Acceptance Rate (GAR), Jaccard Coefficient, and Dice Coefficient. Furthermore, the density maximization approach can be applied to other algorithms of classification such as ensemble classification to develop more efficient anomaly detection system.

References

1. L. Khan, M. Awad, and B. Thuraisingham, "A new intrusion detection system using support vector machines and hierarchical clustering," *The VLDB Journal—The International Journal on Very Large Data Bases,* vol. 16, pp. 507–521, 2007.
2. H.-J. Liao, C.-H. R. Lin, Y.-C. Lin, and K.-Y. Tung, "Intrusion detection system: A comprehensive review," *Journal of Network and Computer Applications,* vol. 36, pp. 16–24, 2013.
3. Y. Bai and H. Kobayashi, "Intrusion detection systems: technology and development," in *Advanced Information Networking and Applications, 2003. AINA 2003. 17th International Conference on*, 2003, pp. 710–715.
4. R. Raut and S. Gawali, "Intrusion detection system using data mining approach," *EXCEL International Journal of Multidisciplinary Management Studies,* vol. 2, pp. 124–138, 2012.

5. W.-C. Lin, S.-W. Ke, and C.-F. Tsai, "CANN: An intrusion detection system based on combining cluster centers and nearest neighbors," *Knowledge-based systems,* vol. 78, pp. 13–21, 2015.
6. I. Butun, S. D. Morgera, and R. Sankar, "A survey of intrusion detection systems in wireless sensor networks," *IEEE Communications Surveys & Tutorials,* vol. 16, pp. 266–282, 2014.
7. L. Koc, T. A. Mazzuchi, and S. Sarkani, "A network intrusion detection system based on a Hidden Naïve Bayes multiclass classifier," *Expert Systems with Applications,* vol. 39, pp. 13492–13500, 2012.
8. S. Joshi and V. S. Pimprale, "Network Intrusion Detection System (NIDS) based on Data Mining," *International Journal of Engineering Science and Innovative Technology (IJESIT) Volume,* vol. 2, 2013.
9. C. N. Modi, D. R. Patel, A. Patel, and M. Rajarajan, "Integrating signature apriori based network intrusion detection system (NIDS) in cloud computing," *Procedia Technology,* vol. 6, pp. 905–912, 2012.
10. Y. Lin, Y. Zhang, and Y.-j. Ou, "The design and implementation of host-based intrusion detection system," in *Intelligent Information Technology and Security Informatics (IITSI), 2010 Third International Symposium on*, 2010, pp. 595–598.
11. S. Axelsson, "Intrusion detection systems: A survey and taxonomy," Technical report 2000.
12. C. Kruegel and T. Toth, "Using decision trees to improve signature-based intrusion detection," in *International Workshop on Recent Advances in Intrusion Detection*, 2003, pp. 173–191.
13. P. Garcia-Teodoro, J. Diaz-Verdejo, G. Maciá-Fernández, and E. Vázquez, "Anomaly-based network intrusion detection: Techniques, systems and challenges," *computers & security,* vol. 28, pp. 18–28, 2009.
14. J. Zhang and M. Zulkernine, "Anomaly based network intrusion detection with unsupervised outlier detection," in *2006 IEEE International Conference on Communications*, 2006, pp. 2388–2393.
15. W. Lee, S. J. Stolfo, P. K. Chan, E. Eskin, W. Fan, M. Miller, *et al.*, "Real time data mining-based intrusion detection," in *DARPA Information Survivability Conference & amp; Exposition II, 2001. DISCEX'01. Proceedings*, 2001, pp. 89–100.
16. M. R. S. Landge and M. A. P. Wadhe, "Review of Various Intrusion Detection Techniques based on Data mining approach," *International Journal of Engineering Research and Applications (IJERA) June,* 2013.
17. M. S. S. Morkhade and M. Bartere, "Survey on Data Mining based Intrusion Detection Systems."
18. S. Zhong, T. M. Khoshgoftaar, and N. Seliya, "Clustering-based network intrusion detection," *International Journal of reliability, Quality and safety Engineering,* vol. 14, pp. 169–187, 2007.
19. R. A. R. Ashfaq, X.-Z. Wang, J. Z. Huang, H. Abbas, and Y.-L. He, "Fuzziness based semi-supervised learning approach for intrusion detection system," *Information Sciences,* 2016.
20. S. Elhag, A. Fernández, A. Bawakid, S. Alshomrani, and F. Herrera, "On the combination of genetic fuzzy systems and pairwise learning for improving detection rates on Intrusion Detection Systems," *Expert Systems with Applications,* vol. 42, pp. 193–202, 2015.
21. N. Sengupta, J. Sen, J. Sil, and M. Saha, "Designing of on line intrusion detection system using rough set theory and Q-learning algorithm," *Neurocomputing,* vol. 111, pp. 161–168, 2013.
22. Y. Xie, D. Feng, Z. Tan, and J. Zhou, "Unifying intrusion detection and forensic analysis via provenance awareness," *Future Generation Computer Systems,* vol. 61, pp. 26–36, 2016.
23. N. Pandeeswari and G. Kumar, "Anomaly detection system in cloud environment using fuzzy clustering based ANN," *Mobile Networks and Applications,* pp. 1–12, 2015.
24. M. Amini, J. Rezaeenour, and E. Hadavandi, "A Neural Network Ensemble Classifier for Effective Intrusion Detection Using Fuzzy Clustering and Radial Basis Function Networks," *International Journal on Artificial Intelligence Tools,* vol. 25, p. 1550033, 2016.

25. B. M. Hosseini, B. Amiri, M. Mirzabagheri, and Y. Shi, "A New Intrusion Detection Approach using PSO based Multiple Criteria Linear Programming," *Procedia Computer Science,* vol. 55, pp. 231–237, 2015.
26. W. Lee, S. J. Stolfo, and K. W. Mok, "Mining Audit Data to Build Intrusion Detection Models," in *KDD,* 1998, pp. 66–72.
27. G. Karypis, E.-H. Han, and V. Kumar, "Chameleon: Hierarchical clustering using dynamic modeling," *Computer,* vol. 32, pp. 68–75, 1999.
28. G. Karypis, "CLUTO-a clustering toolkit," DTIC Document 2002.
29. S. Zhong and J. Ghosh, "A unified framework for model-based clustering," *Journal of machine learning research,* vol. 4, pp. 1001–1037, 2003.
30. J. MacQueen, "Some methods for classification and analysis of multivariate observations," in *Proceedings of the fifth Berkeley symposium on mathematical statistics and probability,* 1967, pp. 281–297.
31. M. Jianliang, S. Haikun, and B. Ling, "The application on intrusion detection based on k-means cluster algorithm," in *Information Technology and Applications, 2009. IFITA'09. International Forum on,* 2009, pp. 150–152.
32. J. D. Banfield and A. E. Raftery, "Model-based Gaussian and non-Gaussian clustering," *Biometrics,* pp. 803–821, 1993.
33. S. C. Johnson, "Hierarchical clustering schemes," *Psychometrika,* vol. 32, pp. 241–254, 1967.
34. I. Davidson, "Understanding K-means non-hierarchical clustering," *Computer Science Department of State University of New York (SUNY), Albany,* 2002.
35. J. A. Hartigan and M. A. Wong, "Algorithm AS 136: A k-means clustering algorithm," *Journal of the Royal Statistical Society. Series C (Applied Statistics),* vol. 28, pp. 100–108, 1979.
36. C. B. Lucasius, A. D. Dane, and G. Kateman, "On k-medoid clustering of large data sets with the aid of a genetic algorithm: background, feasiblity and comparison," *Analytica Chimica Acta,* vol. 282, pp. 647–669, 1993.
37. W. Bul'ajoul, A. James, and M. Pannu, "Improving network intrusion detection system performance through quality of service configuration and parallel technology," *Journal of Computer and System Sciences,* vol. 81, pp. 981–999, 2015.
38. R. Kaur and S. Singh, "A survey of data mining and social network analysis based anomaly detection techniques," *Egyptian Informatics Journal,* 2015.
39. T. Ha, S. Kim, N. An, J. Narantuya, C. Jeong, J. Kim, *et al.*, "Suspicious Traffic Sampling for Intrusion Detection in Software-Defined Networks," *Computer Networks,* 2016.
40. X. Ni, D. He, and F. Ahmad, "PRACTICAL NETWORK ANOMALY DETECTION USING DATA MINING TECHNIQUES," *VFAST Transactions on Software Engineering,* vol. 9, pp. 1–6, 2016.
41. S. Pan, T. Morris, and U. Adhikari, "Developing a hybrid intrusion detection system using data mining for power systems," *IEEE Transactions on Smart Grid,* vol. 6, pp. 3104–3113, 2015.
42. F.-Y. Leu, K.-L. Tsai, Y.-T. Hsiao, and C.-T. Yang, "An Internal Intrusion Detection and Protection System by using Data Mining and Forensic Techniques," 2015.
43. A. L. Buczak and E. Guven, "A survey of data mining and machine learning methods for cyber security intrusion detection," *IEEE Communications Surveys & Tutorials,* vol. 18, pp. 1153–1176, 2015.
44. M. Goyal, "Data Mining Based Classification Technique for Adaptive Intrusion Detection System using Machine learning," *International Journal of Artificial Intelligence and Knowledge Discovery,* vol. 5, pp. 34–37, 2015.
45. M. A. Faisal, Z. Aung, J. R. Williams, and A. Sanchez, "Data-stream-based intrusion detection system for advanced metering infrastructure in smart grid: A feasibility study," *IEEE Systems Journal,* vol. 9, pp. 31–44, 2015.
46. A. Chaudhary, V. Tiwari, and A. Kumar, "Design an anomaly-based intrusion detection system using soft computing for mobile ad hoc networks," *International Journal of Soft Computing and Networking,* vol. 1, pp. 17–34, 2016.

47. J. Kevric, S. Jukic, and A. Subasi, "An effective combining classifier approach using tree algorithms for network intrusion detection," *Neural Computing and Applications,* pp. 1–8, 2016.
48. S.-Y. Ji, B.-K. Jeong, S. Choi, and D. H. Jeong, "A multi-level intrusion detection method for abnormal network behaviors," *Journal of Network and Computer Applications,* vol. 62, pp. 9–17, 2016.
49. M. A. Adibi and J. Shahrabi, "Online Anomaly Detection Based on Support Vector Clustering," *International Journal of Computational Intelligence Systems,* vol. 8, pp. 735–746, 2015.

A Study of Memory Access Patterns as an Indicator of Performance in High-Level Synthesis

Meena Belwal and T.S.B. Sudarshan

Abstract CPU/FPGA hybrid systems have emerged as a viable means to achieve high performance in the field of embedded applications and computing. High-Level Synthesis (HLS) tools facilitate software designers and programmers to utilize the underlying hardware in a hybrid system without requiring deep insights into hardware. HLS tools execute the program in sequential order by default. However, these tools provide mechanisms to parallelize the code wherein the user/programmer can apply constructs such as loop-unrolling, loop-flattening, and pipelining in the form of pragmas. Along with all these constructs in place, it is also important for programmers to understand the memory access pattern used in the program for efficiently utilizing the underlying capabilities of CPU/FPGA hybrid system. Memory access patterns in array references play a major role in deciding the latency and area required for a specific computation. Four typical memory access patterns with growing input sizes in array context were exercised in Vivado HLS with C code as an input and it was observed that change in the memory access pattern leads to a different area and timing requirements and change in the coding style may improve the performance of HLS tools.

Keywords Memory access pattern · CPU-FPGA hybrid systems
HLS · Embedded systems

M. Belwal (✉) · T.S.B. Sudarshan
Department of Computer Science & Engineering, Amrita School of Engineering,
Amrita Vishwa Vidyapeetham, Amrita University, Bengaluru, India
e-mail: b_meena@blr.amrita.edu
URL: http://www.amrita.edu

T.S.B. Sudarshan
e-mail: tsb_sudarshan@blr.amrita.edu

A.K. Somani et al. (eds.), *Proceedings of First International Conference on Smart System, Innovations and Computing*, Smart Innovation, Systems and Technologies 79, https://doi.org/10.1007/978-981-10-5828-8_61

1 Introduction

Memory access pattern for classifying the array reference for affine loop nests has been used as a technique in the literature that enables programmers as well as compilers to exploit multi-threading and to efficiently use memory hierarchy [1]. These patterns play a major role in deciding the execution time of an application executed in software due to the use of DRAM. DRAM latency varies with different DRAM transactions based on the kind of addresses that need to be fetched. Locality-of-reference which is same as a page-hit in DRAM would show low latency, whereas consecutive transactions going to the different page of DRAM would show higher latency than page-hit because of the protocol implications of DRAM. The memory access pattern transactions would have different latency in case a DRAM is used as a target, an off-chip memory.

However, FPGAs would show similar latency independent of any kind of memory access pattern since FPGA has memory built in it, i.e., a silicon memory. In case of FIFO kind of access for FPGA, one data is available per clock in streaming/sequential manner. When single port BRAM is synthesized in FPGA, one data is available based on one input address to the RAM, whereas in the case of dual port BRAM two data are available in one clock. But any kind of address would have same latency in FPGA. Executing the application with less execution time and constant execution time with the growing input size is the basic idea to use FPGA for complex computation to use enormous parallelism available with FPGAs. But since hardware coding is not as easy as writing a high-level language code, HLS tools act as a solution.

HLS tools enable software programmers to utilize the underlying parallelism and resource availability of CPU/FPGA hybrid systems without the burden of hardware coding. HLS tools such as Vivado HLS [2] and LegUp [3] allow programmers to write code in the high-level language of their choice such as C/C++/SystemC. HLS tools also provide various statistics for the code in execution such as clock cycle time, number of clock cycles, and logic utilization in terms of flops, multiplexers, registers, and LUTs. The execution time and the area utilization for a particular loop in high-level language depend on the memory access patterns applied in the computation.

This article explores the impact of the change in input sizes on the performance of different memory access patterns, described in Sect. 2, typically found in C language arrays. Programs were coded in C language corresponding to each memory access pattern classification and executed in Vivado HLS [2]. The synthesis statistics were observed with increasing input sizes. The results indicate that there is a linear increase in area with growing input size for single-location access. Whereas the increase in area is nonlinear with growing input size for other three patterns namely neighbor-location, shared access, and part-of-array access. The execution time observed with increasing input size is constant for single-location, neighbor-location, and part-of-array access. Whereas execution time increases with growing input size for shared-location access due to the inherent dependency in the code used for experimentation. Even after converting the computation for shared access pattern to the other two coding styles the execution time and area changes with the growing input size.

The main objective of this study is:

- To analyze the timing and area requirements for different memory access patterns with an increase in the input size.
- To guide programmers achieve a higher gain in terms of timing and area by utilizing the appropriate memory access pattern as one access pattern can be converted to another. For example, the single-location-access is a kind of part-of-array access, the part-of-array access is a kind of neighbor-location access without overlapping in subsequent accesses [1].
- To suggest memory access patterns as a new parameter to be considered by programmers while writing code for HLS.

The rest of the paper is organized as follows. Section 2 discusses the memory access classification based on array references. The memory optimization techniques applied in the two popular HLS tools, Vivado HLS [2] and LegUp [4] are also explained briefly in this section. Section 3 explores the related work in the field of memory access patterns, computational patterns and CPU/FPGA frameworks based on memory access requirements. Section 4 describes the latency and area changes based on increasing input size for all the memory access patterns and analyzes the impact of growing input size on the various classifications. Finally Sect. 5 derives the conclusions and discusses the possible scope of further exploration.

2 Memory Access Pattern and HLS

The memory access pattern can be classified similar to the classification performed by the algorithmic species theory [1] into four categories as below. The classification is based on identifying the specific kind of memory access performed to achieve a computation.

2.1 Single-Location Access

This classification is mapped to "element species" of algorithmic species theory [1]. A code snippet, Example 1, from matrix addition is shown below as a representation of single-location access:

Matrix addition

```
for (i = 1; i < n; i++)
{
c[i] = a[i] + b[i];
}
```

Example 1: Code listing for single-location access To calculate a specific element of output matrix c, one element from both the input matrices a and b is required. To perform this calculation only one location of both the inputs is accessed at a time. This type of access pattern is classified as single-location access.

2.2 Neighbor-Location Access

This classification is mapped to neighborhood species of algorithmic species theory [1]. The code below, Example 2, represents the neighbor-location access:

```
for (i = 1; i < index-1; i++)
{
b[i-1] = 3 * (a[i-1] + a[i] + a[i+1]);
}
```

Example 2: Code listing for neighbor-location access To calculate one element of output matrix b three accesses of the input matrix a are required and these three accesses are neighbor-locations in memory, a[i − 1], a[i] and a[i + 1]. Therefore the input access in this example can be classified as neighbor-location access whereas the output access can be classified as single-location access.

2.3 Shared Access

This classification is mapped to "shared or full species" of algorithmic species theory [1]. As per the original algorithmic species theory [5], the memory access is classified as "shared species" when only the partial locations of the array are accessed and are classified as "full species" when all the locations of the array are accessed in a computation. The following code represents the case of shared access:

```
for (i = 0; i < size-2; i++)
{
r[0]+ = a[i] + b[i+2];
}
```

Example 3: Code listing for shared access, coding style 1 Array b is accessed starting from location 2 such that the locations b[0] and b[1] are never accessed. It is classified as shared access as the input array b is partially accessed to generate the output r[0].

The following code which is a part of matrix multiplication represents “full access”:

```
for (i = 0; i < m; i++)
{
for(j = 0; j <n; j++)
{
r[i]+ = a[i][j] * b[j];
}
}
```

Example 4: Code listing for full access Matrix a is of size m × n and matrix b is a vector of size n × 1. To obtain one element of the output matrix r, one complete row of matrix a is operated with the full vector b. Therefore, the access pattern for vector b is classified as full and the access pattern for matrix a is classified as chunk (explained further in Sect. 2.4).

Nugteren et al. redefined their original algorithmic species theory [5] by adding more parameters for memory access classification [1]. As per this new theory [1] “shared” and “full” species are differentiated only by the purpose of access (used for reading or writing). Full species is the case of reading and shared species is the case of writing the memory location with a step size of zero. This article maps the shared access to “shared or full species” of algorithmic species theory [1], and does not differentiate between reading and writing as only the memory access is considered for experiments.

2.4 Part-of-Array Access

This classification is mapped to chunk species of algorithmic species theory [1]. As represented in Example 4, to obtain one element of output matrix r, one complete row of input matrix a, i.e., some part of the array a is accessed. This kind of memory access is classified as part-of-array access. It is different from the neighbor-location access as even though the neighbor-locations are accessed in the form of subsequent elements of a row, there is no overlap of the same memory location in each new computation. Each computation fetches the new row of the matrix a.

Another example of part-of-array access is shown below:

```
for (i = 0; i < n-1; i = i +2)
{
temp = 0; for(j = 0; j <2; j++)
{
temp+ = a[i+j];
}
b[i] = temp;
b[i+1] = temp;
}
```

Example 5: Part-of-array access To obtain the values of output matrix b[i] and b [i + 1] two accesses of input matrix a[i] and a[i + 1] are required. Since variable i is incremented by 2 in each iteration the access is non-overlapping in subsequent iterations.

Since the experiments are performed in Vivado HLS [2], the optimization techniques available in the two popular HLS tools, Vivado HLS [2] and LegUp [4] HLS are briefly explained. By default, HLS tools execute the program in sequential order. However, these tools provide mechanisms to parallelize the code for better performance in terms of execution time and area.

2.5 Vivado HLS [2]

Vivado [2] provides various parallelizing constructs in the form of pragmas such as loop-unrolling, loop-flattening, and pipelining. Vivado HLS's [2] compiler creates a memory architecture as per the need of the application. As per Vivado's HLS specification, the memory architecture is decided by the size of memory blocks in the programs and also by data usage throughout the program. Vivado HLS [2] implements the memory for an array as registers, shift registers, FIFOs, or BRAM depending on the computation involved in the algorithm. The specification does not clearly mention whether the memory access patterns are applied in designing the memory architecture for synthesis.

2.6 LegUp HLS [4]

LegUp [4], another HLS tool which is also open source, allows programmers to use OpenMP and Pthreads APIs for enabling parallelism. LegUp [4] facilitates the programmers to select the functions to be executed in hardware and rest of the program to be executed in software, i.e., in Tiger MIPS by default, to make it a hybrid flow for CPU/FPGA hybrid computing. LegUp [4] enables the designers to apply certain constraints related to the array such as LOCAL_RAMS to determine if an array is used only inside one function. In that case, the array is placed in local RAM rather than global memory to enhance memory access performance as local RAMs can be accessed in parallel while the global memory is limited to two ports in LegUp [4]. Another LegUp [4] constraint to be utilized is GROUP_RAMS, to group all arrays in the global memory controller into four RAMs (one for each bit width: 8, 16, 32, 64) to avoid the case where the entire array can take up the whole block of memory. Another useful constraints feature provided by LegUp [4] is NO_ROMS that places constant arrays into RAMs instead of the default read-only memory.

However, most of the HLS tools do not take into consideration the feature of memory access patterns. Sufficient knowledge of the hardware design is required to effectively utilize HLS tools. The programmer needs to understand the syntax as well as the complete synthesis flow of the HLS tool since the high-level language code may not run as is in the HLS [6]. The focus of this study is to analyze the behavior of different memory access patterns with an increase in input size through HLS.

3 Related Work

The classification used in this work is motivated by algorithmic species theory [1]. This theory claims that the classification scheme can be used by programmers and compilers to take parallelization as well as memory optimization decisions. Bones [7], a source-to-source compiler also performs optimizations based on memory access patterns classified in algorithmic species theory [1] and generates code for different target platforms.

The mathematical model AEcute [8] provides Access/Execute specification to classify the algorithm. The specification keeps the information about the iteration space, the memory locations that may be written or read in an iteration, the partition to mention the set of iterations to be executed in the same processing unit and the precedence to set the execution order. The work by Membarth et al. [9] uses AEcute [8] to develop a source-to-source compiler from C++ classes to OpenCL or CUDA code.

Compiler directives applied in various APIs such as OpenMP [10] and OpenACC [11] can also be considered as one kind of classification of the code as they provide the information to the compiler to execute the code in parallel. The pragmas applied in Vivado HLS [2] are also a way to classify the code as they provide extra information to compiler to work in a specific way.

Another parallel stream of work is to optimize based on computational patterns. The work by Nithin George et al. [12] focuses on applying parallel computational patterns such as map, reduce, zipwidth, and foreach that enable designers to optimize based on the specific patterns to generate extremely parallel hardware modules. Their study tries to find out how the data is utilized to parallelize various computations across hardware and also mentions the need to find out the data access patterns involved in the computation to generate a better hardware design.

CoRAM++ [13] creates a programming framework to support complex data structures such as multi-dimensional arrays and linked list. It defines an appropriate application-level interface for each data structure to generate streamlined soft-logic implementations that can selectively instantiate components in order to improve performance. These performance benefits can be especially noticeable for irregular pointer-based structures without penalizing DRAM access performance. CoRAM++ [13] framework also supports the simple and regular memory access patterns such as streaming and block copy.

In this work, various memory access patterns for array references found in C language have been compared with growing input size and analyzed for timing and area using synthesis statistics generated by Vivado HLS [2].

4 Results and Analysis

4.1 Experimental Setup

The case studies were worked out on Vivado HLS tool [2]. The project setup was based on Zynq FPGA family and targeted device used was xc7z030sbv485-3. Following points have been considered while performing high-level synthesis:

- The "main" function represents the "test bench" similar to any HDL verification flow. The "main" function calls the "top" function which implements the core computation in all the cases. The top function is similar to the "design top" of the actual hardware unit. The arguments passed from "test bench" to "top" function define the interface (inputs/outputs) of the hardware which gets synthesized on the target device.
- The inputs provided to the top function are assumed to be in parallel, i.e., the entire array elements are available on ports from outside. The case studies use integer type declaration of variables which creates 32-bit ports on the device for one element of the array. The number of ports can be reduced by creating user defined types of small bit widths, e.g., a short would create less ports than using an integer declaration of I/O ports.
- Directives have been added so that HLS tool creates a combinational logic for each of the cases, e.g., unrolling for-loops. Based on the size of input vector and iterations to be performed in for-loops, the amount of hardware resources of targeted device increases. Synthesis directives have been added to use more resources so that computation occurs in parallel and estimated time of input reaching to output ports after combinational logic is approximately the same. The following synthesis directives have been used:
  - INTERFACE: It synthesizes the arguments to a top-level function into RTL ports. Interface synthesis is used to achieve custom interface.

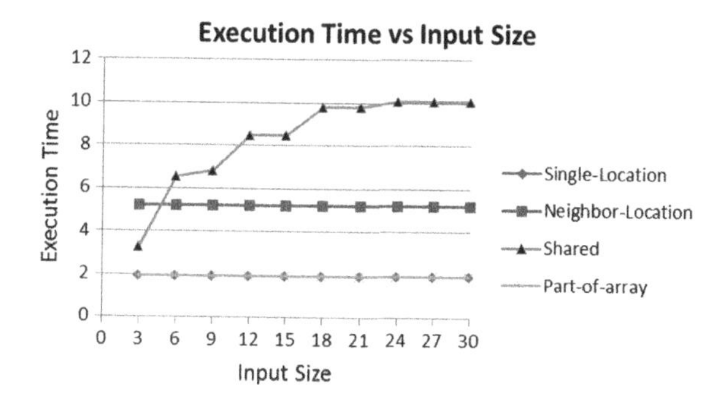

Fig. 1 Impact of input size on (execution time) for various memory access patterns

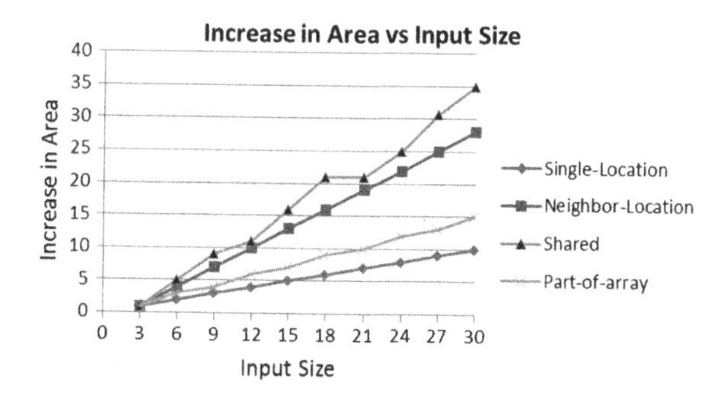

Fig. 2 Impact of input size on (area in terms of LUTs) for various memory access patterns

- ARRAY PARTITION: This synthesis directive improves access to input data by providing parallel access to data of the input array. For array input, this helps to override the default INTERFACE directive.
- UNROLL: It tries to create multiple independent operations rather than a single collection of operations and also improves latency that otherwise could suffer due to a long chain of combinational logic.

Figures 1 and 2 show the plot of execution time and gain in area with respect to the change in input size for all the four memory access patterns identified in Sect. 2. Arrays a, b, and c are single dimensional. The size of input arrays was increased from 3 to 30 with an increase of 3 in each iteration.

4.2 Assumptions

All the cases have been synthesized as a combinational logic with readily available input data for processing. This creates a 0 cycle clock latency but the input suffers from a combinational path delay to reach output ports. For simplicity and ease of study, the synthesis has been done without looking into any upper bound on area.

4.3 Single-Location Access

The matrix addition as described in Example 1, was experimented and observed. The area in terms of LUTs count increased linearly with respect to input size. However, the execution time remains constant as 1.93 ns even as the input size increases due to the for-loop being executed in parallel.

4.4 Neighbor-Location Access

The computation performed to study this case is same as Example 2. The arrays a and b are of size index and index-2, respectively. The size of the array a was increased from 3 to 30 and size of array b was increased from 1 to 28 with an increase of 3 in each iteration. The area increase is nonlinear. However, the execution time remains same as 5.19 ns even as the input size increases due to the for-loop being executed in parallel.

4.5 Shared Access (Coding Style 1)

The computation to test this case is described in Example 3. Arrays a and b are of size-2 and size long, respectively, and array r is of size 1. The size of array b was increased from 3 to 30 and size of the array a was increased from 1 to 28. In this case, the increase in execution time, as well as the area, is nonlinear. The *shared access* case shows the dependency of HLS Csynthesis on the coding style. Since r [0] is part of computation inside for-loop, it gets reassigned in each iteration. r[0] happens to be the output port as well, therefore, it is possible that the logic is routed in the following pattern: input port → combinational logic → output port → back to combinational logic → final output → output port. This could produce additional routing from output port till the LUTs where computational logic resides. This may be one of the reasons for the change in execution time.

4.6 Part-of-Array Access

Example 5 was executed to perform the analysis in this case. Execution time observed was 1.93 ns and was same for all the iterations. However, there was a nonlinear change in the area.

To analyze the case of *shared access* where there is a change in timing another two coding styles were executed as shown below in Example 6 and Example 7, respectively.

```
rtemp = 0;
for (i = 0; i < size-2; i++)
{
rtemp = a[i] + b[i+2] + rtemp;
}
r[0] = rtemp;
```

Example 6: Code listing for shared access, coding style 2

```
for (i = 0; i < n-1; i++)
{
temp[i] = a[i] + b[i+2];
}
t = temp[0];
for(i = 2; i <n-2; i = i+2)
{
t = t + temp[i-1] + temp[i] + temp[i+1];
}
r[0] = t;
```

Example 7: Code listing for shared access, coding style 3 The execution time and change in the area were plotted for the three cases (Examples 5, 6 and 7) of *shared access*, as shown in Fig. 3 and Fig. 4, respectively.

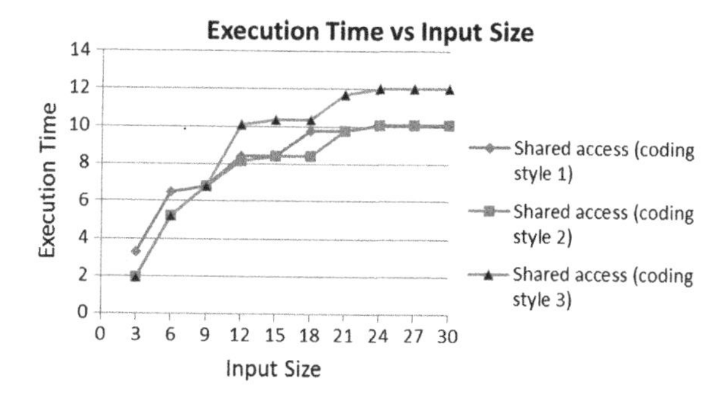

Fig. 3 Impact of input size on (execution time) for shared access with different coding styles

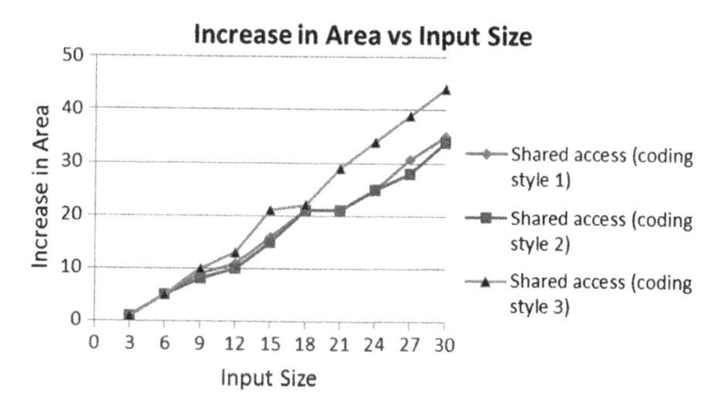

Fig. 4 Impact of input size on (area in terms of LUTs) for shared access with different coding styles

4.7 Shared Access (Coding Style 2)

Example 6, *shared access* (coding style 2), is a modified version of Example 5, *shared access* (coding style 1). In this case, rtemp variable is used to store the entire computation result from each iteration. This may lead to the following distribution or placement of logic in FPGA: input port → combinational logic → temp(gets produced here) → output port. Since there is no path from output port back to combinational logic the same delay as in *shared access* (coding style 1) is not observed even though the computation is same. There is a nonlinear increase in execution time as well as area with growing input size but this increase is less than the increase observed in *shared access* coding style 1.

4.8 Shared Access (Coding Style 3)

Example 7, *shared access* coding style 3, is a modified version of Example 5, *shared access* coding style 1. The array temp is created to store the intermediate results and then the elements of temp are added using for-loop which is a *neighbor-location access*. It is still a *shared access* for array r. This coding style consumes more execution time because of the two dependent for-loops.

The change in timing observed in case of *shared access* is due to synthesis (routing and placement) of small logic that suffers more delay because the output is routed back to combinational logic. But as the size of input grows, the code dependency is removed or minimized and all the three logics experience similar latency, since the combinational logic may reside near to the output port. The plots shown in Figs. 3 and 4 for the three different coding styles of *shared access* represent differences due to different routing and placement possibilities available while synthesis is performed by Vivado HLS tool [2]. Also when the combinational logic increases with the increase in input sizes or for-loop iterations, the latency differences due to such routing and placement appears to merge in all three cases.

This analysis re-enforces the notion that change in the memory access patterns leads to different area and timing requirements and change in coding style may improve the performance in hardware generation through HLS tools.

5 Conclusion and Future Directions

Memory access patterns play a very important role in deciding the execution time as well as the area utilization by an application for hardware generation through HLS. HLS tools perform optimizations by looking into the minimum latency required for a specific computation or operation. Since the computing consumes and produces data, memory access patterns have an impact on computation. For example, if the array is stored in column major form rather than row major form, the memory access calculations will change completely. Therefore optimizing on the basis of a specific computation may not be the sole key parameter to achieve higher performance gains as the performance of the resulting hardware depends on operation latency, data dependency as well as on data access pattern. Various experiments were conducted by executing programs in C language using Vivado HLS [2] to study the impact of growing input sizes on different memory access patterns. This study re-enforces the notion that change in memory access pattern leads to different area and timing requirements and change in coding style may help in working with HLS tools for better performance gains.

There are multiple ways in which hardware synthesis can be performed by keeping same latency. To synthesize hardware at a higher frequency, multiple clock cycle latencies can be added to hardware by registering the input ports and breaking the combinational path through registers before delivering the outputs. For large-sized arrays, there could be a limitation on ports. In such cases, the performance can be enhanced by the use of hardware ports partitioning at acceptable granularity or by using some computational cores. The experiments performed in this study focus on memory access patterns found in one-dimensional arrays. The study of memory access patterns for two-dimensional arrays and the limitations of HLS in Vivado [2] and LegUp [4] HLS tools will be the extension of this work.

References

1. C. Nugteren and R. Corvino and H. Corporaal.: Algorithmic species revisited: A program code classification based on array references. 2013 IEEE 6th International Workshop on Multi-/Many-core Computing Systems (MuCoCoS).
2. Xilinx Vivado HLS, http://www.xilinx.com/support/documentation/sw_manuals/ug998-vivado-intro-fpga-design-hls.pdf.
3. Canis, Andrew and Choi, Jongsok and Aldham, Mark and Zhang, Victor and Kammoona, Ahmed and Czajkowski, Tomasz and Brown, Stephen D. and Anderson, Jason H.: LegUp: An Open-source High-level Synthesis Tool for FPGA-based Processor/Accelerator Systems. ACM Trans. Embed. Comput. Syst., year: 2013, issn: 1539-9087, pages: 24: 1–24: 27.
4. LegUp HLS, http://legup.eecg.utoronto.ca/.
5. Nugteren, Cedric and Custers, Pieter and Corporaal, Henk.: Algorithmic Species: A Classification of Affine Loop Nests for Parallel Programming. ACM Trans. Archit. Code Optim., year: 2013, issn: 1544-3566, pages: 40: 1–40: 25.
6. M. Belwal and M. Purnaprajna and Sudarshan TSB.: Enabling seamless execution on hybrid CPU/FPGA systems: Challenges amp; directions. 2015 25th International Conference on Field Programmable Logic and Applications (FPL).
7. Nugteren, Cedric and Corporaal, Henk.: Bones: An Automatic Skeleton-Based C-to-CUDA Compiler for GPUs. ACM Trans. Archit. Code Optim. January 2015, issn: 1544-3566, pages: 35: 1–35: 25.
8. Howes, Lee W and Lokhmotov, Anton and Donaldson, Alastair F and Kelly, Paul HJ.: Deriving efficient data movement from decoupled access/execute specifications. High Performance Embedded Architectures and Compilers, pages: 168–182, year: 2009, publisher: Springer.

9. Membarth, Richard and Lokhmotov, Anton and Teich, Jürgen: Generating GPU code from a high-level representation for image processing kernels. Euro-Par 2011: Parallel Processing Workshops, pages: 270–280, year: 2011, Organization: Springer.
10. OpenMP, http://www.openmp.org.
11. OpenACC, www.openacc.org.
12. Nithin George, HyoukJoong Lee, David Novo, Muhsen Owaida, David Andrews, Kunle Olukotun and Paolo Ienne.: Automatic support for multi-module parallelism from computational patterns. 25th International Conference on Field Programmable Logic and Applications, FPL 2015, London, United Kingdom, September 2–4, 2015, pages: 1–8, year: 2015.
13. G. Weisz and J. C. Hoe.: CoRAM++: Supporting data-structure-specific memory interfaces for FPGA computing. 2015 25th International Conference on Field Programmable Logic and Applications (FPL), London, 2015, pp. 1–8.

Image Segmentation Using Hybridized Firefly Algorithm and Intuitionistic Fuzzy C-Means

Sai Srujan Chinta, Abhay Jain and B.K. Tripathy

Abstract Fuzzy clustering methods have been used extensively for image segmentation in the past decade. The most commonly used soft clustering algorithm is Fuzzy C-Means. An improvised version of FCM called Intuitionistic Fuzzy C-Means (IFCM) has also gained popularity in the recent past. In this paper, we propose a new hybrid algorithm which combines intuitionistic fuzzy c-means and firefly algorithm to propose Intuitionistic Fuzzy C-Means with Firefly Algorithm (IFCMFA). Experimental analysis confirms that IFCMFA is far more superior to both FCM and IFCM. Several measures like DB-index and D-index are used for this purpose. Also, different types of images like MRI scan, Rice, Lena and satellite image are used as inputs to establish our claim.

Keywords Data clustering · Fuzzy set · Intuitionistic fuzzy set
DB-index · D-index

1 Introduction

Image Segmentation is the process of partitioning an image space into several non-overlapping meaningful homogeneous regions. Segmentation has traditionally relied on data clustering methods. There are essentially two types of clustering algorithms: hard clustering and soft clustering. In hard clustering, every pixel of the

S.S. Chinta (✉) · B.K. Tripathy
School of Computing Science and Engineering, VIT University,
Vellore 632014, Tamil Nadu, India
e-mail: chintasai.srujan2014@vit.ac.in

B.K. Tripathy
e-mail: tripathybk@vit.ac.in

A. Jain
School of Information Technology and Engineering, VIT University,
Vellore 632014, Tamil Nadu, India
e-mail: abhay.jain2014@vit.ac.in

A.K. Somani et al. (eds.), *Proceedings of First International Conference on Smart System, Innovations and Computing*, Smart Innovation, Systems and Technologies 79, https://doi.org/10.1007/978-981-10-5828-8_62

image belongs to only one cluster. Soft clustering revolves around the concept of Fuzzy Sets which was proposed by Zadeh et al. [12] in 1965. In soft clustering, every pixel of the image belongs to every cluster with a certain probability. This probability is recorded in a matrix called the membership matrix. However, in IFCM, an additional property called hesitation degree is added. This further enhances the clustering process. In the recent past, the need for fuzzy clustering has increased tremendously due to unclear grayscale values and the uncertainty associated with class definitions and boundaries. The major shortcoming of most data clustering algorithms is their susceptibility to random initialization of data. In this paper, we have nullified this shortcoming of IFCM by fusing it with an optimization algorithm called firefly algorithm which was invented by Yang [11] in 2009. The rest of the paper is structured as follows: Sect. 2 consists of a brief Literature Survey. We describe the methodology of our algorithm in detail in the third section. Section 4 consists of a brief summary of firefly algorithm followed by a brief summary of IFCM in Sect. 5. Our experimental results have been displayed in Sect. 6 followed by the conclusion in Sect. 7.

2 Literature Review

Feng [5] performed a survey on the feasibility of using genetic algorithms for performing data clustering. K-Means coupled with a traditional Genetic Algorithm is better than existing clustering algorithms like Fuzzy c-Means and k-Means. Thus, Genetic Algorithm is used to provide the clustering algorithm an optimal initial guess about the position of the centroids. Chaira [3] introduced the Intuitionistic Fuzzy C-Means (IFCM) algorithm for data clustering. The main idea is to introduce the concept of non-membership of data. Along with the degree of certainty that a particular data point belongs to a particular cluster, there is also a degree of uncertainty that it does not belong to that cluster. Dash et al. [4] compared the performances of the traditional K-Means and a data clustering algorithm coupled with a Genetic Algorithm. The experimental results show that the performance of the GA-based clustering algorithm is superior to the traditional K-means clustering algorithm. Alsmadi et al. [1] combined Fuzzy c-Means and firefly algorithm. It returns the optimal centroids to FCM which then completes the image segmentation process. Huang et al. [6] improvised on IFCM by considering neighborhood pixel tuning and further optimized the performance by employing a genetic algorithm for providing the clustering algorithm with the optimum initial centroid locations. It is observed that the performance of this algorithm is better than FCM and IFCM. Tripathy et al. [9] have combined Scalable Rough C-means and firefly algorithm. Firefly algorithm is used to compute the initial cluster centroids so as to make the clustering algorithm independent of the sensitivity associated with the random selection of initial centroids.

3 Methodology

It is evident from the literature survey that IFCM is a very powerful data clustering algorithm. However, IFCM is susceptible to random initialisation of data. Attempts have been made in the past to use genetic algorithms in order to provide near-optimal solutions to IFCM. However, they have largely been unsuccessful due to their susceptibility to local optima. In the literature survey, we have also mentioned about the robustness and the efficiency of firefly algorithm. Thus, we propose a hybrid algorithm between firefly and IFCM. In this algorithm, firefly will provide the optimal membership matrix and the cluster centers to IFCM which will then iterate to find the actual cluster centers and the optimal membership matrix. Experimental results prove that IFCMFA outperforms both FCM which was proposed by Ruspini et al. [8] and IFCM (Fig. 1).

The above flowchart clearly depicts the functioning of the firefly algorithm. It is to be noted that there is a degree of randomness associated with the movement of each firefly towards the brightest firefly. Moreover, the fireflies which are geometrically closer to the brightest firefly tend to get more attracted to it than those which are far away. This property of fireflies is called Attractiveness. Thus, in this project, firefly algorithm is being used to return the near-optimal values of the membership matrix and cluster centers. These near-optimal values are then passed to IFCM (Fig. 2).

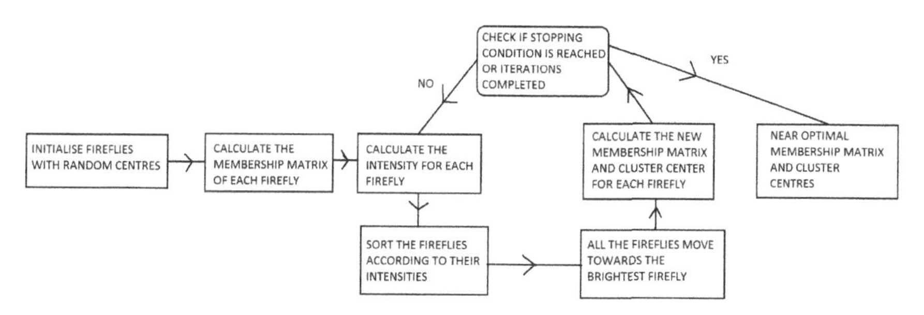

Fig. 1 Functioning of Firefly Algorithm in IFCMFA

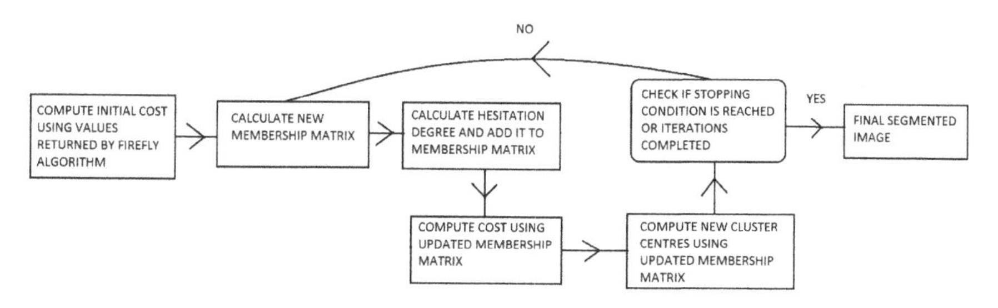

Fig. 2 Functioning of IFCM in IFCMFA

The above flowchart depicts the functioning of IFCM. IFCM is explored in detail in Sect. 5.

4 Firefly Algorithm

Firefly Algorithm is a nature-inspired optimization algorithm which mimics the behavior of fireflies. There are certain predefined attributes associated with each firefly which are problem-specific. In firefly algorithm, there are three idealized rules defined by Yang [10]

(1) Every firefly is attracted to every other firefly irrespective of its gender.
(2) Attractiveness of a firefly is directly proportional to its brightness and inversely proportional to the distance between the two fireflies in question. The brightest firefly will move randomly.
(3) The brightness of a firefly is decided by the objective function that is to be minimized or maximized. The attractiveness function β is determined by

$$\beta(r_{i,j}) = \beta_0 e^{-\gamma r_{i,j}^2}, \quad (1)$$

where β_0 is the attractiveness at $r = 0$, γ is the light absorption coefficient at the source and $r_{i,j}$ is the Euclidean distance between any two fireflies i and j. The movement of firefly i to j is determined by

$$x_i = x_i + \beta_0 e^{-\gamma r_{i,j}^2}(x_i - x_j) + \alpha\left(rand - \frac{1}{2}\right), \quad (2)$$

where, the second term is the attraction and the randomization parameter is α. For most cases in implementation, Yang [10] introduced $\beta_0 = 1$, $\alpha \in [0, 1]$, and $\gamma \in [0.01, 100]$.

5 Intuitionistic Fuzzy C-Means

IFCM is based on the concepts of intuitionistic fuzzy sets [2]. Chaira [3] has taken the intuitionistic fuzzy generator or the fuzzy complement as defined by Yager [7]. Fuzzy complement is defined as

$$M(\mu(x)) = f^{-1}(f(1) - f(\mu(x))), \quad (3)$$

where f is an increasing function between 0 and 1. Yager's intuitionistic fuzzy complement is written as $f(x) = (1 - x^{\alpha})^{\frac{1}{\alpha}}$, $\alpha > 0$, where $M(1) = 0$, $M(0) = 1$.

Non-membership values are calculated from Yager's complement $M(x)$. Thus the hesitation degree of x is

$$\pi_A(x) = 1 - \mu_A(x) - (1 - \mu_A(x)^{\alpha})^{\frac{1}{\alpha}} \tag{4}$$

The intuitionistic fuzzy membership values are obtained as follows:

$$u'_{ik} = u_{ik} + \pi_{ik}, \tag{5}$$

where $u'_{ik}(u_{ik})$ denotes the intuitionistic (conventional) fuzzy membership of the kth data in jth class. The modified cluster center is written as

$$v'_i = \frac{\sum_{k=1}^{n} u'_{ik} x_k}{\sum_{k=1}^{n} u'_{ik}} \tag{6}$$

The second function is written as $\sum_{i=1}^{c} \pi'_i e^{1-\pi'_i}$ where $\pi'_i = 1/n \sum_{k=1}^{n} \pi_{ik}$, k $\in$ [1, n]. π_{ik} is the hesitation degree of the kth element in cluster 'i'. Final cost function

$$J = \sum_{i=1}^{c} \sum_{k=1}^{n} u'^{m}_{ik} d(x_k, v_i)^2 + \sum_{i=1}^{c} \pi'_i e^{1-\pi'_i} \text{ With m} = 2 \tag{7}$$

6 Results

We have compared the performance of IFCMFA with IFCM and FCM by running these clustering algorithms on four different images. All the images were obtained from Math Works. In order to compare the performance of the algorithms, two indices have been used: DB and Dunn.

$$\mathrm{DB} = \frac{1}{\mathrm{m}} \sum_{\mathrm{i}=1}^{\mathrm{m}} \max_{\mathrm{k} \neq \mathrm{i}} \left\{ \frac{\mathrm{Q}(\mathrm{v_i}) + \mathrm{Q}(\mathrm{v_k})}{\mathrm{d}(\mathrm{v_i}, \mathrm{v_k})} \right\} \quad \text{for} \quad 1 < i, k < m \tag{8}$$

$$\mathrm{Dunn} = \min_{\mathrm{i}} \left\{ \min_{\mathrm{k} \neq \mathrm{i}} \left\{ \frac{\mathrm{d}(\mathrm{v_i}, \mathrm{v_k})}{\max_{\mathrm{l}} \mathrm{Q}(\mathrm{v_l})} \right\} \right\} \quad \text{for} \quad 1 < k, i, l < m \tag{9}$$

DB-index is the ratio of intra-cluster distance to inter-cluster distance. Dunn index is essentially the reciprocal of DB-index. Lower the value of DB-index and higher the value of D-index, the better the clustering quality. Stopping condition value is 0.001.

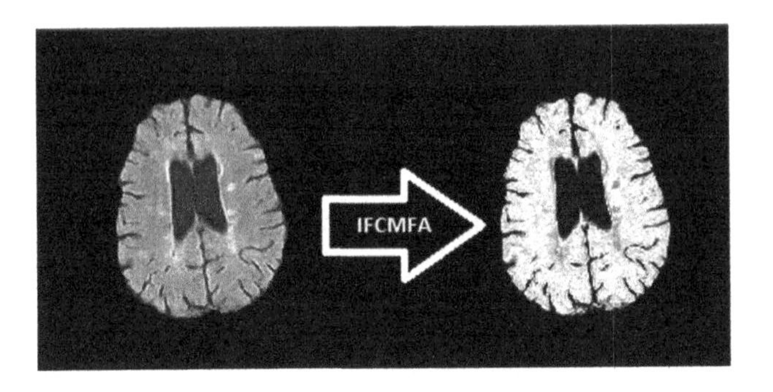

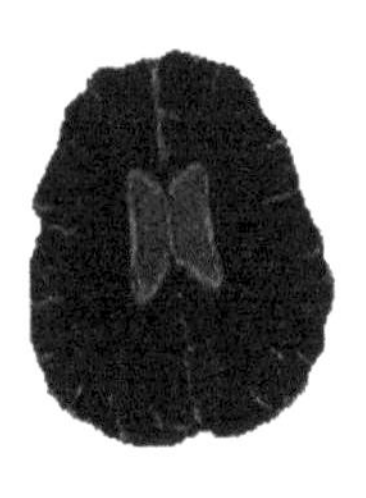

Fig. 3 The Brain MRI image represents the input image and the black part of the other two images show the clustered area after applying IFCMFA

Table 1 DB and Dunn index values for Brain MRI Image

Iterations	FCM		IFCM		IFCMA	
	DB	D	DB	D	DB	D
1	126.3954	0.0157	134.9710	0.0147	0.9179	1.1174
5	13.0985	0.1436	0.9261	1.0947	0.9173	1.1177
10	4.5823	0.3693	0.9172	1.1177	0.9172	1.1177
25	0.9443	1.1070	0.9172	1.1177	0.9172	1.1177
Convergence	0.9443	1.1070	0.9172	1.1177	0.9172	1.1177

6.1 Brain MRI Segmentation

It can be clearly observed that the algorithm has successfully segmented the tumor and the brain as separate clusters. The DB-index value of IFCMFA in its very first iteration is as low as 0.9179 and its D-index value is as high as 1.1174. It can be observed that FCM takes more than 25 iterations and IFCM takes 10 iterations to reach the clustering quality which IFCMFA achieved in its very first iteration (Fig. 3 and Table 1).

6.2 Rice Image Segmentation

See Fig. 4 and Table 2.

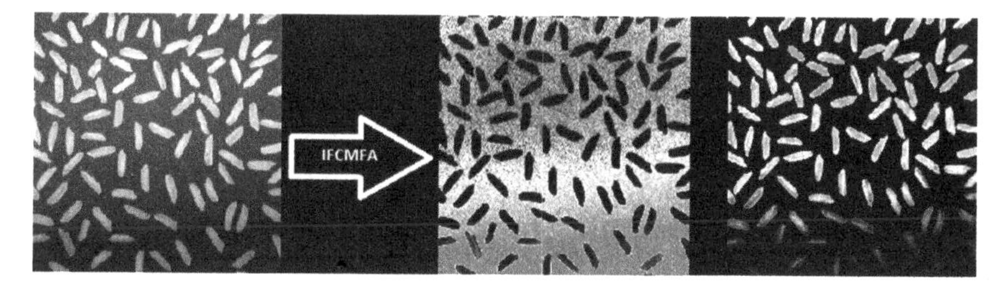

Fig. 4 The Rice input image i.e. the first image on the left is segmented into different clusters represented by the dark areas in the other two figures by IFCMFA

Table 2 DB and Dunn index values for Rice Image

Iterations	FCM		IFCM		IFCMA	
	DB	D	DB	D	DB	D
1	4.183e+03	4.778e−4	225.2375	0.0089	1.5011	0.8920
5	1.0592e+03	0.0019	2.9908	0.5199	1.4469	0.9123
10	4.0830	0.4175	1.4485	0.9116	1.4468	0.9124
25	1.6447	0.8473	1.4468	0.9124	1.4468	0.9124
Convergence	1.6447	0.8473	1.4468	0.9124	1.4468	0.9124

6.3 Lena

It can be clearly seen that IFCMFA is clustering all the relatively dark parts of the image into one cluster and the relatively brighter parts into the other cluster (Fig. 5 and Table 3).

Fig. 5 The Lena input image on the left is segmented into two different clusters as represented by the dark areas in the other two images by IFCMFA

Table 3 DB and Dunn index values for Lena Image

Iterations	FCM		IFCM		IFCMA	
	DB	D	DB	D	DB	D
1	1.7055e+03	0.0012	987.5390	0.0020	1.6372	0.8222
5	506.0081	0.0039	53.4034	0.0368	1.6070	0.8342
10	5.3643	0.3307	1.6212	0.8302	1.6039	0.8358
25	2.0966	0.7059	1.6034	0.8360	1.6034	0.8360
Convergence	2.0966	0.7059	1.6034	0.8360	1.6034	0.8360

6.4 Satellite Image

The effectiveness of IFCMFA is clearly established in this example. IFCMFA achieves a DB-index value of 2.3721 in its first iteration. This value was not achieved by FCM or IFCM even in 20 iterations (Fig. 6 and Table 4).

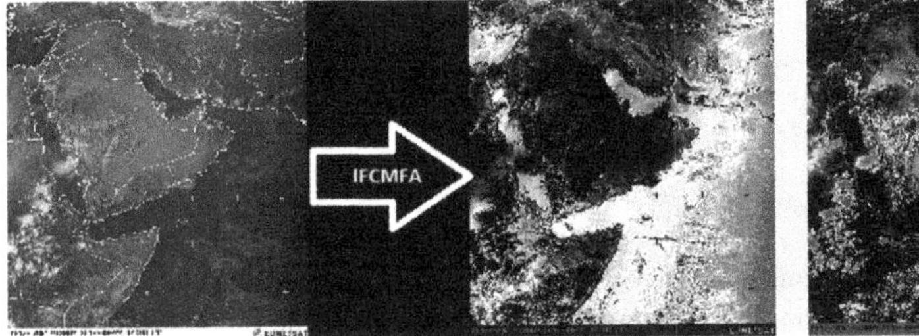

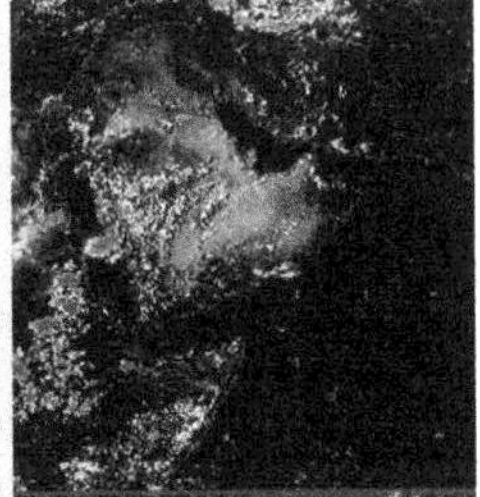

Fig. 6 Segmentation of Satellite Image using IFCMFA

Table 4 DB and Dunn index values for Satellite Image

Iterations	FCM		IFCM		IFCMA	
	DB	D	DB	D	DB	D
1	2.1391e +03	9.3392e −04	2.0392e +03	9.8071e −04	2.3721	0.5802
3	535.4130	0.0037	110.2924	0.0180	2.3398	0.5852
5	343.3901	0.0058	9.3566	0.1945	2.3548	0.5875
10	9.0052	0.1986	2.3796	0.5919	2.3488	0.5868
20	2.7708	0.5180	2.3559	0.5879	2.3505	0.5871

7 Conclusion

IFCMFA retains the advantages of IFCM and further enhances its performance by drastically reducing the number of iterations required for it to converge. This is made possible due to Firefly algorithm which supplies IFCM with a near-optimal solution for it to work with instead of a random membership matrix and cluster centers. By doing so, we are eliminating the problem of being stuck at the local optima. It is to be noted that though IFCMFA is more powerful than IFCM, the clustering quality of both these algorithms is the same. The main advantage in IFCMFA is the reduction in the number of iterations required. This property can prove to be more useful in big data clustering. Our future work includes the usage of firefly with other data clustering algorithms.

References

1. Alsmadi, M. K. A. and Khalil, M.: A hybrid firefly algorithm with fuzzy-C mean algorithm for MRI brain segmentation, Malaysian Journal of Library and Information Science, (2014).
2. Atanassov, Krassimir T. "Intuitionistic fuzzy sets." *Fuzzy sets and Systems* 20.1 87–96 (1986).
3. Chaira, T.: A novel intuitionistic fuzzy C means clustering algorithm and its application to medical images, Applied Soft Computing, 11(2), 1711–1717, (2011).
4. Dash, R. and Dash, R.: Comparative analysis of k-means and genetic algorithm based data clustering, International Journal of Advanced Computer and Mathematical Sciences, 3(2), 257–265, (2012).
5. Feng, Z.: Data clustering using Genetic Algorithms, course project for CSE848-Evolutionary Computation with the instructor of Dr. Bill Punch, fall (2012).
6. Huang, C. W., Lin, K. P., Wu, M. C., Hung, K. C., Liu, G. S., and Jen, C. H.: Intuitionistic fuzzy C-means clustering algorithm with neighborhood attraction in segmenting medical image, Soft Computing, 19(2), 459–470, (2015).
7. R.R. Yager: On the measures of fuzziness and negation part II lattices, Information and Control 44 236–260 (1980).
8. Ruspini, Enrique H. "A new approach to clustering." *Information and control* 15.1 22–32 (1969).
9. Tripathy, B.K, and Namdev, A.: Scalable Rough C-Means clustering using Firefly algorithm, International Journal of Computer Science and Business Informatics, 16(2) (2016).
10. Yang, X. Firefly algorithm, stochastic test functions and design optimization.;In Proceedings of IJBIC, 78–84 (2010).
11. Yang, Xin-She. "Firefly algorithms for multimodal optimization." *International Symposium on Stochastic Algorithms*. Springer Berlin Heidelberg, (2009).
12. Zadeh, Lotfi A. "Fuzzy sets." *Information and control* 8.3 338–353 (1965).

A Review Spoof Face Recognition Using LBP Descriptor

Tanvi Dhawanpatil and Bela Joglekar

Abstract Passwords are normally used for authentication in systems. They have several drawbacks like passwords can be guessed easily, they can be copied. Since biometric authentication is excelling in every field whether it be banking sector, corporate sector, etc., they are considered quite secure and mostly preferred for authentication. But every system has some flaws; therefore biometric authentication can be attacked so as to obtain any confidential information. One of them is face authentication system. Face is a unique characteristic that can be used to authenticate a person. Face authentication systems can be easily spoofed by using Replay and Printed paper attacks. Spoofing means real person's identity is copied and used for harming any type of data. In this review paper, mainly LBP (Local Binary Pattern) descriptor is used, which is considered especially for texture analysis. LBP descriptor divides the captured face into blocks and calculates histogram for each block. Thus each block histograms are concatenated and finally are combined together. The formed histogram of whole face is compared with other face histograms and the similarity between the faces is found out. Spoof faces will not have similar histograms like the real face. And this helps in detecting Spoof face. Different spoof face detection methods are discussed in this review paper. Detection of spoof face is done by considering Moiré patterns, image distortion analysis algorithm. This review paper aims at securing confidential information by providing face unlock mechanism wherein spoof faces are to be detected.

Keywords LBP · Spoof face · Moiré pattern · Image distortion
Replay attack · Printed paper attack

T. Dhawanpatil (✉) · B. Joglekar
Department of Information Technology, Maharashtra Institute of Technology, Pune, India
e-mail: tanvidhawanpatil@gmail.com

B. Joglekar
e-mail: bela.joglekar@mitpune.edu.in

A.K. Somani et al. (eds.), *Proceedings of First International Conference on Smart System, Innovations and Computing*, Smart Innovation, Systems and Technologies 79, https://doi.org/10.1007/978-981-10-5828-8_63

1 Introduction

Biometric authentication is gaining popularity in every field. Whether it be in case of iris recognition or finger print, or face recognition and many more. One of them is face detection. Face recognition has its own challenges that can be unconstrained environment, light effect, occlusion, etc. Face detection techniques can be used in various applications like forensic department in finding criminals, video surveillance, recognizing face expression, identifying age and gender of a person, face tracking, etc. One of the challenges in biometric authentication is that the increasing database must not affect the working of the authentication system.

Nowadays people prefer to store their confidential information in smartphone and laptops. Therefore preserving that information from an unauthorized user is a matter of concern. Face unlock is one of the secure methods to protect the confidential data and to detain unauthorized or fake people from accessing it. With challenges comes attacks that can be performed on face detection. This includes Replay attack and Printed Paper attack. Replay attack Fig. 1b can be done by an illegal person by showing the live video of the authorized user from tablet or any other medium and placing that device in front of the camera of smartphone or laptop [1]. While in case of Printed Paper attack, Fig. 1a, this attack is achieved by placing the printed paper having the image of the authorized user in front of camera [1].

We can crawl through different social media sites like Facebook, Instagram, etc., wherein we can obtain photos of people and through that these types of attacks can be performed so as to get access to confidential data. There are different Android applications developed so far that precisely identify the person, i.e., the authenticated user and unlock the screen in smartphones. Also different face detection and recognition apps have been developed for desktops so far like Face Unlock, Facelock Pro, and Visidon. For laptops, we have Veriface, Luxand Blink, and

Fig. 1 Spoof attacks in smartphones: **a** shows example of Printed paper attack, **b** shows example of Replay attack

FastAccess. But these desktop-based apps are not robust to spoof attacks. Therefore security mechanism in systems must be very strong as they can be highly vulnerable to different attacks. Author K. Jain implemented the Moire Pattern and Image Distortion approach on smartphone. Also the Moire pattern implementation was tested on non facial images [1]. The proposed approach is implemented on desktop providing face authentication as soon as the login screen appears. And the dataset used to test the approaches consist of facial images.

2 Literature Survey

In the past, there were several methods proposed to detect spoof attacks. Face liveness detection methods like eye blink detection, lip movement detection, life sign based detection were used for detection. In eye blink detection, the movement of the eyes was observed as live face has maximum movement of eyes as compared to the printed photo presented in the spoof attack. Same with mouth movements. Randomly system asks user for lip or eye movements and this sequence is not fixed; hence attacker cannot try to crack the system. So the author proposed that the eye movement and lip movement detection will be random and not in sequence. Here Principal Component Analysis descriptor is used for feature extraction. And eye movements and mouth movements are captured using HSV (Hue, Saturation, Value). It needs user collaboration, which is its drawback [2].

Another method used was Dynamic Mode Decomposition (DMD) algorithm to capture dynamic features in a video. Also along with that it can detect Moiré patterns. It does not depend on any cues. In DMD Krylov subspace and Arnoldi algorithm are used to extract dynamic features. For every frame of the video sequence a dynamic mode is created and a dynamic mode image with phase shift 0 is selected. Then using LBP descriptor feature extraction is done and then the result is passed to SVM for classification. Author found that, DMD+LBP+SVM outperformed optical flow, LBP methods that were previously used for spoof face detection [3].

Another method used was Diffusion speed model, it is independent on the medium, i.e., paper or electronic device. The main motivation for diffusion speed model is that the fake face reflects light uniformly whereas the real face reflects randomly. Diffusion speed is calculated to detect illumination characteristic using total variation (TV) flow scheme. TV uses pixel variation and finally we get diffused image. For feature extraction from a diffused image Local speed Pattern (LSP) is used. LSP represents a block in the image in the form of matrix in which the neighboring pixels are compared with the center pixel and values are given as shown in. From that f_{LSP}, LSP image is built and finally histograms are built for each block in the image. SVM is used for classifying the image as spoof or not. It takes total 0.34 ms for execution and performance with respect to accuracy is better as compared to LBP and DoG [4].

Then came Color texture Analysis method. Chrominance information of the image is used to identify the image as spoof or real. To separate the chrominance and luminance information the authors have considered the RGB (Red, Green, Blue), HSV (Hue, Saturation, Value) and YC_bC_r ("Y" is for luminance, "C_b" for chrominance blue, "C_r" for chrominance red). Initially RGB image is converted to other two color spaces and for each color space, histograms are computed using LBP descriptor. And finally the concatenated histogram is fed to SVM classifier, to classify the image as spoof or real. Author has considered $LBP_{(8,1)}$ operator. HSV and YC_bC_r along with LBP outperforms previous methods used for spoof face detection [5].

In this method, Moiré pattern are detected using Fourier spectrum. Initially band pass filter is applied to the image and Fourier transform of that image is built. Then peak detection is done on the Discrete Fourier Transform (DFT). The author found that spoof face has strong peak as compared to real face. And this is used to detect spoof image. The author found that the high frequency changes of a spoof image are less then the real image. This method applies to Printed photo attack. One of the drawbacks in this method is that distance between camera and display device must be small [6].

This method uses Local Derivative Pattern(LDP) descriptor and input is video. Video is divided into frames and from each frame face is extracted using Viola Jones method, also face image is normalized. The LDP operator is applied on three orthogonal planes XY, YT, XT where "T" is time axis. And based on that, histograms of every plane are constructed. SVM is used for classification purpose. This method performs better as compared to LBP, DMD+SVM, and Image Distortion Analysis (IDA) [7].

This method depends on scales; then resolution. Scale the image and extract texture features from the image. For extracting textual features, LBP descriptor is used. Scaling is required to detect the edges of image. For multiscale filtering, Gaussian scale space, Difference of Gaussian (DoG) scale space, Multiscale Retinex method were used. Gaussian scale space helps in reducing noise in an image. DoG is used for edge enhancement and it subtracts original image from blurred image. Multiscale Retinex method eliminates the lower frequency spectrum in an image. Initially image is scaled at different levels, then using LBP the texture information is captured and plotting of histogram is done for each multiscale image. Finally histograms are concatenated and provided to SVM. Results show that, it performs better than LBP, MLBP descriptor, IDA, DMD. The comparison of the results using DoG scale and without using DoG scale are also given [8].

The next method proposed by author is face detection using Haralick descriptor features. It takes the input image, detects face, and normalizes it. After that the face image is divided into 32*32 non overlapping patches. Then every patch in the image is converted to wavelet form using single level Redundant Discrete Wavelet Transform (RDWT). After that as soon as we get the RDWT subbands, Haralick descriptor is used to extract features for four subbands namely horizontal, vertical, approximation, and diagonal. Author found that this method performs better than color texture analysis method [5] when 30 frames are taken into consideration [9].

The proposed method uses Moiré pattern analysis and image distortion analysis for spoof detection. Adding LBP descriptor improves the performance of the proposed system as said by the author [1]. Thus results can be compared with the other methods that either use only Moire pattern or only use image distortion analysis (IDA) method. Also, as a part of addition to the existing work, dataset of different formats like .bmp, .jpeg, .png, can be used to test results. Further, popular descriptor like GIST descriptor can be added along with LBP to improve spoof face detection process. This review paper discusses the previous face spoof detection methods and focuses on the proposed architecture and possible additions to the existing work.

3 System Modules

The diagram shown above shows the main modules that are used to detect spoof faces in the face authentication system in desktop, it consists of following main modules (Fig. 2).

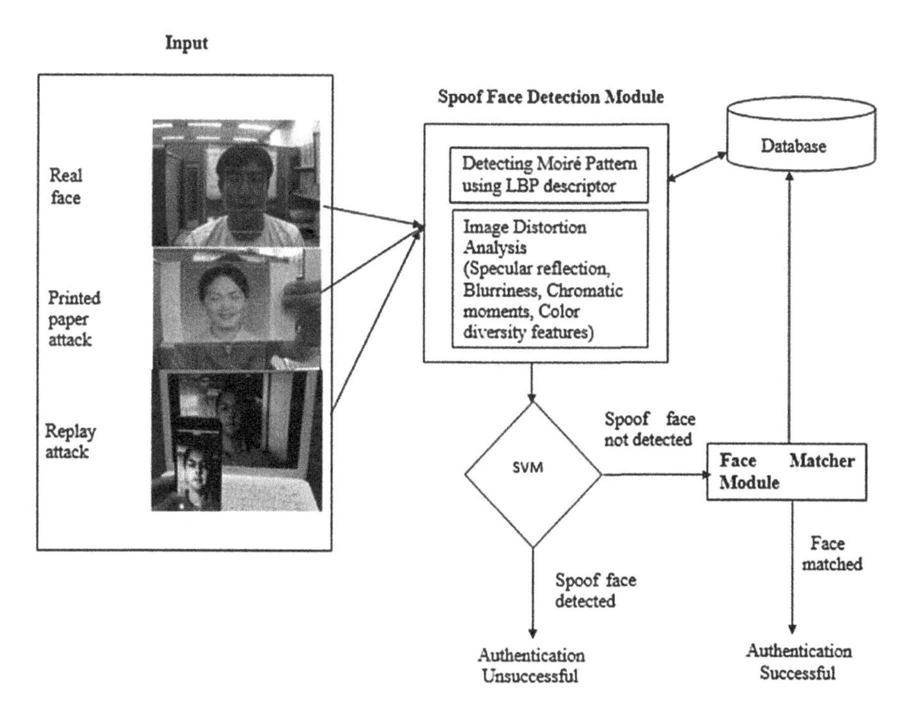

Fig. 2 Block diagram

3.1 Input

Here the input is the face of a person. This face can be the face of authorized user or it can be spoof face. Spoof face attack can be performed by either Replay attack or Printed Paper attack [1].

3.2 Spoof Face Detection and Face Matcher Module

In this, module image is tested for the presence of Moiré pattern [10] and image distortion analysis [11], which helps in identifying whether the face is spoof face or not. Moiré patterns are shown in the Fig. 3. Moiré patterns are produced when two patterns are overlayed on each other. Therefore whenever spoof attack takes place mainly in the case of Replay attack, the light of tablet or any electronic device which shows the video of legitimate user falls on the device's screen which produces the Moiré pattern and which can help to detect the Replay attack. Moiré pattern detection is done with the help of LBP descriptor, wherein LBP histograms of captured face and database faces are compared. Depending on the result, the face is detected as spoof or real face [10].

For Printed Paper attack, surface roughness of photo and real face is observed. When spoof attacks are performed, such Moiré patterns are identified in the image and color histograms of the spoof face and real face are compared by taking Hue, Saturation parameters of an image into consideration. Also image distortion analysis is done by analyzing image regions, intensity channels, and feature descriptors [11]. Finally classifier is used to test whether the image is spoof or not, then the screen is locked otherwise unlocked. Support Vector Machine Classifier (SVM) is used for classifying the face image as spoof face or not.

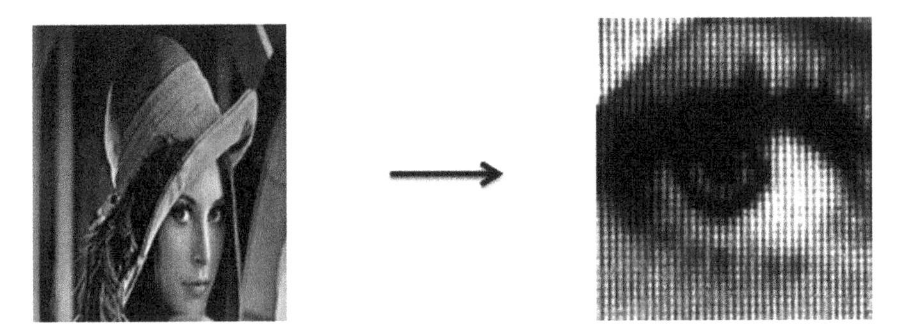

Fig. 3 Moiré patterns while capturing screen of digital devices

4 Applications

This approach can be combined with Internet of Things (IoT) and can be used in various applications like doorbell ringing by identifying the face of the person. Once the face of the person is recognized, doorbell rings automatically. In case of failure in recognition of face of unknown people, additional facilities like voice recognition can be used for blind people. Also opening of locker using face recognition can be used for providing security mechanisms for locker at home or at banks called as Smart Bank Locker. As the face is recognized correctly, the locker opens, else an alert message is sent to authorized user. In this way, proposed approach can be used to provide security to different things in the field of IoT.

5 Conclusion

In this review paper, we have discussed what kind of attacks can be carried out on face authentication systems if face unlock mechanism is used. We have also seen different descriptors, their advantages and disadvantages, their features. And how LBP is more accurate and efficient in case of face recognition. This review paper also discusses different face spoof detection methods used so far. Also Spoof Face Identification using Moiré pattern analysis, image distortion analysis is discussed. Different applications of the proposed approach in the field of IoT are discussed. Therefore face recognition is one of the best techniques for biometric authentication. Also LBP is more robust to illumination changes and poses. This helps in identifying the fake face and thus spoofing can be detected effectively using Moiré pattern using LBP along with image distortion analysis algorithm.

References

1. K. Jain, "Secure Face Unlock: Spoof Detection on Smartphones," in *IEEE Transactions on Information Forensics and Security*, vol. 11, no. 10, pp. 2268–2283, Oct. 2016.
2. A. K. Singh, P. Joshi and G. C. Nandi, "Face recognition with liveness detection using eye and mouth movement," *2014 International Conference on Signal Propagation and Computer Technology (ICSPCT 2014)*, Ajmer, 2014, pp. 592–597.
3. S. Tirunagari, N. Poh, D. Windridge, A. Iorliam, N. Suki and A. T. S. Ho, "Detection of Face Spoofing Using Visual Dynamics," *in IEEE Transactions on Information Forensics and Security*, vol. 10, no. 4, pp. 762–777, April 2015.
4. W. Kim, S. Suh and J. J. Han, "Face Liveness Detection From a Single Image via Diffusion Speed Model," in *IEEE Transactions on Image Processing*, vol. 24, no. 8, pp. 2456 2465, Aug. 2015.
5. Z. Boulkenafet, J. Komulainen and A. Hadid, "Face anti-spoofing based on color texture analysis," *2015 IEEE International Conference on Image Processing (ICIP)*, Quebec City, QC, 2015, pp. 2636–2640.

6. Z. Boulkenafet, J. Komulainen, Xiaoyi Feng and A. Hadid, "Scale space texture analysis for face anti-spoofing," *2016 International Conference on Biometrics (ICB)*, Halmstad, 2016, pp. 1–6.
7. D. C. Garcia and R. L. de Queiroz, "Face-Spoofing 2D-Detection Based on Moiré-Pattern Analysis," in *IEEE Transactions on Information Forensics and Security*, vol. 10, no. 4, pp. 778–786, April 2015.
8. Q. T. Phan, D. T. Dang-Nguyen, G. Boato and F. G. B. De Natale, "FACE spoofing detection using LDP-TOP," *2016 IEEE International Conference on Image Processing (ICIP)*, Phoenix, AZ, 2016, pp. 404–408.
9. A. Agarwal, R. Singh and M. Vatsa, "Face anti-spoofing using Haralick features," 2016 IEEE 8th International Conference on Biometrics Theory, Applications and Systems (BTAS), Niagara Falls, NY, 2016, pp. 1–6.
10. K. Patel, H. Han, A. K. Jain, and G. Ott, "Live face video vs. spoof face video: Use of moir´e patterns to detect replay video attacks," *in Proc.ICB*, 2015, pp. 1–8.
11. D. Wen, H. Han and A. K. Jain, "Face Spoof Detection With Image Distortion Analysis," in *IEEE Transactions on Information Forensics and Security*, vol. 10, no. 4, pp. 746–761, April 2015.

Cube NoC Based on Hybrid Topology: A Thermal Aware Routing

R. Suraj and P. Chitra

Abstract Rolling into modern processor technology, developers are increasing the number of transistors exponentially. NoC is a proficient on-chip communication platform for SoC architecture; partitioning a die into segments and stacking them in 3D fashion significantly reduce latency and energy consumption. A new Cube Network-on-Chip (NoC) based architecture is proposed, which takes the advantage of this exponential increase. In this model, the number of processing elements can be increased exponentially, while reducing the space complexity. In this paper, the thermal impact of the proposed cube NoC model is analyzed. Power and thermal aware hybrid routing method is employed in this model, to improve the reliability and performance. The experimental results reveal that the hybrid routing approach, offers better throughput and failsafe packet delivery, compared to other approaches in the literature.

Keywords Network-on-Chip · Thermal management · Routing protocol Interconnection network

1 Introduction

As the size of transistors have been reduced to nanometer scale, the number of transistors available in a single chip increases to billions or even greater. Multiprocessor System-on-Chip (SoC) technology is becoming an attractive platform for high-performance applications. The performance of the multiprocessor systems is determined not only by the performance of individual processors, but also based on how these processors work together to achieve more performance. As semiconductor technologies continually shrink the feature sizes, metallic interconnects gradually

R. Suraj (✉) · P. Chitra
Thiagarajar College of Engineering, Madurai, Tamil Nadu, India
e-mail: suraj@tce.edu

P. Chitra
e-mail: pccse@tce.edu

A.K. Somani et al. (eds.), *Proceedings of First International Conference on Smart System, Innovations and Computing*, Smart Innovation, Systems and Technologies 79, https://doi.org/10.1007/978-981-10-5828-8_64

become the bottleneck of NoC performance due to the limited bandwidth, high delay, usage of large area, and high power dissipation. The SoC communication architectures have to look to Network-on-Chip (NoC) to overcome the problems of poor scalability, limited bandwidth, and high power consumption in traditional interconnection architectures.

By adding more components to the NoC system, it faces dramatic thermal problem because of continuously increasing over router traffic and power density. To optimize latency, energy leakage, and the overall system reliability NoC thermal problem has to be analyzed for better reliability and thermal distribution should be maintained.

1.1 Network-on-Chip

Network-on-Chip (NoC) integrates the processing element or intellectual property using communication network to satisfy the requirement of bandwidth and delay management. NoC greatly reduces the wire routing issues such as single signal flow, thermal problem, wire delay, and floor planning cost. The NoC systems and its success are highly based on a kind of interfaces between different processing units. The interfacing components reduce the latency introduced by slow bus in response time and limitations like energy and scalability. All these issues are considered while designing the interface between the communication components of Network-on-Chip systems. In Network-on-Chip systems the processor which is hard cored receives the packet from a network and stores them in a storage medium called buffer and schedule them to forward towards any destination.

In general, the NoC architecture is deployed on a 2D space and has scalability in the various factors like routing flits, handling a number of flits, latency, power, etc. To overcome the scalability of 2D-NoC architecture, the architecture can be stacked to form 3D-Cube NoC, where there will be multi layered 2D-NoC integrated in such a way to handle the enormous number of flits to be handled and routed at each millisecond. The architecture of 3D-NoC is shown in Fig. 1.

In Network-on-Chip systems the router receives the packet from any source and stores them in a storage medium called buffer and schedule them to forward towards any destination.

The proposed model can be enhanced by exponentially increasing the number of processing elements that is the IP cores of the NoC system by stacking large number of 2D models to form a cube structure. The overall thermal impact of the proposed cube model is analyzed in this paper by implementing power aware and energy aware routing protocols within the proposed model.

The remainder of the paper is organized as follows: Sect. 2 talks about the various existing methods; Sect. 3 describes the proposed model, its design and working along with the thermal model that is used; Sect. 4 shows the experimental setup and the simulation of the proposed model along with the results and finally Sect. 5 gives the overall conclusion of the paper.

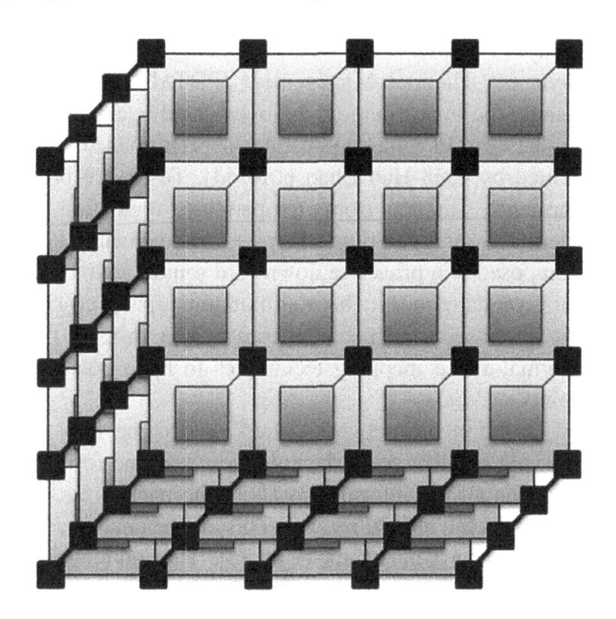

Fig. 1 3D cube NoC architecture

2 Related Work

Thermal problems also arise with the transition from 2D chip systems to 3D stacked system. 3D integrated circuits takes advantage of the dimensional scaling approach and are seen as a natural progression towards the future of large and complex systems. They increase various factors such as the device density, bandwidth, and speed of the network. But because of the increased integration, the amount of heat per unit footprint increases, that result in a much higher on-chip temperatures compared to normal level [1]. Considering thermal on 3D, floorplanning plays a crucial role.

In three dimensional Network-on-Chip (3D NOC) thermal issues is an important concern. 3D NOC typically improves the scalability and bandwidth of the network by using die and chip stacking. Due to high integration of chip density and die stacking, it increases the temperature of the network, and in parallel it degrades the reliability and overall performance of the network. So it is needed to ensure the thermal safeties of the network, for that various thermal technique have been evolved and implemented. NoC also has the ability of multiple clocking which is extensively used in many SoC. In [2], annealing algorithm for thermal and power-aware test scheduling of cores in a NoC-based SoC using multiple clock rate proves about the stability of system.

Due to die stacking, high heat sink is introduced in the packets, to reduce that a Proactive Thermal Dynamic Buffer Allocation (PTDBA) scheme was proposed by Yuan-Sheng et al. [3] that gives information about overheated regions. Based on this information the forwarding of packets is managed in such a way that the overheated regions are not affected further till the temperature drops to an optimal level. To improve the system performance further, the overheated region should be analyzed so that it can be easily managed. For this purpose en effective Proactive Dynamic Thermal Management (PDTM) scheme was proposed that can easily handle the forwarding of the overheated packets.

Increase in temperature affects performance, power, reliability, and cost presented by Chih-Hao Chao et al. [4]. To ensure both thermal safety and less performance impact from temperature regulation, we propose a traffic- and thermal-aware run-time thermal management (RTM) scheme. The scheme is composed of a proactive downward routing and a reactive vertical throttling. Based on a validated traffic-thermal mutual-coupling co-simulator, our experiments show the proposed scheme is effective. The proposed RTM can be combined with thermal-aware mapping techniques to have potential for higher run-time thermal safety.

Traffic in the network occurs due to the flow of unknown packet within the network and this could degrade the system performance. Every node in the network contains all the information about the packets to which node it has to be sent or from which it should receive. This can be analyzed by using the clock frequency as proposed by Chunsheng et al. [5]. The unknown packet flow that occurs in the network due to the congested node will increase the heat. To reduce this increase in temperature, Traffic and Thermal aware Adaptive Beltway Routing (TTABR) was proposed by Kun-Chih et al. [6]. Jieyi Long et al. [7] proposed a idea of reducing the monitoring sensor over the NoC grid without compromising the accuracy.

Capturing traffic patterns and predict future behaviors using table-driven predictor named Network Traffic Prediction Table (NTPT) proposed by Yoshi Shih-Chieh Huang et al. [8] for recording and predicting traffic in NoC, it has ability to predict end-to-end traffic data and simulation offers low area overhead and is very feasible.

Feiyang Liu [9] proposed a Dynamic Thermal-Balance Routing (DTBR) algorithm for Network-on-Chip, which can solve both thermal problems such as hotspot and thermal distribution of the network. DTBR is a minimal adaptive routing algorithm based on an architectural thermal model. Based on the simulation results, the proposed DTBR algorithm can obtain a much better and uniform thermal distribution and occurrence of hotspot nodes is reduced. The hotspot temperature is cut down about 20% in different traffic patterns.

3 Methodology

3.1 Motivation of Work

Increased chip integration density and reduced device size had led to the fast growing and evolvement in technology. It is necessary to maintain the system reliability, scalability, and performance for this purpose thermal management scheme is proposed to maintain thermal balance throughout the network.

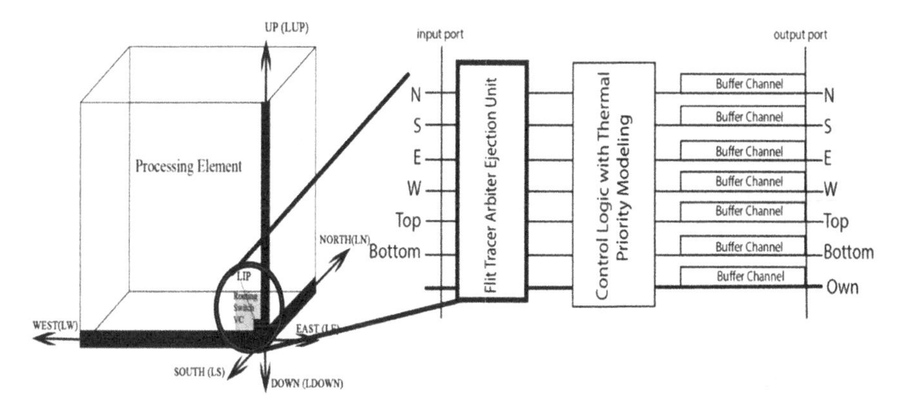

Fig. 2 Processing element with router

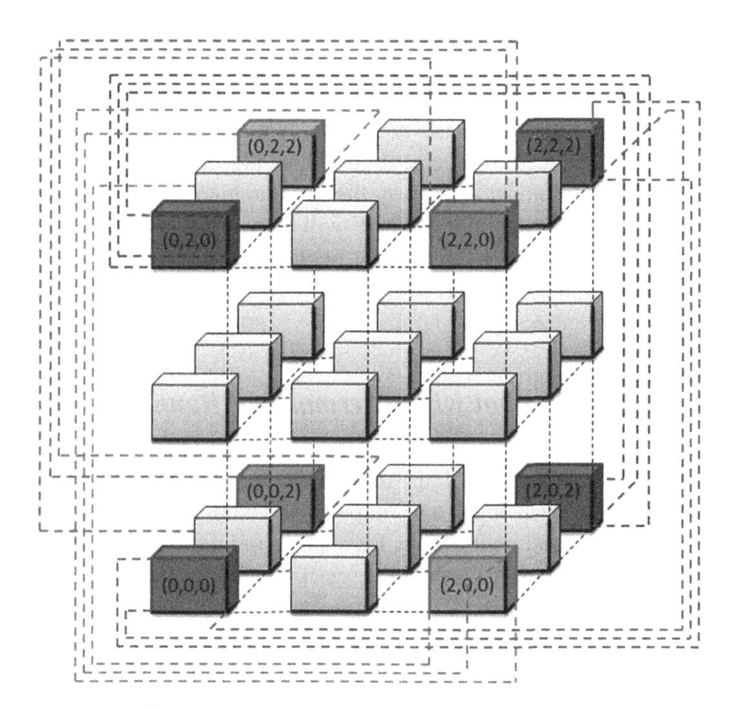

Fig. 3 Topology for n^3 model example $3 * (3 * 3)$

3.2 *Cube NoC Model*

The proposed cube-based NoC model offers a regular structure and non-congested floor planning. For enterprise processing solution, increasing the processing elements linearly will not make much difference. This approach allows the exponential increase of Processing Elements (PEs) and has an effective routing strategy and offers deadlock aware minimal hop network communication, failsafe, and thermal aware execution.

In traditional NoC architecture, the router will be a separate element and attached to a switch in PEs as shown in Fig. 1. In the proposed cube model, each PE is equipped with switching and routing logic. The processing element with the switching and routing logic is shown in Fig. 2. The router is placed within the processing element itself. In case of a 2D NoC layer, the PE can be implemented on one of the physical planes of the system, consisting of $n * n$ PEs in each layer. For the 3D system, n such layers are stacked and hence the total number of PEs will be $n * (n * n)$ which is n^3.

The router contains control logic to monitor the power and thermal impact, it also maintains an access control list which include the neighbor node's power and thermal details. As flit traffic is forwarded across the router the consumption of power is high and this further increases the thermal state of processing element that is the temperature. So to avoid the bad impact of traffic on the nodes the routing logic should be consistent over all traffic conditions.

In the cube-based NoC, every element is arranged in a 3D matrix form and every element of the processing blocks are designed and developed in the form of cubes. Figure 2 shows the processing element cube and Fig. 3 shows the proposed 3D cube NoC architecture model. This way the final architecture of the network will have cube properties. The structure of NoC will be $n * n * n$ matrix model, which grows in order of 3. Such cube-based floor planning offers average equal length of communication path and avoids congested wiring.

3.3 Processing Element with Thermal and Routing

In the proposed cube-based model, the router will be embedded with the processing element. Router has seven input and output ports with selector and arbiter. This model helps to avoid router outside the PE's and avoid unnecessary space and wire delay. The router is designed with the ejection and control logic of thermal modeling.

Figure 3 shows an example cube of $3 * (3 * 3)$ PEs and their interconnection. The interconnection topology makes this model more efficient even during bottleneck conditions during the internal node communication. Mesh is the primary topology over the chip and there are diagonal fail safe link outer of the model to reduce congestion within the center of the cube network. Figure 3 shows the diagonal connection established for the proposed cube model.

The topology switching used here has seven links port in each PEs as shown in Fig. 2, one to connect the IP and the router, and six other to connect to adjacent PEs within the layer or if the router is in edge then out of the six, three ports connect to its diagonal PE router port.

3.4 Thermal Aware Routing Logic

Packet transfer can take place either within the same layer or across the various layers of the cube model. Each and every communication is controlled by the traffic and thermal profile of the subsequent router. The proposed routing logic is defined by considering the probability of failures over a router, which avoids unwanted waits and shifting to alternate path.

The thermal resistance T_{Die} states how long the Die/PE can be able to resist the current situation, which is calculated in accordance with the heat conduction as in Eq. (1)

$$T_{Die} = R_0 + R_n\left(\frac{P}{A}\right) \tag{1}$$

R_0 is the absolute thermal resistance measured by the Fourier law of heat conduction as shown in Eq. (2), R_n is the heat flow value, P is the power dissipation in the network and A is the area in the chip.

$$R_0 = \frac{x}{A * k} \tag{2}$$

Here x denotes the length of the material and k is the thermal conductivity of the material. By considering latency and congestion of communication path, it can be understandable that the nearby router is crowded with heavy traffic and maybe thermally affected. Average throughput of the router defines how much traffic it can handle in a particular time as in Eq. (3)

$$Throughput = \frac{F}{N * C} \tag{3}$$

Here F is the total received flit in the router, N is the total number of nodes connected to the router and C is the total number of cycles of packet transmission in that router. The average packet latency TL is given as in Eq. (4)

$$TL = \frac{1}{M}\sum_{i=1}^{M} L_i \tag{4}$$

Here M is the number of links and L_i is the latency in the link i, where $i = 1, \ldots, M$. Finally the probability of failure P_{fail} is calculated as in Eq. (5):

$$P_{fail} = \frac{\sum_{j=1}^{M} TR + T_{Die} + TL}{Throughput + M} \tag{5}$$

According to the diagram in Fig. 2 there are seven links denoted by LIP, LE, LW, LN, LS, LUP, and LDOWN. The routing algorithm has to determine the communication is within the same layer or with the up/down layer. The use of each of these links is

- LIP—link between the IP and the switching port.
- LE, LW, LN, LS—act as the port for communicating within the layer.
- LUP, LDOWN—act as the port for communicating among the layers.
- LE, LW, LN, LS, LUP, and LDOWN—also used to communicate with diagonal node.

If the local IP core at the bottom layer wants to communicate to the IP core at the top layer, it need not do a multiple hop by jumping layer by layer, instead a jump to the top via diagonal link can be done and then it can be forwarded to the destination IP code there. Figure 6 gives the pseudo code for the proposed routing logic.

Cube NoC Routing Algorithm

Initialization

S: Source node
n: Order of cube
n^3: Number of cube elements
$S(s_x, s_y, s_z)$: Source node
$C(c_x, c_y, c_z)$: Current node
$D(d_x, d_y, d_z)$: Destination node
L_c: Level of current node
L_d: Level of destination node
T: Thermal profile of current node

Algorithm

1. S wants to forward packets to D, take $C = S$
2. Compute L_c and L_d
3. If $L_c == L_d$ go to step 4 else go to step 7
4. If $c_x == d_x$ go to step 5 else if $c_z == d_z$ go to step 6
5. If $c_z > d_z$ then route packet to $-z$ direction else if $c_z < d_z$ then route packet to $+z$ direction else destination reached
6. If $c_x > d_x$ then route packet to $-x$ direction else if $c_x < d_x$ then route packet to $+x$ direction else destination reached
7. Layer difference $diff = c_y \sim d_y$
8. If $(n - c_y) > (n - d_y)$ go to step 9 else go step 10
9. If $c_y + (n - d_y) > diff$ then route packet to $+y$ direction else route packet via diagonal link

10. If $d_y + (n - c_y) > diff$ then route packet to $-y$ direction else route packet via diagonal link
11. Update current node C
12. Repeat steps 2–11 till $C = D$

For packets that are routed across the different layers, the router makes a suitable routing decision based on load and latency on the port. The proposed algorithm also considers the thermal state of the nodes while forwarding the packets. If the node is hotspot then packets are forwarded to a nearby node.

4 Experimental Setup

The proposed routing logic implemented in the proposed cube NoC is evaluated and analyzed by using the flit injection simulation the SystemC simulator with Hotspot Tool. The kernel of the simulator is based on Noxim, which is an open source NoC simulator modified to accommodate the proposed router model and routing logic. The evaluation on the routing logic of the proposed method is compared with existing methods and our proposed method offered better results even during failure of routers.

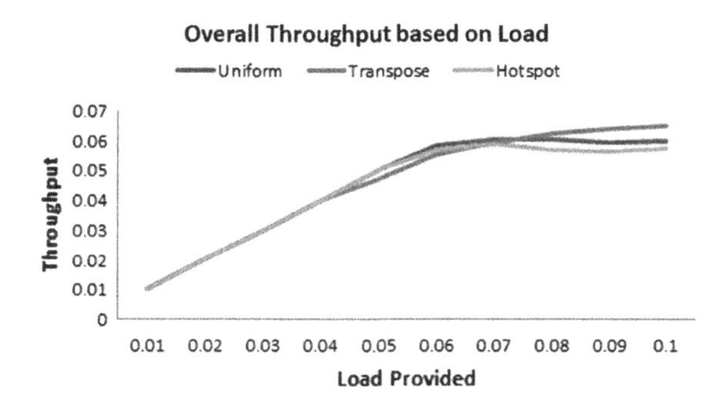

Fig. 4 Throughput of the proposed Cube NoC

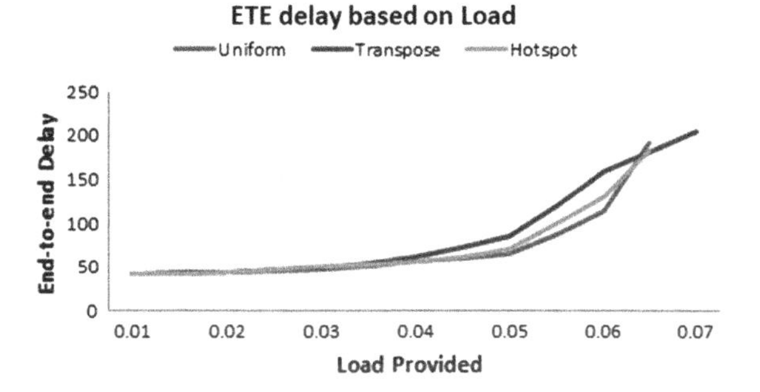

Fig. 5 End-to-end delay of the proposed Cube NoC

5 Performance Evaluation

The proposed approach has been tested with varying number of traffic, payload, and failure of routers. The method has produced efficient results in all the test cases being considered to evaluate the performance of the proposed architecture. The proposed architecture has improved the throughput and increases the flit delivery fraction than the previous designs. Also the delay incurred is comparatively less with respect to the load provided.

The throughput and end-to-end delay are measured with respect to three different traffic conditions such as the uniform traffic, transpose traffic, and hotspot traffic. Uniform traffic means that a source node can forward packets to any destination node. In transpose traffic a given source node can only transfer packets to a pre-defined destination node. Finally hotspot traffic means that some of the nodes within the network are taken as hotspot node initially with a high temperature than normal.

Figures 4 and 5 shows the obtained throughput and end-to-end delay comparison of the proposed cube NoC model with the routing logic for all three traffic patterns. It can be seen that the throughput and delay performance is better in case of uniform traffic and hotspot traffic. But the performance is low in case of transpose traffic. This is because in transpose traffic a particular source node can only send packets to a specific destination node and this will nullify the use of diagonal links in the proposed cube model. Thus the performance here is degraded compared to the other two traffic patterns. Finally the overall throughput ratio is measured based on the overall load that is generated throughout the network.

Next the thermal distribution of the proposed routing logic is analyzed by keeping the initial temperature of the nodes as 30 °C. Figure 6 shows the overall thermal distribution of the proposed Cube NoC model.

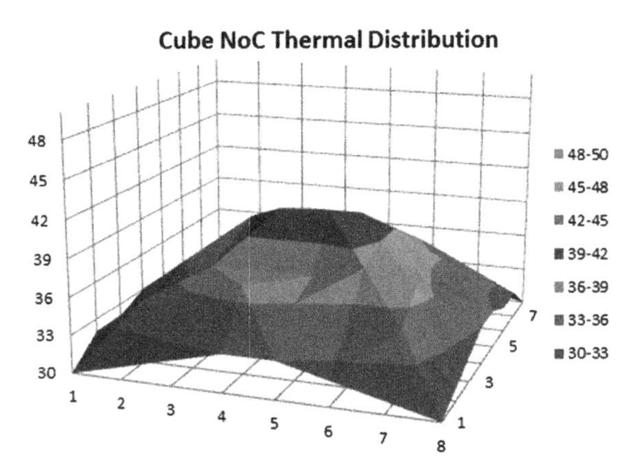

Fig. 6 Thermal distribution in cube NoC

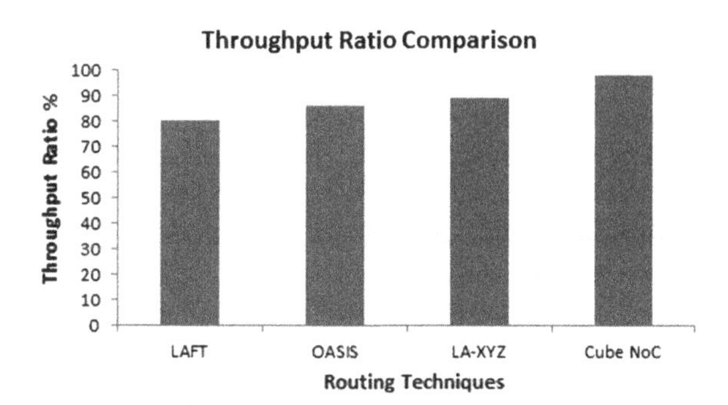

Fig. 7 Throughput ratio comparison

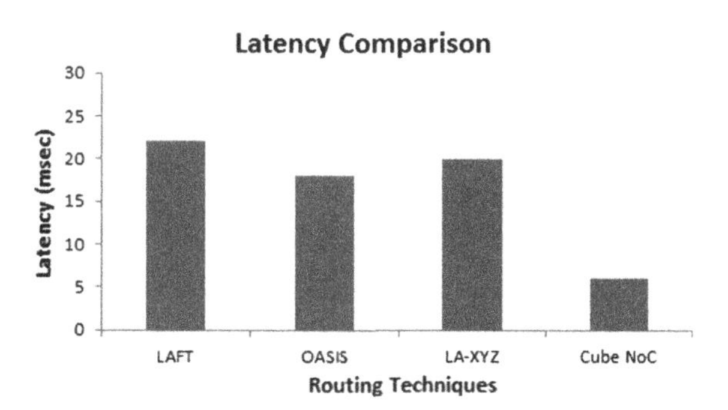

Fig. 8 Latency Comparison

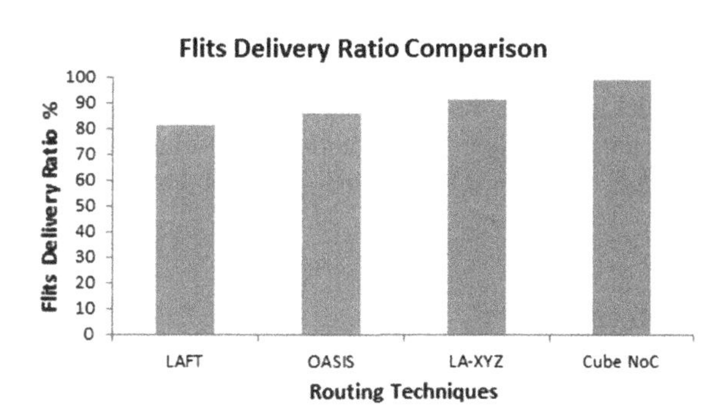

Fig. 9 Flits delivery ratio comparison

5.1 Performance Comparison

The throughput of the cube model and the end-to-end delay is then compared with the other existing protocols based on the throughput ratio and the latency as in Fig. 7 and Fig. 8 respectively. The existing algorithms such as the LAFT, OASIS, and LA-XYZ have been used for comparison.

The graph in Fig. 7 shows the comparative result of different algorithms in throughput ratio, the result shows that the proposed method has produced more throughput ratio than other existing methods.

The graph in Fig. 8 shows the comparative analysis of different methods on latency produced. It shows clearly that the proposed method has produced less latency than other existing methods.

The proposed cube NoC model produces better results in case of throughput obtained and the latency incurred within the network. This is because it makes use of the 3D architecture to obtain high performance and the diagonal links provide better transfer of packets to different layers of the cube model and reduces the overall latency. Apart from these a final comparison is also made based on the amount of packets delivered successfully throughout the network. This is measured as the flits delivery ratio as in Fig. 9.

The graph in Fig. 9 shows flit delivery ratio produced by different methods and it shows clearly that the proposed method has produced more flit delivery than other methods.

6 Conclusion

In this paper, novel cube architecture based Network-on-Chip system is proposed that contains large number of processing elements. The number of processing elements here can be increased exponentially by stacking many 2D NoC layers over one another. This way the processing power of the NoC system improves. With this model a routing logic is also proposed that can maintain the overall throughput and thermal distribution of the network. The thermal impact, throughput, and latency of the proposed method are analyzed. From the results, it is shown that the Cube NoC model improves the performance, reliability, and scalability of the network by large amounts. In future, this method can be further enhanced by providing more efficient routing algorithms for thermal management and hotspot reductions.

References

1. Hamedani. P.K, Hessabi. S, Sarbazi-Azad. H, Jerger. N.E., "Exploration of Temperature Constraints for Thermal Aware Mapping of 3D Networks on Chip", ACM International Journal of Adaptive, Resilient and Autonomic Systems, Volume 4, Issue 3, pp. 42–60, July 2013.
2. Salamy. H. Harmanani. H, "An effective solution to thermal-aware test scheduling on network-on-chip using multiple clock rates", IEEE 55th International Midwest Symposium on Circuits and Systems, pp. 530–533, August 2012.
3. Yuan-Sheng Lee, Hsien-Kai Hsin, Kun-Chih Chen, En-Jui Chang, A.Y.A. Wu, "Thermal-aware Dynamic Buffer Allocation for Proactive routing algorithm on 3D Network-on-Chip systems" IEEE International Symposium on VLSI Design, Automation and Test, pp. 1–4, April 2014.
4. Chih-Hao Chao, Kai-Yuan Jheng, Hao-Yu Wang, Jia-Cheng Wu, An-Yeu Wu, "Traffic-and Thermal-Aware Run-Time Thermal Management Scheme for 3D NoC Systems", 4th ACM/IEEE International Symposium on Network-on-Chip, pp. 223–230, 2010.

5. Chunsheng Liu, V. Iyengar, "Test scheduling with thermal optimization for network-on-chip systems using variable-rate on-chip clocking", IEEE International Conference on Design, Automation and Test in Europe, Volume 1, March 2006.
6. Kun-Chih Chen, Che-Chuan Kuo, Hui-Shun Hung, A.Y.A. Wu, "Traffic-and Thermal-aware Adaptive Beltway Routing for three dimensional Network-on-Chip systems", IEEE International Symposium on Circuits and Systems, pp. 1660–1663, May 2013.
7 Jieyi Long, Seda Ogrenci Memik, Gokhan Memik, Rajarshi Mukherjee, "Thermal Monitoring Mechanisms for Chip multiprocessors", ACM Transactions on Architecture and Code Optimization, Volume 5, Issue 2, Article No. 9, August 2009.
8 Yoshi Shih-Chieh Huang, Kaven-Chun-Kai Chou, Chung-Ta King, Shau-Yin Tseng, "NTPT: On the End-to-End Traffic Prediction in the On-Chip", 47th ACM/IEEE Design Automation Conference, pp. 449–452, June 2010.
9 Shu-Yen Lin, Tzu-Chu Yin, Hao-Yu Wang, An-Yeu Wu, "DTBR: A dynamic thermal-balance routing algorithm for Network-on-Chip", IEEE Computers & Electrical Engineering, Volume 38, Issue 2, pp. 270–281, March 2012.

Real-Time VANET Applications Using Fog Computing

Jyoti Grover, Ashish Jain, Sunita Singhal and Anju Yadav

Abstract The main objective of vehicular ad hoc networks (VANETs) is to improve driver safety and traffic efficiency. Most of VANET applications are based on periodic exchange of safety messages between nearby vehicles and between vehicles and nearby road side communication units (e.g., traffic lights, roadside lights etc.). This periodic communication generates huge amount of data that have typical storage, computation, and communication resources needs. In recent years, there have been huge developments in automotive industry, computing, and communication technologies. This has led Vehicular Cloud Computing (VCC) as a solution to satisfy the requirements of VANETs such as computing, storage, and networking resources. Fog computing is a standard that comprehends cloud computing and related services to the proximity of a network. Since VANET applications have special mobility, low latency, and location awareness requirement. Fog computing plays a significant role in VANET applications and services. In this paper, we present real-time applications of VANET that can be implemented using fog computing.

Keywords Fog computing · VANET · Cloud computing

J. Grover (✉) · A. Jain · S. Singhal · A. Yadav
School of Computing and Information Technology, Manipal University Jaipur, Jaipur, India
e-mail: jyoti.grover@jaipur.manipal.edu

A. Jain
e-mail: ashish.jain@jaipur.manipal.edu

S. Singhal
e-mail: sunita.singhal@jaipur.manipal.edu

A. Yadav
e-mail: anju.yadav@jaipur.manipal.edu

A.K. Somani et al. (eds.), *Proceedings of First International Conference on Smart System, Innovations and Computing*, Smart Innovation, Systems and Technologies 79, https://doi.org/10.1007/978-981-10-5828-8_65

1 Introduction

Due to large-scale improvement in computation and wireless communication technologies, vehicles are being equipped with communication devices in order to communicate with other vehicles and take appropriate actions based upon received messages. Such communication creates new opportunities for enhancing road safety. VANET has emerged as a solution to many road safety problems by providing safety information to drivers on time [1]. VANET is a special class of mobile ad hoc network (MANET) [2, 3]. Main characteristics of VANET that distinguish it from MANET are high mobility of nodes, scalability, and frequent topology changes. VANETs use short range radios [4] in each vehicle, which allows various vehicles to communicate with other vehicles and roadside infrastructure. Safety and traffic management applications require real-time information and can play significant role in life and death decisions.

In VANET, vehicle to vehicle (V2V) and vehicle to roadside (V2R) communication is used to propagate safety information to nearby vehicles on time. Apart from the communication unit, each vehicle is also equipped with high computation and storage unit. But most of the time, these units are underutilized. Vehicular cloud computing (VCC) is the paradigm [5] emerged to utilize VANET resources efficiently by taking advantages of cloud computing. It serves the drivers of VANET with a pay as you go model. In VCC, group of vehicles cooperates with each other to dynamically share computing, sensing, and communication resources for decision-making on the road in order to improve traffic management and road safety. Some examples of VCC applications are:

1. Local traffic condition can be collected from nearby vehicles for route planning.
2. Current transportation system can be improved by big data processing of traffic information by local traffic authorities.
3. Collaborative image of critical events can be reconstructed such as car accident, congestion on the road, etc.

VCC is a very promising solution to share the computation and storage resources among the vehicles and roadside units in order to implement these applications. But VCC is not sufficient for many VANET applications due to mobility of vehicles and the latency sensitive requirements imposed by these. It is difficult to meet the Quality of Service (QoS) requirements using VCC. So, a new approach is designed that comprehends cloud computing with VANET applications named as fog computing. Fog computing leverage computation infrastructure that is closer to the network edge to compliment cloud computing in providing latency sensitive applications and services. Fog computing is similar to cloud computing but the only difference is that time sensitive applications can be implemented at the network edge rather than sending huge amount of data to remote cloud. The idea of fog computing [6, 7] is to place cloud-like facility at the proximity of application users. It provides storage, computation, and application services to the edge of network (nodes within proximity), thereby reduces burden on the cloud and facilitates real-time VANET applications implementation. In VANET, any node having communication, computation, and storage resources can become fog nodes. Fog computing is very useful to implement VANET applications as these are time sensitive and real time which is the prime agenda of Intelligent Transportation system (ITS) [4].

This paper discusses various applications of VANET which can be implemented using fog computing. The rest of the paper is organized as follows. Section 2 provides overview of cloud computing and its emergence with VANET. It also discusses the concept of VCC. Fog computing and its related work in relation with VANET is discussed in Sect. 3. Some of the VANET real-time applications which can be implemented using fog computing are discussed in Sect. 4. Finally, concluding remarks with future work are covered in Sect. 5.

2 Vehicular Cloud Computing

Cloud computing is an Internet-based computing technique that provides shared computer processing resources and data to computers and other devices on demand. It enables the users with various capabilities to store and process their data in privately owned or third party data centers that may be located far from user (in different city or country). Cloud computing provides high computation power, cheap cost of service, high performance, scalability, accessibility, availability, etc.

Inspired by the success of cloud computing integrated with mobile communications, authors [5, 8, 9] have introduced the concept of vehicular cloud computing (VCC). Figure 1 describes vehicular cloud environment in VANET.

Huge advancements in automotive industry, computation, and communication technologies have led the dream of smart cars true. Ford has announced the manufacturing of four million vehicles with SYNC systems (having integrated in-vehicle communication system, GPS and Radar, etc.). Recently, Google has also tested a self-driving car which is equipped with computation and communication device, radar sensors, GPS and video camera, etc.

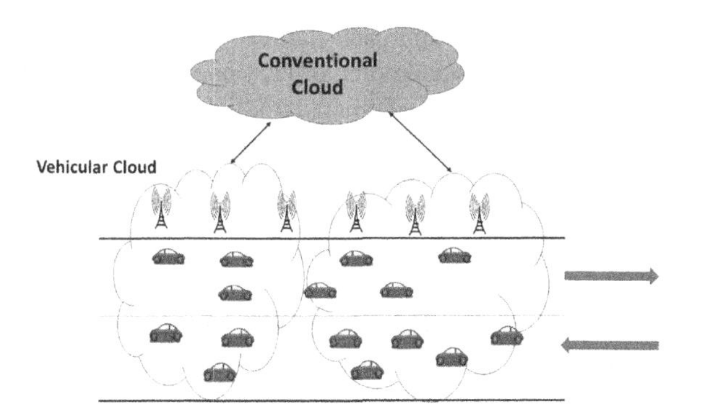

Fig. 1 Vehicular Cloud environment for Vehicular Ad Hoc Network

Nowadays, vehicles are just more than transportation machines because these have high computation and storage capability. People can get different types of services (such as navigation, weather, and entertainment, etc.) when they are driving. Most of the time (especially when vehicles are parked in offices, shopping malls etc.), computing and storage resources of these vehicles are untapped. These vehicles could be the potential resources for vehicular cloud (VC). The main issue in implementing VC is due to the dynamic behavior of vehicles. VC services can be switched to traditional cloud placed nearby in the case of vehicles move in/out a specific area.

3 Fog Computing

Fog computing is defined as a platform where large number of heterogeneous, ubiquitous, and distributed devices communicate and cooperate by forming a network in order to perform storage and processing functions without the intervention of third parties. These functions can be providing support to new applications that can run in sandboxed environment. Most of VANET applications are area specific and needs to be handled locally such as navigation, infotainment and safety information, etc. In these cases, fog computing just fits in because it allows the contents and application services as close as possible to drivers on the road in particular road segment. Fog computing addresses the limitation of location awareness of cloud computing. For example, a person visiting a new city would like to seek information on the places of interest, news, and weather conditions of this city rather than interested in other city information. Fog computing is well suited for these types of applications. However, Firdhous et al. [12] mentioned that fog computing paradigm needs extensive research before its practical deployment.

Cloud computing is the central portal of information and does not have location awareness. Fog computing overcomes this issue by providing localized services to specific deployment sites. Bonomi et al. [13] have presented the basic architecture of storage, computing, and networking for cloud and fog computing. Fog computing is a most suitable approach that can control critical resources like energy, traffic, health care, etc. Vaquero et al. [15] presented a border and integrative view of fog computing. Concept of smart building and software-defined networking (SDN) is also presented using fog computing in [6]. Different authors presented their different perception of fog computing, Yi et al. [14] strongly differentiated the fog computing from related technologies. Sarkar et al. [17] have shown that for a scenario where 25% of applications demand real-time low latency services, means energy expenditure in fog computing is 40% less than conventional cloud computing model. This analysis motivates the research and academia to explore fog computing more extensively.

As can be seen from Fig. 2, fog computing is overlaid onto cloud computing in VANET scenario. It creates a distributed computing platform for VANET components to tackle with data processing and storage services. The role of fog computing

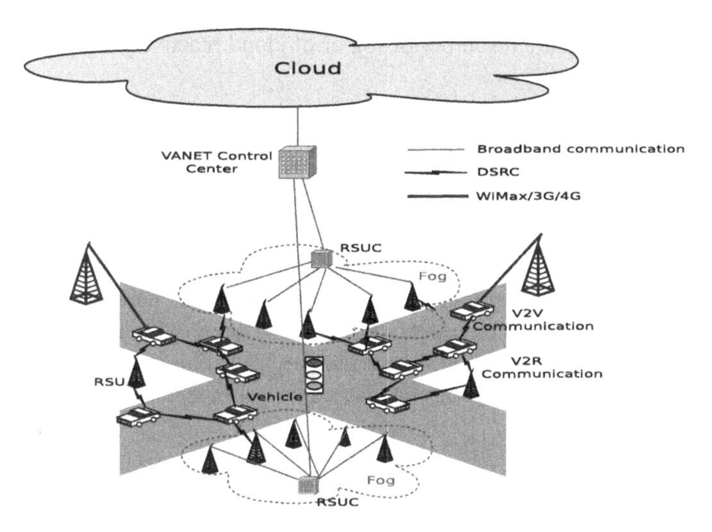

Fig. 2 Architecture of Fog Computing for VANET

is like near-end computing proxies between the front-end devices and the far-end servers. There are three types of connections made by fog server in VANETs:

- **RSU to Fog Server**: In VANET, V2V and V2R communication is used to propagate safety/non-safety information. RSUs are also able to communicate with each other. Hence RSUs act as a backbone of fog. There is wired/wireless connection between RSU and fog server. In Fig. 2, **RSUC** is the fog server.
- **Fog Server to Fog Server**: Fog servers at different locations manage a pool of resources for localized area. There can be direct wired/wireless communication between peer fog servers, or these can be connected via vehicular control center. It allows the collaborative service provision and content delivery among peered fog servers, thereby improving the performance of entire system.
- **Fog Server to Cloud**: The cloud is the central controller of fog servers deployed at different locations. Each fog server provides services to users at specific locations. Cloud aggregates the information received from fog servers and performs centralized computation and fog servers convey the information received from cloud to application users.

Luan et al. [10] summarized the differences between fog and cloud computing. Some of the main features of fog computing compared to cloud computing are:

1. It simplifies the development model for huge number of decentralized heterogeneous devices.
2. Fog computing enhances service quality with increased data rate and reduced service latency and response time.
3. It avoids the back-and-forth traffic between cloud and users. This saves the network bandwidth and reduces the energy consumption.

4. Fog users can use the resources of fog or of cloud reactively based on their demands.

4 Fog Computing for VANET Applications

Applications of fog computing in vehicular scenario are not as much as for VCC. Giang et al. [16] argued that traditional cloud computing is not sufficient for many VANET applications due to mobility of vehicles and their real-time applications. Fog computing leverage computation infrastructure that is closer to the network edge, thereby supports applications with low latency requirements. Some of the applications are listed below:

4.1 Smart Traffic Lights

A Dynamic Traffic Light System (DTLS) is an important application of VANET that optimizes traffic at traffic light junctions efficiently by using the concept of fog computing. As this application is distributed (location proximity specific and time sensitive), centralized solution such as VCC cannot be used. Different fogs can be used at traffic light junctions to compute the duration of each signal based on the traffic around the junction. The current system of traffic lights allots a dedicated time towards each direction, thereby making the whole system static. The limitation of this system is that traffic from one side has to wait for the green light even if there is no vehicle crossing the road in the current direction of green light. A dynamic

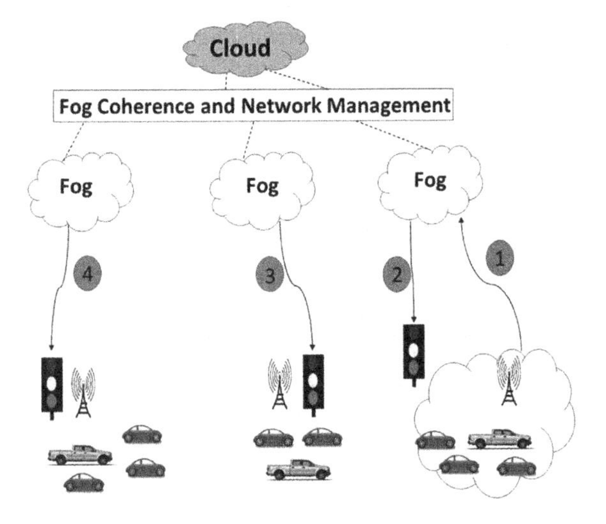

Fig. 3 Smart Traffic Light System using Fog Computing

DTLS allots the time of each signal based on the traffic present around the junction. A sample smart traffic light system is shown in Fig. 3.

Traffic lights can serve as fog devices to send warning messages to approaching vehicles in particular direction. The interaction between vehicle and RSUs is based on Dedicated Short Range Communication (DSRC). RSUs can communicate with each other to form a fog and send the aggregated perceived safety information to traffic signal. Traffic lights can be switched from one mode to another as number or type of vehicles approach towards them.

DTLS is one of the main applications of fog computing. Communication range of each RSU is generally 500–1000 m and it is a usual norm that each vehicle sends beacon packets periodically (Please refer [18]). Hence, RSUs are able to estimate the traffic on particular road segment. Based upon the computed information, RSUs can propagate the information to fog servers and thereby duration of traffic light can be decided accordingly. Now, we discuss different scenarios of DTLS.

1. If there is heavy traffic on one side of road segment and other side of the road is free, traffic signal can be turned green for the heavy traffic side and red for the free road. This can be implemented using fog servers collecting information from RSUs and timely making decision by performing computations.
2. In a scenario, where emergency vehicle such as an ambulance is approaching towards road junction. Nearby RSUs can identify ambulance by information sent by it. Video cameras deployed in RSUs are also able to recognize flashing lights. RSU observing the event spread this information in the fog. Neighboring smart lights serving as fog can coordinate and create green light signal in addition to sending warning signals to approaching vehicles.
3. Roadside lights can also detect the presence of vehicles and measures the speed and distance of nearby vehicles. Sensors deployed on fog servers can facilitate turning on these lights on identification of vehicles and vice versa. Smart lights along the road side serve as fog devices can collaboratively make efficient use of resources.

4.2 Parking System

Finding an empty parking space in rush hour is really a big problem in urban areas. Kim et al. [11] have presented their work on solving parking problems to relieve the traffic congestion, reduce air pollution, and enhance driving experience. They have shown how fog computing and roadside clouds can be used to find a vacant spot. They assume that unused public and private parking slot (e.g., in hotels, parks, universities etc.) can be shared for the vehicles. Matching theory is used to solve the parking problem by bringing profit to the owners of these places.

4.3 Content Distribution

Fog computing can be used for distribution of any type of information (safety or non-safety). For example, if a particular section of a road is blocked due to some accident or natural hazard. This information can be conveyed by the fog servers to vehicles approaching towards this site.

4.4 Decision Support System

Safety of drivers can be improved by using intelligent decision support system (DSS) using fog computing. Different fog servers can exchange their localized conceived information with each other and form DSS in order to monitor traffic violation and safety system for drivers.

VCC has vast number of applications as compared to fog computing. But fog computing fulfills the service requests on localized information, thereby reducing the service latency and response time.

5 Conclusion

In this paper, we have discussed how fog computing can be used in implementing real-time VANET applications. We have also presented the architecture of VCC and fog computing in VANET environment. Fog computing has the potential to handle the data generated by VANET components and take appropriate actions in order to implement desired VANET applications. The features of fog computing like mobility, proximity to end users, low latency, heterogeneity, location awareness make it best suited for implementation of some applications of VANET. We also have discussed different real-time applications of VANET that can be implemented using fog computing. In our future work, we would like to design the layout and implementation of these applications using fog computing in VANET.

References

1. Bai, F., Krishnan, H., Sadekar, V., Holl, G. and Elbatt, T. (2006) Towards characterizing and classifying communication-based automotive applications from a wire- less networking perspective. *IEEE Workshop on Automotive Networking and Applications (AutoNet)*, pp. 1–16.
2. S. Yousefi, M. Fathy (2006) Vehicular Ad-hoc NETworks (VANETs): Challenges and Perspectives. *6th International Conference on ITS Telecommunications, IEEE*, pp. 761–766.
3. S. Biswas, R. Tatchikou and F. Dion (2006) Vehicle-to-Vehicle wireless communication protocols for enhancing highway traffic safety. *IEEE Communication Magazine*, **44(1)**, pp. 74–82.

4. D. Jiang, V. Taliwal, A. Meier, W. Holfelder and R. Herrtwich. (2006) Design of 5.9 GHz DSRC-based vehicular safety communication *IEEE Wireless Communications*, 13(5), pp. 36–43.
5. M. Whaiduzzaman, M. Sookhak, M. Gani. (2014) A survey on Vehicular cloud computing *Journal of Network and Computer applications*, 40(1), pp. 325–344.
6. I. Stojmenovic, S. Wen. (2014) The Fog Computing Paradigm: Scenario and security Issues *Proceedings of the 2014 Federated Conference on Computer Science and Information Systems (FedCSIS'14)*, Warsaw, Poland, pp 8–16.
7. K. Kai, W. Cong, L. Tao (2016) Fog computing for Vehicular Ad hoc Networks: Paradigms, Scenarios and Issues *The Journal of China Universities of Posts and Telecommunications*, 23(2), Elsevier, pp. 56–96
8. M. Eltoweissy, S. Olariu, M. Younis. (2010) Towards autonomous Vehicular clouds *Ad Hoc Networks*, Berlin, Germany, springer, pp 1–16.
9. S. Olariu, T. Hristov, G. Yan (2013) The next paradigm shift: from vehicular networks to vehicular clouds *Basagni S, Conti M., Giordano S. et al. mobile Ad hoc Networking: cutting edge directions, 2nd edition, John wiley sons*
10. T. H. Luan, L. X. Gao, Y. Xiang (2015) Fog computing: focusing on mobile users at the edge arXiv:1502.0181v1
11. O. T. T. Kim, N. Dang. Tri, V. D. Nguyen, N. H. Tran, C. S.Hong (2015) A shared Parking Model in Vehicular Network Using Fog and Cloud environment *17th Asia Pacific Network Operations and Management Symposium (APNOMS)*, Busan, pp. 321–326.
12. M. Firdhous, O. Ghazali, S. Hassan (2014) Fog computing: will it be the future of cloud computing? *Proceedings of the 3rd International Conference on Informatics and Applications (ICIA'14)*, Kuala Terengganu, Malaysia, pp. 8–15.
13. F. Bonomi (2011) Connected Vehicles, Internet of things and fog computing *Proceedings of 8th ACM international Workshop on Vehicular Inter-Networking (VANET'11)*, Las Vegas, USA.
14. S. H. Yi, C. Li, Q. Li (2015) A Survey of fog computing, concepts, applications and Issues *Proceedings of the 2015 workshop on mobile big data (MobiData'15)*, Hangzhou, china, pp 37–42.
15. L. M. Vauero, L. Rodero Merino (2014) Finding your Way in the Fog: towards a Comprehensive Definition of Fog computing *ACM SIGCOMM computer communication Review*, 44(5), pp 27–32.
16. N. K. Giang, c. M. Victor, L. Rodger (2016) On Developing Smart Transportation Applications in Fog Computing Paradigm *Proceedings of the 6th ACM Symposium on development and Analysis of Intellegent Vehicular Networks and Applications (DIVANet'16)*, New York, USA, pp. 91–98.
17. S. Sarkar, s. Misra (2016) Theoretical modelling of fog computing: a Green Computing to support IoT Applications *IET Networks* 5(2), pp. 23–29.
18. D. Jiang, L. Delgrossi (2008) IEEE 802.11p: Towards an International Standard for Wireless Access in Vehicular Environments *Vehicular Technology Conference, VTC Spring 2008*, IEEE, pp 2036–2040.

Simplifying Spaghetti Processes to Find the Frequent Execution Paths

M.V. Manoj Kumar, Likewin Thomas and B. Annappa

Abstract Control-flow discovery algorithms of Process Mining are capable of generating excellent process models until the process is structured (less number of activities and paths connecting between them). Otherwise, process model with Spaghetti structure will be generated. These models are unstructured, incomprehensible and cannot be used for operational support. This paper proposes the techniques for (1) converting Spaghetti (unstructured) process to Lasagna (structured) process, and (2) Identifying the frequent execution paths in the process under consideration.

Keywords Process mining · Control-flow · Spaghetti · Lasagna · Operational support

1 Introduction

Nowadays information systems supporting operational process are in mainstream. A range of information systems are available for supporting a various kind or real-world operational processes. For example, work flow management systems (Staffware) [10], case handling systems (FLOWer) [2], product data management systems (Windchill) [6], customer relationship management systems (Microsoft Dynamics CRM) [9], hospital information systems (Chipsoft) [8], etc. These systems are capable of recording the process execution information in the form of event log. A

M.V. Manoj Kumar (✉) · L. Thomas · B. Annappa
Department of Computer Science and Engineering, National Institute of Technology Karnataka, Surathkal, India
e-mail: manojmv24@gmail.com
URL: http://www.cse.nitk.ac.in

L. Thomas
e-mail: likewinthomas@gmail.com
URL: http://www.cse.nitk.ac.in

B. Annappa
e-mail: annappa@ieee.org
URL: http://www.cse.nitk.ac.in

A.K. Somani et al. (eds.), *Proceedings of First International Conference on Smart System, Innovations and Computing*, Smart Innovation, Systems and Technologies 79, https://doi.org/10.1007/978-981-10-5828-8_66

young data analysis horizon named Process Mining [1] offers a range of techniques for analysing these event logs.

Lion's share of work in Process Mining is related to discovering the control flow model of the process. It depicts the casual relationships between the different steps(activities) in the process. Depending on the structure and complexity of discovered control flow model it can be classified as,

- ***Lasagna:*** relatively structured, readable and easily understandable. Process model shown in Fig. 5 is an example for Lasagna process.[1]
- ***Spaghetti:*** unstructured, cannot be readable and incomprehensible. Process model shown in Fig. 2 is an example for Spaghetti process. Generally, Spaghetti processes consist of huge number of activities and transitions.

Spaghetti and Lasagna processes are exact opposite of each other. Due to the structural complexity of Spaghetti processes, only minuscule of Process Mining techniques can be practised. However, Spaghetti processes are highly interesting from the perspective of Process Mining, as they always allow for continuous improvements. Knowing the frequently carried out paths of execution is always a useful and it helps in path optimization, performance improvements and resource management.

The goal of this paper is to provide the tools and techniques for analysing Spaghetti processes. This paper addresses following two major problems:

- Reducing the structural complexity of Spaghetti process to make it more comprehensible (Reducing Spaghetti structure to Lasagna structure).
- Identifying the most frequently carried out paths of execution in process under consideration.

Upcoming sections of this paper are organised as follows. Complete framework followed in this paper is given in Sect. 3. Section 4 describes a method for converting Spaghetti process to Lasagna process. The method for finding frequent execution paths is described in Sect. 5. Results are illustrated in Sect. 6.

2 Related Work

Attempts have been made for simplifying the structure of Spaghetti process by abstracting loops [3, 4] and clustering [7] the least significant activities. These methods cannot be applied for reducing the structural complexity of process beyond some threshold value. Concept presented in this paper applies fuzzy mining approach [5] for solving this problem and also proposes the effective way of finding frequent execution paths.

[1] All control flow process models appear in this paper are created using process discovery tool Disco (https://fluxicon.com/).

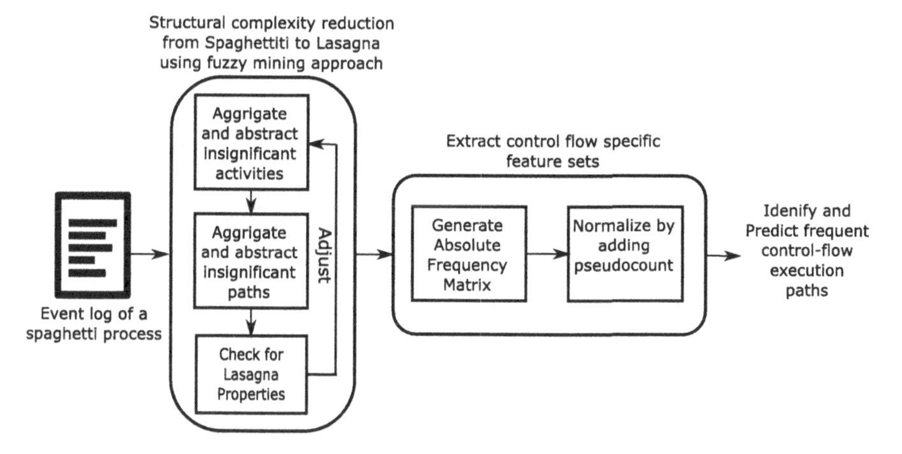

Fig. 1 Framework used for simplifying Spaghetti process and finding frequent paths

3 Framework

Framework for simplifying Spaghetti processes to find the frequent execution paths is given in Fig. 1. Initially a event log of Spaghetti process is read and simplified by applying fuzzy mining technique [5] (fuzzy miner abstracts insignificant paths, aggregates least occurred activities and customises by showing only necessary details). Different variants of control flow will be created with different percentage of activity and path percentage. This step is repeated until obtaining concrete Lasagna structure with fitness[2] value at least equal to 0.8 (i.e. at least 80% of behaviours should match between model and the event log).

The resulting Lasagna model is used for extracting control flow related features. These features serve as the basis for identifying the frequent execution paths in the process at hand.

4 Reducing the Complexity

For illustrating the concepts discussed in this paper, we use a running example related to *road traffic fine collection process*. The control-flow process model of the same is shown in the Fig. 2. It is consists of 27 activities and 45,000 events grouped into 1,000 cases. Traffic fine management event log is taken from the standard Process Mining repository.[3]

[2]Discovered model should allow for the behaviour seen in the event log. A model having a good fitness is able to replay most of the traces in the log.

[3]http://data.4tu.nl/repository/uuid:270fd440-1057-4fb9-89a9-b699b47990f5.

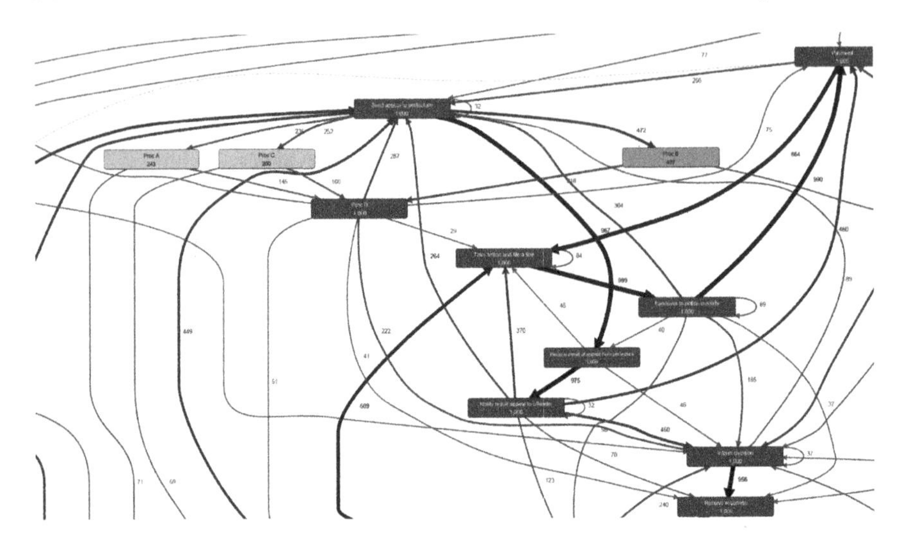

Fig. 2 A Spaghetti process: Excerpt showing 25% of *road traffic fine management process*

4.1 *From Spaghetti to Lasagna*

For reducing the process structure from Spaghetti to Lasagna, this paper follows the premise stated by founder of Process Mining (Wil Van Der Aalst) in [1]. According to [1], "*a process is a Lasagna process if with limited efforts it is possible to create an agreed-upon process model that has a fitness of at least* 0.8, *i.e. more than* 80% *of the events happen as planned*".

With the help of fuzzy miner [5], road traffic fine management process is reconstructed at various percentages of path and activity abstraction and aggregation. This step is repeated until the condition in the premise is satisfied.

For example, Fig. 4 shows three of the many possible control flow structures of road traffic fine management process with different abstractions of activities and paths. It is evident that reducing the structural complexity in terms of activity and path increases fitness value. Here the goal is to find the control flow with the fitness equal to or greater than 0.8. It is achieved by repeatedly calibrating activity and path constituent percentage in the process. Balloon graph depicting the fitness value for various combinations of path and activity percentages is shown in Fig. 3. Resulting process model with fitness of 0.8 is shown in Fig. 5. This fitness level is achieved at 40% of activity and 25% of paths. The simplified Lasagna process shown in Fig. 5 serves as the basis for identifying frequent execution paths.

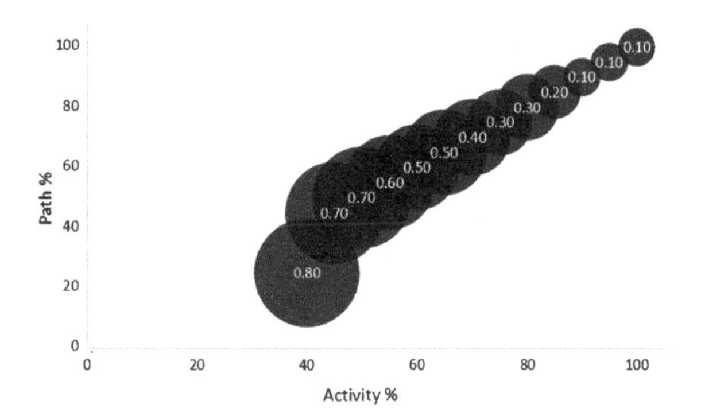

Fig. 3 Fitness increases with decrease in the structural complexity (i.e. with less activity and path percentage)

5 Identifying the Frequent Execution Paths

The Lasagna process shown in Fig. 5 is used to extract the control-flow features.

Feature matrices such as absolute frequency and pseudocounts are extracted and used for determining the frequent execution paths.

We assign single alphabet names for referring each of the 11 different activities in the simplified Lasagana process shown in Fig. 5. The activity name and corresponding short names are—file fine (a), create fine (b), send fine (c), insert fine notification (d), insert data appeal to perfecture (e), add penalty (f), send for credit collection (g), send appeal to perfecture (h), receive the result of appeal from perfecture (i), Notify the result of appeal to offender (j), and payment (k).

5.1 Absolute Frequency Matrix

Absolute Frequency Matrix (AFM) shows the number of times one activity is directly followed by another activity (absolute frequency is inscribed on arcs shown in Fig. 5) of appeal from perfecture (i), Notify the result of appeal to offender (j), and payment (k).

5.2 Absolute Frequency Matrix

Absolute Frequency Matrix (AFM) shows the number of times one activity is directly followed by another activity (absolute frequency is inscribed on arcs shown in Fig. 5). AFM of traffic management process is given in Table 1. For instance, $|a >_{\mathcal{L}} b| = 502$, i.e. in the entire log activity named b has been carried out 502 times after the end of activity a.

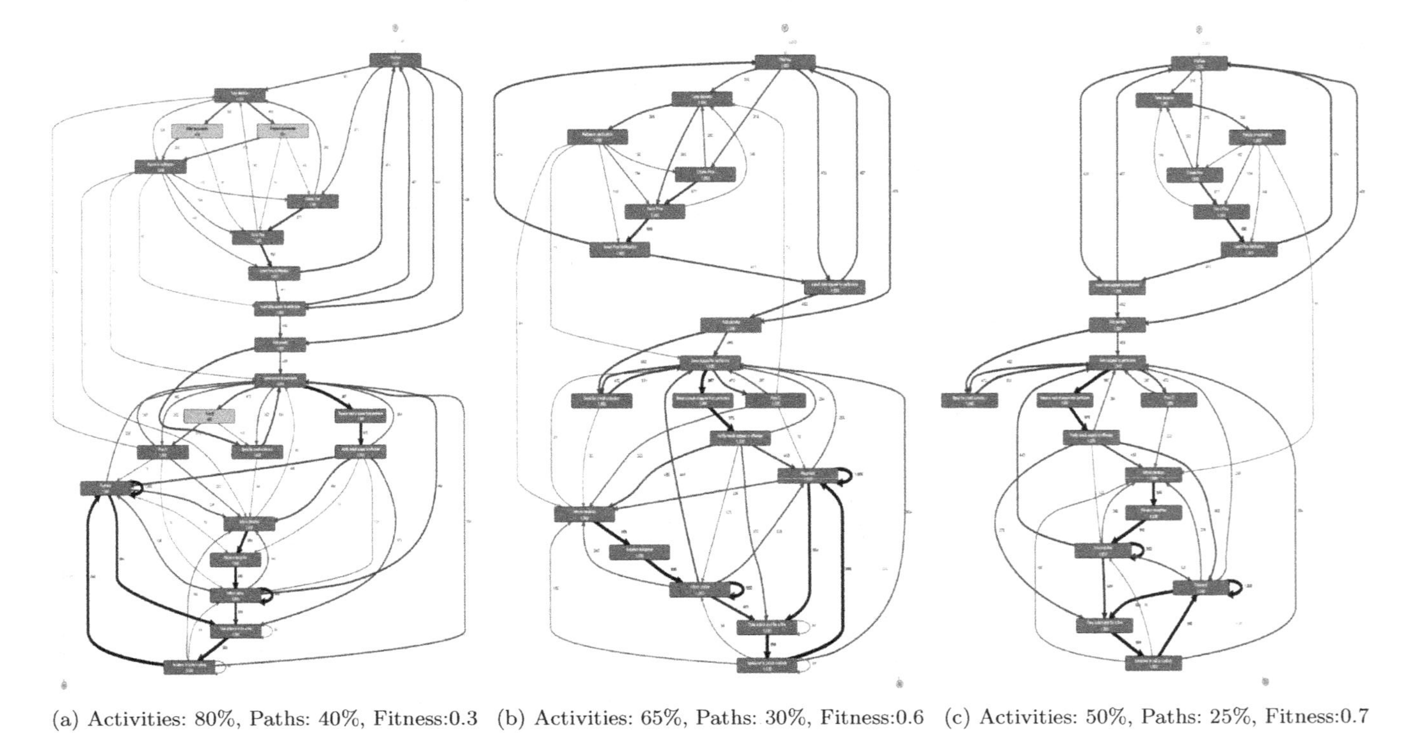

(a) Activities: 80%, Paths: 40%, Fitness:0.3 (b) Activities: 65%, Paths: 30%, Fitness:0.6 (c) Activities: 50%, Paths: 25%, Fitness:0.7

Fig. 4 Control flow models of road traffic fine management process with different levels of aggregation and abstraction

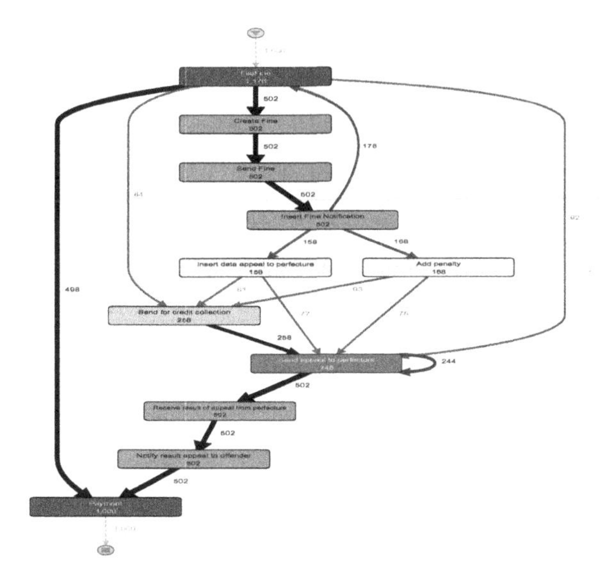

Fig. 5 Lasagna process: Simplified version of road traffic fine management process with fitness of 0.8 (at activity: 40% and path 25%)

Table 1 Frequency of the directly follows relation in event log $\mathcal{L}$: $|x >_{\mathcal{L}} y|$ is the number of times x is directly followed by y in $\mathcal{L}$

$\|>_{\mathcal{L}}\|$	**a**	**b**	**c**	**d**	**e**	**f**	**g**	**h**	**i**	**j**	**k**
a	0	502	0	0	0	0	84	92	0	0	498
b	0	0	502	0	0	0	0	0	0	0	0
c	0	0	0	502	0	0	0	0	0	0	0
d	176	0	0	0	158	168	0	0	0	0	0
e	0	0	0	0	0	0	81	77	0	0	0
f	0	0	0	0	0	0	93	75	0	0	0
g	0	0	0	0	0	0	0	258	0	0	0
h	0	0	0	0	0	0	0	244	502	0	0
i	0	0	0	0	0	0	0	0	0	502	0
j	0	0	0	0	0	0	0	0	0	0	502
k	0	0	0	0	0	0	0	0	0	0	0

AFM is updated with pseudocount for removing the zero values. Each cell with a zero value in the absolute frequency matrix is updated according to Eq. 1. Where, N is the number of activities in Lasagna process, $AFM[i, j]$ is the particular entry in the AFM, $Background(i)$ is the fraction between number of times activity i is executed to the total number of times all activities are executed. Updated AFM values with pseudocounts is shown in Table 2. Values in this table will be used for calculating the scores for individual traces. For example, value in the cell $PC[1, 1]$ is calculated as $log_2(\sqrt{11} - \frac{502}{5016})$ and the value in $PC[1, 2] = \log_2(502)$.

Table 2 Updated values of absolute frequency matrix with pseudocounts

$>\mathcal{L}_s$	**a**	**b**	**c**	**d**	**e**	**f**	**g**	**h**	**i**	**j**	**k**
a	1.634	8.972	1.690	1.690	1.717	1.716	6.392	6.524	1.690	1.690	8.960
b	1.634	1.690	8.972	1.690	1.717	1.716	1.709	1.670	1.690	1.690	1.690
c	1.634	1.690	1.690	8.972	1.717	1.716	1.709	1.670	1.690	1.690	1.690
d	7.459	1.690	1.690	1.690	7.304	7.392	1.709	1.670	1.690	1.690	1.690
e	1.634	1.690	1.690	1.690	1.717	1.716	6.340	6.267	1.690	1.690	1.690
f	1.634	1.690	1.690	1.690	1.717	1.716	6.539	6.229	1.690	1.690	1.690
g	1.634	1.690	1.690	1.690	1.717	1.716	1.709	8.011	1.690	1.690	1.690
h	1.634	1.690	1.690	1.690	1.717	1.716	1.709	7.931	8.972	1.690	1.690
i	1.634	1.690	1.690	1.690	1.717	1.716	1.709	1.670	1.690	8.972	1.690
j	1.634	1.690	1.690	1.690	1.717	1.716	1.709	1.670	1.690	1.690	8.972
k	1.634	1.690	1.690	1.690	1.717	1.716	1.709	1.670	1.690	1.690	1.690

$$PC[i, j] = \begin{cases} \log_2(\sqrt{N} - Background(i)) & \text{if } AFM[i, j] = 0 \\ \log_2(AFM[i, j]) & \text{otherwise} \end{cases} \quad (1)$$

6 Results and Discussion

We use the set of possible traces T_p = {`abcdeghijk`, `abcdehijk`, `abcdfhijk`, `abcdfghijk`, `aghijk`, `ahijk`, `ahhhijk`, `abcdabcdfhijk`, `aghhhijk`} and set of impossible traces T_i = {`kjjihgfdcb`, `adcbjh`, `kjig`

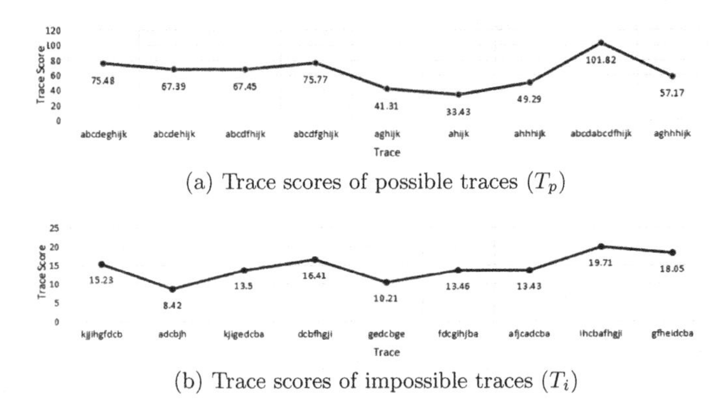

(a) Trace scores of possible traces (T_p)

(b) Trace scores of impossible traces (T_i)

Fig. 6 Traces scores for set of possible and impossible traces. Calculated on T_p and T_i based on the scores given in Table 2

edcba, dcbfhgji, gedcbge, fdcgihjba, afjcadcba, ihcbafh gji, gfheidcba} for evaluation.

The total score obtained by traces in both the set (i.e. T_p and T_i) is visualised in graphs shown in Fig. 6. It is evident from the graph that the set of possible traces have obtained comparatively much higher trace score than impossible traces. None of the traces from impossible trace set have managed to get score greater than 20. This implies that the traces given in set T_p are possible and can be observed frequently.

7 Conclusion

Spaghetti process are unstructured, difficult to comprehend and impossible to use during operational support. This paper proposed a technique for reducing the unstructured (Spaghetti) process to structured (Lasagna) process. Also a method has been proposed for finding the frequently possible and impossible paths of execution in the simplified process. These techniques have been tested on real life event logs and results are validated.

References

1. Van der Aalst, W.M.: Process mining discovery, conformance and enhancement of business processes (2010)
2. Van der Aalst, W.M., Weske, M., Grünbauer, D.: Case handling: a new paradigm for business process support. Data & Knowledge Engineering 53(2), 129–162 (2005)
3. Bose, R.J.C., Van der Aalst, W.M.: Abstractions in process mining: A taxonomy of patterns. In: International Conference on Business Process Management. pp. 159–175. Springer (2009)
4. Bose, R.J.C., Van der Aalst, W.M.: Context aware trace clustering: Towards improving process mining results. In: SDM. pp. 401–412. SIAM (2009)
5. Günther, C.W., Van Der Aalst, W.M.: Fuzzy mining–adaptive process simplification based on multi-perspective metrics. In: International Conference on Business Process Management. pp. 328–343. Springer (2007)
6. Liu, D.T., Xu, X.W.: A review of web-based product data management systems. Computers in industry 44(3), 251–262 (2001)
7. Luengo, D., Sepúlveda, M.: Applying clustering in process mining to find different versions of a business process that changes over time. In: International Conference on Business Process Management. pp. 153–158. Springer (2011)
8. Mans, R., Schonenberg, M., Song, M., Van der Aalst, W.M., Bakker, P.J.: Application of process mining in healthcare–a case study in a dutch hospital. In: International Joint Conference on Biomedical Engineering Systems and Technologies. pp. 425–438. Springer (2008)
9. Mithas, S., Krishnan, M.S., Fornell, C.: Why do customer relationship management applications affect customer satisfaction? Journal of Marketing 69(4), 201–209 (2005)
10. Tsuiki, K., Majima, H., Ono, H., Suga, K., Toge, T., Akifuji, S.: Work flow management system (Aug 17 1999), uS Patent 5,940,829
11. Van Der Aalst, W.M., Adriansyah, A., De Medeiros, A.K.A., Arcieri, F., Baier, T., Blickle, T., Bose, J.C., van den Brand, P., Brandtjen, R., Buijs, J., et al.: Process mining manifesto. In: International Conference on Business Process Management. pp. 169–194. Springer (2011)

Analysis of NIR and VW Light Iris Images on the Bases of Different Statistical Parameters

Harish Prajapati and Rajesh M. Bodade

Abstract To carry out accurate research for person's identification, it is necessary to know about NIR and VW images, by some easy method especially for iris recognition. This work presents method to know diversity and noise in the NIR and VW iris images, on the bases of eight different statistical parameters. For this work four VW databases: UBIRIS.v2, UBIRIS.v1 (season1), UBIRIS.v1 (season2), UPOL and four NIR databases: IIT Delhi.v1, CASIA.v1, CASIA.v4-iris interval, and CASIA.v4-iris twin were chosen. In calculated points, it is found that VW iris databases UBIRIS.v2 and UBIRIS.v1 (season2) posses most diversity. VW images are noisy compared to NIR images. Out of which UPOL images is most noisy. Intensity of VW images are less as compared to the NIR images. Average standard deviation of both NIR and VW images is equal (approx. 13.5).

Keywords NIR (Near infrared rays) · VW (visible wavelength)
Diversity (heterogeneous) · Noisy

1 Introduction

Iris is an annular region between pupil and sclera. It is one of the most important biometric traits useful for personal identification purpose. As it has desirable properties such as uniqueness ($1:10^{31}$), stability (do not change with age), non-invasiveness [1], and reliability (less rejection rate). Moreover, there are fewer chances to cover human iris. Most of the researches in iris recognition are based on NIR images (wavelength range 700–900 nm) [2]. This wavelength region requires cooperation from the user; that is why these NIR images may not be useful for

H. Prajapati (✉)
Chameli Devi Group of Institutions, Indore, India
e-mail: prajapati.harish@gmail.com

R.M. Bodade
Military College of Telecommunication Engineering, Mhow, India
e-mail: rajeshbodade@gmail.com

A.K. Somani et al. (eds.), *Proceedings of First International Conference on Smart System, Innovations and Computing*, Smart Innovation, Systems and Technologies 79, https://doi.org/10.1007/978-981-10-5828-8_67

identification purposes. To focus research toward identification, analysis of VW images (wavelength range 390–700 nm) along with NIR images becomes an essential task. Person's identity includes premises entry, computer login, forensic and police applications, any other transactions in which personal identification relies [3].

Motivation to analyze various databases came from studying research methodology subject. In which we studied sample for the research work, should cover diverse parameters then we can get accurate results. So, thought came in mind how to know diversity of images easily. And, in case of iris images: they are of two types on the bases of illumination of light, that is, NIR and VW iris images. Of them, VW iris could be used for distant identification purpose. So, another thought came in mind to know the difference between NIR and VW iris images by some easy method. These methods would be very useful not only for iris recognition but also for all types of images such as face images, X-ray, and other medical images.

The paper is organized in seven sections. Section 1 presents introduction along with motivation for writing this paper. Section 2 has about related work in the area and different iris image databases used in this paper. Section 3 consists of statistical parameters (used in this paper) and experimental plots with their discussion. Section 4 consists of results and discussion. Section 5 is the conclusion of the work. Section 6, presents the acknowledgement, and Sect. 7, future scope in the work.

2 Related Work and Iris Image Databases

2.1 Related Work

Most of the works of comparing databases is based on the difficulty in segmentation process of images. Comparison of NICE.I and UBIRIS.V2 dataset was done on the bases of specular reflection, occlusion, pixel count, pigmentation, pupil dilation, eccentricity, and other parameters by Samata et al. [1]. Similarly comparison of databases on the bases of difficulty in segmentation was done by Bodade et al. They found that UPOL database is more challenging than CASIA and UBIRIS because of its very low-intensity gradient at boundaries of iris [4]. Vijay et al. used statistical parameters in filtering of image [5]. Roy et al. mentioned that iris has high entropy per unit area [6]. By this new method, diversity of images in the database can be found by knowing only some statistical parameters? Our approach is new, less complex and a less programming skill is required. Persons entering in the image processing field could easily know about the complexity of images. Up to the best of our knowledge, it may be the first paper to know the diversity of iris images in the database. It also differentiates NIR and VW iris images by statistical parameters.

2.2 *Databases*

2.2.1 NIR Databases

(a) CASIA.v1: Preprocessing was done to mask out the specular reflections from the NIR illuminators [7]. It consists of 756 images on 108 subjects [8]. Two sample images are given below.

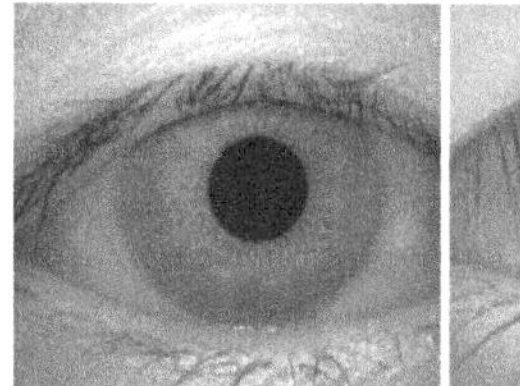 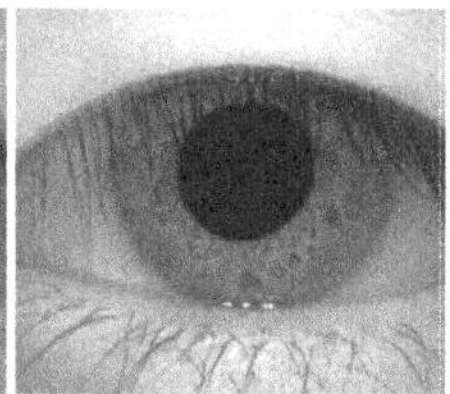

(b) CASIA.v4-iris interval: Images are captured from circular NIR LED array [7]. Two sample images are given below.

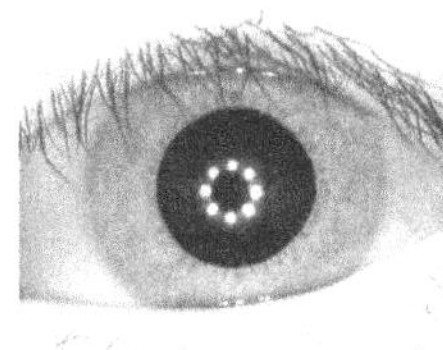 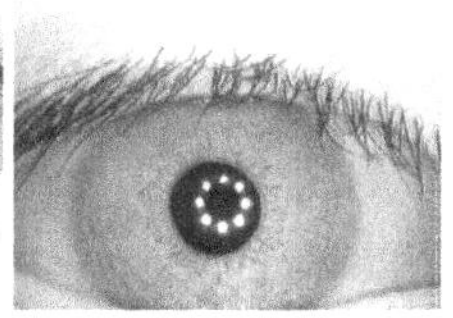

(c) CASIA.v4-iris twin: Database contains iris images of 100 pairs of twins. Images are captured from OKI's IRISPASS-h camera [7]. Two sample images are given below.

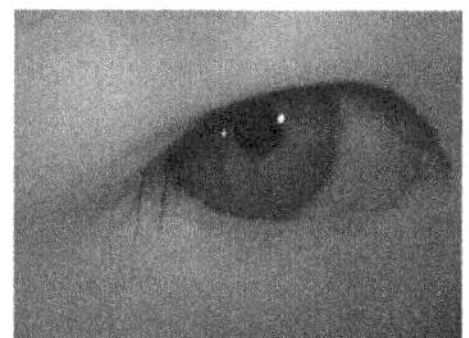 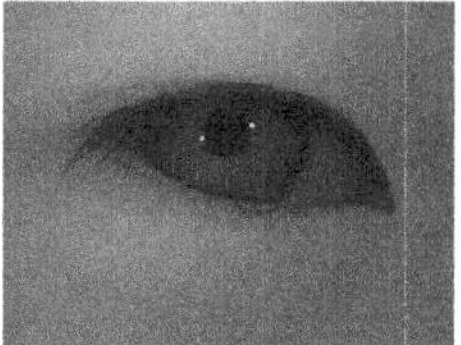

(d) IIT Delhi (V1.0): Database consist 1120 image collected from 224 users [9], the images were acquired in indoor environment and have a resolution of 320 × 240. Two sample images are given below.

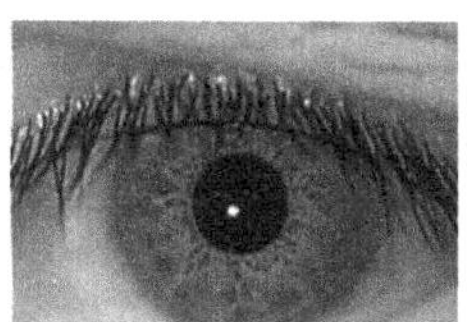 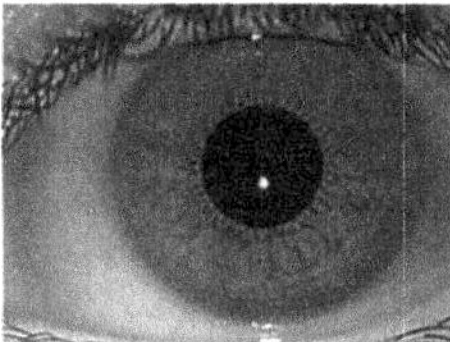

2.2.2 VW Databases

(e) UBIRIS.v1 (season1): It consists of 1214 images with minimized noise factors especially reflections, luminosity, and contrast [10]. Two sample images are given below.

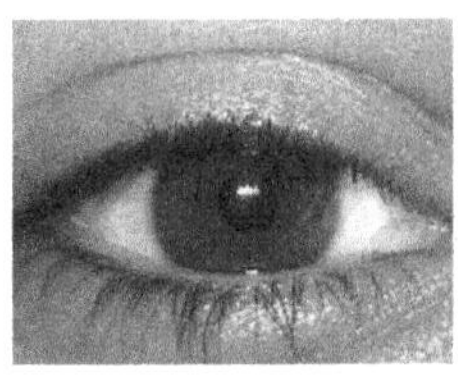 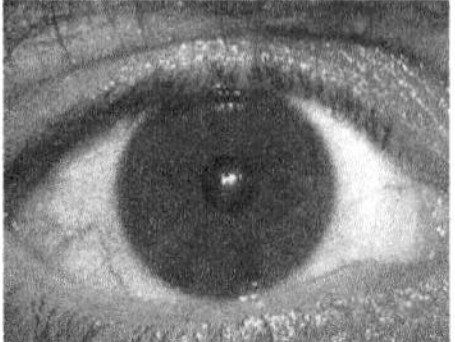

(f) UBIRIS.v1 (season2): It consists of 663 images [11]. Images were captured with minimal active collaboration with the subject. Images are heterogeneous with respect to reflections, contrast, luminosity, and contrast [10]. Two sample images are given below.

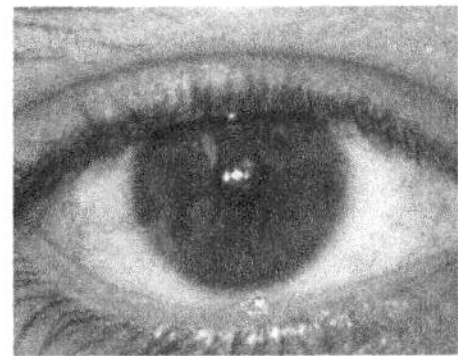 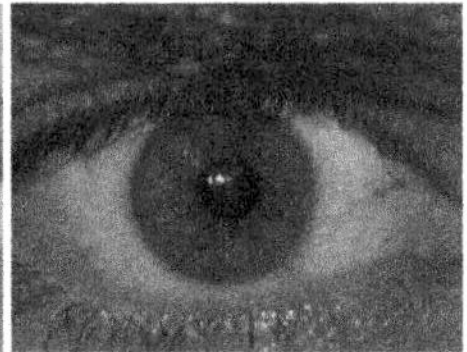

(g) UBIRIS.v2: Database consists of 864 images on 151 subjects. Images are influenced by noisy sources such as motion, occlusions, reflections, off angle, etc. [8]. Two sample images are given below.

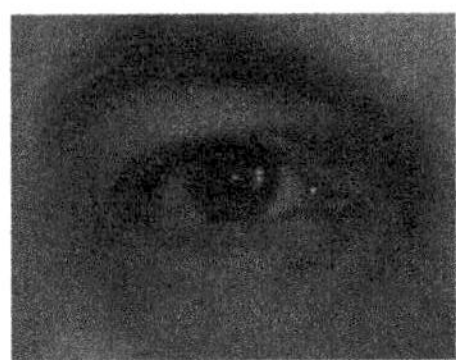 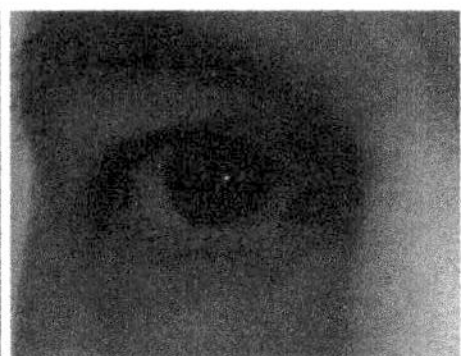

(h) UPOL: Database is of 64 subjects and contains 3 × 64 left and 3 × 64 right iris images [4]. It consists of realistic iris images of 800 × 600 resolution [12]. Database just includes images from the internal part of the eye. Segmentation work almost did [10]. Two sample images are given below.

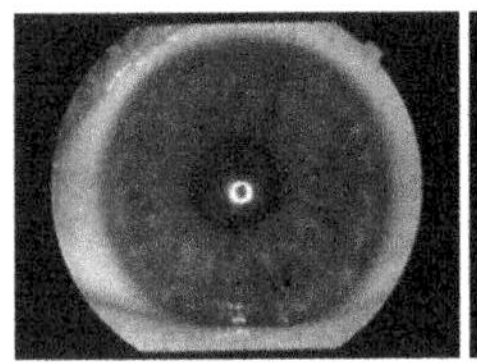 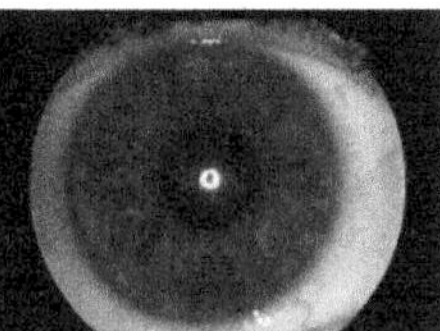

3 Statistical Parameters and Experimental Plots

3.1 Statistical Parameters

1. Mean: It is arithmetic average of set of scores

$$\bar{x} = \frac{\sum_{i=1}^{k} x_i}{N},$$

where,

$\bar{x}$	mean
N	number of scores
$\sum_{i=1}^{k} x_i$	sum of all scores

Here it is used for finding average intensity of images and variation in intensity of images of different databases.

2. Median: It is the value in a set of numbers that divides set of numbers in two equal halves. When median value has been derived, half numbers should be above the score and half numbers should be below the score. Median value of intensity of images was found out. If mean and median values of images are same this signifies that intensity is normally distributed. Otherwise, images have skewed distribution of intensity.
3. Mode: Most frequently occurring number in a set of score. Here, we can find the intensity of image whose frequency of occurrence is more.
4. Standard Deviation: Gives an approximate picture of the average amount, each number in a set varies from the center value. Or how far each number is from the mean. Measure of average contrast in the images.
5. Variance: Average of the square of the deviations of a set of scores from the mean value.
6. Entropy: Measure of randomness [13]. It is useful for texture information.
7. Skewness: It is measure of asymmetry of the probability distribution of real-valued random diverse. It can be positive, negative, or undefined. Darker and glossier surfaces are positively skewed than lighter and matte surfaces [5].
8. Kurtosis: It ensures the shape of the probability distribution of a real-valued random diverse. Kurtosis value is constant for images taken in good light (clean/natural images) and it varies due to noise inherent in the images [14].

3.2 Experimentation

Experiment was performed on eight types of databases, i.e., four of VW images and four of NIR images (choice of the database depends on the availability of database). Here, for reducing the complexity of plots, data is arranged in descending order. From each database 100 equal dimension images were randomly selected. After that mean, median, mode entropy, skewness, variance, standard deviation, and kurtosis of all 100 images of each database were found out. Then averages of the values of the parameters were calculated. Now, to calculate the deviation of parameter value from average value, high average value, and low average values were found out. High and low average values were based on highest and lowest averages of parameters of five images. Thus positive and negative deviation from the average value was found out. Out of positive and negative deviations, maximum deviation of each parameter was found out. Parameter values of all 100 images were plotted in MATLAB software.

For example, plot of mean values of all 100 images of all databases have been shown in Fig. 1. To differentiate databases mean values of different databases are plotted with different colors and markers. For example, red color and star as marker for CASIA v.1, blue color and diamond as marker for CASIA v.4 iris interval database respectively.

Figure 1 plots the graph of mean values of all the databases. We can see that there is highest deviation in the mean value of UBIRIS.v1 (season2) database plotted with cyan color marked by circles, i.e., mean value can vary up to 70.04

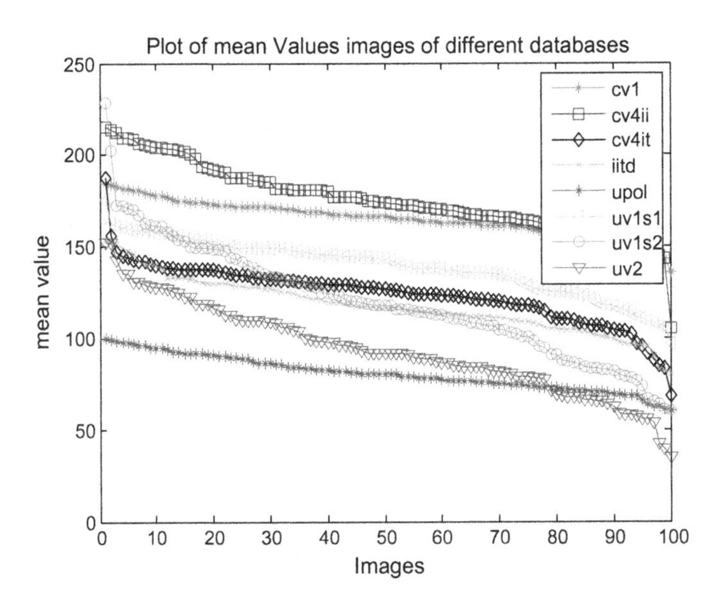

Fig. 1 Plot of mean values of images of different databases

from average value of 119.63 (0.1, red color in mean value column).Second highest, deviation in mean value was found in UBIRIS.v2 database plotted by magenta color and marked by triangles, i.e., mean value can vary up to 50.03 from average value 93.42. And lowest variation in mean value was found in the UPOL database plotted with magenta color marked by stars.

Median values of all the databases are plotted in Fig. 2. We can see that there is highest deviation in the median value of UBIRIS.v1 (season2) database plotted with cyan color marked by circles, i.e., median values can vary up to 77.07 from average value 125.73 (Table 1, red color in median value column).Second highest, deviation in UBIRIS.v2 database plotted with magenta color and marked by triangles, i.e., median values can vary up to 58.86 from average value 93.26. And lowest deviation in median values was found in the CASIA v.1 database plotted with red color marked by stars. In all the databases, average values of mean and median are approximately the same; it suggests that intensity of images is normally distributed.

Figure 3 shows the plots of modes of images. It can be seen that there is highest deviation in the mode values of CASIA v.1 database plotted with red color marked by stars, i.e., mode values can vary up to 161.39 from average value 93.61 (Table 1, red color in mode value column). Second highest, variation in the mode values of images is in UBIRIS.v1 (season2) database plotted with cyan color marked by circles, i.e., mode values can vary up to 131.77 from average value 100.23. And lowest deviation in mode values were found in the UPOL database plotted with magenta color and marked by stars, i.e., zero deviation or constant plot merged with the X-axis.

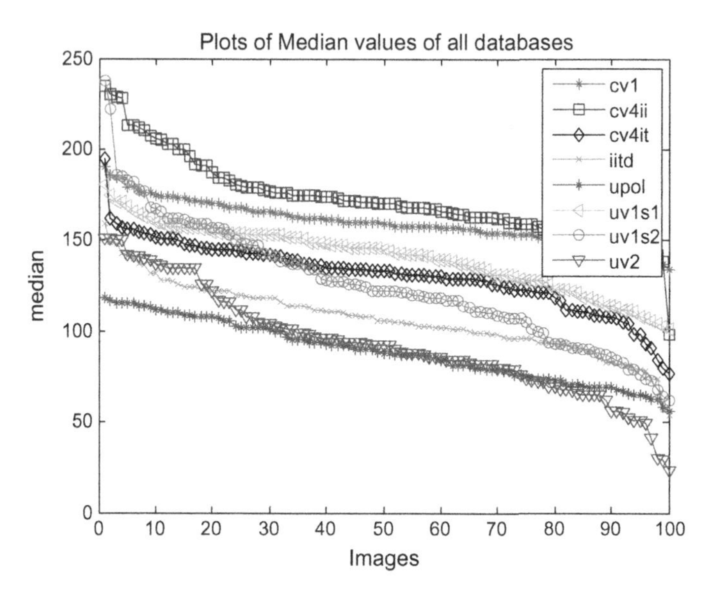

Fig. 2 Plot of median values of images of different databases

Table 1 Statistical values of NIR and VW databases

S.NO	Type of data base	Data base name	value	Mean	Median	Mode	Entropy	Skewness	variance	Standard deviation	Kurtosis
1	NIR	Casia v4it	Avg.	124.0751	130.95	122.47	6.908691	−0.39151	8.61E+05	14.09321	5.654065
			p. dev	31.73306	34.85	86.53	0.519529	1.435192	1.37E+06	5.696571	13.17222
			n.dev	4.08E+01	45.95	103.87	0.611591	0.878468	6.57E+05	6.658429	3.368365
			M.dev	4.08E+01	45.95	103.87	0.611591	1.435192	1369466	6.658429	13.17222
2	NIR	Casia v4iint	Avg.	175.8333	172.36	215.98	7.051979	−0.37157	2.49E+06	14.95634	5.308549
			p. dev	36.16865	54.64	39.02	0.560461	0.787754	4.74E+06	7.873525	11.72847
			n.dev	3.78E+01	39.76	190.18	1.158119	0.691166	1.93E+06	6.272635	2.954109
			M.dev	37.75841	54.64	190.18	1.158119	0.787754	4735204	7.873525	11.72847
3	NIR	IITD	Avg.	117.6281	107	5	7.390637	−0.75639	1.63E+06	1.05E+01	8.050233
			p. dev	33.25825	41	95	0.232763	0.482012	2.77E+06	6.74E+00	46.19570
			n.dev	3.15E+01	39.6	4	0.280577	0.715588	1.32E+06	5.17E+00	5.592313
			M.dev	33.25825	41	95	0.280577	0.715588	2.77E+06	6.74E+00	46.19570
4	NIR	Casia v1	Avg.	164.9814	159.82	93.61	7.020176	−0.35926	2.39E+06	15.01855	4.649632
			p. dev	17.878895	24.78	161.39	0.299084	1.047642	2.35E+06	4.231553	7.993608
			n.dev	2.19E+01	22.02	56.81	0.463936	1.169278	1.60E+06	5.171727	2.615732
			M.dev	21.89096	24.78	161.39	0.463936	1.169278	2351336.	5.171727	7.993608
5	VW	Ubiri sv1s1	Avg.	139.2465401	140.73	142.74	7.445327	−0.21427	6.29E+05	8.945462	3.887614
			p. dev	24.81545	32.67	98.26	0.179153	0.710632	1.05E+06	5.883558	12.44540

			n.dev	3.38 E+01	36.53	106.5 4	0.215 967	0.576 748	4.98 E+05	4.062 962	2.182 494
			M.dev	33.75 238	36.53	106.5 4	0.215 967	0.710 632	1.05 E+06	5.883 558	12.44 540
6	VW	Ubiri sv1s 2	Avg.	119.6 292	125.7 3	100.2 3	7.136 861	0.128 734	4.32 E+05	8.377 289	11.66 36
			p. dev	70.04 324	77.07	131.7 7	0.339 919	1.759 826	1.13 E+06	6.068 131	33.13 223
			n.dev	5.62 E+01	57.53	86.03	1.001 861	1.359 314	3.65 E+05	3.621 809	9.855 015
			M.dev	70.04 324	77.07	131.7 7	1.001 861	1.759 826	1127 301	6.068 131	33.13 223
7	VW	Ubiri sv2	Avg.	93.41 868	93.26	94.72	7.108 273	0.191 416	3.33 E+05	7.993 923	14.85 564
			p. dev	50.03 775	55.54	105.6 8	0.443 407	2.415 144	1.02 E+06	6.508 917	62.20 578
			n.dev	4.82 E+01	58.86	85.92	0.841 653	1.593 836	3.09 E+05	4.595 103	12.87 666
			M.dev	50.03 775	58.86	96.88	0.841 653	2.415 144	1019 602	6.508 917	62.20 578
8	VW	UPOL	Avg.	80.76 054	88.94	0	5.896 851	5.812 992	8.38 E+06	28.61 411	61.30 885
			p. dev	17.62 89	27.06	0	0.198 769	1.579 428	4.94 E+06	3.175 529	70.20 569
			n.dev	1.87 E+01	28.14	0	0.272 611	1.079 672	4.27 E+06	3.772 971	29.35 795
			M.dev	18.73 304	28.14	0	0.272 611	1.579 428	4935 146	3.772 971	38.25 479

Figure 4 shows the plots of entropies of all the databases. We can see that there is highest deviation in the entropy of CASIA.v4-iris interval database plotted with blue color marked by squares, i.e., entropy of images can vary up to 1.16 from average value 7.05 (Table 1, red color in entropy column). Second highest, deviation is in entropy of UBIRIS.v1 (season2) database, plotted with cyan color marked by circles, i.e., entropy can vary up to 1.0 from average value 7.13. And lowest deviation in entropy of images was found in the UBIRIS.v1 (season1) database plotted with yellow color marked by lesser sign.

Figure 5 shows the plots of skewness of all the databases. We can see that there is highest deviation in the skewness of UBIRIS.v2 database plotted with magenta color marked by triangles, i.e., skewness of images can vary up to 2.41 from average value 0.19 (Table 1, red color in skewness column). Second highest,

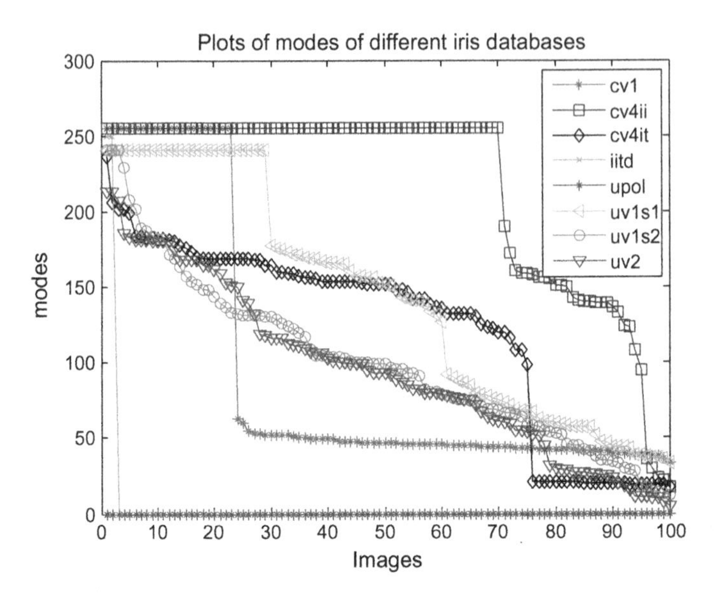

Fig. 3 Plot of mode values of images of different databases

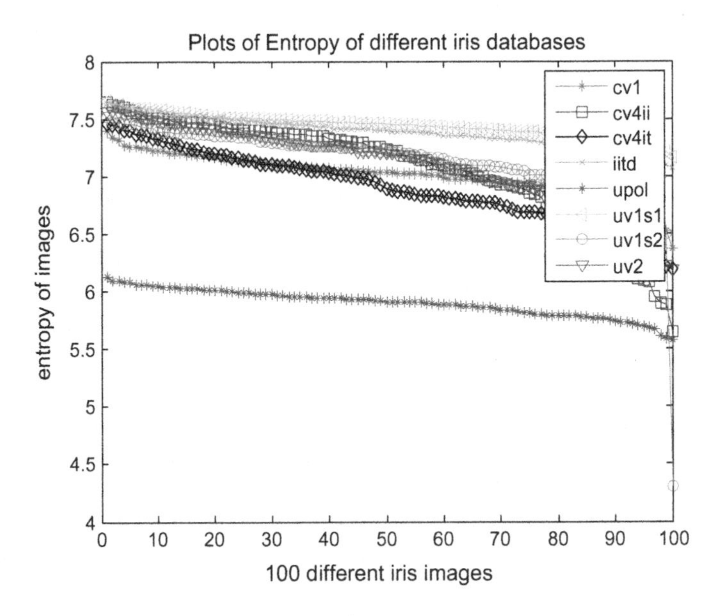

Fig. 4 Plot of entropy of images of different databases

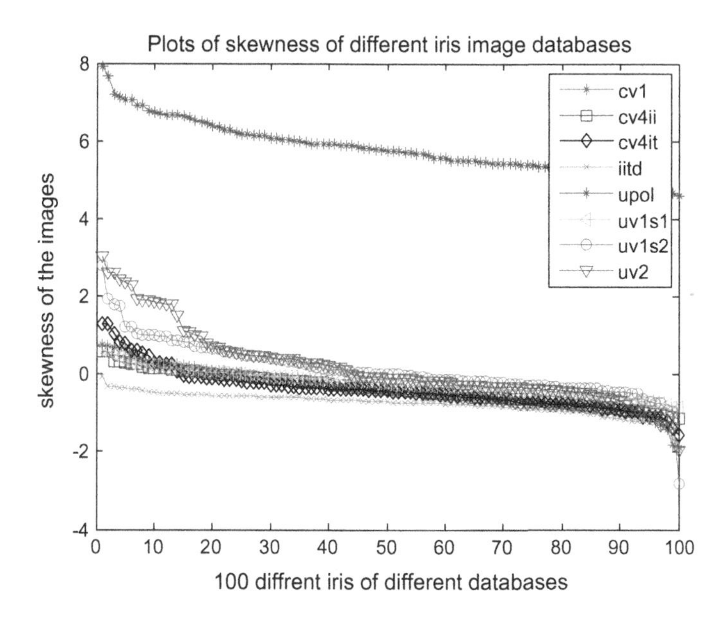

Fig. 5 Plot of skewness of images of different databases

deviation in the skewness of UBIRIS.v1 (season2) database plotted with cyan color marked by circles, i.e., skewness can vary up to 1.76 from average value 0.12. And lowest deviation in skewness of images was found in the UBIRIS.v1 (season1) database plotted with yellow color and marked by lesser sign.

Figure 6 shows the plots of variances of all the databases. In plots we can see that there is highest deviation in the variance of UPOL database plotted with magenta color marked by stars, i.e., variance of images can vary up to 4.93E+06 from average value 8.38E+06 (Table 1, red color in variance column). Second highest, deviation is in variance of CASIA.v4-iris interval database plotted with blue color marked by squares, i.e., variance can vary up to 4.73E+06 from average value 2.49E+06. And lowest variation in the variance of images was found in the UBIRIS.v1 (season1) database plotted with yellow color, marked by lesser sign.

Figure 7 shows the plots of standard deviation of all the databases, we can see that there is highest deviation in the standard deviation of CASIA.v4-iris interval database plotted with blue color marked by squares, i.e., variance can vary up to 7.87 from average value 14.97. Second highest, deviation in the standard deviation of IIT Delhi database plotted with green color marked by dashes, i.e., variance can vary up to 6.74 from average value 10.5. And lowest variation in standard deviation of images was found in the UPOL database plotted with magenta color, marked by stars.

From Fig. 8 plots of kurtosis of all the databases, it can be seen that there is highest deviation in the kurtosis of UBIRIS.v2 database plotted with magenta color marked by triangles, i.e., kurtosis can vary up to 62.20 from average value 14.85.

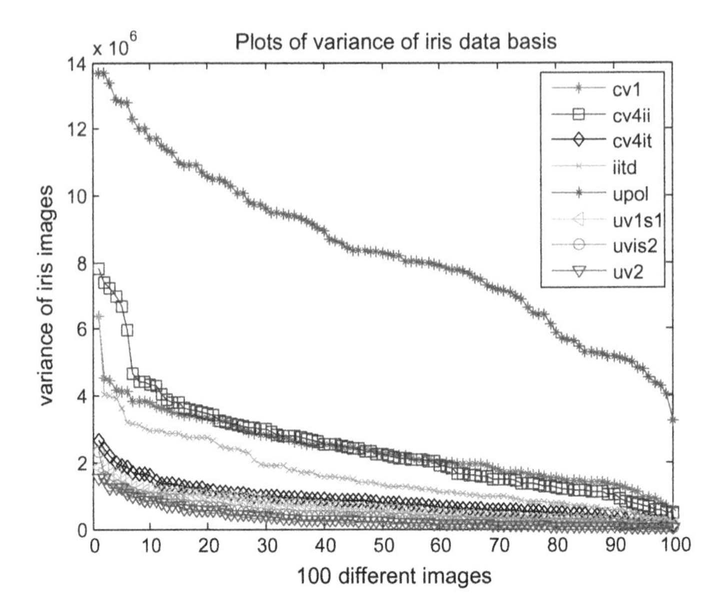

Fig. 6 Plot of variance of images of different databases

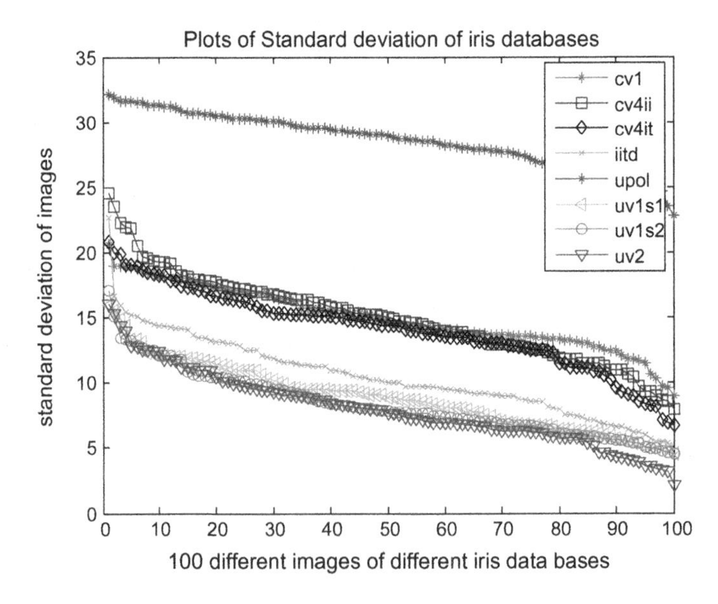

Fig. 7 Plots of standard deviation of images of different databases

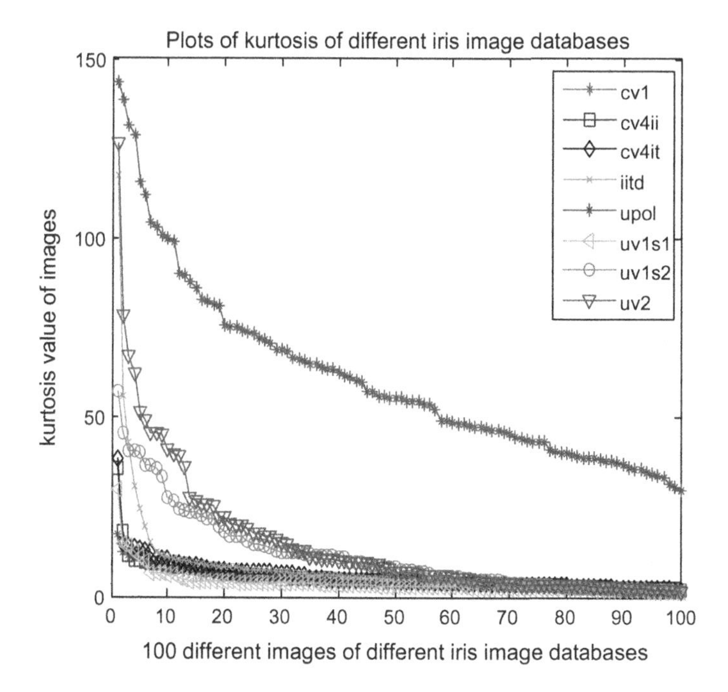

Fig. 8 Plot of kurtosis of images of different databases

Second highest, deviation is in the kurtosis of IIT Delhi database plotted with green color marked by dashes, i.e., variance can vary up to 46.19 from average value 8.05. And lowest, variation in kurtosis of images was found in the CASIAv.1 database plotted with red color, marked by stars.

3.3 *Calculations of Diversity in the Databases*

Different databases were assigned positions on bases of maximum deviation from its average value (Table 2). Points were allotted on the bases of deviation from the average value. Highest deviated database from the average value of any statistical parameter was allotted three points, second highest deviated was allotted two points and for third highest deviated three points. For example, highest deviated database in mean value from the average of the mean value was allotted three points. Second highest deviated database in mean value from average mean value was allotted two points. And, for third highest deviated database in mean value was allotted one point. Likewise, points were allotted on different statistical parameters to different databases. For example, databases CASIA.v1 database was highest deviated from average value of mode so was allotted three points. CASIA.v4-iris interval database was highest deviated in entropy and standard deviation, second highest in variance,

Table 2 Diversity points of NIR and VW databases

Type of database	Databases	Total I positions (each 3 points)	Total II positions (each 2 points)	Total III positions (1 points)	Total points scored
NIR	Casiav4it	–	–	1	1
	Casiav4iint	2	1	1	9
	IITD.v1	–	2	1	5
	Casiav1	1	–	–	3
	Total points NIR databases				18
VW	Ubirisv1s1	–	–	1	1
	Ubirisv1s2	2	3	–	12
	Ubirisv2	2	2	2	12
	UPOL	1	1	–	5
	Total points VW databases				30

third highest in median; thus it was allotted nine points. Likewise, deviation points of all databases were calculated to compare the diversity in the database.

4 Results and Discussion

4.1 Observations from the Result

CASIA.v4-iris interval database stands on nine points (Table 2). Hence it is highest diverse database in NIR iris image databases and second highest diverse database in all eight iris databases. IIT Delhi.v1 database secured five points. Thus, it is second diverse database in NIR images and third diverse database in all databases. Its diversity is equivalent to the UPOL database. In all, total diversity points of NIR databases are 18.

In VW databases: UBIRIS.v1 (season2) database stands on 12 points. This score is highest among all the databases and is equivalent to the UBIRIS.v2 database points. Since, UBIRIS.v2 database also stands on 12 points. Thus, we can say that UBIRIS.v1 (season2) and UBIRIS.v2 are highly diverse databases. UPOL database got five points. All total points for VW image databases were 30; this shows that VW iris images are more heterogeneous as compared to NIR iris images.

In average values of different statistical parameters (Table 3) the highest averages of mean, median, and mode values are of CASIA.V4 iris interval database.

Table 3 Averages of different statistical parameters of NIR and VW iris images

Type base of data	Databases	Mean (Avg.)	Median (Avg.)	Mode (Avg.)	Entropy (Avg.)	Skew-ness (Avg.)	Variance (Avg.)	Standard deviation (Avg.)	Kurtosis (Avg.)
NIR	Casiav4it	124.075	130.95	122.47	6.90869	−0.391	8.61E+05	14.093	5.654
	Casiav4iin	175.833	172.36	215.98	7.05197	−0.3715	2.49E+06	14.956	5.308
	IITD.v1	117.628	107	5	7.3906	−0.756	1.63E+06	1.05E+01	8.0502
	Casiav1	164.9814	159.82	93.61	7.02017	−0.359	2.39E+06	15.01855	4.6496
	Average of NIR images	145.6294	142.532	109.26	7.09287	−0.4696	1.84E+06	13.642	5.91561
VW	Ubirisv1s1	139.24654	140.73	142.74	7.44532	−0.214	6.29E+05	8.945462	3.88761
	Ubirisv1s2	119.6292	125.73	100.23	7.13686	0.12873	4.32E+05	8.377289	11.6636
	Ubirisv2	93.41868	93.26	94.72	7.10827	0.19141	3.49E+05	7.993923	14.8556
	UPOL	80.76054	88.94	0	5.89685	5.81299	8.38E+06	28.61411	61.3088
	Average of VW images	108.26374	112.165	84.422	6.89682	1.47971	2.44E+06	13.48269	22.9289

These averages are 175.83, 172.36, and 215.98, respectively. And lowest average mean, median, and mode values are found in UPOL database which are 80.76, 88.94, and 0, respectively. Highest average entropy is of UBIRIS V1 (season 1) database, i.e., 7.44; which is approximately equal to average entropy CASIA.v4 iris interval database (i.e., 7.05). And lowest average entropy is of UPOL database (i.e., 5.89). Similarly, highest average skewness, variance, standard deviation, and kurtosis are found in UPOL database these are 5.81, 8.38E+06, 28.61, and 61.30, respectively. And lowest average skewness, variance, standard deviation, and kurtosis are found in IIT Delhi.v1, UBIRIS.v2, UBIRIS.v2, and UBIRIS.v1 (season1) databases which are −0.756, 3.49E+05, 7.99, and 3.89, respectively. Average values of skewness, variance, standard deviation and kurtosis show that UPOL database is highest noisy database. Reason for this may resolution of light in the images.

Comparison of NIR images (only NIIR database images) and VW images (only VW database images): Averages of mean, median, and mode in VW images are less, these parameters (Table 3) show that average intensity of light in VW images is less as compared to NIR images. This point is also proved by average skewness value of VW images moving towards positive side which confirms that VW images are darker (having less intensity) than NIR images. Average standard deviation is same for both types of images (Table 3). This shows that both NIR and VW images are equally deviated from the center (approx. 13.5).

The VW images are noisy as compared to the NIR images. Average kurtosis value of VW images (Table 3) varies more (very much greater than 3) as compared to the NIR images. And, in most of the VW databases, kurtosis value is more as compared to NIR images. Reason for this may be that VW images are taken from distance or are noisy as compared to NIR images.

5 Conclusion

From the above outcomes, it can be concluded that VW iris images are more diverse as compared to NIR iris images. And in comparison of databases UBIRIS. v1 (season2) and UBIRIS.v2, are most heterogeneous databases. Second, conclusion which can be drawn from this work is that the intensity of light in VW images is less as compared to NIR images. In other words, VW images are darker then NIR images. Third and most important outcome of the work is that VW images are noisy as compared to the NIR images. Since, these images are captured in uncontrolled environment or at distance from the camera. And in comparison of databases UPOL images are highly noisy images.

6 Acknowledgements

We thank UBIRIS, CASIA, UPOL, and IIT Delhi for providing us iris databases. Without their help this work was impossible.

7 Future Scope

This work can be extended to know the factors such as motion, reflections, etc., those effects VW iris images by calculating certain statistical parameter. This entity would be very useful for all image processing fields such as face images, X-ray, and other medical images.

References

1. Kharul, S., Chaskar, U.: Quality Factors estimation of iris images. Int. J. of recent trends in Engineering and Technology. 276–284 (2014)
2. Wild. P., Radu, P., Ferryman, J.: On Fusion of Multispectral Iris Recognition. Biometrics (ICB), Int. Conf., (2015)
3. Proenca, H., Filipe, S., Santos, R., Oliveira, J., Alexandre, L. A.: The UBIRIS.v2: A Database of Visible Wavelength Iris Imag Captured On-The-Move and At-A- Distance.: IEEE Transactions on Pattern Analysis and Machine Intelligence, (2009)
4. Bodade, R., Talbar, S.: Fake Iris detection: A Holistic Approach.: Int. J. of computer Applications, vol. 19, no. 2, (2011)
5. Kumar, V., Gupta, P., Importance of Statistical Measures in Digital image processing.: Int. J. of Emerging Technology and advanced and advanced Engineering: vol. 2, no. 8, (2012)
6. Roy, S., Biswasr, A.: A Personal Biometric Identification technique based on iris recognition. Int. J. of Computer Science and Information Technologies, 1474–1477 (2011)
7. http://biometrics.idealtest.org/
8. Tan, C.-W., kumar A.: Accurate iris recognition at a distance using stabilized iris encoding and Zernike moments phase features.: IEEE Trans. Image Process., vol. 23, no. 9, pp. 3751–3765, (2014)
9. http://www4.comp.polyu.edu.hk/~csajaykr/IITD/Database_Iris.htm
10. Proenca, H., Alexandre, L. A.: UBIRIS: A noisy iris image database.: springer, pp. 970–977, (2005)
11. Yahya, A. E., Nordin, M. J.: Accurate iris segmentation method for Non-Cooperative iris recognition System.: J. of computer science., vol. 6, no. 5, pp. 492–497, (2010)
12. High contrast iris image database, http://www.sinobiometrics.com
13. Gonzalez, R. C., Woods, R. E., Eddins, S. L.: Digital Image Processing Using.: McGraw Hill Education
14. Zoran, D., Weiss, Y.: Scale Invariance and noise in natural images,.: IEEE 12th International conference on computer vision, pp. 2209–2216, (2009)
15. Proenca, H.: The UBIRIS.v2: Quality assessment of degraded iris images acquired in visible wavelength.: IEEE Transactions on Information forensic and security

Study and Analysis of Demonetization Move Taken by Indian Prime Minister Mr. Narendra Modi

Vijay Singh, Bhasker Pant and Devesh Pratap Singh

Abstract Demonetization is the process of stopping currency in the monetary standard. Mr. Narendra Modi shocked the whole country at live telecast on November 8, 2016 at 8:15 pm by announcing demonetization of 500 and 1000 rupee notes from November 9, 2016, to handle the threat of black money, terrorist funding, and fake currency. In this article, we study and analyzed what the people of India feel about this decision of the government on demonetization and analyzed the effect on the business community. The results show that 51.02% people are supported this while the 34.69% are against it, rest 14.28% are neutral.

Keywords Demonetization · Naïve Bayes · Latent Dirichlet Allocation

1 Introduction

Demonetization is a financial procedure in which the status of the currency's legal tender declared invalid. The demonetization move is taken by the Government when it is required to replace old currencies with the new currencies, to remove black money from the economy, tackle various kind of terrorist funding, Naxalite funding, and fake currency. Generally, Government takes these kinds of decisions periodically to boost the economy. In the Indian context, it is implemented in 1948 and the then Government head by Janata party in 1978 to tackle counterfeit money and black money and in November 2016 by Mr. Narendra Modi. In this demonetization step, government demonetized 500 and 1000 rupee notes and replacing

V. Singh (✉) · B. Pant · D.P. Singh
Computer Science and Engineering Department, Graphic Era University,
566/6 Bell Road, Clement Town, Dehradun, India
e-mail: vijaysingh_agra@hotmail.com

B. Pant
e-mail: pantbhaskar2@gmail.com

D.P. Singh
e-mail: devesh.geu@gmail.com

A.K. Somani et al. (eds.), *Proceedings of First International Conference on Smart System, Innovations and Computing*, Smart Innovation, Systems and Technologies 79, https://doi.org/10.1007/978-981-10-5828-8_68

them with the new 500 and 2000 rupee notes. There are various challenges associated with the demonetization implementation. How to replace old currencies with the new one? In the country like India, where millions of people not having a bank account, there are various other reasons to implement demonetization by the Government, some of them are:

- Interest rate can be low.
- Push economic growth.
- The government wants to change the people's mindset of not using cash money, they promote plastic money.
- Controlling the fake currency in the Indian economy.
- To stop terrorist and Naxalite fundings.

Naïve Bayes and Latent Dirichlet Allocation: In this section, the brief introduction of the Naïve Bayes and Latent Dirichlet Allocation is given to better understanding to the further content of the paper. Twitter is one of the effective media where a users can share their feeling instantly, and this data can be further used to detect nation's reaction on the specific topic or specific event. In this article, we analyzed the emotion of the people into the following six categories: Unknown, Joy, Sadness, anger, surprise, fear, and disgust [1] and Latent Dirichlet Allocation is applied to detect the most concern issues of the people about demonetization [2].

Naïve Bayes: Naïve Bayes classifier is one of the simplest Bayesian classifiers, but it is very effective on a variety of data or unstructured data. Naïve Bayes classifier can be direct applications [3]. Naïve Bayes classifier assumes attributes have the independent distribution shown in Eq. 1.

$$p(C_j|d) = \frac{p(d|C_j)p(C_j)}{p(d)} \tag{1}$$

- $p(C_j|d)$ = Probability of instance d being in class Cj.
- $p(d|C_j)$ = Probability of generating instance d given class Cj.
- $p(C_j)$ = Probability of occurrence of class Cj.
- $p(d)$ = Probability of instance d occurring.

Latent Dirichlet Allocation: Latent Dirichlet Allocation (LDA) is an unsupervised, generative probabilistic method used to determine the topic from the corpus or text collection in the unigram form [2]. In the recent years, it is observed that Latent Dirichlet Allocation outperforms well in the topic mining as well as text classification domain [4]. Suppose there are X independent topics in the corpus which consist of D documents. Here every topic is the polynomial probabilistic distribution of word and every document randomly generated by X topics. In LDA, it is essential to extract the composition information (θ, Z) of hidden topic. If it is given that Dirichlet parameter α and β, the distribution of a random variable θ, z, and W in document d is computed as [5] shown in Eq. 2:

$$p(\theta, Z, W/\alpha, \beta) = p(\theta/\alpha) \prod_{i=1}^{N_m} p(\frac{Z_i}{\theta}) p(W_i/Z_i, \beta) \quad (2)$$

2 Literature Survey

Kaiquan Xu et al. [6] proposed a novel method for identification of online communities with a similar type of sentiments in Social Networks. They implemented polarity and rating-based sentiment analysis. Deepa Anand et al. [7] proposed a two-class classification method for plots and reviews without the need for labeled data. They train the data with minimal supervision and find the promising results. Nur Azizah Vidyaet al. [8] identifies the public perception on the quality of the service toward the customer of telco companies. They compared the results with Naïve Bayes, Support Vector Machine, and Decision Tree, and finds Support Vector Machine outperforms over the Naïve Bayes and Decision tree with respect to processing and accuracy. Asad Ulla Rafiq Khan et al. [9] used comparative opinion mining by Naïve Bayes. They performed Naïve Baye's multilevel classification based on the Naïve assumption that the words around keyword play an important role for extracting opinion. Chenghua Lin et al. [10] proposed a joint sentiment/topic model for sentiment analysis, which detects the sentiment and the topic model simultaneously from textual the data set. The experiments show improvements in the sentiment classification accuracy. Ralf Krestel et al. [11] proposed a method for automatically generating topic-dependent lexica from the huge data set of review contents by exploiting user preferences or ratings. Their technique combines text segmentation and discriminative feature analysis, and Latent topic extraction. They investigate the performance improvements in comparison with existing sentiment lexica. Wanying Ding et al. [12] proposed a novel hybrid Hierarchical Dirichlet Process–Latent Dirichlet Allocation model, which automatically detect the number of aspects, different factual words and opinioned words. The system is efficient to extract the aspect specific sentiment words. Their model efficiently works aspect level to sentiment classification.

Dataset Preparation and Methodology

Twitter is a famous microblogging service and one of the best media to get the pulse of the nation. For demonstration purpose, the data is collected through various twitter hashtags related to demonetization of 500 and 1000 rupees note. There are around 50,000 tweets were collected for study and analysis of demonetization move by the Government of India. Sample tweets are shown in Table 1.

Table 1 Sample tweets

S. No.	Sample tweets
1.	Crossing 2.5 lakh deposit threshold to attract heavy penalty
2.	Dis guy had no money 2 pay Ola cab driver and d driver's reply will win ur heart
3.	Every child in India also knows that all Modi ji is doing is fooling
4.	This is how life changes in 24 h!
5.	Right step to mainstream all economic activity, but traditional business will go into a tizzy for a while
6.	Modi's financial move will hit the illegal drug business hard and help reduce drug addictions
7.	After late-night surge in business, jewellery sale drops 70%
8.	In 24 h two of the greatest democracies have gone from black to white. TrumpPresident BlackMoney DeMonetisation
9.	Rs 3 crore was recovered from the car of a BJP leader in Ghaziabad
10.	Lets all cooperate to make India a more honest and corruption free country which is free from BlackMoney

Methodology: The whole approach is divided into two task, task 1, and task 2. Important Steps are shown in Fig. 1.

Task 1: In task 1, Naïve Bayes algorithm is applied. Further, the task 1 is subdivided into two parts. In the first past tweets are classified into various emotions like unknown, joy, sadness, anger surprise, fear, and disgust shown in Fig. 1. In the second part, the sentiment polarity is calculated on the collected dataset.

Task 2: In task 2, The Latent Dirichlet Allocation is applied. LDA is a topic modeling approach used to determine the discussed topic in the corpus or in the dataset. There is a drawback associated with this Latent Dirichlet Allocation that it

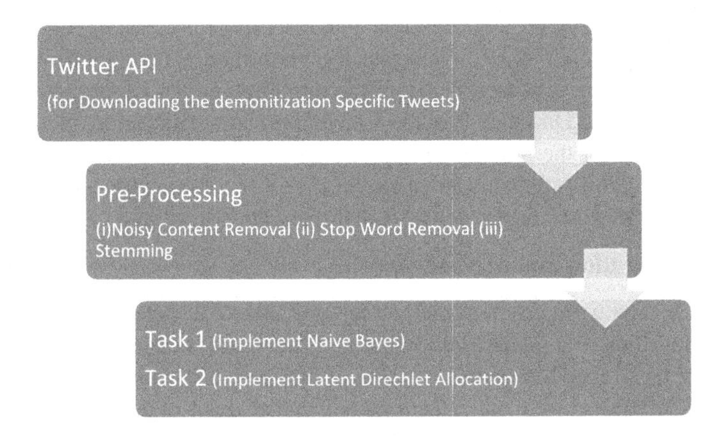

Fig. 1 Important steps

works on Unigram data, so it is difficult to understand in which sense the word is using contextually. But it is efficient to extract the most discussed word from the huge collection of data.

3 Results and Discussions

Emotions categories by the Naïve Bayes algorithm is shown in Fig. 2, and most of the people of India who used twitter are happy (joy) with this demonetization move by the Government of India. Due to the limitation of the Naïve Bayes algorithm

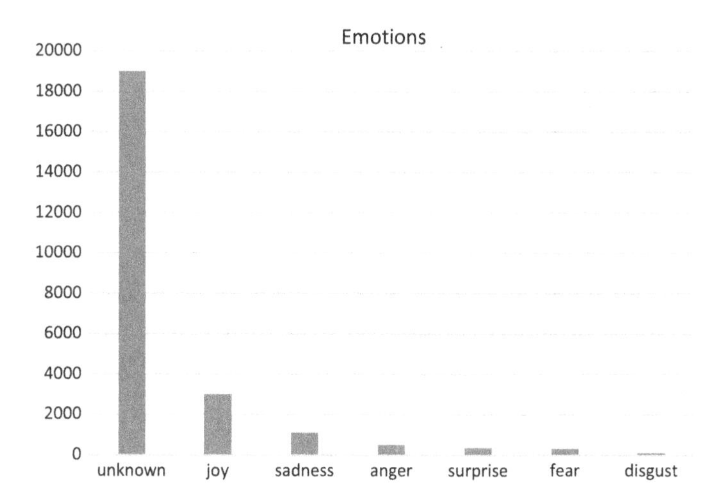

Fig. 2 Emotion categories

Fig. 3 Polarity categories

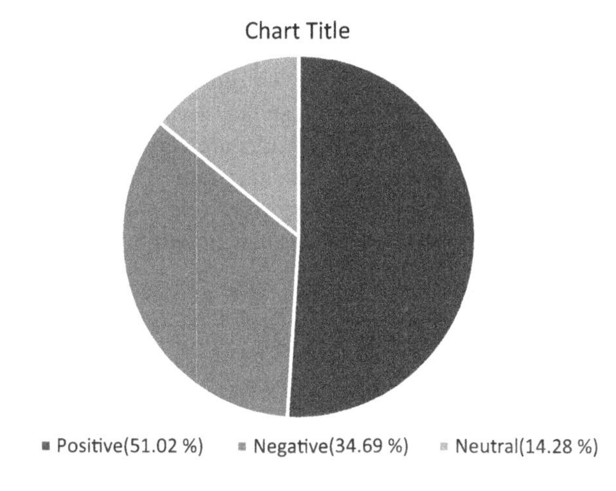

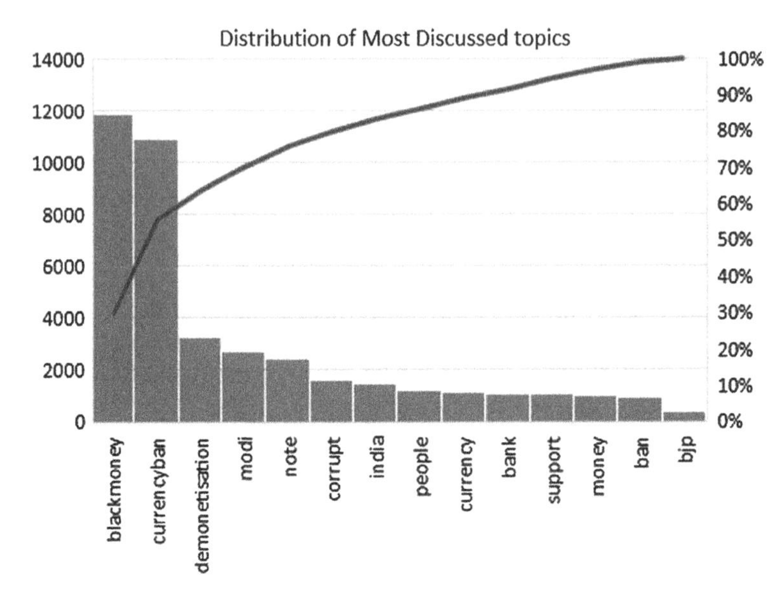

Fig. 4 Distribution of most discussed topics

maximum tweets comes under the category of unknown. In the second part of task 1, polarity analysis is done and the results show that 51.02% consider this move as positive, 34.69% against this decision and 14.28% are neutral on this decision is shown in Fig. 3. In task 2, LDA applied and the most discussed topics are extracted from the huge collection of tweets and the distribution of the most discussed topics is shown in Fig. 4. The LDA output also supports the Naïve Bayes results, that most of the twitter use supports the government decision of Demonetization.

4 Conclusion

In this study and analysis, we demonstrated how Latent Dirichlet Allocation (LDA) and Naïve Bayes can be used to understand people emotion on the demonetization move. The results are very promising and clearly shows that most of the people are happy with this step. This analysis can be applied to another similar domain. In the future work, we would demonstrate the mentioned algorithms in the distributed environment with a huge twitter data.

References

1. Sanchez, Gaston (2015). Sentiment Analysis with "sentiment". Mining twitter with R. 09 November 2015.
2. David M. Blei, Andrew Y. Ng, Michaell. Jordan (2003). Latent Dirichlet Allocation. Journal of machine Learning Research (3) 993–1022.
3. Sona Taheri, Musa Mammadov and Adil M. Bagirov (2011). Improving naïve Bayes Classifier using conditional probabilities. AusDM Ninth Australasian Data Mining Conference, (121) 63–68.
4. Jiguang Liang, Ping Liu, Jianlong Tan and Shuo Bai (2014). Sentiment Classification Based on AS-LDA Model. Information Technology and Quantitative Management, ITQM 2013, Procedia Computer Science (31)511–516.
5. Kunlun Li, Jing Xie, Xue Sun, Yinghui Ma and Hui Bai (2011). Multi-class text categorization based on LDA and SVM. Advanced in Control Engineering and Information Science, Procedia Engineering, (15) 1963–1967.
6. Kaiquan Xu, Jiexun li and Stephen Shaoyi Liao (2011). Sentiment Community detection in Social Networks. Proceedings of the 2011 iConference, 804–805.
7. Deepa Anand and Deepan Naorem (2016). Semi-supervised Aspect Based Sentiment Analysis for Movies using Review Filtering. 7th International Conference on Intelligent Human Computer Interaction. Procedia Computer Science, (84) 86–93.
8. Nur Azizah Vidya, Mohamad Ivan Fanany and Indra Budi (2015). Twitter Sentiment to Analyze Net Brand Reputation of Mobile Phone Providers. The Third Information Systems International Conference. Procedia Computer Science, (72)519–526.
9. Asad Ullah Rafiq Khan, Madiha Khan and Mohammad Badruddin Khan (2016). Naïve Multi-label classification of YouTube comments using comparative opinion mining. Symposium on Data Mining Applications. Procedia Computer Science, (82), 57–64.
10. Chenghua Lin and Yulan He (2009). Joint sentiment/topic model for sentiment analysis. 18th ACM conference on Information and Knowledge Management, 375–384.
11. Ralf Krestel and Stefan Siersdorfer (2013). Generating contextualized sentiment lexica based on latent topics and user ratings. 24th ACM Conference on Hypertext and Social Media, 129–138.
12. Wanying Ding, Xiaoli Song, Lifan Guo, Zunyan Xiong and Xiaohua (2013). A Novel Hybrid HAD-LDA Model for Sentiment Analysis. IEEE/WIC/ACM International Joint Conference on Web Intelligence and Intelligent Agent Technologies, (01)329–336.

Prediction of Gallstone Disease Progression Using Modified Cascade Neural Network

Likewin Thomas, M.V. Manoj Kumar, B. Annappa, S. Arun and A. Mubin

Abstract Prediction of disease severity is highly essential for understanding the progression of disease and initiating an early diagnosis, which is priceless in treatment planning. A Modified Cascade Neural Network (ModCNN) is proposed for stratification of the patients who may need Endoscopic Retrograde Cholangiopancreatography (ERCP). In this study, gallstone disease (GSD) whose prevalence is increasing in India is considered. A retrospective analysis of 100 patients was conducted and their case history was recorded along with the routine investigations. Using ModCNN, the associated risk factors were extracted for the prediction of disease progression toward severe complication. The proposed model outperformed showing better accuracy with an *area under receiver operating characteristic curve (area under ROC curve) of 0.9793, 0.9643, 0.9869, and 0.9768* for choledocholithiasis, pancreatitis, cholecystitis, and cholangitis, respectively, when compared with Artificial Neural Network (ANN) showing an accuracy of *0.884*. Hence, the proposed technique can be used to conduct a nonlinear statistical analysis for the better prediction of disease progression and assist in better treatment planning, avoiding future complications.

L. Thomas (✉) · M.V. Manoj Kumar · B. Annappa
Department of Computer Science and Engineering, National Institute of Technology Karnataka, Surathkal, Mangalore, India
e-mail: likewinthomas@gmail.com

M.V. Manoj Kumar
e-mail: manojmv24@gmail.com

B. Annappa
e-mail: annappa@gmail.com

S. Arun
Department of Surgery, Government Medical College, Manjeri, Malappuram, Kerala, India
e-mail: arunsuthan@gmail.com

A. Mubin
Department of Surgery, Government Medical College, Calicut, Kerala, India
e-mail: Mubin05abdul@gmail.com

A.K. Somani et al. (eds.), *Proceedings of First International Conference on Smart System, Innovations and Computing*, Smart Innovation, Systems and Technologies 79, https://doi.org/10.1007/978-981-10-5828-8_69

Keywords Cascade Neural Network · Artificial neural network
Gallstone disease · Area under curve · Receiver operating curve

1 Introduction

Developing an information model for a structured and nonlinear representation of medical data is a real challenge. The information model supports the decision-making process with statistical analysis. It is a comprehensive and consistent way of representing the clinical data [1]. The information retrieved by analyzing the clinical data assist in building a decision support system. It is very difficult to analyze such clinical data due to its uncertainty and complexity of diagnostic information. This decision supporting system helps in taking appropriate clinical decisions about the patient's management. In this study, we developed an information model which compares and selects the effective statistical tool for analyzing the clinical data and identifying the risk factors associated with them. These significant risk factors can be used as predictors for the prediction of patients who may need an emergency intervention or who may enter into complication. This is a retrospective study on 100 patients with GSD admitted at Calicut medical college, Kerala, India, during the period of 2014–2015.

GSD, which was once considered as the western disease, in last few decades has increased its prevalence in India. A group from All India Institute of Medical Science, Delhi, conducted a study in Kerala, a southern Indian state. On detailed examination, ~126/100,000 population showed the prevalence of acute pancreatitis and ~98/100,000 showed calcific pancreatitis against ~27/100,000 populations in western countries, which is too high [2]. Its prevalence is also increasing to 29% in the north Indian women [3]. This drastic increase is due to changes in the food habits. GSD is also a significant risk factor of gall bladder cancer [4]. Hence, an early detection and management of GSD will prevent the progression toward adverse complications. For this reason, there is a need of an optimal technique to classify and predict the patients who may enter the complicated stage. This inspired us to develop a statistical information model for extracting the information from GSD data and assist in taking critical decisions.

The current study proposes a Modified Cascade Neural Network (ModCNN) and compares its performance of prediction with well-established statistical model: ANN [5]. Jovanovic P et al. used ANN for selecting the patients who may need Endoscopic Retrograde Cholangio-Pancreatography (ERCP). This was a prospective study of 291 patients where ANN showed an accuracy with *area under ROC curve* of *0.884* for the prediction. *92.3%* of patients were accurately classified with positive findings who may need ERCP.

2 Materials and Methods

The proposed ModCNN evaluated the risk level of each patient at the time of admission without getting affected by concomitant disorders. The study design was a retrospective analysis of 100 patients admitted to Government Medical College, Calicut India, from 2014 to 2015. The percentages of observation of the investigations are shown in Fig. 1.

ANNs mimics the human brain and are composed of nonlinear combination of computational elements known as neurons. They are trained with the cases of known outcome, during which the weights interconnecting neurons are adjusted using the back-propagation technique. They have been successfully applied for diagnostic radiology including pulmonary embolism on ventilation–perfusion scans and differentiation of benign from malignant breast lesions [6–8]. They are also seen applied in lung disease and coronary artery disease [9, 10], and for the outcome analysis of pancreatitis [11]. Hence, ANNs are well-established models for classifying the disease severity and pattern recognition. CNN is a form of ANN, with cascade correlation of network and is a supervised learning architecture introduced by *Scott Fahlman* and *Christian Lebiere* in 1989 [12]. Here, the weights are kept frozen by adding hidden units at each iteration. ModCNN is a modified architecture of CNN, where at each iteration, an optimal number of neurons are

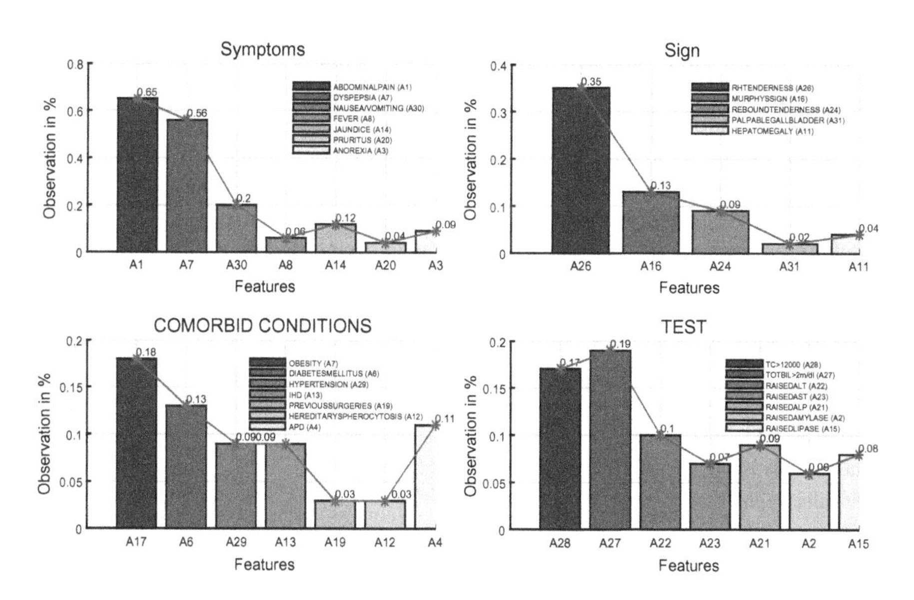

Fig. 1 Anthropometry and demographic characteristic of GSD in the proposed study. Height of bar is the total percentage (mentioned on the top of the bar) of observation the feature observed in 100 patients. Here, USG readings are excluded from the study as they were used for validation of the result. These tests were conducted and recorded at Calicut medical college by the fourth and fifth co-authors of this paper

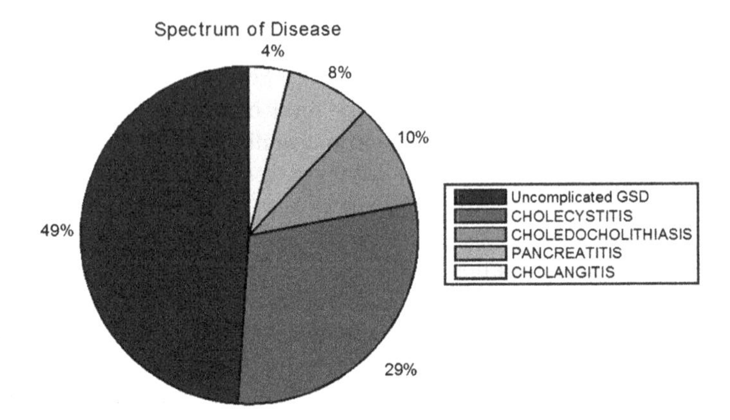

Fig. 2 Spectrum of GSD obtained by ModCNN classifier. Each slice shows the percentage of prevalence of each spectrum. The original observation through USG with error percentage is shown in Table 1

identified for every hidden units. This optimal combination of neurons in hidden unit is obtained using Least Mean Square (LMS) and gradient descent. Due to the nonlinearity in clinical data, a nonlinear activation function is employed at an *output* of hidden and output units.

The classified spectrum of GSD is shown in Fig. 2. An accuracy of ModCNN classification was validated by comparing it with the USG findings. This is shown in Table 1 with the percentage of drift from the actual values. From this comparison, we were able to analyze that ModCNN erroneously identified one of the cases as *cholangitis* and also identified *choledocholithiasis* as *cholecystitis*, which were later ruled out on validation after USG. This observation found its similarity with the California study [13], as shown in Table 2. ModCNN also found *11, 10, 13, and 7* significant risk factors for *Cholecystitis, Choledocholithiasis, Pancreatitis, and Cholangitis,* respectively, among 26 inputs which were fed as its input.

Table 1 The comparison of ModCNN classification with USG findings and percentage of error in the classification

Spectrum of disease	ModCNN classification (%)	USG findings (%)	Error (%)
Uncomplicated GSD	50	51	0.019608
Cholecystitis	28	27	0.037037
Choledocholithiasis	10	11	0.090909
Pancreatitis	8	8	0.000000
Cholangitis	4	3	0.333334
Average percentage error			0.480888

The classifier classified 50% as uncomplicated GSD against 51% of actual findings using USG. The readings in the second and third columns are the result of the classifier and USG, respectively. The USG was conducted and recorded at Calicut medical college. The fourth column is the % of error in the classification. The spectrum of GSD is shown in 3–6 rows

Table 2 Comparison of percentage of observation in the proposed study with hospital-based and community-based California study

Spectrum of disease	Hospital-based California study (%)	Community-based California study (%)	Proposed study (%)
Cholecystitis	29	35.9	27
Choledocholithiasis	5	3.1	11
Pancreatitis	6	4	8
Cholangitis	0	0.2	3

The second and third columns of the study are the observation seen in hospital-based and community-based California study conducted by Robert. E. Glasgow et al. in 1996 [13] and last column is the result of the current study. This comparison shows that the prevalence of GSD in India is almost same as that of California study

3 Evolution of Cascade Neural Network

The synaptic connections in human brain are very unique for every individual and are not completely inherited. As a learning process, this synaptic connection iteratively trains itself. The Carnegie Mellon University conducted an experiment on neurons and studied that every feedback from different parts of human body is purely due to neurons in the brain, which are connected by these synaptic connections. Hence, as per the learning process since birth, this feedback system trains the brain to interpret and understand sensory stimuli. It is highly impossible by any convention physiological experiment to understand the mechanism of pattern recognition inside the brain. But in 1943, the neurophysiologist *Warren* and *Mc Cullon* and a mathematician *Walter Pitts* [14] understood this mechanism. Using this knowledge of neurons, they introduced its functionality and modeled simple neural network. Since then, a lot of research has been done. But neural network found its actual emergence in late 1980s and soon got diminished in late 1990s. Due to the recent advancement in computational units, there is resurgence in neural network.

Scott Fahlman and Christian Lebiere [12] in 1989 introduced a new CNN architecture with cascade correlation of network which learns by experience. Here, the weights are frozen as the hidden units are added to the network. CNN proved to be quicker than traditional ANN as it did not use backpropogation algorithm for adjusting the weights, but instead, weights were adjusted to maximize the sum "S" over the all output units "O". CNN receives the trainable inputs to its pre-existing hidden units, with no output units connected to it. The input units are then processed by adjusting the weights to maximize the sum "S". The output "O" is of the magnitude of correlation between "V" (the candidate unit values) and "E_O" (the residual output error):

$$S = \sum_{O} \left| \sum_{p} (V_p - \bar{V})(E_{p,O} - \bar{E}_O) \right|. \tag{1}$$

In Eq. (1), "P" is the training pattern, and $\bar{V}$ and $\bar{E}$ are averaged over "V" and "E", respectively. The objective here is to identify the best combinations of hidden units which maximizes "S". In order to maximize "S" the backpropogation rule of taking partial derivation of "S" with respect to each combination of input weights θ is applied:

$$\frac{\partial S}{\partial \theta_i} = \sum_{p,O} \sigma_o (E_{p,O} - E_O) f'_O I_{ip}. \tag{2}$$

4 Modified Cascade Neural Network

The current study proposes a modified CNN (ModCNN) for analyzing the patterns of the clinical data and predict the disease progression. The process begins with a minimum number of hidden units and neurons in each hidden unit. After each iteration, error E_O is calculated using LMS algorithm. The objective here is to adjust the weights so that it minimizes MSE. After computing MSE, the gradient descent algorithm is applied to find the minimum error E_O. ModCNN works in two main layers of architecture. *The first layer* is for adding new neurons to each hidden unit and to identify the number of neurons that gives minimized error E_O. On identifying the optimal number of neurons, a new hidden unit is added to the architecture. *The second layer* is for adding new hidden unit by freezing the number of neurons in the previous layer and computing the MSE with this new hidden unit. Obtained MSE of current hidden unit is compared with the MSE of the previous hidden unit. If the MSE has decreased, then that hidden unit is frozen. Process is continued to find the optimal number of neurons for the frozen hidden unit. The architecture of ModCNN

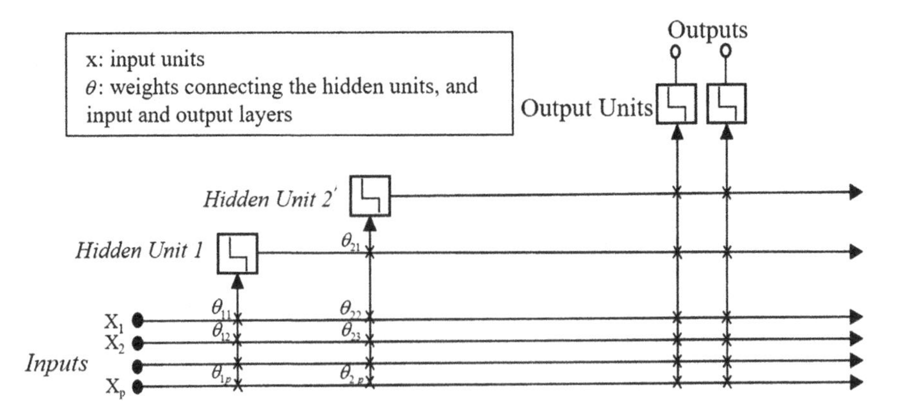

Fig. 3 Schematic diagram of ModCNN architecture. Here, there are two hidden units, p inputs, and two output units. X showing that hidden units and neurons are trained repeatedly. The architecture is a slight modified CNN architecture in [12]

is shown in Fig. 3. The architecture is slightly modified version of CNN architecture in [12].

4.1 Least Mean Square Algorithm (LMS Algorithm)

LMS also known as *Widrow-Hoff algorithm* or *delta algorithm* is used to identify the least MSE E_O^2. Each input $X = x_0, x_1, x_2, \ldots, x_p$ goes to intermediate weights $\theta = \theta_0, \theta_1, \theta_2, \ldots, \theta_{nl}$ (where n is the number of neurons for l number of hidden units) to give the *predicted output* $h_\theta(x) = I^T \times \theta = \sum_{i=0}^{n} I^T \times \theta$. For each input pattern, there are actual output O, predicted output $h_\theta(x)$, and an error $E_o = h_\theta(x) - O$. The error is used to adjust the weights so that MSE is minimum. On squaring the error E, *square of error* equation is obtained $= E^2 = h_\theta(x)^2 - (2 \times h_\theta(x) \times X^T\theta) + (X^T\theta \times \theta^T X)$. Hence, the MSE is obtained by averaging E^2 and is given in Eq. (3):

$$MSE = E[E_o^2] = E[h_\theta(x)^2] - (2E[h_\theta(x) \times X^T]\theta) + (\theta^T E[X \times X^T]\theta).$$

Let $P = E[h_\theta(x) \times X]$ and $R = E[X \times X^T]$.

$$MSE = E[h_\theta(x)^2] - 2 \times P^T \times \theta + \theta^T \times R \times \theta \quad (3)$$

This quadratic Eq. (3) yields a bowl-shaped plane with weights on x-axis and y-axis, and MSE on Z-axis. The gradient descent algorithm is used to reach the local minimum of the surface, where the weights are optimal and give minimum error “E_O”. The ADALINE circuit is generalized and the quadratic equation is shown in Eq. (3).

5 Result

The ModCNN imitates ANN architecture with an *input layer*, an *output layer*, and *hidden layers* between them. The feed-forward weights connecting the input layer to output layer through the hidden layer is adjusted by the back-propagation technique to get minimum MSE. The proposed method dynamically discovers an optimal model by identifying the number of neurons for each hidden unit and total number of hidden units by running gradient descent algorithm. It calculates the MSE for each neuron added to the frozen hidden. In this study, out of the 100 patients with GSD, there were *48%* males and *52%* females. The majority of cases were found in between the age group of 30–60 with 34 females and 35 males. ModCNN identified the different risk factors associated with each spectrum of GSD. On comparison of ModCNN with ANN, it was noted that ModCNN showed

Table 3 Statistical analysis and prevalence of significant risk factors associated with different spectrums of GSD

Spectrum of disease	Factors associated	Area under ROC curve
Cholecystitis	A1(*P* = *8.09e−06*), A3(*0.1568*), A4(*3.26e−05*), A7 (*P* = *0.7350*), A8(*P* = *4.91e−08*), A11(*P* = *0.0002*), A16(*P* = *0.0431*), A17(*P* = *0.2145*), A24 (*P* = *0.0656*), A26(*P* = *0.6537*), A29(*P* = *0.9828*)	0.9348
Choledocholithiasis	A1(*P* = *0.0099*), A3(*P* = *0.9179*), A7(*P* = *0*), A11 (*P* = *0.2689*), A13(*P* = *0.4730*), A14(*P* = *0.2689*), A19(*P* = *0.5364*)	0.9653
Pancreatitis	A1(*P* = *0.0305*), A4(*P* = *0.6994*), A7(*P* = *0.4563*), A8(*1.4e−15*), A9(*P* = *0.0134*), A10(*P* = *1.2e−*), A12 (*P* = *0.6040*), A14(*P* = *0.8876*), A19(*P* = *5.55e −16*), A21(*P* = *0.4855*), A22(*P* = *0.6040*), A26 (*P* = *0.5300*), A28(*P* = *1.6e−10*)	0.9875
Cholangitis	A1(*P* = *0.1969*), A4(*P* = *0.1190*), A7(*P* = *7.04e −06*), A8(*P* = *0.2483*), A9(*P* = *0.0165*), A10 (*P* = *0.8016*), A14(*P* = *0.7571*), A19(*P* = *0.2094*), A21(*P* = *0.5802*), A26(*P* = *0.0038*)	0.9768

The table listed the risk factors shown by the model which showed the highest accuracy in *area under ROC curve*. The proposed ModCNN showed better accuracy when compared to ANN. The third column shows the *area under ROC curve* value of the ModCNN. These risk factors are used for the prediction of disease progression. Hence, having appropriate risk factors is highly essential for giving accurate prediction of disease progression. The comparison result is shown in Fig. 4. (A1, A2, etc., are the factors associated with GSD and is explained in Fig. 1)

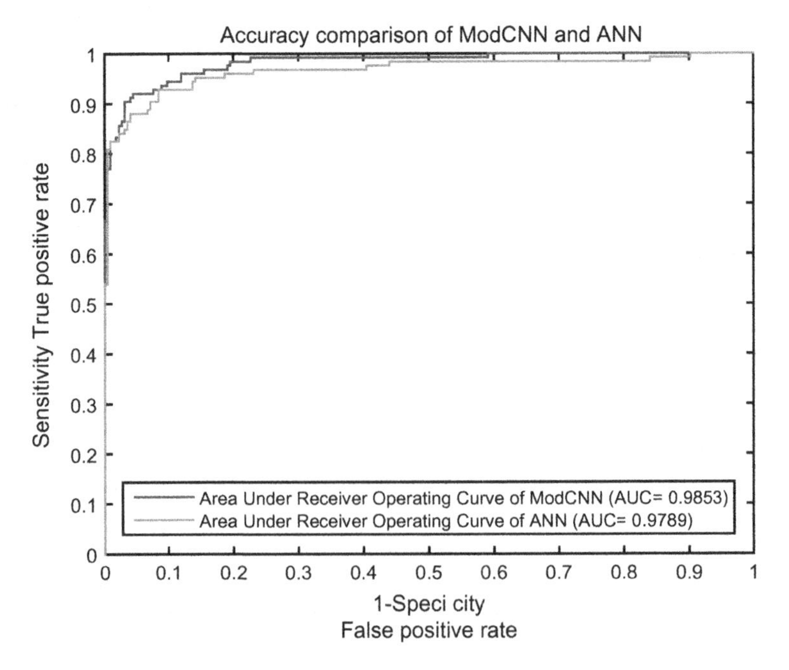

Fig. 4 *Area under ROC curve* of ModCNN for predicting and classifying the patients with severe GSD

better accuracy when tested for *area under ROC curve* for cholangitis, pancreatitis, cholecystitis, and choledocholithiasis with an *area under ROC curve* = 0.9768, 0.9643, 0.9869, and 0.9793, respectively. This is tabulated in Table 3. This analysis shows that ModCNN outperformed ANN and is shown in Fig. 4. The identified risk factors were then used as the predictors and were fed to ModCNN. Using ModCNN was able to predict the patients who may need immediate interventions at the time of admission and avoid future complications.

6 Conclusion

The proposed ModCNN is an attempt to analyze the clinical data and build a decision supporting system. This system predicts the severity and disease progression. Using this prediction an appropriate treatment can be initiated. In the current study, an information model is built to assist in taking clinical decision for GSD. ModCNN proposed showed the better *area under ROC curve* when compared with ANN. It was trained and validated using k-fold cross-validation technique for identifying the best *fold* where the accuracy was higher. It indicated that there are *seven, thirteen, eleven,* and *ten* risk factors significantly associated with *choledocholithiasis*, *pancreatitis*, *cholecystitis*, and *cholangitis,* respectively. These risk factors accurately predicted the patients on whom CECT abdomen had to be performed. They also predicted the patients who may need ERCP. Hence, ModCNN can be used as a routine clinical practice to predict the patients who need immediate interventions at the time of admission and avoid any further complications.

Acknowledgement The Institutional Ethics Committee, Govt. Medical College, Kozhikode has evaluated the protocol of the thesis entitled "Gall stone disease, modes of presentations and its association with gall stone pancreatitis" submitted by Dr. Mubin A, Junior Resident, Dept. of General Surgery, Govt. Medical College Kozhikode.

References

1. Huff S. M., Rocha R. A., Bray B. E., Warner H. R. and Haug P. J. "An event model of medical information representation", Journal of the American Medical Informatics Association, Vol. 2 no. 2, pp. 116–134, Mar. 1995.
2. Garg P. K., editor, "Chronic Pancreatitis-ECAB". Elsevier Health Sciences, 2013 Jun. 17.
3. Khuroo M. S., Mahajan R., Zargar S. A., Javid G. and Munshi S. "Prevalence of peptic ulcer in India: an endoscopic and epidemiological study in urban Kashmir". Gut., Vol. 30, no. 7, pp. 930–934, Jul. 1989.
4. Kapoor V. K. "Cholecystectomy in patients with asymptomatic gallstones to prevent gall bladder cancer–the case against", Indian Society of Gastroenterology, 2006.
5. Jovanovic, Predrag, Nermin N. Salkic, and Enver Zerem. "Artificial neural network predicts the need for therapeutic ERCP in patients with suspected choledocholithiasis", Gastrointestinal endoscopy, Vol. 80, no. 2, pp. 260–268, Aug. 2014.

6. Tourassi G. D., Floyd C. E, Sostman H. D. and Coleman R. E. “Acute pulmonary embolism: artificial neural network approach for diagnosis”, Radiology, Vol. 189, no. 2, pp. 555–558, Nov. 1993.
7. Chan H. P., Sahiner B., Petrick N., Helvie M. A., Lam K. L., Adler D. D. and Goodsitt M. M. “Computerized classification of malignant and benign microcalcifications on mammograms: texture analysis using an artificial neural network”, Phys. in Med. and Biol., Vol. 42, no. 3, pp. 549–567, Mar. 1997.
8. Baker J. A., Kornguth P. J., Lo J. Y., Williford M. E. and Floyd Jr. C. E. “Breast cancer: prediction with artificial neural network based on BI-RADS standardized lexicon”. Radiology, Vol. 196, no. 3, pp. 817–822, Sep. 1995.
9. Fujita H., Katafuchi T., Uehara T. and Nishimura T. “Application of artificial neural network to computer-aided diagnosis of coronary artery disease in myocardial SPECT bull's-eye images”, J. Nucl. Med. Vol. 33, pp. 272–276, 1992.
10. Ashizawa K., Ishida T., MacMahon H., Vyborny C. J., Katsuragawa S. and Doi K. “Artificial neural networks in chest radiography: application to the differential diagnosis of interstitial lung disease”, Acad. Radiol., Vol. 6, no. 1, pp. 2–9, Jan 1999.
11. Keogan M. T., Lo J. Y., Freed K. S., Raptopoulos V., Blake S., Kamel I. R., Weisinger K., Rosen M. P, and Nelson R. C. “Outcome analysis of patients with acute pancreatitis by using an artificial neural network”. Academic radiology, Vol. 9, no. 4, pp. 410–419, Apr. 2002.
12. Fahlman Scott E., and Christian Lebiere, “The cascade-correlation learning architecture”. Vol. 2, 1989.
13. Glasgow R. E., Cho M., Hutter M. M. and Mulvihill S. J. “The spectrum and cost of complicated gallstone disease in California”. Archives of Surgery, Vol. 135, no. 9, pp. 1021–1025, Sep. 2000.
14. Mc. Culloch W. S. and Pitts W. “A logical calculus of the ideas immanent in nervous activity”. The bulletin of mathematical biophysics, Vol. 5, no. 4, pp. 115–33, Dec 1943.

AMIPRO: A Content-Based Search Engine for Fast and Efficient Retrieval of 3D Protein Structures

Meenakshi Srivastava, S.K. Singh, S.Q. Abbas and Neelabh

Abstract Proteins are macromolecules which are virtually involved in all of the life processes. The study of protein structures is of utmost importance in the field of bioinformatics. With the advancement in the field of computational biology, there has been tremendous upsurge in the sequential and the structural data deposition. The structure of a protein depends upon the sequence of the amino acids present in it, although similarity in sequence does not guarantee a similarity in structure. Despite the fact that the three-dimensional structure of protein molecule is very important to predict its functionality, yet the backbone of the searching has been majorly dependent upon the sequences rather than the structures. The leading platforms for searching structural similarity in proteins make use of *sequence-based searching* or *text-based searching* but do not provide the desired results. In the current manuscript, a model has been proposed to perform "content-based searching" on protein images. Content-based searching takes into account the visual/structure-based similarity and the information contained in the data sets rather than the traditional sequence-based searching. Intelligent Vision Algorithm has been applied to extract the visual features from the protein images for determining the similarity between two proteins. The proposed search engine model will result in an efficient and fast retrieval of similar protein structures.

M. Srivastava (✉) · S.K. Singh
Amity Institute of Information Technology, Amity University, Lucknow, UP, India
e-mail: msrivastava@lko.amity.edu

S.K. Singh
e-mail: sksingh1@amity.edu

S.Q. Abbas
Computer Science Department, Ambalika Institute of Information Technology, Lucknow, UP, India
e-mail: sqabbas@yahoomail.com

Neelabh
Department of Zoology (MMV), Banaras Hindu University, Varanasi, UP, India
e-mail: srivastava.neelabh@gmail.com

A.K. Somani et al. (eds.), *Proceedings of First International Conference on Smart System, Innovations and Computing*, Smart Innovation, Systems and Technologies 79, https://doi.org/10.1007/978-981-10-5828-8_70

Keywords Content-based searching · Sequence-based searching
Text-based searching · Visual similarity · Structural similarity

1 Introduction

Access to computational frameworks and organic databases now accessible on the Internet is turning into a major role player for scholars and bioinformaticians. Customary illustrations are the notable databases of arrangements of biomolecules such as nucleic acids (GenBank and the EMBL Data Library) and those for protein sequences SWISSPROT and PIR. In case of biology, proteins are one of the most complex biomolecules. Studying the spatial arrangement of each atom inside the protein and also understanding and analyzing the bonding between the atoms can be a very tedious job. Computational biologists and bioinformaticians have in their arsenal some useful bioinformatics tools which make this tedious job easier. But major problem faced by bioinformaticians is developing efficient and accurate algorithms in order to search similar protein structures. For example, in the case of mutated protein, for knowing the extent to which the structure has changed by a given mutation, one can go for querying biomolecules database for the structural similarity search which is different from the similarity search done in other fields, such as generic databases of complex types like GenBank. Unlike the other complex databases, here the query will not be generated directly over the textual attributes of the sequence, or feature vector but most frequently used alignment algorithm such as FASTA or BLAST will be run and ranked result of similar sequences will be displayed. The PDB data set contains textual annotations followed by a list of coordinates of atoms. In structural databases, like PDB textual fields, such as key words, organism, resolution are filled in the form for querying in database, which in general does not provide the user the desired output. Identifying groups of similar objects is a popular first step in biomedical data analysis, but it is error-prone and impossible to perform manually. Many computational methods have been developed to tackle this problem [1]. That is main reason that different authors make effort to develop more refined and dedicated domain based accessing systems based on the information itself stored in file [2, 3]. In the present manuscript, a web-based search engine "AMIPRO" has been proposed which will retrieve the protein images based on their visual similarity in their 3D structure.

2 Literature Review

Proteins of known function which are structurally similar to proteins of unknown function exhibit a relationship among themselves. Dobson et al. 2004 have discussed many examples where sequence-based searches are insufficient for annotating protein functions. Hence, establishment of structure to structure relationship

Table 1 Comparison of existing structural alignment servers

Parameters	Distance alignment matrix method (DALI)	Combinatorial extension (CE)	Sequential structure alignment program (SSAP)	Mammoth
Breaking into smaller fragments	Breaks the input structures into hexa-peptide fragments and calculates a distance matrix	Breaks structures of the inquiry set into a progression of parts then plans to reassemble into an entire arrangement	No	Disintegrate the protein structure into hepta-peptides which help in the computation of closeness scores.
Matrix	Distance matrix is designed which has two diagonals depending on overlapping sub matrices of size 6 × 6	Similarity matrix is used to store the scores obtained in the process	A series of matrices are constructed during the process	Similarity matrix is used to store the scores obtained in the process
Programming	2 rounds of dynamic programming are performed. The original version used a Monte Carlo simulation for structural alignment	Linear progression through the sequences is applied and extension of the alignment is done with the aid of the next high scoring AFP pair	Atom–Atom based Structural alignment. Double dynamic programming is applied	Hybrid (local/global) dynamic programming applied. Similarity Score calculated through RMS (URMS) method

is of immense importance in the field of Structural Genomics. The majority of the recent web tools for structure comparison are either devoted to a database search [4] or for pairwise structural alignments [5, 6]. A comparative study between some existing structural alignment servers is shown in Table 1. Extracting knowledge from these databases, we need efficient and accurate tools; which is a major goal of computational structural biology.

Extra effort and funds are currently being invested to improve and speed up the processing potential of many computer-based tools that reign in the field of structural bioinformatics [7]. In the current manuscript, a visual similarity based search engine for protein structures has been proposed. To our knowledge, no previous system has been developed which retrieves the protein image based on their visual similarity.

3 Proposed Model

The query-by-content search system involves interfacing among its five main modules, namely, query module, feature extraction module, the database module, the search engine module, and the visual interface which communicate with the user. A diagrammatic representation of the flow of procedures among these components is shown in Fig. 1.

3.1 Query Interface Module

Implementation of query interface has been done as query by example interface. Visualization methods, like backbone, ribbon and rocket may be used for querying the database. Present paper has discussed the query image representation and retrieval based on combination of three visualization method. Various visualizations of the PDB ID 5GHK have been shown in Fig. 2.

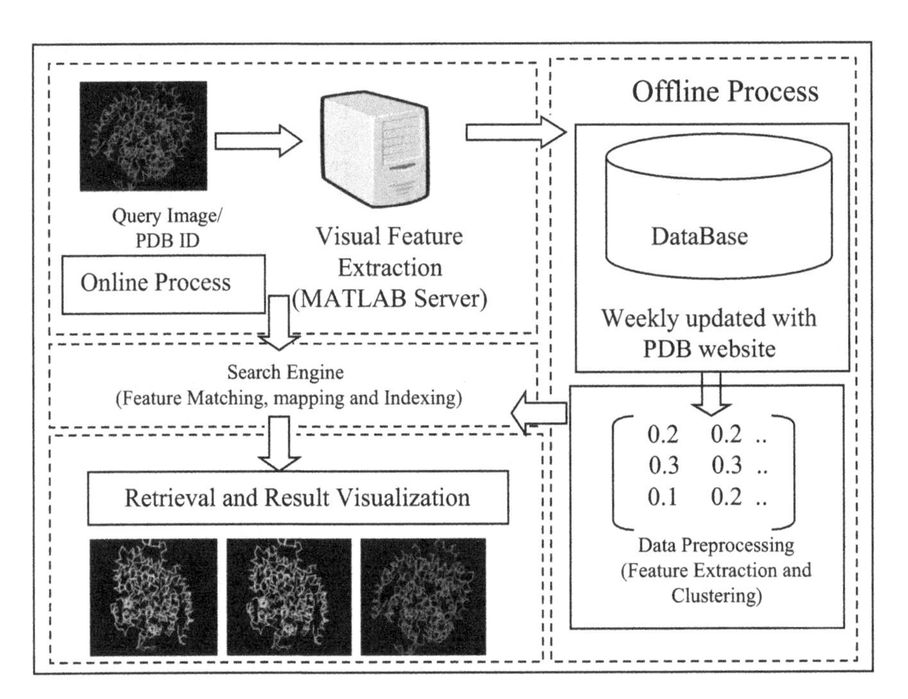

Fig. 1 The system architecture of AMIPRO, containing the five blocks: query image interface block, visual feature extraction block, database block, search engine block, image retrieval, and visualization block

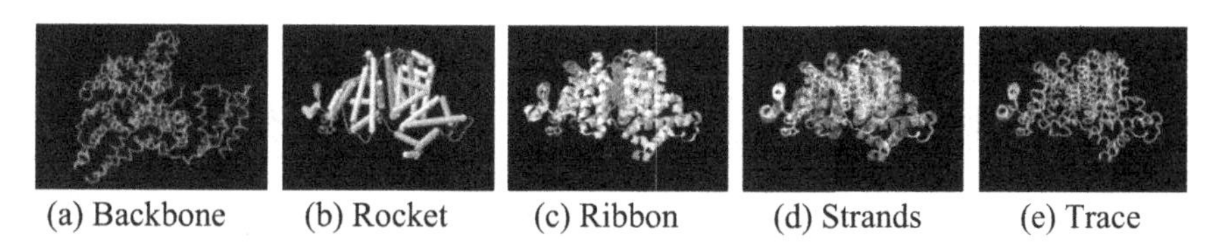

(a) Backbone (b) Rocket (c) Ribbon (d) Strands (e) Trace

Fig. 2 Various visualizations for example query image (PDB ID-5GHK)

3.2 Visual Feature Extraction Module

While designing of automatic feature extraction programs, there is always a trade-off between the precision and computational time involved in presenting a given visual feature. Generally, the representations leading to proficient and fast algorithms are chosen over complex and computational time-consuming approaches even though they might give more accurate results [8]. High-order local autocorrelation features have been extracted from the protein images [9]. Since HLAC features are shift invariant, it is prudent to choose HLAC features for shape-based feature representation [10]. Features are computed by applying the HLAC mask on the binary image. Feature computation involves counting the number of black pixels and white pixels, and the number of pixels is normalized by the image sizes [10]. HLAC feature is derived from *N*th-order autocorrelation functions, defined as:

$$r\binom{N}{f}(a_1, \ldots, a_N) = \int_p f(r)f(r+a_1), \ldots, f(r+a_N)dr, \quad (1)$$

where r is the image coordinate vector, a_i is the displacement vector. The order N is limited to the second order ($N \in \{0, 1, 2\}$). Then, the *N*th-order autocorrelation function counts the number of pixels satisfying the following logical condition: [9]

$$f(r) \bigwedge f(r+a_1) \bigwedge, \ldots, \bigwedge f(r+a_N) = 1 \quad (2)$$

3.3 Database Module

Database module is updated weekly by RCSB PDB. 200 Proteins were randomly collected and were synthesized using MATLAB Bioinformatics toolbox. Jmol [11] was used to acquire 6000 images at 128 × 128 pixels for each protein. The extracted primitive features were combined for pattern recognition and trainability. Principal component analysis is performed on the HLAC feature vector which performs a linear mapping of the data to a lower dimensional space in such a way that the variance of the data in the low-dimensional representation is maximized.

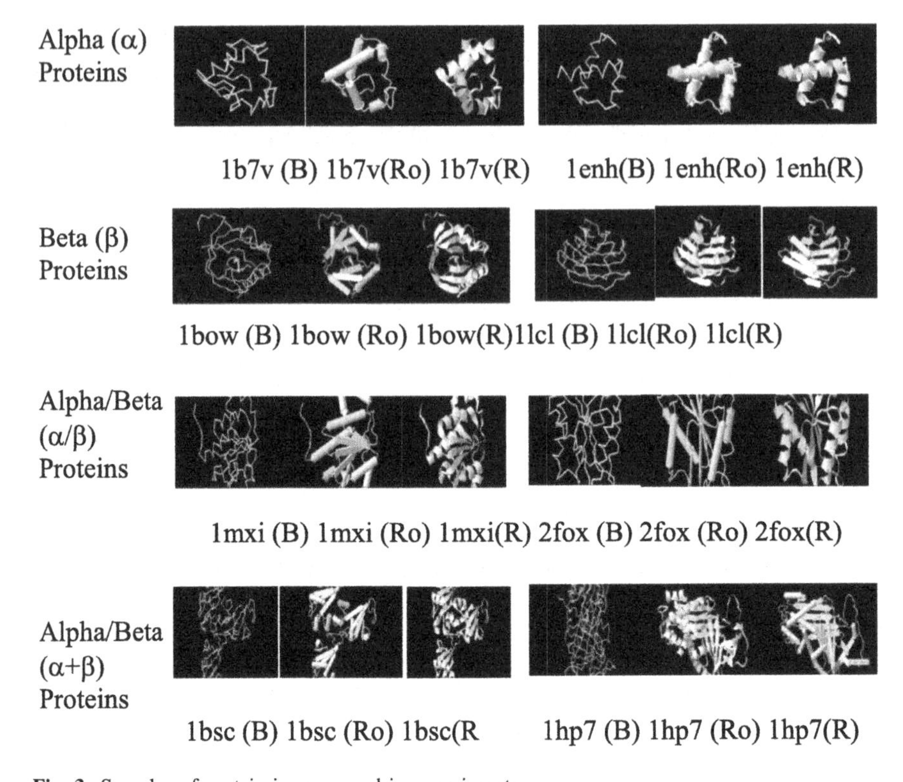

Fig. 3 Samples of protein images used in experiments

Finally, FCM [12] algorithm has been used to cluster the high dimensional data in the database. Ruspini [12] had introduced the concept of fuzzy c-partition which states that (Figs. 3, 4 and 5).

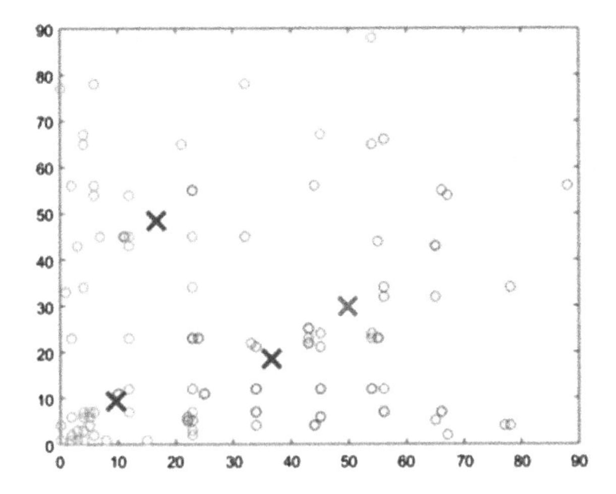

Fig. 4 FCM cluster (No of cluster = 4)

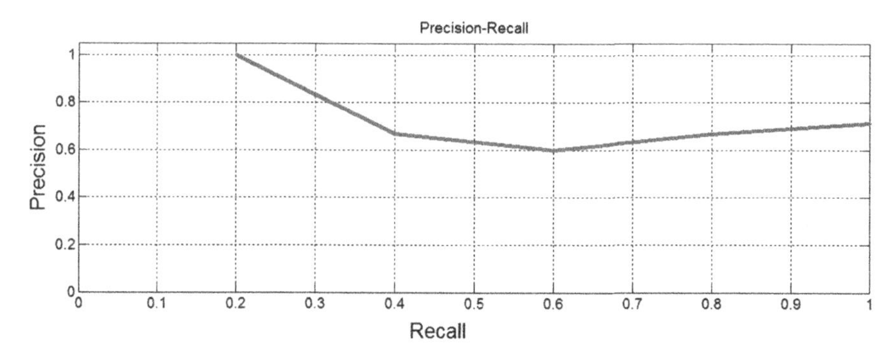

Fig. 5 Image retrieval accuracy

Given a set of data X = $\{x_1, \ldots, x_n\}$, in which each data point is a vector, U_{cn} be a ser of real c x n matrices, and c be an integer, $2 \le c \le n$. Then, the fuzzy c-partition space for X is the set [5]

$$Mfcm = \left\{ \mathrm{U} \in \mathrm{Ucn{:}\ uik} \in [0,\ 1]\ \sum_{i=1}^{k} \mathrm{uik} = 1,\ 0 < \sum_{k=0}^{c} ,\mathrm{uik} < n \right\} \tag{3}$$

The aim of the algorithm is to find fuzzy partitioning which conceded out through an iterative optimization of the objective function.

$$Jm(U, V; X) = \sum_{k=1}^{n} \sum_{i=1}^{n} (uik)||xk - vi||^2. \tag{4}$$

Here, $V = (v_1, v_{2,} \ldots, v_c)$ *is a matrix of unknown* cluster centers.

3.3.1 Search Engine Block

Retrieval speed is an important issue in image retrieval system, while image information is abundantly, and the data of image database is large [6]. The retrieval time is reduced by comparing the query with all the cluster centers rather than comparison with all the images in the database. Distance between the query image and the cluster center is calculated and the clusters are ranked according to their similarity with the query image. Finally, the two clusters from the top of this ranked list are selected and the query image is directly compared with the images in these clusters.

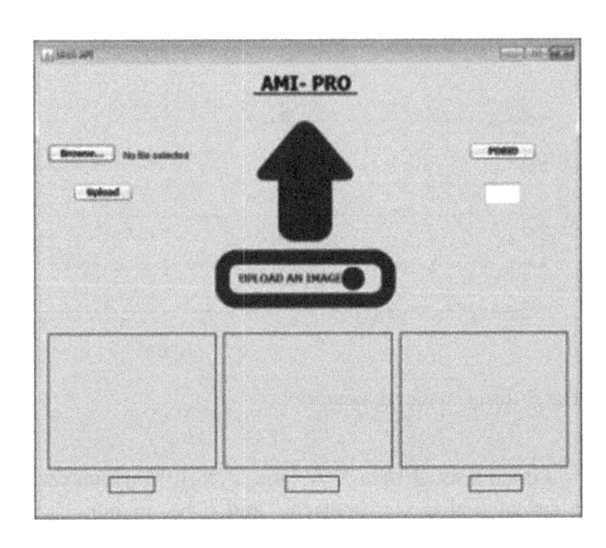

Fig. 6 Main page

3.3.2 Image Retrieval and Visualization Block

Top 3 similar images are shown in decreasing order of their similarity. Euclidian distance has been taken as the measure for calculating the similarity between the query image and center of cluster. Image retrieval is shown in Fig 6.

4 Results and Discussions

Prototype of the model is implemented in MATLAB [13]. Oracle Database Server version 12.1.0.2.0 and MATLAB R2015a have been used. The GUI-based interface is designed using Java (Fig. 5).

To validate the effectiveness of the proposed method, we did experiments on the randomly collected protein images of four classes of protein defined by SCOP database: Alpha (α), Beta (β), Alpha/Beta (α/β), and Alpha + Beta (α/β) [14]. Samples of protein images used in experiments are shown in Fig. 3.

In order to test the validity and efficiency of the proposed algorithm, FCM method is compared with other retrieval methods, including MSM-based k-means clustering [15] and FCM clustering. In mutual subspace based method, similarity is defined by the canonical between them [15]. A comparative study has been depicted in Table 2. Our results show that FCM-based clustering results in better performance in all cases.

The accuracy of the algorithm has been shown in Fig. 5 through precision recall graph. For test purpose, the bucket size of the database has been set to 20 and the best 5 matches for the query image have been retrieved.

The retrieval of proposed model is depicted in Fig. 7 a, b Two cases have been shown, first case is when the query image was present in the database and the

Table 2 Comparison of clustering results for MSM-based k-means clustering and FCM clustering (Number of protein images 200, number of cluster = 4)

Measurements\method\	MSM-based k-means clustering	FCM clustering
Accuracy (%)	84.00	87.00
Specificity (%)	78.46	83.08
Sensitivity (%)	94.29	94.29
F measure	80.49	83.54
Retrieval time (ms)	176.32	88.68

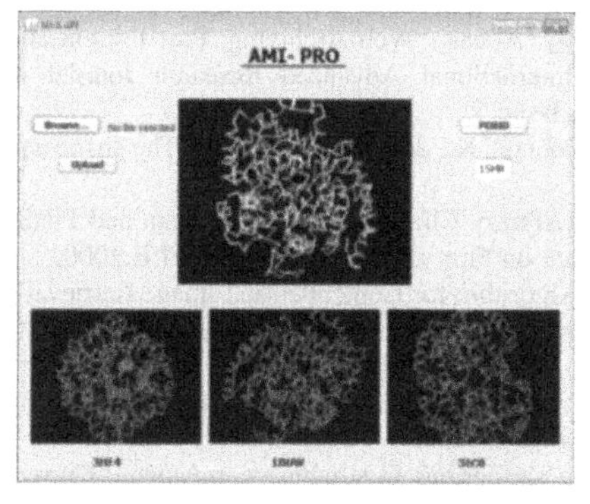

(a) PDBID 1A3N (b) PDBID 1A3N
(c) PDB ID 1GZX (d) PDB 1HBB

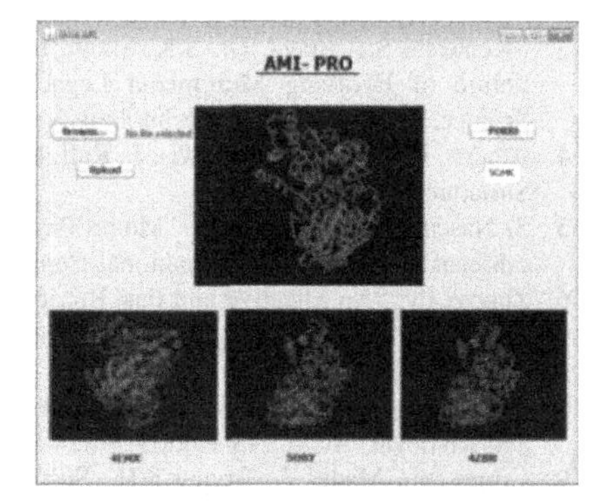

(a) PDBID 1A3N (b) PDBID 1GZX
(c) PDBID 1HBB (d) PDB ID 1MBN

Fig. 7 Results of retrieval

second case is when the query image was removed from the database. Results confirm that proposed model can easily retrieve the structurally similar protein at faster rate.

5 Conclusion and Future Work

The advantage of the proposed method is that no protein alignment is required. The said search engine improves the efficiency of retrieval of biomolecular images tremendously. The given approach has been designed on the basis of structural similarity rather than backbone structure which is the conventional retrieval methodology. Limitations of the traditional algorithms designed for protein structure similarity like nonoptimal alignment due to dependency on backbone length, run time cost of dynamic programming are tackled effectively by the proposed model as the protein images are stored on the basis of precomputed visual similarity between them. Since the proposed model AMIPRO works only on visual similarity, future work would focus on developing an algorithm which would engage both content-based similarity and visual similarity to give better retrieval results.

References

1. Wiwie C, Baumbach J, Röttger R Comparing the performance of biomedical clustering methods, NCBI, Nat Methods. 2015 Nov;12(11):1033-8. doi:10.1038/nmeth.3583.
2. Mussarat Yasmin, Sajjad Mohsin, Muhammad Sharif, Intelligent Image Retrieval Techniques: A Survey, Journal of Applied research and technology, Vol 14 December 16.
3. Meenakshi Srivastava, Dr. S.K. Singh, Dr. S.Q. Abbas, "Web Archiving: Past Present and Future of Evolving Multimedia Legacy", International Advanced Research Journal in Science, Engineering and Technology Vol. 3, Issue 3.
4. Alberts, B., Johnson, A., Lewis, J., Raff, M., Roberts, K., & Walter, P. (2002). The shape and Structure of proteins.
5. S. Nascimento, B. Mirkin, F. Moura-Pires, " AFuzzy Clustering Model of Data and Fuzzy c-Means", Ninth IEEE International Conference on Fuzzy Systems. FUZZ- IEEE 2000.
6. Zhuoyi Lv. "An Effective and Fast Retrieval Algorithm for Content-Based Image Retrieval", 2008 Congress on Image and Signal Processing, 05/2008.
7. www.jmolbiochem.com.
8. Luo, Zhang, Huang, Gao, Tian, Superimage: Packing Semantic-Relevant Images for Indexing and Retrieval, ACM 978-1-4503-2782-4/14/04.
9. Gibrat J.F., Madej T., Bryant S.H.: Surprising similarities in structure comparison. Current Open Struct Biology 6(3), pp. 377–385 (1996).
10. H. Sakano and S. T. Classifiers under continuous observations. *Lecture Notes in Computer Science*, Volume 2396/2002:798, 2002.
11. A. Herr´aez. Biomolecules in the computer: Jmol to the rescue. *Biochemistry and Molecular Biology Education*, 34(4):255–261, 2006.
12. E. Ruspini, "A new approach to clustering", Information and Control, 15, pp. 22–32, 1969.
13. MATLAB and Statistics Toolbox Release 2015a, The Math Works, Inc., Natick, Massachusetts, United States.
14. A. Andreeva, D. Howorth, J. M. Chandonia, S. E. Brenner, T. J. P. Hubbard, C. Chothia, and A. G. Murzin. Data growth and its impact on the scop database: new developments. *Nucleic Acids Research*, 36:D419–D425, 2007.
15. Tetsuo Shibuya, Hisashi Kashima, Jun, Pattern Recognition in Bio-Informatics, 7th IAPR International Conference, PRIB 2012 Proceedings.

A Risk Averse Business Model for Smart Charging of Electric Vehicles

Md. Muzakkir Hussain, Mohammad Saad Alam, M.M. Sufyan Beg and Hafiz Malik

Abstract Smart collaborations among the smart grid, electric vehicles, and aggregators will provide range of benefits to stakeholders involved in an intelligent transportation system (ITS). The EVs, nowadays, are becoming the epicenter of smart power system research towards the electrification of transport. However, massive penetration of EVs will pose management threats to the supporting smart grid in the foreseeable future. This work proposes a risk averse optimization framework for smart charging management of electric vehicles. Adopting conditional value at risk (CVaR) for estimating the risks, the work attempts to propose an optimized bidding strategy for the smart charging stations (SCS) that act on behalf of aggregators for managing the financial risk caused by the uncertainties. Finally, a fuzzified translation model is discussed along with notable methodologies as a solution strategy to the risk averse cost optimization problem.

Keywords Electric vehicles · Smart grid (SG) · Smart charging stations (SCS) · Conditional value at risk

Md.Muzakkir Hussain (✉) · M.M. Sufyan Beg
Department of Computer Engineering, ZHCET, AMU, Aligarh, India
e-mail: md.muzakkir@zhcet.ac.in

M.M. Sufyan Beg
e-mail: mmsbeg@cs.berkeley.edu

M.S. Alam
Department of Electrical Engineering, ZHCET, AMU, Aligarh, India
e-mail: saad.alam@zhcet.ac.in

H. Malik
Department of Electrical and Computer Engineering, University of Michigan, Dearborn, MI, USA
e-mail: hafiz@umich.edu

A.K. Somani et al. (eds.), *Proceedings of First International Conference on Smart System, Innovations and Computing*, Smart Innovation, Systems and Technologies 79, https://doi.org/10.1007/978-981-10-5828-8_71

1 Introduction

There are pressing discussions, arguments, panels, industrial research, and innovations for the overhaul of contemporary power grid infrastructures aided with rollout of smart grid [1]. The latter integrates green power generation options, control and monitors the energy usage and dynamically shuffles the generation, transmission, and distribution procedures in accordance to the power rates and system loads. The power grid is under threat from incapability to scale to the load. Repercussions in the blackouts are growing more severe [2]. Upgrading to smart grid utilities coupled with micro-nano services is not only a vision but the growing sinks in the power utilization has transformed it into an urgent imperative.

Due to the substantial price hike in the crude oil and prolonged dependence on foreign oil over the past decades, power GENCOs as well as the consumer are directly or indirectly forced to opt for alternative energy sources [3]. The bulk of the research is going on to ensure a zero emission mode of transport. Fortunately, positive results have been shown by such inventions and electric vehicles (EVs) which can serve as the best recourse in this storyline [4]. Conceptually, EVs integrate the electrical networks with so-called data and communication infrastructures through smart metering and sensing utilities [5].

Electric vehicles (EVs) possess potentials to guarantee environmental and energy security over conventional gasoline vehicles [6]. The year 2015 saw the global threshold of one million EVs on the road, a symbolic milestone demonstrating noteworthy efforts deployed jointly by governments and automotive industries over the past 10 years [7]. Despite its advantages, the integration of EVs to the power grid poses a serious challenge for electric utilities [8].

2 Risk Analysis for Mission Critical Smart Grid Operations

Switching from traditional power grid to a multitenant smart grid introduces substantial risks to power sector, an issue that needs to get fixed in inception phase. The penetration of EVs into modern road transport manifolds such concerns. The EVs are becoming a basin for multidimensional data production, an asset if mishandled, may befool the execution of whole systems. Moreover, the data generated due to cloud-IoT integrated transportation telematics coupled with advanced metering infrastructures (AMI) can be harmful to its stakeholders, specifically privacy and security [9]. Thus, it is an earnest need for the stakeholders to be assured with stringent protection protocols and be inert from the vulnerabilities. Such scenario necessitates incorporating robust risk analysis procedures that will evaluate and quantify the computational and business risks that persist in such critical infrastructures. In an attempt to have successful penetration of contemporary EVs fleet, this work proposes a risk averse business model for efficient mobility and charging management of electric vehicles. The work adopts conditional value at risk (CVaR) [10], for estimating the risks and proposes an optimized bidding strategy for the smart charging stations (SCS) that act on behalf of aggregators for managing the financial risk caused by the uncertainties.

2.1 Risk Management in Smart Grid Infrastructures Supporting EVs

Concerns for maintaining confidentiality, integrity, and availability in the transportation and metering utilities have stressed the research folks to work for embedding risk analysis into design in a way that ensures transparency and understandability among involves stakeholders. An ideal risk management model is consequence of successive phases namely risk scope definition, risk identification, risk characterization, risk evaluation, and risk mitigation planning [11]. The outputs of such models are iteratively communicated, reviewed, and evolved in course of execution to incorporate resiliency and sustainability in the SG infrastructures.

Since the architecture is leveraged to multiple modes of services and operations like grid to vehicle (G2V), vehicle to grid (V2G), vehicle to cloud (V2C) and mobility services, an efficient and intelligent synchronization mechanism is required at varying levels to enhance the adoption of more EVs [5]. Further, the commercial uncertainties invoked due to heavy penetration of EVs and smart utilities in the foreseeable future will pose threat in smart grid operations [12]. In such scenario, the infrastructure demands careful assignment of tasks to each of the players engaged in the execution. The power sink created by EVs needs to be sustained via optimal bidding strategies that undertake the risks and uncertainties into consideration, at the same time circumvent load on the smart grid. Owing to this, in this work, it is intended to have an optimized bidding scheme coupled with an exhaustive assessment of probable risks and commercial uncertainties.

The smart charging stations (SCS) are tailored to act on behalf of aggregators. In the first part, optimization scheme for SCS's payoff function is proposed considering its participation in varying services like G2V, V2G, V2C, mobility solutions, etc. In the next part, risks and uncertainties have been incorporated into the payoff function through stochastic formulations. Lastly, the whole bidding scheme is mapped to a multiobjective fuzzy logic to obtain a near optimal solution.

3 Problem Formulation

3.1 Optimization of the Payoff

The fluctuating nature of the power market and the uncertainties pertained in the infrastructure are incorporated in the charging station's payoff objective function. The revenue for the CS would be from payments of xEVs users, selling the excess power into the day ahead or into reserve markets and the rental that is procured by availing regulatory services. The cost pertaining to the CS vendors will include cost of electricity procurement from generation sources, cost of hired cloud services, cost involved when V2G mode is enabled, and battery degradation costs.

$$F_{s,t}^{payoff} = P_{s,t}^{revenue} + P_{s,t}^{dayahead} + P_{s,t}^{reg} + P_{s,t}^{V2G} - C_{s,t}^{gen} - C_{s,t}^{cloud} - C_{s,t}^{deg} \quad (1)$$

$P_{s,t}^{revenue}$ is the revenue obtained from the xEVs customers that may vary according to the demand, i.e., it may achieve higher value during rush hours. This fueling charge also depends on the mode of charging adopted by the xEVs owner. Thus, it can be formulated as

$$P_{s,t}^{revenue} = \sum_{p=1}^{N_p} n^p \left\{ \sum_{t=1}^{T} \sum_{p=1}^{N_s} w.\alpha_s (E_{p,t}^{demand} . \delta_{s,t}^{sale}) \right\} \forall p, \forall s, \forall t, \quad (2)$$

where w is the weight factor that decides the mode of charging, α_s is the probability of the occurrence of the *sth* scenario, n^p gives the number of vehicles having similar driving pattern, N_p is the size of vehicle fleet under the control of each charging hub. Each xEVs fleet is assumed to have a second-order aggregated concave demand payoff [3], given by

$$E_{p,t}^{demand} = -a_p (E_t^{p.sale})^2 + b_p (E_t^{p.sale}), \forall p, \forall t \quad (3)$$

Thus, Eq. (2) changes to

$$P_{s,t}^{revenue} = \sum_{p=1}^{N_p} n^p \left\{ \sum_{t=1}^{T} \sum_{p=1}^{N_s} w.\alpha_s ((-a_p (E_t^{p.sale})^2 + b_p (E_t^{p.sale})) . \delta_{s,t}^{sale}) \right\}, \quad (4)$$

where a_p and b_p are nonnegative cost coefficients. Similarly, if δ_t^R is the reserve market price at the *tth* hour (USD/MWh), the $P_{s,t}^{dayahead}$ is given by

$$P_{s,t}^{dayahead} = \sum_{t=1}^{T} \sum_{s=1}^{N_s} \left\{ \frac{\psi}{100} . (\delta_t^R)(E_t^{p,R}) + (\delta_t^R)(E_{s,t}^{p,R}) + (\delta_t^{tariff})(G_t^p) \right\}, \quad (5)$$

where ψ is the percentage of the energy price that is received by players as option fee, δ_t^{tariff} is the predetermined and constant tariffs that the seller facilitates the customer. $P_{s,t}^{reg}$ is the constant regulation cost that is received from the PHEV users due to ancillary services that are provided and is independent of the scenario. When the CS participates in the reserve market and dynamic energy management, it would act as a virtual power plant and inject the excess power in its store and the drained power from PHEV batteries into the grid. Here, a binary variable ε is employed that dictates whether the charging station is called by independent system operator (ISO) for ancillary services like demand management or not. Mathematically,

$$E_t^{p,R}.\varepsilon = E_{s,t}^{p,R}, \quad \varepsilon = [0, 1] \tag{6}$$

when the CS is called by ISO $\varepsilon = 1$ and 0 otherwise. In this mode, the CS imports power from the PHEV batteries $(P_{s,t}^{V2G})^-$ and trades it to the grid system $(P_{s,t}^{V2G})^+$.

$$(P_{s,t}^{V2G}) = (P_{s,t}^{V2G})^- + (P_{s,t}^{V2G})^+, \tag{7}$$

where

$$(P_{s,t}^{V2G})^- = (\delta_{s,t}^{import})(E_{s,t}^{import}) \tag{8}$$

$$(P_{s,t}^{V2G})^+ = (\delta_{s,t}^{export})(E_{s,t}^{export}) \tag{9}$$

From (8) and (9), we get

$$(P_{s,t}^{V2G}) = (\delta_{s,t}^{import})(E_{s,t}^{import}) + (\delta_{s,t}^{export})(E_{s,t}^{export}) \tag{10}$$

The cost of electricity procurement from the generation sources $C_{s,t}^{gen}$ is given by

$$C_{s,t}^{gen} = \sum_{t=1}^{T} \sum_{p=1}^{N_s} \left\{ \alpha_s . \left\{ (E_t^{buy} . \delta_{s,t}^{buy}) + (E_t^{buy} . \delta_{s,t}^{buy}) . \frac{\theta}{100} \right\} \right\}, \tag{11}$$

where θ is the incentive rate. The first term in (11) gives the charge under normal rate while the second on is the price during offer hour. Since the management of the infrastructure is controlled by cloud and the CS hires cloud services on lease, a significant portion of the cost also goes to the cloud providers. This is given by

$$C_{s,t}^{cloud} = (1 + \gamma).E_t^{storage} + (C_{s,t}^E + C^P) + \sum C_A^i . \omega_A \tag{12}$$

The addend in (12) gives storage cost, cost due to power consumed by cloud processing elements and cost of analytics algorithms used, respectively.

$$\sum C_A^i.\omega_A = C_A^{i1} + C_A^{i2} + \cdots + C_A^{ik}, \tag{13}$$

where $i1, i2, i3 \ldots ik$ are the sequence of data analytics modules employed.

The cost due to degradation in PHEV battery due to prolonged charging/discharging process is given by

$$C_{s,t}^{deg} = \sum_{t=1}^{T} \sum_{p=1}^{N_s} \{c_d.(DE_{s,t}^p)\}, \tag{14}$$

where

$$c_d = \frac{c_b.E_b + c_l}{L_c.E_b.DOD} \tag{15}$$

The cost term c_d is a function of battery energy capacity cost E_b, the battery cost per kilowatt-hour c_b, the battery replacement labor cost c_l and the battery cycle life at a specific depth of discharge (DOD) L_c. The overall objective function is to maximize the net profit which can be formulated as:

$$Maximize_{P_t^{sale}, C_t^{cloud}, C_{s,t}^{buy}} \{F_{s,t}^{payoff} = P_{s,t}^{revenue} + P_{s,t}^{dayahead} + P_{s,t}^{reg} + P_{s,t}^{V2G} - C_{s,t}^{gen} - C_{s,t}^{cloud} - C_{s,t}^{\circ}\} \tag{16}$$

Subject to constraints

$$SOC_{\min}^p \le SOC_{s,t}^p \le SOC_{\max}^p \tag{17}$$

$$SOC_{\min}^{CS} \le SOC_{s,t}^{CS} \le SOC_{\max}^{CS} \tag{18}$$

$$\delta_{\min}^{sale} \le \delta_{s,t}^{sale} \le \delta_{\max}^{sale} \tag{19}$$

$$\delta_{\min}^{cloud} \le \delta_{s,t}^{cloud} \le \delta_{\max}^{cloud} \tag{20}$$

$$R_{\min}^p \le R_{s,t}^p \le R_{\max}^p \tag{21}$$

$$E_t^{p.sale} + E_{s,t}^{p,R} = DE_{s,t}^p, \forall p, \forall t \tag{22}$$

$$R_t^p.E_t^{p,R}.E_t^{p.sale} = 0 \quad \forall p, \forall t \quad \text{and} \quad G_{p,t} \neq 0 \tag{23}$$

$$SOC_{s,t}^p \le SOC_{s,t-1}^p + (\varphi_p E_t^p - G_t^p - \varsigma_p.DE_{s,t}^p), \forall p, \forall s, \forall t \tag{24}$$

$$E_t^{p,R} \le SOC_{s,t}^p - SOC_{\min}^p \tag{25}$$

Equations (17) and (18) put the limits on the state of charge of both xEVs and charging station's storage, whereas (19) and (20) maintains the power as well cloud data usage cost within limits. Constraint in (21) defines the bounds on minimum and maximum rates of charging (MWh/hour). Discharged energy during plugged in time is stated in (22), while constraint (23) restricts that PEV can be charged and discharged only during plugged times. The power sold in the market and its SOC at the previous hour are given by Eq. (24), which shows how the SOC of the battery of each PEV would be updated using power which is drawn from the grid E_t^p, the power used for driving G_t^p. Finally, constraint (25) considers the limitation on the rate of SOC change during charging process.

3.2 Optimization of Risk Model for Handling Uncertainties

In this work, conditional value at risk (*CVaR*) measure is utilized for managing the financial risks associated with the charging stations (CS) and xEVs that may arise in uncertain scenarios [3]. *CVaR* is an extension of value at risk (*VaR*) that estimates the financial loss for the decision-making players under the circumstances where the uncertainty parameters fluctuate with a certain confidence level. *CVaR* is the mean excess loss, measured as the difference between the expected profit and average of potential profit values less than *VaR*. *CVaR* assumes to be one of the most efficient and coherent risk measure used in stochastic programming that satisfies properties like monotonicity, sub-additivity, translational invariance, and positive homogeneity. At a confidence level, μ, μ-*CVaR* is the expected profit of the $(1-\mu)$ scenario with lowest profit. If N_s scenarios could completely represent the possible uncertainties then,

$$CVaR(\mu) = \xi - \frac{1}{(1-\mu)}\left(\sum_{s=1}^{N_s} \alpha_s . \phi_s^p\right), \tag{26}$$

where ϕ_s^p is the nonnegative auxiliary variable, measured as the difference between VaR and the scenario payoff and ξ is the VaR at confidence level μ.

$$\phi_s^p = \xi - P_{s,t}^p \tag{27}$$

Hourly charging energy $E_{s,t}^p$ in each of the Monte Carlo (MCS) scenario is the energy purchased in the day ahead and real-time markets and is bounded by capacity of the xEVs battery.

$$E_{s,t}^p = E_{s,t}^o + \Delta E_{s,t}^p \tag{28}$$

Thus, the overall objective function for *CVaR* is formulated as

$$Maximize\, CVaR(\mu) = \xi - \frac{1}{(1-\mu)}\left(\sum_{s=1}^{N_s} \alpha_s.\phi_s^p\right) \tag{29}$$

Subject to

$$\phi_s^p \geq \xi - P_{s,t}^p \tag{30}$$

$$\phi_s^p \geq 0 \tag{31}$$

$$E_{s,t}^p = E_{s,t}^o + \Delta E_{s,t}^p \tag{32}$$

$$E_{s,t}^p \leq SOC_{\max}^P \tag{33}$$

$$E_{s,t}^p = P_{s,t}^p.\tau \tag{34}$$

$$\left|\Delta E_{s,t}^p\right| \leq \left|\Delta E_{s,t}^{\max}\right| \tag{35}$$

$$SOC_{s,t+1}^p \leq SOC_{s,t}^p + \eta.E_{s,t}^p.\tau \tag{36}$$

$$SOC_{\min}^{CS} \leq SOC_{s,t}^{CS} \leq SOC_{\max}^{CS} \tag{37}$$

The constraints (30) and (31) enact the *VaR* to be greater than the scenario payoff. Hourly charging energy in each MCS scenario is depicted in (32) whereas (33) restricts it to the capacity of battery. Equation (35) defines the range of energy deviation while the evolution of SOC over time is defined in (36). Merging Eqs. (16) and (29) in order to formulate the risk averse scheduling problem of a xEVs using *CVaR*, the resulting objective function obtained is given as

Maximize

$$(1-\beta)(P_{s,t}^{revenue} + P_{s,t}^{dayahead} + P_{s,t}^{reg} + P_{s,t}^{V2G} - C_{s,t}^{gen} - C_{s,t}^{cloud} - C_{s,t}^{deg}) + \beta\left(\xi - \frac{1}{(1-\mu)}\left(\sum_{s=1}^{N_s} \alpha_s.\phi_s^p\right)\right)$$

$$\beta \in [0,1] \tag{38}$$

With two additional constraints $\phi_s^p \geq 0$ and

$$\xi - \sum_{s=1}^{N_s} n^p \sum_{t=1}^{T} \left\{(\delta_{s,t}^p)(E_t^{p.sale}) + \frac{\psi}{100}.(\delta_t^R)(E_t^{p,R}) + (\delta_t^R)(E_t^{p,R}) - (\delta_{s,t}^R)(E_t^{p.buy})\right\} \leq \phi_s^p \tag{39}$$

3.3 Fuzzification of the Risk Model

The objective function for the net profit represented by (16) and the objective function for *CVaR* represented by (29) are combined using a linear approximation with weights β and $1-\beta$, respectively. This weighted sum method is incapable of dealing the complex situation which can arise when probabilistic uncertainties exist in the objectives. In multiobjective optimization, we find a compromised optimal solution as achieving an optimal solution is not feasible in conflicting circumstances. Fuzzy techniques can be used to achieve the near optimal solution for such uncertain objectives. It also reflects the satisfaction level of each objective represented by a membership function. The corresponding membership function for objectives (16) and (29) can be defined as follows:

$$\lambda_i(x) = \begin{cases} 0 & \text{If } Z_i \leq L_i \\ \frac{Z_i - Z^*}{L_i - U_i} & \text{If } L_i < Z_i < U_i \\ 1 & \text{If } Z_i \geq U_i \end{cases} \tag{40}$$

Here, L_i and U_i represent the lower and upper bound on objective function and Z^* is the optimal objective value obtained independently. The final compromise optimal solution using fuzzy programming technique can be achieved on solving follows:

$$Max\ \lambda \quad \lambda \leq \frac{Z_1 - Z_1{}^*}{L_1 - U_1}, \lambda \leq \frac{Z_2 - Z_2{}^*}{L_2 - U_2} \tag{41}$$

Subject to: Constraints (17)–(25) and (30)–(37), where

$$Z_1 = p_t^{sale}, C_t^{cloud},\ C_{s,t}^{buy} \left\{ F_{s,t}^{payoff} = P_{s,t}^{revenue} + P_{s,t}^{dayahead} + P_{s,t}^{V2G} - C_{s,t}^{gen} - C_{s,t}^{cloud} - C_{s,t}^{\circ} \right\}$$
$$\text{and} \quad Z_2 = \xi - \frac{1}{(1-\mu)} \left(\sum_{s=1}^{N_s} \alpha_s . \phi_s^p \right) \tag{42}$$

In the proposed model, xEVs charging infrastructure market price uncertainty is modeled using the scenario generation approach. This assessment presents an optimization model that accounts for profit payoff and risk optimization simultaneously. The charging station will act on behalf of the aggregators and will be engaged in the business by participating in bidding, cloud monitoring, demand response (DR), as well as dynamic energy management (DEM), and other ancillary markets. The charging stations may modify their hourly charging schedule while formulating its bidding strategies. The Monte Carlo method can be employed in the proposed scheme to model the uncertainties in the xEVs fleet characteristics and power market. In the proposed model, conditional value at risk (CVaR) is used to manage the charging station's (CS's) financial risk that may arise in uncertain market environments. When the probability distribution functions (PDF) of random parameters are included, the Monte Carlo scenario is employed to calculate the CVaR. There may be deterministic target profiles for xEVs drivers where they declare a target level of SOCs thereby forecasting SOC of vehicles at arrival and departure times, while in case of random xEVs penetration, the requirements are managed through stochastic programming formulation. The uncertainties like forecast errors of xEV's hourly electricity consumptions, arrival and departure schedules and the size of xEVs fleet will be depicted by truncated normal distribution functions (NDF), where the aggregate values are the forecasts and percentages of these aggregate values give standard deviations.

Stochastic mixed integer linear programming can be used to formulate the proposed model and will be solved by commercial mixed integer programming (MIP) solver. The trade-off between maximizing expected payoff and minimizing risk due to the physical and commercial uncertainties is modeled by considering the expected downside risk as a constraint. For forecasting, the hourly reserve probabilities of smart grid artificial neural network (ANN) will be applied. The installation of the proposed architecture also introduces unique economic and environmental opportunities in electricity business market. The smart charging stations (SCS) act as an entity that has a function similar to that of a distribution system operator and will participate in the bulk energy market operation by submitting such bidding strategies.

4 Conclusion

The uncoordinated influx of EVs offers unique economic and environmental opportunities and also presents diverse challenges in smart grid and electricity market operations. This work proposes a risk averse, commercially viable optimization framework for enabling smart charging management of EVs. The work introduces a risk assessment model for managing the financial risks in uncertain environments thereby formulating a bidding strategy for coordinating smooth operation of EVs as well as charging stations. The technical, social, and commercial uncertainties of the power market are exhaustively assessed to generate probable scenarios. Adopting conditional value at risk (*CVaR*) for estimating the risks, the work attempts to propose an optimized bidding strategy for the smart charging stations (SCS) that act on behalf of aggregators for managing the financial risk caused by the uncertainties.

References

1. W. Jing, Y. Yan, I. Kim, and M. Sarvi, "Electric vehicles: A review of network modelling and future research needs," *Adv. Mech. Eng.*, vol. 8, no. 1, pp. 1–8, 2016.
2. D. Thoshitha Gamage, David Anderson, C. H. Bakken, Kenneth Birman, Anjan Bose, and and R. van R. Ketan Maheshwari, "Mission-Critical Cloud Computing for Critical Infrastructures," pp. 1–16.
3. V. Razo, S. Member, and C. Goebel, "Vehicle-Originating-Signals for Real-Time Charging Control of Electric Vehicle Fleets," *IEEE Trans.Transportation Electrification* vol. 1, no. 2, pp. 150–167, 2015.
4. N. G. Omran, S. Member, S. Filizadeh, and S. Member, "Location-Based Forecasting of Vehicular Charging Load on the Distribution System," *IEEE Trans. Smart Grid*, vol. 5, no. 2, pp. 632–641, 2014.
5. Y. He, B. Venkatesh, and L. Guan, "Optimal scheduling for charging and discharging of electric vehicles," *IEEE Trans. Smart Grid*, vol. 3, no. 3, pp. 1095–1105, 2012.
6. L. Pieltain Fernández, T. Gómez San Román, R. Cossent, C. Mateo Domingo, and P. Frías, "Assessment of the impact of plug-in electric vehicles on distribution networks," *IEEE Trans. Power Syst.*, vol. 26, no. 1, pp. 206–213, 2011.
7. EVI, "Global EV Outlook 2016: Beyond one million electric cars," 2016.
8. "smart charging : steering the charge, driving the change," no. March, 2015.
9. D. Alahakoon and X. Yu, "Smart Electricity Meter Data Intelligence for Future Energy Systems : A Survey," *IEEE Trans. Industrial Informatics*, vol. 12, no. 1, pp. 425–436, 2016.
10. H. Wu and M. Shahidehpour, "A Game Theoretic Approach to Risk-Based Optimal Bidding Strategies for Electric Vehicle Aggregators in Electricity Markets With Variable Wind Energy Resources," *IEEE Trans. Sustainable Energy*, vol. 7, no. 1, pp. 374–385, 2016.
11. R. Y. and R. Clarke, "Framework for Risk Analysis in Smart Grid Perspective Based Approach," *Proc. 8th Int'l Conf. Crit. Inf. Infrastructures Secur. (CRITIS 2013), Amsterdam, 16–18 Sept. 2013.*
12. S. Deilami *et al.*, "Real-Time Coordination of Plug-In Electric Vehicle Charging in Smart Grids to Minimize Power Losses and Improve Voltage Profile,"*IEEE Trans. Smart Grid,* vol. 2, no. 3, pp. 456–467, 2011.

Counting and Classification of Vehicle Through Virtual Region for Private Parking Solution

Mahesh Jangid, Vivek K. Verma and Venkatesh Gauri Shankar

Abstract This paper presents a new and efficient way to track the movement of vehicles and counting them for the purpose of better and economical vehicle parking system. The numbers of the vehicles are growing very rapidly which produced the problem of the parking. There are diverse techniques that have been introduced by many researchers those are used according to the need and scale. This paper is addressed the problem of vehicle parking for small scale by using the surveillance camera. A background subtraction is used to detect moving vehicle. The moving vehicles are tracked using the Gaussian mixture model and a foreground mask is created. Dilation is applied to remove inconsistent and noise particles. The variable intensity distribution is plotted on a histogram and changes are identified to count vehicles. Each parking spot can accommodate only certain numbers of vehicle. If the count has reached the maximum limit then the display unit guides the driver to next available parking spot.

Keywords Vehicle counting and tracking · Intensity distribution
Gaussian mixture model · Vehicle classification

1 Introduction

Vehicle parking is an important aspect of the smart city. The number of vehicles is rapidly growing worldwide and the usages of the personal vehicle are also increasing for the transportation day by day that has produced a lot of vehicle parking issues. The vehicle parking spaces can be divided into public and private

---

M. Jangid (✉) · V.K. Verma · V.G. Shankar
Department of Computer Science and Engineering, Manipal University Jaipur, Jaipur, India
e-mail: mahesh.jangid@jaipur.manipal.edu

V.K. Verma
e-mail: vivekkumar.verma@jaipur.manipal.edu

V.G. Shankar
e-mail: venkateshgauri.shankar@jaipur.manipal.edu

A.K. Somani et al. (eds.), *Proceedings of First International Conference on Smart System, Innovations and Computing*, Smart Innovation, Systems and Technologies 79, https://doi.org/10.1007/978-981-10-5828-8_72

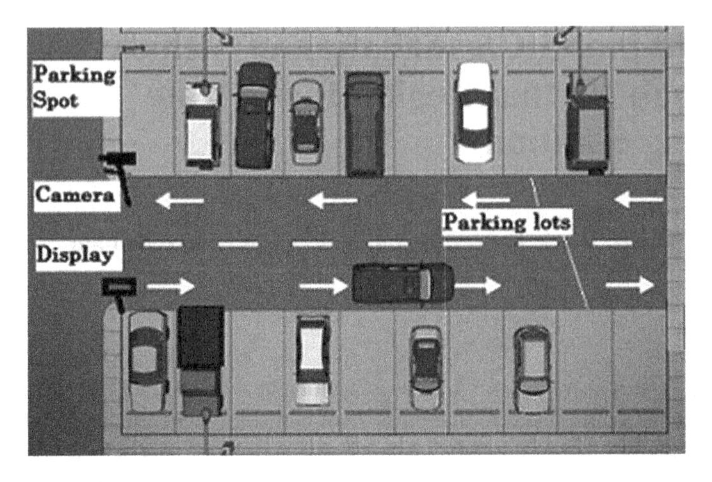

Fig. 1 Structure of parking site

parking space categories. Where the private parking space means the parking space for the employees of a particular organization/university/offices, etc., for which no cost has to pay but only authorized person can park the vehicles and remaining parking space in city is public parking space for which a cost has to pay as per usage but anyone can use the space. The private parking space issues are considered in this paper. Parking spaces are further divided into spots and the spots are divided into the lot to manage properly as shown in Fig. 1. The aim of the smart private parking systems gives the benefit to the authorized driver to find the parking lot with less time [1]. The following are the issues with private parking space.

1. Only the authorized vehicles can be entered.
2. The system should be work in real time.
3. Guide the driver for available parking space.
4. Each parking spot can accommodate only the fix numbers of vehicle.
5. Each parking spot can also accommodate only the specify types of vehicle ex-car, bicycle, etc.

An extensive survey of the technologies available for smart car parking system can be found in [2, 3]. There are different technologies and methodologies can be observed. Most of the technologies are for public parking space like smart parking system based on agent model [4], based on fuzzy logic [5], wireless sensor network based systems [6], based on vehicular to infrastructure [7], based on global positioning system [8], based on RFID technology [9], computer vision [10], and others M2M, IoT systems. This technology has their own advantages and disadvantages and also the cost is also a very important factor. In spite of that, a lot of sensors and good infrastructure are also needed. Our problem was to make a smart private parking system for our university which allows the authorized cars and guides them towards the available parking spots with minimum cost, and also maintains the in and out log information of cars/vehicles. Mostly, entire offices/universities are having the surveillance camera which can be used for the smart parking solution without any addition cost and only authorized car can be allowed by the car number plate recognition.

Background subtraction is used to track the movement of vehicles. A foreground mask is then calculated to track the moving object in subsequent frames. This is done with the help of Gaussian mixture model. The method helps in analyzing consecutive frames and tracking the consistent particles and their movement with respect to the static camera. Inconsistent and secluded particles are removed using a threshold value and dilation. A threshold value is set and if the value of the pixel is less than or equal to the threshold, the value is set to 0 (black). This process helps in erasing noise and thus increasing the efficiency of the system. For tracking the moving object, a rectangular contour is drawn on the moving vehicle. For better efficiency, contours are drawn only on moving objects having an area greater than an assigned value. This prevents tracking of unwanted objects and greatly reduces the error of the system. A virtual region of interest is created and the variable intensity distribution of each pixel in that region is plotted on a histogram. The histogram shows minimal change when no vehicle is passing through it but there is appreciable change when a tracked vehicle passes through. This is used to count the vehicles and guide the driver for next available parking spot. The rest of the paper comprises of various phases use to detect the vehicle and counting them.

2 Our Proposed Vehicle Tracking and Counting Method

Our proposed method for the smart private parking solution has depicted in Fig. 2. The first step is to locate the vehicle in the video that has been done by background subtraction method and followed by some preprocessing techniques to remove the noise. The moving vehicles are tracked using the Gaussian mixture model and a foreground mask is created. A snap is captured to recognition the number plate and search in the database to authorize the vehicle. Further, the vehicle is classified into car and bike to count them and guide the driver towards the available parking lot to park the vehicle.

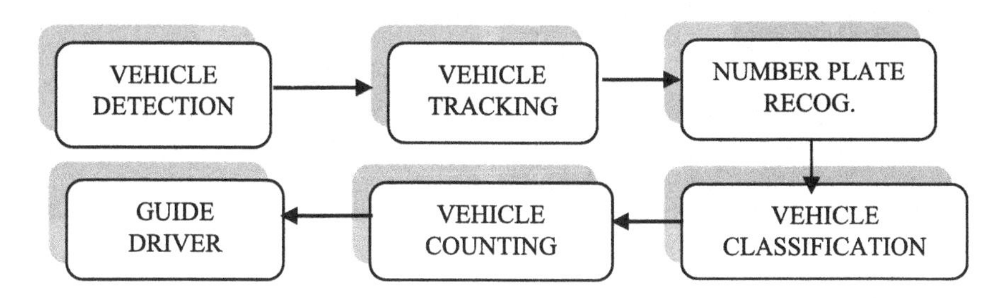

Fig. 2 Flow diagram depicting the process followed

2.1 Vehicle Detection

The smart private parking system starts with the vehicle detection with the specified area. A motion detection algorithm [11] has used to detect the presence of vehicle in front of the surveillance camera. The easiest way to detection the presence of motion is to subtract the current image *I (t1)* taken at time *t1* with the background image *I (t0)* taken at time *t0* and the resultant image would have the intensity values for location where the motion is observed. Figure 3b has shown the resultant image.

$$\mathrm{Im}(t1) = \mathrm{I}(t1) - \mathrm{I}(t0) \tag{1}$$

The above way is the only work where the foreground pixels are only in moving and the background pixels are static. But in reality, there is always noise in the resultant image as shown in Fig. 3b. The foreground mask extracted after background subtraction is error prone and requires further processing for better vehicle tracking. The mask contains lots of secluded white portions and a lot of noise. Applying thresholding and then dilating the foreground mask results in a better image with minimum noise. The foreground mask obtained is shown in Fig. 3b. To achieve this, a threshold value is assigned and each pixel of the frame is compared. The pixels having a value less than or equal to the threshold are assigned the value 0

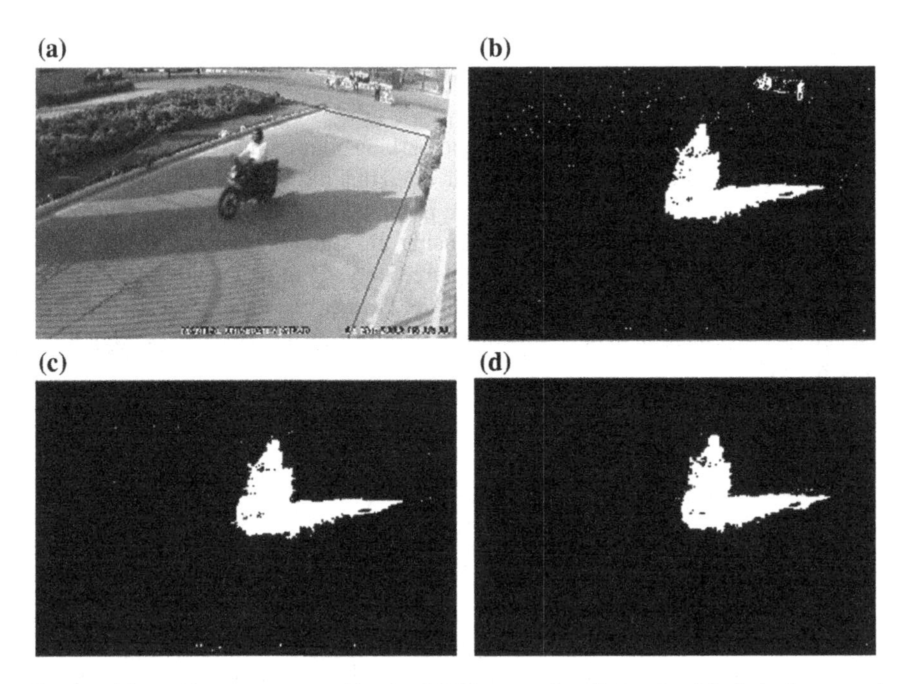

Fig. 3 **a** The real image as captured by the CCTV camera installed at the sight **b** the foreground mask extracted after background subtraction **c** Extracted foreground before thresholding and dilation **d** Foreground after dilation and reduced noise

(black) and the pixels having a value higher than the threshold are assigned white. The region fell out of the specified area which was shown in Fig. 3a, of the resultant image is considered as background and the impact of that is shown in Fig. 3c.

During this process, each pixel in the image matrix is compared to its surrounding pixels. Suppose a pixel has value 255, if at least one pixel surrounding it has the same value then its value is not changed, else it is changed to zero. After applying dilation, the noise white particles and white dots which are far apart are erased. This process greatly enhances the efficiency of the system and results in better object tracking. The comparison is described in Figs. 3c, d. The dilation procedure greatly helps in eliminating unimportant and inconsistent white areas of the foreground mask and thus greatly improves the performance of the system.

2.2 *Vehicle Tracking*

The proposed method also includes drawing rectangular contours on the moving vehicle for better tracking. This also helps in only tracking the vehicles of interest and ignoring vehicles in other lanes or other non-vehicle entities. The contour is drawn on the extracted foreground image after dilation is applied. The extracted background image is taken as input and the white areas are searched. An area threshold is defined, white portions of the image having area greater than this threshold only are further considered for contours. This helps the system in ignoring small non-vehicle particles and only draws contours on vehicles. The coordinates of the top-left and bottom-right edges of the white portion are returned and a rectangular contour is drawn with the above-mentioned points being the opposite vertices of the rectangle. The contours are shown in Fig. 4. Many advanced techniques can be used for the purpose of object tracking in real-time environments.

Fig. 4 Rectangular contour draw on moving vehicle

Particle filtering [12] and non-drifting mean shift [13] can be used for better object tracking.

2.3 Number Plate Recognition

Car number plate recognition was done by Tesseract [14] which is an open-source OCR engine that was developed at HP and maintained by Google from 2006. After detection of the vehicle, a grayscale image of vehicle is passed through the Tesseract OCR to get the registration number of the vehicle. There is no need to include page numbers or running heads; this will be done at our end. If your paper title is too long to serve as a running head, it will be shortened. Your suggestion as to how to shorten it would be most welcome.

2.4 Vehicle Classification

Furthermore, the counted vehicle will be further analyzed and counted as a two-wheeler or four-wheeler. For this process, a snap of the vehicle is taken while it enters the virtual region of interest. Background subtraction is applied again and the area of white pixels is calculated if the value greater than a threshold value than the vehicle being counted as a four-wheeler and otherwise it is counted as a two-wheeler.

2.5 Vehicle Counting

Vehicle counting is an important part of our system. Whether the specific parking spot has available lots for the vehicle or not that has been decided by the maximum count of the vehicle for that particular parking spot. As it has been already discussed that there are no sensors being used to detect the status of parking lots. Because sensors introduced addition cost to install and maintain them. A camera and display unit are installed at the entry point of parking spot. The display unit shows the number of available lot in that particular parking spot and if no lot is available then it guide the driver for next available parking spot.

A virtual region of interest is defined for vehicle counting as shown in Fig. 5. The vehicle counting is only performed at user-defined virtual region, which may be defined at any position of the frame. Each pixel in a frame consists of values which represent the color of that pixel. These values are plotted as a histogram to identify changes when the vehicle passes through the region. The histogram [15] shows minimal change when there is no object passing through it but changes appreciably when a vehicle does pass through it as shown in Fig. 6a, b.

Fig. 5 Virtual ROI with no vehicle passing through

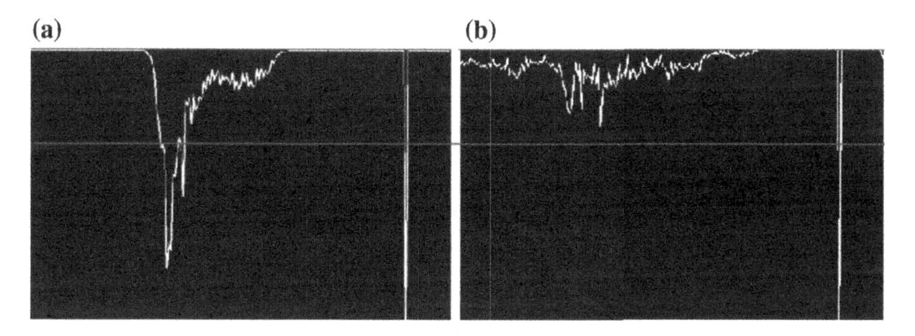

Fig. 6 **a** Histogram representing the variable intensity distribution when no vehicle is passing through it **b** Histogram representing the variable intensity distribution in the region of interest when a vehicle is passing through it

$$F_{Change} = Hist_{F1} - Hist_{F0} \quad (2)$$

$$\begin{cases} F_{change} > threshold(t) \cdots VehicleCount \\ F_{change} \leq threshold(t) \cdots NotCount \end{cases}, \quad (3)$$

where $Hist_{F1}$ is the histogram of frame at time *t1* and $Hist_{F0}$ is the histogram of frame at time *t0*. F_{Change} is the record of changes between two consecutive frames. A threshold (t) has been fixed empirically.

Table 1 Vehicle count

Video samples →	I	II	III	IV	V
Visual inspection count	31	37	34	42	32
Correct classification	29	35	31	39	30
Incorrect classification	2	2	3	3	2
Accuracy	93.5	94.5	91.1	92.8	93.75

3 Experimental Results

Our goal is to count tracked vehicles on user-defined virtual region of interest. Five experimental videos were taken for testing the system. These videos were captured from the main gate of Manipal University Jaipur in different times. The testing was divided into two main categories, namely vehicle counting and vehicle classification. The videos were processed on Intel(R) i5 processor at a clock speed of 1.60 GHz. The video was of 480 × 720 resolutions with 15 fps. The vehicle counting test checks whether the vehicle is being counted or not. FN describes the false negative scenario, where the vehicle passes through the virtual region of interest but is not counted by the system. FP describes the false positive scenario wherein the system counts a non-vehicle element. Both are calculated and mentioned in Table 1. The accuracy is also calculated for vehicle count method. It can be observed from Table 1 that most of the case the accuracy is more than 96% and in one case for video V, it is observed low. The vehicle classification checks whether the counted vehicle is being classified correctly or not, mentioned in Table 2. Hence, the accuracy is determined.

Table 2 Vehicle classification

Video samples →	I	II	III	IV	V
Visual inspection count	31	37	34	42	32
Proposed method count	30	36	34	41	30
FN	3	4	2	3	4
FP	2	3	2	2	2
FN + FP	5	7	4	5	6
Accuracy	96.77	97.29	100	97.61	93.75

4 Conclusion

This paper presents an economical and efficient method for detecting and counting vehicles for smart private parking system and also verifies the vehicle by the recognition of the vehicle registration number. The efficiency of this method was good but due to the fast motion of the vehicle some time it's missed the count them and also in classification them otherwise this method don't include addition cost due to the already installed surveillance camera at site. For future research, we work on the more robust algorithm which can count and classify the vehicle correctly.

Acknowledgements Authors would like thank Manipal University Jaipur, India to support this project and provide the necessary infrastructure and the database (video) to test this project.

References

1. Li, Chieh-Chang, Shuo-Yan Chou, and Shih-Wei Lin. "An agent-based platform for drivers and car parks negotiation." In Networking, Sensing and Control, 2004 IEEE International Conference on, vol. 2, pp. 1038–1043. IEEE, 2004.
2. Mahmud, S. A., G. M. Khan, M. Rahman, and H. Zafar. "A survey of intelligent car parking system." Journal of applied research and technology 11, no. 5 (2013): 714–726.
3. Fraifer, Muftah, and Mikael Fernström. "Investigation of Smart Parking Systems and their technologies." Thirty Seventh International Conference on Information Systems, Dublin 2016
4. Khoukhi, Amar. "An intelligent multi-agent system for mobile robots navigation and parking." In Robotic and Sensors Environments (ROSE), 2010 IEEE International Workshop on, pp. 1–6. IEEE, 2010.
5. Mohammadi, Sasan, Mostafa Tavassoli, and Abolfazl Rajabi. "Authoritative Intelligent Perfect Parallel Parking Based on Fuzzy Logic Controller for Car-Type Mobile Robot." In Information Technology: New Generations (ITNG), 2011 Eighth International Conference on, pp. 135–138. IEEE, 2011.
6. Kianpisheh, Amin, Norlia Mustaffa, Pakapan Limtrairut, and Pantea Keikhosrokiani. "Smart parking system (SPS) architecture using ultrasonic detector." International Journal of Software Engineering and Its Applications 6, no. 3 (2012): 55–58.
7. Geng, Yanfeng, and Christos G. Cassandras. "A new "smart parking" system infrastructure and implementation." Procedia-Social and Behavioral Sciences 54 (2012): 1278–1287.
8. Hanif, Noor Hazrin Hany Mohamad, Mohd Hafiz Badiozaman, and Hanita Daud. "Smart parking reservation system using short message services (SMS)." In Intelligent and Advanced Systems (ICIAS), 2010 International Conference on, pp. 1–5. IEEE, 2010.
9. Šolić, Petar, Ivan Marasović, Maria Laura Stefanizzi, Luigi Patrono, and Luca Mainetti. "RFID-based efficient method for parking slot car detection." In Software, Telecommunications and Computer Networks (SoftCOM), 2015 23rd International Conference on, pp. 108–112. IEEE, 2015.
10. Bong, D. B. L., K. C. Ting, and K. C. Lai. "Integrated approach in the design of car park occupancy information system (COINS)." IAENG International Journal of Computer Science 35, no. 1 (2008): 7–14.
11. Zivkovic, Zoran. "Improved adaptive Gaussian mixture model for background subtraction." In Pattern Recognition, 2004. ICPR 2004. Proceedings of the 17th International Conference on, vol. 2, pp. 28–31. IEEE, 2004.

12. Bouttefroy, P. L. M., Bouzerdoum, A., Phung, S. L., & Beghdadi, A. (2008). Vehicle tracking by non-drifting mean-shift using projective kalman filter. In 2008 11th International IEEE conference on intelligent transportation systems (ITSC 2008) (pp. 61–66). IEEE.
13. Bouvie, C., Scharcanski, J., Barcellos, P., & Escouto, F. (2013). Tracking and counting vehicles in traffic video sequences using particle filtering. In 2013 IEEE international instrumentation and measurement technology conference (I2MTC) (pp. 812–815). IEEE.
14. Smith, Ray. "An overview of the Tesseract OCR engine." (2007).
15. Barcellos, Pablo, Christiano Bouvié, Fabiano Lopes Escouto, and Jacob Scharcanski. "A novel video based system for detecting and counting vehicles at user-defined virtual loops." Expert Systems with Applications 42, no. 4 (2015): 1845–1856.

Towards Incorporating Context Awareness to Recommender Systems in Internet of Things

Pratibha and Pankaj Deep Kaur

Abstract Technological innovation in communication and embedded systems has led to a new paradigm called Internet of Things (IoT). Context in association with the things in IoT plays a crucial role in effective data communication and processing. Contextual information if integrated with recommender systems (RS) can recommend more appropriate things to users. In this paper, context-aware recommender system (CARS) from an IoT perspective has been explored and its various aspects have been considered in detail. In addition, correlation between context and IoT as well as perquisite of making context-aware recommendations in IoT has been discussed. Moreover, the context flow depicting the movement of context in context-aware systems and an application scenario has been delineated. Finally, the paper concludes with highlighting various challenges and open issues in CARS.

Keywords Context · Context-aware recommender system (CARS)
Context-aware systems · Internet of things (IoT)

1 Introduction

Recent advancements in computing technologies have made it possible to identify, locate, and tract almost all the routine objects across the Internet. With numerous objects attached and communicating over the Internet, there is a compelling requirement to build an efficacious approach to search, recommend, and categorize among the collection of things in order to extract some patterns which are of interest to the user.

The traditional (2D) RS focused on suggesting meaningful items to users by understanding their preferences without considering any additional contextual

Pratibha (✉) · P.D. Kaur
GNDU Regional Campus, Jalandhar, India
e-mail: pratibha.cse05@gmail.com

P.D. Kaur
e-mail: pankajdeepkaur@gmail.com

A.K. Somani et al. (eds.), *Proceedings of First International Conference on Smart System, Innovations and Computing*, Smart Innovation, Systems and Technologies 79, https://doi.org/10.1007/978-981-10-5828-8_73

information for instance location and time [1]. It is not adequate to consider only users and items for applications like tourism and e-learning which require additional information regarding user's context.

Therefore, it is essential to integrate contextual information into RS under certain situations [1]. As a consequence of adding contextual information to 2D recommender systems, the new system so devised is known as the Multidimensional (MD) RS or CARS.

Recommending things is one of the important measures taken to possess full privilege of IoT; it not only benefits individuals but also businesses and organizations. If things recommendations in IoT can be made context-aware, it can help provide more relevant things to users according to their interest, current context, and preferences. Furthermore, it can benefit user in searching information and reducing search time in an IoT network. In this paper, the use of CARS in making things recommendations in IoT has been explored. The main contributions are presented as follows:

- Correlation between context and IoT has been discussed.
- Circulation of context in context-aware systems has been presented.
- Use of context awareness while making recommendation in IoT environment along with an application scenario has been highlighted.

Rest of the paper is structured in the following sections.

Section 2 provides an introduction to IoT. Section 3 presents context fundamentals such as definitions and categories of context. Section 4 describes relationship between context and IoT. Section 5 discusses context flow and identifies its five stages: discovery, acquisition, representation, processing, and distribution. Section 6 discusses context-aware. This section explains architecture and types of context-aware systems. Section 7 presents CARS. This section discusses different ways to obtain contextual information followed by incorporation of context into recommendation process at different pivots. Section 8 discusses scenario of using CARS in an IoT-based environment. Section 9 identifies open issues and challenges in CARS. Section 10 presents conclusion remarks.

2 Internet of Things (IoT)

IoT has attained considerable acknowledgement in industry and academia during the last decade. The phrase "Internet of Things" was primarily introduced by Ashton [2] in 1998. Further, a formal definition was given by ITU (International Telecommunication Union) [3].

Research into IoT is still in its early stage. Therefore, IoT has been defined differently by distinct researchers in [4–6]. In general, IoT can be defined as worldwide network (wired or wireless) of interrelated objects that are provided with distinctive identifiers over the Internet to share or exchange information with each other without human intervention.

3 Contextual Fundamentals

3.1 *Definition of Context*

Individual inventors have defined the term "context" in different ways by considering several aspects of user's context. Views of some of them are described in [7–9] and generalized below.

Any aspect of user's context (like current place, time, etc.) can be utilized to identify the current state of an individual, location, or thing.

3.2 *Classification of Context*

Context has been classified into different classes by various researchers [1, 10, 11].

After reviewing the above classifications of context proposed by different researchers, this article derived appropriate categories (minimum but essential) as shown in Fig. 1.

- User's profile, location of the user, community condition, people nearby constitutes **user context**.
- Temperature, traffic conditions, weather, and noise level add up to **physical context**.
- Resources close by such as work stations, printers, and network connectivity form **computing context**.
- Homologous profiles, common interest, etc., comprise **social context**.
- Geographical information, nationality, etc., represent **location context**.

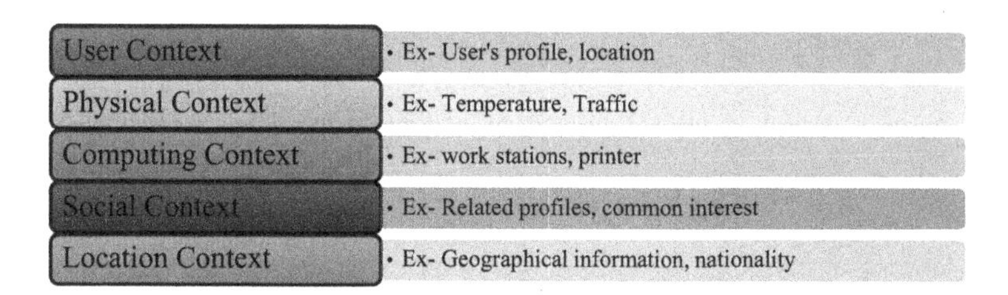

Fig. 1 Classification of context

4 Correlation Between Context and IoT

Context awareness and context-aware computing have become important aspects of ubiquitous computing systems. Many scientists and researchers have chosen these areas for research in various application domains, starting from desktop and web-enabled applications to the IoT. The term "context-aware" was first introduced by Schilit and Theimer in 1994 [7].

Since then, a large number of solutions have been proposed by different inventors using context-aware computing approaches. Even though the basis of each project is different, one issue that remained common is the number of data sources. For instance, most of the suggested solutions collect data from a limited number of sources making it possible to analyze the collected data. On the other hand, in IoT, huge numbers of sensors are interconnected, making it difficult to handle the data acquired from various data sources. Therefore, context awareness will significantly contribute in finding the appropriate data which is required to be processed and analyzed.

Due to innovations in sensor technology, large number of sensors can be planted over the same area because of its compact size. It is speculated that this number will grow speedily over the next decade. Eventually, big data will be produced by these sensors. This raw data will not be of any use unless it is analyzed and interpreted. It is trusted that context-aware computing would be proved successful in tackling this challenge in IoT as it has been proved effective in other areas like mobile and ubiquitous. Contextual information linked to sensor can be stored using context-aware computing, so that more meaningful interpretation can be performed.

5 Context Flow

Context flow considers the flow of context in systems that can adapt themselves with the change in context. Managing context is a necessary component in an application and it is believed that this trend will further grow in the IoT.

The context flow consists of five phases as depicted in Fig. 2. First, the context is detected in "context discovery" phase. Second, context is acquired from different

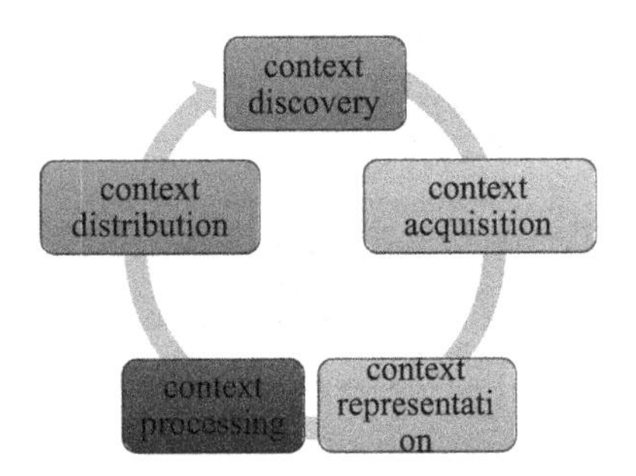

Fig. 2 Context movement in context-aware system

data sources in "context acquisition" phase. Third, the acquired data is represented in worthwhile way in "context representation" phase. Fourth, data represented in phase three is then managed to obtain contextual information from unprocessed data is "context processing" phase. Finally, both context information and unprocessed data are dispersed to the users in "context distribution" phase.

6 Context-Aware System

Context-aware systems provide new occasions for users by bringing together data related to context and adapting the behavior of the system at the same time. Schilit et al. [12] defined those systems as context-aware systems that can adapt themselves to context. Another scientist, Abowd et al. [9] tried to pacify both views "adaptation to context" and "using context" in a single definition: A system is said to be context-aware if it uses context information to provide relevant recommendations to the user with respect to the current task being performed by the user.

6.1 Layered Architecture of Context-Aware System

From an architectural point of view, it is composed of three layers: sensor (top layer), middleware (middle layer), and end user application (bottom layer) as shown in Fig. 3, where top and bottom layers are compulsory layers as they produce and consume contextual information.

In the middle of these two layers, there is an optional layer, a middleware layer; to handle coordination and communication issues between components.

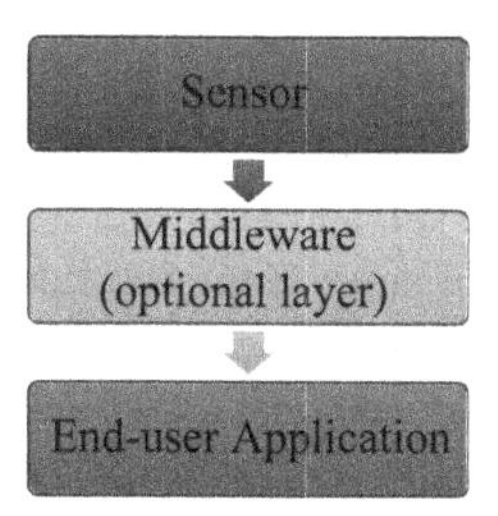

Fig. 3 Layered architecture of context-aware system

Table 1 Forms of context-aware system

Forms of context-aware systems	Further classification	Features
Local context-aware system	None	Direct physical connection between sensors and end user applications
Distributed context-aware system	Distributed collaborative context-aware system	Two or more humans working to attain same goal
	Distributed non-collaborative context-aware system	Supports only individual goals

6.2 Forms of Context-Aware System

Two forms of context-aware systems are local context-aware system and distributed context-aware system. In local context-aware systems, top and bottom layers are connected with a direct physical connection.

On the contrary, distributed context-aware systems do not have a direct physical connection between top and bottom layer. As a consequence, multiple end user applications can receive information from the same sensor or vice versa.

Two further classifications of distributed context-aware systems are—collaborative and non-collaborative context-aware system as shown in Table 1. Systems that help group members (dispersed at different locations) attain a common goal are distributed collaborative context-aware systems.

7 Context-Aware Recommender System (CARS)

The RS captures user preferences for different items to measure R (Rating function). After primary set of ratings is collected from users' previous browsing history, a RS assesses the function R, (R → User × Item) for the (user, item) unrated sets. Here, rating is a totally ordered set, and User and Item are the domains of users and items, respectively. The RS can suggest the top-ranked k items for each user after the function R is estimated for the whole (User × Item) space. Such systems are known as traditional or two-dimensional (2D) RS since they consider only two dimensions (User and Item) in the process of generating recommendations.

While suggesting items to users if we analyze additional contextual information such as location, place, and time as additional categories of data such type of systems are known as CARS. Preferences of users are formed as the function of not only (user, item) sets, but context is included as well (R → User × Item × Context). Such contextual information can be attained in following ways:

- **Explicitly**—Context can be captured explicitly from the input entered by the users. For instance, entries are made during registration process.
- **Implicitly**—These methods capture contextual information automatically, for instance current location is frequently discovered via implied activities accompanied by Location-Based Services for example Wi-Fi, GPS.
- **Contextual information**—These methods can be concluded by interpreting user communications with devices for example, by assessing the ongoing activity of a user like searching for hotels.

7.1 Incorporation of Context in Recommender System

There are three different ways in which contextual information can be incorporated into recommendation process as depicted in Fig. 4:

- **Context pre-filtering approach**—the relevant contextual data is selected with the help of current context and rating function is predicted by the traditional RS (users × items) on selected data.

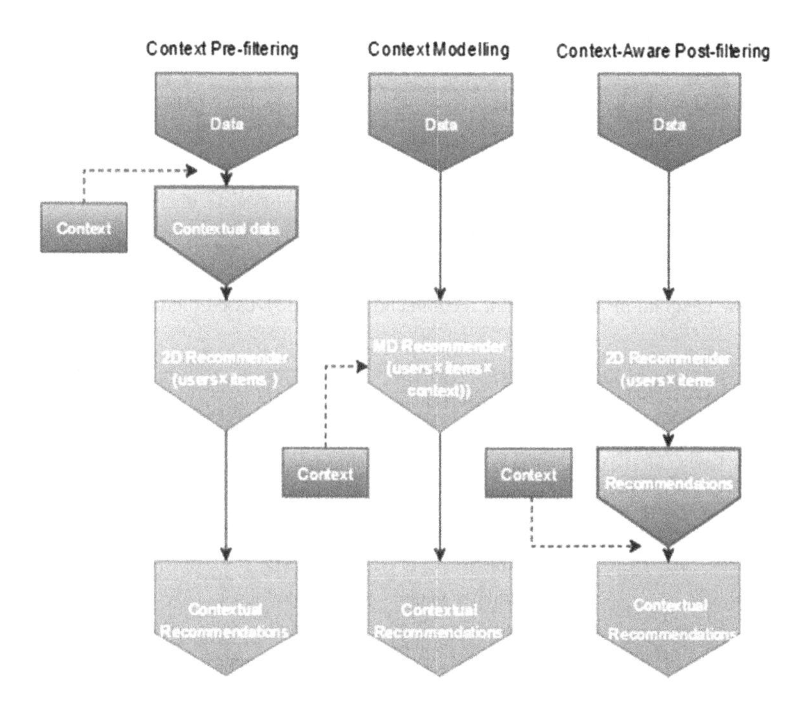

Fig. 4 Different ways to incorporate context in RS

- **Context-aware post-filtering approach**—anticipation of ratings on raw data is done using 2D RS (users × items), then the resultant recommendation list is filtered by reviewing the context values.
- **Context modeling approach**—context information is added in the 2D recommendation system. Thus, this algorithm makes use of multidimensional recommendation system.

8 Motivation: A Use Case Scenario

This section presents a context-aware advertisement recommendation scenario on IoT-based digital signage. Many of the digital signage products are fitted with the digital cameras to capture the demographic features of its viewer such as gender and age bracket as depicted in Fig. 5.

Suppose, below is the advertisement policy used by the digital signage in Table 2 which shows which ad will be displayed to males or females viewers considering their age brackets.

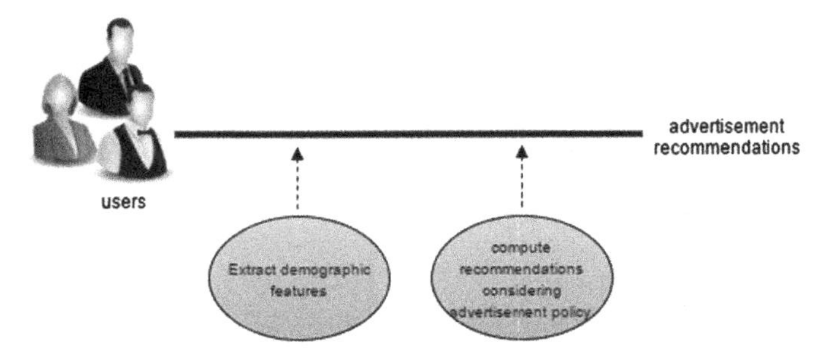

Fig. 5 Flow Diagram of advertisement recommendation on IoT-based digital signage

Table 2 Advertisement policy for displaying recommendations on digital signage

Advertisement	Gender	Age bracket
Boating	Male/female	0–100
Camera	Male/female	25–40
Clothing	Male/female	0–100
Gardening	Male/female	50–70
Cosmetics	Female	18–35
Computer game	Male/female	15–25
Handbags	Female	20–50
Bicycle	Male/female	8–25
Bike	Male	18–35

Table 3 (a) Recommendations to a male between age 35 and 40 (b) recommendations to a female between 20 and 25

Recommended advertisements (a)	Recommended advertisements (b)
Boating	Boating
Camera	Camera
Clothing	Clothing
Bike	Cosmetics
	Computer game
	Handbags
	Bicycle

As a viewer watches an ad on the digital signage, it recognizes their demographic features in real time and displays the advertisement in which the viewer would be most interested in. Suppose the viewer John is recognized as a man with age between 35 and 40, the list of advertisements recommended to him based on the advertisement policy of the digital signage will be (Table 3).

More user context like time, location, and weather can also be included to generate more relevant recommendations.

9 Discussion and Open Issues

Various researches have proved that for improving the efficiency and accuracy of predictions in RS, contextual information can be used. Following are some of the open issues in CARS:

- Accumulating effective and adequate data, and pinpointing the elements that have unusual influence on the subsequent steps.
- Discovering context factors and predicting recommendations are major issues in designing more efficient CARS. These issues can be handled by finding correct approximations about the effect of contextual aspect before accumulating data from real environment [1].
- Shortage of publicly accessible datasets is one of the principal problems in conducting empirical analysis of context-aware systems. This problem may be induced by the trouble of accumulating contextual information for CARS and the preservation for user privacy.
- Security and privacy are other two aspects which need attention as data from user's profile and other personal information are used in contextual information.
- CARS can also be integrated with proactivity for generating recommendations without submitting any query [13].

10 Conclusion

Recently, CARS have received significant attention in recommending items or services to users based on the contextual information. Contextual information was integrated with 2D RS to increase the effectiveness of traditional 2D system. In this article, it has been recognized that context awareness is a fundamental IoT research demand for generating more relevant things recommendations to users by understanding the user's current context.

References

1. Asabere, N. Y.: Towards a Viewpoint of Context-Aware Recommender Systems (CARS) and Services. Int'l J. of Comp. Sci. and Telecom. vol 4, pp. 19–29. (2013)
2. Ashton, K.: That' internet of things' thing in the real world, things matter more than ideas. J. RFID (2009)
3. International Telecommunication Union: The internet of things. Workshop Report, International Telecommunication Union (2005)
4. Lu, T., Neng, W.: Future internet: The internet of things. In: 3rd International Conference on Advanced Computer Theory and Engineering (ICACTE), vol. 5, pp. 376–380. (2010)
5. Guillemin, P., Friess, P.: Internet of things strategic research roadmap. Technical Report, The Cluster of European Research Projects (2009)
6. Gubbi, J. et al.: Internet of Things (IoT): A vision, architectural elements, and future directions. Future Generation Computer Systems. vol. 29, issue 7, pp. 1645–1660. (2013)
7. Schilit, B., Theimer, M.: Disseminating Active Map Information to Mobile Hosts. IEEE Network, pp. 22–32. (1994)
8. Brown P.J., Bovey J.D., Chen X.: Context-Aware Applications: From the Laboratory to the Marketplace. IEEE Personal Communications, pp. 58–64. (1997)
9. Abowd, G. D., Dey, A. K.et al.: Towards a better understanding of context and context-awareness. Handheld and ubiquitous computing. pp. 304–307. Springer (1999)
10. Yeung, K. F., Yang, Y.: A Proactive Personalized News Recommender System. Presented at: Development in E-systems Engineering (DESE), London. (2010)
11. Gallego, D., Barra, E., et al.: Incorporating Proactivity to Context-Aware Recommender Systems for ELearning. Presented at: IEEE. (2013)
12. Schilit, B., Adams, N., Want, R.: Context-aware computing applications. In: 1st Workshop on Mobile Computing Systems and Applications, WMCSA'94, pp. 85–90. IEEE Computer Society, Washington (1994)
13. Woerndi, W., Huebner, J. et al.: A Model for Proactivity in Mobile, Context-aware Recommender Systems. Presented at: RecSys, USA (2011)

Latency Factor in Bot Movement Through Augmented Reality

Suman Bhakar and Devershi Pallavi Bhatt

Abstract Augmented Reality (AR) is a new and developing technology capable of unlimited future prospects. It has gained much attention in recent years and is in its initial phases of research and development. Slowly and steadily, AR is spreading into day-to-day lives through audiovisual media and in other fields such as sports, entertainment, business, designs, etc. This paper provides an overview of the concept of AR including its definition, architecture, and the different latency factors which affect the rotation of arm of machine and limitations of AR.

Keywords Augmented reality · Virtual reality · Head mounted display Latency · Glyph

1 Introduction

The technological advancement has a direct impact on our personal life and personal behavior routine. Many companies are trying to attract different users with new emerging techniques. Amongst technical advancements, AR (Augmented Reality) is the most recent development. Basically, AR is a combination of a real scene viewed by a camera through user and a virtual scene generated by a computer that superimposed the scene with additional images/video [1]. In other words, augmented reality is the technique that adds virtual scene/audio generated by computer and superimposes it on real image [2].

In last few years, AR technology is accessible through smart phones, where user can get additional information about any application [3]. AR is audiovisual media, and it is used for entertainment, news, and gaming; additionally used in tourism, electronics, commerce, and shopping. It is fundamentally a combination of the

S. Bhakar (✉) · D.P. Bhatt
Department of Computer Science, Manipal University, Jaipur, India
e-mail: Sumanbhakar2016@gmail.com

D.P. Bhatt
e-mail: Pallavi25.bv@gmail.com

A.K. Somani et al. (eds.), *Proceedings of First International Conference on Smart System, Innovations and Computing*, Smart Innovation, Systems and Technologies 79, https://doi.org/10.1007/978-981-10-5828-8_74

real-world and virtual environment. In this survey, the main task is understanding the existing and future applications in augmented reality [4]. In such a way, AR has no limit to sense of sight. It can apply in all senses including smell, hearing, and touch. Augmented reality applications are also used in vehicles for a safer drive. During manufacturing of car, developed HUD (head-up display) [5] is installed and this technology offers many key information relating to velocity/speed, warning messages, and driving directions. Augmented reality applications have a significant impact on images as with the help of AR special effects can be added in images which improve the images processing technique [6].

1.1 Definitions: Augmented Reality and Virtual Reality

It is a combination of a real scene viewed by a camera through user and a virtual scene generated by a computer that superimposed the scene with additional images/video. Then it said to be augmented reality.

On the other hand, virtual reality ignores the real world and transports you somewhere else entirely, it emerged you fully in a virtual world, as shown in Fig. 1.

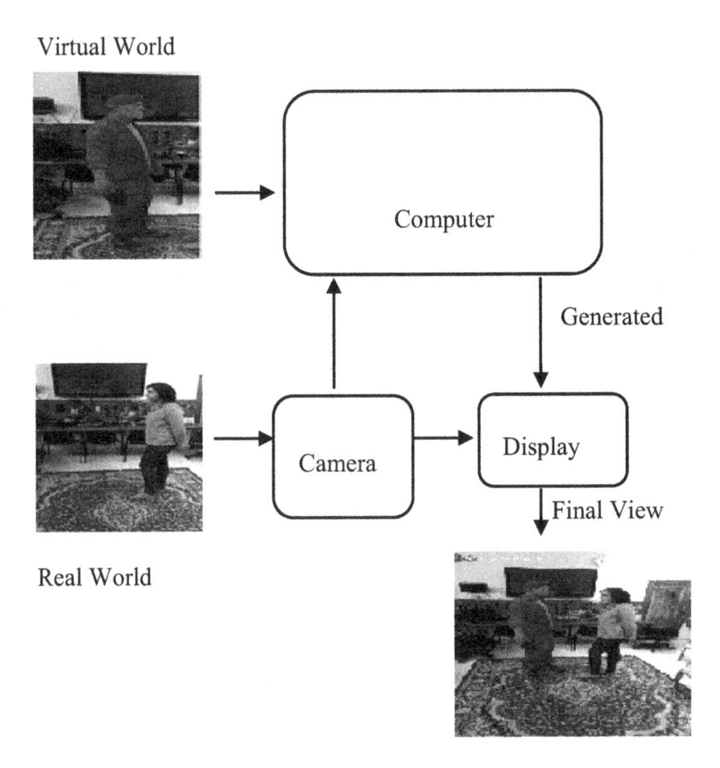

Fig. 1 Example of augmented reality

2 Architecture of Augmented Reality System

The concept of Augmented Reality revolves mainly around four steps: choose a real world image—add your virtual image/data in it—remove real-world environment—not a virtual reality since the environment is real [6, 7], as shown in Fig. 2.

The tasks performed by an Augmented Reality system are as follows:

- Scene Capture
- Scene Identification
- Scene Processing
- Visualization of the Augmented Scene.

2.1 *Scene Capture*

There are two types of devices which are used in scene capture:

Video-Through Devices: Video cameras and smartphones are the classic examples of video-through devices [8].
See-Through Devices: Head-mounted displays are an example of see-through devices. These devices give a picture of captured reality along with the information of the augmented object [9].

2.2 *Scene Identification*

The main function in this stage is to classify the scenes. There are two types of techniques used for scene identification:

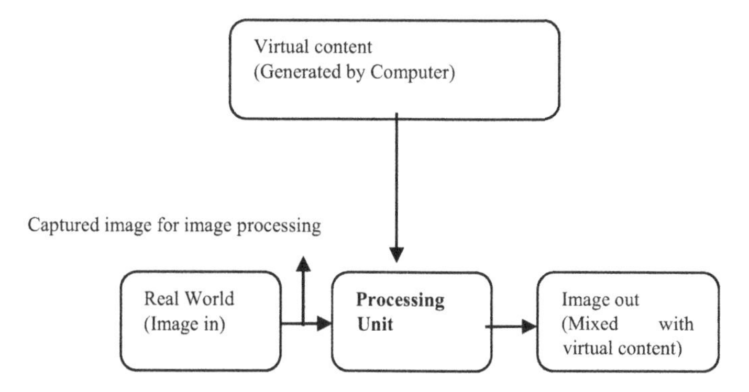

Fig. 2 Augmented reality block diagram

Marker-based: Marker-based techniques use the visual tags as markers within a real scenario which can be retained by the AR system.

Non-marker-based: In this technique, devices are used instead of markers for scene identification. For example, AR browser helps the users in surfing virtual data in real world environment [10].

2.3 Scene Processing

At this stage, the inner and outer parameters of camera spot specific marker in a real environment and thereafter projects virtual model of marker in 3D [11].

2.4 Visualization Scene

This is the last stage where the virtual scene is generated by a computer that superimposed the scene with additional images/video [12].

3 Glyph (Marker) Specification

Table 1 shows different markers used in the project which rotates the arm of the machine.

4 Different Factors Which Affects the Rotation of Arm of Machine

Following are the factors which affect the rotation of arm of machine:

1. Camera response time
2. Converting Glyph in black and white and adding the threshold to remove the noise
3. Distance from camera to Glyph.

Table 1 Glyph specification

Glyph (marker)	Action
Marker "A"	Hello
Marker "B"	Rotation from 0° to 30°
Marker "C"	Rotation from 30° to 90°
Marker "D"	Rotation from 90° to 135°
Marker "E"	Rotation from 130° to 180°
Marker "F"	Stop

Table 2 Camera specifications

Resolution hardware	500 K pixels
Image quality	RGB24 or I420
Exposure	Auto or manual
Interface	USB2.0
Frame rate	30 fps (MAX)
Lens	f = 6.0, F = 2.0
Focus range	4 cm to infinity
Weight	175 g

4.1 Camera Response Time

If the work is done by camera then camera response time is very important factor. In the project, glyph recognized by camera and arms of machine rotated from 0° to 180° through glyphs. Here specifications of camera such as focus range, image quality, frame rate, resolution affect the rotation of arm of machine. In Table 2, the specifications of camera used in the project are shown as below.

4.2 Converting Glyph in Black and White and Adding Threshold to Remove Noise

Image quality is very important factor in latency. Here the original glyph contains noise (Fig. 3) so to remove noise, first converts the original image into black and white image (Fig. 4) and even after that if noise is present then apply the threshold to minimize the noise (Fig. 5). Here different black and white and threshold images with noise are shown below:

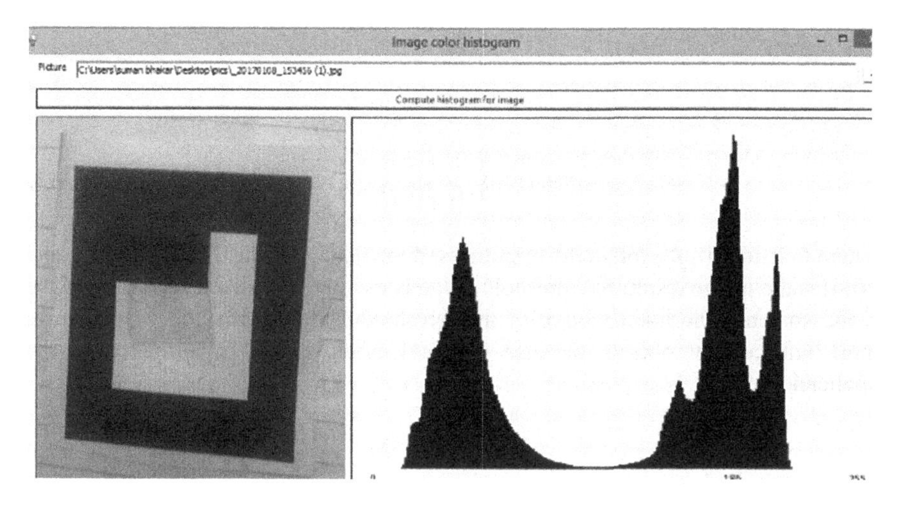

Fig. 3 Original image with noise

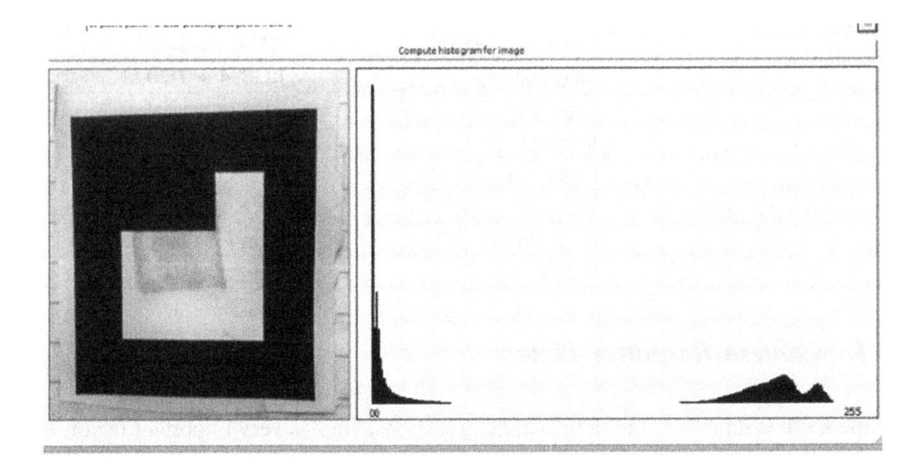

Fig. 4 Convert black and white but noise detected

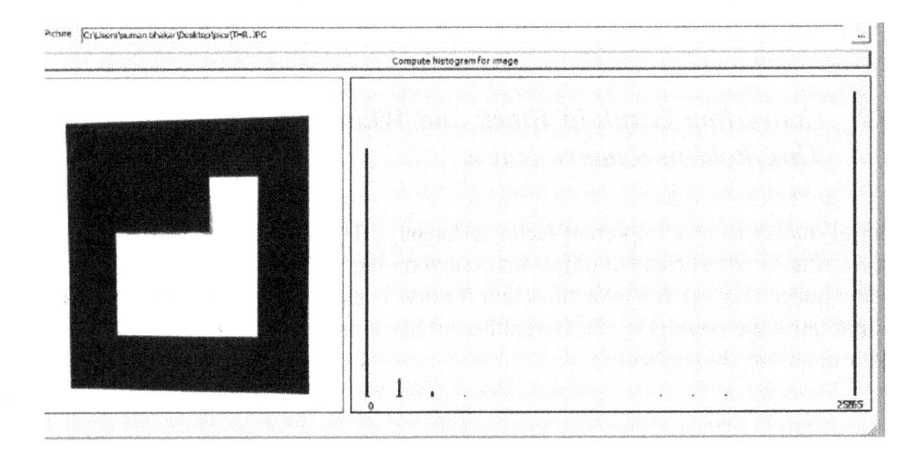

Fig. 5 Apply threshold noise minimized but still present

$$g(i,j) = \begin{cases} 0, & f(i,j) \leq p \\ 1, & f(i,j) > p \end{cases} \quad (1)$$

where function f(i, j) explain the brightness of the pixel, p is the threshold point, and g(i, j) is the image result, the threshold value is usually considered according to the scene content. The main motive of the thresholding transformation is to remove most unnecessary objects because it could improve the performance of the application.

Fig. 6 Distance from camera to glyph (approximate 35 cm)

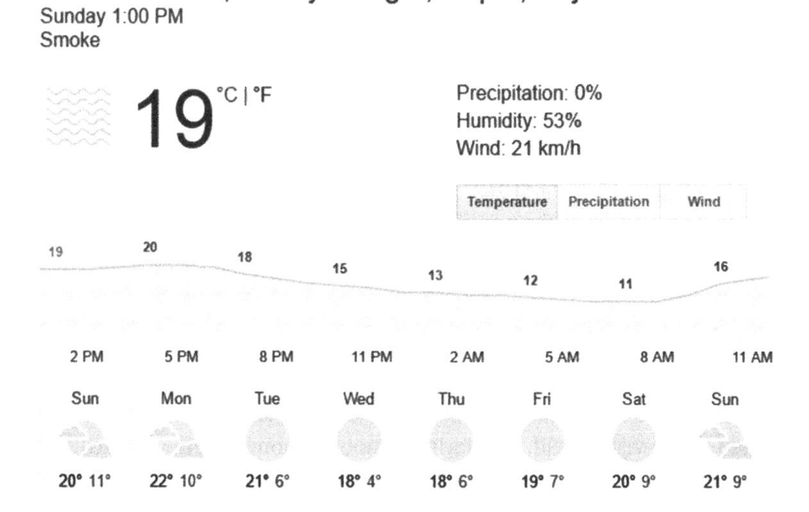

Fig. 7 Cloudy time at 01:00 PM

Table 3 Distance from camera to glyph

Glyphs (marker)	Distance from camera (cm)	Visibility (action perform)	Camera detection
A (rotation from 0–30)	~35	Visible	Yes
A (rotation from 0–30)	~58	Visible	Yes
A (rotation from 0–30)	~78	Visible	Yes
A (rotation from 0–30)	~125	Visible	Yes
A (rotation from 0–30)	~132	Starts blinking	Yes
A (rotation from 0–30)	~135	Blinking	No
A (rotation from 0–30)	~136	Blinking	No
A (rotation from 0–30)	~138	Invisible	No

4.3 Distance from Camera to Glyph (in Centimeter)

There must be approximately fixed distance from camera to glyph, otherwise the arm of machine will not rotate in a desired manner (Figs. 6 and 7).

In the cloudy environment (8/1/2016) at 1.00 pm, one glyph which is having height and width (5.5 × 5.5 cm) and its focus range from camera is approximately 132 cm in ambient light then it works properly otherwise if the distance is more than 132 cm then it doesn't work. At 135 cm approximate it starts blinking and response time is very less and the machine does not work properly (Table 3).

5 Limitations of Augmented Reality

There have been great advancements in the technology of augmented reality and through constant innovations, it has evolved itself from bulky and awkward head mounted devices to mobile phones. However, there are still many challenges lying in its development path as the technique is too much dependent on the innovation of computers and digital media [13].

Some of the challenges/limitations of augmented reality are as follows.

5.1 Technological Limitations

The main challenge of augmented reality is to convert digital data into graphics and scaling it to fit the visual field. Similarly, in mobile phones augmented reality has to function with less storage, limited processing power, and very small memory.

5.2 Negative Marketing and Advertising

Marketing and advertising are a grave concern for the future of augmented reality. Financial motives and excessive consumerism may lead augmented reality in the hands of few powerful companies who can pay hefty amounts to the augmented reality developers so as to exclude other small and less financially sound companies out of the loop. For, e.g., in search features of smart phones instead of searching all the city based restaurants, only few big companies like subways or pizza huts will pop up [14].

5.3 Privacy

With the development of augmented reality, another area of concern is privacy. Developers are working in a direction where merely by pointing a device at a person all the information such as date of birth, marital status, etc., can be gathered [15]. This is still a distant dream however through facial or image search in near future augmented reality can become a tool which is capable of invading the privacy of individuals.

Apart from these augmented reality faces, the challenges of social acceptance and becoming profitable for businesses to use it [16]. Augmented reality has still to cover a lot of distance for becoming an everyday tool. These are minor challenges and may be with the kind of developing augmented reality is seeing it will become a part of digital lives in coming years.

6 Conclusion and Future Trends

Augmented Reality has seen a tremendous expansion in recent years, but still it is considered in its nascent stage and the upcoming possibilities for AR are endless. There has been a remarkable rise in the field of AR products which are presented in different forms throughout the world. AR has played a significant role in the broader transition of computing from desktop to smartphones. However, apart from the research and development of AR certain limitations have also surfaced that need to be addressed. In this paper, we have analyzed some latency factors which affects the rotation of arm of the machine. Experiments of computing response time and methodology will be undertaken in next proposed paper.

References

1. Yang, R.: The Study and Improvement of Augmented Reality Based on Feature Matching. 2011 IEEE 2nd International Conference on Software Engineering and Service Science (ICSESS), Beijing, 15–17 July 2011, pp. 586–589.
2. Yuen, S. C.-Y., Yaoyuneyong, G., Johnson, E.: Augmented Reality: An Overview and Five Directions for AR in Education. Journal of Educational Technology Development and Exchange, 2011, pp. 119–140.
3. Krevelen, D., Poelman, R.: A Survey of Augmented Reality Technologies, Applications and Limitations. The International Journal of Virtual Reality, 2010, pp. 1–20.
4. Wu, H., Wen-Yu, S., Chang, H., Liang, J.: Current Status, Opportunities and Challenges of Augmented Reality in Education. Computers & Education, 2013, pp. 41–49.
5. Carmigniani, J., Furht, B., Anisetti, M., Ceravolo, P., Damiani, E., Ivkovic, M.: Augmented Reality Technologies, Systems and Applications. Multimedia Tools and Applications, 2011, pp. 341–377.
6. Carmigniani, J., Furht, B.: Augmented Reality: An Overview. In: J. Carmigniani and B. Furht, Eds., Hand-book of Augmented Reality, Springer, 2011, pp. 3–46.
7. Cai, S., Wang, X., Gao, M., Yu, S.: Simulation Teaching in 3D Augmented Reality Environment. 2012 IIAI International Conference on Advanced Applied Informatics (IIAIAAI), Fukuoka, 20–22 September 2012, pp. 83–88.
8. López, H., Navarro, A., Relaño, J.: An Analysis of Augmented Reality Systems. 2010 Fifth International Multi-Conference on Computing in the Global Information Technology (ICCGI 2010), 20–25 September 2010, pp. 245–250.
9. Jackson, T., Angermann, F., Meier, P.: Survey of Use Cases for Mobile Augmented Reality Browsers. In: T. Jackson, F. Angermann and P. Meier, Eds., Handbook of Augmented Reality, 2011, pp. 409–431.
10. Grasset, R., Langlotz, T., Kalkofen, D., Tatzgern, M., Schmalstieg, D.: Image-Driven View Management for Augmented Reality Browsers. 2012 IEEE International Symposium on Mixed and Augmented Reality (ISMAR), Atlanta, 5–8 November 2012, pp. 177–186.
11. Mohana, Z., Musae, I., Tahir, M.A., Parhizkar, B., Ramachandran, A., Habibi, A.: Ubiquitous Medical Learning Using Augmented Reality Based on Cognitive Information Theory. Advances in Computer Science, Engineering & Applications, Vol. 167, 2012, pp. 305–315.
12. Echeverry, J., Méndez, D.: Augmented Reality Training. 2010. Information Science, 2011, pp. 62–71.
13. Juan, C., Toffetti, G., Abad, F., Cano, J.: Tangible Cubes Used as the User Interface in an Augmented Reality Game for Edutainment. 2010 IEEE 10th International Conference on Advanced Learning Technologies (ICALT), Sousse, 5–7 July 2010, pp. 599–603.
14. Furata, H., Takahashi, K., Nakatsu, K., Ishibashi, K., Aira, M.:A Mobile Application System for Sightseeing Guidance Using Augmented Reality. 2012 Joint 6th International Conference on Soft Computing and Intelligent Systems (SCIS) and 13th International Symposium on Advanced Intelligent Systems (ISIS), Kobe, 20–24 November 2012, pp. 1903–1906.
15. Haugstvedt, A., Krogstie, J.: Mobile Augmented Reality for Cultural Heritage: A Technology Acceptance Study. 2012 IEEE International Symposium on Mixed and Augmented Reality (ISMAR), Atlanta, 5–8 November 2012, pp. 247–255.
16. Januszka, M., Moczulskia, W.: Augmented Reality for Machinery Systems Design and Development. New World Situation: New Directions in Concurrent Engineering, 2010, pp. 79–86.

Enhancing Web Search Through Question Classifier

Gaurav Aggarwal, Neha V. Sharma and Kavita

Abstract Question answering field has evolved alongside the Natural Language processing. There are several small-scale applications that use the linguistic, semantic, and syntactic interpretations of text and consume it in further processing. In case of a web search, knowing what the query is intended for can save hours of CPU processing and decrease response time tremendously. We have taken a small step in this direction by treating the query as a question and classifying it with the best suited classification algorithm. In this paper, we have tried to find out that when a perfect-informer of the question (knowing what is asked) is provided as an input for classification algorithms like SVN, Naïve Base, and decision tree, we want to observe their accuracy on the same data set of questions. In our experiment, we have used the concept of CRF to find question features that are relevant. CRF is a probabilistic model that treats features as observation sequence and emits all sequence labels with probability values.

Keywords Text mining · Question classification · CRF · SVM
NB · DT · Web search

1 Introduction

To be able to enhance text search, we require smart NLP techniques. In this paper, we do not focus on document classification which is the primary task of a search engine. We focus more on the information extraction part from a question or a user

G. Aggarwal (✉) · N.V. Sharma · Kavita
School of Computing and Information Technology, Manipal University Jaipur, Jaipur, Rajasthan, India
e-mail: gaurav.aggarwal@jaipur.manipal.edu

N.V. Sharma
e-mail: nehav.sharma@jaipur.manipal.edu

Kavita
e-mail: kavita.jhajharia@jaipur.manipal.edu

A.K. Somani et al. (eds.), *Proceedings of First International Conference on Smart System, Innovations and Computing*, Smart Innovation, Systems and Technologies 79, https://doi.org/10.1007/978-981-10-5828-8_75

entered query. This indeed can rule out processing of unwanted documents and increase computation power of the CPU. The final result is quicker and relevant response to the user. NLP involves the linguistic, semantic, and syntactic interpretations of the question or query in our case. It is still a developing field. Its uses range from speech processing to generating opinion from common social networking sites.

Big Data analysis or data analytics is in huge demand today. Techniques like Hadoop lessen the burden on the CPU but still the question remains which is the most useful data in this chunk. What is it that I am missing? What is it that I don't know and would be very helpful to me if I knew it?

Today, business runs on data analysis and the fact that internet has brought so many opportunities its all about who knows it first. There is a marginal difference now between small and huge firms; no doubt today data experts are in more demand than IT people.

2 Concept and Classification

2.1 Recent Work

By taking advantage of machine learning algorithms, we are now able to cater to larger datasets and more complex classifications. As the learning techniques advance, will the classification systems. UIUC is used as standard classification. It has 6 coarse and 50 semantically linked classes. It has 5500 training and 500 test questions. Question classification has somewhat benefitted less till now from machine learning.

Li and Roth achieved more than 78% accuracy for 50 classes by including question tokens such as parts of speech, chunks, named entity. After the work of Li and Roth, SVM was applied to question classification [1]. SVM two class classification was extended to multiclass by [2]. Zee and Lee [3] obtained 79% accuracy using linear SVM. Zhang and Lee designed a kernel on question parse trees, which yielded visible gains for the 6 coarse labels in the UIUC classification system. The accuracy gain for the 50 fine-grained classes was surprisingly small. The authors explain this in the following terms: "the syntactic tree does not normally contain the information required to distinguish between the various fine categories within a coarse category."

2.2 Features

Features are a subset of original features selected to make the learning more powerful. In feature selection, we reduce the number of features and in construction we create new meaningful features [4]. Any amount of data that needs to be mined must first be preprocessed before any data mining algorithm can be applied. There are two approaches in this matter: In the first approach, the mining algorithm can itself preprocess the data or in other words convert the data into a format on which the mining algorithm can function. This approach has several disadvantages, the main being the mining algorithm may not be able to handle large data which becomes a limitation. To overcome this limitation, we separate the preprocessing from inside the mining algorithm, which has several other advantages. For this matter, a very good concept is the introduction of features. Features basically are the characteristics of the data that distinguish one set of data from another. Features are used in text mining, speech processing, etc. The concept of features reduces the dimensionality of the same data in the feature space. Therefore, it is a very important step in data preprocessing because the learning time and predicting accuracy of the mining algorithm depend on it. In text mining features, can be in any granular level, they can be a word or at higher granularity can be a text layout. When we train the machine learning algorithm, we provide it with the near to a real dataset. There is an emerging concept in this field called continuous random fields (CRF). According to [5], this is a probabilistic framework for segmenting and labeling sequence data and is better than HMM in reducing redundant features. In case of text mining, the process of text annotation and information extraction can be sort as the same problem as segmenting and labeling a sequence of data. Many concepts have been developed in this area. In [6], the authors have used CRF to extract the informer span of the question.

HMM, stochastic grammar, MEMM, and CRF find use in scientific computing and numerical analysis; typically in biological sequence identification like in case of finding RNA secondary structure. Typically computational linguistics and in computer science, these have been applied to text processing and speech processing. The area they have been applied are topic segmentation, part of speech tagging, information extraction, etc.

While in HMM joint probability is attached to observation and label sequence, CRF gives joint probability to all possible label sequences, hence reducing burden on the observation sequence which is anyway fixed. According to [6] CRF has been proved better than HMM in face of redundant features. In our model, we extend on the model depicted in [6] and give the input of CRF-informer segment to all three algorithms SVM, Naïve bayes, and DT.

In [6], the author suggests using a hierarchical classification system. In this system, he uses the UIUC standard classification system. In the first layer, he uses CRF to generate the informer segment. They say that for a question with one question token two to three words are enough for classification and they call it the informer segment. They trained CRF to identify the informer segment in the first layer. And in the second layer, they have used SVM to classify the question based on the informer segment. Accuracy obtained is 93.2% for 6 class and 86.2 for 50 classes in UIUC standard. Features are important to remove unwanted data in a text mining process. According to [7], the detailed process is described in Fig. 1.

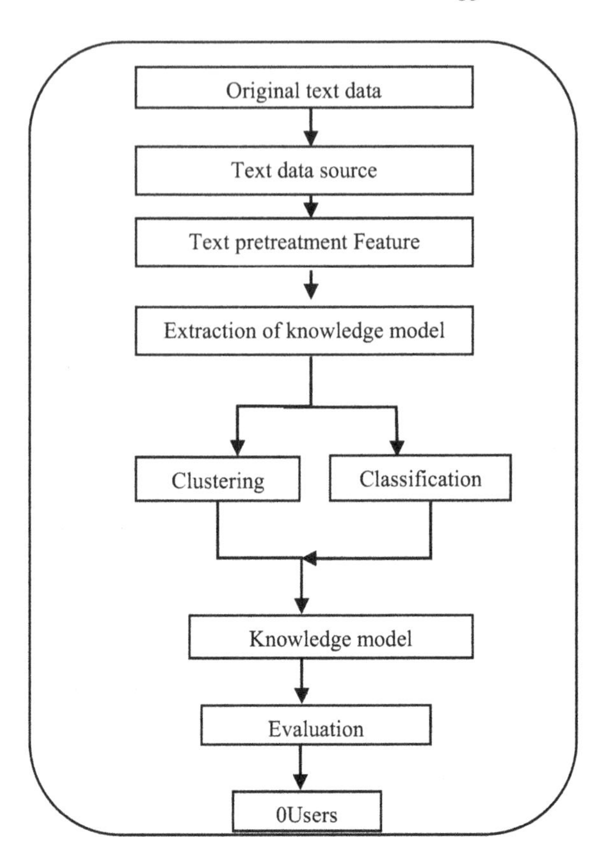

Fig. 1 Text mining process and order of the feature extraction in the process

2.3 Question Classification

As in [8], question classification is the task in which we map the given question into one of the x classes which specify a semantic constraint on the sought after answer. They have suggested the use of hierarchical classifiers. First, classifier segments the data into coarse categories and second classifier taking these as input again classifies into finer segmentation groups. In both cases, the attributes or features of the input data are same. The distribution of the coarse and fine-grained classes is shown below reference is TREC. In [8], the acceptable standard for the UIUC classification with 6 coarse and 50 fine classes is described according to Table 1.

In [8], however, no evident performance improvement was noticed between hierarchical classifier and multiclass classifier. As in case of former, there is a chance of wrongly categorizing in coarse grained class. Our below model is based on question hierarchy.

Table 1 Showing the UIUC classification with 6 coarse and 50 fine classes

Abbrev.	Description
abb	Manner
exp	Reason
Entity	Human
Animal	Group
Body	Individual
Color	Title
Creative	Description
Currency	Location
Dis.med.	City
Event	Country
Food	Mountain
Instrument	Other
Lang	State
Letter	Numeric
Other	Code
Plant	Count
Product	Date
Religion	Distance
Sport	Money
Substance	Order
Symbol	Other
Technique	Period
Term	Percent
Vehicle	Speed
Word	Temp
Definition	Weight

2.4 Our Model

Figure 2 describes the common learning model for SVN, NBC, and DT.

2.5 Experiment

In most of the feature generating models, develop unnecessary features because they only relate to either semantic or syntactic or both. These are developing NLP systems. In Chap. 10 [6], the author develops a logic which can find our informer segment from a question and here after meaning full features can be identified. Even the training becomes easy and so is the accuracy of the algorithm. Another problem is feature ambiguity which seems to have lowered with CRF. But a better alternative can be sought for. What I propose in this paper is an experiment where if the

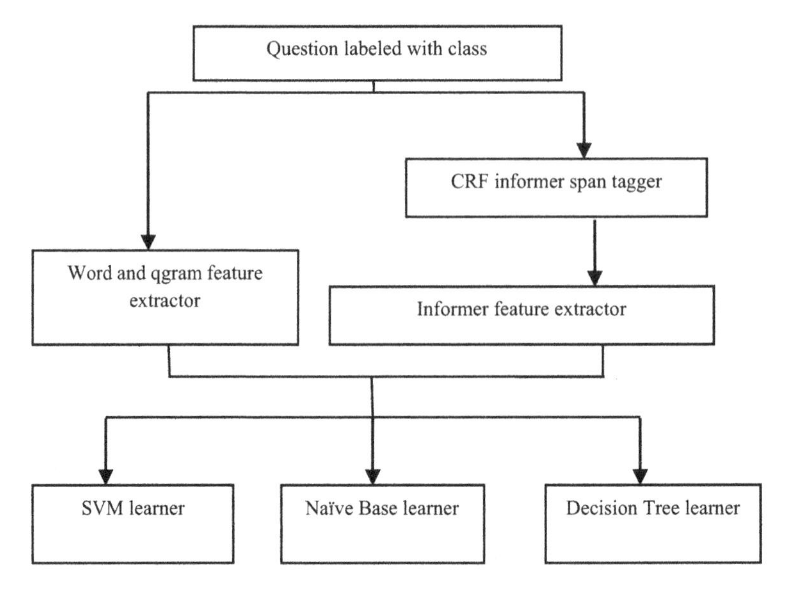

Fig. 2 Common learning model for SVM, NB and DT

informer span is correctly identified with the former technique using CRF, then in case of one line of text/question with only one identifying token (this is a constraint) which classification algorithm performs better—linear SVM, Naïve Base which is based on conditional probability distribution [9] or decision tree in which we split the instance space [9]. Data set being used here is the standard developed by TREC which has 6 coarse and 50 fine classes. In this paper we have studied the accuracy obtained by SVM, Naïve Base and Decision tree. Accuracy here would simply mean how many questions are correctly classified in the first attempt. Both the CRF section and the algorithm are to be trained separately but with the same dataset of questions. The result obtained here is combined with the result obtained in [6]. Results are summarized as below (Table 2).

In Table 3, columns represent the dataset and the rows represent the algorithms. We can easily compare the results of our experiment for each algorithm corresponding to each dataset.

In our experiments, we see that SVM is most accurate with one line of question (having one question token). We have yet to determine the accuracy of the three

Table 2 Result of experiment on the algorithms SVM, NBC and DT

Algorithm	6-class	50-class
SVM+ECOC	–	80.2–82
Linear SVM	87	79.2
SVM, perfect-informer	94.2	88
SVM, CRF-informer	93.4	86
NBC, CRF-informer	82	75
C4.5, CRF-informer	90	82

Table 3 Accuracy results with different datasets taken from the TREC data

Algo	1000	2000	3000	4000	5500
NB	52	58	62	65	72
DT	62	67	72	75	79
SVM	64	69	74	77	81

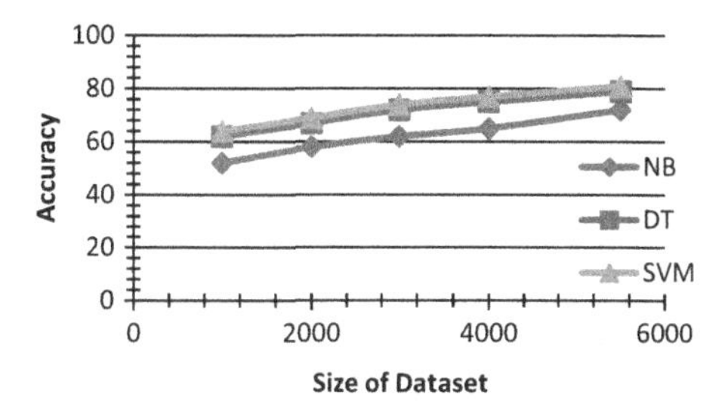

Fig. 3 Diagram depicting relative accuracy of NB, DT and SVM

algorithms when there are two or more informer spans in the question. We are yet to see the classification accuracy in such a case. Below we can see the graphical representation of the tabular data which is in Table 2. The graphical representation clearly depicts the marginal difference between Decision Tree and the SVM. It shows that the classification is based on numerical analysis it is far more effective than probabilistic approach. But where numerical data cannot be computed we benefit greatly from this probabilistic approach (Fig. 3).

3 Conclusion

Referring to the above results, it can be concluded that question class categorization depends on the choice of algorithm, classification categories and quality of training data. It can be further concluded that the comparison algorithms such as SVM do surpass the commutative algorithms such as DT in this case.

4 Future Work

Based on the experimental results, we would further like to extend the model which can be trained to choose which classification algorithm to use for particular set of questions. With more sophisticated and efficient natural language processing, the applications of question answering can also be widened. Clustering techniques also can be studied for accuracy in question classification.

References

1. Xin Li and D. Roth.: Learning question classifiers: The role of semantic information. Natural Language Engineering. 12(03):229–249 (2006).
2. Vapnik. V, Golow1ich. S, and Smola. A.J.: Support vector method for function approximation, regression estimation, and signal processing. Advances in Neural Information Processing Systems. MIT Press (1996).
3. Zhang, D. and Lee. W.: Question classification using support vector machines. SIGIR, pages 26–32, (2003).
4. Motoda, Hiroshi, and Huan Liu. "Feature selection, extraction and construction." Communication of IICM (Institute of Information and Computing Machinery, Taiwan) Vol 5, pp: 67–72, (2002).
5. Lafferty, J., McCallum, A. and Pereira. F.: Conditional random fields: Probabilistic models for segmenting and labeling sequence data. ICML (2001).
6. Srivastava, Ashok N., and Mehran Sahami, eds. Text mining: Classification, clustering, and applications. CRC Press, (2009).
7. Shi, Guoliang, and Yanqing Kong. "Advances in theories and applications of text mining." Information Science and Engineering (ICISE), 2009 1st International Conference on. IEEE, (2009).
8. Lowd, D. and Domingos. P.: Naive Bayes Models for Probability Estimation. Proceedings of the 22nd International Conference on Machine Learning, Bonn, Germany, pp: 529–536. (2005).
9. Cohen. S., Rokach. L, and Maimon. O.: Decision-Tree Instance-Space Decomposition with Grouped Gain-Ratio. International Journal of Information Science. Volume 177 Issue 17, pp: 3592–3612. (2007).

A Chaotic Genetic Algorithm for Wireless Sensor Networks

Anju Yadav, Vipin Pal and Ketan Jha

Abstract Today, clustering for Sensor Node (SN) is a method in Wireless Sensor Networks (WSNs) to diminish the energy consumption of the SN by avoiding long distance communication between the SNs. This will protract the lifetime of sensor networks. However, a cluster head has to perform various tasks such as collection of data from member nodes, aggregation of the collected data, and send that data to the BS. Network load balance is a challenging issue in WSNs for the clustering schemes. Genetic algorithms (GA) with clustering schemes are implemented for better cluster formation. The GA run through again over a large no of iterations to find the optimal solution that leads to premature convergence. The chaotic GA (CGA) will solve this problem by avoiding local convergence, i.e., by choosing a chaotic map to generate the random values instead of traditional random function and improves the performance of the traditional GA. A chaotic GA (CGA) based clustering algorithm for WSNs has been proposed in the proposed work that has better convergence rate for cluster head selection and consequently improves the performance of sensor network.

Keywords Logistic map · WSNs · Genetic algorithm · Clustering

1 Introduction

An enormous number of SNs deployed over the region of interest (RoI) are the prime constituent of WSNs (WSN) [1, 2]. Spatially distributed SNs work cooperatively to communicate the sensed information of the RoI to the sink\Base Station

A. Yadav (✉) · V. Pal
Manipal University, Jaipur, India
e-mail: anju.anju.yadav@gmail.com

V. Pal
e-mail: vipinrwr@gmail.com

K. Jha
Central University of Rajasthan, Ajmer, India
e-mail: jha.ketan555@gmail.com

A.K. Somani et al. (eds.), *Proceedings of First International Conference on Smart System, Innovations and Computing*, Smart Innovation, Systems and Technologies 79, https://doi.org/10.1007/978-981-10-5828-8_76

(BS) through wireless links. SNs, the central point of WSNs, are resource constraints and are embedded with low computational circuitry, less memory, and limited battery power. In most of the applications, harsh or hostile environment is monitored by sensor nodes; consequently, recharge or replace the battery unit of nodes is nearly impossible, see Fig. 1. Consequently, efficient and economical energy utilization of nodes in RoI is the essential design prospect for WSNs [3]. In the literature, clustering approach is examined as the energy efficient that also provide load balancing and scalability to the network [4, 5].

The basic idea of clustering is to group the nodes in independent sets, known as clusters, and electing one representative of each set, called as a cluster head. Member nodes of a cluster communicate only to the respective cluster head. The sensed data of each member nodes is received by cluster head (CH) and may apply data aggregation or fusion technique [6] to condense the collected information to relevant information and send the aggregated data to BS directly or multi-hop relaying other cluster heads. The hierarchical natures of clustering approaches reduce the communication distance, the prime contributor of energy consuming of nodes [7]. To achieve the better load balancing, the role of cluster head is shared by all nodes or few highly resource rich nodes are deployed with other nodes that act as cluster heads all the time. Clustering may be applied as distributed approach—nodes locally perform clustering or may be operated as centralized—a centralized point performs the clustering and provides information to the nodes of RoI.

Performance of a clustering approach depends on various parameters of clustering like—selection of cluster head, cluster formation, size of clusters, number of cluster heads, round-time and many more. Cluster head selection is the vital for the performance of clustering algorithms as cluster head is accountable for the respective cluster. Location and remaining energy are two prime factors for

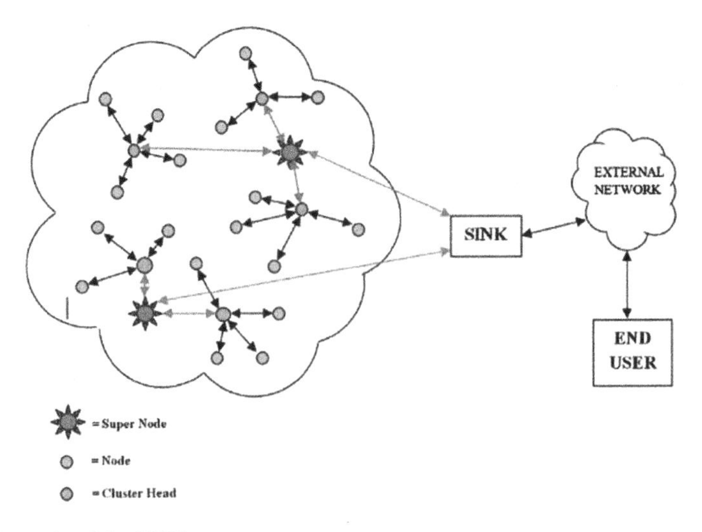

Fig. 1 Scenario of the WSNs

selection of a node as a cluster head. In the literature, various optimization techniques are implemented with centralized clustering techniques to select efficient cluster heads. Genetic algorithms (GA) [8, 9] are extensively implemented in WSNs with clustering approaches to select cluster heads and cluster formation [10–12]. However, the traditional GA has some disadvantages, like it will take more no of iteration to reach optimal solution and also having premature convergence [13]. These key problems are solved by introducing chaos in GA.

In 1975, Chaos was proposed by Li et al. [14]. Chaos has a dynamic behavior in nonlinear systems and it has three important properties: Randomness, Ergodicity, and Sensitivity to the initial condition. Recently nonlinear dynamics, especially chaos due to its properties is applied on various fields such as cryptography [15, 16], image processing [17, 18], secure transmission [19], nonlinear circuits [20], etc. Even the chaos may be used for solving non optimization algorithm [5, 21, 22]. In [23], the logistic model, i.e., a population model is used to generate the initial population in GA, but still not able to deal the mutation diversity in various complicated cases. The above problem is solved by the author Mohammad et al. in 2015 by considering two types of chaotic map in GA. Further, some more authors used chaos in GA or hybrid GA using chaos and particle swarm optimization to solve the key problem of GA [13, 24].

In proposed work, we have considered chaotic GA to optimize the CH selection with cluster formation to reduce the energy utilization and to improve the performance of the WSNs. In Sect. 2, we have explained our proposed approach with the help of flow chart and we have concluded our work in Sect. 3.

2 Chaotic Genetic Algorithm for WSNs

2.1 Chaos Theory

In nonlinear system, chaos shows dynamic behavior in deterministic conditions and it is also extremely sensitive to the initial state. A small perturbation in initial value can lead to a big change in dynamical system behavior. Three main properties of chaos are: Ergodicity, Randomness, and Sensitive to the initial state. The property randomness of the chaos avoids the search trapped in local minimum solution [24]. This can be achieved with the help of ergodicity property as it results in non-repeated values in the specific range [25]. The third major property of the chaotic system is sensitivity to initial condition which ensures that no two new population obtained are identical even the two best-fit solutions obtained are very close to each other [26].

This unpredictable effect (i.e., sensitive dependence) on initial states yields extensively diverging results for the chaotic systems rendering to a long-term prediction, and it is called as butterfly effect [13]. So, the chaotic behavior depends on the initial values of dynamical systems. To understand the chaotic map and the properties of chaos, we have taken a logistic map.

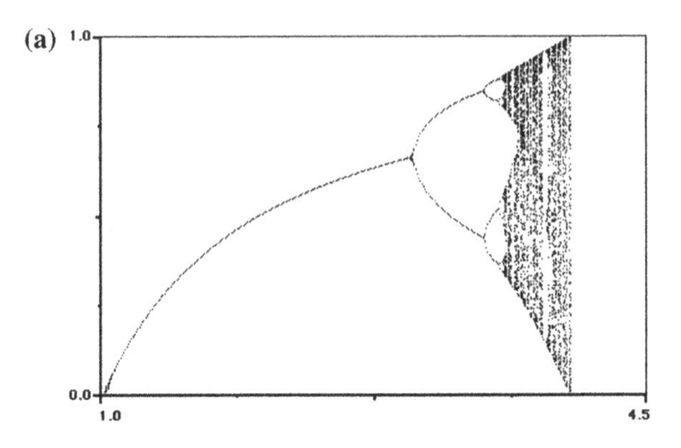

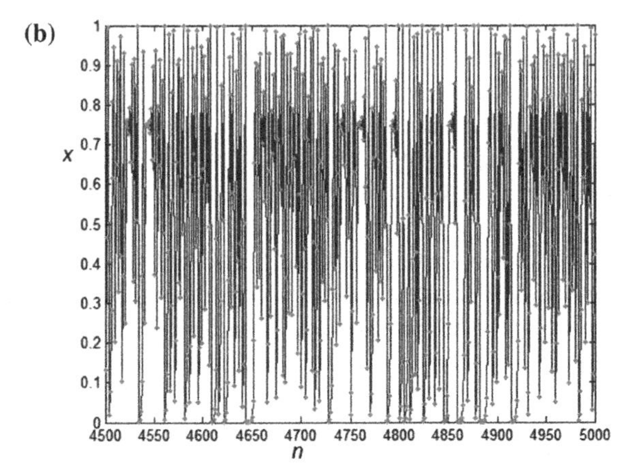

Fig. 2 **a** Bifurcation diagram and **b** Time series (at $r = 4$) of logistic map

The logistic map is a model for population dynamics, which was given by Belgique mathematician Pierre F. Verhulst during his studies in 1844 and 1847. Indeed, it has always been interesting to study the chaotic behavior of the logistic model $t(x) = rx(1 - x)$, $x \in [0, 1]$, r is the control parameter. Following behavior of the logistic model is observed:

For $0 < r \leq 2.75$, the logistic model is convergent for all $x \in [0, 1]$. For $2.75 < r \leq 3.2$, the logistic map shows cyclic behavior, which does not converge but still stable. For $3.2 < r < 4$, the logistic model is unstable. At $r = 4$ iterates span the entire space, i.e., the map becomes chaotic in nature [13]. The results for chaotic values are explained with the help of bifurcation diagram, See Fig. 2a and the time series at $r = 4$. See Fig. 2b.

2.2 Chaotic Genetic Algorithm

As we know that GA has various applications in many fields. The traditional GA has some disadvantages as discussed in introduction section which is solved in the chaotic genetic algorithm. Chaos theory is used with the genetic algorithm (CGA) to generate set of the random initial population with the help of chaotic map [23]. Further, chaos method is used in selection method for global searching capability. Wang et al. gave a new hybrid GA that is the combination of PSO and chaos to improve the convergence [24, 27].

Finally, this will create a population which reserves best-fit chromosome and also maintain the population diversity to avoid the premature convergence or trapping in a local optimum. In 2015, [13] used logistic and tent map to generate the sequence of random numbers instead of random process to generate the more diversity in the population. In traditional GA, random sequences generated by random function are used at mutation, crossover, and initial population. But, in chaotic GA by using the tent map we can generate uniformly distributed random numbers and the logistic map is used in crossover and mutation to solve the problems of traditional GA.

2.3 Proposed Approach: Chaotic Genetic Algorithm in WSNs

The Genetic algorithm is a popular evolutionary approach to find the most efficient method for solving the clustering in WSNs and it is worth proved [8, 10–12]. In this paper, an algorithm based on a chaotic genetic algorithm for clustering of SNs in a network to reduce the energy consumption and to maximize the lifetime of the network is proposed. The clustering algorithm is explained with the help of block diagram, see Fig. 3.

Now, in Fig. 4, we will explain the working of chaotic GA used in above clustering algorithm.

The proposed solution of Chaotic Genetic Algorithm based clustering scheme has wider diversity in the initial population and the population maintains the best fit chromosome by this it avoids the trapping in local optimum because of that it results in better convergence than the traditional GA. Consequently, the proposed solution has a better aspect of cluster head selection and cluster formation that leads to well load balanced network.

3 Performance Results

We have implemented CGA in MATLAB and compared its performance with GA, for generating a set of random numbers we have considered two maps logistic and tent map. Initially, we have evaluated the performance of CGA in comparison to GA for two different mathematical functions, i.e., Zakharov, Rosenbrock. The mathematical equation for these two functions is given in Table 1. For the purpose

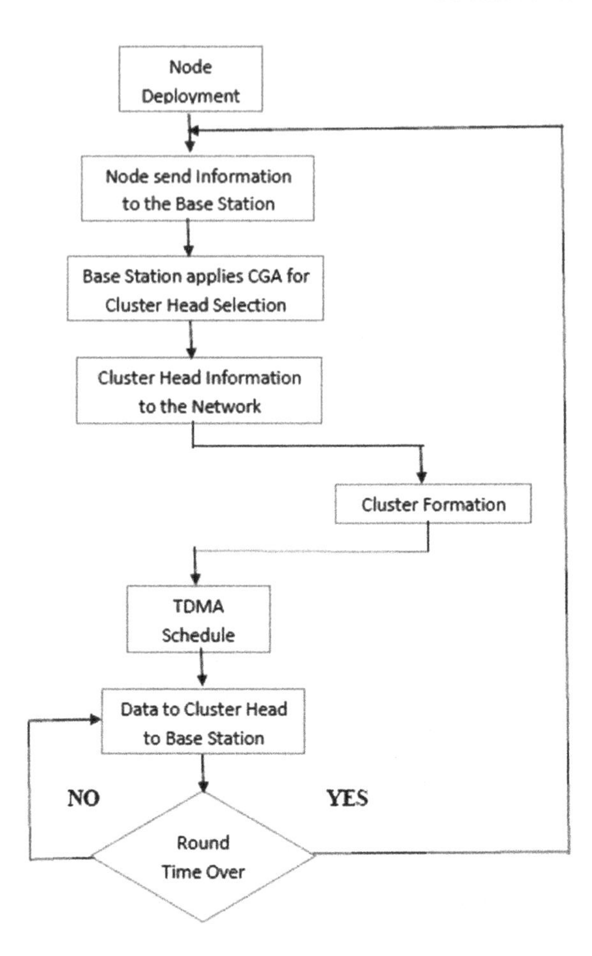

Fig. 3 Flow chart diagram for the clustering based on the chaotic GA in WSN

of simulation following parameters are considered population size as 100, the mutation rate is 0.2 and crossover rate is 0.8. And from simulation we have observed that the CGA has converged for less no of iterations, i.e., its convergence rate is faster than GA (see Table 2). The above simulation result shows that the CGA has reduced the no of iterations to find the global optimum solution and also it avoids the local convergence.

Further, the convergence and correctness of both the GA and CGA are compared for cluster based WSN's to find the optimal number of CH's in a round in MATLAB. A network of 100 SN's (N) deployed over 100 × 100 m^2 area is considered for experimental results. The population size of the initial population is 100 (N) and size of each chromosome is (N). A '1' in chromosome represents the status of a node as CH while a '0' means member nodes. Selection type for creating new population is Roulette Wheel while crossover rate and mutation rate are 0.7 and 0.01 respectively. Fitness function of the GA and CGA is same that consider

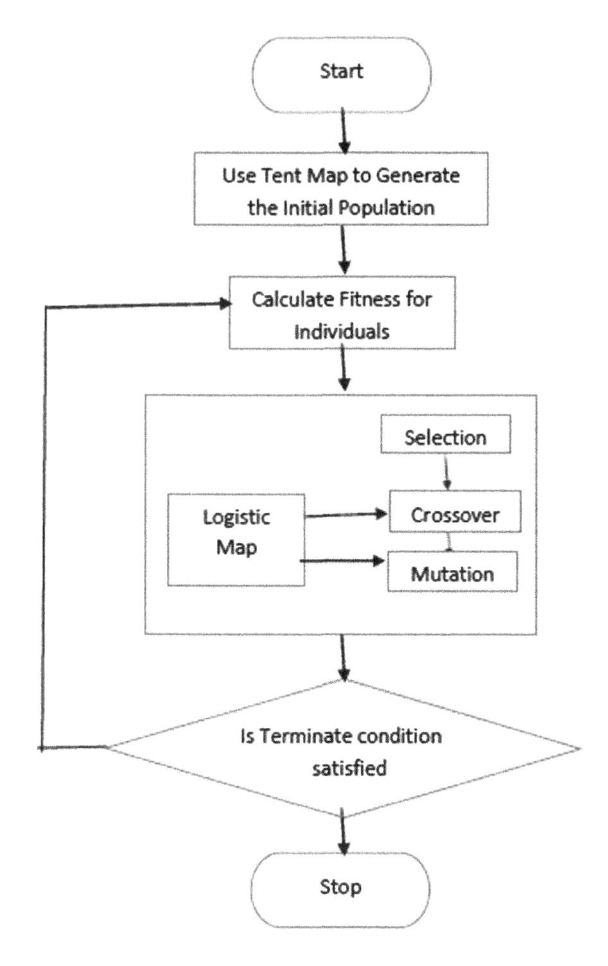

Fig. 4 Flow chart diagram of chaotic genetic algorithm

Table 1 Mathematical functions as benchmark for simulation

Function	Mathematical expressions
Zakharov	$h(x) = \sum_{i=1}^{n} x_i^2 + (\sum_{i=1}^{n} 0.5x_i)^2 + (\sum_{i=1}^{n} 0.5x_i)^4$
Rosenbrock	$h(x) = \sum_{i=1}^{n-1} (100(x_i^2 - x_{i+1})^2 + (x_i - 1)^2)$

Table 2 CGA and GA average no of iteration

Bioinspired algorithms	Zakharov	Rosenbrock
GA	406	351
CGA	336	317

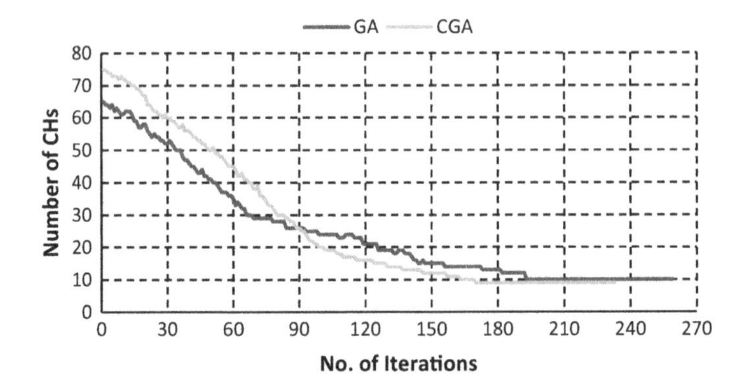

Fig. 5 Performance analysis of CGA in comparison to GA

the number of CH's (NCH), total intra-cluster communication distance (IC) and total distance from CH's to BS (BSD). The normalized fitness function (t) is as

$$t = (N - NCH) + \frac{IC}{N} + \frac{BSD}{N} \tag{1}$$

Figure 5 anatomizes the comparison of convergence and correctness of GA and CGA. It can be analyzed from the figure that the initial search area within the initial population is much higher for CGA as compared to GA. Initial average number of CH's in case of CGA is 75 while it is 65 for GA. It can also be analyzed that near 90 iterations an average number of CH's for CGA goes down to that of GA. Convergence of CGA is better than GA; GA starts converging after 193 iterations for final solution while CGA start converging after 170 iterations

4 Conclusion

We have proposed an algorithm using evolutionary algorithms, i.e., CGA for clustering in WSNs. In our approach, logistic and tent map is used in GA to produce the set of random sequences to avoid the trap in local optimum or premature convergence. Then the chaotic GA is used in clustering of the network to diminish the energy utilization and to increase the network performance.

References

1. Akyildiz I., Su W., Sankarasubramaniam Y., and Cayirci E., "WSNs: a survey," Computer Networks, vol. 38, no. 4, pp. 393–422, 2002.
2. H. Karl and A. Willig, "Protocols and architectures for WSNs" John Wiley & Sons, 2007.

3. Anastasi G., Conti M., Francesco Di M., and Passarella A., "Energy conservation in WSNss: A survey," Ad Hoc Network, vol. 7, no. 3, pp. 537–568, May 2009.
4. Abbasi A. A. and Younis M., "A survey on clustering algorithms for WSNs," Computer Communication, vol. 30, no. 14–15, pp. 2826–2841, Oct. 2007.
5. Afsar M. M. and Tayarani-N M.-H., "Clustering in sensor networks: A literature survey," Journal of Network and Computer Applications, vol. 46, pp. 198–226, 2014.
6. Rajagopalan R. and Varshney P. K., "Data aggregation techniques in sensor networks: A survey," Communication Surveys and Tutorials, IEEE, vol. 8, pp. 48–63, 2006.
7. Rappaport T., "Wireless Communications: Principles and Practice," 2nd ed. Upper Saddle River, NJ, USA: Prentice Hall PTR, 2001.
8. Gen M. and Cheng R., "Genetic Algorithms and Engineering Optimization (Engineering Design and Automation)" Wiley-Interscience, 1999.
9. Goldberg D. E., "Genetic Algorithms in Search and Machine Learning" Pearson Education, 2006.
10. S. Hussain, A. W. Matin, and O. Islam, "Genetic algorithm for energy efficient clusters in WSNs," in ITNG. IEEE Computer Society, 2007, pp. 147–154.
11. Liu J.-L. and Ravishanka C. V., "LEACH-GA: Genetic Algorithm-Based Energy-Efficient Adaptive Clustering Protocol for WSNs," International Journal of Machine Learning and Computing, vol. 1, no. 1, pp. 79–85, December 2011.
12. Pratyay k., Gupta S. K. and Jana P.K., " A novel evolutionary approach for load balancing clustering problem for WSNs", Swarm Evolv. Comput., 12, pp. 48–56, 2013.
13. Javidi M. and Hosseinpourfard R., "Chaos genetic algorithm instead genetic algorithm", Int. J. Inf. Tech., 12(2), pp. 163–168, 2015.
14. Li S., Chen G., and Zheng X., Chaos-based Encryption for Digital Images and Videos, Multimedia Security Handbook, Internet and Communications Series, vol. 4, CRC Press, USA, 2004.
15. Suneel M., "Chaotic Sequences for Secure CDMA," available at: http://arxiv.org/ftp/nlin/papers/0602/0602018.pdf, last visited 2006.
16. Wong K., Man P., Li S., and Liao X., "More Secure Chaotic Cryptographic Scheme Based on Dynamic Look-up Table Circuits," System Signal Process Journal, vol. 24, no. 5, pp. 571–84, 2005.
17. Faraoun K., "Chaos-based Key Stream Generator Based on Multiple Maps Combinations and its Application to Images Encryption," the International Arab Journal of Information Technology, vol. 7, no. 7, pp. 231–240, 2010.
18. Gao H., Zhang Y., Liang S., and Li A., "New Chaotic Algorithm for Image Encryption," Chaos, Solitons and Fractals, vol. 29, no. 2, pp. 393–399, 2006.
19. Heidari-Bateni G. and McGillem A., "Chaotic Direct-sequence Spread Spectrum Communication System," IEEE Transaction on Communication, vol. 42, no. 2, pp. 1524–1527, 1994.
20. Arena P., Caponetto R., Fortuna L., Rizzo A. and Rosa M., "Self-Organization in Non-Recurrent Complex System," Applied Sciences and Engineering, vol. 10, no. 5, pp. 1115–1125, 2000.
21. Li B. and Jiang W., "Optimizing Complex Functions by Chaos Search," Cybernetics and System Journal, vol. 29, no. 4, pp. 409–419,1998.
22. Manganaro G. and Pineda J., "DNA Computing Based on Chaos," in Proceedings of the IEEE International Conference on Evolutionary Computation, Indianapolis, pp. 255–60, 1997.
23. Cheng T., Wang C., Xu M., and Chau W., "Optimizing Hydropower Reservoir Operation using Hybrid Genetic Algorithm and Chaos," Water Resources Management, vol. 22, no. 7, pp. 895–909, 2008.
24. Wang Y. and Yoo M., "A New Hybrid Genetic Algorithm Based on Chaos and PSO," in Proceedings of the IEEE International Conference on ICIS, Shanghai, China, pp. 699–703, 2009.
25. Tan D., "Application of Chaotic Particle Swarm Optimization Algorithm in Chines Documents Classification," in Proceedings of IEEE International Conference on Granular Computing, CA, USA, pp. 763–766, 2010.

26. Juan L., Zi-xing C., and Jian-qin L., "A Novel Genetic Algorithm Preventing Premature Convergence by Chaos Operator," the Journal of Central South University of Technology, vol. 7, no. 2, pp. 100–103, 2000.
27. Alatas B., Akin E., and Ozer A., "Chaos Embedded Particle Swarm Optimization Algorithms," Chaos Soliton and Fractals, vol. 40, no. 4, pp. 1715–1734, 2009.

An Empirical Investigation of Impact of Organizational Factors on Big Data Adoption

Mrunal Joshi and Pradip Biswas

Abstract Big data and Analytics (BDA) is one of the most talked about technology trend having a widespread impact on organizational value chain. The objective of the study is to explore and examine the key organizational factors that impact the big data adoption in service organizations. A research framework—grounded in organizational theories and IT adoption—examines the impact of four organizational variables on big data adoption and finds that three of them have a strong positive impact. The survey instrument is developed by employing rigorous measurement scales. The study targeted around 500 service organizations headquartered at Mumbai; of which 109 suitable responses are received. Structural equation modeling using the variance based, prediction-oriented PLS model estimation—SmartPLS is applied for testing. The precision estimation and standard errors are evaluated using bootstrapping with 109 cases and 300 samples (resamples).

Keywords Big data adoption · Organizational factors · Service organizations SmartPLS SEM · Organizational commitment · Organizational data environment · Organizational size

1 Introduction

Organizations are getting bigger and amassing with ever-increasing amounts of data. However, it is scattered across multiple sources and exists in multiple formats. The process of data consolidation at a centralized location happens through data warehousing. Google's chief executive, Eric Schmidt, believes the world creates

M. Joshi (✉)
IES Management College and Research Centre, Mumbai, India
e-mail: mrunal.joshi@ies.edu

M. Joshi · P. Biswas
National Institute of Industrial Engineering (NITIE), Mumbai, India
e-mail: pradip2468@gmail.com

A.K. Somani et al. (eds.), *Proceedings of First International Conference on Smart System, Innovations and Computing*, Smart Innovation, Systems and Technologies 79, https://doi.org/10.1007/978-981-10-5828-8_77

5 EB of data every 2 days. Its value is increasing since the generated information leads to organizational decisions. It gets recorded and has structure. It has the potential meaning which makes it potentially useful. Data is in a form suitable for subsequent interpretation and processing. Information can be inferred and deduced from it and can be associated with other data [1]. It can be consolidated and stored in a data warehouse across the organizations over a period of time where it can be used for trend analysis, comparisons, and forecasting. This way, the enterprise level data warehouse is allowed to enjoy the benefit of a strategic system. Further to the data warehouse, today organizations are exploring the ways and means of generating value from the data, scattered and captured across different verticals in the organizations. This data, fondly called as "big data," is not readily suitable for shifting to the warehouse but may have a lot of potentials which is required to be reconnoitered for better business decisions [2].

2 Literature Review

2.1 Big Data

According to Gartner, the high-volume, high-velocity data which has high-variety information assets and which demands cost-effective and innovative forms of information processing is defined as big data. It is used for enhanced insights and decision making. It is the voluminous amount of structured, semistructured and unstructured data that has potential to be mined for information [3]. The "bigness" of data attracts researchers' attention to the size of the dataset, but at the same time the emergent discussion also points out that "big" cannot be a defining parameter, rather, how "smart" insights the data can provide is important. Social networks and social engagement behaviors of individuals can be analyzed by mapping mobility patterns onto physical layouts of workspaces. For this, the data captured by sensors that track entry and exit patterns can be used or even the frequency of meeting room usage using remote sensors can be useful. This could provide information on communication and coordination needs based on project complexity and approaching deadlines. This data provides a richness of individual behaviors and actions that has not been fully tapped in management research.

2.2 Implication of Big Data on the Organizations

It is undeniably big to fit on a single server and continuously flowing to fit into a static data warehouse. Further, the unstructured nature makes it unsuitable for a two-dimensional database. The literature indicates that the world used over 2.8 ZB of data in the year 2012 [4]. Big Data takes data from a variety of systems and combines it into significant and actionable insights. For example, in retail business the lagging indicators with the current and predictive data may be used to understand the performance of a product and its contribution to the revenues. A variety of data like the customer-satisfaction measures, real-time sales data, competitor advertising, campaign-management reports, customer surveys, blogs, website comments, etc., can be used to convey the performance of each product in each market over time. Based on the performance data, companies can identify and focus on the high return improvement opportunities in product development. This may help them to improve the revenue performance too. For a customer-focused business like banking, big data solution can integrate structured customer data with unstructured reports and social media. It can also be useful in developing the strategies to maximize share pricing by identification and prioritization of global opportunities and threats. Big data can also address the need for predictive analytics and collaborative decision making around various opportunities like countries, industries, competitors, etc. It can improve decision-making team's ability to prioritize their actions to mitigate enterprise risks. In a manufacturing environment where process complexity, process variability, and capacity restraints are vital, the big data can be a critical tool for realizing the improvements in yield. Such companies can differentiate themselves by building their capabilities by conducting quantitative assessments [5]. Big Data challenges managers for becoming more fact based and savvy about their decisions.

2.3 Organizational Need for Big Data Readiness

Studies infer that the data driven companies perform better on financial and operational front. They are found to be 5% more productive and 6% more profitable than their competitors. The studies also illustrate that with statistical and economic significance, they reflected a better stock market valuation [4, 6]. The traditional analytics focuses on to sharing of data, technology, and people to achieve the analytical objectives across the organization. The startups and online firms who were the early adopters of big data were concerned about something new and hardly cared about the integration of big data and traditional analytics. Even in many large organizations where very few big data projects are running, there was not a strong need to coordinate with those initiatives. But in future, Davenport claims that many organizations would be following multiple data initiatives across various functions and units and they will feel a greater need to coordinate those initiatives. The basic value does not come from raw data, but comes from its processing and the insights, products, and services emerging from it [7].

2.4 Insights for Organizational Big Data Adoption

Big Data signifies a vivacious shift in business decision making. Organizations are familiar to analyzing internal operational data like sales, shipments, inventory, etc. [8]. Recently, they have started recognizing the existence of other sources of data which need to be tracked and harness apart from operational databases. The newer data sources like sensor data, click-stream-logs, location data from mobile devices, chat transcripts and customer support emails, surveillance videos, etc., are becoming prevalent and important. Big data systems help to handle these new sources of data and allow enterprises to evaluate and extract business value [9]. It is helpful but challenging to get insights into customers, markets, supply chains, and operations. The study suggests that out of 2.8 ZB of data, 25% has potential value but only 0.5% is analyzed by various ways. Imposing a structure to unstructured data is the greatest barrier for further analysis [4]. For many enterprises, big data has brought big headaches with it. The storage cost and complexity have increased but the organizations are struggling for assigning a "positive" value to the data. The traditional technologies are unlikely to meet future needs for storage and the IT professionals will have to become accustomed to the diverse data driven skills [10].

3 Research Gaps

"Big data and analytics"-is considered to be a major innovation in the field of analytics and data storage technologies. Organizations are trying to invest and get returns from it. The big data market is still immature and lot many avenues are to be explored [11]. It is the future for analytics and can help the organizations to gain competitive advantage [7]. So it is important to consider organizational characteristics in information systems' adoption and acceptance [12–14]. The availability of resources to support the innovation acceptance defines the organizational context. The organizational silos and a dearth of data specialists are the main obstacles to use big data to work effectively for decision making. Literature indicates that being a new technology innovation—big data and organizational context is a less researched area. This leads to a query to examine the organizational factors affecting the big data adoption.

4 Research Objective

The extant literature lead to the identification of the research gaps and research problem leading to the identification of following objectives:

- To identify the key organizational factors of big data adoption in service organizations in India.
- To develop a conceptual framework for understanding the impact of organizational factors on big data adoption in service organizations in India.
- To validate the conceptual framework with empirical testing.

5 Theoretical Background

Big data has a potential to provide the considerable organizational benefits, but it may trigger extensive changes to business processes. This leads to a need for exploration of key organizational factors that influence its adoption. Further, it would be useful in developing evaluation guidelines and formulating strategies to overcome the adoption constraints. Extensive literature review has been conducted and various theories are studied to derive the theoretical background [15–23]. Based upon the organizational theory/behavioral research, the following model has been suggested along with the associated propositions (Fig. 1).

5.1 Big Data Adoption

The extent of big data adoption describes the level of practices related to handling, storage, and use of big data in an organization or business unit to support the business decisions. Various assessment scales related to big data have been examined to determine an organizational fitness and extent of adoption. Extent of big data adoption is a dependent variable in the study to analyze the impact of various organizational factors identified by reviewing literature.

5.2 Organizational Size

Big data adoption requires varied skills and competency and is an expensive investment. Various studies have identified organizational size to be an influential factor in technology adoption. Many of them suggest that the smaller organizations may be more efficient in generating and adopting innovations than large organizations. Studies also suggest that post-adoption large organizations are able to utilize it more effectively because of their capability of mobilizing resources and offering economies-of-scale/scope. They can also justify the cost justification for the big data adoption. Some of the studies have found organizational size is positively related to adoption of innovation [24], whereas others could not. The findings are inconsistent. Organizational size being a key variable in innovation literature the study undertaken proposes that:

H1 Organizational size is positively related to the extent of big data adoption.

5.3 *Organizational Culture*

Culture at various levels such as national, organizational, or group, can affect success of information technology. It is one of the predictors with highest predictive power of "IT innovation adoption" in information systems and computer science [25]. It plays a great role in approach of employees and managerial processes that may influence adoption [26]. Organizational openness or innovativeness to new ideas is an aspect of organizational culture [27]. Impatience with the status quo and a sense of urgency attribute to big data culture. It enables the organizations to provide a strong focus on innovation and exploration with a belief in technology as a source of disruption. The organizations with a culture that emphasizes analytical and fact-based decisions are more likely to succeed with big data [7]. So it can be postulated that:

H2 Organizational culture is positively related to the extent of big data adoption

5.4 *Organizational Data Environment*

Organizational data resource management contributes to its data environment. It is responsible for the database operations, data planning, and data policy functions

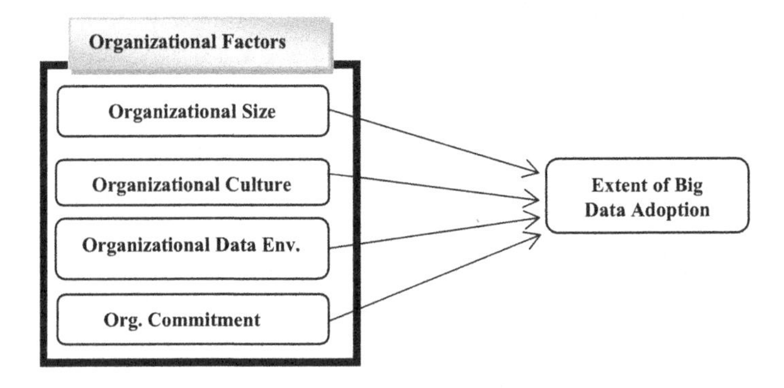

Fig. 1 Proposed framework

including data naming conventions, data dictionary standards, data modeling and design, data integrity, security policies, etc. [24]. If organized effectively, it can offer a number of benefits like reduced error rate and information agility [28, 29]. Data is collected in various formats and context across the organizations. They are extracted, cleaned, transformed, combined and formatted before being used for spawning analytics. With interorganizational environments and distributed systems, the challenge becomes many-fold. Organizations with proper data resources management guidelines make the adoption of innovation easier. Therefore, it can be proposed that:

H3 Organizational data environment is positively related to the extent of big data adoption.

5.5 Organizational Commitment

Organizational commitment refers to the support from senior management and key stakeholders. It also depends upon their participation and involvement. Management support has been a key variable in many innovation studies [30, 31] and IT implementation literatures [13, 24, 32, 33]. Similarly users/stakeholders' participation and involvement have been seen as an important factor in creating their buy-in [34]. Support and positive involvement of top-level management enables a strategic vision and direction for technology/innovation adoption and sends appropriate signals to different parts of organizations leading to a favorable attitude. Thus, organizational commitment is a key element for an innovation adoption and its subsequent use. Big data as a strategic tool requires high investment and resources. It can turn around the organizations and requires proper rollout of technology and experts to extract value out of voluminous data. This demands proper attention, backing, active engagement, and commitment across the organization [7]. It may influence the organizational competitive positioning and business relationships creating the need for involvement of senior management and stakeholders. So with a support from Swanson's tri-core model of IT innovation, the study proposes that

H4 Organizational commitment is positively related to the extent of big data adoption.

6 Research Methodology

The individual organization or strategic business unit (SBU) are considered as unit of analysis and the research looks at the data gathered just once (i.e., cross-sectional) from each individual. The responses are solicited from the senior executives working in the field of information systems (CIO, CTO, Head-Data Analytics, VP-BDA, BD Architect, etc.) or at strategic or tactical level; having familiarity to the new technology adoption initiative in the organization or unit's approaches to existing data environment and big data projects. It is administered by the service organizations in information technology services, banking and finance, and telecom industries headquartered in the city Mumbai. These industries are selected since they rely on heavy data usage where a variety of data has outstripped the capacity of manual analysis. The expert sampling method [35] was used since this is a very niche and specialized area and lacks the observational evidence. After the pilot study, the stratified random sampling method was used for data collection. The questionnaire was mailed to the companies sourced from Indian Merchant's Chamber (IMC) and CMIE databases. By seeking an appointment, the senior executives were interviewed and their responses were marked. Finally, the data was collected from 115 respondents out of which 6 were discarded as the responses were not suitable for various reasons like the organizations did not see any need of big data adoption for their business because of the area of business they operate into, incomplete questionnaires, etc. Structural equation modeling using the variance based, prediction oriented PLS model estimation—SmartPLS [36] was found to be best suited for testing the complex structural model used in the study [35, 37]. To evaluate the precision of the estimates and the standard errors, bootstrapping is used with 109 cases and 300 samples (resamples) to generate the standard error of the estimate and t-value.

7 Study Results

After evaluation of the data quality, the PLS algorithm was run to calculate the model parameter's estimates to evaluate the proposed hypothesis with the collected set of data and validate the quality criteria required for the empirical work. A two-step process covering the assessment of the measurement and structural model was followed.

7.1 The Assessment of the Measurement Models

The suggested model is a reflective measurement model so the measures like uni-dimensionality, internal consistency reliability, indicator reliability, convergent validity, and discriminant validity are required to be verified [38]. When the measurement items converge to its corresponding latent variable/construct better than any other variables, it is called as unidimensionality [39]. It is difficult to measure directly but can be evaluated using an exploratory factor analysis (EFA). The loading coefficient above 0.600 is considered as high and below 0.400 is considered as low [40] (ref: Table 1).

Table 1 Cross loading

	BD_Extent	OC	OCult	ODE	O_Size
Ext1	0.7393	0.4739	0.4188	0.4838	0.2729
Ext2	0.8313	0.4687	0.4484	0.5252	0.3302
Ext3	0.7150	0.4834	0.4696	0.5260	0.2831
Ext4	0.6441	0.4356	0.3633	0.3455	0.1556
Ext5	0.7809	0.4379	0.3976	0.5048	0.4365
Ext6	0.7882	0.4893	0.3950	0.3814	0.2598
Ext7	0.7495	0.5514	0.3921	0.4327	0.2846
Ext8	0.7109	0.5322	0.3431	0.3103	0.4064
OC1	0.5353	0.8436	0.5680	0.3851	0.3037
OC2	0.5975	0.8502	0.5887	0.4823	0.2141
OC3	0.4282	0.7891	0.4064	0.3419	0.2704
OC4	0.5829	0.8906	0.5191	0.4538	0.3171
OC6	0.5874	0.7586	0.5171	0.5137	0.2987
OC7	0.4225	0.7380	0.3846	0.2491	0.3033
OC8	0.4749	0.7558	0.5584	0.4736	0.2908
OC9	0.4507	0.7441	0.4738	0.4990	0.2046
O_Cult1	0.4975	0.5115	0.8638	0.5349	0.2791
O_Cult2	0.3965	0.5647	0.8011	0.3845	0.2547
O_Cult3	0.4246	0.4886	0.7811	0.5616	0.1655
O_Dev2	0.5896	0.5062	0.5170	0.9143	0.3009
O_Dev3	0.4264	0.4283	0.5595	0.8285	0.0343
Osize_1	0.3878	0.2641	0.2017	0.1955	0.9342
Osize_2	0.3822	0.3792	0.3357	0.2062	0.9321

In the factor analysis, the cross-loading of the factors is required to be checked. The cross-loading indicates how strongly each item loads on the other (non-target) factors. There should be a gap of at least ~0.2 between the primary target loadings and each of the cross-loadings. After eliminating the items showing strong corelation with other factors unidimensionality is achieved (ref: Table 1).

The internal consistency reliability is assessed using Cronbach's alpha (CA). The high alpha value indicates that all item scores with one construct have the same range and meaning [41]. Composite Reliability (CR) is another measure for internal consistency [42]. Chin [37] recommends use of CR; saying it overcomes some of CA's deficiencies. CA assumes that all indicators are equally reliable and underestimates the internal consistency reliability of latent variables (LVs), whereas CR takes into account the different loadings of indicators [43]. Research indicates values above 0.600 are acceptable but 0.700 is desirable for exploratory research and values above 0.800 or 0.900 are required in more advanced stages of research [44]. The composite reliability (CR) of all the constructs was found to be above 0.7 in the acceptable range (ref: Table 2).

Table 2 Reliability validation for LV and results for outer reflective model

Latent variables (LV)	Indicators	Outer loading	Indicator's reliability	Composite reliability (CR)	AVE	R square	Cronbachs alpha (CA)
Extent_BDA	Ext1	0.74	0.55	0.91	0.56	0.54	0.89
	Ext2	0.83	0.69				
	Ext3	0.72	0.51				
	Ext4	0.64	0.41				
	Ext5	0.78	0.61				
	Ext6	0.79	0.62				
	Ext7	0.75	0.56				
	Ext8	0.71	0.51				
Organizational commitment	OC1	0.84	0.71	0.93	0.64	0	0.92
	OC2	0.85	0.72				
	OC3	0.79	0.62				
	OC4	0.89	0.79				
	OC6	0.76	0.58				
	OC7	0.74	0.54				
	OC8	0.76	0.57				
	OC9	0.74	0.55				
Organizational culture	O_Cult1	0.86	0.75	0.86	0.67	0	0.75
	O_Cult2	0.8	0.64				
	O_Cult3	0.78	0.61				
Organizational data environment	O_Dev2	0.91	0.84	0.86	0.76	0	0.69
	O_Dev3	0.83	0.69				
Organizational size	Osize_1	0.93	0.87	0.93	0.87	0	0.85
	Osize_2	0.93	0.87				

A single or set of variables consistent in what it intends to measure is referred to as Indicator Reliability. At least 50% of each indicator's variance should be explained by a latent variable. So, indicator loadings should be substantial at least at the 0.050 level and value should be greater than 0.707 (≈0.500) [37]. Indicator reliability is the square of each of the outer loadings. The value 0.70 or higher is preferred; whereas if it is an exploratory research 0.4 or higher is acceptable [45].

The degree to which individual items of a construct converge in comparison to items measuring different constructs is called as Convergent validity. Fornell and Larcker had proposed AVE (Average Variance Extracted) as the criteria for determining the convergent validity [46]. An AVE of 0.500 and above indicates the sufficiency of value and indicates that an LV is able to explain more than half of the variance of its indicators (ref: Table 2).

The degree to which different construct's measure differ from one another is the discriminant validity. It examines if the items do not unintentionally measure

Table 3 Fornell-Larcker criterion analysis for discriminant validity

	BD_Extent	OC	OCult	ODE	O_Size
BD_Extent	0.75				
OC	0.65	0.80			
OCult	0.54	0.64	0.82		
ODE	0.59	0.54	0.61	0.87	
O_Size	0.41	0.34	0.29	0.22	0.93

something else. It is measured by verifying cross-loadings [37] and the Fornell–Larcker criterion [46]. It is found that each of the indicator's loading is higher for its designated construct and each of the constructs loads highest with its assigned items than other constructs (ref: Table 1). The AVE of each LV is greater than the LV's highest squared correlation with any other LV (ref: Table 3).

Based on these tests, it can be concluded that the research variables exhibit satisfactory psychometric properties.

7.2 *Assessment of the Structural Model*

The assessment of PLS structural equation model is done with the coefficient of determination (R^2). The value of R^2 indicates the relationship of a latent variable's explained variance to its total variance. The R^2 test identifies predictability of test data through PLS regression model. Values approximately around 0.670 are considered to be substantial; 0.333 as average; and values less than 0.190 and lower as weak [37]. Overall, value more than 0.50 for R^2 is considered to be a good value. The indicated R^2 value for the study variables infers the higher probability of acceptance. The evaluation of the path coefficients between latent variables of the model is another criterion for assessing the structural model. In the proposed study the paths coefficient's signs are found to be in line with the postulated assumptions. The strength of the relationship between two latent variables is analyzed by the coefficient's magnitude. In the proposed study, the variables like Organizational Size (O_Size), Organizational Commitment (OC), and Organizational Data Environment (O_Dev) have shown the strong relationship with the extent of big data adoption in organizations. For determining the significance in PLS, resampling techniques such as bootstrapping [47, 48] or jackknifing [49] should be used. In the present study, significance is determined using bootstrapping with 300 resamples. The proposed hypotheses in the research study are established by examining the

Table 4 Summary of hypotheses tests (path coefficients and hypothesis testing)

Total effect (mean, STDEV, T-value)			
	Original sample (O)	T statistics (|O/STERR|)	Hypothesis supported
H1: O_Size → BD_Extent	0.20000	2.07970	Yes
H2: OCult → BD_Extent	0.05450	0.41990	No
H3: ODE → BD_Extent	0.31490	3.06800	Yes
H4: OC → BD_Extent	0.37510	3.73460	Yes

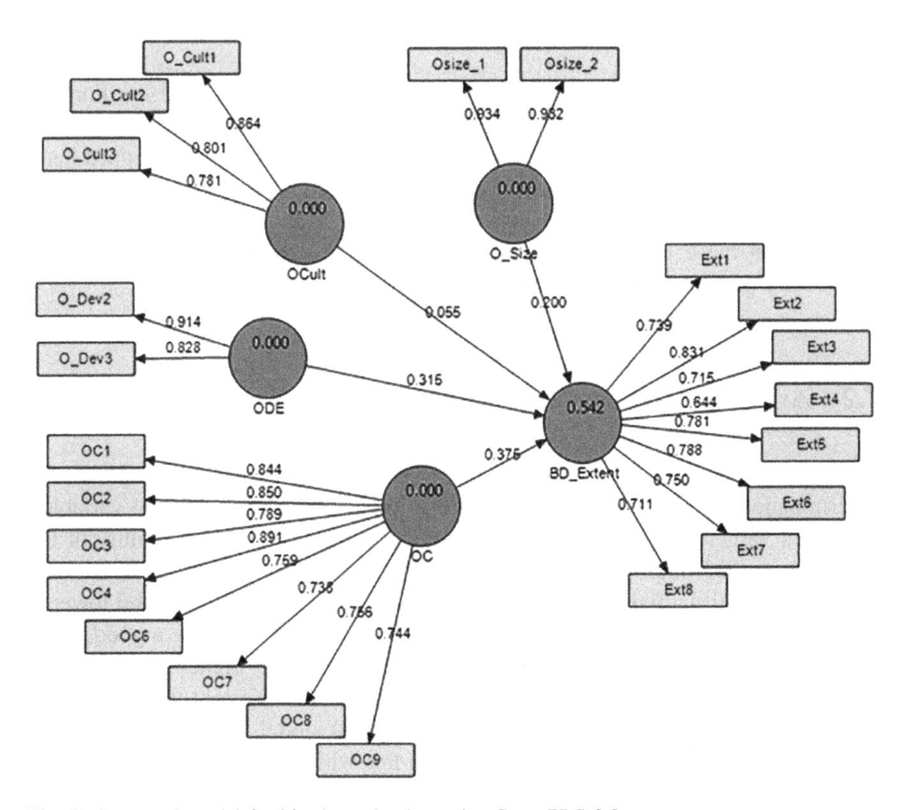

Fig. 2 Structural model for big data adoption using SmartPLS 2.0

structural model using SmartPLS software. The corresponding t-values show the level of significance using the magnitude of the standardized parameter estimates between the constructs. The results of the structural model are summarized in Table 4 (Figs. 2 and 3).

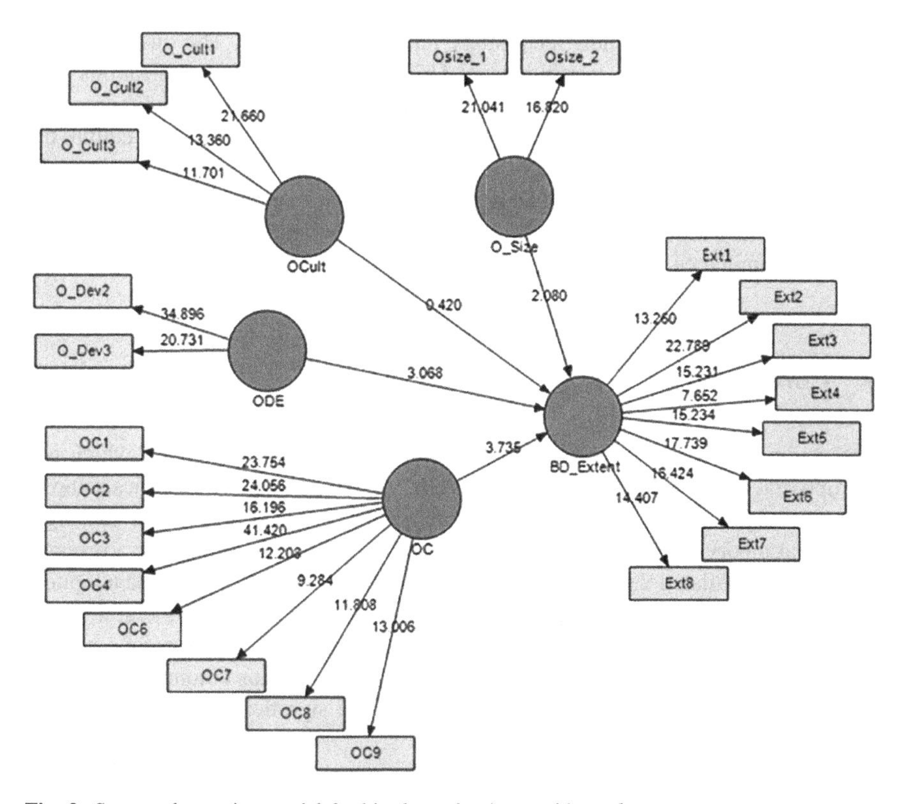

Fig. 3 Structural equation model for big data adoption—with t-value

8 Discussion

The study explored and examined the impact of four organizational variables—organizational size, commitment, culture, and data environment—on big data adoption. Three out of four variables are found to have a positive impact on the big data adoption decision. The emergence of organizational size as a key factor is in line with the previous literature. It affirms the fact that the larger organizations are able to afford and more importantly capitalize on the economies-of-scale and scope in organizational environment following the technology implementation [6]. The affordability of the technology is not an issue with the smaller organizations but the larger organizations have higher adoption rate because of the availability of organizational resources and scope. The organizational adoption research also affirms the results of organizational commitment [14, 18]. The study covered sustained commitment of all stakeholders including the c-suit support. It affirms the fact that big data adoption is not only the technology decision but an organizational decision for differentiation and the c-suit role becomes important in leading the decision from front. The organizational data environment is also found to be an influencer in the adoption decision of new technology initiative—big data. The quality of data environment suggests its maturity level [17, 20]. The bigger organizations with matured setup for data warehousing and analytics are suitably seen to be proactive in embracing big data technology. It is surprising that the organizational culture did not emerge as a determinant. It can be inferred that in larger organizations the strategic direction with supportive organizational data environment and commitment leads the adoption decision. It is also possible that multicollinearity of the data items or dominance of other factors might be affecting the results.

9 Conclusions

The study undertaken examined the determinants of big data adoption which is a disruptive technology innovation and has challenged the existing analytics setup of organizations. The existing body of research in IT innovation adoption and organizational theory helped in drawing the necessary inferences to propose a research model that postulated the impact of four organizational factors on big data adoption. The proposed model was empirically validated as a Smart PLS structural model with 109 organizations in service sectors located in Mumbai, the financial capital of India. The data analysis supported the suggested relationships among the variables and three out of four hypotheses were proved to be significant from the collected data.

Considerations and future work: The findings of adoption decision make it essential and fascinating to examine the organizational benefits that accrue from subsequent use/diffusion of innovation within the organization. The documentation of use cases and success stories can really be advantageous as it can be a motivational factor for non-adopters. It would also be interesting to examine if there exists any difference in big data usage among the early and late adopters; proactive, and reactive adopters.

References

1. Checkland, P., & Holwell, S. (1997). Information, Systems and Information Systems: Making Sense of the Field. Wiley.
2. Beyer, M. A., & Friedman, T. (2013). Big Data Adoption in the Logical Data Warehouse. Gartner Inc.
3. Rouse, M. (2015, June). Big Data. Retrieved June 16, 2015, from searchcloudcomputing.techtarget.com: http://searchcloudcomputing.techtarget.com/definition/big-data-Big-Data.
4. Brynjolfsson, E., & McAfee, A. (2013). Is Your Company Ready for Big Data? Retrieved Jan 17, 2015, from http://hbr.org: http://hbr.org/web/2013/06/assessment/is-your-company-ready-for-big-data.
5. Auschitzky, E., Hammer, M., & Rajagopaul, A. (2014, July). How big data can improve manufacturing. McKinsey & Company.

6. Gopalkrishnan, S., & Damanpour, F. (2000). The impact of organizational context on innovation adoption in commercial bank. IEEE Transactions on Engineering Management, 47 (1), 14–25.
7. Davenport, T. H. (2014). big data @ work. Boston: Harvard Business School Publishing Corporation.
8. Economist Intelligence Unit. (2012). The Deciding Factor: Big data and decision-making. Capgemini.
9. Oracle Corporation, W. H. (2014, September). Oracle Database 12c for Data Warehousing and Big Data. CA, USA.
10. (2013). The Big Data Readiness Study. Vodafone.
11. Finos, R. (2015). Wikibon Big Data Analytics Survey: Adoption Maturity by Vertical Market. Wikibon.
12. Rogers, E. (1995). Diffusion of Innovation (5th Edition ed.). New York: The Free Press.
13. Thong, J. Y., Yap, C.-S., & Raman, K. S. (1996). Top Management Support, External Expertise and Information Systems Implementation in Small Businesses. Information Systems Research, 7(2), 248–267.
14. Tornatsky, L., & Fleischer, M. (1990). The Process of Technology Innovation. Lexington, MA.: Lexington Books.
15. Ajzen, I. (1991). The Theory of Planned Behaviour. Organizational Behaviour and Human Decision Processes, 179–211.
16. Davis, F. D. (1986). A Technology Acceptance Model for Empirically Testing new end-user information systems: Theory and Results. Doctoral Dissertation. Cambridge, MA: Sloan School of Management, Massachusetts Institute of Technology.
17. Humphrey, W. (1988). Characterizing the software process: A maturity framework. 5(2).
18. Kwon, T., & Zmud, R. (1987). Unifying the fragmented models of information systems implementation. Critical Issues in Information Systems Research, 227–251.
19. Oliveira, T., & Martins, M. F. (2011). Literature Review of Information Technology Adoption Models at Firm Level. The Electronic Journal Information Systems Evaluation, 14(1), 110–121.
20. Paulk, M., & Weber, B. C. (1995). The Capability Maturity Model: Guidelines for Improving the Software Process. MA: Addison-Wesley.
21. Premkumar, G., Ramamurthy, K., & Crum, M. (1997). Determinants of EDI Adoption in the Transportation Industry. European Journal of Information Systems, 11, 157–186.
22. Scott, W. (2001). Institutions and organizations, 2 ed. Thousand Oaks, CA: Sage Publications.
23. Scott, W., & Christensen, S. (1995). (1995) The institutional construction of organizations: International and longitudinal studies. Thousand Oaks, CA: Sage Publications.
24. Ramamurthy, K., Sen, A., and Sinha, A. P. (2008). An empirical investigation of the key determinants of data warehouse adoption. Decision Support Systems 44, 817–841.
25. Basole, R., Seuss, D., & Rouse, W. B. (2013). IT innovation adoption by enterprises: Knowledge discovery through text analytics. Decision Support System, 54, 1044–54.
26. Leidner, D. E., & Kayworth, T. (2006, June). A Review of Culture in Information Systems Research: Toward a Theory of Information Technology Culture Conflict. MIS Quarterly, 30 (2), 357–399.
27. Venkatesh, V., & Bala, H. (2012). Adoption and impacts of inter-organizational business process standards: Role of partnering synergy. Information Systems Research, 23(4), 1131–1157.
28. Goodhue, D. L., Wybo, M. D., & Kirsch, L. J. (1992). The impact of data integration on the costs and benefits of information systems. MIS Quarterly, 293–311.
29. Jain, H., Ramamurthy, K., Ryu, H.-S., & Yasai-Ardekani, M. (1998). Success of Data Resource Management in Distributed Environments: An Empirical Investigation. MIS Quarterly, 22(1).
30. Boynton, A. C., & Zmud, R. W. (1984). An assessment of critical success factors. Sloan Management Review, 17–27.

31. Ettlie, J. (1986). Implementing manufacturing technologies: Lessons from experience. In D. D. (Eds.), Managing technological innovation. San Francisco: Jossey-Bass.
32. Hwang, H.-G., Ku, C.-Y., Yen, D. C., & Cheng, C.-C. (2004). Critical factors influencing the adoption of data warehouse technology: a study of the banking industry in Taiwan. Decision Support Systems, 37(1), 1.
33. Sanders, G., & Courtney, J. (1985). A Field Study of Organizational Factors Influencing DSS Success. MIS Quarterly, 9(1).
34. Hartwick, Jon, & Barki, H. (1994). Explaining the Role of User Participation in Information Systems Use. Management Science, 40(4), 440–465.
35. Chin, W. W., Marcolin, B., & Newsted, P. (2003). A partial least squares latent variable modeling approach for measuring interaction effects: Results from Monte Carlo simulation study and an electronic mail emotion/ adoption study. Information Systems Research, 14(2), 189–217.
36. Ringle, C. M., Wende, S., and Will, A. (2005). SmartPLS 2.0. Hamburg: University of Hamburg.
37. Chin, W. (1998b). The partial least squares approach to structural equation modeling. Modern Methods for Business Research, Marcoulides, G.A. (ed.), pp. I295–1336.
38. Straub, D., Boudreau, M., & Gefen, D. (2004). Validation guidelines for IS positivist research. Communications of the AIS, 380–427.
39. Gerbing, D., & Anderson, J. (1988). An updated paradigm for scale development incorporating unidimensionality and its assessment. Journal of Marketing Research (JMR), 2, pp. 186–192.
40. Gefen, D., & Straub, D. (2005). A practical guide to factorial validity using PLS-Graph: Tutorial and annotated example. Communications of the AIS, 16, pp. 91–109.
41. Cronbach, L. (1951). Coefficient alpha and the internal structure of tests. Psychometrika, 16 (3), pp. 297–334.
42. Werts, C. E., Linn, R. L., & Joreskog, K. (1974). Intra class reliability estimates: Testing structural assumptions. Educational and Psychological Measurement, 34, 25–33.
43. Henseler, J., Ringle, C., & Sinkovics, R. (2009). The use of partial least squares path modelling in international marketing. In R. Sinkovics, & P.N. Ghauri, Advances in International Marketing (pp. 277–320). Bingley: Emerald.
44. Nunnally, J., & Bernstein, I. (1994). Psychometric Theory. New York: McGraw-Hill.
45. Hulland, J. (1999). Use of partial least squares (PLS) in strategic management research: A review of four recent studies. Strategic Management Journal, 20, pp. 195–204.
46. Fornell, C., & Larcker, D. (1981). Evaluating structural equation models with unobservable variables and measurement error. Journal of Marketing Research, 18(1), 39–50.
47. Efron, B. (1979). Bootstrap methods: Another look at the jackknife. Annals of Statistics, 7(1), pp 1–26.
48. Efron, B., & Tibshirani, R. (1993). An Introduction to the Bootstrap. New York: Chapman Hall.
49. Miller, R. (1974). The jackknife—A review. Biometrika, pp. 1–15.

User Feedback-Based Test Suite Management: A Research Framework

Akhil Pillai and Varun Gupta

Abstract Delivering better product value on each revision is the principal task for an organization's existence in a competitive market. The customer's base needs to be continuously increased to earn more revenue. Feedback is collected from customers to decide functionalities that need to be included in the future versions of the software. A software company gets flooded with feedback, which makes it difficult to analyze, process, and prioritize. Feedback may result in addition of new code in order to implement new functionality or to delete/modify existing one. Each task which alters the source code to implement a feedback requires modifying test suites which enhance the testing effort thereby impacting project schedules, budget, and plans. In this paper, we highlight a research framework that attempts to lower the testing effort by reducing the number of test cases by compounding the feedback which is similar to existing requirements of same or similar projects.

Keywords Test cases · Customer feedback · Requirements

1 Introduction

A software company delivers software incrementally which gives its users enough time to use a particular increment and record their expectations as feedback, which shapes the next increment. This feedback oriented development is followed till the software is taken off the market. Thus, feedback is used by firms to make decisions about the functionality of next increment and various improvements/modification.

Each requirement is mapped to certain test case/s, which are always increasing. In other words, requirements expressed implicitly or explicitly through user feedback increases the size of test suite. Feedback is an integral part of the development

A. Pillai (✉) · V. Gupta
Amity School of Engineering and Technology, Amity University, Noida, Uttar Pradesh, India
e-mail: akhilpillai18@gmail.com

V. Gupta
e-mail: vgupta7@amity.edu

A.K. Somani et al. (eds.), *Proceedings of First International Conference on Smart System, Innovations and Computing*, Smart Innovation, Systems and Technologies 79, https://doi.org/10.1007/978-981-10-5828-8_78

of evolutionary software, it also aims at reducing testing effort as it is one of the objectives of a software firm.

This paper highlights the motivation and rationale behind the research, i.e., focusing on building a test suite on the basis of user feedback for the current version. Once some feedback is obtained from customers, it is analyzed to check whether it requests for new requirement/s in which case new test case/s need to be designed or whether it requests for deletion of feature/s, accordingly test case/s are required to be removed completely. Feedback received can also indicate modification of an existing requirement, thus either a change is made in the existing test cases or new test cases are created. Minimization of the test suite is challenging in mass market development, because in such development the number of requirements (feedbacks) is always increasing [1].

2 Research Framework

Let R_i be a requirement in set R with T_i, T_j test cases associated with it. Let F_i be feedback relaying either new functionality, deletion request or modification request. F_i is associated with a single requirement, R_{Fi} and T_{Fi}, T_{Fj}, ... T_{Fn} test cases. For new functionality, the test cases associated with F_i say T_f prompts an increase in the size of the test suite. Whereas for deletion, the test cases associated with F_i say T_f will decrease the size of test suite by maximum value of N (where N is the number of test cases associated with feedback F_i), it is important to note that not all test cases associated with feedback F_i be removed, since some test cases may be required to be repeated as they might be associated with requirements from set R. While for modification, the cardinality of test suite will increase since few test cases will be modified and few more might be added (Fig. 1).

For the first increment, effort is given by

$$E_1 = E_{R1} + E_{R2} + \cdots + E_{Rn} \tag{1}$$

For the second increment, effort is given by feedback

$$E_2 = E_{regression} + E_{Rn1} + \cdots + E_{Rmod1} + E_{Rmod2} + \cdots + E_{Rmodn} \tag{2}$$

Now $E_2 > E_1$, E_2 is too high.

The objective of the research is to handle feedback in such a manner that E_2 is minimized. This is possible if we can consider high priority feedbacks and attempt to minimize the number of test cases associated with these feedbacks.

Thus,

- E is directly proportional to $E_{Feedback}$
- E is directly proportional to $E_{Regression\ Testing}$
- $E_{Feedback}$ is directly proportional to the number of feedback
- The number of test cases is directly proportional to the number of feedback.

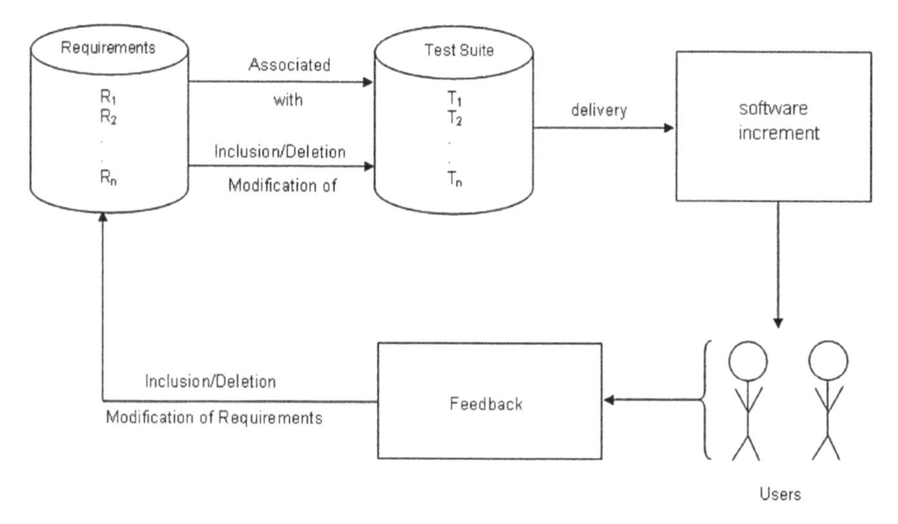

Fig. 1 Suggested framework for user feedback based software testing

So by reducing the number of feedback, the number of test cases are reduced, which in turn reduces the effort (E) required. Reduction of feedback is possible by careful prioritization of feedback and reduction of test cases by analysis of the similarity of new feedback with existing requirement of the same project.

2.1 *Impact of Feedback*

During the initial phase of a development cycle, test cases are created on the basis of requirements. Each requirement or a group of requirements are referred to create a test case. After the first iterative release of the software product, feedback is collected from customers. It is important to note that the feedback received from the customer might contradict an existing requirement, complement an existing requirement or have no effect on the existing requirement. The nature of feedback is pivotal in ensuring elimination of redundancy in test cases as well as minimization of test cases [2–4].

The feedback received from a customer might request a new feature, change in some existing feature or even removal of a feature. This nature of feedback has an impact on the existing set of requirements. This impact is to be measured before creating a new set of test cases based on the feedback and starting the development process [5].

If the nature of feedback contradicts the existing set of requirements, then it has the highest impact since it indicates a new set of requirements which would require the creation of completely new test cases. For instance, the username field in a login webpage might initially allow numeric characters but after the first increment is

Table 1 Measuring impact of feedback on existing requirement

	R_1	R_2	R_n
F_1	0	2	1
...	.	.	.
Fn	2	1	0

delivered the feedback collected from the client might specify for alphanumeric characters in the username field.

If the nature of feedback complements the existing set of requirements, it has a moderate impact on the existing set of requirements since the previously created test cases can be used or modified. For instance, the username field in a login web page might initially allow a length of 4 numeric characters but after the first increment is delivered the feedback collected from the client might specify to allow a length of 32 numeric characters in the username field. The least impact is caused by the feedback which is similar to the existing set of requirements, in which case we only have to reuse the existing set of test cases.

The impact values of the feedback are numerically graded, least impact as 0, moderate impact as 1 and one with the highest impact as 2. Table 1 can be created where columns correspond to requirements and rows correspond to feedback obtained from customers.

3 Proposed Solution

In this section, we propose an algorithm that attempts to reduce the number of test cases on the basis of user feedback. The feedback generated is input to the algorithm. Feedbacks are generated in natural language. It is difficult to process natural language, since the processing and conversion of natural language is out of scope of the presented research. It is assumed that the feedback provided will be in a structured manner such that it matches the input conditions of the algorithm. Feedback is generally about requirements the user wants to add, remove or edit. So the feedback should be represented as requirements are represented.

3.1 Proposed Algorithm

Step 1. Procure feedback from the team, and process and format the feedback similar to that of the requirement form.

Step 2. Choose one feedback from the list and compare it with the existing requirements.

Step 3. Determine whether this feedback is contradictory to the original requirements and assign a maximum weight (02) per each of the requirement it contradicts, sum all the weights $W_{contradictory}$.

Let us suppose that feedback F_1 contradicts with R_1 to R_N.

$$W_{contradictory} = \sum N*2 \tag{3}$$

Step 4. Determine whether this feedback supplements the existing requirements and assign moderate weight (01) to it, sum all the weights $W_{supplement}$. Let us suppose that feedback F_1 supplements with R_1 to R_M.

$$W_{supplement} = \sum M*1. \tag{4}$$

Step 5. Determine the similarity of the feedback to the existing requirements and assign minimum weight (0), if it is almost similar to the existing feedback. Sum all the weights, $W_{similar}$.

Step 6. Find the net weight of the feedback using Eq. (5)

$$W_{net} = W_{contradictory} - W_{supplement} \tag{5}$$

Step 7. Repeat from Step 2 to step 6 for all feedbacks.

Step 8. The feedback with the maximum net weight (W_{net}) is considered of high priority and all test cases (new or old) are considered of high priority. However, these test cases could also be further prioritized.

4 Conclusion and Future Work

The feedback given by users may cause an increase in the number of test cases in the test suite, which ultimately increases the total effort. If the feedback is analyzed and prioritized, it can be linked with the existing test cases which enable a firm to alter cardinality of the test suite.

In future, techniques addressing feedback oriented test suite management, i.e., based on the presented research framework will help a software company to test complete software with minimal effort. Analysis of requirements stated through customer feedback, will be used for reduction of the test suite. The proposed algorithm needs to be validated on a live case study. The number of the test cases associated with high priority feedback could also be further reduced, which is considered as future work.

References

1. Karlsson, L., Dahlstedt, Å. and Dag, J.: Challenges in market-driven requirements engineering–an industrial interview study. In: Eighth International Workshop on Requirements Engineering: Foundation for Software Quality (2002)

2. Karambir and Kuldeep Kaur: Survey of Software Test Case Generation Techniques. In: International Journal of Advanced Research in Computer Science and Software Engineering, Volume 3, Issue 6, June (2013)
3. Kohsuke Yatoh, Kazunori Sakamoto, Fuyuki Ishikawa, Shinichi Honiden: Feedback-Controlled Random Test Generation. In: Proceedings of the International Symposium on Software Testing and Analysis, Pages 316–326, (2015)
4. Carlos Pacheco, Shuvendu K. Lahiri, Michael D. Ernst, and Thomas Ball: Feedback directed Random Test Generation. In: Proceedings of the 29th International Conference on Software Engineering, ICSE'07, pages 75–84 (2007)
5. Helena Holmström Olsson and Jan Bosch From Requirements To Continuous Re-prioritization Of Hypotheses. In: International Workshop on Continuous Software Evolution and Delivery, pages 63–69 (2016)

Mobile Big Data: Malware and Its Analysis

Venkatesh Gauri Shankar, Mahesh Jangid, Bali Devi and Shikha Kabra

Abstract The quick extension of mobile big data by telecoms vendors, has presented a flexible worldwide platform that creates a user interface for many app data bases or app stores. Big data is a tremendously well-known idea, however what are we truly talking about? From a security point of view, there are two particular issues: securing the app stores or app databases with its source of information in big data context and utilizing big data procedures to break down, and even anticipate, security flaws. The main issue arise that many hackers or attackers are targeting mobile big data in the form of signaling big data, mobile traffic big data, location-based big data, and heterogeneous data in app store. In this paper, we are taking Android-based mobile operating system for experimental setup. This paper contains an extraction technique to extract the malware in different big data context and also analysis of these malware. We have worked with many Mallarme family (as approx 40 K malware) in mobile big data and result of the whole analysis is approx 90% to identify the current malware in mobile big data.

Keywords Big data · Mobile big data · Malware analysis · Android malware Security

V.G. Shankar (✉) · M. Jangid · S. Kabra
SCIT, Manipal University Jaipur, Jaipur, India
e-mail: venkateshgaurishankar@gmail.com

M. Jangid
e-mail: mahesh_seelak@yahoo.co.in

S. Kabra
e-mail: kabra.shikha1990@gmail.com

B. Devi
CSE, Jayoti Vidyapeeth Women's University, Jaipur, India
e-mail: baligupta03@gmail.com

A.K. Somani et al. (eds.), *Proceedings of First International Conference on Smart System, Innovations and Computing*, Smart Innovation, Systems and Technologies 79, https://doi.org/10.1007/978-981-10-5828-8_79

1 Introduction

There is no rigid lead about precisely what measure a database should be all together, for the information within it to be viewed as "big." Instead, what ordinarily characterizes mobile big data is the requirement for new procedures and mobile devices keeping in mind the end goal to have the capacity to process it [1]. Keeping in mind the end goal to utilize big data, you require programs which traverse different physical as well as virtual mobile devices cooperating with each other, so as to process the greater part of the information in a sensible traverse of time [1, 2].

The feature of mobile big data are practically as changed as they are in big size. Unmistakable illustrations you are likely officially acquainted with including online networking system breaking down their individuals information to take in more about them and associate them with substance and publicizing significant to their interests, or web search tools taking an important role at the relationship among inquiries and results to give better solutions for client's inquiries [2].

Be that as it may, the potential uses go much further! Two of the biggest wellsprings of mobile big data in extensive amounts are value-based information, including everything from stock costs to mobile-related bank's data to individual dealer's buy histories; and mobile sensor data, quite a bit of it originating from what is ordinarily alluded to as the Internet of Things (IoT) [3]. This mobile sensor data may be anything from estimations taken from robots on the assembling line of an automaker, to area information on a wireless system, to prompt electrical use in homes and organizations, to traveler boarding data gone up against a travel management system.

Mobile big data having same data storage as big data like social media data, bank transactional data, enterprise data, user activity generated data. These kind of mobile big data also shows three V's of big data as velocity, variety, and volume in Fig. 1. The main aim towards these mobile big data is to secure from malware. We are

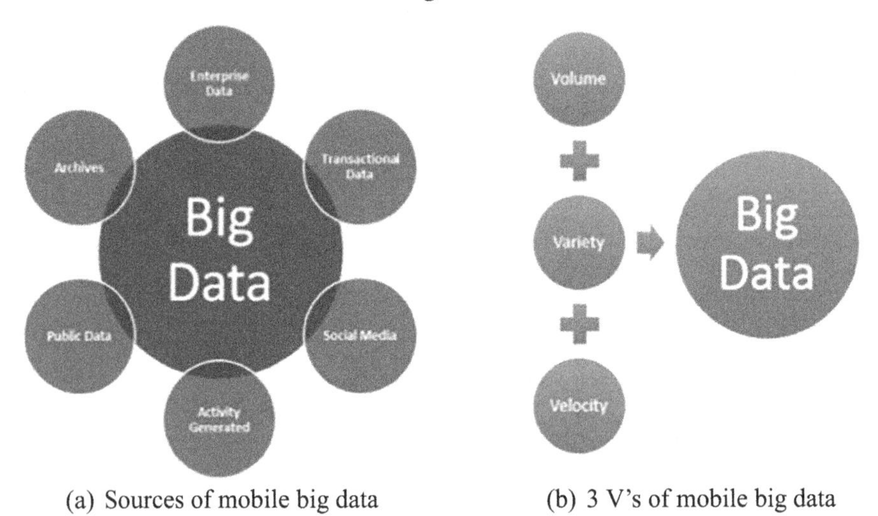

(a) Sources of mobile big data (b) 3 V's of mobile big data

Fig. 1 Concept of mobile big data

creating a execution script to analyze the behavior of malware and also give the study of some malware in Android operating system. We are taking the platform for experimental setup as Android, due to its popularity among the users [4].

2 Related Work

Nada Elegendy et al. [5] described the review of big data in all scenario and also its sources of big data with properties of big data in different phases. They also explained some security issues in big data with the future possibilities. Demetrious et al. [6] explained the research on big data with its opportunities in mobile big data. They also presented many future issues in security as well as privacy. They proposed good practices of big data in different context. Yeng He et al. [7] showed his view on cellular mobile network with the possible security issues, vulnerabilities and future cellular malware on mobile big data. They also explained network-related data usage and mobile data storage. Elgendy N. et al. [8] Proposed a decision-making process on big data analytics and its use in big data resources. They also created a support system for decision-making structure of big data. EMC [9] gives an idea for data science environment and its possibilities in future. It also processes an data science environment on big data analytics. He Y. et al. [10] implemented a map reduce based big data structure for fast and Space efficient Data Placement Structure in MapReduce. They also presented many warehousing resources in big data. Herodotou H. et al. [11] created a tool named Starfish, which is a self-tuning system for big data environment. They also structured huge amount of data presentation in big data. Kubick et al. [12] signified a role of big data in current scenario with his idea about data processing and information processing. Lee R. et al. [13] proposed an intrinsic translator for data base system and they are using map reduce to create a big data based translator. Manyika, J. et al. [14] presented an emerging role of big data in the productivity, innovation and also for data extraction strategy. Venkatesh G.S. et al. [15] proposed an idea as Anti-Hijack, which capture current session and intent-based malware using honey pot technology. They are also collected a large set of Android malware and its analysis. Arzt Steven et al. [16] created an idea for Android Malware named FlowDroid. They also justified many sources of Android malware and its analysis to prevent them.

Our novel idea is based on extraction based analysis using Strace [17] as tool in mobile big data. We identified many malware and its family based on their extraction. We also identified many recent malware in mobile big data. We are taking the experimental platform on Android operating system, due to its popularity.

3 Android and Its Malware

Android operating system is in market with many security features. Many hacker or attacker are prominent to find data vulnerabilities in mobile big data and misuse the user sensitive information. Hacker or attacker is very smart to create malware in mobile big data, which is difficult to catch up. Few malware creation strategy such as analysis vulnerabilities, code obfuscation, piggybacking already exist in Android operating system. Hands-on based analysis is not feasible to detect the malware because of big data storage as App store [18].

Malware alludes to an infected program that tries to contaminate a PC, mobile, or mobile big data. Programmers utilize malware for any number of reasons, for example, separating individual data or passwords, stealing money, or preventing proprietors from getting to their mobile device. You can ensure yourself against malware by utilizing anti- malware programming [19, 20].

Malware most regularly accesses your mobile big data device through the Internet and by means of email, however it can likewise get access through hacked sites, gaming environment, music documents, toolbars, programming packages, free memberships and subcriptions, or whatever else you download from the web on to a mobile big data device which is not secured with anti-malware programming [4].

There are many types of malware in mobile big data which are spyware, worms, adware, rootkits, ransomware, session hijacker, intent hijacker, bot nets, etc. Here we are presenting some more malware families in Android mobile big data find in the Q3-Q4 of the year 2016 (Table 1). The families type are same but many anti-malware techniques unable to identified these families [19, 21].

We have implemented the extraction algorithm for extracting these mobile big data malware and find its behavior on the basis of signatures in the next section as experimental setup.

4 Experimental Setup

We are maintaining an extraction script and we are taking different malicious app as input in the extraction script. These malicious apps are fetching from many mobile big data stores as Google play store [21], SlideMe app stores [22] and some GNOME malware data sets [23].

The process flow of working of our model is given below

- Taking any malicious app from any one of mobile big data.
- Extract the manifest file for human readable form.
- Extract the present intent of the manifest file.
- Extract the services, permission, activities of the manifest file.
- Store the extracted data in Hadoop Distributed File System (HDFS) and apply MapReduce

Table 1 Android mobile big data malware family in Q3-Q4 of the year 2016

Malware name	Description
Switcher trojan	It is just for fun or some times demand as some monetary charges (Trojan)
Hijack router DNS	It collects user sensitive information such as contacts list browsing details, etc. (Session Hijacker)
Swift mailer	It captures user credential such as bank information, credit card, or debit card information's (Spyware)
Adware adbox	It contains phishing mails, some unusable advertisements and phishing web site also (Adware)
Sprovider	It generates automated malicious SMS and malicious calls to capture user sensitive information's (Adware)
AccuTrack	Track the GPS and get the user control on remote server (Rootkits)
AdSMS	It generates automated malicious SMS and malicious calls to capture user-sensitive information's (Adware)
AVPass	Pass The antivirus and taking root control on server (BotNet)
BeanBot	Take the whole control on remote server by making robot network (BotNet)
Asroot	Gaining root access and steal user-sensitive information (RootKit)

- Compare the extracted file with the existing malicious signature after applying MapReduce and report them.

The extraction script is applied on the family of BeanBot Malware in mobile big data and Asroot Malware also in mobile big data. When we are executing extraction script in one slot (we are taking three apps Parallel). After that we are extracting the beanbot family malicious app and check them for packages and permissions. Same process execute for all apps in the process and reporting the following same process to all components as activities, services, and intents (Table 2).

In the below Table 3, we are also extracted some special intent, which supported the behavior of executed app in runtime. In the given table, we are extracting Asroot big data malware with the extraction of manifest file. This manifest file gives some special data extraction as browser activity, media storage, and bluetooth activity. These intent are some time big exploit of data.

Table 2 BeanBot family permission and package in extraction script

Family: BeanBot (BotNet)			
Input:	1 App	2 App	3 App
Package:	com.keji.danti607	com.keji.danti704	com.apkbook.fengchen
Permissions:	ACCESS_WIFI_LOGS	ACCESS_WIFI_LOGS	READ_PHONE_LOGS
	WRITE_SMS	WRITE_SMS	WRITE_SMS
	RECEIVE_SMS_NOTIFIED	RECEIVE_SMS_NOTIFIED	RECEIVE_SMS_NOTIFIED
	READ_SMS	READ_SMS	READ_SMS
	RECEIVE_SMS_LOGS	RECEIVE_SMS _NOTIFIED	RECEIVE_SMS_LOGS
	SEND_SMS_NOTIFIED	SEND_SMS_NOTIFIED	SEND_SMS_NOTIFIED
	DISABLE_KEYGUARD	DISABLE_KEYGUARD	DISABLE_KEYGUARD
	INTERNET	INTERNET	INTERNET
	ACCESS_NET_LOGS	ACCESS_NET_LOGS	ACCESS_NET_LOGS
	RESTART_PKG_SETTINGS	RESTART_PKG_SETTINGS	WRITE_APN_SETTINGS
	READ_APN_LOGS	WRITE_APN_LOGS	READ_LOGS

Table 3 Asroot family activity in extraction phase

Family: Asroot (Rootkits)			
Input:	1 Browser	2 Media Storage	3 Bluetooth
Activity:	Browser Activity	RingtonePickerActivity	BluetoothOppLauncherActivity
	Shortcut Activity		BluetoothOppLiveFolder
	BrowserPreference Page		BluetoothBpp Activity
	Bookmark Search		BluetoothBppSetting
	Add Bookmark Search		BluetoothBppPrintPrefActivity
	Bookmark Widget Configure		BluetoothBppStatusActivity
			BluetoothBppAuthActivity
			BluetoothBppIntentActivity

Table 4 BeanBot family analysis

Family: BeanBot			
Input:	1 App	2 App	3 App
Attack points:	1 Activity exported	1 Activity exported	1 Activity exported
	1 Broadcast receiver exported	1 Broadcast receiver exported	1 Broadcast receiver exported
	0 Content Provider exported	0 Content Provider find	1 Content Provider find
	1 Services find	0 Services exported	0 Services exported
Activity:	Main Activity	Main Activity	Start Activity
Broadcast receiver:	BaseABroadcastReceiver	BaseABroadcastReceiver	BaseABroadcastReceiver
Native library:	lib/armeabi/libandroidterm.so	lib/armeabi/libandroidterm.so	lib/armeabi/libandroidterm.so

Table 5 Asroot malware family analysis

Family: Asroot			
Input:	1 App	2 App	3 App
Package:	com.liteningrom.liteningconfig	uk.co.neilandtheresa.InstantRoot2	org.zenthought.flashrec
Permissions:	WRITE_EXTERNAL_STORAGE	BLUETOOTH	BLUETOOTH
	READ_EXTERNAL_STORAGE	BLUETOOTH_ADMIN	BLUETOOTH_ADMIN
		WRITE_EXTERNAL_STORAGE	INTERNET
		READ_PHONE_STATE	WRITE_EXTERNAL_STORAGE
		READ_EXTERNAL_STORAGE	READ_PHONE_STATE
			READ_EXTERNAL_STORAGE
Attack points:	1 Activity exported	0 Activity exported	1 Activity exported
	0 Broadcast receiver exported	1 Broadcast receiver find	1 Broadcast receiver find
	1 Content Provider find	0 Content Provider find	0 Content Provider exported
	0 Services exported	0 Services exported	0 Services exported
Activity:	Main Activtiy		FlashRec
Broadcast receiver:		Instant Root Receiver	

Table 6 Result of different malware families in mobile big data

Mobile big data malware categories	# Of apps tested	# Of app found malign	% of result
Adware	11000	9730	88.45
Rootkits	9000	8400	93.33
Botnets	6000	5400	90
Spyware	8000	6500	81.25
Ransomware	6000	5860	81
Total	40000	35890	89.72

5 Result and Discussion

As per the result point of view, the family of BeanBot Malware in mobile big data, after the extraction of data gives result at run time. This mobile big data BeanBot family gives some attack points which are activity, broadcast, intent services, and content. This also shows the point of attack where the malware is exported. Some times these exported activities are different as per the apps but after comparing with the signature it will identified that it is a malicious activity or its a benign mobile big data. We are also tracking the network broadcast as well as broad cast receiver properties at run time. We have also reported many of bot family malign app at rum time. This will give the 90% of accuracy with the mobile big data (Table 4).

The tested result of Asroot malware in mobile big data, after the extraction of data gives result at runtime. As per the result point of view Asroot family gives some attack points which are bluetooth, activity, services, permission, content provider, broadcast reciever. The extraction of these intents also compared with the signature then reported as malign big data or benign big data. As same as the Bean bot family, Asroot family also reported in this experiment as malign app. Approx 90% of malign apps are identified in this experimental script (Table 5).

Here, we are giving a result of different families identified in the experimental execution and this will consist of percentage result on the basis of malign and benign apps (Table 6).

6 Conclusion and Future Work

This approach is a powerful approach to identify malware or malicious app in big data. We have tested approx 40 k apps and the result with the respect of these app is approx 90%. We are also working on some recent network-based malware and in future we will work on recent coming Nougat and Marshmallow version of Android for malware detection as well as prevention. We will also use new android API as 5.1, 5.2, etc., for our research work. There are many more malware in Q4 Of 2016 also targetting the mobile big data in form of proximity sensor, NFC, etc. We will work on same upcoming malware analysis with new version of API.

References

1. Venturebeat Big Data Analytics. www.venturebeat.com/2015/01/22/big-data-and-mobile-analytics-ready-to-rule-2015/, Online; Accessed October 15 2016.
2. Digital Innovation Mobile Big Data. www.digitalinnovationgazette.com/mobile_big_data/, Online; Accessed November 27, 2016.
3. Google Cloud and Big Data. https://cloud.google.com/bigquery/, Online; Accessed October 20 2016.
4. Google Android with its Security. http://googlemobile.blogspot.com/2012/02/androidmobile-world-congress-its-all.html, Online; Accessed October 22 2016.
5. Nada Elgendy, Ahmed Elragal. Big Data Analytics: A Literature Review Paper. In *Article in Lecture Notes in Computer Science, August 2014*. August, 2014.
6. Demetrios Zeinalipour Yazti, Shonali Krishnaswamy. Mobile Big Data Analytics: Research, Practice, and Opportunities. In *Proceeding MDM '14 Proceedings of the 2014 IEEE 15th International Conference on Mobile Data Management - Volume 01, Pages 1–2*, 2014.
7. YING HE, FEI RICHARD YU NAN ZHAO 1, HONGXI YIN, HAIPENG YAO, AND ROBERT C. Big Data Analytics in Mobile Cellular Networks. In *IEEE access 2016 (volume:4)*. https://doi.org/10.1109/ACCESS.2016.2540520.
8. Elgendy, N. Big Data Analytics in Support of the Decision Making Process. In MSc Thesis, German University in Cairo, p. 164 (2013) 2013.
9. EMC. Data Science and Big Data Analytics. In *EMC Education Services, pp. 1–508 (2015)*, 2015.
10. He, Y., Lee, R., Huai, Y., Shao, Z., Jain, N., Zhang, X., Xu, Z.: RCFile. A Fast and Space efficient Data Placement Structure in MapReduce-based Warehouse Systems. In *IEEE International Conference on Data Engineering (ICDE), pp. 1199–1208 (2011).*
11. Herodotou, H., Lim, H., Luo, G., Borisov, N., Dong, L., Cetin, F.B., Babu, S. Starfish: A Self-tuning System for Big Data Analytics. In *Proceedings of the Conference on Innova- tive Data Systems Research, pp. 261?272 (2011)*, pp. 261–272 (2011).
12. Kubick, W.R. Big Data, Information and Meaning. *In: Clinical Trial Insights*, pp. 26–28 (2012).
13. Lee, R., Luo, T., Huai, Y., Wang, F., He, Y., Zhang, X. Ysmart: Yet Another SQL-to-MapReduce Translator. *In IEEE International Conference on Distributed Computing Systems (ICDCS), pp. 25–36 (2011).*
14. Manyika, J., Chui, M., Brown, B., Bughin, J., Dobbs, R., Roxburgh, C., Byers, A.H. Big Data: The Next Frontier for Innovation, Competition, and Productivity. *In: McKinsey Global Institute Reports, pp. 1–156 (2011).*
15. Venkatesh Gauri Shankar, Gaurav Somani. Anti-Hijack: Runtime Detection of Malware Initiated Hijacking in Android. In (Elsevier) Procedia Computer Science 78 (2016): 587–594. https://doi.org/10.1016/j.procs.2016.02.105.
16. Arzt Steven, Rasthofer Siegfried, Fritz Christian, Bodden Eric. Flowdroid: Precise context, flow, field, object-sensitive and lifecycle-aware taint analysis for android apps. *In Proceedings of the 35th International Conference on Security, ACM, Edinburg, United Kingdom, 2014*, United Kingdom, 2014.
17. Strace. System Call Trace. http://linux.die.net/man/1/Strace, Online; Accessed September, 2016.

18. Bruno P.S. Rocha, Mauro Conti, Sandro Etalle, Bruno Crispo. Hybrid Static-Runtime Information Flow and Declassification Enforcement. In (IEEE) Transactions on Information Forensics & Security, 8(8): 1294–1305, 2013. https://doi.org/10.1109/TIFS.2013.2267798, ISSN: 1556-6013.
19. Amazon Inc. Amazon Android App Stores. http://amazon.com/mobile-apps/b?ie=UTF8Ź node=2350149011, Online; Accessed September 12 2016.
20. Chinese Apps Malware Data Set. http://appinchina.com/the-market, Online; Accessed October 16 2016.
21. Google App-Store Android Apps in App Stores. https://play.google.com/store?hl=en, Online; Accessed December 10 2016.
22. SlideME App Store for Smartphone. https://slideme.org/applications, Online; Accessed December 02 2016.
23. GNOME Android Malware Data Set: GNOME. http://malgenomeproject.org/mobile-apps, Online; Accessed September 05 2016.

Toward Smart Learning Environments: Affordances and Design Architecture of Augmented Reality (AR) Applications in Medical Education

Arkendu Sen, Calvin L.K. Chuen and Aye C. Zay Hta

Abstract Medical education places emphasis on situational learning of real-life clinical contexts, while simultaneously focusing on human body three-dimensional (3D) visualization. However, classrooms and laboratories being the main learning environment in early years of medical education, there is limited exposure to real clinical environments to adequately meet such objectives. This study proposes that augmented reality (AR) applications can provide both affordances. Such applications fulfill many of the criteria for smart learning environments (SLE). A systematic review aiming to identify the affordances of AR applications, their design architectures, and impact evaluations was conducted. This review evaluated 25 studies and, with model case studies, analyzed how the different AR applications provided situational learning and visualization in medical education and how their design architecture provided such affordances toward contextualized, interactive, and personalized SLEs. It was found that AR affords facilitation of situational learning and visualization individually. Their integrated educational impact, however, needs to be evaluated further.

Keywords Augmented reality · Basic medical sciences education
Medical education · Design architecture · Situational learning
Visualization

A. Sen (✉) · C.L.K. Chuen
Jeffrey Cheah School of Medicine and Health Sciences, Monash University Malaysia, Bandar Sunway, 47500 Subang Jaya, Selangor, Malaysia
e-mail: arkendu.sen@monash.edu

C.L.K. Chuen
e-mail: ckleo8@student.monash.edu

A.C. Zay Hta
Faculty of Science & Technology, Department of Computing & Information Systems, Sunway University, Bandar Sunway, 47500 Subang Jaya, Selangor, Malaysia
e-mail: 14085468@imail.sunway.edu.my

A.K. Somani et al. (eds.), *Proceedings of First International Conference on Smart System, Innovations and Computing*, Smart Innovation, Systems and Technologies 79, https://doi.org/10.1007/978-981-10-5828-8_80

1 Introduction

Augmented reality (AR) is a technology that "allows the user to create sensory-motor activities in a new space by combining the real environment and the virtual environment" [1]. Ubiquitous use and availability of smart phones with inbuilt camera have renewed interest in the use of AR in mobile learning that can capture the real environment against the AR interface. With widespread public engagement in novel AR games such as Pokémon Go, affordances of AR in various educational applications are increasingly being the subject of educational research.

Affordances are "the qualities or properties of an object that define its possible uses or make clear how it can or should be used" [2]. Such affordances of AR technology have been used in learning of three-dimensional (3D) concepts in various educational settings whether to teach undergraduate Geography regarding Earth–Sun relationships [3] or 3D geometric construction tools for mathematics and geometry education [4]. Affordances of such AR applications have mainly been on *visualization* of complex 3D concepts for various STEM learnings. Visualization requires a student to be able to visualize objects or processes. Medical education, especially in the beginning of the course, also relies heavily on visualization of the 3D structures of the human body. The uniqueness of medical education is the added emphasis on simultaneous *situational learning* within a clinical context. Situational learning requires a student to be in the real or at least, simulated environment to gain familiarity with the hospital or clinic setting. Recent studies have focused on creating 3D resources with such focus especially in relation to Anatomy or Pathology [5, 6].

In this study, we propose two important affordances of AR technology and illustrate such affordances through model and case studies. AR can provide one of the best platforms for superimposing real clinical or laboratory environments (as background) into the classroom activity thus aiding situational learning. Simultaneously, AR intrinsically, by projecting a high-resolution 3D image (as overlay), helps to envision specimens/medical models for visualization. Such an advantage is critical, as only limited actual clinical environment experience can exist, while students are in the classroom in early years of basic medical sciences education.

Both such affordances within an interactive mobile interface can satisfy the criteria of model smart learning environments (SLE). SLE is characterized by being (1) context-aware, i.e., the real-world situation as the context of learning, (2) adaptive (guidance, feedback, and learning tool based upon the student's situation/topic of interest), and (3) personalized (interactive interface suited to learners needs) [7].

2 Problem Statement

Considering the versatility of AR, very few studies have considered the affordances of AR especially on its value in context situational learning and personalized visualization in medical sciences education toward an SLE [7]. Since approaches to AR vary greatly in terms of design, how the design architecture of the AR applications subserves the needs of situational learning and interactive personalized interface for visualization needs to be considered.

Through literature reviews and case studies, the goal of this study was to identify the affordances and functionalities in line with SLE of AR applications and devices in the field of Basic Medical Sciences and Allied Health Sciences education including Nursing.

3 Methodology

A comprehensive search of the databases was undertaken, including Cochrane Library, Google Scholar, and PubMed to evaluate peer-reviewed journal articles. The search terms used included "Augmented reality AND Basic Medical Sciences OR Anatomy OR Pathology OR Physiology OR Biochemistry OR Physiotherapy OR Nursing OR Allied Health Sciences". The inclusion criteria included papers published after 2000 to reflect current developments, those in English, and "Peer Reviewed Journals". The exclusion criteria were clinical sciences, those not for education or solely virtual reality interventions.

Search results which satisfied the inclusion criteria and were relevant to the problem statement are presented in Table 1. Considerable heterogeneity between AR approaches and diversity in the integration of other functionalities in the literature review prevented statistical pooling or subgroup analysis. This has led to the incorporation of a breadth analysis rather than focusing on the level of evidence.

Table 1 Overview of published studies on "Augmented Reality" resources for basic medical education and their affordances

No.	References	Subject	Description of AR used	Functionality	Situation simulated	Evaluation tools	Equipment/Operating System used	Design architecture/-*Augmented Reality Toolkits used*
1	Ai et al. [8]	Medicine	Device for AR visualization of veins	Vein image into AR illumination of veins	Hospital setting	Not evaluated for teaching	Infrared camera and active structured light	AR illumination of veins after image capture by IR camera; Noninvasive near-IR light (NIR), AccuVein, vein contrast enhancers, VueTek Veinsite, VascuLuminator, Vein Viewer Vision, head-mounted vascular imaging system, 3D blood vessel search system, imaging, and augmented system (IAS) *- Microsoft Visual Studio C ++ with OpenCV library and Qt; CorelDraw, MeshLab, 3D reconstruction of the skin*
2	Al Hamidy Hazidar [21]	Anatomy	AR application for AR visualization of human heart	Mobile usable.Easy customizable	Classroom	Not Evaluated	Android operating mobile smart phone	3D model after camera vision, capture, tracking, and identification *-AR bone, Vuforia, Image-target tracking system, Xcode, java IDE and SDK, Unity 3D and Maya, interactive application*

(continued)

Table 1 (continued)

No.	References	Subject	Description of AR used	Functionality	Situation simulated	Evaluation tools	Equipment/Operating System used	Design architecture/-*Augmented Reality Toolkits used*
3	Albrecht, Folta-Schools, Behrends, and von Jan [22]	Forensic Science	AR application for AR visualization of forensic pathologies	Mobile usable. Easy customizable	Live Peers	Significant knowledge gain and "Profile of Mood States" (POMS) questionnaires. n = 6 students	iOS operating mobile smart phone	An overlay image (i.e., pathology) after image acquisition and processing
4	Berry and Board [31]	Biochem	AR display of macromolecules	Easily customizable	Classroom	Not evaluated	Webcam-enabled PC/Macintosh	Formation of a visual display after camera image acquisition and processing; -*ARToolkit*
5	Bifulco et al. [9]	Medicine (ECG operation)	Head-mounted display—enhanced simulation of ECG recording	Easy to use	Hospital setting	Error of precordial electrode placement. Average errors: <3 mm (mannequin) and <7 mm (real patient)	Head-mounted display with webcam	An overlay image (i.e., instructions) after image thresholding, marker detection and perspective computation, Marker-based system, Head-mounted display; -*ARToolkit*
6	Blum et al. [10]	Anatomy	AR "magic mirror" for AR display of CT dataset on user	Interactive	Live Peers	Not evaluated	Color camera, depth camera and visual display utilizing Kinect system	An overlay image (i.e., CT dataset) after pose tracking of user and augmentation of image; -*NITE skeleton tracking software*

(continued)

Table 1 (continued)

No.	References	Subject	Description of AR used	Functionality	Situation simulated	Evaluation tools	Equipment/Operating System used	Design architecture/-*Augmented Reality Toolkits used*
7	Botden et al. [11]	Surgery	AR laparoscopic suturing simulator with haptic feedback	Realistic actual haptic feedback	Hospital setting	Time spent in correct area and strength of knot —positive significant difference; n = 18 participants	ProMIS v2.0 Augmented reality (AR) simulator	An overlay image and haptic feedback after tracking distal end of the laparoscopic instrument shaft
8	Coles et al. [12]	Medicine	AR femoral palpation and needle insertion simulator	Realistic actual haptic feedback	Hospital setting	face and content validation questionnaire. n = 7 participants	AR workstation: LCD display, camera, needle hub coupled to haptic device	A virtual display and haptic feedback after tracking of the user's hand
9	Ferrer-Torregrosa et al. [23]	Anatomy	Visualization of 3D anatomy models	Portable	Classroom	Course exam scores: Notes group (gp) 5.6, Video gp 6.54, AR gp 7.19; n = 171 students	AR software	Not described
10	Jamali, Shiratuddin, Wong, and Oskam [13]	Anatomy	Visualization of 3D anatomy models	Portable	Classroom	Pre- and post-intervention knowledge questionnaire; n = 30 students	Android operating tablet	3D model after camera vision, capture, tracking, and identification interactive, marker-based AR system using Waterfall methodology, -*Vuforia AR toolkit, 3D Studio Max, Unity 3D*

(continued)

Table 1 (continued)

No.	References	Subject	Description of AR used	Functionality	Situation simulated	Evaluation tools	Equipment/Operating System used	Design architecture/-*Augmented Reality Toolkits used*
11	Juanes et al. [24]	Anatomy	Visualization of 3D anatomy models	Portable	Classroom	Not evaluated	Android or iOS operating device	Formation of a visual display (i.e., the 3D model) after camera vision, capturing, tracking and identification. *-Vuforia AR toolkit, AR bone, Image-target tracking system, Xcode, java IDE & SDK, Unity3D & Maya, interactive app*
12	Keri et al. [14]	Medicine	AR ultrasound-guided lumbar puncture simulator	Realistic	Hospital setting	Intervention group -significant less tissue damage; n = 24 participants	Perk Tutor computerized training platform	Formation of a visual display based on the position of the needle and input from the ultrasound *-Perk Tutor software*
13	Küçük et al. [25]	Anatomy	3D video animations and 3D anatomy models	Portable	Classroom	Academic achievement test and cognitive load scale questionnaire. n = 70 students	Android or iOS operating device	Formation of a visual display (i.e., the 3D model) after camera vision, capturing, tracking and identification

(continued)

Table 1 (continued)

No.	References	Subject	Description of AR used	Functionality	Situation simulated	Evaluation tools	Equipment/Operating System used	Design architecture/-*Augmented Reality Toolkits used*
14	Leitritz et al. [15]	Medicine	Simulation of ophthalmoscope	Realistic	Hospital/clinic setting	Questionnaire and skills assessment. n = 19 students	Eyes indirect system. (PC, 2 monitors, headset with built-in binocular display, and camera)	Formation of a visual display (i.e., a simulated fundus image imposed on the dummy 20 D lens)
15	Ma et al. [16]	Anatomy	AR "magic mirror" for AR display of CT dataset on user	Interactive	Live Peers	questionnaire n = 72 students	Color camera, depth camera, and visual display using Kinect system	Overlay (i.e., CT dataset) after pose tracking of user and augmentation of image; -*NITE skeleton tracking software*
16	McKenzie et al. [17]	Medicine	AR auscultation simulator	Interactive	Hospital setting	Performance of n = 53 students	Electronic stethoscope and magnetic sensor	Pathological sound (i.e., bruit) played when the stethoscope is placed over the magnetic sensor on the simulated patient
17	Rahn and Kjaergaard [26]	Nursing	Visualization of internal organs	Portable	Live Peers	Student feedback	iOS and standard shirt	Tracking of shirt location and overlay of augmented image
18	Rodriguez-Pardo et al. [27]	Anatomy	Visualization of hand anatomy	Portable	Live Peers	Evaluated with questionnaire n = 37 participants	Any iOS mobile device	Tracking of user's hand and overlay of augmented image, virtual buttons used to detect hand;*Vuforia AR toolkit, Unity 3d, Javascript & C#*

(continued)

Table 1 (continued)

No.	References	Subject	Description of AR used	Functionality	Situation simulated	Evaluation tools	Equipment/Operating System used	Design architecture/-*Augmented Reality Toolkits used*
19	Saenz et al. [28]	Anatomy	Visualization of 3D anatomy models	Portable	Classroom	Not evaluated	Any iOS mobile device, desktop or smart glasses	Overlay after tracking of markers to determine position of model; *Vuforia AR toolkit*
20	Samosky et al. [18]	Medicine	AR endotracheal tube (ET) simulator	Interactive	Hospital setting	Not evaluated	Full-body human simulator augmented with digital light processing (DLP) projector and noncontact position	Overlay on the mannequin based on the location of the ET tube; -*Development with LabVIEW*
21	Samosky et al. [19]	Anatomy	AR anatomy simulator	Interactive	Hospital setting	Not evaluated	Full-body human simulator augmented with DLP projector and noncontact position sensing. IR pen and Wiimote.	Dynamic overlay image formation on the mannequin based on the movement of the IR pen as captured by the Wiimote *-Development with LabVIEW*
22	Tanaka, et al. [29]	Nursing	3D video animations and 3D anatomy model	Portable	Classroom	Not evaluated	Android or iOS operating mobile device	3D model after camera vision, capture, tracking, and identification

(continued)

Table 1 (continued)

No.	References	Subject	Description of AR used	Functionality	Situation simulated	Evaluation tools	Equipment/Operating System used	Design architecture/-*Augmented Reality Toolkits used*
23	Thomas et al. [32]	Anatomy	3D MRI anatomy models	Easily customizable	Live Peers	Questionnaire; n = 34 participants	Webcam-enabled PC	3D model after camera vision, capture, tracking, and identification -Development with visualization toolkit
24	Vaughn et al. [20]	Medicine	Clinical simulation	Easily customizable	Hospital setting	Simulation design scale (SDS) and self-confidence in learning scale (SCLS) questionnaire; n = 12 students	Google glass	Playing of a video showing a simulated patient during the simulation;
25	Wang et al. [30]	Anatomy	3D animations and anatomy models	Portable	Classroom	Not evaluated	Android or iOS operating mobile device	3D model after camera vision, capture, tracking, and identification *-Development with Layer 3D Model Converter*

4 Results and Discussion

Of the 5268 studies on the initial search term, 60 studies were retrieved for full text analysis and 31 studies complied with the inclusion criteria (Fig. 1). Six (6) studies were excluded as they were either literature review themselves or were overlapping articles (Fig. 1). After filtering, the analyses of 25 studies and cases are presented in Table 1.

This review identified a variety of approaches in the use of AR in medical education with a potential of creating SLE. Of the target devices, dedicated devices were the most common (n = 13) [8–20], followed by Android/iOS mobile devices (n = 10) [21–30] and a few (n = 2) used a desktop application [31, 32].

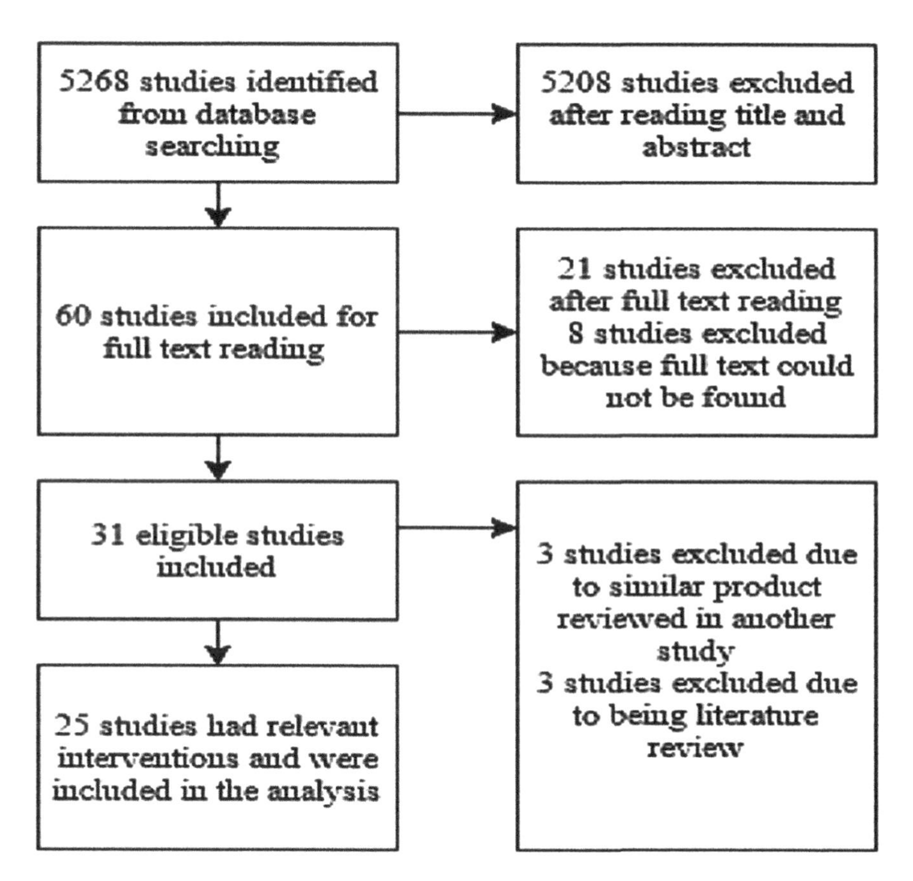

Fig. 1 Flowchart of study design and results of literature search

4.1 Situational Learning

The review identified ten studies that utilized AR in creating situational learning [8, 9, 11, 12, 14, 15, 17, 18, 20, 22], thereby improving competencies in clinical examination or procedures as well as meeting the context-aware criteria of SLE [7].

A few studies reported AR in the context of surgical or procedural skills. An innovative device that allowed for visualization of veins for intravenous procedures [8] that performed relatively well in terms of accuracy of vein imaging was noteworthy and the applicability of the AR device in similar clinical training needs to be investigated. An interactive AR laparoscopic simulator [11] which used haptic and visual feedback for training in appropriate depth of the suture "bite" and knot tying leading to situational learning of laparoscopic suturing was also found to significantly increase competence of trainees (n = 18) in such basic surgical skills. Another interactive, personalized AR simulator, connected to a haptic device to simulate the actual "feel" during a femoral artery palpation and needle insertion [12], readily met all the three criteria of SLE especially as it received general positive feedback in terms of realism. An AR ultrasound lumbar puncture simulator [14] that resembled conventional equipment and realistically simulated actual operation, thus conducive for situational learning was developed. It was tested on students (n = 24) and was found to significantly increase skills, outperforming the control group in terms of individual needle tracking steps, though the overall success rates did not differ.

Some AR applications were developed to ensure clinical competencies in smart learning (clinical) environment. One such simulated recording of electrocardiogram (ECG) [9] on either a life size mannequin or a simulated patient was found to increase competence of untrained personnel (n = 10) in performing a complete ECG test. Another AR employing the use of an AR system in ophthalmoscope training [15] created a smart learning environment of a real patient simulation in a real clinical setting by augmenting a dummy lens in the eyepiece in the background of the viewer's own hand during the procedure, as is the case when using an actual ophthalmoscope. Though questionnaire evaluation found no difference between the control and intervention groups, the intervention group's better performance in their examinations skills, i.e., accurate drawing of the patient's optic disk, highlighted such AR applications efficacy as an SLE.

Some earlier studies reported technical issues while implementing AR to aid situational learning. One such study [17] used a modified electronic stethoscope which played a pathological sound (i.e., a bruit in the carotid artery), when the stethoscope is placed in the correct location. Such technical issues are not highlighted in more recent papers, probably reflecting the recent improvements in AR technology.

As a context-aware interactive SLE, an AR endotracheal tube (ET) simulator [18], though not formally evaluated, afforded situational learning. Using a modified mannequin and Digital Light Processing (DLP) projector [18], a dynamic image of the ET projected on the mannequin was used and was context-aware to change according to the location of the external part of the ET. Another similar simulator used a novel idea of playing a video of a simulated patient in Google Glass, during a simulation to replicate the actual clinical setting [20]. The video which played during the simulation created a setting that there was an actual patient instead of a static mannequin. They tested the system (n = 12) through the simulation design scale (SDS) and the self-confidence in learning scale (SCLS) questionnaires. The students who completed the augmented reality headset (ARH) enhanced simulation exercise scored the simulation design positively and the results of the SCLS indicated that ARH was a favorable addition to the simulation-learning environment.

4.2 Visualization

Majority of the studies (n = 15) analyzed and encompassed interventions that employed the use of AR in visualization. Evaluations were yet to be conducted in the majority probably because the products were still in the proof-of-concept stage [10, 13, 16, 21, 23–25, 31, 33]. Furthermore, few applications were evaluated as a teaching tool (Table 1). One such AR desktop application could project a 3D model of a protein or other macromolecule [31]. This application allowed the user to easily customize the macromolecule visualized rather than only presets from the developer. Thus, as an SLE, there were effective elements of interactivity and personalization in this [7]. Another AR application developed as a “magic mirror” augmented CT datasets [10]. The system used several cameras and the Kinect system (i.e., an add-on to the Microsoft X-Box console) to ascertain the posture of the user in order to correctly project the user as well as the CT dataset in the screen. An improvement on the “magic mirror” concept [16] from Blum et al. (2012) [10] allowed students to visualize CT data on themselves, eliciting (n = 72) generally favorable feedback toward the device in terms of device accuracy, ease of use, fun to use, and applicability for medical education. For visualization of internal organs, an AR tablet application was developed which used a predetermined shirt as a trigger [26]. The application was evaluated qualitatively and was positive as the application gave a higher degree of realism compared to a textbook and the cognitive load was less.

In a distance learning setting, an AR application was developed that could project 3D models [23]. The AR application was tested in a group alongside a “traditional notes” group and a “video” group (n = 130). A course exam, to evaluate the effectiveness, found that the AR group performed the best among the three interventions. An AR mobile application that could project a 3D model of the heart [21] was similarly designed as an AR tablet application that could project 3D anatomy models [13]. This application was also tested (n = 30) with pre- and post-intervention knowledge and usability questionnaires. They found significantly better knowledge gain and the feedback toward the application was generally favorable. This was similar to the AR mobile application that could project 3D anatomy models [24]. They also explored the use of an AR book as means to project the models from, though not evaluated as a teaching tool.

A unique AR mobile application that could project 3D anatomy models as well play 3D animations [25] was tested (n = 70) using the academic achievement test and cognitive load scale questionnaire. The intervention group was found to be statistically significantly more successful and had lower cognitive loads in comparison to the control group students. Another novel AR mobile application for the visualization of hand anatomy [27] was tested (n = 37) using a questionnaire. The participants generally were satisfied with the application. However, the application did not score well in "spatial complexity" which reflects a limitation in the application's ability to project complex spaces like those in actual anatomy. Similarly, though not evaluated for teaching, Saenz et al. (2015) designed an AR mobile application for the visualization of anatomy models [28]. Samosky et al. (2012) used the modified mannequin and digital light processing (DLP) projector from their earlier study to create an AR device for the visualization of internal anatomy and physiology [19]. Each layer of viscera could be removed with gestures using an infrared (IR) pen which was a unique feature in this device and physiological processes such as a beating heart could be modified. These functions aided especially spatial visualization and visualization of processes. The device was not evaluated for teaching. Similar AR mobile application and AR book in the teaching of anatomy by visualization of 3D anatomy models [29], though not quantitatively evaluated, was found to increase motivation. A model case of an AR as an SLE that had the affordance in situational learning to include actual skin pathologies as overlay of an injury or a bullet wound on a subject, simulating the exact position of such medico-legal wounds [22], is discussed below as a case study.

Albrecht et al. (2013) developed a novel mobile AR application for the teaching of criminal pathology. The application, mARble, or "mobile Augmented Reality blended learning environment" can be installed on an iOS mobile device, specifically an iPhone. The base application was developed at the Peter L. Reichertz Institute for Medical Informatics (PLRI) at the Hannover Medical School.

For this study, Albrecht et al. (2013) adapted an existing textbook for "gunshot wounds" used in Germany into the application as the base for context of real-life application. Descriptions and labels were added to the application which was based on the textbook. Using the application, gunshot wounds would be "overlayed" onto a consenting peer where interactive labels and questions would be displayed alongside the gunshot wound thus aiding in visualization. Two (2) groups of students, one group utilizing mARble and one control group, were recruited to evaluate the application. A pre- and post-knowledge questionnaire was used as was the Profile of Mood States (POMS) questionnaire.

The study found that the improvement was higher in the mARble group with 4.7 questions (SD 2.9) compared to the control group showing an improvement of three questions, but also with smaller variability (SD 1.5). The difference in improvement within the mARble group was statistically significant (Wilcoxon, z = 2.232, P =0.03). The study also found a statistically significant decrease of fatigue (z = 2.214, P =0.03) and numbness (z = 2.07, P = .04) for the mARble group in the pre- and post-POMS questionnaire. This decrease was not statistically significant in the control group.

A desktop AR application for visualization of 3D MRI anatomy models [32] was similarly tested (n = 34) using a questionnaire and was found to help participants understand the shape and location of the ventricular system. Such anatomical visualization was also used in an AR mobile application consisting of 3D video animations and 3D anatomy models that could be triggered by QR codes in an AR book [30].

4.3 Design Architecture

Majority of the studies were not very detailed in their explanations of the design architecture. The software generally was already available and seldom was developed from the ground up. Most AR mobile applications utilized available developer software toolkits such as ARToolkit and Vuforia. Some studies described the process from camera image acquisition, tracking, processing to overlaying the augmented image, commonly using toolkits such as Vuforia AR to set the image target and the Unity3D game engine to render the visual graphics to display the 3D models [13]. Another study [27] which also used the Vuforia SDK (Software Development Kit) took advantage of virtual buttons arranged to form a hand shape to trigger the virtual contents when light is obstructed from the virtual buttons [27] using Javascript and C# scripts.

Thus, existing AR toolkits offer readymade architecture allowing 3D renderation linked to GPS location or object-based real context triggers for interactive, adaptive AR applications to be developed as learning tool as part of SLE.

4.4 AR Web Resources on Medical Education

Aside from the literature on AR applications as highlighted in this review, some AR applications developed as interactive resources do fit the criteria of SLE [7] but remains to be researched. To supplement this study, exemplars of such web resources on AR applications were retrieved, analyzed, and reported in Table 2.

Table 2 AR web resources on medical education

Reference	Subject	Description of AR application	Situation simulated	Tools used
Microsoft [34]	Anatomy	Visualization of human body anatomy	Classroom	Microsoft Hololens
Santoso [35]	Anatomy	Visualization of human heart	Classroom	Android mobile
Forsythe et al. [36]	Nursing	Simulation of clinical scenarios on mannequin	Hospital/ clinical setting	iOS tablet and mannequin

One notable application is the Microsoft Hololens Holoanatomy application [34] which allows users to visualize anatomy in a story telling method through the heads-up display (i.e., Microsoft Hololens) while allowing the user to interact with the models through finger gestures (Table 2). Another notable application is the HeART application [35], similar to those described earlier, augmenting a 3D anatomy model for visualization using a heart image as a trigger. The interior of the heart can also be viewed. Simultaneous situational learning is provided through a trigger of real-life heart model with AR app of annotations/labels overlays (Table 2) making it a good SLE as is the one used to train nurses [36] using a live mannequin as trigger to display a video with various clinical scenarios. Such a mannequin, which could also be remotely controlled to continue interacting in tandem with the simulation incorporates many features of an SLE [37] in providing a level of realism and developing skills of empathy toward patients (Table 2).

5 Conclusion

This study has highlighted various AR applications to have affordances toward facilitation of context-aware real-life situational learning and also of 3D visualization especially in medical education. Majority of them incorporated interactive as well as context-aware (as integrated real context presentation) applications in the back drop of mobile learning providing effective and supportive learning environments [37, 38]. The affordances of a smart phone allow, even in a classroom/laboratory setting for early years, basic sciences students, to be engaged in real-life clinical situations like drawing blood, conducting an ECG or skin suturing which could also be personalized depending upon their existing knowledge of a skill or manipulating objects to be visualized and receiving appropriate feedback—a major step toward a model smart learning environment (SLE). The design architecture seems to be simple as most of such applications were developed by using third part AR rendering tool kits rather than developing from ground up. Some of the applications have focused on interactive visualization of 3D Anatomy objects but only a few on presenting it within a clinical environment the most notable being the HEART application [35]. However, what would be most pertinent would be a real integration of AR of basic anatomy (visualization) overlayed on a clinical patient (situational learning), e.g., the heart structure with its valve while being superimposed on a patient being examined for heart valvular disease. However, as most studies have not emphasized on integrating both such affordances, there is a need for future studies to relate the design architecture of the AR applications to such affordances and effect on pedagogies, especially for increasing clinical competencies in a smart (clinical) learning environment.

References

1. Hugues O., Fuchs, P., Nannipieri, O.: New Augmented Reality Taxonomy: Technologies and Features of Augmented Environment. In: Furht B, editor. Handbook of Augmented Reality. New York, NY: Springer New York; p. 47–63 (2011)
2. Merriam Webster Online. Affordance
3. Shelton, B.E., Hedley, N.R., Using augmented reality for teaching Earth-Sun relationships to undergraduate geography students. The First IEEE International Workshop Agumented Reality Toolkit (2002)
4. Kaufmann, H., Schmalstieg, D.: Mathematics and geometry education with collaborative augmented reality. Computers & Graphics.27(3):339–45 (2003)
5. Sen, A., L. Selvaratnam, K.L. Wan, J.J. Khoo, P.A. Rajadurai.: Virtual histopathology—essential education tools to ensure pathology competence for tomorrow's medical interns. In INTED2016 Proceedings. eds L. Gómez Chova et al. p. 4633–4640. International Association of Technology, Education and Development, Spain (2016)
6. Wan, K.L., Sen, A., Selvaratnam, L., Khoo, J.J., Rajadurai, P.A.: Addressing the 'Pathology Gap' in Clinical Education and Internship: The Impetus to Develop Digital (3D) Anatomic Pathology Learning Resources. Proceedings of the 2016 IEEE International Conference On Teaching and Learning In Education (ICTLE'16), Kuala Lumpur, Malaysia, pp. 101–106 (2016)
7. Hwang, G.J.: Definition, framework and research issues of smart learning environments—a context-aware ubiquitous learning perspective. Smart Learn Environ **1** (1), 1–14 (2014)
8. Ai, D., Yang, J., Fan, J., Zhao, Y., Song, X., Shen, J., et al.: Augmented reality based real-time subcutaneous vein imaging system. Biomedical optics express.7(7):2565–85 (2016)
9. Bifulco, P., Narducci, F., Vertucci, R., Ambruosi, P., Cesarelli, M., Romano, M.: Telemedicine supported by Augmented Reality: an interactive guide for untrained people in performing an ECG test. Biomedical engineering online.13:153 (2014)
10. Blum, T., Kleeberger, V., Bichlmeier, C., Navab, N.: Mirracle: An augmented reality magic mirror system for anatomy education. 2012 IEEE Virtual Reality Workshops (VRW) IEEE Press (2012)
11. Botden, S.M., de Hingh, I.H., Jakimowicz, J.J.: Suturing training in Augmented Reality: gaining proficiency in suturing skills faster. Surgical endoscopy. 23(9), 2131–2137 (2009)
12. Coles, T.R., John, N.W., Gould, D.A., Caldwell, D.G.: Integrating Haptics with Augmented Reality in a Femoral Palpation and Needle Insertion Training Simulation. IEEE transactions on haptics. 4(3), 199–209 (2011)
13. Jamali, S.S., Shiratuddin, M.F., Wong, K.W., Oskam, C.L.: Utilising mobile-augmented reality for learning human anatomy. Procedia-Social and Behavioral Sciences.197, 659–668 (2015)
14. Keri, Z., Sydor, D., Ungi, T., Holden, M.S., McGraw, R., Mousavi, P., et al.: Computerized training system for ultrasound-guided lumbar puncture on abnormal spine models: a randomized controlled trial. Canadian Journal of Anesthesia. 62(7), 777–784 (2015)

15. Leitritz, M.A., Ziemssen, F., Suesskind, D., Partsch, M., Voykov, B., Bartz-Schmidt, K.U., et al.: Critical evaluation of the usability of augmented reality ophthalmoscopy for the training of inexperienced examiners. Retina (Philadelphia, Pa). 34(4), 785–91 (2014)
16. Ma, M., Fallavollita, P., Seelbach, I., Von Der Heide, A.M., Euler, E., Waschke, J., et al.: Personalized augmented reality for anatomy education. Clinical anatomy (New York, NY). 29 (4), 446–453 (2016)
17. McKenzie, F.D., Hubbard, T.W., Ullian JA, Garcia HM, Castelino RJ, Gliva GA.: Medical student evaluation using augmented standardized patients: preliminary results. Studies in health technology and informatics. 119, 379–384 (2006)
18. Samosky, J.T., Baillargeon, E., Bregman, R., Brown, A., Chaya, A., Enders, L., et al.: Real-time "x-ray vision" for healthcare simulation: an interactive projective overlay system to enhance intubation training and other procedural training. Studies in health technology and informatics.163, 549–551 (2011)
19. Samosky, J.T., Wang, B., Nelson, D.A., Bregman, R., Hosmer, A., Weaver, R.A.: BodyWindows: enhancing a mannequin with projective augmented reality for exploring anatomy, physiology and medical procedures. Studies in health technology and informatics. 173, 433–439 (2012)
20. Vaughn, J., Lister, M., Shaw, R.J.: Piloting Augmented Reality Technology to Enhance Realism in Clinical Simulation. Computers, informatics, nursing: CIN. 34(9), 402–405 (2016)
21. Al Hamidy Hazidar, R.S.: Visualization Cardiac Human Anatomy using Augmented Reality Mobile Application (2014)
22. Albrecht, U.V., Folta-Schoofs, K., Behrends, M., von Jan, U.: Effects of mobile augmented reality learning compared to textbook learning on medical students: randomized controlled pilot study. Journal of Medical Internet Research.15(8), e182 (2013)
23. Ferrer-Torregrosa, J., Jimenez-Rodriguez, M.A., Torralba-Estelles, J., Garzon-Farinos, F., Perez-Bermejo, M., Fernandez-Ehrling, N.: Distance learning ects and flipped classroom in the anatomy learning: comparative study of the use of augmented reality, video and notes. BMC medical education.16(1), 230 (2016)
24. Juanes, J.A., Hernández, D., Ruisoto, P., García, E., Villarrubia, G., Prats, A.: Augmented reality techniques, using mobile devices, for learning human anatomy. Proceedings of the Second International Conference on Technological Ecosystems for Enhancing Multiculturality; ACM (2014)
25. Küçük, S., Kapakin, S., Göktaş, Y.: Learning anatomy via mobile augmented reality: effects on achievement and cognitive load. Anatomical sciences education (2016)
26. Rahn, A., Kjaergaard, H.:Augmented Reality As A Visualizing Facilitator In Nursing Education. INTED2014 Proceedings. 6560–6568 (2014)
27. Rodriguez-Pardo, C., Hernandez, S., Patricio, M.Á., Berlanga, A., Molina, J.M.: An Augmented Reality Application for Learning Anatomy. International Work-Conference on the Interplay Between Natural and Artificial Computation; Springer (2015)
28. Saenz, M., Strunk, J., Maset, K., Seo, J.H., Malone, E.: FlexAR: anatomy education through kinetic tangible augmented reality. ACM SIGGRAPH 2015 Posters; ACM (2015)
29. Tanaka, M., Uchida, A., Kanda, A., Matsuo, T.: Effects of Digital Textbook-Based Nursing Education in Professional School. Computer Application Technologies (CCATS), 2015 International Conference on; IEEE Press (2015)
30. Wang, L.L., Wu, H.H., Bilici, N., Tenney-Soeiro, R.: Gunner Goggles: Implementing Augmented Reality into Medical Education. Studies in health technology and informatics. 220, 446–9 (2016)
31. Berry, C., Board, J.:A Protein in the palm of your hand through augmented reality. Biochemistry and molecular biology education: a bimonthly publication of the International Union of Biochemistry and Molecular Biology. 42(5), 446–9 (2014)
32. Thomas, R.G., John, N.W., Delieu, J.M.: Augmented reality for anatomical education. Journal of visual communication in medicine. 33(1), 6–15 (2010)
33. Rahn, A., Buhl, M.: Augmented Reality as Wearable Technology in Visualizing Human Anatomy. Is the adaptive researcher the road to success in design-based research? 30 Students

as Math Level Designers: How students position themselves through design of a math. 90 (2016)
34. Microsoft: HoloAnatomy, https://www.microsoft.com/en-us/store/p/holoanatomy/9nblggh4ntd3#
35. Santoso, M.: HeARt: Augmented and Virtual Reality (AR / VR) for Medical School Learning Tool Enhancement (2016) https://www.youtube.com/watch?v=dWf7oEwZKbs&ab_channel=MarkusSantoso
36. Forsythe, R., Lewis, P., Brailsford, M., Clark, D., Bishop, P.: Introducing Patient ScenARios – SimMan gets personality! (2014) http://campus-interactive-media.com/work/test-work/
37. Kamarainen, A.M., Metcalf, S., Grotzer, T., Browne, A., Mazzuca, D., Tutwiler MS, Dede C.: EcoMOBILE: Integrating augmented reality and probeware with environmental education field trips. Comput. Educ. 68, 545–556. 10.1016/j.compedu.2013.02.018 (2013)
38. Andujar, J.M., Mejias, A., Marquez, M.A.:Augmented reality for the improvement of remote laboratories: an augmented remote laboratory. IEEE Trans. Educ. 54(3), 492–500. 10.1109/TE.2010.2085047 (2011)

Safe Sole Distress Alarm System for Female Security Using IoT

Parth Sethi, Lakshey Juneja, Punit Gupta and Kaushlendra Kumar Pandey

Abstract The paper presents development of a sole for protection of female using Arduino microcontroller which is named as 'SAFE SOLE'. The research goal of safe sole is to develop a safety device/tool for females in the event they might face any danger. The advantages of automaticity of safe sole are hands-free user independence; it uses GPS and GSM to ping user's location automatically. The device is programmed in such a way that it recognizes defined movements as tapping forcefully three times on ground or any abnormal/vigorous movement/gait of user instantaneously and pings the distress signal and the location to relevant authorities and saved contacts, and additionally taking countermeasures—triggering of siren. In view of this, aforementioned system is a source of developing a product for general population which helps in maintaining the safety of the user by eliminating the involvement of user to initiate any procedural activities against any situation which user might feel in danger.

P. Sethi (✉) · L. Juneja
Department of Electronics and Communication Engineering,
Jaypee University of Information Technology, Waknaghat, District Solan,
Himachal Pradesh, India
e-mail: parthsethi16.ps@gmail.com

L. Juneja
e-mail: juneja.lakshey@live.co.uk

P. Gupta
Department of Computer Science, Jaypee University of Information Technology,
Waknaghat, District Solan, Himachal Pradesh, India
e-mail: punitg07@gmail.com

K.K. Pandey
Department of ECE, Central Institute of Technology, Kokrajhar, Assam, India
e-mail: kaushlendre@gmail.com

A.K. Somani et al. (eds.), *Proceedings of First International Conference on Smart System, Innovations and Computing*, Smart Innovation, Systems and Technologies 79, https://doi.org/10.1007/978-981-10-5828-8_81

1 Introduction

Women safety, today, is a very real issue and has major implications on the lifestyle of women. Even in this day and age when technology is rapidly advancing and scientific temperament is being promoted, women still feel insecure. Every day our news feed includes news on women being sexually or physically assaulted and there are many more incidents that go unreported or do not make it to the news. The corporate world is evolving and women are forming a higher percentage of the work force than they have ever before. In order to promote equality and promote their participation, we need to make sure they are safe. Women need to be able to go to work, to the grocery store, or to a party whenever they want to and not be held back by the thought that it is not a safe time to go out. The Council of Europe Convention on preventing and combating violence against women and domestic violence defines violence against women as 'Violence against women' is described as an infringement of basic human rights or any other discrimination against women and includes any gender-based violence that can result in physical, emotional, psychological, sexual, or economic harm to women. So, security has become a crucial part of society to provide a healthy and safe environment, especially for female.

To aid the security for women, several steps are being taken by various authorities and these days 'Internet of Things' (IoT) is leading to a new era and opening a new dimension in security. IoT is the huge number of interconnected devices over the internet capable of managing automated tasks smartly without any direct human involvement. As time is passing, more and more devices are being interconnected and IoT is becoming an integral part of our lives which allows us to perform the daily tasks faster and efficiently, and in ways we could not before. Ideally, IoT will optimize the future routines with intelligent and strong systems that will make our lifestyle stress-free and also adapt to the ever-growing demands and priorities. It will enable us to have completely automated systems that will have its applications in making our security arrangements intelligent, by removing the user interference with any security device. In a nutshell, IoT has the power to meet our every need before we even realize what we want and will need. Security and interconnectedness are the real power of IoT solutions. But still, several things are yet to enter this dimension of IoT and the prevailing security systems have yet to be modernized with the application of IoT, which can reduce the overhead from the user to still be the only control in any system.

Literally, security is a set of precautions or countermeasures taken to protect from harm or malicious intent. It is applicable to anything from a person to an organization which might be valuable or vulnerable. The situation is asymmetric since the 'defender' must cover all the weak points from the attacker, and on the other hand only needs to focus on a single weak point. The goal of security is to deter the attacker or someone with malicious intent from causing harm to potential victim.

Personal safety has become an issue of importance for everyone, especially for women. This issue is rising due to the increment of their work to promote equality. So having confidence to move freely with an assured safety asset on your side will promote women even more. A smart security will not even take any space but would just provide a guardian to you at all times. This is made possible by fitting the basic security services into a sole. This sole will enable the user to feel secure and protected without inconvenience of carrying a load of safety equipment. This paper proposes a design of an autonomous security system known as safe sole to fit the basic security requirements like alarm, calling, messaging and location ping autonomously in case of danger.

2 Related Work

Wearable security equipment is a new domain that need more research to have such advanced services to be fitted in normal wear but the solution may exist in IoT [2, 5, 6] which inspired us to find a new Innovative solution to track the human foot steps. Most existing related work is in the form of mobile applications and devices which are completely manual. Some examples of such existing work are as follows.

2.1 Locca

Locca [3] uses AGPS, GSM, cell triangulation, Wi-Fi, Bluetooth Low Energy and FSK for tracking. With tracking possible across international borders, it has a SIM card in it to allow it to use this wide array of tracking options. It can be used to keep track of valuables using IoT. It suffers from poor battery life compared to its Bluetooth counterparts. Using IoT, it can track different methods when needed. It can be used to track your pet if it has run away or your bike if it gets stolen.

2.2 Safer

SaferWalk [7] lets you connect with one guardian who can see the beginning and destination of your journey, and where you are when you are ‘on the way’. Its key features are as follows:

- Press the button twice to send an alert message to the guardian.
- SaferWalk: Nominate your guardians to monitor your movement while travelling.
- Press the button once to click a selfie.
- It uses Bluetooth Low Energy, with a 7-day battery life and 15 min charging.

2.3 Charm Alarm

There is also the anti-theft-focused charm alarm, which is a connected necklace that keeps tabs on your purse and wallet. It warns you if you walk too far away from your belongings or if someone tries to grab your bag and walk away. It also makes noise to attract attention to the theft.

2.4 Athena

Athena [1] provides a simple device to protect women at the click of a button. Once pressed, it emits a loud alarm and messages friends and family with the current GPS location. Working of athena is as follows:

- It is activated by holding down the button for 3 s which results in a loud alarm emitted by the device.
- Text messages the location to the saved contacts.
- The emergency contacts receive a distress text with the current location.

2.5 Spot N Save

SpotNSave is a safety wristband which pairs with the SPOT N SAVE app via LE Bluetooth 4.0. The SPOT 'N' SAVE wristband broadcasts an SMS with the user's location data to loved ones. The SPOT N SAVE band helps the wearer reach a wide network so they know where the wearer is and needs help. The user can choose up to five people to call for help and they will receive your location and turn by turn drive instructions to reach you. It also sends them the user's last known location when the user goes out of coverage or the battery runs out [8].

Functions:

- SOS
- Notifications alert
- Fitness tracker
- Anti-lost device.

Specifications:

- Bluetooth 4.0 Device
- Bluetooth range: 50 m
- Battery life: 14 days
- Time to charge: 1–2 h.

Support:

- iPhone 4S onward
- Android: All Android
- Devices that are powered by Bluetooth 4.0/Android 4.3+. Various other Existing solution for human safety are proposed by [4, 9].

3 Proposed System

As mentioned above, there are many wearable safety devices in the market. But as observed, nearly all rely on manual user input to send out an SOS which can be hard to do in situations where the user might be in distress. So we have proposed a system that in the form of a shoe sole provides a fully autonomous wearable security device that performs the necessary security functions all the while being discretely hidden in the shoe. It is a handy tool that can fit in any footwear and pair up with any modern Android mobile device. The proposed system can work autonomously, where on detecting unusual movements by the user it gets triggered or even manually using a gesture where the user can tap her heel thrice on the ground to trigger the SOS. Triggering the SOS leads execution of the distress algorithm. The algorithm flowchart as shown in Fig. 1 constantly polls the sensor readings (from the accelerometer and the pressure sensor), processes the signals and passes the processed information through a set of conditions. If the conditions are met, the device gets triggered into sending distress signals and other countermeasures—Sending the location of the user to the list of saved contacts and activating a

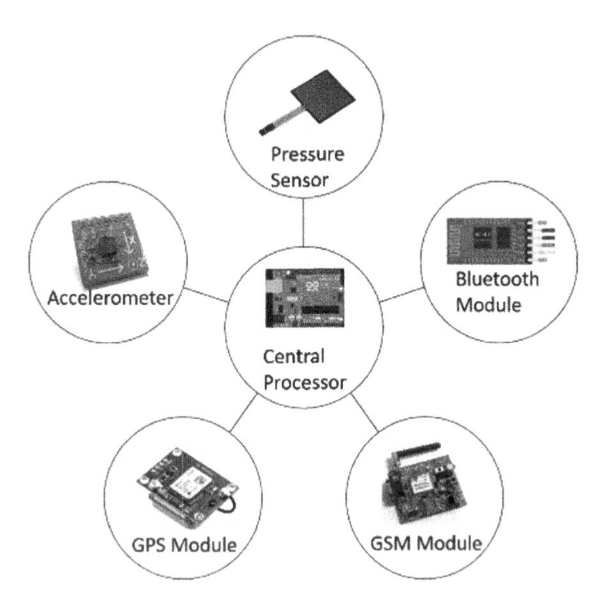

Fig. 1 Showing block diagram of safe sole

Table 1 Illustrates the hardware specifications of safe sole

Hardware specifications	
Controller	Arduino
GSM module	SIM 900A
GPS module	NEO6MV2
Pressure sensor	Pressure sensing resistor square
Accelerometer	ADXL 3-axis
Bluetooth module	HC-02

loud siren in the sole. The complete system consists of various subsystems as shown in (Table 1):

- The central controller
- The GSM module
- The GPS module
- The gesture control system
- The smartphone connecting.

3.1 Gesture Control

The gesture control works on the basis of 3-axis accelerometer built in the safe sole. Gesture sensing is done by

1. Three consecutive heel taps on the ground
2. Vigorous/abnormal movements of foot.

The tapping is obviously differentiated from any accidental taps such as while running or jumping. This is done by definitive taps on the ground from the heel which produces distinctive spike in the signal from the pressure sensor. The threshold value for triggering the distress is selected after testing for many test cases in various environments and calibrated so as to prevent an accidental triggering.

The abnormal movement in the user's feet is sensed with the help of 3-axis accelerometer. Here also the threshold rate is selected after analysing different test cases in controlled and simulated environments. The design and model of project is shown in (Fig. 2).

3.2 The Sole

The hardware consists of a removable sole that can fit in any regular size shoe with ease. It mainly consists of a battery pack that can support its functioning for a day long with a comfortable cushion padding over it, and then it consists of a three-layer structure to accommodate various modules, namely, layer 1 consists of all the

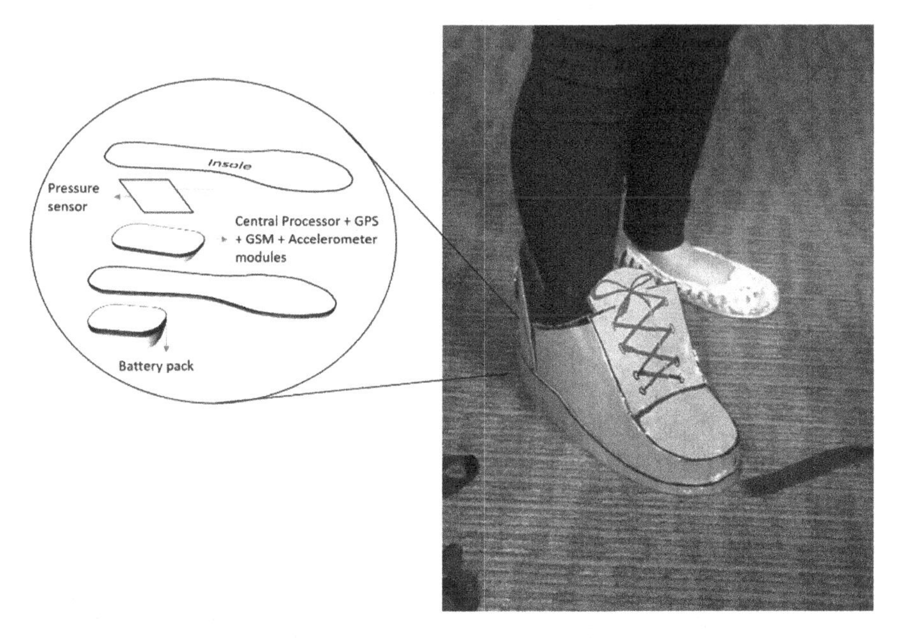

Fig. 2 Shows safe sole's deployment model

sensors (accelerometer, GSM, GPS modules), the Bluetooth module; layer 2 consists of the pressure sensor which is being put at the heal rather than the paw due to the reason that the tapping algorithm proves best if the pressure sensor is put at the heal so as to distinguish the tapping from running or jumping; and the last layer gives it a final coating sealing of all sensors and hardware, hence completing the sole structure.

The pressure sensor that is put beneath the sole is calibrated with the accelerometer to give a proper orientation and adequately evaluate the 3D motion of the foot. When tapped, readings for pressure sensor increase exponentially showing an anomaly, and the first step for this algorithm is initiated. Now, reading from the accelerometer is read and seen if the motion was due to running or any other accidental case or was it actually a tap to trigger the safety mechanisms; this step will repeat itself two more times till three taps are confirmed. Once the count reaches three, the microprocessor flags that the user is in distress and initiates the emergency procedure. The emergency procedure has three main features: sending GPS location, sound alarm and call contacts (Fig. 3).

3.3 The Distress Algorithm

The algorithm flowchart is shown in Fig. 4 constantly polling the readings from pressure sensor and accelerometer; those readings and the signals are passed to the

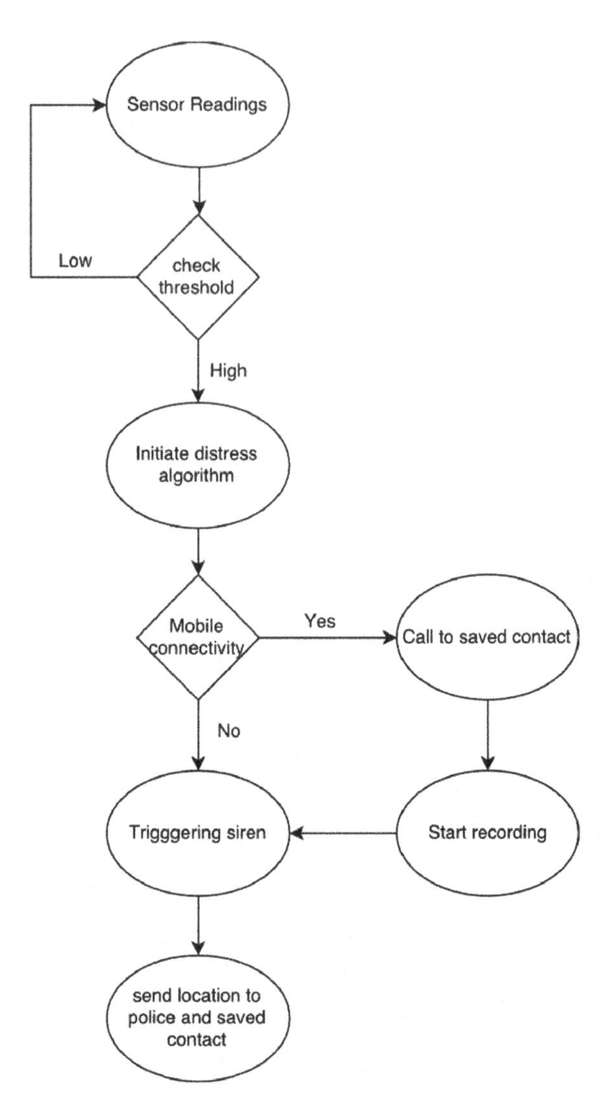

Fig. 3 Shows safe sole's flow chart

controller Arduino which processes that information through a set of conditions—intentional tapping thrice on ground or the abnormal gait of user's foot has value greater than the threshold set. If the conditions are met, the device gets triggered into sending the GPS location to saved contacts and the police station, producing a loud sound, and sole has an optional feature to pair up with an android device via Bluetooth and if paired up gives the user-added functionality of calling up the saved contacts and audio recording of the incident taking place. If there is no android device paired up, these two functionalities will be bypassed.

Fig. 4 Android app for safe sole

4 Experimental Results

The proposed smart automated system to provide a hands-free and completely autonomous security tool is being deployed and tested in numerous controlled test cases that sense the real-time circumstances where the user might feel distressed which can be differentiated with any running or jumping based on the algorithm designed specifically for this system. The system used here is Arduino board, and the sensor used for sensing the real-time occupancy is pressure sensing resistor and a 3-axis accelerometer. The system architecture of the proposed model is explained by the given below figures which include the additional phone's connectivity.

Fig. 5 Page after receiving the distress

The main page over the app is designed to connect to the sole having the buttons to add caller's ID to be called and enable or disable the recording option. The recording starts automatically triggered if the recording of the respective radio button is selected. And the saved contact is automatically called. The call button asks to enter a contact whom to call (Fig. 5).

Figure 6 shows the call page where user has to enter the distress number. Record button confirms whether to record or not. The final product was set up in shoes and was checked while jogging and skipping which did not trigger any false alarm.

Fig. 6 Contact settings button interface

5 Conclusion

Safe sole as described in the proposed system has proved quite successful in order to provide the required set of facilities to the user. The main idea of this system is to provide basic security facilities to any user and at the same time it should be hidden from the attacker, and this proposed system meets all those arguments. The arrangements made in context to trigger the alarm send the distress message proved to be working instantaneously with the different variations of inputs like the pre-planned situations where the user might feel distress. And there was almost no deviation from the actual gesture described and any accidental tap or movement of feet in any sense. All the false alarms if triggered could be cancelled easily with the help of the application.

References

1. Athena safety jewelry: http://www.roarforgood.com/.
2. A. Gupta, P. Gupta, and J. Chhabra, "IoT based power efficient system design using automation for classrooms", pp. 285–289.
3. Locca: http://www.locca.com/.
4. Meyersfeld, Bonita C. "Council of Europe Convention on preventing and combating violence against women and domestic violence–Introductory note by". International legal materials 1 (2012): 106–132.
5. Pham Hoang Oat, Micheal Drieberg, Nguyen ChiCuong, "Development of Vehicle Tracking System using GPS and GSM Modem", 2013 IEEE Conference on Open Systems (ICOS), pp. 89–95.
6. M.F. Saaid, M.A. Kamaludin, M.S.A. Megat Ali, "Vehicle Location Find Using Global Position System and Global System for Mobile", 2014 IEEE 5th Control and System Graduate Research Colloquium, pp. 279–285.
7. SAFER: https://www.leafwearables.com/.
8. Spot n Safe: http://www.spotnsave.com/.
9. G. Toney, F. Jabeen, S. Puneeth, "Design and Implementation of Safety Armband for Women and Children using ARM7", International Conference on Power and Advanced Control Engineering (ICPACE), pp. 300–303.

QoS Aware Grey Wolf Optimization for Task Allocation in Cloud Infrastructure

Punit Gupta, S. P. Ghrera and Mayank Goyal

Abstract Cloud computing is a reliable computing platform for large computationally intensive or data-intensive tasks. This has been accepted by many industrial giants of the software industry for their software solutions; companies like Microsoft, Accenture, Ericson, etc. have adopted cloud computing as their first choice for cheap and reliable computing. By increasing the number of clients adopting this, there is a requirement of much more cost-efficient and high-performance computing for more trust and reliability among the client, and the service providers to guarantee cheap and more efficient solutions. So the tasks in cloud need to be allocated in an efficient manner to provide high resource utilization and least execution time for high performance, and at the same time provide least computational execution cost. We have proposed a learning-based grey wolf optimization algorithm for task allocation to reduce the request time and scheduling time to improve QoS (Quality of Service). Proposed algorithm has been inspired from behaviour of wolfs hunting in real world with a unique technique which they have evolved from long evolution and learning cycle

Keywords Cloud computing · Reliability · Learning-based algorithm
Grey wolf · Cloud IaaS · Makespan

P. Gupta (✉) · S. P. Ghrera
Department of Computer Science Engineering, Jaypee University of Information Technology, Waknaghat, Himachal Pradesh, India
e-mail: Punitg07@gmail.com

S. P. Ghrera
e-mail: s.ghrera@redifmail.com

M. Goyal
Department of Computer Science Engineering, Amity University, Noida 201301, Uttar Pradesh, India
e-mail: mayankrkgit@gmail.com

A.K. Somani et al. (eds.), *Proceedings of First International Conference on Smart System, Innovations and Computing*, Smart Innovation, Systems and Technologies 79, https://doi.org/10.1007/978-981-10-5828-8_82

1 Introduction

Cloud computing has been a new trend in problem-solving and providing a reliable computing platform for big and high computational tasks. This technique is used for business industries like banking, trading and many e-commerce businesses to accommodate high request rate, high availability for all time without stopping system and system failure. In case of failure, the requests are migrated to a different reliable server with letting the user knowing about it, providing fault-tolerant behaviour of system. Other application zone includes scientific research like computing weather forecasting report, satellite imaging and many more applications which require high computation, which is now possible without creating a private infrastructure. Cloud deals with various kinds of application which can have different request types and computation servers based on their capability, i.e. hardware and software configurations. However, to compute these large requests, count datacenters consume high power, and load over the datacenters is also very high which may be stored as computational or network load. This may lead to reduce QoS (Quality of Service) provided by a datacenter that may be due to deadline failure or a fault over a datacenter due to various reasons. Survey in 2006 shows the power consumption of servers around 4.5 billion kWh, equivalent to a large portion of total power consumed by USA, which may increase by 18% yearly [1].

Cloud computing, in general, has various issues listed below:

(1) With adoption of cloud computing techniques by industries and corporate users, the user count is increasing rapidly with increase in a number of cloud computing services. This increases the datacenter count and power consumption. (2) Resource allocation is among data centres, i.e. allocation of virtual machine on datacenters to provide high-quality resource keeping in mind the behaviour and characteristics of datacenters to provide high QoS at resource level. (3) Current task allocation algorithms focus on static load balancing algorithm to balance the load when request load increases but does not take into consideration previous behaviour of the datacenter under high load which leads us to design efficient learning-based algorithms. (4) High loaded data centre has high failure probability to compute requests, and with an increase in load over data centre request completion time increases, which degrades the performance of cloud. This leads to SLA (Service Level Agreement) failure promised to the client.

Proposed algorithm emphases on learning base strategy to find a global appropriate solution which cannot be achieved using static algorithm like max–min, min–min and ant colony optimization to improve request failure count, make span and overall reliability of the system with increase in QoS and SLA promised to users/clients.

2 Related Work

Heuristic-based task allocation in cloud environment is very new and requires a lot of refinement to optimize the system performance and quality of service provided to the user. Various task allocation algorithms have been proposed to solve and optimize the problem of task allocation. Many of these algorithms are inspired by basic computational algorithm and physical phenomena, animal behaviour or universe evolution concepts. In this section, we have discussed some of those existing algorithms.

Nathani [2] has proposed a deadline-based resource allocation and load balancing algorithm for cloud Infrastructure to fulfil the request within the deadline without deadline failure. Author has proposed a method to schedule the task taking into care the deadline and start time of request. Migration of tasks is done by rescheduling the task and swapping the task to idle server known as back filling. Gao [3] proposed an ant colony-based task allocation algorithm, which is inspired by ant behaviour in the real world. Author has proposed a method to update the pheromone value on the basis of resource available and power consumption of the host. The proposed system improves the makespan of system. Dasgupta [4] proposed a load balancing algorithm for task allocation in cloud environment inspired by the genetic evolution of the universe to find the fittest survivor and only the fittest will survive the next generation of evolution. Proposed algorithms perform a mutation for set of iteration to evolve and find a global best solution. Algorithm fitness value is based on the task deadline and task size. Quang-Hung [5] has improved and proposed a genetic algorithm for task scheduling in cloud environment. Proposed algorithm has used a power model based on utilization of system and MIPS of virtual machine. Experimental studies show that the proposed algorithm has improved the power efficiency over existing static scheduling algorithm. PSO (particle swarm optimization)-based task scheduling algorithm is another algorithm evolved from animal behaviour to design a scheduling algorithm for cloud [6]. Abrishami [7] has presented a technique to schedule a task workflow based on deadline in cloud infrastructure environment. The importance of this work is to take the various deadline parameters into consideration to evaluate final deadline value. The parameters are as follows: Earliest Start Time, Latest Finish Time and execution time. The evaluation of deadline at the real time helps to schedule the task in much efficient manner with least deadline failure. Suraj [8] proposed an adaptive genetic algorithm for task allocation with least execution time and high resource utilization. Fitness function used in this algorithm is a function of utilization of online request under execution and the execution cost of upcoming request to find the fittest host for requests. So to find a global best solution rather than sticking in local minima and improve the scheduling time, we have proposed a grey wolf-based task scheduling algorithm to improve the quality of service of system by improving makespan and scheduling time. Since GA can also provide global solution, the execution time to find global best solution is too high. To study grey wolf behaviour and their hunting style to achieve high optimization, we have used a study by Mirjalili [9] and [13] to study the importance of algorithm in software engineering. Many other proposed huristic based algorithms [11–13] are also been reviewed. The solution is being used to optimize many problems in software engineering like searching algorithm, finding fittest solution and much more. In the next section, we have proposed an algorithm, which will deal with such problems, and requests have to be completed with higher resource utilization, least make span and least scheduling delay.

3 Proposed Model

In the above section, we have discussed various existing algorithms from the field of task allocation and load balancing algorithm. But the existing algorithm is static or dynamic in nature, and they may suffer from local minima solution considering that as the best solution. But a better solution for task allocation may be possible. So to overcome these learning-based algorithms, genetic algorithm and PSO (particle swarm optimization) were proposed. The issue with these algorithms is that they have very large time complexity, moreover, they depend upon the iteration and the initial population size, which affects their solution. If the population size of the iterations/generation is less, then there is less probability to get best solution. Moreover, if we talk about genetic-based algorithm, they may provide you global solution but at a cost of high scheduling time due to high count of iterations involved while scheduling. To overcome these problems and find the same best solution of less time complexity, we have proposed a task allocation algorithm based on grey wolf optimization algorithm (GWO). The algorithm is inspired by hunting behaviour of wolfs to let only the fittest wolf which can hunt the animal with highest success probability. Grey wolfs are considered to be a predator who lies high in the food chain and an animal which hunts in group that can have a group size of 10–12 on average. They follow a social chain of hierarchy for their hunting and living. There are four types of wolfs in the hierarchy as shown in Fig. 1; most dominating is α “alpha” that can be a male or a female wolf. They are responsible for decision-making and hunting. The dominance decreases the hierarchy to move from β –> δ to δ –> ω. Beta wolf is responsible for helping alpha leader in decision-making indirectly. Beta is also responsible for reinforcing the alpha leader command to the group and also to send the feedback in case of improvement to alpha. Delta and omega wolves are the wolves which are responsible to define the range or the boundary for the hunting animal. They also

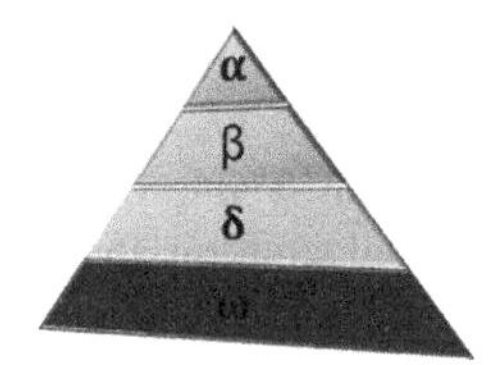

Fig. 1 Grey wolf hierarchy

act as scout to find and watch the food, watching the boundary of territory. The animal to be hunted down is said as "Prey". Finally, all wolves surrounding the prey help alpha and beta wolves to find the weak location to attack and the weak point of prey. Attack is always executed by alpha wolves, because they are considered to be the best in hunting. To find the best location, alpha wolf changes location to attack the prey and changing location is a set of iteration in order to find the fittest and best of them. Hunting process of grey wolf is a social behaviour and takes into consideration the hunting strategy and social hierarchies of wolves. Hunting strategy is divided into three phases as follows:

1. Tracking, chasing and catching the prey.
2. Pursuing, inclosing and harassing the prey until it stops any activity.
3. Hunting and killing the prey.

These phases are used to design the task allocation algorithm for cloud infrastructure.

Proposed algorithm uses Poisson probability distribution for random request at virtual machine, i.e. at host and datacenter level. Based on computing capability of a system, we have proposed a task allocation policy to minimize the total span over the system and reduce time complexity of solution. According to algorithm, collect the information of data centre resources and its capability. Proposed algorithm is similar to genetic algorithm (GA), but the problem size reduces after each phase and will give you a single point solution, i.e. the global solution. But in existing GA the population size remains the same and there is no guarantee that the global best is achieved.

Proposed algorithm is divided into four phases which are as follows:

I. Initialization phase,
II. Tracking, chasing the prey,
III. Harassing the prey, and
IV. Hunting the prey.

I. Initialization

In this phase, we have a set of tasks (T1, T2, T3, T4, T5, T6, …, Tn) and a set of resources in terms of virtual machine (VM1, VM2, VM3, VM4, VM5, …, VMm) which are pre-allocated on hosts in distributed datacenters. Here, we initialized a set of sequences or schedules allocated randomly, each sequence acts as a wolf for proposed algorithm. The complete set of wolfs or schedules acts as an input for algorithm. The prey is to find the best makespan to schedule the task set.

II. Tracking, chasing the prey

In this phase, the grey wolf encircles the prey for hunting. The wolf includes a set of alpha, beta and delta wolves. To simulate this behaviour mathematically, the equations are as follows:

$$D = |C \cdot Xp(t) - X(t)| \tag{1}$$

$$X(t+1) = X_p(t) - A \cdot D, \tag{2}$$

t	The current iteration.
A & C	Coefficient vectors.
Xp	Fitness value of the wolf with respect to its position from prey.
X	The final fitness value of the wolf in the next iteration.

The vector A and C can be evaluated using the following equations:

$$A = 2a \cdot r1 - a \tag{3}$$

$$C = 2 * r_2, \tag{4}$$

r1 and r2	Random vectors ranging from 0 to 1.
a	Linear vector with decreasing value from 2 to 0, each iteration linearly.

Here, Xp can be defined as a function of total execution time of schedule:

$$Xp(t) = \sum_{i=1-n} \frac{T\_Length_i}{VM\_MIPS_i}, \tag{5}$$

VM_MIPS i	MIPS of ith virtual machine.
T_Leng i	Length of ith Task.

III. Harassing the prey

In this phase, the wolf encircles the prey because they have the ability to find the approximate location of prey. The hunting process starts with the alpha, beta and delta wolves. The process is dominated by alpha wolf. Wolf starts a search since in starting no wolf has the best solution. The wolf starts their hunting strategy using three steps.

1. Evaluate the distance from the prey.
2. Evaluate first three best solutions, given alpha, beta and delta wolves.
3. Alpha wolf changes the position with a wolf having least distance from the prey. The wolf whose position can be alpha or any other wolf.
4. The wolf whose position is interchanged is moved out from hunting.
5. The process repeats until the least distance is reached or <1.

To simulate this behaviour mathematically, we need to evaluate the fitness value of all wolves. The top three best solutions are with least fitness value. They are stored as $X\alpha(t)$, $X\beta(t)$ and $X\delta(t)$. X(t) is the solution from previous iteration which for the first iteration is the minimum fitness value. The steps are followed by evaluating $D\alpha$, $D\beta$ and $D\delta$ using Eqs. 6–8 with C1, C2 and C3 and random value between (0, 2). Evaluate X1, X2 and X3 and evaluate X(t + 1) using Eqs. 9–11. Evaluate the fitness value for new evolution x(t + 1). Mutate the alpha, beta and delta solutions with the wolf having the fitness value near X(t + 1) with $X\alpha(t)$, $X\beta(t)$ and $X\delta(t)$ wolves. Mutation refers to interchanging few random task allocations between two schedules and finding a new better solution. New $X\alpha1(t)$, $X\beta1(t)$ and $X\delta1(t)$ are used to iterate new solution:

$$D\alpha = |C1 \cdot X\alpha(t) - X(t)| \quad (6)$$

$$D\beta = |C2 \cdot X\beta(t) - X(t)| \quad (7)$$

$$D\delta = |C3 \cdot X\delta(t) - X(t)| \quad (8)$$

$$X1 = X\alpha(t) - A1 \cdot (D\alpha) \quad (9)$$

$$X2 = X\beta(t) - A2 \cdot (D\beta) \quad (10)$$

$$X3 = X\delta(t) - A3 \cdot X(D\delta) \quad (11)$$

$$X(t+1) = \frac{(X1 + X2 + X3)}{3}. \quad (12)$$

The process repeats unless only alpha wolf is left or the there is no improvement in fitness value.

IV. Hunting the prey

In this phase after evaluating the stem III with several iterations equals to a number of wolves, in each iteration, alpha wolf replaces with best beta or delta wolf and the total wolf count reduces by one. When the wolf count remains one, i.e. the alpha wolf has reached the best solution and attacks the prey with the highest probability of prey being killed, in theoretical view, it is said that the wolf stops when prey stops moving in the simulation model, and it is simulated that when fitness value after each iteration remains same, no improvement is achieved stop changing positions. If they are still achieving improvement in that, we can keep on iterating until only alpha wolf is left.

Proposed algorithm

Proposed algorithm provides a benefit over existing static scheduling algorithm, which it can search for the best global solution rather than assuming the local best solution as the best solution. Moreover, the proposed algorithm takes into consideration the faulty behaviour of cloud, which helps in finding a solution with similar high utilization, least complexity, high reliability and less time complexity as compared to genetic algorithm (Figs. 2, 3 and 4).

Fig. 2 Proposed algorithm phase 1

Initialization

```
Algorithm:-GWA (VM List VM i , Task list Ti, Wolf Count size
Po)
//Input : Po, VM i, Itr and Ti
1. VMi← VM List() ;
2. i ←No. of VM
3. Ti← Task List();
4. C ← GWA_algo(Vmi ,Ti, Po);
5. Allocate Resource( C);   // processing the client request.
7. End
```

Fig. 3 Tracking, chasing the prey phase

Chasing the Prey

```
GWA algo(VMi ,Ti, Po)
//Input : Po, VM i, Ti
1. Po ← Initiate ( Ti,) ;
2. Xp ← Evaluate fitness(Po);
3. Xα(t) ← getFittest1();
4. Xβ(t) ← getFittest2();
5. Xδ(t) ← getFittest3();
6. Harassing()
7. End
```

Fig. 4 Harassing the prey phase get fittest

EHarassing the prey

```
1.  Harassing(){
2.  If( round==1)
3.  { X(t)= Xδ(t) ;}
4.  Else{
5.  Dα =mod (C1·Xα(t)−X(t))
6.  Dβ =mod(C2·Xβ(t)−X(t))
7.  Dδ =mod(C3·Xδ(t)−X(t))
8.  X1 = Xα(t) − A1·(Dα)
9.  X2 = Xβ (t) − A2·(Dβ)
10. X3 = Xδ (t) − A3·X(Dδ)
11. X(t+1)=x1+x2+x3/3
12. If ((A1 || A2 || A3) <1)
13.  {Allocate Resource(x(t+1));
14.  }
15. Else
16.  { mutate(Xα(t) , x(t+1));
17.    mutate(Xβ (t) , x(t+1));
18.    mutate(Xδ (t) , x(t+1));
19.    X(t)=X(t+1);
20.    Harassing();
21.  }
22. } } }
```

4 Experimental Result

For simulation, we used CloudSim 3.0 [10] cloud simulation toolkit. We have simulated and proposed grey wolf task allocation algorithm in CloudSim which is compared with the existing genetic algorithm. The proposed solution is simulated and tested over various simulation environments with 10 servers D0–D9 and Poisson distribution model for random request in distributed environment. Proposed algorithm is compared with a basic genetic algorithm proposed by Suraj [8]. Simulation is done for various size requests with population size of 100. Iteration for simulation of each simulation is 100. Figures 5 and 6 compare the improvement in execution time with increase in a number of requests over the system and population size in terms of genetic algorithm and count of wolfs attacking for grey wolf algorithm. Execution time has reduced over the proposed system in increase in completed requests over the system. Figure 7 discourses the improvement in

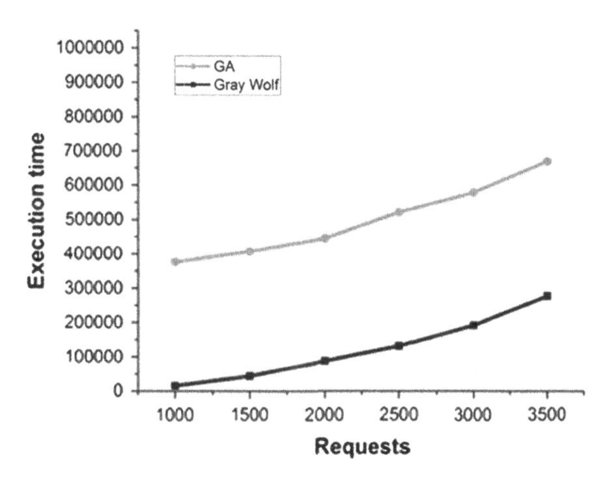

Fig. 5 Comparison of improvement in scheduling time

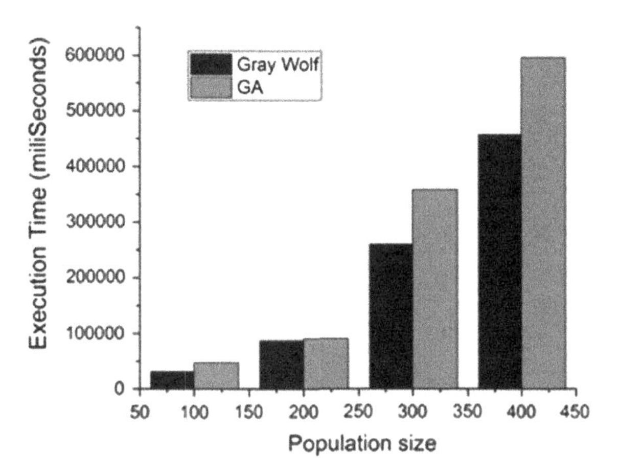

Fig. 6 Comparison of improvement in execution time with changes in population size

average start time with an increase in a number of request over the system, which shows that the proposed algorithm proves to provide better start time than the conventional genetic algorithm. Figure 8 discourses the improvement in average finish time which reduces with increasing requests over the system; the experiment has been performed over 1000, 1500, 2000, 2500, 3000 and 3500 request counts. Proposed algorithm proves to provide reduced finish time as compared to existing genetic algorithm.

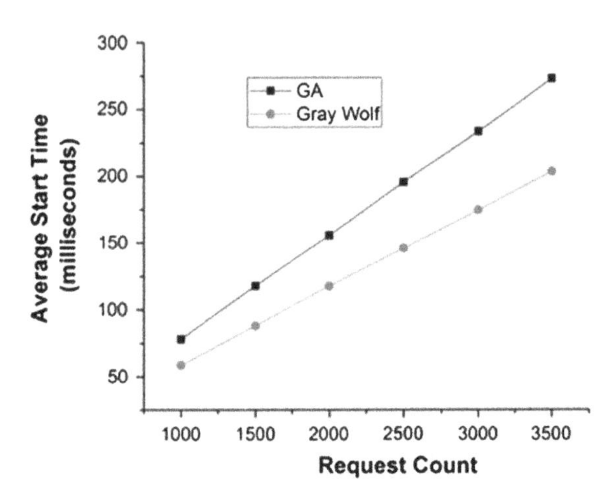

Fig. 7 Comparison of average start time of system with increase in request count

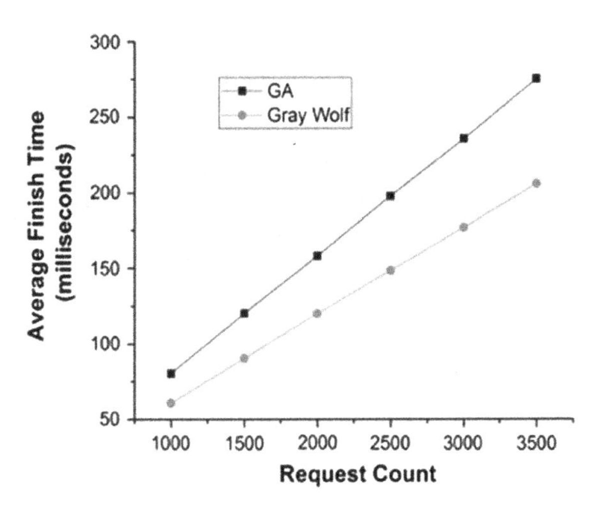

Fig. 8 Comparison of average finish time of system with increase in request count

Table 1 Simulation environment

Server	RAM (Mb)	MIPS	Storage (Gb)	Core	PE	HOST
D0	2000	10000	100000	4	6	2
D1	2000	10000	100000	4	6	2
D2	2000	10000	100000	4	6	2
D3	2000	10000	100000	4	6	2
D4	2000	10000	100000	4	6	2
D5	2000	10000	100000	4	6	2
D6	2000	10000	100000	4	6	2
D7	2000	10000	100000	4	6	2
D8	2000	10000	100000	4	6	2
D9	2000	10000	100000	4	6	2

5 Conclusion

From the above results, it is clear that proposed grey wolf optimization algorithm provides better QoS (Quality of service) as compared to the previously proposed GA algorithm. The aim of this algorithm is to complete the maximum number of requests with least execution time; proposed algorithm showed that it can provide better execution time over large requests which reduces average start time and average finish time over the system. Proposed algorithm reduces the number of iteration required to achieve a global best solution with least scheduling time. This strategy has proven that it provides better QoS in terms of high reliability with increase in a number of requests and resources with least scheduling time with decrease in execution time with increase in population size and number of requests. Proposed algorithm ensures that the schedule achieved is global best solution (Table 1).

References

1. B. Richard. "Report to congress on server and data center energy efficiency: Public law 109-431". Lawrence Berkeley National Laboratory, pp. 0–10, 2008.
2. Nathani, Amit, Sanjay Chaudhary, and Gaurav Somani. "Policy based resource allocation in IaaS cloud". Future Generation Computer Systems 28.1 (2012): 94–103.
3. Gao, Y., Guan, H., Qi, Z., Hou, Y., & Liu, L. (2013). A multi-objective ant colony system algorithm for virtual machine placement in cloud computing. Journal of Computer and System Sciences, 79(8), 1230–1242.
4. Dasgupta, K., Mandal, B., Dutta, P., Mandal, J. K., & Dam, S. (2013). A genetic algorithm (GA) based load balancing strategy for cloud computing. Procedia Technology, 10, 340–347.
5. Quang-Hung, N., Nien, P. D., Nam, N. H., Tuong, N. H., & Thoai, N. (2013). A genetic algorithm for power-aware virtual machine allocation in private cloud. In Information and Communication Technology (pp. 183–191). Springer Berlin Heidelberg.
6. Jena, R. K. (2015). Multi Objective Task Scheduling in Cloud Environment Using Nested PSO Framework. Procedia Computer Science, 57, 1219–1227.

7. Abrishami, S., & Naghibzadeh, M. (2012). Deadline-constrained workflow scheduling in software as a service cloud. Scientia Iranica, 19(3), 680–689.
8. Suraj, S. R., and R. Natchadalingam. “Adaptive Genetic Algorithm for Efficient Resource Management in Cloud Computing”. International Journal of Emerging Technology and Advanced Engineering 4.2 (2014): 350–356.
9. Mirjalili S, Mirjalili SM, Lewis A. Grey wolf optimizer. Advances in Engineering Software. 2014 Mar 31;69:46–61.
10. Calheiros, R. N., Ranjan, R., Beloglazov, A., De Rose, C. A., & Buyya, R. (2011). CloudSim: a toolkit for modeling and simulation of cloud computing environments and evaluation of resource provisioning algorithms. Software: Practice and Experience, 41(1), 23–50.
11. Bhoi, U., & Ramanuj, P. N. (2013). Enhanced max-min task scheduling algorithm in cloud computing. International Journal of Application or Innovation in Engineering and Management, 2(4), 259–264.
12. Gupta, P., & Ghrera, S. P. (2015). Load and Fault Aware Honey Bee Scheduling Algorithm for Cloud Infrastructure. In Proceedings of the 3rd International Conference on Frontiers of Intelligent Computing: Theory and Applications (FICTA) 2014 (pp. 135–143). Springer International Publishing.
13. Komaki, G. M., & Kayvanfar, V. (2015). Grey Wolf Optimizer algorithm for the two-stage assembly flow shop scheduling problem with release time. Journal of Computational Science, 8, 109–120.

Stable Period Enhancement for Zonal (SPEZ)-Based Clustering in Heterogeneous WSN

Pawan Singh Mehra, M.N. Doja and Bashir Alam

Abstract A lot of energy is consumed in organizing the wireless sensor network. Several cluster-based protocols have been proposed to increase the lifespan of the network. The objective of WSN is the lifetime enhancement through energy efficiency. Clustering not only enhances the lifetime but also makes the network scalable and energy efficient. In this proposed work (SPEZ), we have divided the network into zones and randomly deployed the sensor nodes. Extraneous energy is supplied to some of the nodes resulting in heterogeneity. Energy-efficient approach is applied so as to increase the stability span which plays a vital role in many scenarios. Simulation and evaluation of performance validate the elongated lifetime in terms of the stable period in comparison with ZSEP, SEP, LEACH, and DEEC protocols.

Keywords Enhanced lifetime · Wireless sensor network (WSN) Heterogeneity · Clustering · Energy consumption

1 Introduction

WSN has a huge number of sensors placed in deterministic or random fashion. These small sensor devices are capable enough to collect information from the surrounding with each other cooperation. A small sensor consists of a microcontroller for processing, transceiver for transmission and reception of data, and battery-based power supply and memory for storage. The size of these sensor nodes

P.S. Mehra (✉) · M.N. Doja · B. Alam
Computer Engineering, Faculty of Engineering & Technology, Jamia Millia Islamia, Jamia Nagar, Delhi 110025, India
e-mail: pawansinghmehra@gmail.com

M.N. Doja
e-mail: ndoja@yahoo.com

B. Alam
e-mail: babashiralam@gmail.com

A.K. Somani et al. (eds.), *Proceedings of First International Conference on Smart System, Innovations and Computing*, Smart Innovation, Systems and Technologies 79, https://doi.org/10.1007/978-981-10-5828-8_83

depends upon the application, e.g., tiny for surveillance and large for environmental monitoring. The cost of sensor nodes is dependent on storage, processing power, and power supply [1]. Sensor nodes gather the physical parameters from surrounding through sensing and forward to base station. This forwarding can be single hop or multiple hops. Mostly, single hop is least preferred as it dissipates sensor nodes energy faster. Clustering considerably dissipates less energy of individual sensor nodes as compared to single-hop transmission [2]. The mechanism of clustering reduces the decision-making task of individual sensor node to great extent. The clustering technique can be considered as hierarchical routing which enhances efficiency of the WSN in terms of scalability as well as communication. Hierarchical routing can be single hop or multiple hops. Both the communication mechanisms have their merits and demerits. For example, increase in distance leads to more energy loss of individual sensor node in single-hop transmission [3], whereas multiple hops can lead to energy hole problem. In every case, the network lifetime relies on the limited power supply be it network lifetime or lifetime of individual sensor node. Thus, the pertinent concern is well-organized use of energy of the sensor nodes. Some of the popular schemes in WSN are [4–10]. Although several researches have been proposed in energy efficiency issues, there are many parameters with which the analysis of wireless sensor network can be done, for example, deployment of sensor nodes, packet delivery ratio, scalability, stability, connectivity, lifetime, reliability, latency, energy consumption minimization cost, etc. [11]. The categorization of wireless sensor network can be non-cluster and cluster-based. The latter can be further divided into homogeneous and heterogeneous environments. When we talk about homogeneous environment, we mean all the sensors have same capability in terms of energy, link-communication, and processing power, whereas in heterogeneous environment, sensor nodes differ from one another's capability. A variety of sensors can be deployed to meet the requirement of particular WSN to perform diverse tasks, e.g., to gather sound waves, images from the area, and physical phenomena in the surrounding. The ultimate aim of the network is to accumulate and fuse the data at one level and forward it to next level for requisite action.

In this proposed work, we have chosen the heterogeneity-based sensor nodes which differ from each other in terms of energy level. The network is divided into three zones comprising sensor nodes with two energy levels. The sensor is deployed randomly in each zone. Since long-distance transmission dissipates more energy, the sensor nodes far from base station are equipped with more level of energy in comparison to sensor nodes nearby the base station. The ultimate aim is to maximize the stability period. Rest of the literature is organized as follows. In Sect. 2, pertinent proposals are discussed. In Sect. 3, the heterogeneity-based network model is discussed. Section 4 discusses the proposed protocol. Evaluation of the proposed protocol is done in Sect. 5. Conclusion is discussed in Sect. 6.

2 Pertinent Work

Clustering in LEACH [4] algorithm minimizes the expenditure of energy of nodes in the field. LEACH employs random rotation of the role of cluster head to balance the load distribution in homogenous environment. Every sensor node has equal probability to be chosen to play the role of coordinator of a cluster. PEGASIS [6] protocol is chain based on avoiding the formation of cluster. In this protocol, one of the nodes in the chain forwards the data to sink/base station instead of involvement of multiple nodes thereby conserving energy. TEEN [7] protocol is based on two thresholds: soft threshold and hard threshold. Cluster head broadcasts this threshold to member nodes. If the value sensed by the node is higher than the hard threshold, then the data is forwarded, else no transmission takes place. TEEN protocol is unsuitable for periodic data collection application. Adaptive threshold sensitive energy-efficient sensor network [8] gathers data periodically and reacts to events which are time critical. Complexity of cluster formation in multiple levels and complexity of implementation of threshold-based functions make it cumbersome. SEP [9] is a heterogeneity-based protocol with two energy level nodes, i.e., normal and advanced nodes. This proposed work discusses the effects of instability and heterogeneity in WSN. In this scheme, each sensor node is selected by its weight for each round. The author ensures load balancing, and energy level of sensor node is put under consideration. Additional energy is provided to advanced nodes which in turn is considered during the calculation of weighted probability of each node. This protocol prolongs the lifetime but remnant energy is not under consideration. Qing et al. proposed DEEC [12] for multiple levels of energy of sensor nodes for WSN. In his work, cluster head selection depends upon network's remnant energy and average energy of nodes in network. This proposed work emphasizes on load balancing. ZSEP [13] is based on SEP. In ZSEP, the targeted area is divided into three zones. Normal nodes are kept nearby the base station, and for those zones which are farther from base station, advanced nodes are placed. It incorporates direct transmission for normal nodes and cluster-based transmission for advanced nodes which are far away from the base station.

3 Heterogeneity-Based Network Model

In this protocol, the field size is 100×100 and hundred sensor nodes are placed randomly. The sensor nodes location is not known and can be computed with RSSI. The sensor nodes deployed in the field are presumed to be immobile, i.e., the topology is fixed. Optimum probability for a count of cluster heads (p_{opt}) is ten percent of the total sensor nodes alive in the network.

As discussed in [14], heterogeneity in WSN is classified as energy, link, and computational. Link heterogeneity is concerned with the connection for reliable and broader bandwidth for communication. Computational heterogeneity deals with higher processing capability and huge data storage. Energy-based heterogeneity provides extraneous power to the sensor nodes. As compared to homogeneous model, the energy exhaustion in heterogeneity-based model is minimum and communication links are more reliable [15].

In this paper, we have used bi-level energy model. There are some nodes (m %) which have additional power and are named as advanced nodes. Rest of the sensor nodes in the field are termed as normal sensor nodes. The energy level of the nodes during the deployment is kept E_o and the advanced nodes are provided with α times additional power as compared to normal nodes, i.e., $[E_o + \alpha]$. The total energy available with the network with J nodes at the time of deployment can be determined by

$$E_{total} = (1-m)JE_o + JmE_o(1+\alpha) = JE_o(1+m\alpha). \quad (1)$$

For analysis of dissipation of energy, energy model used in [9] is considered. In order to transmit 1 bit to other node placed at distance d, the amount of energy expenditure by the transceiver of the node is calculated by

$$E_{Tx}(l,d) = \begin{cases} lE_{elec} + l\varepsilon_{fs}d^2, & d < d_o \\ lE_{elec} + l\varepsilon_{mp}d^4, & d \geq d_o \end{cases} \quad (2)$$

E_{elec} is the amount of energy exhausted per bit by the receiver and transmitter circuitry, and $\varepsilon_{mp}d^4$ and $\varepsilon_{fs}d^2$ depend upon the in-between distance. The amount of energy expenditure by cluster head is determined by

$$E_{CH} = n(lE_{elec} + l\varepsilon_{fs}d^2 + E_{DA}), \quad (3)$$

where E_{DA} is the energy needed for data aggregation and n is the total count of members of the clusters. The base station position is presumed to be at the center of the field. The nodes in the field generate reports on regular basis for the same event. Every node in the field is capable of adjusting the transceivers communication range to minimize the unwanted long transmission and each node has the capability to communicate every other node deployed in the field. As in [14], sensor nodes are capable of directional propagation for more energy efficiency.

4 SPEZ Protocol

In this section, the proposed protocol SPEZ is discussed. The clustering algorithm has huge impact on network lifetime and performance. Premature death of sensor nodes at distant place from base station decreases the efficiency of network. This issue was addressed in [13]. In ZSEP protocol, the field is divided into three zones, viz., Zone0 where normal nodes are placed randomly, and Head Zone1 and Head Zone2 where advanced nodes are placed randomly. Sensor nodes in Zone 0 directly transmit the data, whereas nodes in Zone 1 and Zone 2 form cluster and then transmit the data.

In this protocol, there are two stages: topology setup stage and forwarding stage. A BS_MSG packet which contains the position of the base station is broadcasted to the field so as to provide the position of the base station to the deployed nodes. In the topology setup phase, to group the nodes into clusters, each alive node in the field finds its probability to proclaim its candidature for cluster head by the given formula:

$$CHSF_i = \frac{E_i(r) \times \omega_i(r) \times \delta_i}{Ch \times \Psi i}, \tag{4}$$

where $Ei(r)$ is the residual energy level, $\omega_i(r)$ is the average reachability transmission power to neighbor nodes within transmission range (Rc), and node density is denoted by δ_i. Ψi is the distance from sensor node to the base station, and Ch is the count of sensor node already chosen as cluster head. Each node broadcasts its Cluster Head Selection Factor ($CHSF_i$) within its vicinity and the sensor node with highest $CHSF_i$ is chosen to be the cluster head. Soon after being chosen as cluster head, the node broadcasts a join message to every node inside the transmission range (Rc) of the sensor node. The nodes having lower $CHSF_i$ send a message to the cluster head node to join the cluster. In this manner, p_{opt} clusters are formed. Once all the nodes of the network are assigned a CH, the topology setup stage comes to an end. After the first stage, the second stage, i.e., forwarding stage, comes into play. The physical phenomena in the field are sensed by the deployed sensors and the data is forwarded to their cluster head as per the TDMA slot allotted to the nodes. The cluster head after collection of data from every member fuses the data so as to minimize the redundancy and communication cost. The base station now collects the fused data from every cluster head in accordance with the TDMA schedule provided to them by the base station so that collision-free communication can take place. In this way, one round gets completed by the proposed algorithm.

5 Evaluation of Performance

We will discuss comparative study of the proposed work (SPEZ) with ZSEP, SEP, DEEC, and LEACH. More than hundred simulations have been carried out to find the mean values of results. Evaluation of the proposed work is done on the following factors: stability period, half node death, packets transmitted to base station, and number of dead nodes (nodes with zero energy level) in the field in each round. The more the number of successful transmission, the more the information collected at the base station. Enlarged stability period depicts balanced energy squandering by the sensor nodes in the network. For the simulation work, the parameters and their values used are shown in Table 1.

Table 1 Radio model parameters that are used in simulation [13]

Parameters	Symbol	Value
Energy dissipation when d < d_o (d is shorter distance)	ε_{fs}	10 pJ/bit/m^2
Energy dissipation when d > d_o (d is longer distance)	ε_{mp}	0.0013 pJ/bit/m^4
Energy consumption for reception/transmission	E_{elec}	50 nJ/bit
Energy dissipation for data aggregation	E_{DA}	5 nJ/bit/report

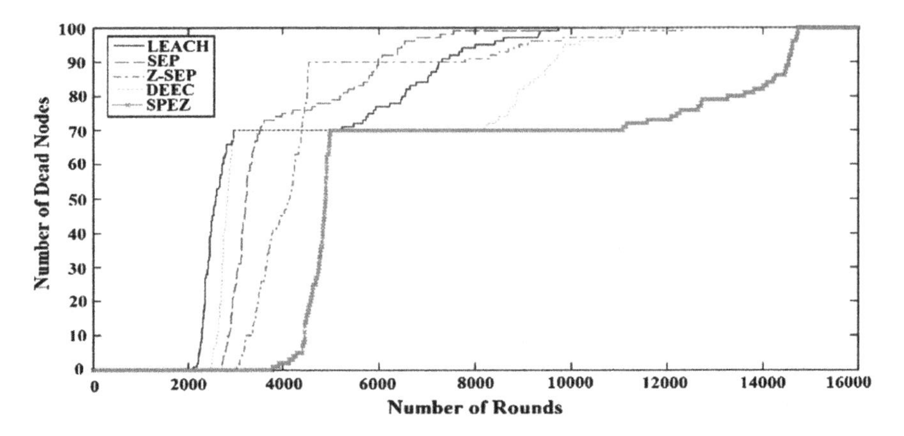

Fig. 1 Evaluation parameters: E_o = 1 J, α = 2 J, m_o = 0.3, base station = (50, 50)

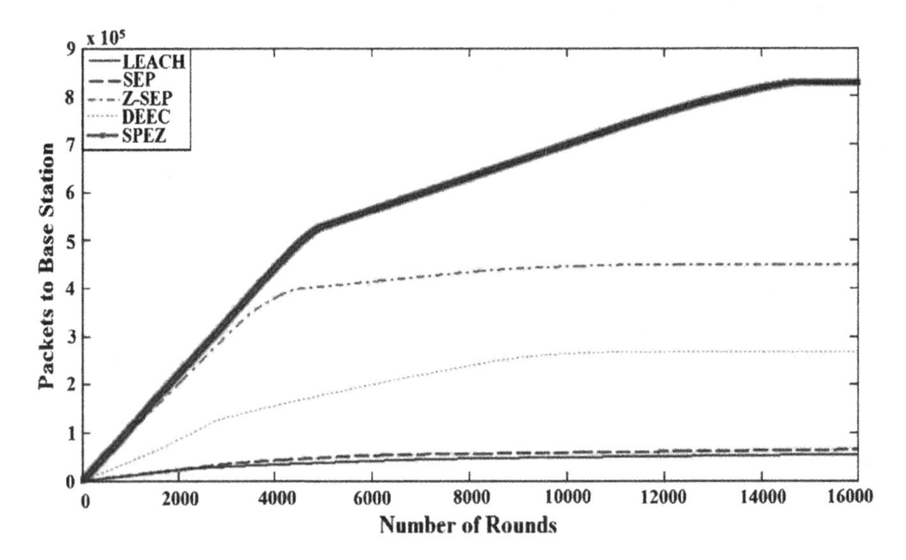

Fig. 2 Evaluation parameters: E_o = 1 J, α = 2 J, m_o = 0.3, base station = (50, 50)

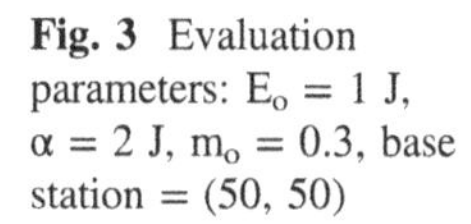

Fig. 3 Evaluation parameters: $E_o = 1$ J, $\alpha = 2$ J, $m_o = 0.3$, base station = (50, 50)

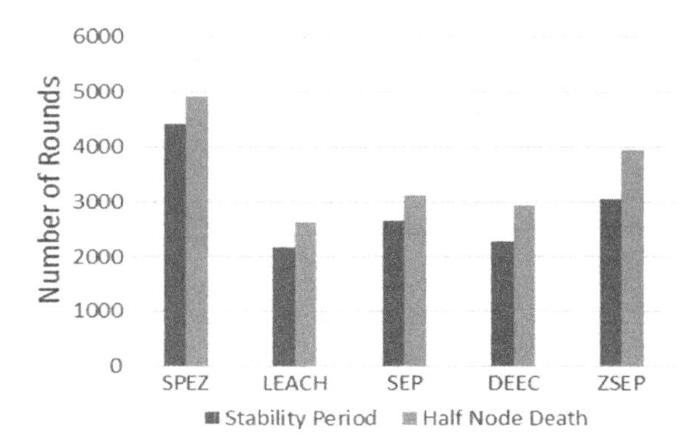

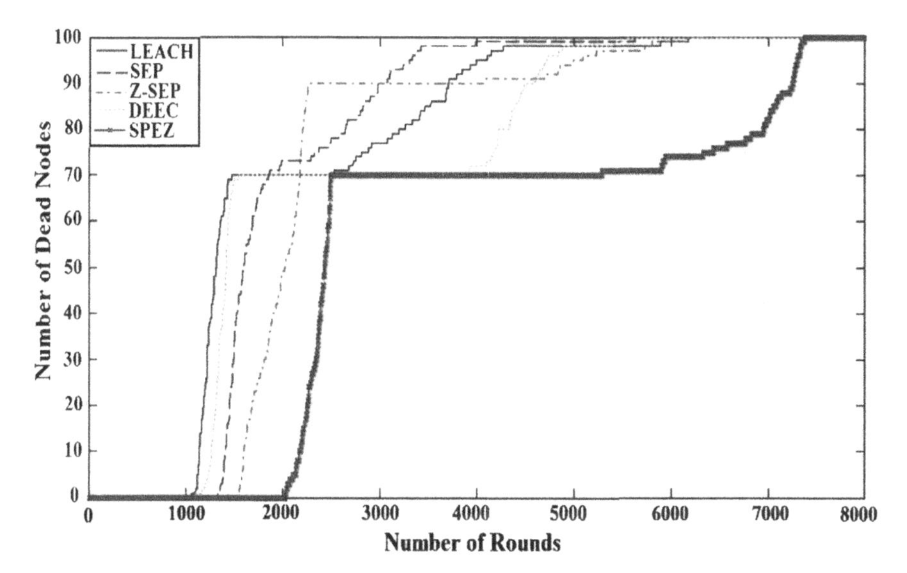

Fig. 4 Evaluation parameters: $E_o = 0.5$ J, $\alpha = 2$ J, $m_o = 0.3$, base station = (50, 50)

For the proposed work, we have simulated the network by varying initial energy levels of the sensor nodes in the field. In Fig. 1, the lifetimes of the SPEZ are 85, 64.4, 34.54, and 24.4% more than SEP, LEECH, DEEC, and ZSEP, respectively.

In Fig. 2, the number of successful packets delivered by SPEZ is around 400, 550, 700, and 750 K more than ZSEP, DEEC, SEP, and LEACH. In Fig. 3, the stability period and half node death are 50.8 and 28.20% better than ZSEP, 106 and 72.4% better than DEEC, 73.6 and 61.3% better than SEP, and 109 and 92.3% better than LEACH which depicts extended life time and load balanced energy dissipation by the sensor nodes in proposed work.

In Fig. 4, we can see that the lifetime of SPEZ is around 7500 rounds as compared to 6200 for LEACH, 6000 for ZSEP, 5700 for SEP, and 5500 for DEEC which validates extended lifetime of proposed work from their comparatives with

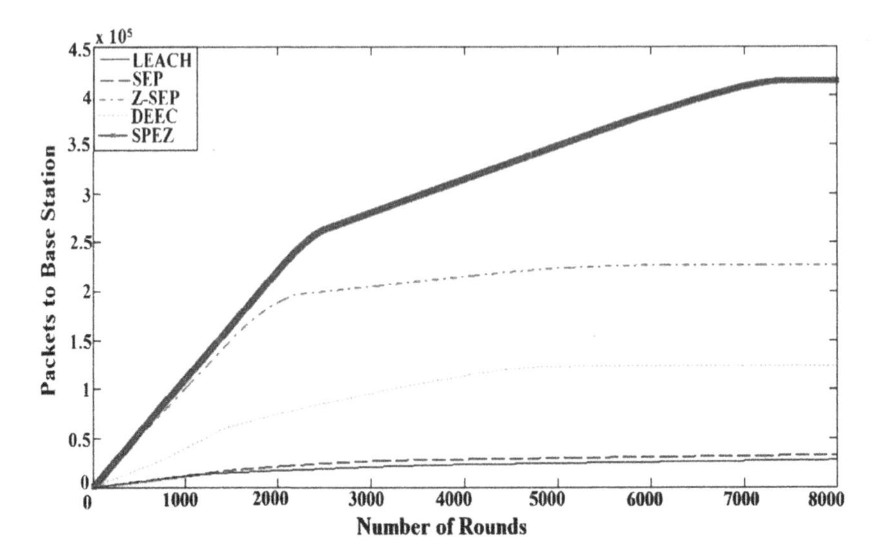

Fig. 5 Evaluation parameters: $E_o = 0.5$ J, $\alpha = 2$ J, $m_o = 0.3$, base station = (50, 50)

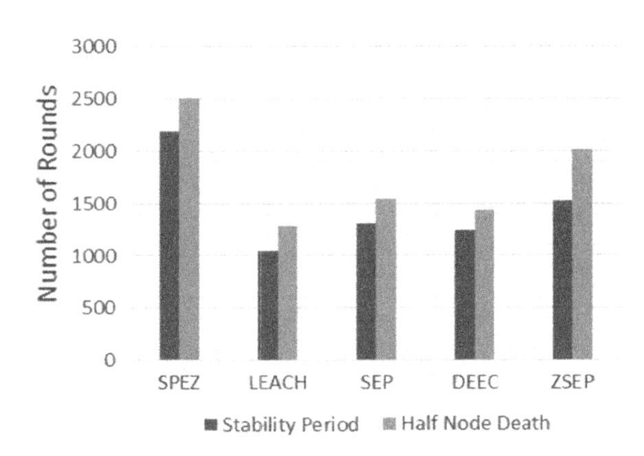

Fig. 6 Evaluation parameters: $E_o = 0.5$ J, $\alpha = 2$ J, $m_o = 0.3$, base station = (50, 50)

initial energy level $E_o = 0.5$ of the sensor nodes. As we can see, the lifetime of the network for SPEZ is extended for the same energy level with its comparatives in Fig. 5. Figure 6 establishes longer stability span and balanced energy distribution of the proposed work.

6 Conclusion

We proposed SPEZ, a Zone-based protocol, by deploying specific type of energy level nodes from the location of the BS with cluster formation within rectangular zones, and the cluster head election depends on the probabilistic election based on weighted remnant energy and local parameters of the sensor nodes in their respective zones. Dense clusters are formed locally in the zones. The key idea proposed here is that the network field is divided into rectangular zones, and the total number of CH elected is determined in proportion to the number of nodes and the size of the zone. Furthermore, unlike ZSEP, our protocol increases the stability period by the intelligent selection of cluster heads by avoiding the direct transmission technique of ZSEP. Simulation results show balanced consumption of energy, thereby enhancing network lifetime and stability of WSNs.

References

1. Metadata K. Romer and F. Mattern, "The design space of wireless sensor networks," in *IEEE Wireless Communications*, vol. 11, no. 6, pp. 54–61, Dec. 2004.
2. Sudeep Tanwar, Neeraj Kumar, Joel J.P.C. Rodrigues, A systematic review on heterogeneous routing protocols for wireless sensor network, Journal of Network and Computer Applications, Volume 53, July 2015, pp. 39–56, ISSN 1084-8045.
3. Houngbadji T, Pierre S. QoSNET: an integrated QoS network for routing protocols in large scale wireless sensor networks. Computer Communication 2010; 33 (13):341–2.
4. W. Heinzelman, A. Chandrakasan and H. Balakrishnan., "An Application-Specific Protocol Architecture for Wireless Microsensor Networks", *IEEE Trans. Wireless Communications,* Vol. 1, No. 4, October 2002, pp. 660–670.
5. S. Lindsey, C. Raghavendra and K. Sivalingam, "Data gathering in sensor networks using the energy*delay metric," *Parallel and Distributed Processing Symposium., Proceedings 15th International*, San Francisco, CA, USA, 2001, pp. 2001–2008.
6. S. Lindsey and C. S. Raghavendra, "PEGASIS: Power-efficient gathering in sensor information systems," *Aerospace Conference Proceedings, 2002. IEEE*, 2002, pp. 3–1125-3-1130 vol.3.
7. A. Manjeshwar and D. P. Agrawal, "TEEN: a routing protocol for enhanced efficiency in wireless sensor networks," *Parallel and Distributed Processing Symposium., Proceedings 15th International*, San Francisco, CA, USA, 2001, pp. 2009–2015.
8. Manjeshwar A. and Agrawal D., "APTEEN: A Hybrid Protocol for Efficient Routing and Comprehensive Information Retrieval in Wireless Sensor Networks," *in Proceedings of International Parallel and Distributed Processing Symposium*, pp. 195–202, 2002.
9. Smaragdakis G., Matta I., and Bestavros A., "SEP: A Stable Election Protocol for Clustered Heterogeneous Wireless Sensor Networks," *in Proceedings of 2nd International Workshop on Sensor and Actor Network Protocols and Applications (SANPA' 2004)*, 2004.
10. Wang X., and Zhang G., "DECP: A Distributed Election Clustering Protocol for Heterogeneous Wireless Sensor Networks," *in Proceedings of the 7th International Conference on Computational Science (ICCS)*, Part III, LNCS, pp. 105–108, 2007.
11. Wang Y, Shi PZ, Li K, Chen ZK. An energy efficient medium access control protocol for target tracking based on dynamic convey tree collaboration in wireless sensor networks. International Journal of Communication System 2012;25(9):1139–59.
12. Qing, L., Qingxin, Z., Mingwen, W.: Design of a distributed energy-efficient clustering algorithm for heterogeneous wireless sensor networks, Computer Communications, Volume 29, Issue 12, 4 August 2006, pp. 2230–2237, (2006).
13. S Faisal, N Javaid, A Javaid, M A Khan, S H Bouk, Z A Khan, Z-SEP: Zonal–Stable Election Protocol for Wireless Sensor Networks, *Journal of Basic and Applied Scientific Research (JBASR),* 2013.

14. Chen, Y.-C. Wen, C.-Y., "Distributed Clustering With Directional Antennas for Wireless Sensor Networks," *Sensors Journal, IEEE*, vol. 13, no. 6, pp. 2166, 2180, June 2013.
15. D. Kumar, Trilok C. Aseri, R.B. Patel, EEHC: Energy efficient heterogeneous clustered scheme for wireless sensor networks, Computer Communications, Volume 32, Issue 4, March 2009, pp. 662–667, (2009).

Evaluation of the Contribution of Transmit Beamforming to the Performance of IEEE 802.11ac WLANs

N.S. Ravindranath, Inder Singh, Ajay Prasad, V. Sambasiva Rao and Sandeep Chaurasia

Abstract Transmit beamforming in MIMO technology is employed to improve receiver SNR. It was added to 802.11n WLAN as an optional feature but not implemented. In 802.11ac, as this technique is highly simplified and also standardized, it is foreseen as a major contributor in improving performance and is expected to be extensively used in 802.11ac devices. In this paper, this feature is studied in depth through simulations in MATLAB and the performance improvement is measured (using parameters like received power, EVM, and constellation diagrams) comparing the results to another MIMO mechanism—Spatial Expansion.

Keywords 802.11ac · 802.11n · MIMO · Spatial expansion
EVM · Constellation diagrams · Transmit beamforming
Transmit opportunity

N.S. Ravindranath (✉) · I. Singh · A. Prasad
UPES, Dehradun, India
e-mail: nethins@yahoo.co.in

I. Singh
e-mail: inder@ddn.upes.ac.in

A. Prasad
e-mail: aprasad@ddn.upes.ac.in

V. Sambasiva Rao
PES University, Bangalore, India
e-mail: vssrao@gmail.com

S. Chaurasia
Manipal University Jaipur, Jaipur, India
e-mail: sandeep.chaurasia@jaipur.manipal.edu

A.K. Somani et al. (eds.), *Proceedings of First International Conference on Smart System, Innovations and Computing*, Smart Innovation, Systems and Technologies 79, https://doi.org/10.1007/978-981-10-5828-8_84

1 Introduction

Transmit beamforming (TBF), also called as MIMO beamforming or co-phasing, is an optional feature on the transmit side and is achieved by introducing complex signal processing features in the WLAN chipset. An array of antennas is used to transmit with high gain to the 802.11 client, resulting in higher downlink signal-to-noise ratio (SNR), higher data rate over a longer range, and hence better overall system performance.

If the same data is transmitted from multiple antennas through a wireless channel, data received by the receive antennas has varied attenuations and phases due to reflection by different objects and also due to the different paths traversed along the channel. Before 802.11n standard was established, network designers selected omnidirectional antennas having a fixed radiation patterns as internal antennas and antennas with different radiation patterns as external antennas for their APs. With the introduction of beamforming, starting with 802.11n, an array of antennae (smart antenna) can be used to actively change the AP's transmit radiation pattern and that too as frequently as every frame.

The contents of this paper are structured as follows: Sects. 2 and 3 introduce theory behind TBF in 802.11n and 802.11ac, respectively, while Sect. 4 discusses test setup for simulating TBF and spatial expansion; results and analysis are discussed in Sect. 5. The conclusions derived in this paper are in Sect. 6.

2 Transmit Beamforming in 802.11n

TxBF in IEEE 802.11n [1, 2] can be either implicit or explicit.

Implicit beamforming is simple and assumes that the RF channel between the beamformer and beamformee is reciprocal. Hence, the transmit beam is formed by the beamformer using the received weights. Channel state information (CSI) is not sent by the beamformee to the beamformer. The constraint in the implicit method is that though the channel between the transmitter and the receiver is reciprocal, the transmitter and receiver are not so. In explicit beamforming (which is not mandatory in IEEE802.11n) with NTx number of transmit antennas and NRx number of receive antennas, a special sounding packet is transmitted through the NTx antennas and it is received by the receiver through the NRx receive antennas. The beamformee derives information regarding the channel from this sounding packet and sends back this information (Channel State Information or CSI) to the beamformer thus incurring overheads.

There are three methods of explicit beamforming with the basic procedure being the same as described above. In the first method, the CSI from beamformee indicates how each NRx antenna has heard the sounding pattern. In the second and third explicit methods, the beamformee itself calculates the weights to optimize the beam and sends the weights in the form of "V" matrix to beamformer. The CSI and V matrices are large as they include precise data for each OFDM subcarrier, for each NTx stream, and for each NRx antenna. The burden on the channel can be reduced by using compression techniques and hence the "compressed V matrix" (method 3).

Hence, TxBF can be effectively used to improve SNR when NRx = 1 and not for broadcast/multicast messages as optimizing phases and calculating the phases for (NTx, NRX) antenna pairs is very complex and data intensive. Also, TxBF can improve data rates and range but not the field-of-view as the beams are made to point to the specific receiver.

3 Transmit Beamforming in 802.11ac

3.1 Introduction

TxBF in 802.11ac [3–5] is much simpler than in 802.11n as the various TBF techniques in 802.11n namely implicit and explicit, and then again CSI matrix and V matrix methods are all replaced by a single method of TBF in 802.11ac, i.e., the null data packet (NDP) sounding which is an explicit mechanism. Beamforming can be theoretically implemented in the dual directional link, but is practically realized in the AP-to-client direction due to the availability of memory, DC power, computational power, and NTx antennas in the AP.

3.2 Channel Measurement (Sounding) Procedures

Figure 1 shows the data transfer mechanism when beamforming is enabled. Channel sounding consists of the following major steps:

Step 1. The Null Data Packet Announcement (NDPA) frame from the beamformer is to access the channel and discover beamformees. By responding to the NDPA,

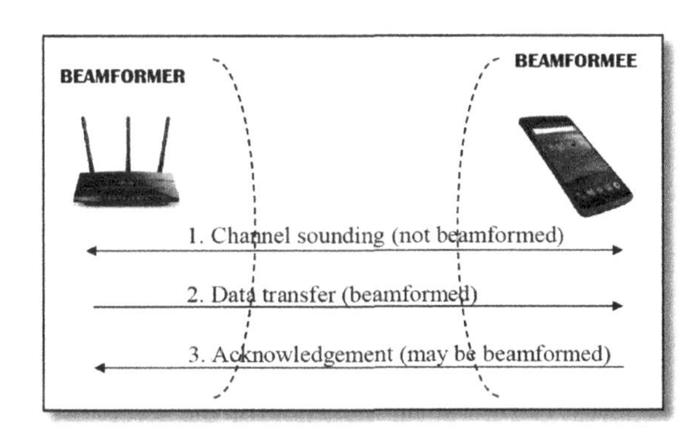

Fig. 1 Data transfer with beamforming

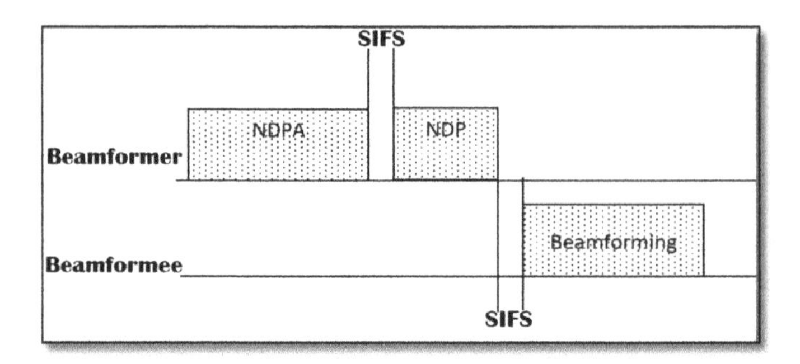

Fig. 2 First two steps of channel sounding

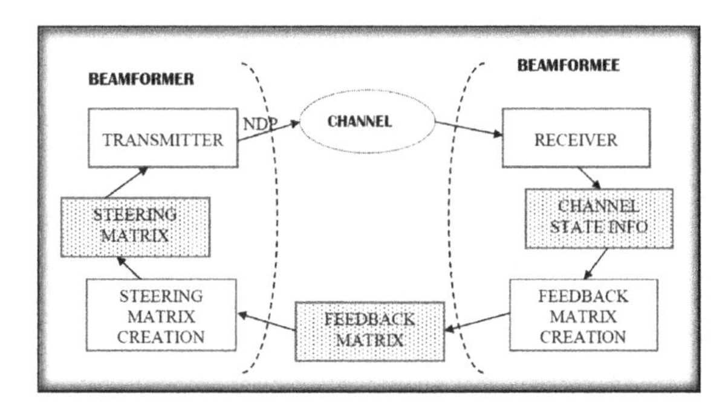

Fig. 3 Last two steps of channel sounding

the beamformee ensures that channel is not accessed by others until the end of the sounding sequence.
Step 2. An NDP from the beamformer follows the NDPA.
Steps 1 and 2 are explained through Fig. 2.
Step 3. On receiving the NDP, the beamformee analyzes the OFDM training fields, processes the individual OFDM subcarrier associated with each (NTx, NRX) antenna pair, and forms the V feedback matrix based on the amplitude and phase of each signal. The feedback matrix is transformed by a series of matrix and mathematical operations to calculate the angles by the beamformee and transmitted to the beamformer as a string of bits in the format specified by the IEEE 802.11ac standard. One of the operations performed is compression which results in a smaller frame and hence lesser airtime. The size of the feedback matrix V is dependent on channel bandwidth which decides the number of OFDM subcarriers and hence the amount of data contained in the matrix. It also depends on the number of (NTx, NRx) pairs.
Step 4. The beamformer forms the steering matrix "Q" based on the contents of the "V" feedback matrix. The effect of the steering matrix on the data to be transmitted is to create a pointed beam from an omnidirectional beam. The beam, before TBF, with approximately same energy in all directions is manipulated after TBF to have maximum energy in the direction of the receiver (by constructive addition) and

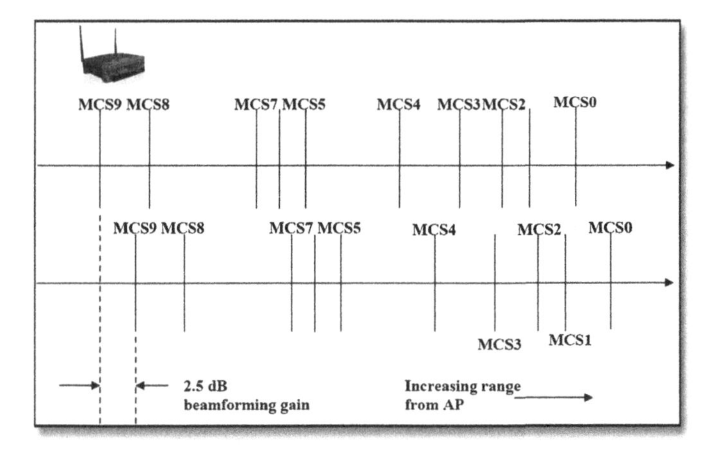

Fig. 4 Relationship between range and MCS

weak energy in all other directions. The phase data required to converge or cancel the energy is contained in the steering matrix. Figure 3 explains the steps 3 and 4. The local maximum at the receive antenna improves the signal strength and hence SNR, which in turn ensures a sustainable RF link rate. TBF can thus be thought of as directing a beam using phase shifts toward a particular receive antenna. As the TBF technique can converge or cancel a beam by introducing appropriate phases, it can apply the same concept to cancel interfering signals in the direction of the receiver. On the flip side, if the amount of data exchanged between the beamformer and beamformee to converge the beam and increase the data rate is so much that speed advantage achieved is countered by the time taken to transmit the huge volume of the steering matrix data, then TBF is not attractive as the overall speed will reduce. Also, the advantage of TBF is more pronounced when communication is between devices separated by a distance in the mid- to long range. At short range, the advantage gained by TBF is countered by the degradation of the signal due to multipath effects and incorrect estimation of the signal properties. At long ranges, TBF gain is insufficient to increase the speed of transmission. To summarize, TBF gain (of about 3-dB) in the direction of the receiver is best in mid-range.

Figure 4 shows the explanation of the increasing range for a sustainable link from an AP with reducing orders of MCS. The two horizontal lines indicate the effect of range improvement at the client device due to gain offered by TBF. Also, there is at least 5-dB change between 256-QAM rates (MCS8) and the next set of lower data rates (i.e., MCS7). At any range distance, the data rate after TBF will be higher and it is most efficient at pulling in the 64-QAM (MCS5 to MCS7) rates in the middle order ranges.

4 Test Setup

MATLAB R2016a [6] version has incorporated a WLAN System Toolbox with provision to design, configure the physical layer in IEEE 802.11ac WLAN standard, and to simulate, analyze, and test the performance of WLAN communications systems. Various parameters of the transmitter, channel, and receiver are adjusted to study the performance parameters like constellation diagrams, EVM, and SNR.

4.1 Transmission with Spatial Expansion

As a single user beamformee is only optional in 802.11ac standard, a transmitter with multiple antennas uses a feature called "Spatial Expansion" to communicate with a receiver that is not a beamformee. Using this feature, many STSs are transmitted on even more transmit antennas to a receiver which is incapable of performing as beamformee.

As shown in Fig. 5, custom spatial mapping scheme is selected which generates a custom spatial mapping matrix that contains details to map a subcarrier for a space time stream (STS) to all transmit antennas leading to a matrix size Nst × Nsts × Nt, where Nst, Nsts, and Nt are the number of occupied subcarriers, STSs, and transmit antennas, respectively. Some STSs may be replicated to match the number preferred to be transmitted. At the receiver, demodulation is performed to get back the data symbols and BER is measured.

The transmit diversity gain is further enhanced more by spatial expansion in channels with flat fading than by directly mapping STSs to transmit antennas.

To demonstrate the benefits of TBF specifically, the data packet is transmitted using TBF without changing the channel conditions.

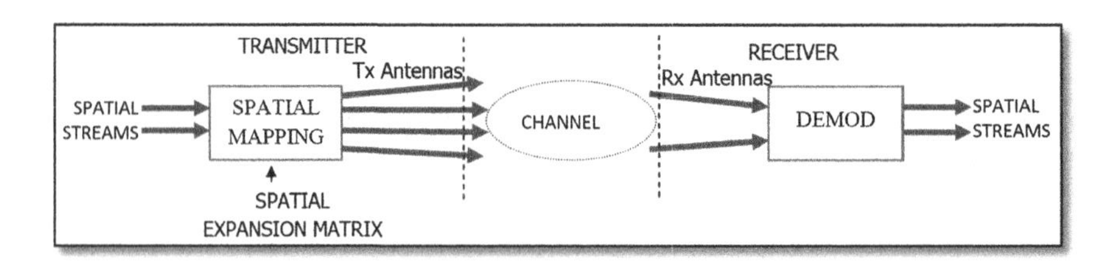

Fig. 5 Transmission with spatial expansion

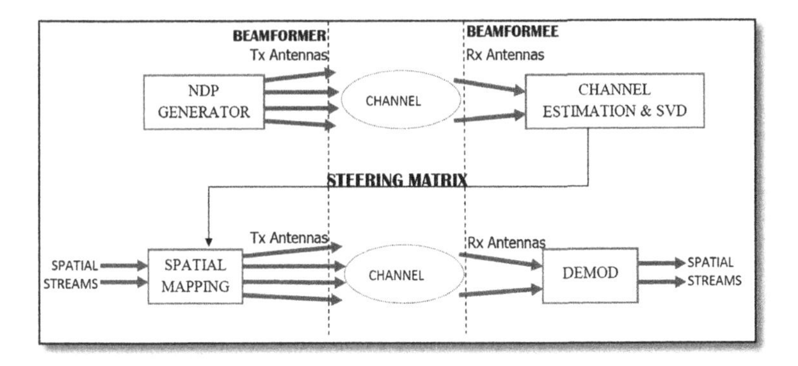

Fig. 6 Transmission with beamforming

4.2 Transmission with Beamforming

When the receiver can function as a beamformee, the achieved SNR is higher as compared to that achieved with spatial expansion. An NDP is passed through the channel using "direct" spatial mapping and is used to create the beamforming steering matrix. The number of STSs selected for transmission matches the number of antennas used for transmission, thus enabling the VHT-LTF to be utilized to sound channels between the transmit–receive antenna pairs. The beamforming matrix is used to form a transmit beam in the channel.

It may be noted that the same channel is used for transmitting sounding information and data. Also, as the feedback information from the beamformee is not compressed, beamforming can be considered as perfect (Fig. 6).

In Fig. 5, NDP is first transmitted to sound the channel and create the steering matrix. The steering matrix is then used to transmit, receive, and demodulate the data packet.

5 Results and Discussion

5.1 Constellation Diagram

The sample constellation diagrams are simulated for a single STS single receive antenna, 2- 3- and 4-transmit antennas with CBW 160 and 40 MHz and MCS values 1, 9. Figure 7a–e depicts the constellation for MCS 9 and Fig. 8a–f for MCS 1.

Comparing TBF scenarios independently, in cases where highest throughput is achieved at the expense of poor constellation, for example, CBW = 160 MHz and MCS = 9, TBF improves performance significantly. This is visible with four-transmit antenna TBF case which is much better than the two-transmit antenna

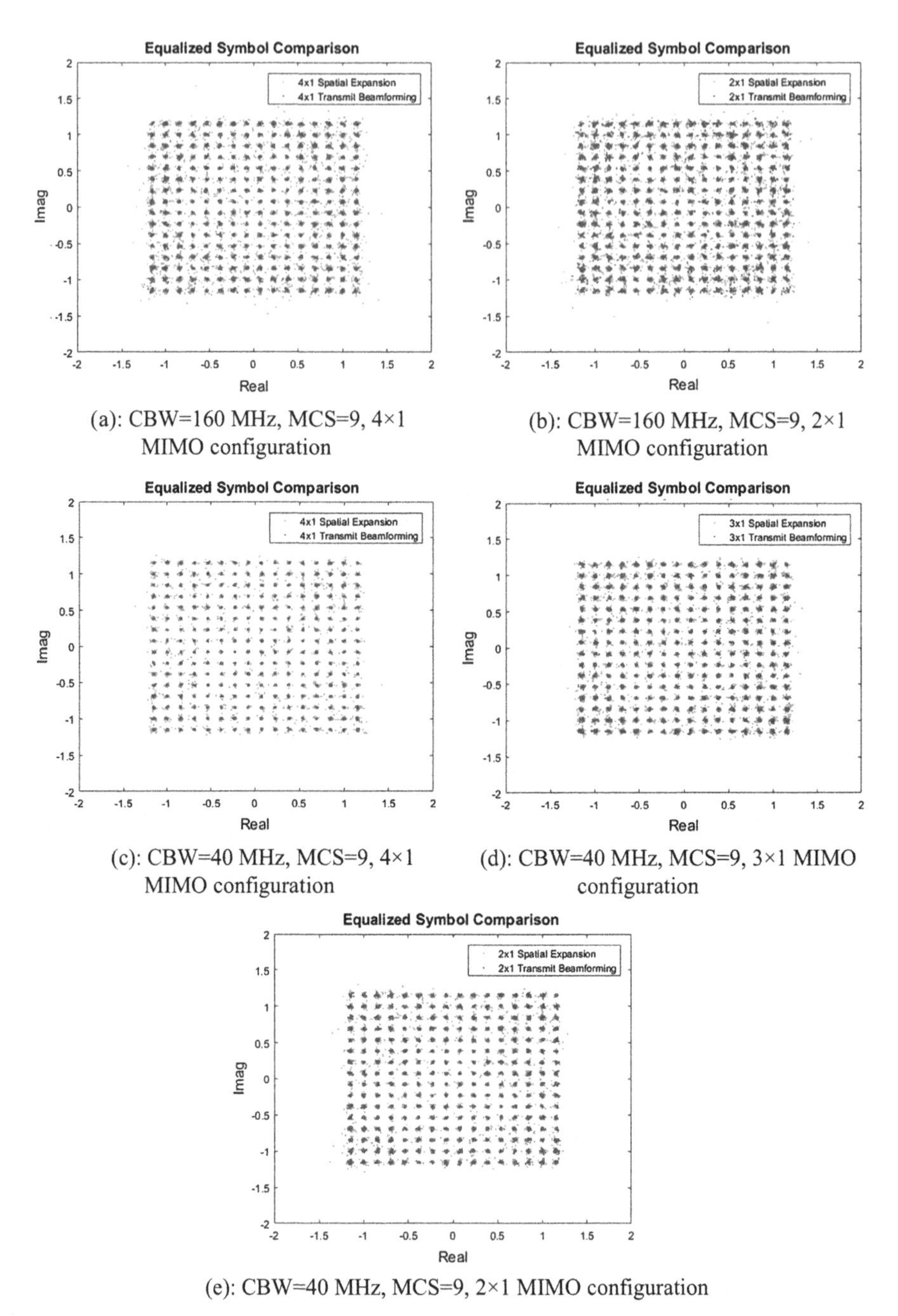

(a): CBW=160 MHz, MCS=9, 4×1 MIMO configuration

(b): CBW=160 MHz, MCS=9, 2×1 MIMO configuration

(c): CBW=40 MHz, MCS=9, 4×1 MIMO configuration

(d): CBW=40 MHz, MCS=9, 3×1 MIMO configuration

(e): CBW=40 MHz, MCS=9, 2×1 MIMO configuration

Fig. 7 **a** CBW = 160 MHz, MCS = 9, 4 × 1 MIMO configuration. **b** CBW = 160 MHz, MCS = 9, 2 × 1 MIMO configuration. **c** CBW = 40 MHz, MCS = 9, 4 × 1 MIMO configuration. **d** CBW = 40 MHz, MCS = 9, 3 × 1 MIMO configuration. **e** CBW = 40 MHz, MCS = 9, 2 × 1 MIMO configuration

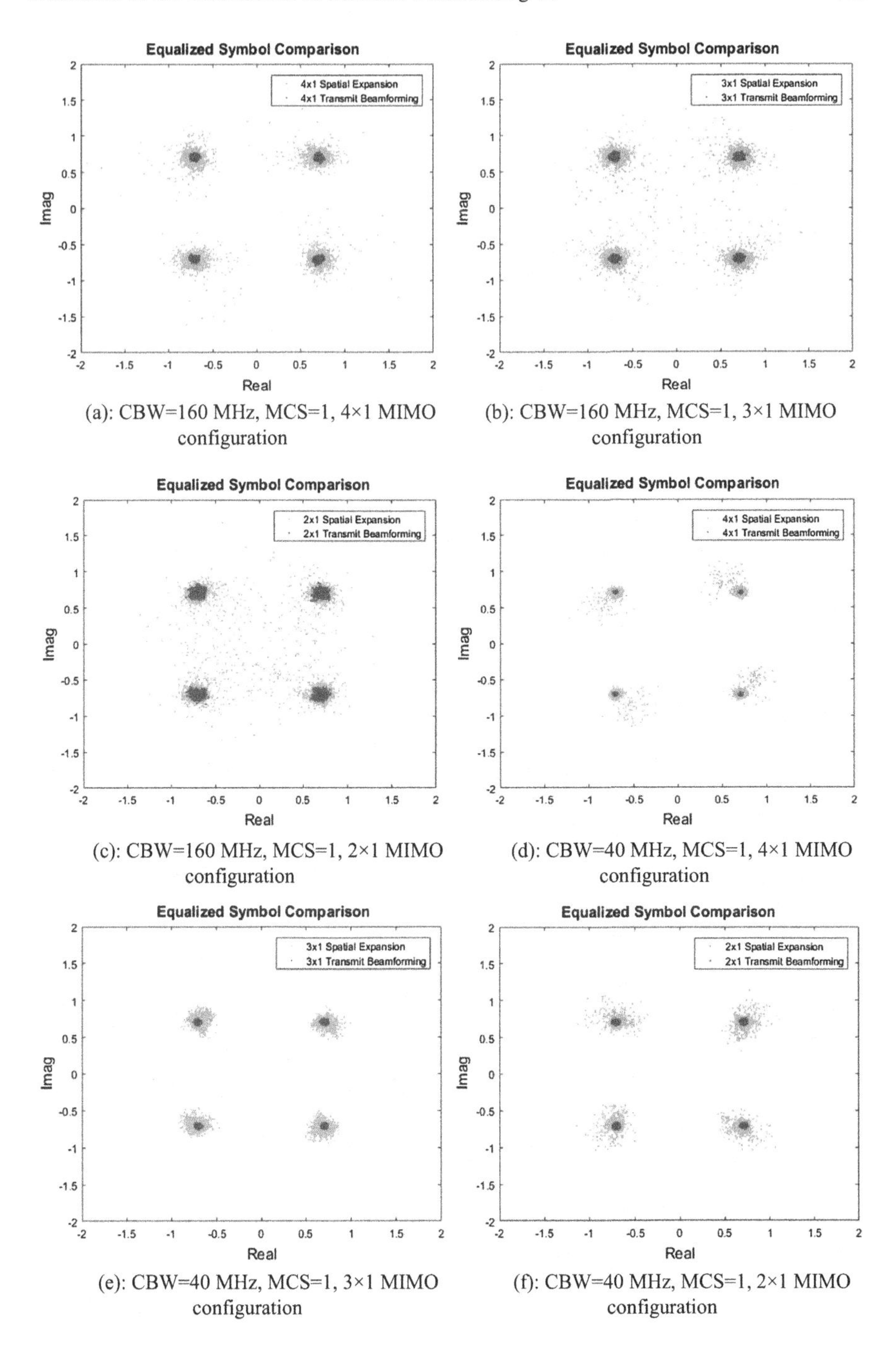

(a): CBW=160 MHz, MCS=1, 4×1 MIMO configuration

(b): CBW=160 MHz, MCS=1, 3×1 MIMO configuration

(c): CBW=160 MHz, MCS=1, 2×1 MIMO configuration

(d): CBW=40 MHz, MCS=1, 4×1 MIMO configuration

(e): CBW=40 MHz, MCS=1, 3×1 MIMO configuration

(f): CBW=40 MHz, MCS=1, 2×1 MIMO configuration

Fig. 8 **a** CBW = 160 MHz, MCS = 1, 4 × 1 MIMO configuration. **b** CBW = 160 MHz, MCS = 1, 3 × 1 MIMO configuration. **c** CBW = 160 MHz, MCS = 1, 2 × 1 MIMO configuration. **d** CBW = 40 MHz, MCS = 1, 4 × 1 MIMO configuration. **e** CBW = 40 MHz, MCS = 1, 3 × 1 MIMO configuration. **f** CBW = 40 MHz, MCS = 1, 2 × 1 MIMO configuration

TBF scenario. Similar results are noticed in lower throughput cases (MCS = 1, CBW = 40 MHz) with two, three, and four antennas also.

5.2 Error Vector Magnitude

The error vector magnitude (EVM) is a tool to measure the performance of digital communication systems. The IQ points are used to estimate EVM of an ideal transmitted signal and, hence, the quality of the demodulated signal. The following figures show the EVM (max) values for CBW = 40 and 160 MHz, MCS = 1–9, and for MIMO configurations 2 × 1, 3 × 1, and 4 × 1.

From Fig. 9a–c, the following observations are deduced:

- 40 MHz CBW exhibits lesser EVM as compared to 160 MHz CBW.
- Dispersion of the EVM values reduces greatly after TBF irrespective of the number of transmit antennas, modulation scheme, or CBW.

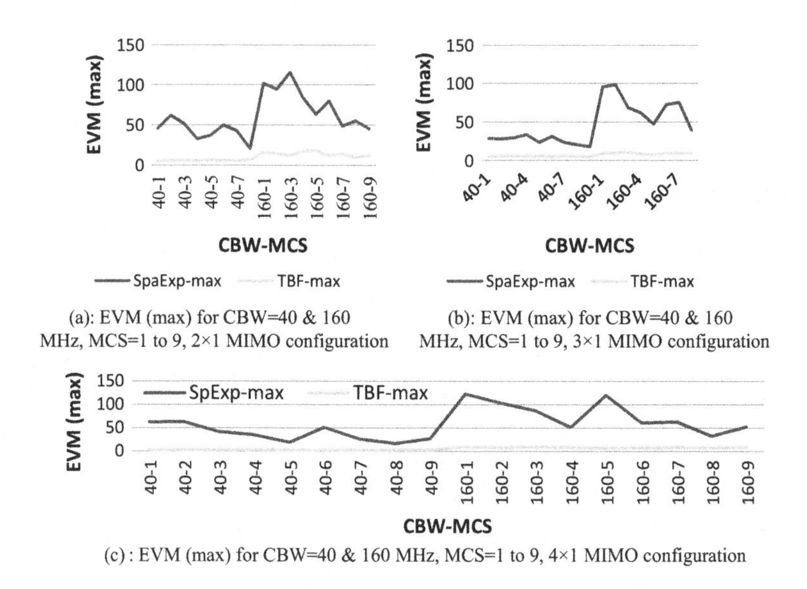

Fig. 9 **a** EVM (max) for CBW = 40 and 160 MHz, MCS = 1–9, 2 × 1 MIMO configuration. **b** EVM (max) for CBW = 40 and 160 MHz, MCS = 1–9, 3 × 1 MIMO configuration. **c** EVM (max) for CBW = 40 and 160 MHz, MCS = 1–9, 4 × 1 MIMO configuration

EVM (rms) with Spatial Expansion & TBF with CBW 40 MHz

	40-1	40-2	40-3	40-4	40-5	40-6	40-7	40-8	40-9
Spatial Expansion- 2x1	7.4	5.1	3.4	3.2	3.9	3.4		7.5	6
TBF- 2x1	2.3	2.3	1.5	1.5	1.5	1.4		2.8	2.9
Spatial Expansion- 3x1	6.5	5	3.8	3.5	4.1	4	3.3	5.4	4.7
TBF- 3x1	1.8	1.8	1.6	1.5	1.5	1.5	1.5	1.4	1.5
Spatial Expansion- 4x1	6.1	4.8	2	2.1	3.5	2.5	2.4	4.9	4.5
TBF- 4x1	1.4	1.5	0.6	0.6	0.7	0.7	0.7	1.3	1.3

(a): EVM (rms) for CBW=40 MHz, MCS=1 to 9, 2×1, 3×1 & 4×1 MIMO configuration with Spatial Expansion and TBF

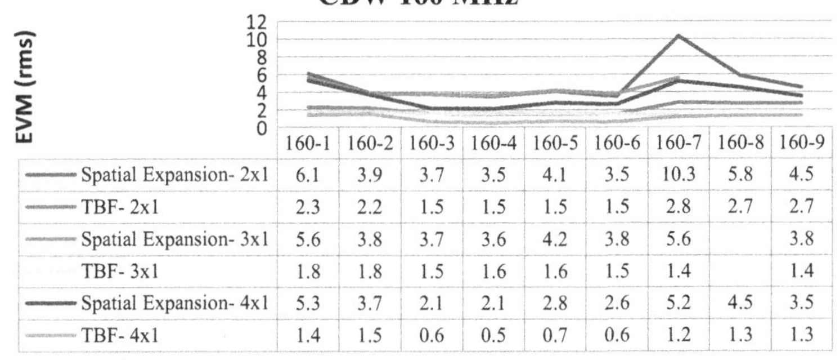

	160-1	160-2	160-3	160-4	160-5	160-6	160-7	160-8	160-9
Spatial Expansion- 2x1	6.1	3.9	3.7	3.5	4.1	3.5	10.3	5.8	4.5
TBF- 2x1	2.3	2.2	1.5	1.5	1.5	1.5	2.8	2.7	2.7
Spatial Expansion- 3x1	5.6	3.8	3.7	3.6	4.2	3.8	5.6		3.8
TBF- 3x1	1.8	1.8	1.5	1.6	1.6	1.5	1.4		1.4
Spatial Expansion- 4x1	5.3	3.7	2.1	2.1	2.8	2.6	5.2	4.5	3.5
TBF- 4x1	1.4	1.5	0.6	0.5	0.7	0.6	1.2	1.3	1.3

(b): EVM (rms) for CBW=160 MHz, MCS=1 to 9, 2×1, 3×1 & 4×1 MIMO configuration with Spatial Expansion and TBF

Fig. 10 **a** EVM (rms) for CBW = 40 MHz, MCS = 1–9, 2 × 1, 3 × 1 and 4 × 1 MIMO configuration with spatial expansion and TBF. **b** EVM (rms) for CBW = 160 MHz, MCS = 1–9, 2 × 1, 3 × 1 and 4 × 1 MIMO configuration with spatial expansion and TBF

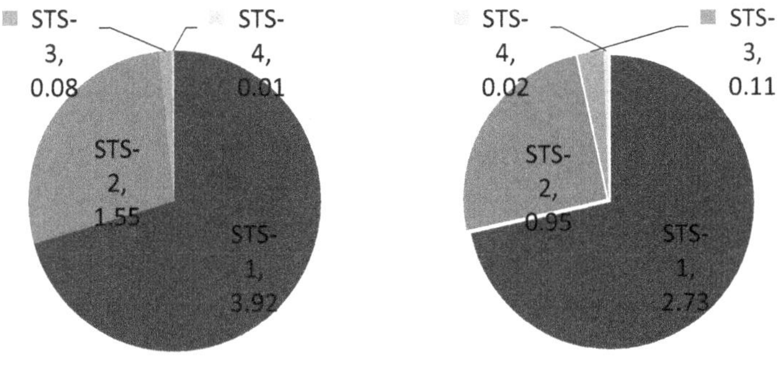

(a): Mean received channel power in Watts per STS with TBF for CBW 40 MHz

(b): Mean received channel power in Watts per STS with TBF for CBW 160 MHz

Fig. 11 **a** Mean received channel power in watts per STS with TBF for CBW 40 MHz. **b** Mean received channel power in watts per STS with TBF for CBW 160 MHz

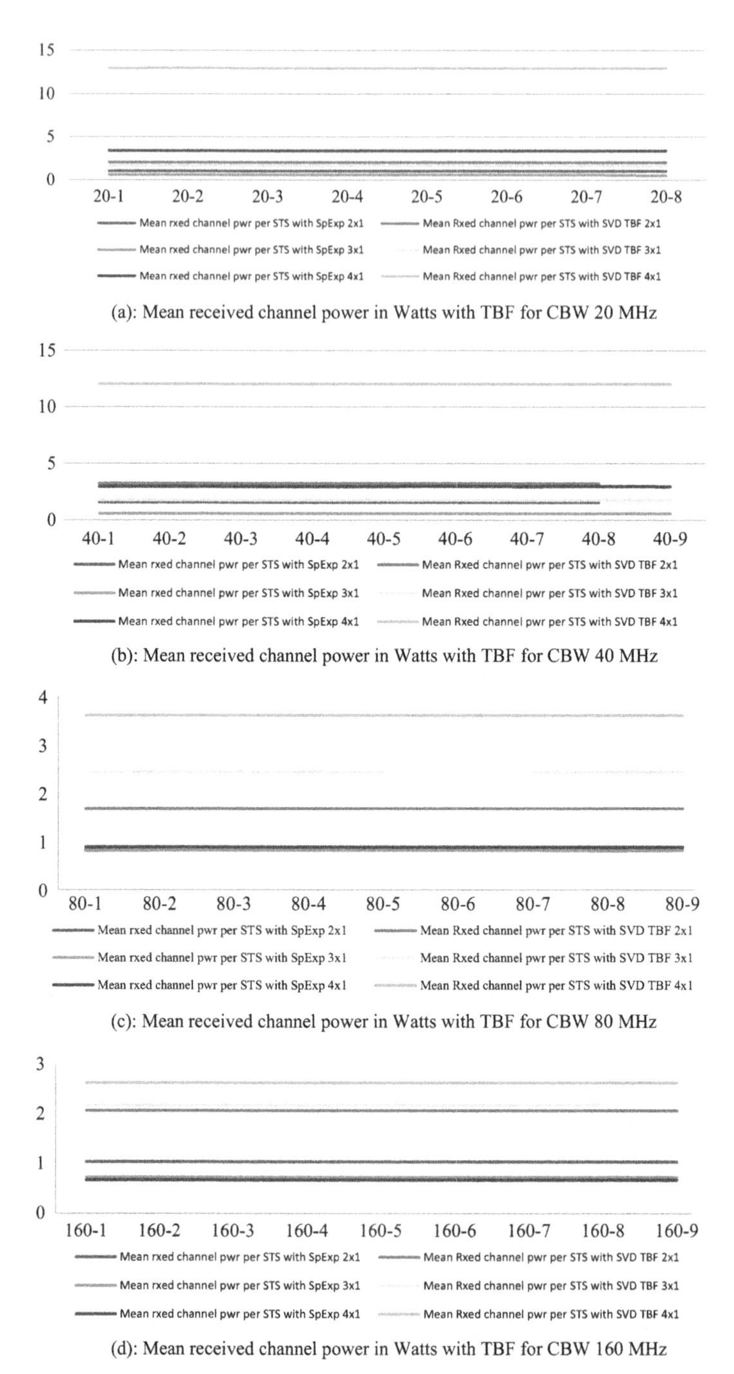

(a): Mean received channel power in Watts with TBF for CBW 20 MHz

(b): Mean received channel power in Watts with TBF for CBW 40 MHz

(c): Mean received channel power in Watts with TBF for CBW 80 MHz

(d): Mean received channel power in Watts with TBF for CBW 160 MHz

Fig. 12 **a** Mean received channel power in watts with TBF for CBW 20 MHz. **b** Mean received channel power in watts with TBF for CBW 40 MHz. **c** Mean received channel power in watts with TBF for CBW 80 MHz. **d** Mean received channel power in watts with TBF for CBW 160 MHz

After TBF 4 × 1 MIMO configuration displays, a lower EVM (max) value is compared to 2 × 1 MIMO configuration. For example, the EVM (max) values are 15.7, 9.5, and 11.9 for MCS1, MCS8, and MCS9, respectively, for the 2 × 1 MIMO configuration, whereas they are reduced to 10, 7.6, and 8.8 for the 4 × 1 MIMO configuration. As observed in Fig. 10a, b, after TBF, EVM (rms) values are also lower for higher transmit antenna MIMO configurations and also as compared to the numbers obtained by spatial expansion alone for CBW 40 and 160 MHz.

5.3 Received Power and Space Time Streams

A 4 × 4 MIMO configuration with four spatial streams is used to study the power received in each stream after TBF. The distribution of power in STS-1, 2, 3, and 4 for 40 MHz channel bandwidth is shown in Fig. 11a and for 160 MHz channel bandwidth in Fig. 11b. The simulation is done for all MCS values between 1 and 9. The mean received channel power per STS with TBF in watts for a particular STS which is independent of MCS or the number of space streams. After TBF, power is maximum in the STS1 and reduces rapidly in STS-2, 3, and 4.

It may also be noted that the power received per STS, say in STS-1, does not vary whether the number of STSs is 1, 2, 3, or 4. Same observation holds for STS-2, 3, and 4 also. After TBF, power is maximum in STS1 and reduces rapidly in STS-2, 3, and 4. As in the case of 40 MHz channel bandwidth, with 160 MHz CBW, also, the power received per STS, say in STS-1, does not vary whether the number of STSs is 1, 2, 3, or 4 and similar observation holds for STS-2, 3, and 4.

With a single STS, and different MIMO configurations, power received is plotted for various MCS values when CBW is 20, 40, 80, and 160 MHz as shown in Fig. 12a–d, respectively.

Mean received channel power is always maximum for the 4 × 1 MIMO configuration.

6 Example Scenarios Where Beamforming Can Be Applied

6.1 Scenario-1

While the spatial division multiplexing (SDM) technique boosts the throughput, TBF achieves better SNR. As the two end results are contradictory, the two mechanisms can be implemented only with a compromise. The following example discusses the implication of implementing both the mechanisms [7]. A major difference between SDM and TBF is that in SDM data has to be transmitted in multiple streams, whereas in TBF a single stream is used. Hence, in a 2 × 2:2 AP/Client MIMO configuration, both SDM and TBF cannot be simultaneously implemented. If the configuration is 3 × 3:3 AP/Client, theoretically two SDM streams and one TBF stream can be configured. This enhances the SDM stream but the achieved data rate is 300 Mbps only against the expected 450 Mbps as the third antenna is utilized to transmit the other antennas spatial streams in SDM instead of independent data of its own. Also, a 4 × 4:3 AP MIMO configuration, theoretically, can communicate three SDM streams and one TBF stream. This boosts the throughput of the SDM streams but reduces the data that could have been transmitted on the SDM stream, and hence the maximum data rate achievable with this mixed implementation is 450 Mbps only.

6.2 Scenario-2

An example to depict the performance of TBF in interiors is discussed. In a medium big-sized house, i.e., in medium ranges, TBF enhances the signal strength even in edges of the home or in a closet. Short ranges do not require this enhancement as the signal strength/SNR is sufficiently high to support the maximum data rate. Hence, the effect of TBF on the performance of WLANs is not well exhibited in short ranges. And at long ranges, TBF gain is unable to support increase in data rates.

7 Conclusion

On studying the various scenarios in which TBF is used, it can be deduced that TBF is contributing to performance improvement. Even in cases where highest throughput is achieved at the expense of poor constellation, effect of TBF is observed and specifically more in the four-transmit antenna case than the two-transmit antenna TBF scenario. The response of TBF is similar in cases of lower throughput with 2, 3, and 4 transmit antennas. TBF also aids in reducing the magnitude of EVM. With reference to CBW, 40 MHz has reduced EVM as compared to 160 MHz CBW. Also, 4 × 1 MIMO configuration displays a lower EVM (max) values compared to lower transmit antenna MIMO configuration. The same is seen for the rms value of EVM too.

On simulating the received power, the first STS always has the maximum share of power irrespective of the number of STS, CBW, or MCS. The same parameter is maximum for the 4 × 1 MIMO configuration as compared to other configurations. It is demonstrated that if a receiver is capable of being a beamformee, the SNR can potentially be improved when a transmission is beamformed compared to a spatial expansion transmission. The increase in received power when using beamforming can lead to more reliable demodulation or potentially even a higher order modulation and coding scheme to be used for the transmission.

References

1. IEEE standard 802.11n-2009.
2. Elda Perahia, Robert Stacey, 2008, "Next Generation Wireless LANs Throughput, Robustness and Reliability in 802.11n", Cambridge University Press.
3. IEEE standard 802.11ac-2013.
4. Elda Perahia, Robert Stacey, 2013, "Next Generation Wireless LANs: 802.11n and 802.11ac", Cambridge University Press.
5. Matthew S. Gast, 2015, "Wi-Fi at Gigabit and Beyond 802.11ac A Survival Guide", o Reilly Media.
6. www.mathworks.com.
7. XIRRUS White Paper, "802.11ac Demystified-High Performance Wireless Networks" examples.

Author Index

A.K. Somani et al. (eds.), *Proceedings of First International Conference on Smart System, Innovations and Computing*, Smart Innovation, Systems and Technologies 79, https://doi.org/10.1007/978-981-10-5828-8